Natural Polymers and Biopolymers II

Natural Polymers and Biopolymers II

Editor

Sylvain Caillol

MDPI • Basel • Beijing • Wuhan • Barcelona • Belgrade • Manchester • Tokyo • Cluj • Tianjin

Editor
Sylvain Caillol
CNRS Research Director, Institute Charles Gerhardt
France

Editorial Office
MDPI
St. Alban-Anlage 66
4052 Basel, Switzerland

This is a reprint of articles from the Special Issue published online in the open access journal *Molecules* (ISSN 1420-3049) (available at: https://www.mdpi.com/journal/molecules/special_issues/biopolymer_II).

For citation purposes, cite each article independently as indicated on the article page online and as indicated below:

LastName, A.A.; LastName, B.B.; LastName, C.C. Article Title. *Journal Name* **Year**, *Volume Number*, Page Range.

ISBN 978-3-0365-0404-9 (Hbk)
ISBN 978-3-0365-0405-6 (PDF)

Contents

About the Editor

Sylvain Caillol is Research Director at CNRS. He was born in 1974. He graduated engineer from the National Graduate School of Chemistry of Montpellier in 1998 and then received his M. Sc. Degree in Chemistry from the University of Montpellier. He received his PhD degree in 2001 from the University of Bordeaux. Subsequently he joined Rhodia Company. Later, promoted Department Manager, he headed the Polymer Research Department in the Research Center of Aubervilliers. In 2007 he joined the CNRS at the Institute Charles Gerhardt of the University of Montpellier where he started a new research topic dedicated to Green Chemistry and Biobased Polymers. He is co-author of more than 150 articles, patents and book chapters. He is Chairman of "Oleochemistry, Molecule and Polymer Science" division of European Federation of Lipids. He won the Innovative Techniques for Environment Award in 2010, the Green Materials Prize in 2018 and 2020 and was nominated Pioneering Investigator by the Royal Society of Chemistry in 2019.

Preface to "Natural Polymers and Biopolymers II"

BioPolymers could be either natural polymers – polymer naturally occurring in Nature, such as cellulose or starch..., or biobased polymers that are artificially synthesized from natural resources. Since the late 1990s, the polymer industry has faced two serious problems: global warming and anticipation of limitation to the access to fossil resources. One solution consists in the use of sustainable resources instead of fossil-based resources. Hence, biomass feedstocks are a promising resource and biopolymers are one of the most dynamic polymer area. Additionally, biodegradability is a special functionality conferred to a material, bio-based or not. Very recently, facing the awareness of the volumes of plastic wastes, biodegradable polymers are gaining increasing attention from the market and industrial community.

This special issue of Molecules deals with the current scientific and industrial challenges of Natural and Biobased Polymers, through the access of new biobased monomers, improved thermo-mechanical properties, and by substitution of harmful substances.

This themed issue can be considered as collection of highlights within the field of Natural Polymers and Biobased Polymers which clearly demonstrate the increased interest in this field. We hope that this will inspire researchers to further develop this area and thus contribute to futures more sustainable society.

Sylvain Caillol
Editor

 *molecules*

Editorial

Special Issue "Natural Polymers and Biopolymers II"

Sylvain Caillol

ICGM, Univ Montpellier, CNRS, ENSCM, 34296 Montpellier, France; sylvain.caillol@enscm.fr

Received: 23 December 2020; Accepted: 25 December 2020; Published: 29 December 2020

BioPolymers could be either natural polymers (polymer naturally occurring in Nature, such as cellulose or starch …), or biobased polymers that are artificially synthesized from natural resources. Since the late 1990s, the polymer industry has faced two serious problems: Global warming and the anticipation of limits in accessing fossil resources. One solution involves the use of sustainable resources instead of fossil-based resources. Hence, biomass feedstocks are a promising resource because of their sustainability. The current production of biopolymers is around 15 Mt/y, but biopolymers are one of the most dynamic polymer area.

Natural polymers are materials that widely occur in nature, or are extracted from plants or animals. Some examples of natural polymers are proteins, cellulose, natural rubber, silk, and wool, starch or natural rubber.

Biobased polymers are synthesized from renewable resources (vegetal, animal or fungal) but it does not mean that they are biodegradable polymers. Hence, biodegradability is a special functionality conferred to a material, bio-based or not, and biobased sourcing does not entail biodegradability. Very recently, due the awareness of the volumes of plastic wastes, biodegradable polymers have gained increasing attention from the market and both scientific and industrial communities.

This special issue of Molecules deals with the current scientific and industrial challenges of Natural and Biobased Polymers, through the access of new biobased monomers, improved thermo-mechanical properties, and by substitution of harmful substances.

Firstly, concerning the renewable resources, this issue proposes the use of innovative biobased derived from terpenes. Indeed, Nishida et al. reports a series of *exo*-methylene 6-membered ring conjugated dienes, which are directly or indirectly obtained from terpenoids, such as β-phellandrene, carvone, piperitone, and verbenone, were radically polymerized [1]. New terpene-based epoxy monomers were also synthesized by Couture et al. for the synthesis of high properties epoxy networks [2]. Additionally, Mora et al. reported the synthesis of vanillin-derived amines for the curing of epoxy thermosets [3]. Moreover, Della Vacche et al. studied the photocuring of cardanol-based monomers for biobased composites [4]. Photopolymerization was also studied on innovative eugenol-based methacrylates by Molina-Gutiérrez et al. for elaboration of biobased coatings [5]. Additionally, Montané et al. proposed an original mechanical study on the synthesis of humins-based epoxy resins [6]. Vegetable oils remain one of the most used resources for the synthesis of biobased polymers. Kohut et al. reported a feature article dedicated to plant oil-based monomers (POBM) in emulsion polymerization [7]. Hence, POBMs with different unsaturations in copolymerization reactions with conventional vinyl monomers allows for obtaining copolymers with enhanced hydrophobicity, provides a mechanism of internal plasticization and control of crosslinking degree.

Then, natural polymers were also studied in order to improve or propose new properties. Hence, curcumin loaded biobased films were studied from alginate/cellulose/gelatin by Chiaoprakobkij et al. These films showed non-cytotoxicity to human keratinocytes and human gingival fibroblasts but interestingly exhibited potent anticancer activity in oral cancer cells [8]. Moreover, Shirosaki et al.

reported chitosan microfibers with properties thanks to silane coupling agents [9]. Furthermore, Saremi et al. proposed a feature article on adhesion and stability of nanocellulose coatings on films and textiles [10]. Additionally, Bernal-Ballen et al. reported chitosan/PVA-based system with antibacterial activity against bacterial strains without adding a high antibiotic concentration [11]. The findings of this study suggest that the system may be effective against healthcare-associated infections, a promising view in the design of novel antimicrobial biomaterials potentially suitable for tissue engineering applications. Chitosan/PVA-based systems were also reported as a promising alternative for bone tissue regeneration by Pineda-Castillo et al. [12]. A feature article presenting new analytical strategy dedicated to multiscale structure of starches grafted with hydrophobic groups was also reported by Volant et al. [13]. Furthermore, Allegretti et al. studied the tuning of lignin characteristics by fractionation, based on solvent extraction and membrane-assisted ultrafiltration [14]. Rebiere et al. managed the characterization of non-derivatized cellulose by SEC in Tetrabutylammonium Fluoride/Dimethylsulfoxide [15]. Then, Klapiszewski et al. reported the preparation and comprehensive characterization of innovative additives to abrasive materials based on functional, pro-ecological lignin-alumina hybrid fillers [16]. Miscibility is also an important issue in biopolymer blends for analysis of the behaviour of polymer pairs through the detection of phase separation and improvement of the mechanical and physical properties of the blend. Ghaeli et al. reported a study dedicated to the formulation of a stable and one-phase mixture of collagen and regenerated silk fibroin (RSF), with the highest miscibility ratio between these two macromolecules [17]. Ghica et al. studied the development and optimization of some topical collagen-dextran sponges with flufenamic acid, designed to be potential dressings for burn wounds healing [18]. Bundled actin structures play an essential role in the mechanical response of the actin cytoskeleton in eukaryotic cells. Although responsible for crucial cellular processes, they are rarely investigated in comparison to single filaments and isotropic networks. Strehle et al. presented a new method to determine the bending stiffness of individual bundles, by measuring the decay of an actively induced oscillation [19]. Their experiments revealed that thin, depletion force-induced bundles behave as semiflexible polymers and obey the theoretical predictions determined by the wormlike chain model.

Biobased polyurethanes (PUs) are a very dynamic class of polymers that are gaining increased attention in both academic and industrial communities. Hence, Carrico et al. proposed castor oil and glycerol based formulations for the synthesis of thermal insulation PU foams [20]. Moreover, Peyrton et al. reported an original kinectic study on the synthesis of oleo-based Pus [21]. Other polymers are also very attractive for scientific communities, such as poly(lactic acid) (PLA) and derivatives. Hence, Bensabeh et al. proposed the acrylate functionalization of butyl-lactate for the synthesis of biobased ABA copolymers which demonstrated competitive performance when compared with commercial pressure-sensitive tapes [22]. PLA was also studied for three-dimensional (3D) printing with cellulose nanofibrils by Wang et al. [23]. Additionally, Duchiron et al. proposed a feature article dedicated to the enzymatic synthesis of amino acids endcapped polycaprolactone in order to propose functional polyesters [24]. End of life of polymers are also a real challenge and reversible polymers are becoming more and more attractive. Hence, Durand et al. proposed an original article on bio-based thermo-reversible furan-based polycarbonate networks [25].

This themed issue can be considered as collection of highlights within the field of Natural Polymers and Biobased Polymers which clearly demonstrate the increased interest in this field. We hope that this will inspire researchers to further develop this area and thus contribute to futures more sustainable society.

Funding: This research received no external funding.

Conflicts of Interest: The author declares no conflict of interest.

References

1. Nishida, T.; Satoh, K.; Kamigaito, M. Biobased Polymers via Radical Homopolymerization and Copolymerization of a Series of Terpenoid-Derived Conjugated Dienes with *exo*-Methylene and 6-Membered Ring. *Molecules* **2020**, *25*, 5890. [CrossRef] [PubMed]
2. Couture, G.; Granado, L.; Fanget, F.; Boutevin, B.; Caillol, S. Limonene-Based Epoxy: Anhydride Thermoset Reaction Study. *Molecules* **2018**, *23*, 2739. [CrossRef] [PubMed]
3. Mora, A.-S.; Tayouo, R.; Boutevin, B.; David, G.; Caillol, S. Synthesis of Pluri-Functional Amine Hardeners from Bio-Based Aromatic Aldehydes for Epoxy Amine Thermosets. *Molecules* **2019**, *24*, 3285. [CrossRef] [PubMed]
4. Dalle Vacche, S.; Vitale, A.; Bongiovanni, R. Photocuring of Epoxidized Cardanol for Biobased Composites with Microfibrillated Cellulose. *Molecules* **2019**, *24*, 3858. [CrossRef] [PubMed]
5. Molina-Gutiérrez, S.; Dalle Vacche, S.; Vitale, A.; Ladmiral, V.; Caillol, S.; Bongiovanni, R.; Lacroix-Desmazes, P. Photoinduced Polymerization of Eugenol-Derived Methacrylates. *Molecules* **2020**, *25*, 3444. [CrossRef] [PubMed]
6. Montané, X.; Dinu, R.; Mija, A. Synthesis of Resins Using Epoxies and Humins as Building Blocks: A Mechanistic Study Based on In-Situ FT-IR and NMR Spectroscopies. *Molecules* **2019**, *24*, 4110. [CrossRef] [PubMed]
7. Kohut, A.; Voronov, S.; Demchuk, Z.; Kirianchuk, V.; Kingsley, K.; Shevchuk, O.; Caillol, S.; Voronov, A. Non-Conventional Features of Plant Oil-Based Acrylic Monomers in Emulsion Polymerization. *Molecules* **2020**, *25*, 2990. [CrossRef]
8. Chiaoprakobkij, N.; Suwanmajo, T.; Sanchavanakit, N.; Phisalaphong, M. Curcumin-Loaded Bacterial Cellulose/Alginate/Gelatin as A Multifunctional Biopolymer Composite Film. *Molecules* **2020**, *25*, 3800. [CrossRef]
9. Shirosaki, Y.; Okabayashi, T.; Yasutomi, S. Silane Coupling Agent Modifies the Mechanical Properties of a Chitosan Microfiber. *Molecules* **2020**, *25*, 5292. [CrossRef]
10. Saremi, R.; Borodinov, N.; Laradji, A.M.; Sharma, S.; Luzinov, I.; Minko, S. Adhesion and Stability of Nanocellulose Coatings on Flat Polymer Films and Textiles. *Molecules* **2020**, *25*, 3238. [CrossRef]
11. Bernal-Ballen, A.; Lopez-Garcia, J.; Merchan-Merchan, M.-A.; Lehocky, M. Synthesis and Characterization of a Bioartificial Polymeric System with Potential Antibacterial Activity: Chitosan-Polyvinyl Alcohol-Ampicillin. *Molecules* **2018**, *23*, 3109. [CrossRef] [PubMed]
12. Pineda-Castillo, S.; Bernal-Ballén, A.; Bernal-López, C.; Segura-Puello, H.; Nieto-Mosquera, D.; Villamil-Ballesteros, A.; Muñoz-Forero, D.; Munster, L. Synthesis and Characterization of Poly(Vinyl Alcohol)-Chitosan-Hydroxyapatite Scaffolds: A Promising Alternative for Bone Tissue Regeneration. *Molecules* **2018**, *23*, 2414. [CrossRef]
13. Volant, C.; Gilet, A.; Beddiaf, F.; Collinet-Fressancourt, M.; Falourd, X.; Descamps, N.; Wiatz, V.; Bricout, H.; Tilloy, S.; Monflier, E.; et al. Multiscale Structure of Starches Grafted with Hydrophobic Groups: A New Analytical Strategy. *Molecules* **2020**, *25*, 2827. [CrossRef] [PubMed]
14. Allegretti, C.; Boumezgane, O.; Rossato, L.; Strini, A.; Troquet, J.; Turri, S.; Griffini, G.; D'Arrigo, P. Tuning Lignin Characteristics by Fractionation: A Versatile Approach Based on Solvent Extraction and Membrane-Assisted Ultrafiltration. *Molecules* **2020**, *25*, 2893. [CrossRef] [PubMed]
15. Rebière, J.; Rouilly, A.; Durrieu, V.; Violleau, F. Characterization of Non-Derivatized Cellulose Samples by Size Exclusion Chromatography in Tetrabutylammonium Fluoride/Dimethylsulfoxide (TBAF/DMSO). *Molecules* **2017**, *22*, 1985. [CrossRef] [PubMed]
16. Klapiszewski, Ł.; Jamrozik, A.; Strzemiecka, B.; Koltsov, I.; Borek, B.; Matykiewicz, D.; Voelkel, A.; Jesionowski, T. Characteristics of Multifunctional, Eco-Friendly Lignin-Al2O3 Hybrid Fillers and Their Influence on the Properties of Composites for Abrasive Tools. *Molecules* **2017**, *22*, 1920. [CrossRef] [PubMed]
17. Ghaeli, I.; De Moraes, M.A.; Beppu, M.M.; Lewandowska, K.; Sionkowska, A.; Ferreira-da-Silva, F.; Ferraz, M.P.; Monteiro, F.J. Phase Behaviour and Miscibility Studies of Collagen/Silk Fibroin Macromolecular System in Dilute Solutions and Solid State. *Molecules* **2017**, *22*, 1368. [CrossRef]

18. Ghica, M.V.; Albu Kaya, M.G.; Dinu-Pîrvu, C.-E.; Lupuleasa, D.; Udeanu, D.I. Development, Optimization and In Vitro/In Vivo Characterization of Collagen-Dextran Spongious Wound Dressings Loaded with Flufenamic Acid. *Molecules* **2017**, *22*, 1552. [CrossRef]

19. Strehle, D.; Mollenkopf, P.; Glaser, M.; Golde, T.; Schuldt, C.; Käs, J.A.; Schnauß, J. Single Actin Bundle Rheology. *Molecules* **2017**, *22*, 1804. [CrossRef]

20. Carriço, C.S.; Fraga, T.; Carvalho, V.E.; Pasa, V.M.D. Polyurethane Foams for Thermal Insulation Uses Produced from Castor Oil and Crude Glycerol Biopolyols. *Molecules* **2017**, *22*, 1091. [CrossRef]

21. Peyrton, J.; Chambaretaud, C.; Avérous, L. New Insight on the Study of the Kinetic of Biobased Polyurethanes Synthesis Based on Oleo-Chemistry. *Molecules* **2019**, *24*, 4332. [CrossRef] [PubMed]

22. Bensabeh, N.; Jiménez-Alesanco, A.; Liblikas, I.; Ronda, J.C.; Cádiz, V.; Galià, M.; Vares, L.; Abián, O.; Lligadas, G. Biosourced All-Acrylic ABA Block Copolymers with Lactic Acid-Based Soft Phase. *Molecules* **2020**, *25*, 5740. [CrossRef] [PubMed]

23. Wang, Q.; Ji, C.; Sun, L.; Sun, J.; Liu, J. Cellulose Nanofibrils Filled Poly(Lactic Acid) Biocomposite Filament for FDM 3D Printing. *Molecules* **2020**, *25*, 2319. [CrossRef] [PubMed]

24. Duchiron, S.W.; Pollet, E.; Givry, S.; Avérous, L. Enzymatic Synthesis of Amino Acids Endcapped Polycaprolactone: A Green Route Towards Functional Polyesters. *Molecules* **2018**, *23*, 290. [CrossRef] [PubMed]

25. Durand, P.-L.; Grau, E.; Cramail, H. Bio-Based Thermo-Reversible Aliphatic Polycarbonate Network. *Molecules* **2020**, *25*, 74. [CrossRef]

Article

Biobased Polymers via Radical Homopolymerization and Copolymerization of a Series of Terpenoid-Derived Conjugated Dienes with *exo*-Methylene and 6-Membered Ring

Takenori Nishida [1], Kotaro Satoh [1,2] and Masami Kamigaito [1,*]

[1]   Department of Molecular and Macromolecular Chemistry, Graduate School of Engineering,
     Nagoya University, Furo-cho, Chikusa-ku, Nagoya 464-8603, Japan;
     nishida@chiral.apchem.nagoya-u.ac.jp (T.N.); satoh@cap.mac.titech.ac.jp (K.S.)
[2]   Department of Chemical Science and Engineering, Tokyo Institute of Technology, School of Materials and
     Chemical Technology, 2-12-1-H120 Ookayama, Meguro-ku, Tokyo 152-8550, Japan
*   Correspondence: kamigait@chembio.nagoya-u.ac.jp; Tel.: +81-52-789-5400

Academic Editor: Silvia Destri
Received: 9 November 2020; Accepted: 9 December 2020; Published: 12 December 2020

Abstract: A series of *exo*-methylene 6-membered ring conjugated dienes, which are directly or indirectly obtained from terpenoids, such as β-phellandrene, carvone, piperitone, and verbenone, were radically polymerized. Although their radical homopolymerizations were very slow, radical copolymerizations proceeded well with various common vinyl monomers, such as methyl acrylate (MA), acrylonitrile (AN), methyl methacrylate (MMA), and styrene (St), resulting in copolymers with comparable incorporation ratios of bio-based cyclic conjugated monomer units ranging from 40 to 60 mol% at a 1:1 feed ratio. The monomer reactivity ratios when using AN as a comonomer were close to 0, whereas those with St were approximately 0.5 to 1, indicating that these diene monomers can be considered electron-rich monomers. Reversible addition fragmentation chain-transfer (RAFT) copolymerizations with MA, AN, MMA, and St were all successful when using *S*-cumyl-*S'*-butyl trithiocarbonate (CBTC) as the RAFT agent resulting in copolymers with controlled molecular weights. The copolymers obtained with AN, MMA, or St showed glass transition temperatures (T_g) similar to those of common vinyl polymers ($T_g \sim 100\ °C$), indicating that biobased cyclic structures were successfully incorporated into commodity polymers without losing good thermal properties.

Keywords: terpenoid; *exo*-methylene; conjugated diene; renewable monomer; biobased polymer; radical polymerization; copolymerization; living radical polymerization; RAFT polymerization; heat-resistant polymer

---

1. Introduction

Polymeric materials, such as plastics, rubbers, and fibers, are indispensable products in our daily life and are produced from a variety of low-molecular-weight molecules as building blocks. Radical polymerization is one of the most effective and widely used techniques to produce a wide variety of polymeric materials from various vinyl monomers due to the high reactivity of radical species and its robustness toward polar functional groups in the monomer substituents as well as aqueous compounds in the reaction mixture [1]. In addition, recent progress in controlled/living or reversible deactivation radical polymerization (RDRP) has dramatically expanded the scope of radical polymerization to synthesize precision polymers with various well-defined structures that can be used as high-performance materials [1–10].

Sustainable industrial development is now indispensable in view of environmental issues, such as global warming, depletion of fossil resources, and plastic waste. One of the promising solutions to these issues is to use renewable resources, in particular plant-derived molecules, as an alternative to the petroleum-derived monomers for synthetic polymers [11–21]. Moreover, the unique structures of naturally occurring molecules can bestow biobased synthetic polymers with characteristic features that are difficult to attain using simpler petroleum-derived molecules, resulting in novel high-performance materials.

A number of vinyl compounds are produced in nature and are found in a family of abundant natural compounds, such as terpenoids and phenylpropanoids [22–25]. As we have proposed, these biobased vinyl compounds can be classified into similar vinyl monomer families, such as nonpolar olefins, styrenes, and acrylic monomers, to those of petroleum-derived monomers, depending on their substituents and structures [26–29]. For example, renewable polar monomers with conjugated carbonyl groups, such as tulipalin A and itaconic acid, are considered to be acrylic monomers [30–41], whereas phenylpropanoids and their derivatives, such as anethole and vinylguaiacol, are considered to be styrenes [42–48]. These conjugated monomers are generally reactive in radical polymerizations due to the C=C bonds conjugated with carbonyl or phenyl groups, resulting in relatively stable propagating radical species.

In contrast, among various terpenes [22–25], the most abundant terpenes are pinene and limonene, which are unconjugated nonpolar olefins without carbonyl or aromatic groups and have difficulty in undergoing radical homopolymerization [49–55]. However, these unconjugated terpenes are radically copolymerized with electron-deficient polar monomers, such as acrylates and maleimides, resulting in biobased copolymers with a relatively high terpene incorporation ratio of up to 50 mol%. In contrast, a linear conjugated diene, such as myrcene, is another nonpolar olefin terpene and is radically homopolymerized, resulting in rubbery materials [56–62]. However, linear conjugated dienes undergo concurrent Diels-Alder reactions with electron-deficient vinyl monomers via their *s-cis* forms during radical copolymerization.

Another nonpolar conjugated diene observed in terpenoids is a *transoid* cyclic conjugated diene with a reactive *exo*-methylene group, such as β-phellandrene (β-Phe) (Scheme 1). As we briefly reported in a patent [63], β-Phe possesses high reactivity in cationic polymerization resulting in heat-resistant cycloolefin polymers after subsequent hydrogenation. More recently, we synthesized another *exo*-methylene 6-membered ring conjugated diene (HCvD: hydrogenated carvone-derived diene) from carvone, which is an abundant naturally occurring α,β-unsaturated carbonyl compound, via simple chemical transformations consisting of hydrogenation and Wittig reactions [64]. We have also found that HCvD is readily polymerized in cationic polymerization, resulting in alicyclic hydrocarbon polymers with good thermal properties. In addition to HCvD, other *exo*-methylene *transoid* 6-membered ring conjugated dienes (PtD and VnD) have been prepared from a series of naturally occurring α,β-unsaturated carbonyl compounds, i.e., piperitone and verbenone, respectively, via the Wittig reaction [65].

The *exo*-methylene-conjugated dienes can be radically homo- and copolymerized with petroleum-derived common vinyl monomers. However, there have been no comprehensive studies on radical homo- and copolymerization of these terpenoid-derived *exo*-methylene 6-membered ring conjugated dienes except for a few results on the homo- and copolymerizations with some specific monomers, such as styrene, acrylonitrile, and *N*-phenylmaleimide [65–67]. These results suggest that some of the *exo*-methylene 6-membered ring conjugated dienes do not seem highly reactive in radical homopolymerization and that they can be copolymerized with electron-deficient monomers. Similar results were reported for 3-methylenecyclopentene, which is an *exo*-methylene 5-membered ring conjugated diene obtained from the ring-closing metathesis of myrcene [68,69].

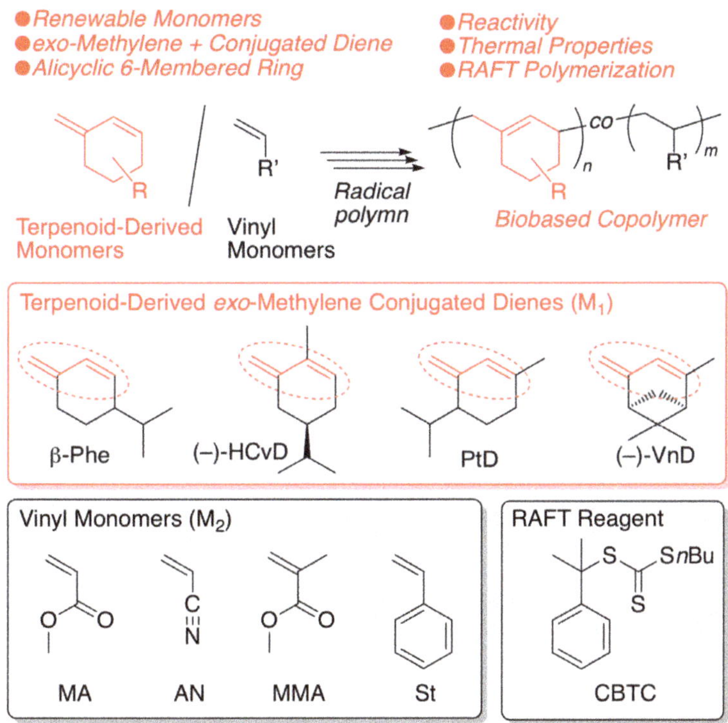

Scheme 1. Radical homo- and copolymerization of a series of terpenoid-derived *exo*-methylene 6-membered-ring conjugated dienes.

In this work, we fully examined radical homo- and copolymerizations of a series of terpenoid-derived *exo*-methylene 6-membered ring conjugated dienes (β-Phe, HCvD, PtD, and VnD) using various common vinyl monomers, including methyl acrylate (MA), acrylonitrile (AN), methyl methacrylate (MMA), and styrene (St), as comonomers to evaluate their reactivity in racial polymerization and the thermal properties of the resulting biobased polymers as heat-resistant materials with unique alicyclic skeletons originating from naturally occurring terpenoid molecules.

2. Results and Discussion

2.1. Radical Homopolymerization

Radical homopolymerizations of a series of terpenoid-derived *exo*-methylene 6-memered ring conjugated dienes, β-Phe, (−)-HCvD, PtD, and (−)-VnD, which were obtained as previously reported [63,64], were examined in toluene at 60 and 100 °C using 2,2′-azobisisobutyronitirle (AIBN) and 2,2′-azobis(*N*-butyl-2-methylpropionamide) (VAm-110), respectively, as radical initiators (Table 1). The polymerizations of all monomers at 60 °C were totally sluggish, resulting in a trace amount of polymers with number-average molecular weights (M_n) less than 10^4. Upon increasing the temperature to 100 °C, although the polymerizations were still slow, they visibly proceeded to result in soluble polymers with relatively high molecular weights ($M_n > 10^4$) except for (−)-VnD, which resulted in only low molecular weight oligomers ($M_n < 10^3$) at 60 and 100 °C. All size-exclusion chromatography (SEC) curves of the obtained polymers showed unimodal shapes (Figure S1), suggesting that cross-linking reactions with the unconjugated olefins formed in the polymer chains rarely occurred.

Table 1. Radical polymerization of a series of terpenoid-derived *exo*-methylene 6-membered ring conjugated dienes [a].

Entry	Monomer	Temp. (°C)	Conv. (%) [b]	M_n(SEC) [c]	M_w/M_n [c]	1,4-/1,2- [d]	T_g (°C) [e]
1	β-Phe	60	2	3700	2.82	93/7	n.d. [f]
2	β-Phe	100	27	17,700	1.79	83/17	66
3	HCvD	60	0	10,800	3.03	>99/0	n.d. [f]
4	HCvD	100	32	34,400	1.89	>99/0	105
5	PtD	60	0	3200	2.47	>99/0	n.d. [f]
6	PtD	100	29	10,700	1.69	>99/0	113
7	VnD	60	0	240	1.10	n.d.	n.d. [f]
8	VnD	100	2	300	1.15	n.d.	n.d. [f]

[a] Polymerization condition: $[M]_0/[\text{radical initiator}]_0$ = 5000/30 mM in toluene at 60 (AIBN) or 100 °C (VAm-110), reaction time = 280 h. [b] Determined by ^{1}H NMR of reaction mixture. [c] Determined by size-exclusion chromatography (SEC). [d] Determined by ^{1}H NMR of the obtained polymers. [e] Determined by DSC. [f] Not determined.

The ^{1}H NMR spectra of the obtained polymers showed that the polymerization of (−)-HCvD and PtD proceeded via regioselective 1,4-conjugated addition (Figure 1). There are no olefinic protons at approximately 5–6 ppm for poly((−)-HCvD), whereas no methyl groups attached to olefins were observed at approximately 1.5–2.0 ppm for poly(PtD). Furthermore, for poly(PtD), the peak intensity ratios of methyl protons (0.6–1.0 ppm), which are attached to saturated carbons, to olefin (5.1–5.3 ppm) protons were 8.9:1, indicating not only regioselective 1,4-conjugated addition but also no cross-linking reactions. However, the 1,4-regioselectivity was not complete for poly(β-Phe), which consisted of 83/17 of 1,4-/1,2-units as estimated from the peak intensity ratios of the olefinic to alkyl protons, in contrast to the complete 1,4-regioselective polymerization of β-Phe via the cationic mechanism as briefly reported previously [63].

The stereoregularity is another interesting issue of the obtained polymers. The poly((−)-HCvD) obtained via radical polymerization of the chiral monomer was easily soluble in common organic solvents, such as hexane and tetrahydrofuran, whereas an insoluble polymer in these solvents was obtained via the regioselective and stereospecific cationic polymerization of the same chiral monomer [64]. The ^{13}C NMR spectrum of poly((−)-HCvD) obtained via the radical process showed a larger number of peaks than that obtained in the cationic process, in which the chiral center attached to the isopropyl group can dictate the stereospecificity at low temperature (Figure S2).

Thus, both regioselectivity and stereospecificity in the polymerizations of a series of terpenoid-derived *exo*-methylene conjugated dienes were influenced by the propagating species, among which the radical species resulted in lower selectivities due to the free radical propagating species as well as the higher reaction temperature.

The thermal properties of poly(β-Phe), poly((−)-HCvD), and poly(PtD) obtained in radical polymerization at 100 °C were evaluated by differential scanning calorimetry (DSC) (Figure S3). The glass transition temperatures (T_g) were observed for all polymers. The T_gs of poly((−)-HCvD) and poly(PtD) were approximately 110 °C due to the cyclic structures incorporated into the main chain via regioselective 1,4-conjugation, whereas the T_g of poly(β-Phe) was 66 °C, which was lower than that of poly(β-Phe) (T_g = 87 °C) obtained in cationic polymerization [63] due to the decreased 1,4-regioselectivity in radical polymerization. In addition, the poly((−)-HCvD) obtained in radical polymerization showed a slightly lower T_g than that obtained in the cationic polymerization of (−)-HCvD (T_g = 105 vs. 114 °C) and showed no melting peak due to the lack of stereospecificity in radical polymerization [64].

These results show that a series of the *exo*-methylene 6-membered ring conjugated dienes are slowly homopolymerized by radical species and that the regioselectivity and stereospecificity are lower for some monomers than those in cationic polymerization.

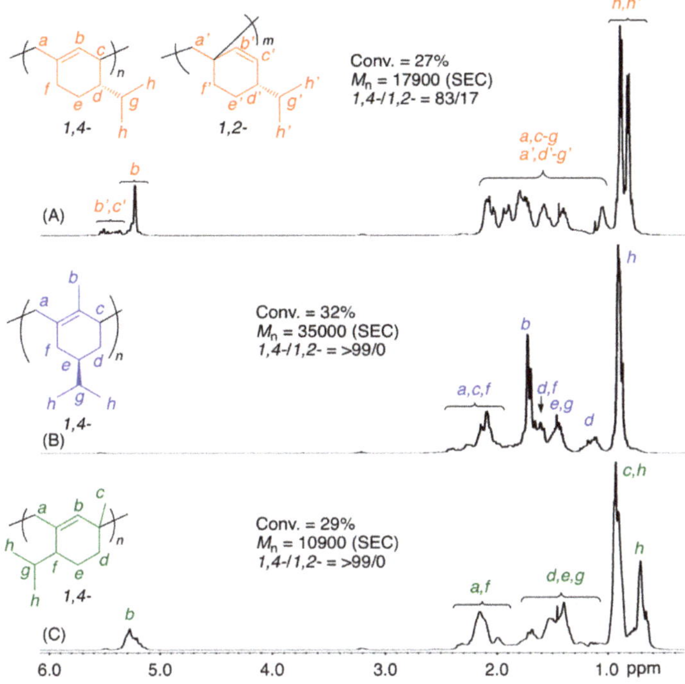

Figure 1. ^{1}H NMR spectra (in CDCl$_3$ at 55 °C) of poly(β-Phe) (A), poly((–)-HCvD) (B), and poly(PtD) (C) obtained in the radical polymerization of the *exo*-methylene conjugated dienes: [M]$_0$/[VAm-110]$_0$ = 5000/30 mM in toluene at 100 °C.

2.2. Radical Copolymerization with Various Common Vinyl Monomers

A series of terpenoid-derived *exo*-methylene 6-membered ring conjugated dienes were then copolymerized with various common vinyl monomers, such as MA, AN, MMA, and St, at a 1:1 feed ratio in toluene at 60 °C using AIBN (Figure S4 and Table S1). Radical copolymerizations occurred in almost all cases except for copolymerization with (–)-VnD, in which conversions of both (–)-VnD and comonomers were very low and the obtained products were only oligomers ($M_n \leq 1 \times 10^3$). Therefore, (–)-VnD is only slightly reactive in radical homo- and copolymerization, probably because the bulky structure with a bicyclic structure and a methyl group at the 4-position makes propagation very difficult in the radical process. Although the other *exo*-methylene-conjugated dienes were all copolymerized, the reactions at 60 °C were generally slow and resulted in polymers with relatively low molecular weights ($M_n = 1 \times 10^3 - 2 \times 10^4$).

Therefore, the radical copolymerizations of β-Phe, (–)-HCvD, and PtD with MA, AN, MMA, and St at a 1:1 feed ratio were investigated at 100 °C in toluene using VAm-110 as the radical initiator (Table 2 and Figure 2 and Figure S5). The polymerization rates, monomer conversion, and molecular weight of the resulting copolymers ($M_n = 6 \times 10^3 - 4 \times 10^4$) were drastically improved, indicating that a higher reaction temperature is favorable for not only homo- but also copolymerization of a series of the *exo*-methylene 6-membered ring conjugated dienes.

Table 2. Radical copolymerization of terpenoid-derived *exo*-methylene 6-membered ring conjugated dienes (M_1) and various common vinyl monomers (M_2) in toluene at 100 °C [a].

Entry	M_1	M_2	Time (h)	Conv.(%) [b] M_1/M_2	M_n (SEC) [c]	1,4-/1,2- [d]	M_1/M_2 (NMR) [d]	M_1/M_2 (Calcd) [e]	r_1 [f]	r_2 [f]
1	β-Phe	MA	90	38/27	7900	83/17	61/39	58/42	0.66	0.04
2	HCvD	MA	20	44/36	26,000	95/5	52/48	55/45	0.14	0.08
3	PtD	MA	20	43/35	23,800	96/4	51/49	55/45	0.11	0.08
4	β-Phe	AN	40	51/45	14,600	84/16	52/48	53/47	0.17	0.03
5	HCvD	AN	10	44/50	42,100	95/5	53/47	47/53	0.09	0.03
6	PtD	AN	10	45/50	42,300	99/1	50/50	47/53	0.02	0.05
7	β-Phe	MMA	150	36/31	7800	89/11	60/40	54/46	0.68	0.09
8	HCvD	MMA	50	44/51	12,900	95/5	47/53	46/54	0.17	0.31
9	PtD	MMA	120	55/62	13,500	>99/0	46/54	47/53	0.19	0.25
10	β-Phe	St	120	38/29	6700	85/15	60/40	57/43	1.53	0.43
11	HCvD	St	100	40/54	19,000	95/5	43/57	43/57	0.40	1.09
12	PtD	St	50	29/35	10,900	>99/0	41/59	45/55	0.48	0.96

[a] Polymerization condition: $[M_1]_0/[M_2]_0/[\text{VAm-110}]_0$ = 1500/1500/30 mM in toluene at 100 °C. [b] Determined by ^{1}H NMR of reaction mixture. [c] Determined by SEC. [d] Determined by ^{1}H NMR of the obtained polymers. [e] Determined by the monomer feed ratio and monomer conversion. [f] Determined by Kelen–Tüdõs method for the copolymerizations at varying monomer feed ratios.

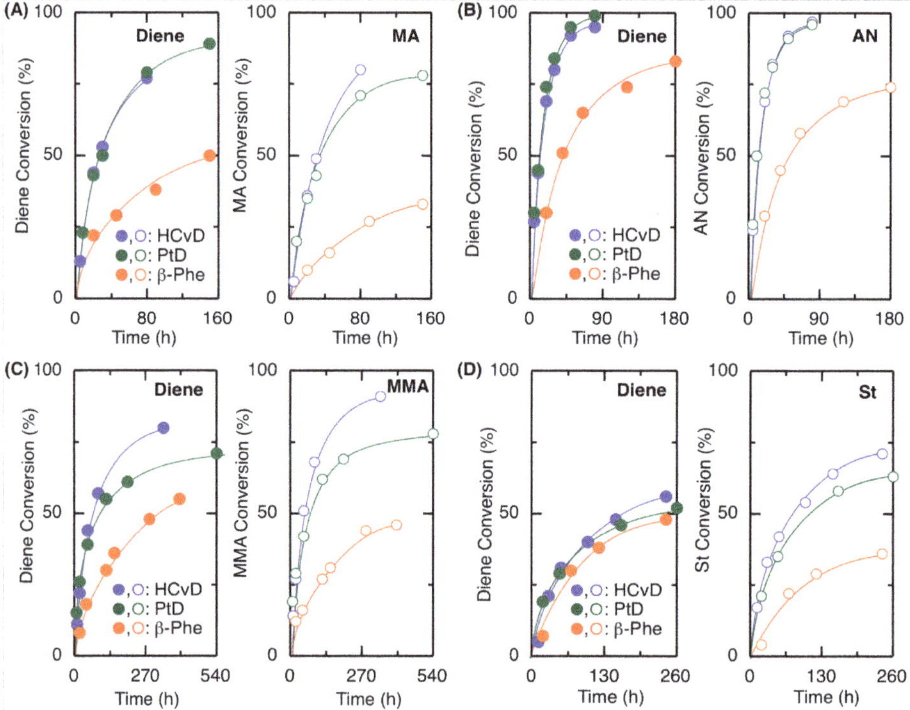

Figure 2. Time-conversion curves for the radical copolymerization of *exo*-methylene 6-memered-ring conjugated dienes with MA (**A**), AN (**B**), MMA (**C**), or St (**D**) as a comonomer: $[\text{diene}]_0/[\text{comonomer}]_0/[\text{VAm-110}]_0$ = 1500/1500/30 in toluene at 100 °C.

In general, electron-deficient comonomers were suitable for copolymerization, where the reaction rates, conversions, and molecular weights increased in the following order: AN > MA > MMA > St, suggesting that these *exo*-methylene cyclic conjugated dienes are electron-rich monomers. Among the cyclic conjugated dienes, (−)-HCvD and PtD showed similar reactivities, resulting in polymers with similar molecular weights in all cases, while β-Phe resulted in slower polymerizations and lower molecular weights. However, the incorporation ratio of β-Phe was generally greater than that of

(−)-HCvD and PtD for any comonomer (Table 2). These results suggest that relatively high incorporation of β-Phe leads to a high probability of the β-Phe-derived propagating radical species, by which the propagation is slow to result in copolymers with relatively low molecular weights. More details on the monomer reactivity ratio will be studied and discussed in the next section.

All the obtained copolymers were analyzed by ¹H NMR to clarify the copolymer structures, particularly regarding the incorporation ratio of comonomers and the regiospecificity of the diene monomers during the copolymerizations (Figure 3, Figures S6–S8). For example, Figure 3 shows spectra of a series of copolymers obtained with β-Phe, (−)-HCvD, or PtD and MA. In addition to the characteristic signals of the conjugated diene units, all the spectra showed additional signals of MA units, including not only main-chain methylene and methine protons but also methyl ester protons (*k*). In particular, olefinic protons (*b*) were observed at approximately 5–6 ppm for the copolymers of β-Phe or PtD (Figure 3A,C), although the peaks were slightly broader than those of the homopolymers of β-Phe or PtD (Figure 1A,C). In addition, almost no olefinic protons were observed for the copolymers of (−)-HCvD (Figure 3B), indicating that the high 1,4-conjugated addition polymerization also proceeded in the copolymerization. For the other various copolymers with AN, MMA, or St (Figures S6–S8), β-Phe or PtD olefinic protons were similarly observed at approximately 4–6 ppm, although the chemical shifts were dependent on the comonomers. Furthermore, for the copolymers of (−)-HCvD with all comonomers, almost no olefinic protons were observed.

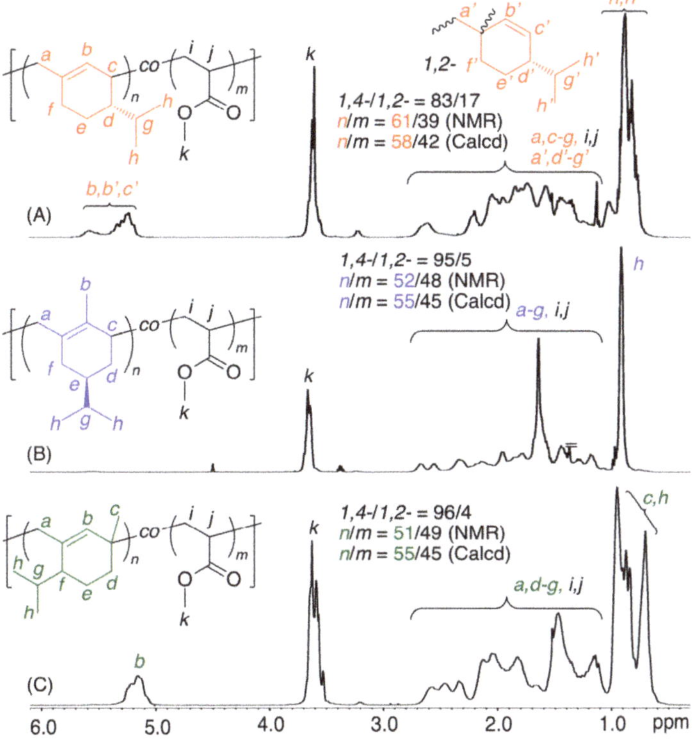

Figure 3. ¹H NMR spectra (in CDCl₃ at 55 °C) of copolymers obtained in the radical copolymerization of β-Phe (**A**), (−)-HCvD (**B**), or PtD (**C**) with MA: [diene]₀/[MA]₀/[VAm-110]₀ = 1500/1500/30 mM in toluene at 100 °C.

The incorporation ratios of the conjugated dienes (M_1) and common vinyl monomers (M_2) were calculated by the peak intensity ratios of the diene and vinyl monomer units in the ^1H NMR spectra. The incorporation ratios (M_1/M_2(NMR)) obtained by NMR of the resulting copolymers were close to the calculated values (M_1/M_2(Calcd)) based on the feed ratio and monomer conversions, indicating that the consumed monomers were efficiently incorporated into the copolymers. Furthermore, the regioselectivities of the diene units were calculated by the olefinic protons and other protons for all copolymers. The regioselectivities in the copolymers were basically the same as those for the homopolymers. Namely, high 1,4-selectivity was attained for (−)-HCvD and PtD, whereas the 1,4-selectivity for β-Phe was approximately 80%.

Thus, a series of terpenoid-derived *exo*-methylene 6-membered ring conjugated dienes are efficiently copolymerized with various common vinyl monomers, including MA, AN, MMA, and St, via mainly 1,4-conjugated additions resulting in copolymers with biobased monomer incorporation ratios ranging from 40 to 60 mol% at a 1:1 feed ratio.

2.3. Monomer Reactivity Ratio

To further clarify the copolymerizability of a series of the *exo*-methylene 6-membered ring conjugated dienes in radical copolymerizations, the monomer reactivity ratios were determined by analyzing the comonomer compositions of the copolymers obtained in the initial stages (total conversion < 10%) of the copolymerizations at various monomer feed ratios ($[M_1]_0 + [M_2]_0 = 5000$ mM, $[VAm-110]_0 = 30$ mM in toluene at 100 °C). Figure 4 shows the copolymer composition curves, in which the diene unit content in the resulting copolymers was plotted against the diene content in the monomer feeds. The data were analyzed using the terminal model and were fitted well with the solid lines obtained using the Kelen–Tüdõs method. The obtained r_1 and r_2 values are summarized in Table 2.

In general, all the plots and fitted curves for HCvD and PtD were close to each other, whereas those for β-Phe were located above the others, indicating that the incorporation of β-Phe units in the resulting copolymers was consistently greater than that of HCvD and PtD at the same feeds. Correspondingly, the r_1 values of β-Phe were consistently larger than those of HCvD and PtD. In addition, the r_1 values of HCvD and PtD were relatively close. These results indicate that HCvD and PtD, which both undergo selective 1,4-regioselective propagation, possess similar reactivities, while β-Phe with a decreased 1,4-regioselectivity shows a higher copolymerizability than HCvD and PtD.

When focusing on the copolymerizations with electron-deficient comonomers (MA, AN, and MMA), both r_1 and r_2 were less than 1, and in particular, those for HCvD and PtD with AN were less than 0.1, indicating that alternating copolymerization predominantly occurs, as also suggested by the almost constant plot for the various monomer feed ratios (Figure 4B). This is because the *exo*-methylene-conjugated diene can be considered an electron-rich monomer, whereas AN is the most electron-deficient monomer among the comonomers. For the copolymerizations with St, the sequence of the copolymer is more random because both of the r values are relatively close to 1, although both monomers are slightly more reactive than the other monomers.

When the r values of the *exo*-methylene cyclic conjugated dienes are compared with those of a linear conjugated diene, such as butadiene (Bd), they are partially different. The r_1 values, using Bd as M_1 and the other common vinyl monomers as M_2, are 1.09 (MA), 0.36 (AN), 0.75 (MMA), and 1.44 (St) [70], which are greater than those of the *exo*-methylene cyclic conjugated dienes due to the low homopolymerizability of the *exo*-methylene cyclic dienes most likely caused by the steric hindrance around the *exo*-methylene group. In contrast, the r_2 values for comonomers M_2 with Bd as M_1 are 0.07 (MA), 0.05 (AN), 0.25 (MMA), and 0.84 (St) [70], which are similar to those obtained in copolymerizations with the *exo*-methylene cyclic conjugated dienes, indicating that the propagating radical species derived from common comonomers can react with cyclic conjugated dienes with reactivity similar to that of the Bd monomer.

Thus, a series of terpenoid-derived *exo*-methylene 6-membered ring conjugated dienes possess sufficiently high copolymerizability with various common vinyl monomers to be efficiently incorporated into the polymer main chains via 1,4-conjugated additions, which results in cyclic structures in the main chain of the copolymers.

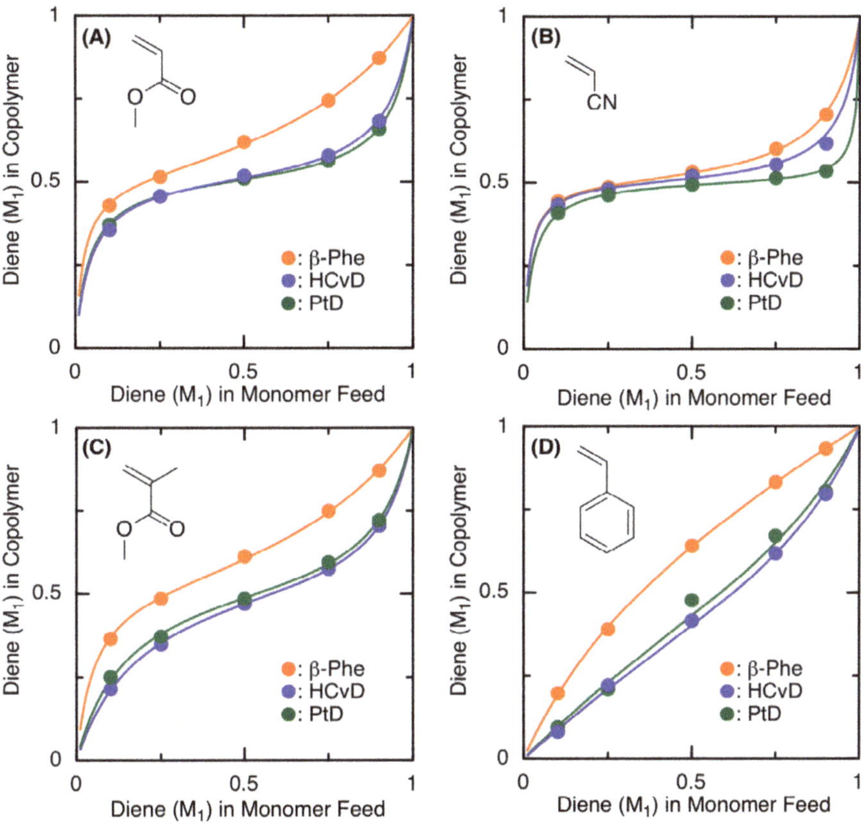

Figure 4. Copolymer composition curves for the radical copolymerization of terpenoid-derived *exo*-methylene 6-membered-ring conjugated dienes as M_1 and MA (**A**), AN (**B**), MMA (**C**), or St (**D**) as M_2 with VAm-110 obtained at varying monomer feed ratios ([M_1]$_0$/[M_2]$_0$ = 1/9, 1/3, 1/1, 3/1, 9/1): ([M_1]$_0$ + [M_2]$_0$)/[VAm-110]$_0$ = 5000/30 mM in toluene at 100 °C. The solid lines were fitted by the Kelen–Tüdős method for the terminal model (see r_1 and r_2 values in Table 2).

2.4. Thermal Properties of Copolymers

The thermal properties of the various obtained copolymers were evaluated by DSC (Figure 5). All the obtained copolymers showed their characteristic T_gs. For the copolymers of MA, the T_g values were much higher than that of the homopolymer of MA ($T_g \sim 10$ °C) [70] due to the incorporated terpenoid-derived alicyclic structure, and the T_g values ranged from 50 to 80 °C depending on the structure of the cyclic conjugated dienes. Furthermore, the T_g values of the copolymers of MMA, AN, and St with (−)-HCvD and PtD were approximately 100 °C, which is similar to the T_g values of the homopolymers, indicating that the biobased units can be introduced into common vinyl polymers without losing their good thermal properties. This could be a unique benefit of the biobased *exo*-methylene 6-membered ring conjugated dienes as comonomers for MMA, AN, and St, in contrast to linear olefins and dienes, which generally have lower T_gs due to the soft hydrocarbon units. Thus, terpenoid-derived alicyclic *exo*-methylene conjugated dienes are promising monomers as novel rigid hydrocarbon units that can be introduced at a high incorporation ratio in radical copolymerization.

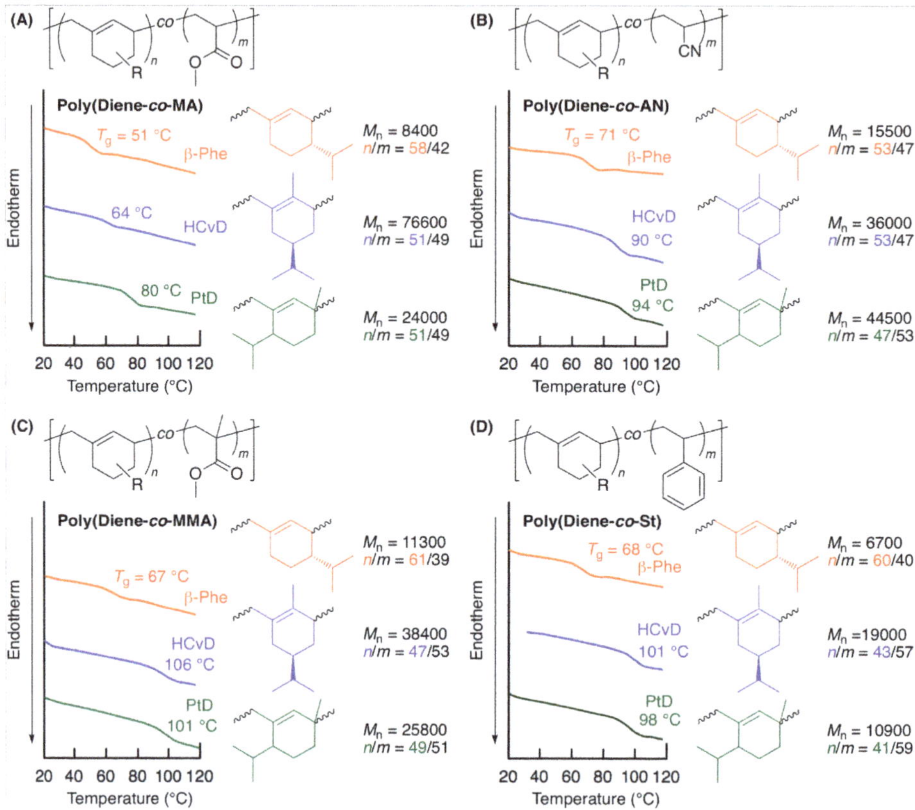

Figure 5. Differential scanning calorimetry (DSC) curves of copolymers obtained in the radical copolymerization of terpenoid-derived *exo*-methylene 6-membered ring conjugated dienes with MA (**A**), AN (**B**), MMA (**C**), and St (**D**).

2.5. Reversible Addition Fragmentation Chain-Transfer (RAFT) Copolymerization of (−)-HCvD and Various Common Vinyl Monomers

To control the copolymerization for precision synthesis of the novel biobased polymers, the RAFT copolymerization of (−)-HCvD and various common vinyl monomers (MA, MMA, AN, and St) was investigated at a 1:1 feed ratio using *S*-cumyl-*S'*-butyl trithiocarbonate (CBTC) as the RAFT agent and VAm-110 as the radical initiator in toluene at 100 °C (Figure 6 and Figure S9). In all cases, the copolymerizations proceeded smoothly even in the presence of the RAFT agent at conversion ratios similar to those in radical copolymerization without the RAFT agent (Figure S9).

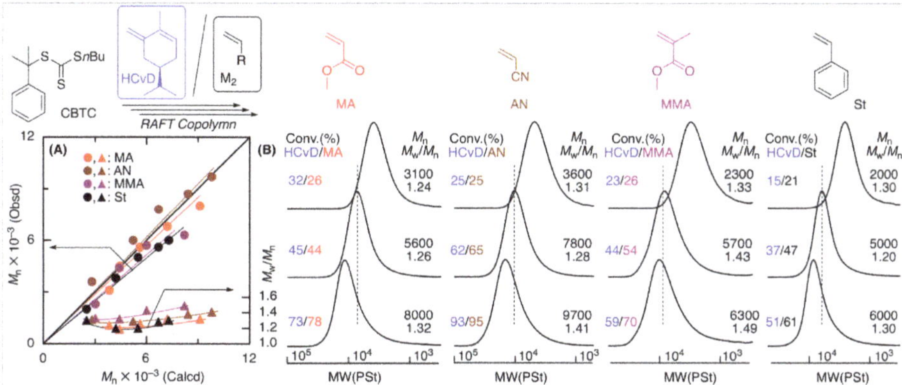

Figure 6. (**A**) M_n, M_w/M_n, and (**B**) SEC curves of the copolymers obtained in the RAFT copolymerization of (−)-HCvD with various common vinyl monomers (M_2): [(−)-HCvD]$_0$/[M_2]$_0$/[CBTC]$_0$/[VAm-110]$_0$ = 1500/1500/30/10 mM in toluene at 100 °C.

The M_n values of the resulting copolymers increased in direct proportion to total monomer conversion in all cases and were close to the calculated values assuming that one molecule of CBTC generates one polymer chain (Figure 6A). In addition, as the polymerization proceeded, the SEC curves shifted to higher molecular weights while maintaining relatively narrow molecular weight distributions (M_w/M_n = 1.2–1.5) (Figure 6B).

The ^{1}H NMR spectra (Figure S10) of the copolymers obtained in the RAFT copolymerization showed similar large absorptions originating from the terpenoid-based cyclic conjugated dienes and common vinyl monomer units and small signals of both the α- and ω-terminals derived from the RAFT agent, for which signals at 7.3–7.4 and 3.3–3.4 ppm can be attributed to the aromatic and methylene protons originating from CBTC, respectively. The incorporations of (−)-HCvD units into the copolymers were similarly calculated by the peak intensity ratios and were almost the same as those for the copolymers obtained in radical copolymerization without the RAFT agent. The M_n values calculated from the peak intensity ratios of the α- and ω-RAFT terminals to the repeating units were close to both M_n(SEC) and M_n(Calcd), suggesting that one CBTC molecule generates one polymer chain. These results indicate that CBTC is suitable for controlling the RAFT copolymerization of (−)-HCvD and various common vinyl monomers to produce copolymers with controlled molecular weights and chain-end groups, which can be used for synthesis of the novel biobased functional polymers.

3. Materials and Methods

3.1. Materials

Toluene (KANTO, >99.5%; H_2O < 10 ppm) and tetrahydrofuran (THF) (KANTO, >99.5%; H_2O < 0.001%) was dried and deoxygenized by passage through Glass Contour Solvent Systems columns before use. (Methyl)triphenylphosphonium bromide (KANTO, >98%) and potassium *tert*-butoxide (Tokyo Chemical Industry, >97%) were used as received. Piperitone (mixture of enantiomers, predominantly (*R*)-form, Tokyo Chemical Industry, >94%), angelica seed oil (Aldrich, containing approximately 80% β-phellandrene), methyl acrylate (MA) (Tokyo Chemical Industry, >99%), acrylonitrile (AN) (Tokyo Chemical Industry, >99%), methyl methacrylate (MMA) (Tokyo Chemical Industry, >98%), styrene (St) (Kishida, 99.5%), and 1,2,3,4-tetrahydronaphthalene (Wako, 97%) were distilled under reduced pressure before use. (−)-Verbenone (Tokyo Chemical Industry, >95%) was purified by column chromatography on silica gel with *n*-hexane/ethyl acetate (6/4) as the eluent to result in the pure compound (>99%). *N*-Cyclohexylmaleimide (Aldrich, 97%) was purified by recrystallization from toluene. 2,2′-Azobisisobutyronitirle (AIBN) (Kishida, >99%) and

2,2′-azobis(*N*-butyl-2-methylpropionamide) (VAm-110) (Wako, >95%) were purified by recrystallization from methanol and hexane. (−)-HCvD [63] and *S*-cumyl-*S*′-butyl trithiocarbonate (CBTC) [71] were synthesized according to the literature.

3.2. Purification of β-Phellandrene (β-Phe)

β-Phellandrene (β-Phe) was obtained from angelica seed oil. *N*-Cyclohexylmaleimide (24.3 g, 0.14 mol) was added to distilled angelica seed oil (75.9 g), which contained β-Phe (92%) and a small amount of linear and *cisoid*-conjugated dienes, such as myrcene and α-phellandrene as impurities, and the linear and *cisoid* compounds were removed using the Diels-Alder reaction. The reaction mixture was stirred at 90 °C for 1 h. After cooling the reaction mixture to 0 °C for quenching, the crude product was purified by column chromatography on silica gel with *n*-hexane as the eluent to yield β-Phe with a higher purity (27 g, 35% yield, 99% purity). Further purification was conducted by distillation to produce a colorless β-Phe oil (14 g, 19% yield, >99% purity, bp 65 °C/2700 Pa. $[\alpha]_D^{25}$ = +27°).

3.3. Synthesis of (−)-VnD

(−)-VnD was prepared by the Wittig reaction of purified (−)-verbenone. (Methyl)triphenylphosphonium bromide (140.2 g, 0.392 mol) was suspended in dry THF (700 mL) at 0 °C in a three-necked 2 L round-bottom flask equipped with three-way stopcocks. Potassium *tert*-butoxide (44.1 g, 0.392 mol) was added resulting in a yellow suspension. After 30 min, (−)-verbenone (40 mL, 0.261 mol) was added to the yellow suspension resulting in an orange product, and the reaction mixture was stirred at room temperature. After 16 h, the conversion of verbenone reached >99% by ^{1}H NMR. Then, water (400 mL) was added to quench the reaction. The organic layer was concentrated by rotary evaporation. Hexane (300 mL) was added to the obtained residue. After the mixture was filtered to remove the precipitated triphenylphosphine oxide, the obtained solution was concentrated. The obtained residue was similarly treated again to yield the crude (−)-VnD product (32.1 g, 82% yield). Further purification was conducted by distillation to produce a colorless oil of pure (−)-VnD (28.9 g, 74% yield, >99% purity, bp 55 °C/1600 Pa, $[\alpha]_D^{25}$ = −44°). ^{1}H NMR (CDCl$_3$, r.t.): δ 0.82 (s, 3H, *CH*$_3$), 1.35 (s, 3H, *CH*$_3$), 1.43-1,46 (d, 1H, *CH*$_2$), 1.78 (s, 3H, CH$_2$=C-CH=C-*CH*$_3$), 2.09–2.12 (t, 1H, CH$_2$=C-CH=C-*CH*), 2.48–2.53 (m, 1H, *CH*$_2$), 2.58–2.61 (s, 1H, CH$_2$=C(*CH*)-CH=C), 4.56 (s, 2H, *CH*$_2$=C-CH=C), 5.77 (s, 1H, CH$_2$=C-*CH*=C). ^{13}C NMR (CDCl$_3$, r.t.): δ 21.9 (*CH*$_3$), 22.9 (CH$_2$=C-CH=C-*CH*$_3$), 26.3 (*CH*$_3$), 35.5 (*CH*$_2$), 43.6 (*C*(CH$_3$)$_2$), 48.3 (CH$_2$=C-CH=C-*CH*), 51.4 (CH$_2$=C(*CH*)-CH=C), 104.9 (*CH*$_2$=C-CH=C), 121.0 (CH$_2$=C-*CH*=C), 147.8 (CH$_2$=*C*-CH=C), 149.7 (CH$_2$=C-CH=*C*). ^{1}H, ^{13}C, COSY, and HMQC NMR spectra are shown in Figures S11–S13.

3.4. Synthesis of PtD

PtD was prepared by the Wittig reaction of distilled piperitone. (Methyl)triphenylphosphonium bromide (131.6 g, 0.369 mol) was suspended in dry THF (700 mL) at 0 °C in a three-necked 2 L round-bottom flask equipped with three-way stopcocks. Potassium *tert*-butoxide (41.3 g, 0.369 mol) was added to the suspension resulting in a yellow suspension. After 30 min, piperitone (40 mL, 0.246 mol) was added to the yellow suspension to produce an orange product, and the reaction mixture was stirred at room temperature. After 16 h, the conversion of piperitone reached >99% by ^{1}H NMR. Then, water (400 mL) was added to quench the reaction. The organic layer was concentrated by rotary evaporation. Hexane (300 mL) was added to the obtained residue. After the mixture was filtered to remove the precipitated triphenylphosphine oxide, the obtained solution was concentrated. The crude product after the Wittig reaction (30.3 g, 82% yield, 81% purity) was purified by column chromatography on silica gel with *n*-hexane as the eluent to yield the product with a higher purity (19.9 g, 36% yield, 94% purity). Further purification was conducted by distillation to produce the colorless PtD oil (16.1 g, 29% yield, 95% purity, bp 70 °C/1600 Pa, $[\alpha]_D^{25}$ = +16°). ^{1}H NMR (CDCl$_3$, r.t.): δ 0.88–0.92 (dd, 6H, cy-CH(*CH*$_3$)$_2$), 1.62–1.71 (m, 2H, CH(*i*Pr)-*CH*$_2$-CH$_2$, cy-*CH*(CH$_3$)$_2$), 1.73 (s, 3H, CH$_2$=C-CH=C-*CH*$_3$), 1.78–1.94 (m, 3H, CH(*i*Pr)-CH$_2$-*CH*$_2$, CH(*i*Pr)-CH$_2$-*CH*$_2$, CH(*i*Pr)-*CH*$_2$-CH$_2$),

2.03–2.19 (m, 1H, CH(*i*Pr)-CH$_2$-C*H*$_2$), 4.63 (s, 1H, *trans*, C*H*$_2$=C-CH=C), 4.74 (s, 1H, *cis*, C*H*$_2$=C-CH=C), 5.85 (s, 1H, CH$_2$=C-C*H*=C). ^{13}C NMR (CDCl$_3$, r.t.): δ 19.9 (cy-CH(CH$_3$)$_2$), 21.9 (cy-CH(CH$_3$)$_2$), 23.6 (CH$_2$=C-CH=C-CH$_3$), 24.7 (CH(*i*Pr)-CH$_2$-CH$_2$), 27.0 (cy-CH(CH$_3$)$_2$), 27.7 (CH(*i*Pr)-CH$_2$-CH$_2$), 46.2 (CH(*i*Pr)-CH$_2$-CH$_2$), 109.4 (CH$_2$=C-CH=C), 125.1 (CH$_2$=C-CH=C), 137.6 (CH$_2$=C-CH=C), 146.4 (CH$_2$=C-CH=C). ^{1}H, ^{13}C, COSY, and HMQC NMR spectra are shown in Figures S11, S14, and S15.

3.5. RAFT Copolymerization

RAFT copolymerization was carried out using the syringe technique under dry nitrogen in sealed glass tubes. A typical example for (−)-HCvD and MA copolymerization with CBTC and VAm-110 in toluene is given below. (−)-HCvD (0.60 mL, 3.3 mmol), MA (0.30 mL, 3.3 mmol), CBTC (0.13 mL, 508 mM solution in toluene, 0.066 mmol), and VAm-110 (1.19 mL of 18.6 mM solution in toluene, 0.022 mmol) were placed in a baked 25 mL graduated Schlenk flask equipped with a three-way stopcock at room temperature. The total volume of the reaction mixture was 2.2 mL. Immediately after mixing, aliquots (0.4 mL each) of the solution were distributed via syringe into baked glass tubes, which were then sealed by flame in a nitrogen atmosphere. The tubes were immersed in a thermostatic oil bath at 100 °C. At predetermined intervals, the polymerization was terminated by cooling the reaction mixtures to −78 °C. The monomer conversion was determined from the concentration of residual monomer measured by ^{1}H NMR with toluene as an internal standard (20 h, 32% for (−)-HCvD and 26% for MA). The quenched reaction solutions were evaporated to dryness to give the product copolymer (M_n = 3100, M_w/M_n = 1.24).

3.6. Measurements

The ^{1}H NMR spectra for the monomer conversion and product polymer were recorded on a JEOL ECS-400 spectrometer operating at 400 MHz. The number-average molecular weight (M_n) and molecular weight distribution (M_w/M_n) of the product polymer were determined by SEC in THF at 40 °C on two polystyrene gel columns (Tosoh Multipore H$_{xL}$-M (7.8 mm i.d. × 30 cm) × 2; flow rate: 1.0 mL/min) connected to a JASCO PU-2080 precision pump and a JASCO RI-2031 detector. The columns were calibrated with standard polystyrene samples (Agilent Technologies; M_p = 580-3,242,000, M_w/M_n = 1.02–1.23). The glass transition temperature (T_g) of the polymers was recorded on a Q200 differential scanning calorimeter (TA Instruments, Inc.). Samples were first heated to 150 °C at 10 °C/min, equilibrated at this temperature for 10 min, and cooled to 0 °C at 5 °C/min. After being held at this temperature for 5 min, the samples were then reheated to 200 °C at 10 °C/min. All T_g values were obtained from the second scan after removing the thermal history.

4. Conclusions

In conclusion, radical polymerizations of a series of terpenoid-derived *exo*-methylene 6-membered ring conjugated dienes were investigated with the aim of developing novel biobased polymeric materials. Although their reactivities for homopolymerization in radical processes were lower than those in cationic processes, copolymerizations with various common vinyl monomers efficiently occurred, resulting in unique biobased copolymers. The substituents on the 6-membered ring of *exo*-methylene conjugated dienes affected the regiospecificity and monomer reactivity. The obtained copolymers contained relatively high biobased units, ranging from 40 to 60 mol% even at a 1:1 feed ratio, and the glass transition temperatures were comparable to those of the homopolymers obtained from common vinyl monomers. In addition, biobased units can be incorporated into copolymers in a controlled fashion via the RAFT process. We believe that these *exo*-methylene-conjugated dienes can be new building blocks to synthesize copolymers with terpenoid-derived alicyclic structures, which serve as hydrocarbon units with good thermal properties.

Supplementary Materials: The following are available online Figure S1. SEC curves of the homopolymers obtained in the radical polymerization of β-Phe, (–)-HCvD, and PtD: $[M]_0/[VAm-110]_0$ = 5000/30 mM in toluene at 100 °C; Figure S2. ^{13}C NMR spectra (in $C_2D_2Cl_4$ at 100 °C) of poly((–)-HCvD) obtained in the cationic (A) and radical (B) polymerization: $[(-)-HCvD]_0/[CEVE-HCl]_0/[SnCl_4]_0/[nBu_4NCl]_0$ = 100/1.0/5.0/4.0 mM in toluene/CH_2Cl_2 (50/50 vol%) at –78 °C (M_n(Calcd) = 15,200) (A) or $[(-)-HCvD]_0/[VAm-110]_0$ = 5000/30 mM in toluene at 100 °C (B); Figure S3. Differential scanning calorimetry (DSC) curves of poly(β-Phe), poly((–)-HCvD), and poly(PtD) obtained in the radical polymerization: $[M]_0/[VAm-110]_0$ = 5000/30 mM in toluene at 100 °C; Figure S4. Time-conversion curves for the radical copolymerization of terpenoid-derived *exo*-methylene 6-membered ring conjugated dienes with MA (A), AN (B), MMA (C), and St (D) as a comonomer: $[diene]_0/[comonomer]_0/[AIBN]_0$ = 1500/1500/30 mM in toluene at 60 °C; Figure S5. SEC curves of the copolymers obtained in the radical copolymerization of terpenoid-derived *exo*-methylene 6-membered-ring conjugated dienes (M_1) with various common vinyl monomers (M_2): $[M_1]_0/[M_2]_0/[VAm-110]_0$ = 1500/1500/30 mM in toluene at 100 °C; Figure S6. ^{1}H NMR spectra (in $CDCl_3$ at 55 °C) of copolymers obtained in the radical copolymerization of β-Phe (A), (–)-HCvD (B), or PtD (C) with AN: $[diene]_0/[AN]_0/[VAm-110]_0$ = 1500/1500/30 mM in toluene at 100 °C; Figure S7. ^{1}H NMR spectra (in $CDCl_3$ at 55 °C) of copolymers obtained in the radical copolymerization of β-Phe (A), (–)-HCvD (B), or PtD (C) with MMA: $[diene]_0/[MMA]_0/[VAm-110]_0$ = 1500/1500/30 mM in toluene at 100 °C; Figure S8. ^{1}H NMR spectra (in $CDCl_3$ at 55 °C) of copolymers obtained in the radical copolymerization of β-Phe (A), (–)-HCvD (B), or PtD (C) with St: $[diene]_0/[St]_0/[VAm-110]_0$ = 1500/1500/30 mM in toluene at 100 °C; Figure S9. Time-conversion curves for the RAFT copolymerization of (–)-HCvD with MA (A), AN (B), MMA (C), and St (D) as a comonomer: $[(-)-HCvD]_0/[comonomer]_0/[CBTC]_0/[VAm-110]_0$ = 1500/1500/30/10 mM in toluene at 100 °C; Figure S10. ^{1}H NMR spectra (in $CDCl_3$ at 55 °C) of copolymers obtained in the RAFT copolymerization of (–)-HCvD with MA (A), AN (B), MMA (C), or St (D) as a comonomer: $[(-)-HCvD]_0/[comonomer]_0/[CBTC]_0/[VAm-110]_0$ = 1500/1500/30/10 mM in toluene at 100 °C; Figure S11. ^{1}H NMR spectra of (–)-VnD (A) and PtD (B) in $CDCl_3$ at 25 °C; Figure S12. ^{13}C NMR and DEPT spectra of (–)-VnD in $CDCl_3$ at 25 °C; Figure S13. ^{1}H-^{1}H COSY and HMQC spectra of (–)-VnD in $CDCl_3$ at 25 °C; Figure S14. ^{13}C NMR and DEPT spectra of PtD in $CDCl_3$ at 25 °C; Figure S15. ^{1}H-^{1}H COSY and HMQC spectra of PtD in $CDCl_3$ at 25 °C; Table S1. Radical copolymerization of terpenoid-derived exo-methylene 6-membered ring conjugated dienes (M_1) and various common vinyl monomers (M_2) in toluene at 60 °C.

Author Contributions: Conceptualization, K.S. and M.K.; methodology, T.N., K.S. and M.K.; investigation, T.N., K.S. and M.K., writing—original draft preparation, T.N.; writing—review and editing, K.S. and M.K.; supervision, K.S. and M.K.; project administration, M.K.; funding acquisition, M.K. All authors have read and agreed to the published version of the manuscript.

Funding: This work was supported by JSPS KAKENHI Grant Number JP20H04809 in Hybrid Catalysis for Enabling Molecular Synthesis on Demand, Japan-France Research Cooperative Program between JSPS and CNRS (Grant number JPJSBP120192907), a project (JPNP18016) commissioned by the New Energy and Industrial Technology Development Organization (NEDO), and the Program for Leading Graduate Schools "Integrative Graduate School and Research Program in Green Natural Sciences".

Conflicts of Interest: The authors declare no conflict of interest.

References

1. Moad, G.; Solomon, D.H. *The Chemistry of Radical Polymerization: Second Fully Revised Edition*; Elsevier: Oxford, UK, 2006.

2. Hawker, C.J.; Bosman, A.W.; Harth, E. New Polymer Synthesis by Nitroxide Mediated Living Radical Polymerizations. *Chem. Rev.* **2001**, *101*, 3661–3688. [CrossRef] [PubMed]

3. Matyjaszewski, K.; Xia, J.H. Atom Transfer Radical Polymerization. *Chem. Rev.* **2001**, *101*, 2921–2990. [CrossRef] [PubMed]

4. Kamigaito, M.; Ando, T.; Sawamoto, M. Metal-Catalyzed Living Radical Polymerization. *Chem. Rev.* **2001**, *101*, 3689–3745. [CrossRef] [PubMed]

5. Moad, G.; Rizzardo, E.; Thang, S.H. Toward Living Radical Polymerization. *Acc. Chem. Res.* **2008**, *41*, 1133–1142. [CrossRef] [PubMed]

6. Matyjaszewski, K. Atom Transfer Radical Polymerization (ATRP): Current Status and Future Perspectives. *Macromolecules* **2012**, *45*, 4015–4039. [CrossRef]

7. McKenzie, T.G.; Fu, Q.; Uchiyama, M.; Satoh, K.; Xu, J.; Boyer, C.; Kamigaito, M.; Qiao, G.G. Beyond Traditional RAFT: Alternative Activation of Thiocarbonylthio Compounds for Controlled Polymerization. *Adv. Sci.* **2016**, *3*, 1500394. [CrossRef]

8. Ouchi, M.; Sawamoto, M. 50th Anniversary Perspective: Metal-Catalyzed Living Radical Polymerization: Discovery and Perspective. *Macromolecules* **2017**, *50*, 2603–2614. [CrossRef]

9. Kamigaito, M.; Sawamoto, M. Synergistic Advances in Living Cationic and Radical Polymerizations. *Macromolecules* **2020**, *53*, 6749–6753. [CrossRef]

10. Nothling, M.D.; Fu, Q.; Reyhani, A.; Allison-Logan, S.; Jung, K.; Zhu, J.; Kamigaito, M.; Boyer, C.; Qiao, G.G. Progress and Perspective Beyond Traditional RAFT Polymerization. *Adv. Sci.* **2020**, *7*, 2001656. [CrossRef]

11. Yao, K.; Tang, C. Controlled Polymerization of Next-Generation Renewable Monomers and Beyond. *Macromolecules* **2013**, *46*, 1689–1712. [CrossRef]

12. Iwata, T. Biodegradable and Bio-Based Polymers: Future Prospects of Eco-Friendly Plastics. *Angew. Chem. Int. Ed.* **2015**, *54*, 3210–3215. [CrossRef] [PubMed]

13. Zhu, Y.; Romain, C.; Williams, C.K. Sustainable polymers from renewable resources. *Nature* **2016**, *540*, 354–362. [CrossRef] [PubMed]

14. Froidevaux, V.; Negrell, C.; Caillol, S.; Pascault, J.-P.; Boutevin, B. Biobased Amines: From Synthesis to Polymers; Present and Future. *Chem. Rev.* **2016**, *116*, 14181–14224. [CrossRef] [PubMed]

15. Llevot, A.; Dannecker, P.-K.; von Czapiewski, M.; Over, L.C.; Söyer, Z.; Meier, M.A.R. Renewability is not Enough: Recent Advances in the Sustainable Synthesis of Biomass-Derived Monomers and Polymers. *Chem. Eur. J.* **2016**, *22*, 11510–11521. [CrossRef]

16. Thomsett, M.R.; Storr, T.E.; Monaghan, O.R.; Stockman, R.A.; Howdle, S.M. Progress in the sustainable polymers from terpenes and terpenoids. *Green Mater.* **2016**, *4*, 115–134. [CrossRef]

17. Llevot, A.; Grau, E.; Carlotti, S.; Grelier, S.; Cramail, H. From Lignin-Derived Aromatic Compounds to Novel Biobased Polymers. *Macromol. Rapid Commun.* **2016**, *37*, 9–28. [CrossRef]

18. Schneiderman, D.K.; Hillmyer, M.A. There is a Great Future in Sustainable Polymers. *Macromolecules* **2017**, *50*, 3733–3749. [CrossRef]

19. Nguyen, H.T.H.; Rostagno, P.; Qi, M.; Feteha, A.; Miller, S.A. The quest for high glass transition temperature bioplastics. *J. Mater. Chem. A* **2018**, *6*, 9298–9331. [CrossRef]

20. Tang, X.; Chen, E.Y.-X. Toward Infinity Recyclable Plastics Derived from Renewable Cyclic Esters. *Chem* **2019**, *5*, 284–312. [CrossRef]

21. O'Dea, R.M.; Willie, J.A.; Epps, T.H., III. 100th Anniversary of Macromolecular Science Viewpoint: Polymers from Lignocellulosic Biomass. Current Challenges and Future Opportunities. *ACS Macro Lett.* **2020**, *9*, 476–493.

22. Erman, W.F. *Chemistry of the Monoterpenes: An Encyclopedia Handbook*; Marcel Dekker, Inc.: New York, NY, USA, 1985.

23. Connolly, J.D.; Hill, R.A. *Dictionary of Terpenoids*; Chapman & Hall: London, UK, 1991.

24. Breitmaier, E. *Terpenes: Flavors, Fragrances, Pharmaca, Pheromones*; Wiley-VCH: Weinheim, Germany, 2006.

25. *Handbook of Essential Oils: Science, Technology, and Applications*; Başer, K.H.C.; Buchbouer, G. (Eds.) CRC Press: Boca Baton, FL, USA, 2016.

26. Satoh, K.; Kamigaito, M. New Polymerization Methods for Biobased Polymers. In *Bio-Based Polymers*; Kimura, Y., Ed.; CMC: Tokyo, Japan, 2013; pp. 95–111.

27. Satoh, K. Controlled/Living Polymerization of renewable vinyl monomers into bio-based polymers. *Polym. J.* **2015**, *47*, 527–536. [CrossRef]

28. Kamigaito, M.; Satoh, K. *Bio-Based Hydrocarbon Polymers, In Encyclopedia of Polymeric Nanomaterials.*; Kobayashi, S., Müllen, K., Eds.; Springer: Heidelberg, Germany, 2015; Volume 1, pp. 109–118.

29. Kamigaito, M.; Satoh, K. Sustainable Vinyl Polymers via Controlled Polymerization of Terpenes. In *Sustainable Polymers from Biomass*; Tang, C., Ryu, C.Y., Eds.; Wiley-VCH: Weinheim, Germany, 2017; pp. 55–90.

30. Akkapeddi, M.K. Poly(α-methylene-γ-butyrolactone) Synthesis, Configurational Structure, and Properties. *Macromolecules* **1979**, *12*, 546–551. [CrossRef]

31. Akkapeddi, M.K. The free radical copolymerization characteristics of α-methylene γ-butyrolactone. *Polymer* **1979**, *20*, 1215–1216. [CrossRef]

32. Mosnacek, J.; Matyjaszewski, K. Atom Transfer Radical Polymerization of Tulipalin A: A Naturally Renewable Monomer. *Macromolecules* **2008**, *41*, 5509–5511. [CrossRef]

33. Mosnacek, J.; Yoon, J.A.; Juhari, A.; Koynov, K.; Matyjaszewski, K. Synthesis, morphology and mechanical properties of linear triblock copolymers based on poly(α-methylene-γ-butyrolactone). *Polymer* **2009**, *50*, 2087–2094. [CrossRef]

34. Zhang, Y.; Miyake, G.M.; Chen, E.Y.-X. Alane-Based Classical and Frustrated Lewis Pairs in Polymer Synthesis: Rapid Polymerization of MMA and Naturally Renewable Methylene Butyrolactones to High Molecular Weight Polymers. *Angew. Chem. Int. Ed.* **2010**, *49*, 10158–10162. [CrossRef]

35. Schmitt, M.; Falivene, L.; Caporaso, L.; Cavallo, L.; Chen, E.Y.-X. High-speed organocatalytic polymerization of a renewable methylene butyrolactone by a phospazene superbase. *Polym. Chem.* **2014**, *5*, 3261–3270. [CrossRef]

36. Pfelifer, V.F.; Vojnovich, C.; Heger, E.N. Itaconic Acid by Fermentation with Aspergillus Terreus. *Ind. Chem. Eng.* **1952**, *44*, 2975–2980. [CrossRef]

37. Kertes, A.S.; King, C.J. Extraction chemistry of fermentation product carboxylic acids. *Biotechnol. Bioeng.* **1986**, *28*, 269–282. [CrossRef]

38. Tate, B.E. Polymerization of itaconic acid and derivatives. *Adv. Polym. Sci.* **1967**, *5*, 214–232.

39. Ishida, S.; Saito, S. Polymerization of Itaconic Acid Derivatives. *J. Polym. Sci. Part. A-1 Polym. Chem.* **1967**, *5*, 689–705. [CrossRef]

40. Satoh, K.; Lee, D.-H.; Nagai, K.; Kamigaito, M. Precision Synthesis of Bio-Based Acrylic Thermoplastic Elastomer by RAFT Polymerization of Itaconic Acid Derivatives. *Macromol. Rapid Commun.* **2014**, *35*, 161–167. [PubMed]

41. Gowda, R.R.; Chen, E.Y.-X. Synthesis of β-methyl-α-methylene-γ-butyrolactone from biorenewable itaconic acid. *Org. Chem. Fornt.* **2014**, *1*, 230–234. [CrossRef]

42. Nonoyama, Y.; Satoh, K.; Kamigaito, M. Renewable β-methylstyrenes for bio-based heat-resistant styrenic copolymers: Radical copolymerization enhanced by fluoroalcohol and controlled/living copolymerization by RAFT. *Polym. Chem.* **2014**, *5*, 3182–3189.

43. Terao, Y.; Satoh, K.; Kamigaito, M. Controlled Radical Copolymerization of Cinnamic Derivatives as Renewable Vinyl Monomers with Both Acrylic and Styrenic Substituents: Reactivity, Regioselectivity, Properties, and Functions. *Biomacromolecules* **2019**, *20*, 192–203. [CrossRef]

44. Terao, Y.; Satoh, K.; Kamigaito, M. 1:3 ABAA sequence-regulated substituted polymethylenes via alternating radical copolymerization of methyl cinnamate and maleic anhydride followed by post-polymerization reactions. *Eur. Polym. J.* **2019**, *120*, 109225. [CrossRef]

45. Imada, M.; Takenaka, Y.; Hatanaka, H.; Tsuge, T.; Abe, H. Unique acrylic resins with aromatic side chains by homopolymerization of cinnamic monomers. *Commun. Chem.* **2019**, *2*, 109. [CrossRef]

46. Takeshima, H.; Satoh, K.; Kamigaito, M. Bio-Based Functional Styrene Monomers Derived from Naturally Occurring Ferulic Acid for Poly(vinylcatechol) and Poly(vinylguaicol) via Controlled Radical Polymerization. *Macromolecules* **2017**, *50*, 4206–4216. [CrossRef]

47. Takeshima, H.; Satoh, K.; Kamigaito, M. Scalable Synthesis of Bio-Based Functional Styrene: Protected Vinyl Catechol from Caffeic Acid and Controlled Radical and Anionic Polymerizations Thereof. *ACS Sustainable Chem. Eng.* **2018**, *6*, 13681–13686. [CrossRef]

48. Takeshima, H.; Satoh, K.; Kamigaito, M. Bio-Based Vinylphenol Family: Synthesis via Decarboxylation of Naturally Occurring Cinnamic Acids and Living Radical Polymerization for Functionalized Polystyrenes. *J. Polym. Sci.* **2020**, *58*, 91–100. [CrossRef]

49. Satoh, K.; Matsuda, M.; Nagai, K.; Kamigaito, M. AAB-Sequence Living Radical Chain Copolymerization of Naturally-Occurring Limonene with Maleimide: An End-to-End Sequence-Regulated Copolymer. *J. Am. Chem. Soc.* **2010**, *132*, 10003–10005. [CrossRef] [PubMed]

50. Matsuda, M.; Satoh, K.; Kamigaito, M. Periodically Functionalized and Grafted Copolymers via 1:2-Sequence-Regulated Radical Copolymerization of Naturally Occurring Functional Limonene and Maleimide Derivatives. *Macromolecules* **2013**, *46*, 5473–5482. [CrossRef]

51. Matsuda, M.; Satoh, K.; Kamigaito, M. Controlled Radical Copolymerization of Naturally-Occurring Terpenes with Acrylic Monomers in Fluorinated Alcohol. *KGK Kaut. Gummi Kunstst.* **2013**, *66*, 51–56.

52. Matsuda, M.; Satoh, K.; Kamigaito, M. 1:2-Sequence-Regulated Radical Copolymerization of Naturally Occurring Terpenes with Maleimide Derivatives in Fluorinated Alcohol. *J. Polym. Sci. Part. A Polym. Chem.* **2013**, *51*, 1774–1785. [CrossRef]

53. Miyaji, H.; Satoh, K.; Kamigaito, M. Bio-Based Polyketones by Selective Ring-Opening Radical Polymerization of α-Pinene Derived Pinocarvone. *Angew. Chem. Int. Ed.* **2016**, *55*, 1372–1376. [CrossRef]

54. Ojika, M.; Satoh, K.; Kamigaito, M. BAB-*random*-C Monomer Sequence via Radical Terpolymerization of Limonene (A), Maleimide (B), and Methacrylate (C): Terpene Polymers with Randomly Distributed Periodic Sequences. *Angew. Chem. Int. Ed.* **2017**, *56*, 1789–1793. [CrossRef]

55. Hashimoto, H.; Takeshima, H.; Nagai, T.; Uchiyama, M.; Satoh, K.; Kamigaito, M. Valencene as a naturally occurring sesquiterpene monomer for radical copolymerization with maleimide to induce concurrent 1:1 and 1:2 propagation. *Polym. Degrad. Stab.* **2019**, *161*, 183–190. [CrossRef]

56. Runckel, W.J.; Goldblatt, L.A. Inhibition of Myrcene Polymerization during Storage. *Ind. Eng. Chem.* **1946**, *38*, 749–751. [CrossRef]

57. Johanson, A.J.; McKennon, F.L.; Goldblatt, L.A. Emulsion Polymerization of Myrcene. *Ind. Eng. Chem.* **1948**, *401*, 500–502. [CrossRef]

58. Marvel, C.S.; Hwa, C.C.L. Polymyrcene. *J. Polym. Sci.* **1960**, *45*, 25–34. [CrossRef]

59. Cawse, J.L.; Stanford, J.L.; Still, R.H. Polymers from Renewable Resources. III. Hydroxy-Terminated Myrcene Polymers. *J. Appl. Polym. Sci.* **1986**, *31*, 1963–1975. [CrossRef]

60. Trumbo, D.L. Free radical copolymerization behavior of myrcene I. Copolymers with styrene, methyl methacrylate or *p*-fluorostyrene. *Polym. Bull.* **1993**, *31*, 629–636. [CrossRef]

61. Sarkar, P.; Bhowmick, A.K. Synthesis, characterization and properties of a bio-based elastomer: Polymyrcene. *RSC Adv.* **2014**, *41*, 61343–61354. [CrossRef]

62. Hilschmann, J.; Kali, G. Bio-based polymyrcene with highly ordered structure via solvent free controlled radical polymerization. *Eur. Polym. J.* **2015**, *73*, 363–373. [CrossRef]

63. Kamigaito, M.; Satoh, K.; Suzuki, S.; Kori, Y.; Eguchi, Y.; Iwasa, K.; Shiroto, H. β-Phellandrene polymer, production method for same, and molded article. WO 2015/060310 A1; filed 21 October 2014 and issued 30 April 2015,

64. Nishida, T.; Satoh, K.; Nagano, S.; Seki, T.; Tamura, M.; Li, Y.; Tomishige, K.; Kamigaito, M. Biobased Cycloolefin Polymers: Carvone-Derived Cyclic Conjugated Diene with Reactive *exo*-Methylene Group for Regioselective and Stereospecific Living Cationic Polymerization. *ACS Macro Lett.* **2020**, *9*, 1178–1183. [CrossRef]

65. Matsumoto, A.; Yamamoto, D. Radical Copolymerization of *N*-Phenylmaleimide and Diene Monomers in Competition with Diels–Alder Reaction. *J. Polym. Sci., Part. A Polym. Chem.* **2016**, *54*, 3616–3625. [CrossRef]

66. Li, Y.; Padias, A.B.; Hall, H.K. Evidence for 2-Hexene-1,6-diyl Diradicals Accompanying the Concerted Diels-Alder Cycloaddition of Acrylonitrile with Nonpolar 1,3-Dienes. *J. Org. Chem.* **1993**, *58*, 7049–7058. [CrossRef]

67. Trumbo, D.L. Synthesis and Polymerization of 1-Methyl-4-isopropenyl-6-methylene-1-cyclohexene. *J. Polym. Sci.: Part. A Polym. Chem.* **1995**, *33*, 599–601. [CrossRef]

68. Kobayashi, S.; Lu, C.; Hoye, T.R.; Hillmyer, M.A. Controlled Polymerization of a Cyclic Diene Prepared from the Ring-Closing Metathesis of a Naturally Occurring Monoterpene. *J. Am. Chem. Soc.* **2009**, *131*, 7960–7961. [CrossRef]

69. Yamamoto, D.; Matsumoto, A. Controlled Radical Polymerization of 3-Methylenecyclopentene with *N*-Substituted Maleimides To Yield Highly Alternating and Regiospecific Copolymers. *Macromolecules* **2013**, *46*, 9526–9536. [CrossRef]

70. *Polymer Handbook*, 4th ed.; Brandrup, J.; Immergut, E.H.; Grulke, E.A. (Eds.) John Wiley & Sons: New York, NY, USA, 1999.

71. Thang, S.H.; Chong, Y.K.; Mayadunne, R.T.A.; Moad, G.; Rizzardo, E. A novel synthesis of functional dithioesters, dithiocarbamates, xanthates, and trithiocarbonates. *Tetrahedron Lett.* **1999**, *40*, 2435–2438. [CrossRef]

Publisher's Note: MDPI stays neutral with regard to jurisdictional claims in published maps and institutional affiliations.

Article

Biosourced All-Acrylic ABA Block Copolymers with Lactic Acid-Based Soft Phase

Nabil Bensabeh [1], Ana Jiménez-Alesanco [2], Ilme Liblikas [3], Juan C. Ronda [1], Virginia Cádiz [1], Marina Galià [1], Lauri Vares [3], Olga Abián [2,4,5,6,7] and Gerard Lligadas [1,*]

[1] Laboratory of Sustainable Polymers, Department of Analytical Chemistry and Organic Chemistry, University Rovira i Virgili, 43007 Tarragona, Spain; nabil.bensabeh@urv.cat (N.B.); juancarlos.ronda@urv.cat (J.C.R.); virginia.cadiz@urv.cat (V.C.); marina.galia@urv.cat (M.G.)

[2] Institute for Biocomputation and Physics of Complex Systems (BIFI), Joint Units IQFR-CSIC-BIFI, and GBsC-CSIC-BIFI, Universidad de Zaragoza, 50018 Zaragoza, Spain; ajimenez@bifi.es (A.J.-A.); oabifra@unizar.es (O.A.)

[3] Institute of Technology, University of Tartu, Nooruse 1, 50411 Tartu, Estonia; ilme.liblikas@ut.ee (I.L.); lauri.vares@ut.ee (L.V.)

[4] Instituto Aragonés de Ciencias de la Salud (IACS), 50009 Zaragoza, Spain

[5] Instituto de Investigación Sanitaria de Aragón (IIS Aragon), 50009 Zaragoza, Spain

[6] Centro de Investigación Biomédica en Red en el Área Temática de Enfermedades Hepáticas Digestivas (CIBERehd), 28029 Madrid, Spain

[7] Departamento de Bioquímica y Biología Molecular y Celular, Universidad de Zaragoza, 50005 Zaragoza, Spain

* Correspondence: gerard.lligadas@urv.cat; Tel.: +34-977-55-8286

Academic Editor: Sylvain Caillol
Received: 26 October 2020; Accepted: 3 December 2020; Published: 5 December 2020

Abstract: Lactic acid is one of the key biobased chemical building blocks, given its readily availability from sugars through fermentation and facile conversion into a range of important chemical intermediates and polymers. Herein, well-defined rubbery polymers derived from butyl lactate solvent were successfully prepared by reversible addition–fragmentation chain transfer (RAFT) polymerization of the corresponding monomeric acrylic derivative. Good control over molecular weight and molecular weight distribution was achieved in bulk using either monofunctional or bifunctional trithiocarbonate-type chain transfer agents. Subsequently, poly(butyl lactate acrylate), with a relative low T_g (−20 °C), good thermal stability (5% wt. loss at 340 °C) and low toxicity was evaluated as a sustainable middle block in all-acrylic ABA copolymers using isosorbide and vanillin-derived glassy polyacrylates as representative end blocks. Thermal, morphological and mechanical properties of copolymers containing hard segment contents of <20 wt% were evaluated to demonstrate the suitability of rubbery poly(alkyl lactate) building blocks for developing functional sustainable materials. Noteworthy, 180° peel adhesion measurements showed that the synthesized biosourced all-acrylic ABA copolymers possess competitive performance when compared with commercial pressure-sensitive tapes.

Keywords: block copolymers; renewable resources; RAFT; alkyl lactate; PSA

---

1. Introduction

Nature uses molecular self-assembly to create precision nanostructures and build large constructs through hierarchical assembly [1,2]. Inspired by these motifs, considerable efforts have been undertaken to recreate such concepts using synthetic block copolymers fashioned from two or more chemically dissimilar components that are covalently-bonded into a single molecule [3]. Linear ABA block copolymers with a soft middle block and hard minority end blocks are of high utility and interest for a wide variety of applications, ranging from adhesives to clothing, automotive and biomedical components due to their remarkable (re)processable structures [4]. Among them, all-acrylic systems offer advantages to the current gold-standard systems based on poly(styrene-*block*-isoprene-*block*-styrene) and poly(styrene-*block*-butadiene-*block*-styrene) copolymers, in terms of wider service temperature range, improved optical transparency and stability to UV light [5,6]. In addition, acrylate-based triblocks may exhibit excellent performance as pressure-sensitive adhesives (PSAs), even without the incorporation of additives such as tackifiers and plastisizers [7].

Making the most of reversible deactivation radical polymerizations techniques [8], the rich assortment of (meth)acrylate monomers from biosourced feedstocks offers an attractive palette of rubbery and glassy polymers for the design of innovative sustainable all-acrylic ABA-type thermoplastic elastomers (TPEs) with competitive properties. For instance, poly(lauryl methacrylate) ($T_g \approx -46\,°\mathrm{C}$) as the rubbery segment and poly(acetylsalicylic ethyl metacrylate) ($T_g \approx 53\,°\mathrm{C}$) as the glassy segment were combined to build up ABA copolymer architectures exhibiting elastomeric behavior at room temperature [9]. Biosourced acrylic monomers derived from glucose [10], isosorbide [11], itaconic acid imides [12] and rosin [13], as well as aromatic lignin derivatives [14], have been shown to provide useful glassy components for developing sustainable elastomeric and adhesive materials based on ABA block copolymers. In most of these studies, poly(butyl acrylate) ($T_g \approx -50\,°\mathrm{C}$) was chosen as the rubbery midblock, since it is a conventional component of acrylic ABA copolymers with tunable properties according to block lengths and molecular composition. Although it is feasible to derive nBA from bio-sourced acrylic acid and butan-1-ol [15], the search for alternative rubbery blocks from biobased feedstocks represents a key step forward towards improving the material's sustainability. In this regard, fatty acids acrylic derivatives [16], dialkyl itaconates [12] and tetrahydrogeraniol derivatives [17] are only selected examples on how to create renewable triblock copolymers with biobased soft phase without compromising performance.

Lactic acid is one of the top value added biomass derived platform chemicals, given its ready availability from sugars through fermentation and its facile conversion into a number of important derivatives [18–20]. Among them, polylactide (PLA), most commonly synthesized by the ring-opening polymerization of the cyclic dimer of lactic acid, is a good alternative replacement for the traditional petroleum-sourced polystyrene glassy end-blocks in ABA TPEs, due to their high modulus characteristics [21]. The potential of alkyl lactate esters, e.g., methyl, ethyl and butyl, goes beyond their well-recognized use as eco-friendly solvents [22], as their secondary alcohol offers a simple access to monovinyl derivatives prone to polymerization by radical mechanisms [23–26]. In this regard, our group reported that poly(alkyl lactate ester) (meth)acrylates with controlled molecular weight, narrow molecular weight distribution and high end-group fidelity are also easily accessible by Cu(0) wire-mediated single-electron transfer living radical polymerization (SET-LRP) [27,28]. In a more recent effort, the different water solubility of poly(ethyl lactate acrylate)s (hydrophobic and water-insoluble) and poly(*N,N*-dimethyl lactamide acrylate) (hydrophilic and water-soluble) was exploited to design amphiphilic AB block copolymers from neoteric lactic acid-based solvents, i.e., ethyl lactate and *N,N*-dimethyl lactamide, that could self-assemble in aqueous solution to form nanoparticles with different morphologies including large-compound micelles and vesicles [29].

Herein, we report on the preparation of all-acrylic ABA copolymers with a lactic acid acrylic derivative as an elastomeric building block. A biobased acrylic monomer, ʟ-butyl lactate acrylate (BuLA), has been synthesized from the corresponding alkyl lactate, which is used as a solvent and a dairy-related flavoring agent approved as a food additive by the US Food and Drug Administration. This monomer was conceived to impart a lactic acid-based sustainable soft segment into ABA-type block copolymers. RAFT polymerization of the BuLA monomer allowed the preparation of fine-tuned all-acrylic triblock copolymers combining soft poly(BuLA) segments with isosorbide and vanillin-derived glassy end blocks. The ABA architecture were investigated for the fidelity of phase-separation in the bulk, and were initially examined for their mechanical and adhesion properties. The results reported here demonstrate that biosourced and non-toxic acrylic derivatives of alkyl lactate esters can complement the classic alkyl acrylate palette for developing sustainable materials with assembly properties that can be utilized in a wide variety of applications.

2. Results and Discussion

2.1. BuLA Monomer Synthesis and RAFT Polymerization Model Studies

As reported in previous publications, alkyl lactate solvents are appealing chemicals to prepare biorenewable acrylic/methacrylic polymers [23–29]. Here, BuLA was prepared from L-butyl lactate (BuL) by acylation with acryloyl chloride in the presence of base using a solvent with low environmental impact such as Me-THF (Figure 1). After vacuum distillation in the presence of hydroquinone to minimize polymer formation, BuLA was isolated as a colorless liquid in high yield (75%). The structure of BuLA was verified by NMR spectroscopy. Figure 2a shows ^{1}H NMR spectrum of BuLA, in which the three acrylic protons appear between 6.5 and 5.8 ppm. The characteristic BuL methine proton [$\underline{\text{CH}}$(CH$_3$)O], at around 4.3 ppm, shifted to 5.6 ppm after the formation of the ester group. The ^{13}C NMR spectrum was also consistent with that expected for BuLA (see experimental section).

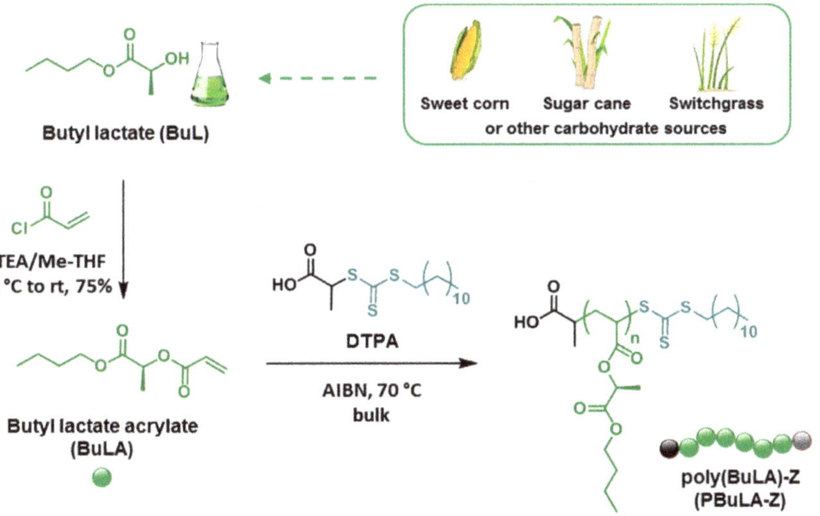

Figure 1. Synthesis of BuLA from biorenewable BuL solvent and their bulk reversible addition–fragmentation chain transfer (RAFT) polymerization using a monofunctional chain transfer agent (CTA), namely 2-(dodecylthiocarbonothioylthio)propionic acid (DTPA) to yield PBuLA-Z.

After monomer synthesis, we investigated its controlled polymerization by using the RAFT technique, on account of its ability to form polymers with predicted molecular weight and relatively low molecular weight distribution (MWD) in a green fashion [30]. Figure 1 depicts the RAFT polymerization of BuLA at 70 °C in bulk, using 2,2′-azobisisobutyronitrile (AIBN) and 2-(dodecylthiocarbonothioylthio)propionic acid (DTPA) as radical initiator and chain transfer agent (CTA), respectively. A series of polymerizations were conducted at different [BuLA]$_0$/[DTPA]$_0$ ratios (50 to 400) to target various polymer chain lengths. PBuLAs grown in one direction, defined as PBuLA-Z, with polydispersity (M_w/M_n) ranging from 1.22–1.27 and gel permeation chromatography (GPC)-determined molecular weight ($M_{n,GPC}$) as high as $M_{n,GPC}$ = 80,300 g/mol could be obtained after 2 h (Figure 3a). In all cases, the bulk RAFT process proceeded to high monomer conversions (>90%) and yielding polymers with $M_{n,GPC}$ close to the theoretical values ($M_{n,th}$), calculated from monomer conversion assuming complete consumption of CTA. The small discrepancy in the $M_{n,GPC}$

and $M_{n,th}$ data may be attributed to differences in hydrodynamic volume between PBuLA-Z and PMMA standards used for calibration.

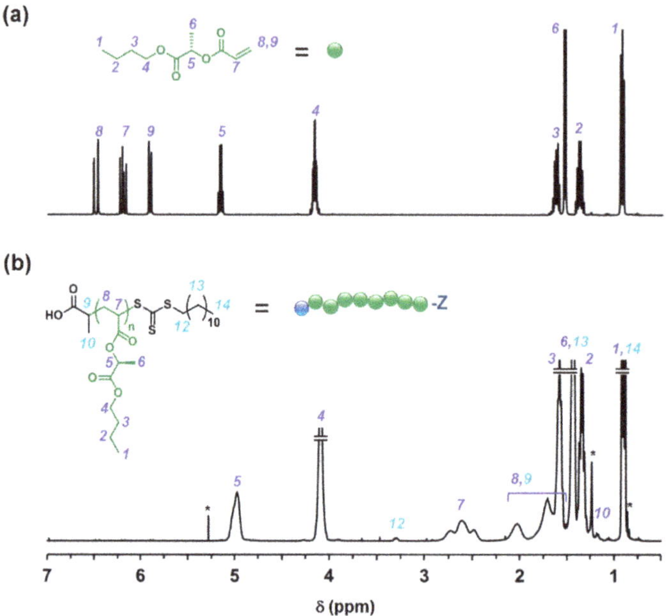

Figure 2. ^{1}H NMR spectra of (**a**) BuLA and (**b**) PBuLA-Z ($M_{n,GPC}$ = 10,900, M_w/M_n = 1.24) in CDCl$_3$. ^{1}H NMR resonances from "grease" impurities are indicated with *.

As a representative example, purified low molar mass PBuLA-Z prepared at degree of polymerization (DP) of 50 ($M_{n,GPC}$ = 10,900 Da, M_w/M_n = 1.24) was characterized by ^{1}H NMR (Figure 2b). After polymerization, the olefinic protons, i.e., 5.8–6.5 ppm, in the BuLA monomer were incorporated into the backbone of the new vinylic polymer (i.e., 1.9–3.0 ppm). Moreover, the peak at ≈ 3.3 ppm, attributed to the CH$_2$-S protons of RAFT CTA groups located at the ω-chain end of the polymer, supports the "livingness" of the synthesized polymer. Next step, we further investigated the kinetics of the polymerization at a DP of 200. As can be seen in Figure 3b (left panel), after a 10 min induction period attributed to the difficulty in deoxygenating a highly viscous reaction mixture, the polymerization achieved ≈ 90% conversion in less than 1 h. The linear evolution of ln[M]$_0$/[M] versus reaction time suggested a constant concentration of active species throughout the entire RAFT process. Moreover, as expected for a controlled polymerization, molecular weights increased linearly with conversion retaining narrow MWD (M_w/M_n ≈ 1.25) even at high values (right panel). Prior to conducting block copolymerization studies, we evaluated the cytotoxic effect of PBuLA-Z RAFT homopolymer by treating HeLa, HT-29 and MIA-Pa-Ca-2 cell lines with different doses of the homopolymer dissolved in the culture medium (Figure 4). Figure 4a depicts the CellTiter plates for different concentrations of PBuLA-Z ($M_{n,GPC}$ = 11,000 Da, M_w/M_n = 1.23) for HeLa cells culture. CellTiter colorimetric assay was used for determining cell viability and cellular proliferation after the addition of test RAFT polymer. This assay determines the number of viable cells due to the bioreduction of the MTS tetrazolium compound (Owen's reagent) into a colored formazan product accomplished by NADPH or NADH produced by dehydrogenase enzymes in metabolically active cells. Therefore, color formation can be a useful marker of viable cells. Notably, the cytotoxicity analysis using the CellTiter assay showed that 24 h exposure of the cells to up to 100 mg·mL^{-1} at 37 °C did not reduce cell viability. Hence, these results further support the use of PBuLA polymers as a soft and hydrophobic building block in well-defined copolymer synthesis employed in a wide range of applications, including biomedical devices and adhesives [31].

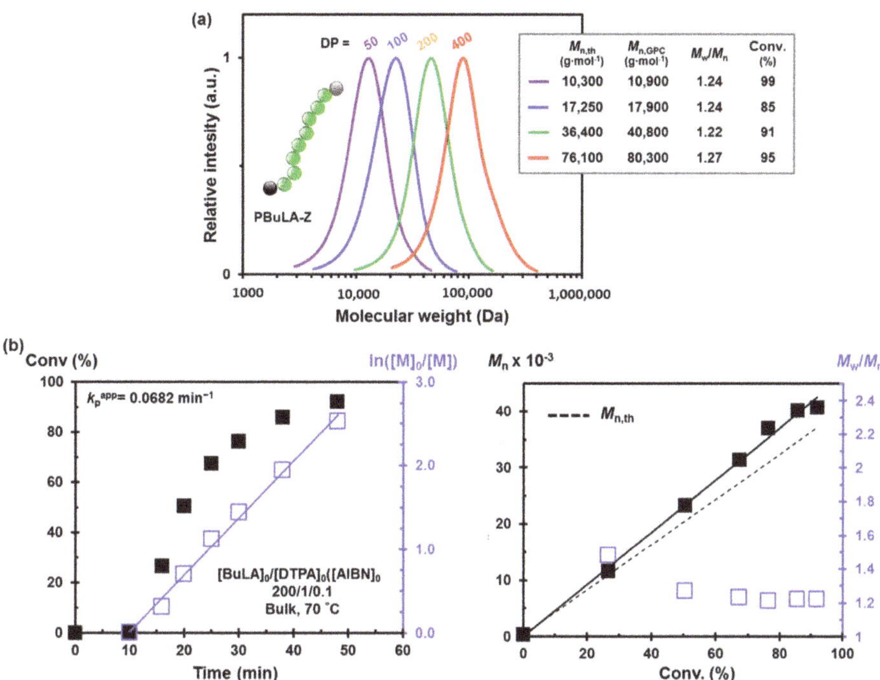

Figure 3. Bulk RAFT polymerization of BuLA with DTPA at 70 °C. (**a**) GPC traces of PBuLA-Z prepared at different degrees of polymerization (DPs), ranging from 50 to 400. The inset contains $M_{n,th}$, $M_{n,GPC}$, M_w/M_n, and monomer conversion values; and (**b**) evolution of monomer conversion and semilogarithmic kinetic plot of $\ln([M]_0/[M])$ versus time (left panel) and evolution of experimental $M_{n,GPC}$ and M_w/M_n, based on the calibration by PMMA standards, with the monomer conversion (right panel). Reaction conditions for (b): BuLA = 1 mL, $[BuLA]_0/[DTPA]_0/[AIBN]_0 = 200/1/0.1$ at 70 °C.

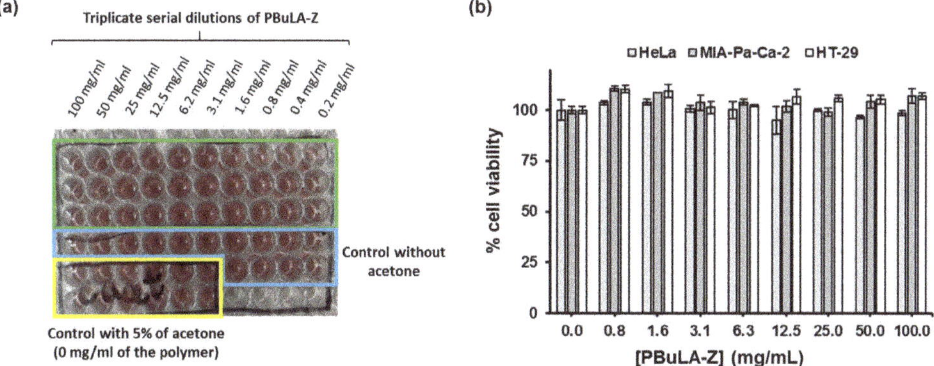

Figure 4. (**a**) The plates for cytotoxic assays of PBuLA-Z ($M_{n,GPC}$ = 11,000 Da, M_w/M_n = 1.23) tested in HeLa cells; and (**b**) CellTiter cell viability assay results for the same polymer at different concentrations assessed in HeLa, MIA-Pa-Ca-2 and HT-29 cell lines.

2.2. Bifunctional PBuLA-2Z macro-CTA via bulk RAFT Polymerization

With the final goal to prepare ABA block copolymers with PBuLA as the soft middle block, we used a commercially available bifunctional CTA, namely 3,5-bis(2-dodecylthiocarbonothioylthio-1-oxopropoxy)benzoic acid (BTCBA), to prepare bifunctional PBuLA macroCTAs (defined as PBuLA-2Z) through divergent RAFT polymerization of BuLA at 70 °C in bulk (see first reaction step in Figure 5a).

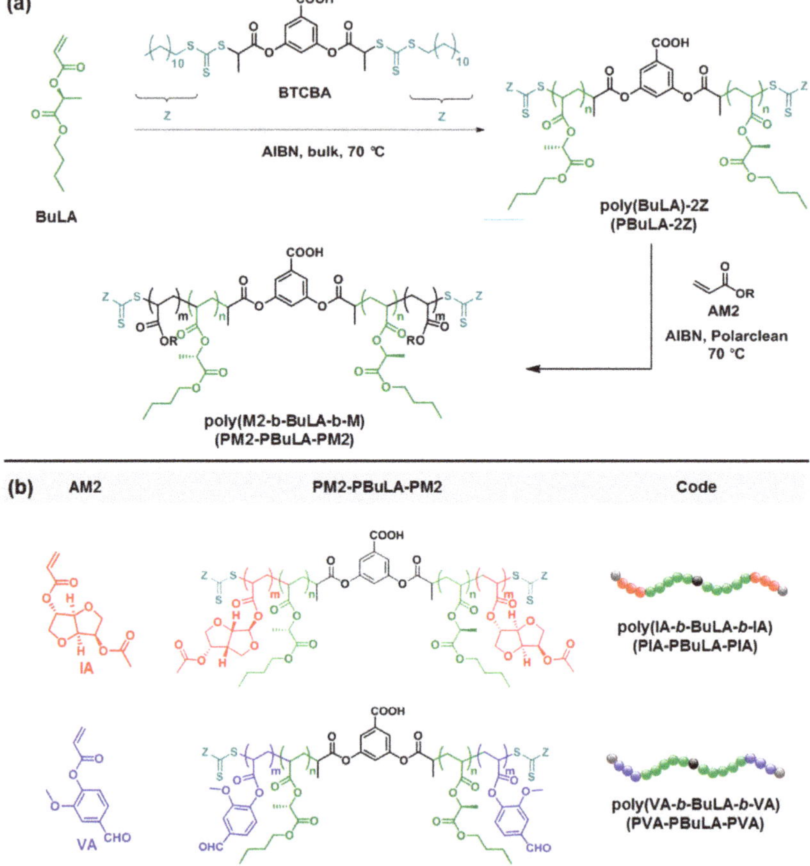

Figure 5. (**a**) Two-step RAFT synthesis of ABA block copolymers with lactic acid-based middle block using AM2 monomer to build up end blocks and BTCBA as a CTA. (**b**) Chemical structures of AM2 acrylic monomers (IA and VA) and ABA block copolymers (PIA-PBuLA-PIA and PVA-PBuLA-PVA) studied herein.

Initially, we targeted a low molar mass polymer (DP = 50, $M_{n,th}$ = 10,830 g·mol^{-1}) to collect structural information about the "livingness" of the system when performing divergent chain growth. Under these conditions, the RAFT polymerization of BuLA achieved 99% monomer conversion in 2 h. BTCPA controlled the polymerization of BuLA and yielded PBuLA-2Z macroinitiator with $M_{n,GPC}$ of 10,500 and M_w/M_n of 1.24 (Supplementary Materials Figure S1). ^{1}H NMR analysis of the isolated polymer clearly revealed the presence of CTA residues at both the polymer core and chain ends, e.g., signals 7.68 and 7.13 ppm for aryl mid group protons and 3.33 ppm for the α methylene protons of the trithiocarbonate end groups (Figure 6). Hence, the ratio of integration of these signals to the methylene (CH$_2$-O) peak of the butyl lactate repeating units enabled to determine $M_{n,NMR}$ (11,200 g·mol^{-1}) which was in good agreement with $M_{n,th}$ and $M_{n,GPC}$ (11,700 g·mol^{-1} and 11,000 g·mol^{-1}, respectively). Note that the $M_{n,NMR}$ was only about 5% above $M_{n,th}$. Thus, within the experimental error of ^{1}H NMR measurement, these results support good control of the bulk RAFT polymerization. Next, we pushed this polymerization system to higher DPs (100–400) under strictly identical reaction conditions, as higher molar mass polymers are required to prepare ABAs with competitive performance. Also in this case, high conversions (>90%) were achieved.

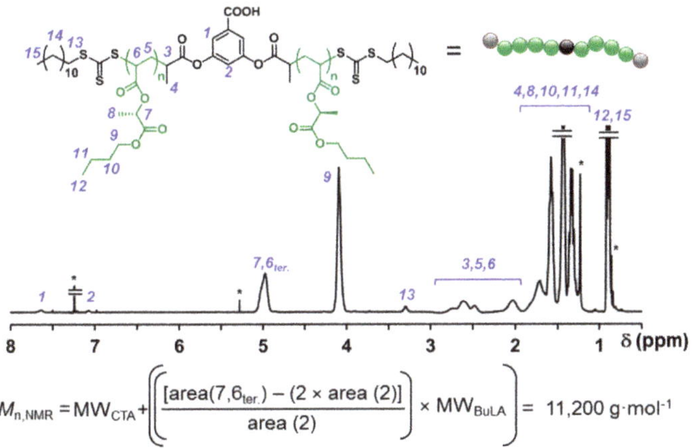

$$M_{n,NMR} = MW_{CTA} + \left(\left[\frac{[area(7,6_{ter.}) - (2 \times area\,(2)]}{area\,(2)} \right] \times MW_{BuLA} \right) = 11,200 \text{ g·mol}^{-1}$$

Figure 6. ^{1}H NMR spectrum PBuLA-2Z ($M_{n,GPC}$ = 10,900, M_w/M_n = 1.24) in CDCl$_3$. ^{1}H NMR resonances from "grease" impurities and residual solvents are indicated with *.

As shown in Figure 7a, the successful synthesis of PBuLA-2Z macro-CTAs with molecular weight as high as $M_{n,GPC}$ = 78,150 g·mol^{-1} was demonstrated by unimodal GPC traces and $M_{n,GPC}$ values very close to the theoretical values and narrow M_w/M_n values (<1.25). Good candidates for the soft middle block in triblock TPEs must have a glass transition temperature (T_g) lower than their application temperature, in order to provide flexibility or tacky characteristics to the final materials.

Differential scanning calorimetry (DSC) analysis of PBuLA-2Z ($M_{n,GPC}$ = 78,150) revealed a T_g value (indicated by dashed line in the green trace) well below room temperature (T_g = −21 °C) (Figure 7b). The reduction of molar mass from $M_{n,GPC}$ = 78,150 g·mol^{-1} to $M_{n,GPC}$ = 20,000 g·mol^{-1} did not significantly influence the value of T_g (data not shown). Note that the T_g value for PBuLA-2Z is 23 °C above that of poly(butyl acrylate) (PBA) of approximately the same M_n prepared by RAFT (Figure 7b). This result can be attributed to the presence of two polar carbonyl units per repeating unit, which decreases segmental motion. Thermogravimetric analysis (TGA) also revealed that PBuLA-2Z is also highly competitive in terms of thermal stability (Figure 7b). The thermal stability of PBuLA-2Z was slightly higher, with a 5% degradation temperature of 340 °C (327 °C for PBA). Overall, these results support the use of PBuLA as an innovative candidate soft block in triblock copolymer-based ABA-type TPEs.

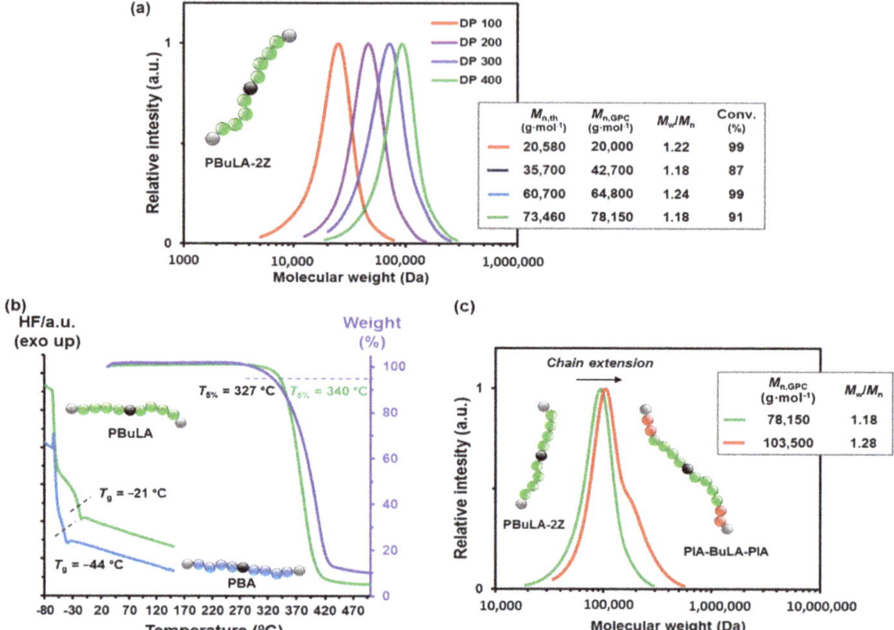

For panel (a) inset table:

	$M_{n,th}$ (g·mol⁻¹)	$M_{n,GPC}$ (g·mol⁻¹)	M_w/M_n	Conv. (%)
	20,580	20,000	1.22	99
	35,700	42,700	1.18	87
	60,700	64,800	1.24	99
	73,460	78,150	1.18	91

For panel (c) inset table:

	$M_{n,GPC}$ (g·mol⁻¹)	M_w/M_n
	78,150	1.18
	103,500	1.28

Figure 7. (**a**) Gel permeation chromatography (GPC) traces of PBuLA-2Z prepared at different DPs ranging from 50 to 400 by the bulk RAFT polymerization of BuLA with BTCBA at 70 °C (The inset contains $M_{n,th}$, $M_{n,GPC}$, M_w/M_n, and monomer conversion values); (**b**) thermal characterization by differential scanning calorimetry (DSC) and thermogravimetric analysis (TGA) analyses of PBuLA-2Z ($M_{n,GPC}$ = 78,150 g·mol⁻¹) and PBA ($M_{n,GPC}$ = 97,900 g·mol⁻¹); and (**c**) GPC traces of the chain extension of PBuLA-2Z macro-RAFT agent with IA monomer at 70 °C in Rhodiasolv® PolarClean solvent. The inset contains $M_{n,GPC}$ and M_w/M_n values.

2.3. ABA Block Copolymers via RAFT in Rhodiasolv® PolarClean Solvent

Using PBuLA-2Z ($M_{n,GPC}$ = 78,150 g·mol⁻¹) as macroCTA, we subsequently studied the incorporation of glassy external segments at the chain ends of PBuLA in a divergent fashion via RAFT polymerization with a second acrylic monomer AM2 (see the second reaction step in Figure 5a,b). Two sustainable triblock copolymers were targeted using poly(vanillin acrylate) (PVA) and a poly(isosorbide acrylate) (PIA) as hard blocks. The chemical structure of both biosourced acrylic monomers (IA, i.e., isosorbide 2-acrylate-5-acetate, and VA, i.e., vanillin acrylate) and the targeted ABA block copolymers (PVA-PBuLA-PVA and PIA-PBuLA-PIA) is shown in Figure 5b. IA and VA were prepared from isosorbide and vanillin, respectively, using previously reported synthetic procedures [32–34]. Unfortunately, the chain extension step required the use of solvent to build the pursued triblocks, as reaction mixtures were not homogeneous at the polymerization temperature (70 °C). Although recent advances in waterborne processes are highly attractive [35], the search for eco-friendly solvents as an alternative to conventional petroleum-derived solvents is also desirable. For this purpose, after control experiments that will be reported in a forthcoming publication, block copolymer syntheses were performed in ethyl-5-(dimethylamido)-2-methyl-5-oxopentanoate (Rhodiasolv® PolarClean) [36]. To our knowledge, this eco-friendly water-soluble and polar solvent was never used as green reaction media in controlled radical polymerization. Hence, chain extension of PBuLA-2Z ($M_{n,GPC}$ = 75,340 g·mol⁻¹) with IA at 70 °C under the reaction conditions of [IA]₀/[PBuLA-2Z]₀/[AIBN]₀ = 109/1/0.2 in PolarClean was afforded the PIA-PBuLA-PIA triblock.

The conversion of IA monomer was calculated to be 86%, using ^{1}H NMR analysis prior to purification of the polymer (Figure S2). A clear shift to higher molar masses in the SEC trace indicated successful chain extension (Figure 7c). Note that a high molar mass shoulder was present in the GPC trace of the ABA copolymer, likely due to termination by combination at high conversion. Nevertheless, the molar mass distribution was relatively narrow (M_w/M_n = 1.28) suggesting that the chain extension occurs, retaining substantial control. As depicted in Figure S3, chain extension PBuLA-2Z with VA also resulted in success to furnish PVA-PBuLA-PVA triblock (78% conversion, $M_{n,GPC}$ = 110,150 g·mol^{-1}). Both triblocks were subsequently characterized by ^{1}H NMR to determine their compositions (Figure 8). For instance, as indicated in Figure 8a, the composition of block copolymer PIA-PBuLA-PIA could be determined from the integration of signals at 4.3–4.5 ppm from H_{17} corresponding to PIA units and the integration of the signals at 4.7–5.3 ppm from PBuLA units, after subtracting the peak area contributed by the PIA units (H_{15}, H_{18}, and H_{21}) in this region. Thus, the obtained PIA-PBuLA-PIA triblock had a PIA (hard block) weight percentage of 15 wt%, as determined via ^{1}H NMR spectroscopy, which is comparable to the theoretical 18 wt% value determined from monomer conversion. As depicted in Figure 8b, the hard block content for the PVA-PBuLA-PVA copolymer was slightly higher (19 wt% PVA), but also comparable with the theoretical value calculated from conversion (21 wt%). Note that the hard block content in these representative copolymers is consistent with typical hard block composition for triblock copolymers used for pressure-sensitive adhesive (PSA) applications, where high tacky materials are pursued [10,14].

2.4. ABA Block Copolymers Characterization: Thermal, Mechanical, and Ahesive Properties

The microstructure and viscoelastic properties of the synthesized triblock copolymers were studied by DSC and dynamic mechanical thermal analysis (DMTA). DSC analysis of both synthesized triblocks revealed a low T_g for the PBuLA midblock at about −20 °C, although no discernible signals corresponding to the T_g of the glassy end blocks were observed, likely due to the lack of a DSC signal based on the low weight fraction of hard block (Figure S4) [9–11,14]. However, the presence of midblock T_g near that of the homopolymer suggests that PBuLA segment is not mixed with either PIA or PVA segments. Note that the T_g of PIA and PVA homopolymer glassy blocks was expected to appear around 80 °C, as indicated DSC analysis of model homopolymers also prepared by RAFT (dashed lines in Figure S4). Next, both copolymers were submitted to DMTA (Figure 9a and Figure S5). DMTA revealed a relaxation process at around 0 °C, corresponding to the glass-rubber transition of the PBuLA soft phase in PIA-PBuLA-PIA and PVA-PBuLA-PVa triblocks, as indicated by a stepwise decrease in the storage (E′) and loss (E″) moduli, as well as a peak in tan δ. The presence of a second tan δ peaks at around 100 °C in the case of PIA-PBuLA-PIA triblock, and at 130 °C for the copolymers integrating PVA, which confirmed the soft-hard-soft microstructure configuration of the synthesized triblock copolymers [37]. Note also that E′ (25 °C) for PVA-PBuLA-PVA triblock was slightly higher than that observed for PIA-PBuLA-PIA (1.3 MPa versus 1.0 MPa). This can be attributed to the higher weight percentage of hard block for PVA-PBuLA-PVA (19 wt% of PVA) in comparison to PIA-PBuLA-PIA (15 wt% of PVA).

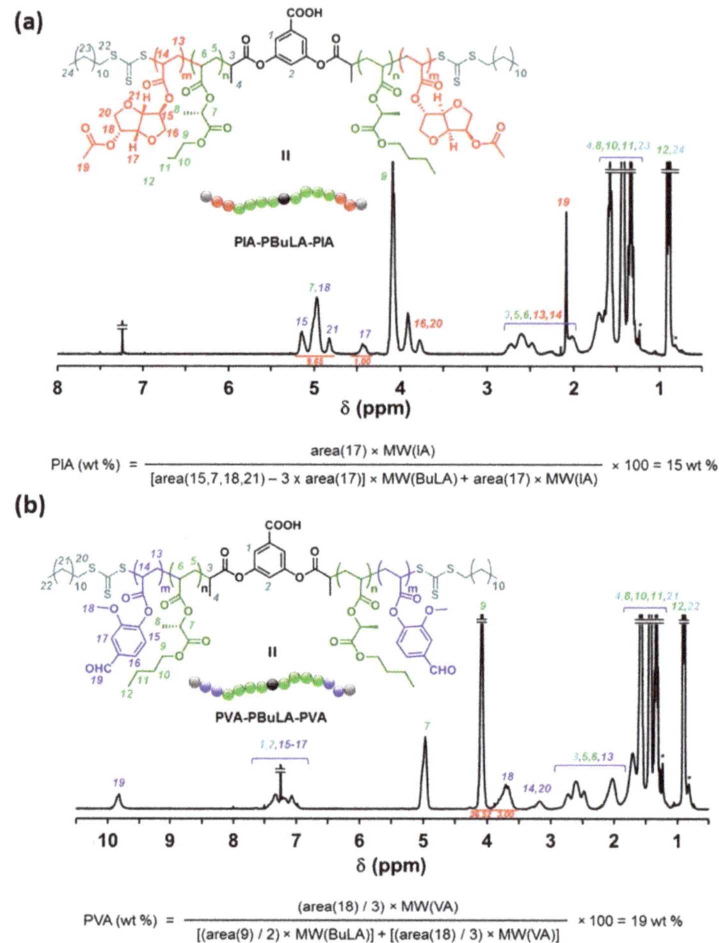

Figure 8. ^{1}H NMR spectra of: (**a**) PIA-PBuLA-PIA ($M_{n,GPC}$ = 103,500 g·mol^{-1}, M_w/M_n = 1.28); and (**b**) PVA-PBuLA-PVA ($M_{n,GPC}$ = 110,150 g·mol^{-1}, M_w/M_n = 1.38) triblocks in CDCl$_3$. ^{1}H NMR resonances from "grease" impurities are indicated with *.

Next step, mechanical properties of both triblocks were measured by monotonic tensile tests on polymer films prepared by a hot-press technique (Figure 9b). Both systems exhibited comparable elongation at a break at around 250%. However, PVA-PBuLA-PVA with 19 wt% hard block content demonstrated improved tensile strength (0.83 versus 0.33 MPa). Nevertheless, it is noteworthy that the relatively low tensile strength (<1 MPa) of these materials points toward good adhesion performance (*vide infra*) [10]. TGA was used to evaluate the thermal stability of the synthesized copolymers (Figure S6), and no significant differences were found between them. TGA under an inert atmosphere showed 5 wt% loss temperatures of PVA-PBuLA-PVA and PIA-PBuLA-PIA to be 386 °C and 389 °C, respectively. Hence, these polymers can be processes at elevated temperatures without degradation, and could therefore be used in practical applications demanding high temperatures, e.g., hot melt adhesives [37]. Finally, the adhesion properties of the synthesized all-acrylic ABA block biopolymers with a lactic acid-based elastomeric phase were evaluated. A 30 wt% solution of PVA-PBuLA-PVA and PIA-PBuLA-PIA was spread on a polyethylene terephthalate (PET) film [10,14].

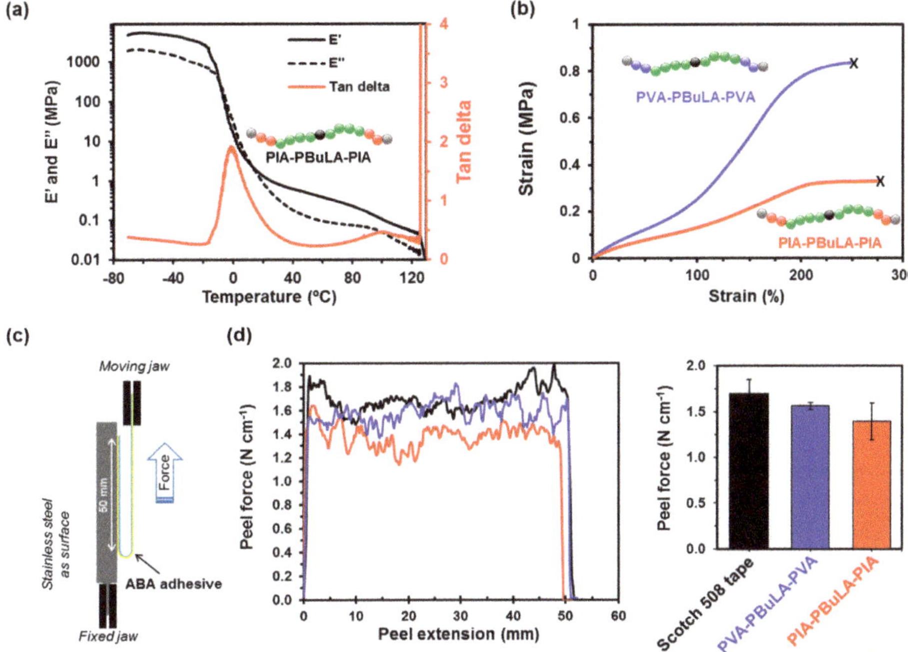

Figure 9. (**a**) Dynamic tensile storage (E′) and loss (E″) moduli and tan (δ = E″/E′) as a function of temperature of PIA-PBuLA-PIA triblock; (**b**) stress-strain curves of PIA-PBuLA-PIA and PVA-PBuLA-PVA; (**c**) schematic design of 180° peel test; and (**d**) 180° peel force of PIA-PBuLA-PIA, PVA-PBuLA-PVA and Scotch 508 tape.

After complete evaporation of the solvent at room temperature, the coated PET films were adhered onto a stainless steel plate, and the peel resistance was measured by pulling the adhered films off the plate at an angle of 180° (Figure 9c). Data for both polymers is presented in Figure 9d, and compared with a commercial Scotch 508 tape. The 180° peel strength for PVA-PBuLA-PVA and PIA-PBuLA-PIA was about 1.5 N·cm^{-1}, which is comparable to the tested commercial tape. All the investigated samples had adhesive failure, leaving no residue on the adherend. It is important to mention that both synthesized triblock copolymers were tested without the addition of tackifier or additive. Hence, taking into account that no optimization studies were conducted, these results suggest that the all-acrylic ABA copolymers with a lactic acid-based elastomeric building block are promising materials for PSA applications, and open the door to further investigations in this line.

3. Materials and Methods

3.1. Materials

3,5-Bis(2-dodecylthiocarbonothioylthio-1-oxopropoxy)benzoic acid (BTCBA, 98%), 2-(Dodecylthiocarbonothioylthio)propionic acid (DTPA, 98%), acryloyl chloride (≥97%), triethylamine (TEA, ≥99%), 2-methyltetrahydrofuran (Me-THF, 99.5%), n-Butyl acrylate (BA, 99%), butyl ʟ-lactate (98%) were all purchased from Merck KGaA (Darmstadt, Germany) and used as received unless otherwise specified. Thus, BA was passed through a short column of basic Al_2O_3 prior to use in order to remove the radical inhibitor and both TEA and Me-THF were distilled prior to use from CaH_2 and sodium/benzophenone, respectively. 2,2′-Azobis(2-methylpropionitrile) [AIBN, ≥98%, Fluka (Buchs, Switzerland)] was recrystallized three times from methanol before further use. Deuterated chloroform ($CDCl_3$) was purchased from Eurisotop, and ethyl-5-(dimethylamido)-2-methyl-5-oxopentanoate

(Rhodiasolv® PolarClean) was kindly donated by Solvay (Aubervilliers, France). Isosorbide acetate, i.e., isosorbide 2-acrylate-5-acetate and vanillin acrylate monomers were prepared following previously reported synthetic procedures [32–34].

3.2. Methods

400 MHz (for ^{1}H) and 100.6 MHz (for ^{13}C) NMR spectra were recorded on a Varian VNMR-S400 NMR instrument at 25 °C in CDCl$_3$ with tetramethylsilane (TMS) as an internal standard. The number-average molecular weight (M_n) and polydispersity (M_w/M_n) of the synthesized polymers were obtained via GPC using an Agilent 1200 series system (Agilent Technologies Inc., Santa Clara, CA, USA) equipped with three serial columns (PLgel 3 μm MIXED-E, PLgel 5 μm MIXED-D and PLgel 20 μm from Polymer Laboratories) and an Agilent 1100 series refractive-index detector. THF HPLC grade (Panreac Química S.L.U., Barcelona, Spain) was used as eluent at a flow rate of 1.0 mL/min. The calibration curves for GPC analysis were obtained with poly(methyl methacrylate) (PMMA) standards purchased from American Polymer Standards (Mentor, OH, USA). The molecular weights were calculated using the universal calibration principle and Mark-Houwink parameters. The glass transition temperature (T_g) of all polymers was determined on the second heating/cooling ramps using DSC measurements conducted on a Mettler DSC3+ thermal analyzers (Mettler-Toledo S.A.E., Barcelona, Spain) using N$_2$ as a purge gas (50 mL/min) at scanning rate 20 °C/min in the −80 to 150 °C temperature range. Calibration was based on an indium standard (heat flow calibration) and an indium-lead-zinc standard (temperature calibration). Thermal stability studies were carried out on a Mettler TGA2/LF/1100 (Mettler-Toledo S.A.E., Barcelona, Spain) with N$_2$ as a purge gas at flow rate of 50 mL/min. The studies were performed in the 30–600 °C temperature range at a heating rate of 10 °C/min. ESI MS were run on a chromatographic system Agilent G3250AA liquid chromatography coupled to 6210 Time of Flight (TOF) mass spectrometer from Agilent Technologies Inc. (Santa Clara, CA, USA) with an ESI interface. Dynamic mechanical analysis of all polymers was performed on a DMA Q800 (TA Instruments, New Castle, DE, USA) in tension configuration. Rectangular-shaped specimens (5 mm length, 5 mm width, 0.12 mm thickness) were prepared by compression molding into a rectangular steel mould (20 mm 5 mm 1.2 mm) between two parallel steel plates on a Specac Atlas Manual Hydraulic Press (Specac Ltd, Orpington, UK) at 100 °C, with an applied load of 4 tons for 30 min. Then, the rectangular specimen was further heated in the range of −70 to 200 °C using a heating rate of 3 °C/min and a fixed frequency of 1 Hz. For tensile testing, the lower grip was stationary, and the upper grip was raised at a force ramp rate of 1 N/min to obtain tensile strength and elongation at break of polymer at 27 °C. The peel strength analysis was performed on an Instron 5965 (Instron, Bucks, UK) at a peel rate of 300 mm·min^{-1}.

3.3. Synthesis of BuLA Monomer

Butyl ʟ-lactate (20.0 g, 0.14 mol) and anhydrous triethylamine (21.4 g, 0.21 mol) were dissolved in dry 2-methyltetrahydrofuran (Me-THF; 50 mL) under a constant flow of argon. After the solution was stirred for 30 min at 0–5 °C, acryloyl chloride (14.8 g, 0.16 mol) dissolved in dry Me-THF (50 mL) was added dropwise. The reaction was allowed to proceed for 24 h at room temperature. The reaction mixture was then filtered and Me-THF was removed under reduced pressure. The resulting residue was dissolved in 150 mL of diethyl ether, and the solution was subsequently washed with HCl 1M (150 mL) and a saturated NaHCO$_3$ solution (150 mL). The organic layer was rinsed with a brine solution and dried over anhydrous MgSO$_4$. The final residue was purified by vacuum distillation in the presence of 5 (*w/w* %) of hydroquinone to afford BuLA (20.5 g, 75%) as a colorless liquid. ^{1}H NMR (400MHz, CDCl$_3$, δ): 6.48 (dd, 1H), 6.19 (dd, 1H), 5.90 (dd, 1H), 5.16 (q, 1H), 4.16 (m, 2H), 1.63 (m, 2H), 1.53 (d, 3H), 1.38 (m, 2H), 0.93 (t, 3H); ^{13}C NMR (100.6 MHz, CDCl$_3$, δ): 170.72, 165.34, 131.74, 127.71, 68.78, 65.13, 30.51, 18.99, 16.96, 13.62. HRMS-TOF for BuLA: [M + NH$_4$]$^+$ calcd for C$_{10}$H$_{20}$NO$_4{}^+$, 218.1392 found, 218.1393 [M + H]$^+$ calcd for C$_{10}$H$_{17}$O$_4{}^+$, 201.1127, found, 201.1119; [M + Na]$^+$ calcd for C$_{10}$H$_{16}$NaO$_4{}^+$, 223.0946, found, 223.0942.

3.4. General Procedure for the RAFT Polymerization of BuLA Monomer

The synthesis of poly(L-butyl lactate acrylate) (PBuLA) macro-CTA was performed in bulk. A representative procedure is described as follows: AIBN (2.07 mg, 0.01 mmol) 3,5-Bis(2-dodecylthiocarbonothioylthio-1-oxopropoxy)benzoic acid (BTCBA) (or DTPA) (103.3 mg, 0.13 mmol) and L-butyl lactate acrylate (10.0 g, 50.40 mmol) were mixed in a 25 mL Schlenk tube equipped with a teflon stirring bar. The reaction mixture was degassed by bubbling argon for 30 min and subsequently submerged into a preheated, stirring oil bath maintained at 70 °C. To monitor the monomer conversion, the side arm of the tube was purged with argon before it was opened to remove two drops of sample using an airtight syringe. Samples were quenched by immediately placing them into liquid nitrogen exposed to air. After that, samples were dissolved in CDCl$_3$ and the monomer conversion was measured by ^{1}H NMR spectroscopy. The M_n and M_w/M_n values were determined by GPC calibrated with Poly(methyl methacrylate) (PMMA) standards. Finally, to stop the reaction, the Schlenk flask was placed into nitrogen liquid and opened to air. The final polymerization mixture was dissolved in the minimal volume of DCM and precipitated in 400 mL of cold hexane and isolated by decanting off the supernatant fluid. The procedure was repeated three times and the final polymer was dried under a vacuum for 24 h at room temperature to lead a viscous yellowish liquid. Model homopolymers prepared from butyl, isosorbide acetate and vanillin acrylates were prepared following the same procedure.

3.5. General Procedure for RAFT Block Copolymerization Experiments

An illustrative example is provided. CTA-poly(BuLA)-CTA (1.0 g, 0.013 mmol), isosorbide acetate acrylate (0.26 g, 1.06 mmol), AIBN (0.47 mg, 2.86 μmol) (from a dilute stock solution), and 3 mL of ethyl-5-(dimethylamido)-2-methyl-5-oxopentanoate (Rhodiasolv® PolarClean) were added to a Schlenk flask equipped with a teflon stirring bar. The flask was sealed and degassed by bubbling argon for 45 min, and then submerged in a preheated oil bath at 70 °C. After stirring vigorously for 24 h, the flask was immediately submerged into liquid nitrogen and opened to air. Two drops of sample were dissolved in CDCl$_3$, and a monomer conversion was determined by ^{1}H NMR. The polymerization mixture was dissolved in the minimum of DCM, and precipitated three times in cold methanol (3 × 200 mL). The final copolymer was then dried in a vacuum at room temperature for 24 h.

3.6. Cell Cultures

HT-29 (human colon adenocarcinoma), HeLa (human cervix epithelioid carcinoma) and MIA-Pa-Ca-2 (human Caucasian pancreatic carcinoma) cells were obtained from ATCC and maintained in Dulbecco's Modified Eagle's Medium (DMEM) (PAN-Biotech GmbH, Germany) supplemented with 10% FBS (Fetal Bovine Serum), 1% P/S (Penicillin/Streptomycin) and 1% NEAAs (Non-Essential Amino Acids) at 37 °C with 5% CO$_2$.

3.7. Cell Viability Assays

The cellular cytotoxicity of selected polymers was assessed in three cell lines: HT-29, HeLa and MIA-Pa-Ca-2 cells. The cell viability experiments were carried out by serial dilutions of PBuLA-Z, which was diluted in acetone.

Cells were plated in 96-well plates (8000 cells/95 μL/well in HeLa and MIA-Pa-Ca-2 cells; 9000 cells/95 μL/well in HT-29 cells) with supplemented DMEM without phenol red. Forty-eight hours later, 5 μL of each serial dilutions of the polymer were added to the cells. 100 mg/mL was the maximum tested concentration of the polymer in the cells (poly(butyl lactate acrylate) stock: 2 g/mL). The borders of the plates were in presence of 100 μL of PBS, in order to prevent evaporation-related problems caused by acetone.

The cells were in the presence of the polymer during 24 h and after this period the cytotoxicity was checked by adding the CellTiter reagent. The CellTiter 96® AQueous One Solution Cell Proliferation Assay Kit (Promega, Madrid, Spain) provides a convenient and sensitive procedure for determining the number of viable cells in cytotoxicity assays. 20 μL of CellTiter reagent (1/4 dilution in DMEM without phenol red) were added into each well of the 96-well assay plates containing the samples. The plates were incubated at 37 °C for 2 h in a humidified, 5% CO_2 atmosphere and later the absorbance was recorded at 490 nm (background correction at 800 nm). Each experiment was performed in triplicate and repeated at least twice.

3.8. General Procedure for 180° Peel Adhesion Tests

The samples for 180° peel adhesion tests were prepared by dissolving the polymer in ethyl acetate to obtain a 30% solid content solution. As a representative example, 200 mg of triblock copolymer was dissolved in 520 μL ethyl acetate. The solution was solvent-cast on a polyethylene terephthalate (PET) film of 19 mm width using a standard laboratory drawdown rod. The film was allowed to dry under ambient conditions open to air in a chemical fume hood for 4 h. The resultant coated film strips were approximately 5 cm long. The peel strength analyses were performed on an Instron 5965 at a peel rate of 300 mm.min^{-1}. 5 cm-wide strips of the coated PET films were placed on a clean stainless steel panel, as an adherend. The coated film was gently pressed against steel plate by manually rolling in order to develop good contact between the adhesive and the steel plate. The strip was then peeled from the stainless steel panel. The reported average peel force and standard deviation values were acquired from at least three replicates. Adhesion testing on commercially available scotch 805 PSA was also performed for comparison under identical conditions.

4. Conclusions

In summary, our study demonstrates that L-butyl lactate solvent can be upgraded to well-defined tacky triblock copolymers, suitable for applications as PSAs. The functionalization of BuL with an acrylate moiety generated a monomer well behaved under RAFT polymerization. Good control over molecular weight and molecular weight distribution was achieved in bulk, using either monofunctional and bifunctional trithiocarbonate-type chain transfer agents. Subsequently, poly(BuLA), with a relative low T_g (−20 °C), good thermal stability (5% wt. loss at 340 °C) and low toxicity in the cell cultures assessed was evaluated as sustainable elastomeric core block in all-acrylate ABA copolymers using isosorbide and vanillin-derived glassy polyacrylates as representative end blocks. Materials with a low content of hard building block (<20 wt%) were targeted, pursuing the preparation of materials with high tack. The thermal and mechanical properties of the prepared copolymers were evaluated to demonstrate suitability of rubbery poly(alkyl lactate) building blocks for developing innovative sustainable materials. As is noteworthy, 180° peel adhesion measurements showed that the synthesized biosourced all-acrylic ABA copolymers possess competitive performance when compared with commercial pressure-sensitive tapes. The significance of developing innovative adhesives based on renewable resources combining good thermal stability and low cytotoxicity is that these materials are promising for a broad range of applications, including both biomedical and technological fields.

Supplementary Materials: The following are available online: Figure S1: GPC analysis PBuLA-2Z; Figure S2: ^{1}H NMR spectrum of the reaction mixture after the chain extension of PBuLA-2Z macro-RAFT agent with IA; Figure S3: GPC traces of the chain extension of PBuLA-2Z macro-RAFT agent with VA; Figure S4: DSC analysis of model homopolymers and block copolymers; Figure S5: DMTA analysis of PVA-PBuLA-PVA triblock; and Figure S6: TGA analysis for PIA-PBuLA-PIA and PVA-PBuLA-PVA under nitrogen and air atmosphere.

Author Contributions: Conceptualization, N.B. and G.L.; methodology, N.B., A.J.-A., I.L., L.V., O.A. and G.L.; investigation, N.B., A.J.-A. and I.L.; writing—original draft preparation, G.L.; writing—review and editing, N.B., A.J.-A., J.C.R., V.C., M.G., L.V. and O.A.; project administration, J.C.R., V.C., M.G. and G.L.; funding acquisition, J.C.R., V.C., M.G. and G.L. All authors have read and agreed to the published version of the manuscript.

Funding: This research was funded by Spanish Ministerio de Ciencia, Innovación y Universidades through project MAT2017-82669-R (to G.L. and J.C.R.), Serra Hunter Programme of the Government of Catalonia (to G.L.), University Rovira i Virgili (DL003536 grant to N.B.), Miguel Servet Program from Instituto de Salud Carlos III (CPII13/00017 to O.A.), Fondo de Investigaciones Sanitarias from Instituto de Salud Carlos III, and European Union (ERDF/ESF, 'Investing in your future') (PI18/00349 to O.A.), Diputación General de Aragón (Predoctoral Research Contract 2019 to A.J.-A., 'Digestive Pathology Group' B25_20R to O.A.), and Centro de Investigación Biomédica en Red en Enfermedades Hepáticas y Digestivas (CIBERehd). I.L. and L.V. gratefully acknowledge support from the European Regional Development Fund (No. MOBTT21), the Estonia-Russia Cross Border Cooperation Programme (ER30) and the Baltic Research Programme in Estonia (EEA grant no. EMP426).

Acknowledgments: The authors would like to thank Rhodia-Solvay for kindly donating Rhodiasolv® PolarClean solvent.

Conflicts of Interest: The authors declare no conflict of interest.

References and Note

1. Bates, C.M.; Bates, F.S. 50th Anniversary Perspective: Block Polymers-Pure Potential. *Macromolecules* **2017**, *50*, 3–22. [CrossRef]
2. Epps, T.H., III; O'Reilly, R.K. Block Copolymers: Controlling Nanostructures to Generate Functional Materials—Synthesis, Characterization, and Engineering. *Chem. Sci.* **2016**, *7*, 1674–1689. [CrossRef]
3. Ruzette, A.V.; Leibler, L. Block Copolymers in Tomorrow's Plastics. *Nat. Mater.* **2005**, *4*, 19–31. [CrossRef] [PubMed]
4. Holden, G.; Kricheldorf, H.R.; Quirk, R.P. *Thermoplastic Elastomers*, 3rd ed.; Hanser Publishers: Munich, Germany, 2004.
5. Shipp, D.A.; Wang, J.L.; Matyjaszewski, K. Synthesis of Acrylate and Methacrylate Block Copolymers using Atom Transfer Radical Polymerization. *Macromolecules* **1998**, *31*, 8005–8008. [CrossRef]
6. Jeusette, M.; Leclère, P.; Lazzaroni, R.; Simal, F.; Vaneecke, J.; Lardot, T.; Roose, P. New "All-Acrylate" Block Copolymers: Synthesis and Influence of the Architecture on the Morphology and the Mechanical Properties. *Macromolecules* **2007**, *40*, 1055–1065. [CrossRef]
7. Nakamura, Y.; Adachi, M.; Tachibana, Y.M.; Sakai, Y.; Nakano, S.; Fujii, S.; Sasaki, M.; Urahama, Y. Tack and Viscoelastic Properties of an Acrylic Block Copolymer/Tackifier System. *Int. J. Adhes. Adhes.* **2009**, *29*, 806–811. [CrossRef]
8. Corrigan, N.; Jung, K.; Moad, G.; Hawker, C.J.; Matyjaszewski, K.; Boyer, C. Reversible-Deactivation Radical Polymerization (Controlled/Living Radical Polymerization): From Discovery to Materials Design and Applications. *Prog. Polym. Sci.* **2020**, *111*, 101311. [CrossRef]
9. Wang, S.; Ding, W.; Yang, G.; Robertson, M.L. Biorenewable Thermoplastic Elastomeric Triblock Copolymers Containing Salicylic Acid-Derived End-Blocks and a Fatty Acid-Derived Midblock. *Macromol. Chem. Phys.* **2016**, *217*, 292–303. [CrossRef]
10. Nasiria, M.; Reineke, T.M. Sustainable Glucose-based Block Copolymers Exhibit Elastomeric and Adhesive Behavior. *Polym. Chem.* **2016**, *7*, 5233–5240. [CrossRef]
11. Gallagher, J.J.; Hillmyer, M.A.; Reineke, T.M. Acrylic Triblock Copolymers Incorporating Isosorbide for Pressure Sensitive Adhesives. *ACS Sustain. Chem. Eng.* **2016**, *4*, 3379–3387. [CrossRef]
12. Satoh, K.; Lee, D.H.; Nagai, K.; Kamigaito, M. Precision Synthesis of Bio-Based Acrylic Thermoplastic Elastomer by RAFT Polymerization of Itaconic Acid Derivatives. *Macromol. Rapid Commun.* **2014**, *35*, 161–167. [CrossRef] [PubMed]
13. Ding, W.; Wang, S.; Yao, K.; Ganewatta, M.S.; Tang, C.; Robertson, M.L. Physical Behavior of Triblock Copolymer Thermoplastic Elastomers Containing Sustainable Rosin-Derived Polymethacrylate End Blocks. *ACS Sustain. Chem. Eng.* **2017**, *5*, 11470–11480. [CrossRef]
14. Wang, S.; Shuai, L.; Saha, B.; Vlachos, D.G.; Epps, T.H. From Tree to Tape: Direct Synthesis of Pressure Sensitive Adhesives from Depolymerized Raw Lignocellulosic Biomass. *ACS Cent. Sci.* **2018**, *4*, 701–708. [CrossRef] [PubMed]
15. Vendamme, R.; Schüwer, N.; Eevers, W. Recent Synthetic Approaches and Emerging Bio-Inspired Strategies for the Development of Sustainable Pressure-Sensitive Adhesives derived from Renewable Building Blocks. *J. Appl. Polym. Sci.* **2014**, *131*, 40669. [CrossRef]
16. Sajjad, H.; Tolman, W.B.; Reineke, T.M. Block Copolymer Pressure-Sensitive Adhesives Derived from Fatty Acids and Triacetic Acid Lactone. *ACS Appl. Polym. Mater.* **2020**, *2*, 2719–2728. [CrossRef]
17. Noppalit, S.; Simula, A.; Ballard, N.; Callies, X.; Asua, J.M.; Billon, L. Renewable Terpene Derivative as a Biosourced Elastomeric Building Block in the Design of Functional Acrylic Copolymers. *Biomacromolecules* **2019**, *20*, 2241–2251. [CrossRef]
18. Wang, Y.; Deng, W.; Wang, B.; Zhang, Q.; Wan, X.; Tang, Z.; Wang, Y.; Zhu, C.; Cao, Z.; Wang, G.; et al. Chemical Synthesis of Lactic Acid from Cellulose Catalysed by Lead(II) Ions in Water. *Nat. Commun.* **2013**, *4*, 2141. [CrossRef]

19. Cubas-Cano, E.; González-Fernández, C.; Ballesteros, M.; Tomás-Pejó, E. Biotechnological Advances in Lactic Acid Production by Lactic Acid Bacteria: Lignocellulose as Novel Substrate. *Biofuels Bioprod. Bioref.* **2018**, *12*, 290–303. [CrossRef]

20. Yee, G.M.; Hillmyer, M.A.; Tonks, I.A. Bioderived Acrylates from Alkyl Lactates via Pd-Catalyzed Hydroesterification. *ACS Sustain. Chem. Eng.* **2018**, *6*, 9579–9584. [CrossRef]

21. Kim, H.J.; Jin, K.; Shim, J.; Dean, W.; Hillmyer, M.A.; Ellison, C.J. Sustainable Triblock Copolymers as Tunable and Degradable Pressure Sensitive Adhesives. *ACS Sustain. Chem. Eng.* **2020**, *8*, 12036–12044. [CrossRef]

22. Galaster™ EL 98.5 FCC, Galasolv™ 003, PURASOLV® ELECT/BL/ML, and VertecBio™ EL are examples of commercial ethyl lactate solvents produced by Galactic, Corbion, or Vertec Biosolvents.

23. Purushothaman, M.; Krishnan, P.S.G.; Nayak, S.K. Poly(alkyl lactate acrylate)s Having Tunable Hydrophilicity. *J. Appl. Polym. Sci.* **2014**, *131*, 40962. [CrossRef]

24. Purushothaman, M.; Krishnan, P.S.G.; Nayak, S.K. Tunable Hydrophilicity of Poly(ethyl lactate acrylate-co-acrylic acid). *J. Renew. Mater.* **2015**, *3*, 292–301. [CrossRef]

25. Purushothaman, M.; Krishnan, P.S.G.; Nayak, S.K. Effect of Isoalkyl Lactates as Pendant Group on Poly(acrylic acid). *J. Macromol. Sci. A* **2014**, *51*, 470–480. [CrossRef]

26. Purushothaman, M.; Krishnan, P.; Santhana, G.; Nayak, S.K. Effect of Nano Sepiolite Fiber on the Properties of Poly(ethyl lactate acrylate): Hydrophilicity and Thermal Stability. *Mater. Focus* **2018**, *7*, 101–107. [CrossRef]

27. Bensabeh, N.; Moreno, A.; Roig, A.; Monaghan, O.R.; Ronda, J.C.; Cádiz, V.; Galià, M.; Howdle, S.M.; Lligadas, G.; Percec, V. Polyacrylates Derived from Biobased Ethyl Lactate Solvent via SET-LRP. *Biomacromolecules* **2019**, *20*, 2135–2147. [CrossRef]

28. Moreno, A.; Bensabeh, N.; Parve, J.; Ronda, J.C.; Cádiz, V.; Galià, M.; Vares, L.; Lligadas, G.; Percec, V. SET-LRP of Bio- and Petroleum-Sourced Methacrylates in Aqueous Alcoholic Mixtures. *Biomacromolecules* **2019**, *20*, 1816–1827. [CrossRef]

29. Bensabeh, N.; Moreno, A.; Roig, A.; Rahimzadeh, M.; Rahimi, K.; Ronda, J.C.; Cádiz, V.; Galià, M.; Percec, V.; Rodriguez-Emmenegger, C.; et al. Photoinduced Upgrading of Lactic Acid-Based Solvents to Block Copolymer Surfactants. *ACS Sustain. Chem. Eng.* **2020**, *8*, 1276–1284. [CrossRef]

30. Semsarilar, M.; Perrier, S. 'Green' Reversible Addition-Fragmentation Chain-Transfer (RAFT) Polymerization. *Nat. Chem.* **2010**, *2*, 811–820. [CrossRef]

31. Holmberg, A.L.; Reno, K.H.; Wool, R.P.; Epps, T.H., III. Biobased Building Blocks for the Rational Design of Renewable Block Polymers. *Soft Matter* **2014**, *10*, 7405–7424. [CrossRef]

32. Matt, L.; Parve, J.; Parve, O.; Pehk, T.; Pham, T.H.; Liblikas, I.; Vares, L.; Jannasch, P. Enzymatic Synthesis and Polymerization of Isosorbide-Based Monomethacrylates for High-Tg Plastics. *ACS Sustain. Chem. Eng.* **2018**, *6*, 17382–17390. [CrossRef]

33. Zhang, P.; Liu, T.; Xin, J.; Zhang, J. Biobased Miktoarm Star Copolymer from Soybean Oil, Isosorbide, and Caprolactone. *J. Appl. Polym. Sci.* **2020**, *137*, 48281. [CrossRef]

34. Zhou, J.; Zhang, H.; Deng, J.; Wu, Y. High Glass-Transition Temperature Acrylate Polymers Derived from Biomasses, Syringaldehyde, and Vanillin. *Macromol. Chem. Phys.* **2016**, *217*, 2402–2408. [CrossRef]

35. Noppalit, S.; Simula, A.; Billon, L.; Asua, J.M. Paving the Way to Sustainable Waterborne Pressure-Sensitive Adhesives Using Terpene-Based Triblock Copolymers. *ACS Sustain. Chem. Eng.* **2019**, *7*, 17990–17998. [CrossRef]

36. Cseri, L.; Szekely, G. Towards Cleaner PolarClean: Efficient Synthesis and Extended Applications of the Polar Aprotic Solvent Methyl 5-(dimethylamino)-2-methyl-5-oxopentanoate. *Green. Chem.* **2019**, *21*, 4178–4188. [CrossRef]

37. Lu, W.; Wang, Y.; Wang, W.; Cheng, S.; Zhu, J.; Xu, Y.; Hong, K.; Kang, N.G.; Mays, J. All Acrylic—based Thermoplastic Elastomers with Upper Service Temperature and Superior Mechanical Properties. *Polym. Chem.* **2017**, *8*, 5741–5748. [CrossRef]

Sample Availability: Samples of the monomeric compounds, macroinitiators and block copolymers are available from the authors.

Publisher's Note: MDPI stays neutral with regard to jurisdictional claims in published maps and institutional affiliations.

Article

Silane Coupling Agent Modifies the Mechanical Properties of a Chitosan Microfiber

Yuki Shirosaki *, Toshinobu Okabayashi and Saki Yasutomi

Faculty of Engineering, Kyushu Institute of Technology, 1-1 Sensui-cho, Tobata-ku, Kitakyushu, Fukuoka 804-8550, Japan; o109011t@mail.kyutech.jp (T.O.); s-yasutomi@che.kyutech.ac.jp (S.Y.)
* Correspondence: yukis@che.kyutech.ac.jp

Academic Editors: Sylvain Caillol and Pietro Russo
Received: 8 October 2020; Accepted: 12 November 2020; Published: 13 November 2020

Abstract: Chitosan microfibers are widely used in medical applications because they have favorable inherent properties. However, their mechanical properties require further improvement. In the present study, a trimethoxysilane aldehyde (TMSA) crosslinking agent was added to chitosan microfibers to improve their tensile strength. The chitosan microfibers were prepared using a coagulation method. The tensile strength of the chitosan microfibers was improved by crosslinking them with TMSA, even when only a small amount was used (less than 1%). TMSA did not change the orientation of the chitosan molecules. Furthermore, aldehyde derived from TMSA did not remain, and siloxane units were formed in the microfibers.

Keywords: chitosan; silane coupling agent; microfiber; crosslinking; mechanical strength

---

1. Introduction

Microfiber-based materials are widely used in medical applications, such as sutures, textiles, and scaffolds [1–3]. Natural polymers are good candidates for the preparation of microfibers because they are biocompatible, non-toxic, biodegradable, and have low immunogenicity. Chitin and chitosan are candidates for biofibers, and their inherent properties have been reported [4]. Chitosan fibers are easier to prepare than chitin fibers because they are soluble in dilute acids.

Chitosan is often crosslinked with reagents, such as glutaraldehyde and epoxy compounds, to improve its mechanical properties and control its biodegradability [4–6]. However, such compounds are highly cytotoxic [7] and reduce the biocompatibility of chitosan. Shirosaki et al. already used the silane coupling agent γ-glycidoxypropyltrimethoxysilane (GPTMS) to crosslink chitosan and demonstrated improvements in the mechanical properties of a chitosan membrane [8]. The epoxide groups of GPTMS can react with the amino groups of chitosan, and each GPTMS molecule polycondenses to form −Si−O−Si− networks. The crystalline properties of chitin and chitosan are also important for their mechanical characteristics. The addition of GPTMS can disturb the orientation of the crystalline domains of chitosan [8,9], reducing its mechanical strength.

In the present study, we used trimethoxysilane aldehyde (TMSA) shown in Figure 1 as a crosslinked agent to improve the tensile strength of the chitosan microfibers. It is to be expected that, compared with GPTMS, a smaller amount of TMSA can effectively react with the amino groups in chitosan owing to crosslinking by the aldehyde groups and the polycondensation of the silanol groups derived from the three methoxysilane groups of TMSA. The reaction between the amino groups of chitosan and the aldehyde groups of glutaraldehyde results in the formation of a Schiff base [10]. The aldehyde groups of TMSA can react as shown in Figure 2.

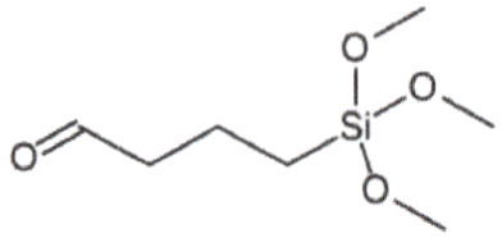

Figure 1. Trimethoxysilane aldehyde (TMSA).

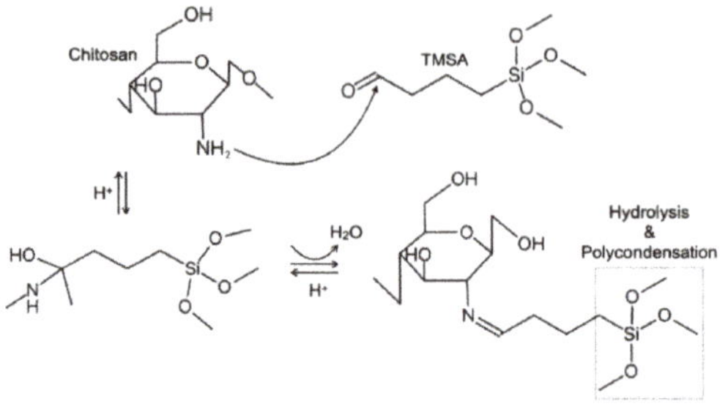

Figure 2. Expected reaction between chitosan and TMSA forming a Schiff base.

2. Results

2.1. Structural Characterization

The fibers were white and flexible, and the diameter was around 200 μm, regardless of the amount of TMSA added (Figure 3). Figure 4 shows the X-ray diffraction (XRD) patterns of the fibers. Only the peaks attributable to chitosan ($2\theta = 20°$, PDF #00-039-1894) and chitin ($2\theta = 10°$, PDF #00-035-1974) were detected. The half width at $2\theta = 20°$ is shown in Table 1. The half width value did not change after the addition of TMSA.

Table 1. The half width values at $2\theta = 20°$ from XRD patterns.

Ch	ChTMSA0001	ChTMSA0005	ChTMSA001
1.0	1.0	1.0	1.0

Figure 3. Photograph of the chitosan−TMSA microfiber (ChTMSA001).

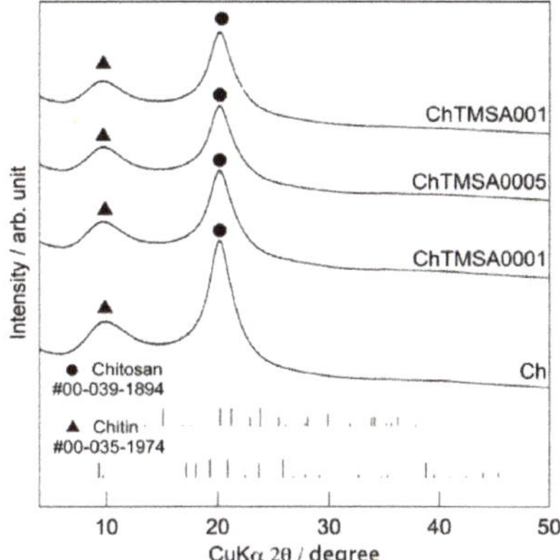

Figure 4. XRD patterns of the chitosan–TMSA microfibers.

The intensity of birefringence in the microscope images obtained using crossed nicols (Figure 5) confirms the orientation of the chitosan molecules in the fibers. Figure 6 shows the scanning electron microscope (SEM) images and energy-dispersive X-ray spectroscopy (EDS) spectra of the fibers. The fibers had flattened surfaces and no deposits were discernable. Only fiber ChTMSA001 had silicon on its surface. The results of the ninhydrin test are shown in Table 2. The number of remaining amino groups decreased as the amount of TMSA increased, and fibers ChTMSA0001, ChTMSA0005, and ChTMSA001 had $102 \pm 4\%$, $92 \pm 2\%$, and $76 \pm 1\%$ amino groups, respectively.

Figure 7 shows the Fourier-transform infrared spectroscopy (FT-IR) spectra of the fibers. The characteristic bands of chitosan were detected in the spectra of all the fibers, as demonstrated in a previous paper [11]. Bands attributable to stretching in amide II (NH_2 str.) were detected at approximately 1563 cm^{-1}, and their intensity decreased after TMSA was added. The peaks in the spectra of the ChTMSA fibers at approximately 800 cm^{-1} can be attributed to symmetrical Si–O–Si stretching vibrations [12–14]. C=O derived from TMSA was not detected in the fibers.

Table 2. Ninhydrin test results indicating the number of amino groups remaining in the chitosan–TMSA microfibers.

Ch	ChTMSA0001	ChTMSA0005	ChTMSA001
100	102 ± 4	92 ± 2	76 ± 1

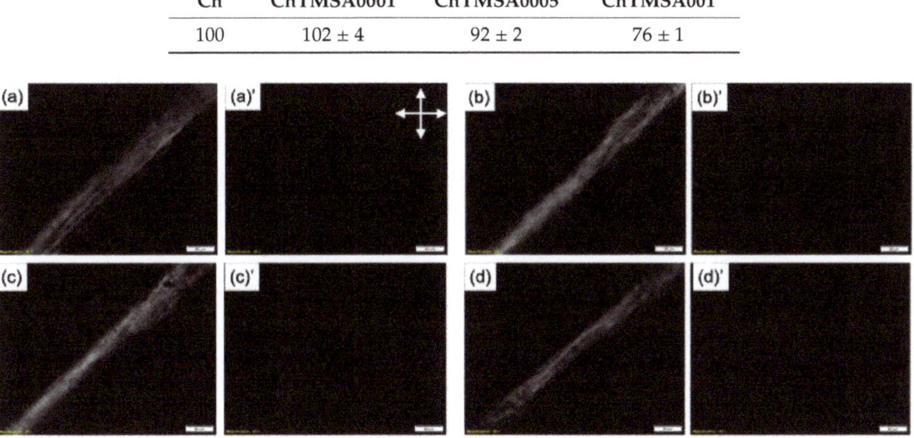

Figure 5. Optical microscope images of the chitosan–TMSA microfibers obtained with crossed nicols at 45° (**a–d**) and 90° (**a**)′–(**d**)′. (**a**) and (**a**)′ Ch; (**b**) and (**b**)′ ChTMSA0001; (**c**) and (**c**)′ ChTMSA0005; and (**d**) and (**d**)′ ChTMSA001. The arrows indicate the direction of the nicols.

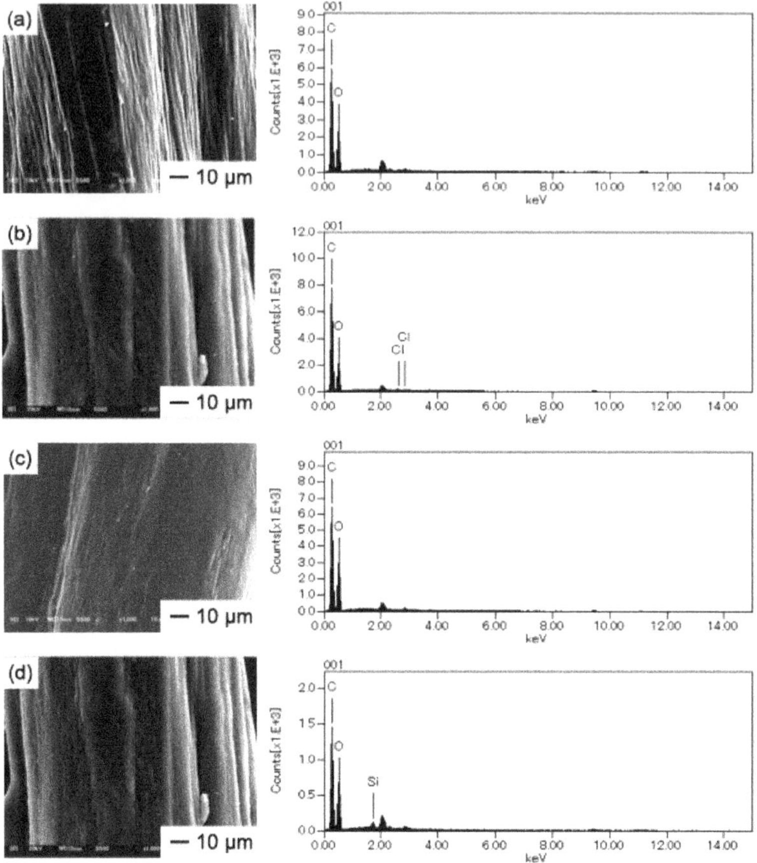

Figure 6. SEM images and EDS spectra of the chitosan−TMSA microfibers. (**a**) Ch; (**b**) ChTMSA0001; (**c**) ChTMSA0005; and (**d**) ChTMSA001.

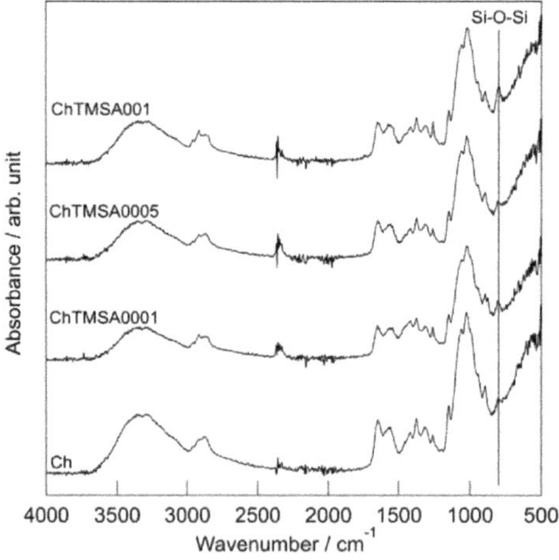

Figure 7. FT-IR spectra of the chitosan−TMSA microfibers.

2.2. Mechanical Strength

The results of the tensile strength tests are shown in Figures 8 and 9. The maximum tensile strengths of Ch, ChTMSA0001, ChTMSA0005, and ChTMSA001 were 121.3 ± 26.7, 151.0 ± 23.5, 184.5 ± 30.0, and 133.8 ± 8.9 MPa, respectively (i.e., ChTMSA0005 had the highest value). The maximum strain values of Ch, ChTMSA0001, ChTMSA0005, and ChTMSA001 were 12.3 ± 5.2, 4.6 ± 1.5, 5.6 ± 1.7, and 5.4 ± 0.8%, respectively (i.e., the ChTMSA fibers had lower maximum strain values than Ch). The Young's modulus values of Ch, ChTMSA0001, ChTMSA0005, and ChTMSA001 were 9.9 ± 5.1, 32.6 ± 15.7, 32.9 ± 17.3, and 24.6 ± 11.3 MPa, respectively (i.e., the ChTMSA fibers had higher Young's modulus values than Ch).

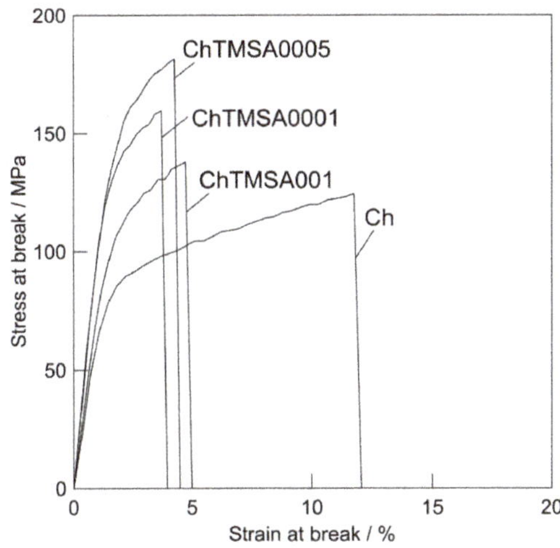

Figure 8. Stress–strain curves of the chitosan–TMSA microfibers.

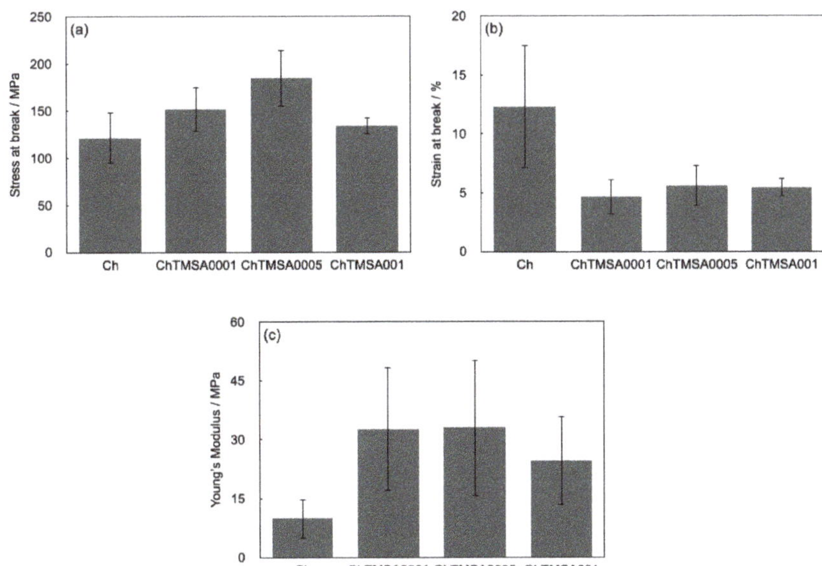

Figure 9. Maximum tensile strength values (**a**), maximum strain values (**b**), and Young's modulus of the chitosan–TMSA microfibers (**c**). Both sets of values were obtained at the breaking point.

3. Discussion

Generally, chitosan molecules have crystalline and amorphous domains derived from the original structures of the chitin nanofibrils [15,16]. The XRD results reveal that the crystallinity of the chitosan molecules was retained even after the addition of TMSA. This suggests that the crystalline chitosan domains were not affected by crosslinking with TMSA. The ninhydrin tests insisted that the number of crosslinked amino groups was too small an amount even in ChTMSA001. FT-IR and the results of the ninhydrin tests demonstrated the existence of crosslinking between chitosan and TMSA. In the present study, the pH of the chitosan solution was approximately 4.0. Therefore, the amino groups probably formed $-NH_3^+$, thereby decreasing the nucleophilicity of the nitrogen atoms and the extent of the Schiff base reaction following mixing with TMSA [10]. However, the remaining TMSA was able to react completely with the amino groups of chitosan during washing with NaOH solution. According to the ninhydrin results, the crosslinked ration of ChTMSA001 was 20%. ChTMSA001 fibers refer to the addition of TMSA to produce 1% crosslinking of a chitosan unit. In addition to aldehyde groups, silanol groups interacted with the amino groups [9]. FT-IR showed symmetrical $-Si-O-Si-$ stretching vibrations, which suggests that the $-Si-OH$ groups formed by the hydrolysis of TMSA polycondensed in the chitosan fibers like silica [12–14]. However, the SEM images confirmed that there are no micro-sized deposits on the surfaces of the fibers. The EDS results also suggested that the addition of TMSA resulted in very little crosslinking within the chitosan molecules, in particular ChTMSA0001 and ChTMSA0005. Eventually, 19% of the amino groups in ChTMSA001 interacted with Si-OH and 1% were reacted with aldehyde. The orientation of the chitosan molecules in the fibers was also retained after the addition of TMSA. Figure 10 represents the expected molecular structure of the ChTMSA001 fiber according to the following characterization: (1) chitosan has crystalline and amorphous domains; (2) each bundle is oriented along the stretching direction; (3) just 1% of the amino groups in the fibers reacted with TMSA, and most of the TMSA reacted outside the bundles; and (4) the silanol groups derived from TMSA interacted with the chitosan molecules or polycondensed. The mechanical propertied of the fibers were improved by crosslinking. The increased Young's modulus values insist that the interaction between chitosan become stronger and chitosan–TMSA fibers have more stiffness.

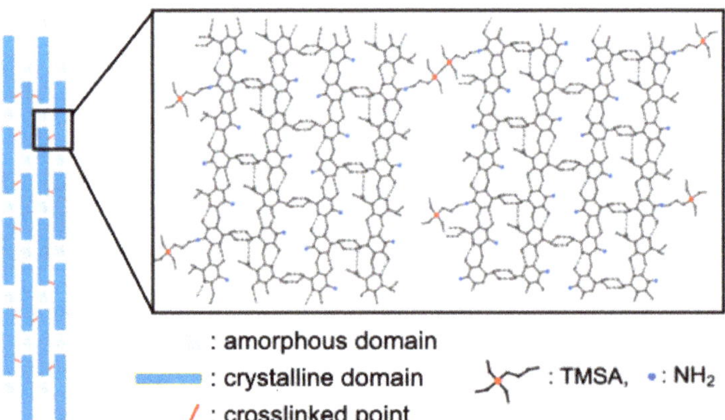

Figure 10. Structural model of the chitosan–TMSA (ChTMSA001) fiber.

Silanol groups ($-Si-OH$) and siloxane networks ($-Si-O-Si-$) derived from GPTMS favored cell attachment and proliferation [8,9]. In the present study, FT-IR indicated that no aldehyde groups remained after the crosslinking reaction and siloxane networks formed. It is to be hoped that TMSA in the fibers is non-toxic and that chitosan–TMSA microfibers can be used in medical applications.

4. Materials and Methods

4.1. Preparation of Chitosan Monofibers with Silane Coupling Agent

Chitosan powder (high-molecular weight, molecular weight = 310,000—375,000 Da, degree of acetylation (DA) = 76.0%; Sigma-Aldrich®, St. Louis, MO, USA) was dissolved into 0.2 M aqueous acetic acid and mixed using a planetary centrifuge (ARE-310, Thinky Corp., Tokyo, Japan) to obtain a homogeneous 3.5% (*w/v*) chitosan homogeneous solution. The appropriate amount of TMSA (UCT, Bristol, TN, USA) was stirred in dimethyl sulfoxide to be hydrolyzed at room temperature, and the hydrolyzed precursor sols were obtained. The obtained precursor sols were added to a 3.5% (*w/v*) chitosan solution, and the mixture was mixed in a planetary centrifuge for 20 min. Mono-fibers were produced using a coagulation method, as described in a previous paper [17]. After washing with ethanol and a 0.2 M aqueous solution of sodium hydroxide, the fibers were stretched by rolling at 30/45 rpm with two rollers. Finally, the stretched fibers were dried at room temperature in a desiccator. The starting material composition of the fiber and sample code of each fiber are presented in Table 3.

Table 3. Starting composition (molar ratio) of the fibers and sample code of each fiber.

	Ch	ChTMSA0001	ChTMSA0005	ChTMSA001
TMSA/Chitosan	0	0.001	0.005	0.01

4.2. Characterization and Mechanical Properties of the Fibers

The crystal structures of the fibers were examined using an automated multipurpose X-ray diffractometer (CuKα, 45 kV, 200 mA, 0.01°/step; Rigaku smartLab, Rigaku, Tokyo, Japan).

The orientation of the fibers was examined using an inverted microscope (IX73; Olympus Co., Tokyo, Japan) under crossed nicols. Each fiber was placed on a glass substrate and sandwiched between crossed nicols at 45° and 90° (Figure 11).

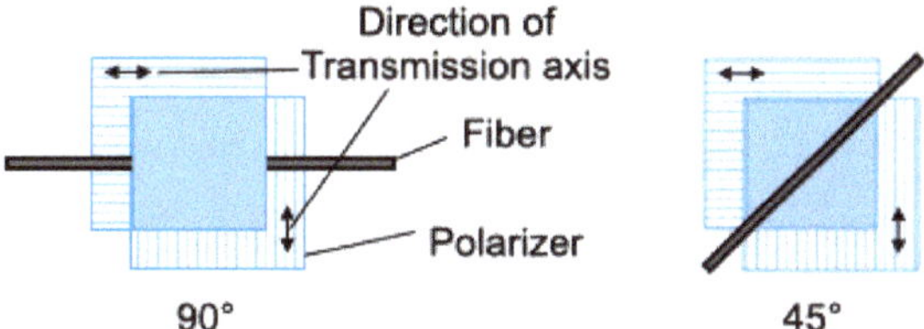

Figure 11. Examination of the molecular orientation of the fibers using crossed nicols. If the molecules were oriented, the image became dark at 90° and bright at 45°.

The surface morphology of each fiber was examined using an SEM (JMS-6010 PLUS/LA; JEOL, Tokyo, Japan) equipped with an EDX. Before examination, the fibers were coated with Pt/Pd to a thickness of approximately 20 nm (MSP-1S Magnetron Sputter; Vacuum Device Inc., Mito, Japan).

The molecular structures of the pulverized fibers were examined by FT-IR (FT/IR-6100; JASCO Co., Tokyo, Japan) using attenuated total reflectance at a resolution of 4 cm^{-1} and the accumulation of 100 scans.

The number of amino groups remaining in the fibers after crosslinking was evaluated by a ninhydrin assay [8]. The pulverized fibers (0.02 g) were suspended in 4 mL of ninhydrin solution (Ninhydrin Kit, L-8500 Set; Wako Chemicals, Osaka, Japan) and incubated at 80 °C for 40 min while shaking at 100 rpm. After cooling to room temperature, the optical densities of the sample (A_{sample}) and chitosan (A_{ch}) solutions were recorded at 570 nm using a spectrophotometer (DS-11+/w; DeNovix,

Wilmington, DC, USA). The ninhydrin test was performed with replicates ($n = 3$). The number of the remaining amino groups was calculated using the following Equation (1):

$$\text{Remaining amino groups (\%)} = \text{Asample/Ach} \times 100 \tag{1}$$

As described in a previous paper [17], the mechanical properties of the fibers were examined by determining their tensile strengths at 0.5 mm/s using a creep meter (RE2-3305C; YAMADEN Co., Ltd., Tokyo, Japan).

5. Conclusions

Flexible chitosan microfibers were prepared by crosslinking them with TMSA. Results of XRD analysis showed that addition of TMSA did not change the crystallinity. Crossed nicols images also showed the orientation of chitosan. TMSA (molar ratio = 0005/amino groups chitosan) crosslinked only 1% of chitosan, but effectively improved the tensile strength of the microfibers (maximum strength around 184.5 MPa) and increased their stiffness. These mechanical properties depend on the crosslinking between chitosan and TMSA, as well as the –Si–O–Si– bonds of TMSA, which was confirmed by FT-IR.

Author Contributions: Y.S. conceived and designed the experiments and prepared the draft of the manuscript; T.O. performed the preparation and characterization; S.Y. analyzed the data and wrote a part of the initial draft of the manuscript. All authors have read and agreed to the published version of the manuscript.

Funding: This research was funded by the Ministry of Education, Culture, Sports, Science and Technology (MEXT), Japan (Promotion and Standardization of the Tenure-Track System, "Kojinsenbatsu"), and by the Japan Society for the Promotion of Science (JSPS) KAKENHI (grant Number JP16K12898).

Acknowledgments: We gratefully acknowledge Katsumi Yamamoto, Center for Instrumental Analysis, Equipment Sharing Sector, Organization for Promotion of Open Innovation, Kyushu Institute of Technology, for conducting XRD analysis of samples. We also would like to thank Frank Kitching, Msc., from Edanz Group (www.edanzediting. com/ac) for editing a draft of this manuscript.

Conflicts of Interest: The authors declare no conflict of interest.

References

1. Gimbel, H.V.; Raanan, M.G.; DeLuca, M. Effect of suture material on postoperative astigmatism. *J. Cataract. Refract. Surg.* **1992**, *18*, 42–50. [CrossRef]
2. Onoe, H.; Takeuchi, S. Cell-laden microfibers for bottom-up tissue engineering. *Drug Discov. Today* **2015**, *20*, 236–246. [CrossRef] [PubMed]
3. Tuzlakoglu, K.; Reis, R.L. Biodegradable Polymeric Fiber Structures in Tissue Engineering. *Tissue Eng. Part B Rev.* **2009**, *15*, 17–27. [CrossRef] [PubMed]
4. Pillai, C.; Paul, W.; Sharma, C.P. Chitin and chitosan polymers: Chemistry, solubility and fiber formation. *Prog. Polym. Sci.* **2009**, *34*, 641–678. [CrossRef]
5. Ngah, W.W.; Hanafiah, M.; Yong, S. Adsorption of humic acid from aqueous solutions on crosslinked chitosan–epichlorohydrin beads: Kinetics and isotherm studies. *Colloids Surf. B. Biointerfaces* **2008**, *65*, 18–24. [CrossRef] [PubMed]
6. Poon, L.; Wilson, L.D.; Headley, J.V. Chitosan-glutaraldehyde copolymers and their sorption properties. *Carbohydr. Polym.* **2014**, *109*, 92–101. [CrossRef] [PubMed]
7. Zeiger, E.; Gollapudi, B.; Spencer, P. Genetic toxicity and carcinogenicity studies of glutaraldehyde? A review. *Mutat. Res. Mutat. Res.* **2005**, *589*, 136–151. [CrossRef] [PubMed]
8. Shirosaki, Y.; Tsuru, K.; Hayakawa, S.; Osaka, A.; Lopes, A.; Santos, J.D.; Costa, M.A.C.; Fernandes, M.H. Physical, chemical and in vitro biological profile of chitosan hybrid membrane as a function of organosiloxane concentration. *Acta Biomater.* **2009**, *5*, 346–355. [CrossRef] [PubMed]
9. Shirosaki, Y.; Okayama, T.; Tsuru, K.; Hayakawa, S.; Osaka, A. In vitro bioactivity and MG63 cytocompatibility of chitosan-silicate hybrids. *Int. J. Mater. Chem.* **2013**, *3*, 1–7. [CrossRef]
10. Wang, T.; Turhan, M.; Gunasekaran, S. Selected properties of pH-sensitive, biodegradable chitosan–poly(vinyl alcohol) hydrogel. *Polym. Int.* **2004**, *53*, 911–918. [CrossRef]

11. Shirosaki, Y.; Tsuru, K.; Hayakawa, S.; Osaka, A.; Lopes, M.A.; Santos, J.D.; Fernandes, M.H. In vitro cytocompatibility of MG63 cells on chitosan-organosiloxane hybrid membranes. *Biomaterials* **2005**, *26*, 485–493. [CrossRef] [PubMed]

12. Yan, W.; Liu, D.; Tan, D.; Yuan, P.; Chen, M. FTIR spectroscopy study of the structure changes of palygorskite under heating. *Spectrochim. Acta Part A Mol. Biomol. Spectrosc.* **2012**, *97*, 1052–1057. [CrossRef] [PubMed]

13. Mozgawa, W.; Król, M.; Dyczek, J.; Deja, J. Investigation of the coal fly ashes using IR spectroscopy. *Spectrochim. Acta Part A Mol. Biomol. Spectrosc.* **2014**, *132*, 889–894. [CrossRef] [PubMed]

14. Sanaeishoar, H.; Sabbaghan, M.; Mohave, F. Synthesis and characterization of micro-mesoporous MCM-41 using various ionic liquids as co-templates. *Microporous Mesoporous Mater.* **2015**, *217*, 219–224. [CrossRef]

15. Oh, D.X.; Cha, Y.J.; Nguyen, H.-L.; Je, H.H.; Jho, Y.S.; Hwang, D.S.; Yoon, D.K. Chiral nematic self-assembly of minimally surface damaged chitin nanofibrils and its load bearing functions. *Sci. Rep.* **2016**, *6*, 23245. [CrossRef] [PubMed]

16. Lertworasirikul, A.; Noguchi, K.; Ogawa, K.; Okuyama, K. Plausible molecular and crystal structures of chitosan/HI type II salt. *Carbohydr. Res.* **2004**, *339*, 835–843. [CrossRef] [PubMed]

17. Okada, T.; Nobunaga, Y.; Konishi, T.; Yoshioka, T.; Hayakawa, S.; Lopes, A.; Miyazaki, T.; Shirosaki, Y. Preparation of chitosan-hydroxyapatite composite mono-fiber using coagulation method and their mechanical properties. *Carbohydr. Polym.* **2017**, *175*, 355–360. [CrossRef] [PubMed]

Sample Availability: The datasets analysed during the current study available from the corresponding author on reasonable request.

Publisher's Note: MDPI stays neutral with regard to jurisdictional claims in published maps and institutional affiliations.

 *molecules*

Article

Curcumin-Loaded Bacterial Cellulose/Alginate/Gelatin as A Multifunctional Biopolymer Composite Film

Nadda Chiaoprakobkij [1,2], Thapanar Suwanmajo [3], Neeracha Sanchavanakit [4,*] and Muenduen Phisalaphong [2,*]

[1] Biomedical Engineering Program, Faculty of Engineering, Chulalongkorn University, Bangkok 10330, Thailand; naddakij@hotmail.com

[2] Department of Chemical Engineering, Faculty of Engineering, Chulalongkorn University, Bangkok 10330, Thailand

[3] Centre of Excellence in Materials Science and Technology, Department of Chemistry, Faculty of Science, Chiang Mai University, Chiang Mai 50200, Thailand; thapanar.s@cmu.ac.th

[4] Center of Excellence for Regenerative Dentistry, Department of Anatomy, Faculty of Dentistry, Chulalongkorn University, Bangkok 10330, Thailand

* Correspondence: neeracha.r@chula.ac.th (N.S.); muenduen.p@chula.ac.th (M.P.)

Academic Editor: Sylvain Caillol
Received: 14 July 2020; Accepted: 19 August 2020; Published: 21 August 2020

Abstract: Multifunctional biopolymer composites comprising mechanically-disintegrated bacterial cellulose, alginate, gelatin and curcumin plasticized with glycerol were successfully fabricated through a simple, facile, cost-effective mechanical blending and casting method. SEM images indicate a well-distributed structure of the composites. The water contact angles existed in the range of 50–70°. Measured water vapor permeability values were 300–800 g/m^2/24 h, which were comparable with those of commercial dressing products. No release of curcumin from the films was observed during the immersion in PBS and artificial saliva, and the fluid uptakes were in the range of 100–700%. Films were stretchable and provided appropriate stiffness and enduring deformation. Hydrated films adhered firmly onto the skin. In vitro mucoadhesion time was found in the range of 0.5–6 h with porcine mucosa as model membrane under artificial saliva medium. The curcumin-loaded films had substantial antibacterial activity against *E. coli* and *S. aureus*. The films showed non-cytotoxicity to human keratinocytes and human gingival fibroblasts but exhibited potent anticancer activity in oral cancer cells. Therefore, these curcumin-loaded films showed their potential for use as leave-on skin applications. These versatile films can be further developed to achieve desirable characteristics for local topical patches for wound care, periodontitis and oral cancer treatment.

Keywords: bacterial cellulose; alginate; gelatin; curcumin; biomaterials

---

1. Introduction

Over the past years, various types of biopolymers have been studied and proposed as alternatives for biomedical approaches. However, there are still attempts to develop functional biomaterials to meet the increasing demands for practical and safe options for clinical uses. Antimicrobial biomaterial obtained by incorporation of antimicrobial agents into biocomposites shows promise as a useful and environmentally friendly method; however, antimicrobial agents such as antiseptics might cause some drawbacks as they can lead to bacterial resistance with prolonged use, allergic reactions and side effects with high dose usage. Recently, silver nanoparticles (AgNPs) have been one of the prevalent metallic materials being studied, owing to their antimicrobial properties. Despite potential as antimicrobial agents, AgNPs have been reported to possibly be harmful to human health. The cytotoxic characteristics of AgNPs against mammalian cells have been documented [1–3]. For example, silver and gold nanoparticles were found to reduce fibroblasts migration in vitro [4]. Long-term studies are required to assess the safety for humans of metal nanoparticle-based biomaterials. On the other hand, natural-based compounds have been popular in modern therapy due to the low cost, limited adverse effects and high efficacy. Bacterial cellulose (BC) is an extracellular polysaccharide excreted by bacteria. Despite the identical chemical formula to that of cellulose from plants, BC is free of lignin and hemicellulose. Several bacterial cellulose-based biocomposites have been studied as alternative biomedical materials. Disintegrated BC (in gel suspension form) is an alternative form for producing BC composite films for large-scale production. Disintegrated BC can behave as a reinforcing agent or binder to the composites; the resulting composite showed better performance and improved mechanical strength and stiffness [5]. Alginate (A) is a natural polysaccharide that is biocompatible, biodegradable and non-toxic. Alginate is not only biocompatible and gellable under mild conditions but also able to improve compatibility, homogeneity, binding and dispersibility of polymer blends [6]. However, it does not present binding sites for cell attachment to which mammalian cells can bind. Gelatin (G) is one of the promising biomaterials because of its biological origin from collagen with excellent biocompatibility. Gelatin contains Arg-Gly-Asp sequences that are crucial for cell proliferation. Curcumin (C) is a yellow, non-water-soluble bioactive agent extracted from *Curcuma longa*. In recent years, it has been well documented in many in vitro and in vivo studies that curcumin possess a wide range of beneficial properties including wound healing, along with anti-inflammatory, antioxidant, antimicrobial and anticancer properties [7–9]. Recently, there have been reports on loading curcumin into polymers to produce functional polymer composites for various approaches. Cellulose/curcumin films were developed by using 1-allyl-3-methylimidazolium chloride (AMIMCl) for the dissolution of cellulose and curcumin and fabricated via by casting method [10]. Nanocurcumin prepared by ultrasonic process were impregnated into gelatin cellulose fibers to form microbial resistant fibers [11]. Chitosan/cellulose microcrystal films incorporated with curcumin were fabricated by vapor induced phase inversion (VIPI) [12]. Antibacterial curcumin/PVA/cellulose nanocrystal films were fabricated as antimicrobial wound dressing film [13]. Curcumin were entrapped in nanocellulose/chitosan hydrogel by swelling equilibrium method [14]. Curcumin was incorporated into nanocellulose by using Tween 20 as a surfactant for delivery of curcumin to stomach and upper intestinal tract [15]. Curcumin loaded bacterial cellulose films was developed and reported for anticancer against malignant melanoma skin cancer cells [16]. Despite a number of benefits, instability under physiological condition could limit the efficacy of curcumin [17].

In our previous study, homogenized BC/alginate/gelatin (BCAGG) composite films were successfully fabricated [18]. This current work is intended to broaden the implementation feasibility of multifunctional BCAGG composites by directly loading curcumin into the polymer matrix during the film fabrication via a casting method. Film mechanical properties, fluid uptake ability in various media, surface wettability, bioadhesive residence time and antibacterial activity were thoroughly evaluated. In addition, evaluation as an oral wound dressing, cytocompatibility and proliferation of human keratinocytes (HaCaT), human gingival fibroblasts (GF) and oral cancer cells (CAL-27) on the film surface were preliminary assessed.

2. Results

2.1. Morphology

The morphology of BC, homogenized BC and plant cellulose (Whatman filter paper No. 1) fibers were depicted by SEM, as presented in Figure 1A–C. The micrometer scale SEM images show no significant difference between mechanically-disintegrated BC fibers and native BC fibers. The fibrillated BC appeared highly entangled and formed a web-like network structure (Figure 1B) with fiber diameter in nanometers and length in micrometers. Figure 1D–G shows the surface morphology of curcumin-loaded films of BCAGG-C1, BCAGG-C2, BCAGG-C3 and BCAGG-C4 (containing curcumin at 0.17 ± 0.3, 0.34 ± 0.2,

0.55 ± 0.2 and 0.75 ± 0.1 mg/cm^3, respectively). It was observed that the entire surface was rather rough, with uniform appearance and no cracking. No significant microstructural differences existed for any film samples in this study.

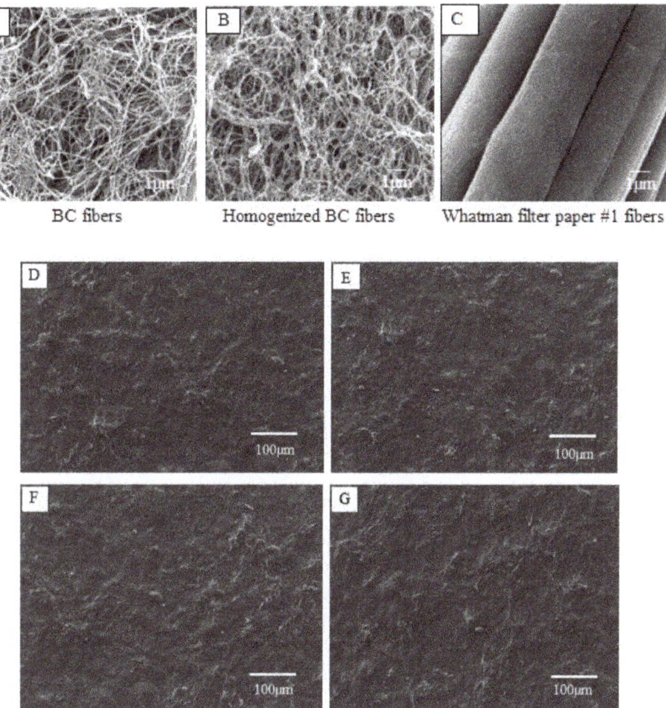

Figure 1. SEM images of: (**A**) BC fibers; (**B**) mechanically-disintegrated BC fibers; and (**C**) Whatman filter paper No. 1 fibers at 10,000× magnification. Surface morphology of: (**D**) BCAGG-C1; (**E**) BCAGG-C2; (**F**) BCAGG-C3; and (**G**) BCAGG-C4.

2.2. FT-IR Analysis

The interactions among the components of samples were investigated by FT-IR spectra (Figure 2). The typical absorption bands of BC were –OH groups at 3392 cm^{-1}, H-O-H bending of absorbed water at 1647 cm^{-1} and C-O-C stretching at 1061 cm^{-1} [18]. For FT-IR spectrum of alginate, the peaks at 3411, 1607 and 1423 cm^{-1} were attributed to –OH stretching, -COO- asymmetric stretching and -COO- symmetric stretching, respectively. For gelatin, the peak at 3408 cm^{-1} was attributed to the partially overlapped stretching vibrations of O-H and N-H groups. The amide I (C=O (stretching, amide II (N-H) bending and amide III (C-H) stretching, characteristic peaks of gelatin (protein), were clearly observed at 1643, 1535 and 1239 cm^{-1}, respectively. The major peaks of interest for curcumin were at around 3400 and 1601 cm^{-1}. FT-IR spectra of pure curcumin showed a peak at 3413 cm^{-1}, denoted as phenolic O-H stretching, and sharp peaks at 1601 cm^{-1}, indicative of C=C benzene stretching vibrations.

The FTIR results of all of curcumin-loaded films displayed all the characteristic bands of BC, alginate, gelatin and curcumin components. The result indicates their bonding interactions. Broad absorption bands between 3200 and 3550 cm^{-1} were due to overlapping of –OH groups and -NH$_2$ groups stretching vibrations [19]. The peaks of –OH stretching in the spectrum of curcumin-loading composites were obviously shifted and intensity changed in the range of 3200–3500 cm^{-1} as compared to that of BCAGG. This is an indication of intermolecular interaction between phenolic groups of curcumin and BCAGG matrix; therefore, curcumin likely interacted via hydrogen bonding interaction in the polymer matrix. The peaks at 1650, 1628, 1625 and 1616 cm^{-1} of all curcumin-loaded composites were assigned to the C=C stretching vibration of benzene rings, regarded as a characteristic peak of curcumin [20]. The major peaks of interest for curcumin are at around 3500 and 1601 cm^{-1}. FT-IR spectra of curcumin showed a peak at 3513 cm^{-1}, denoted as phenolic O-H stretching, and sharp peaks at 1601 cm^{-1}, indicative of C=C benzene stretching vibrations [21].

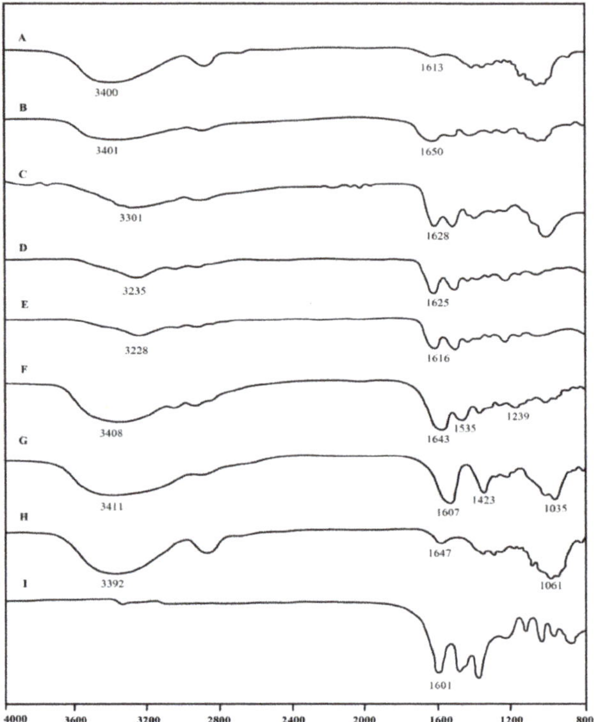

Figure 2. FT-IR spectra of: (**A**) BCAGG; (**B**) BCAGG-C1; (**C**) BCAGG-C2; (**D**) BCAGG-C3; (**E**) BCAGG-C4; (**F**) gelatin; (**G**) alginate; (**H**) BC; and (**I**) curcumin.

2.3. Fluid Absorption Capacity

The samples were completely submerged into PBS and artificial saliva in an effort to simulate clinical conditions. Fluid uptake capacity is an important factor in maintaining the optimum moist environment at the target site. Figure 3A,B shows that the fluid uptake ability decreased with an increase in curcumin contents in both PBS and artificial saliva. During the first 6 h, the fluid uptake ability in artificial saliva solution was slightly higher than that of PBS buffer uptake ability. The fluid uptakes increased rapidly and reached maximum after 24 h of incubation. After 24 h of incubation, BCAGG and BCAGG-C1 showed relatively higher PBS absorption compared to artificial saliva (Figure 3B). The fluid absorptions were in the range of 100–700%. All samples were able to maintain integrity throughout the 48 h of immersion (Figure 3C). The release of curcumin was not observed in the testing mediums during the experimental study. The amount of released curcumin was determined by UV-Vis spectrophotometry (Shimadzu UV-2550, Tokyo, Japan) at the wavelength of 420 nm. All tests were done in triplicate. The results show that there was no curcumin in the testing mediums. This is similar to observations of the previous report of curcumin loaded in gelatin/curcumin films, in which curcumin did not diffuse out of the polymer matrix [22]. Due to the high entanglement of mechanically-disintegrated BC fibers and good colloidal properties of

BC/alginate/gelatin blend [23], curcumin was well embedded into the matrix. In this study, curcumin was entrapped inside the cross-linked BCAGG network. During the drying process, water evaporated, and a dense and stable structure of films was formed and entrapped curcumin within the matrix. The FTIR result also indicate an intermolecular interaction between phenolic groups of curcumin and BCAGG matrix. Because curcumin has poor solubility in water, it hardly dissolves in water-based mediums. Therefore, curcumin was not released from the BCAGG-C films into PBS and artificial saliva.

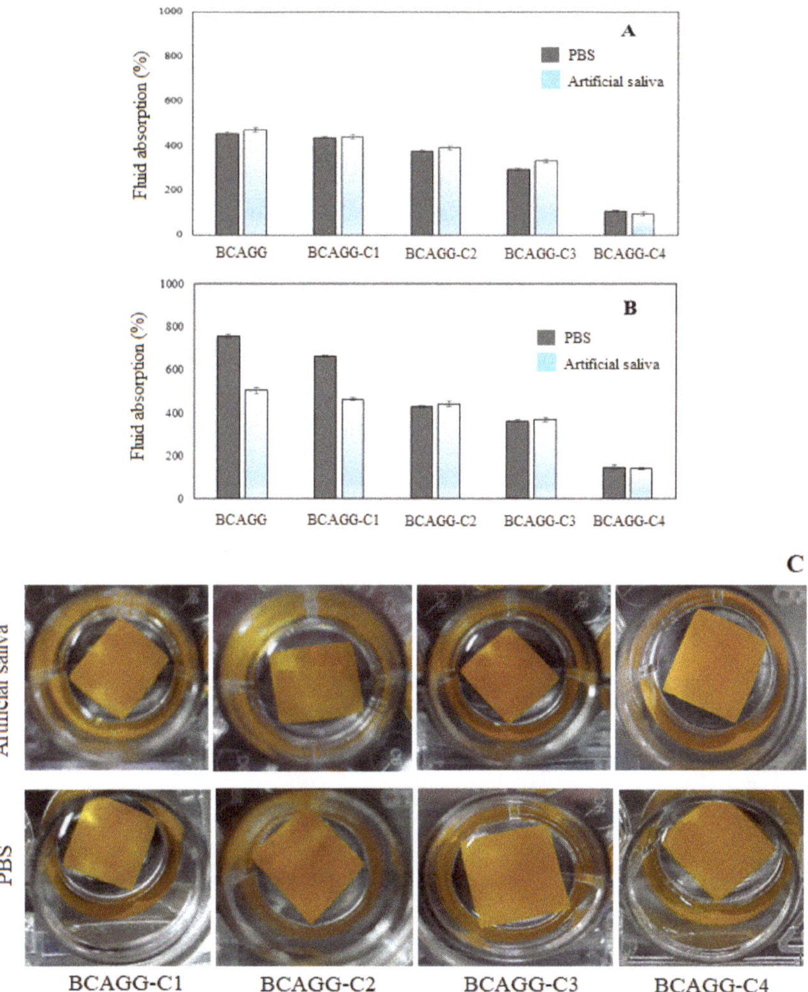

Figure 3. Fluid absorption after the immersion at 37 °C in PBS (pH 7.4) and artificial saliva (pH 6.2) for (**A**) 6 h and (**B**) 24 h; and (**C**) film samples after the immersion for 48 h.

2.4. Surface Wettability

The water contact angle of BCAGG was 49.5° [23], whereas those of BCAGG-Cs were 54.7–73.3°. The water contact angle of curcumin-loaded films increased with increasing curcumin content, as shown in Figure 4A–D. It was found that all hydrated films provided perfect coverage and neatly adhered to the skin with good flexibility, as shown in Figure 4E,F, respectively.

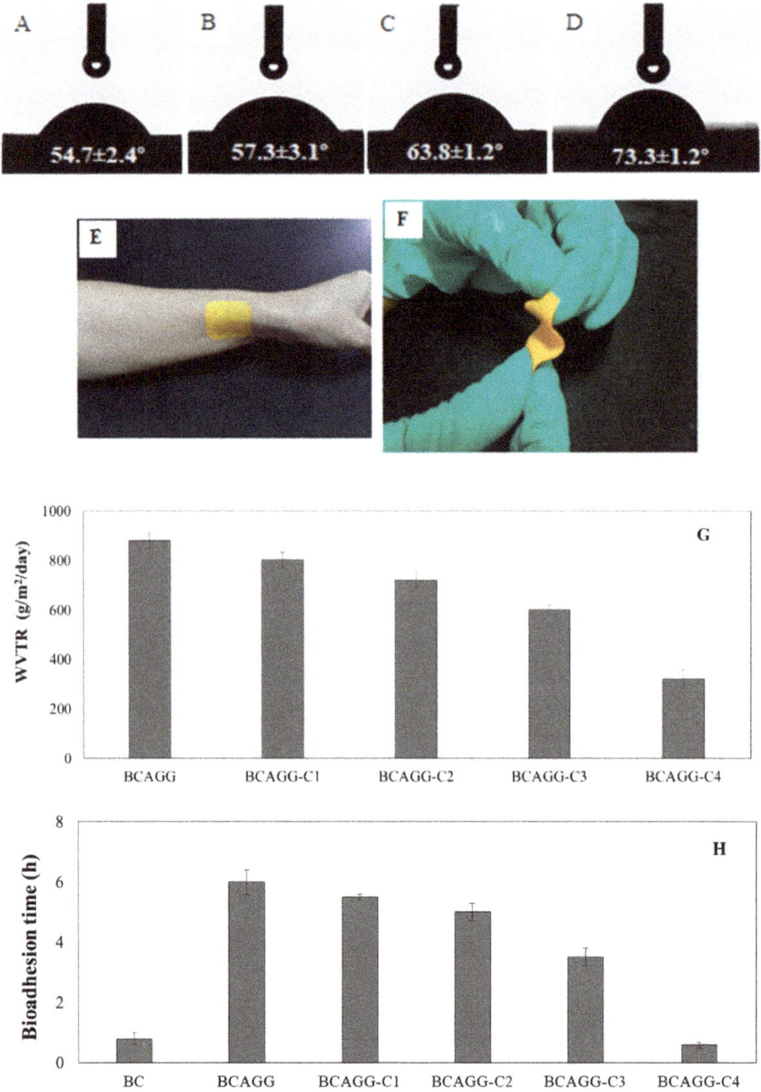

Figure 4. Water contact angles of: (**A**) BCAGG-C1; (**B**) BCAGG-C2; (**C**) BCAGG-C3; and (**D**) BCAGG-C4; (**E**) film attachment to skin; (**F**) film flexibility; (**G**) WVTR; and (**H**) bioadhesion time.

2.5. Water Vapor Transmission Rate (WVTR)

Figure 4G shows that WVTR of curcumin- loaded films decreased with the addition of curcumin. WVTRs of BCAGG-C films were around 300–800 $g/m^2/24$ h.

2.6. Bioadhesion Time

Bioadhesion times of BCAGG and BCAGG-C films are shown in Figure 4H. BCAGG exhibited more hydrophilic surface, leading to a faster rate of water absorption and higher adhesive performance compared with the curcumin-loaded films. It was observed that the adhesion time decreased with the addition of curcumin. The adhesion of the composite films lasted 0.5–6 h.

2.7. Mechanical Properties

The mechanical properties of dry and hydrated samples are shown in Figure 5. Tensile strength and Young's modulus of the dry films were in the range of 80–170 and 7000–8000 MPa, respectively, and the mechanical properties of the films decreased with increase in curcumin content. The mechanical strengths were much lower in the wet state; however, it was observed that tensile strength and Young's modulus of hydrated curcumin-loaded films relatively increased in comparison to non-curcumin loaded ones. Elongation at break of the films in both dry and wet states (Figure 5D) decreased with curcumin loading content. Elongation at break of the hydrated films was in the range of 15–40%.

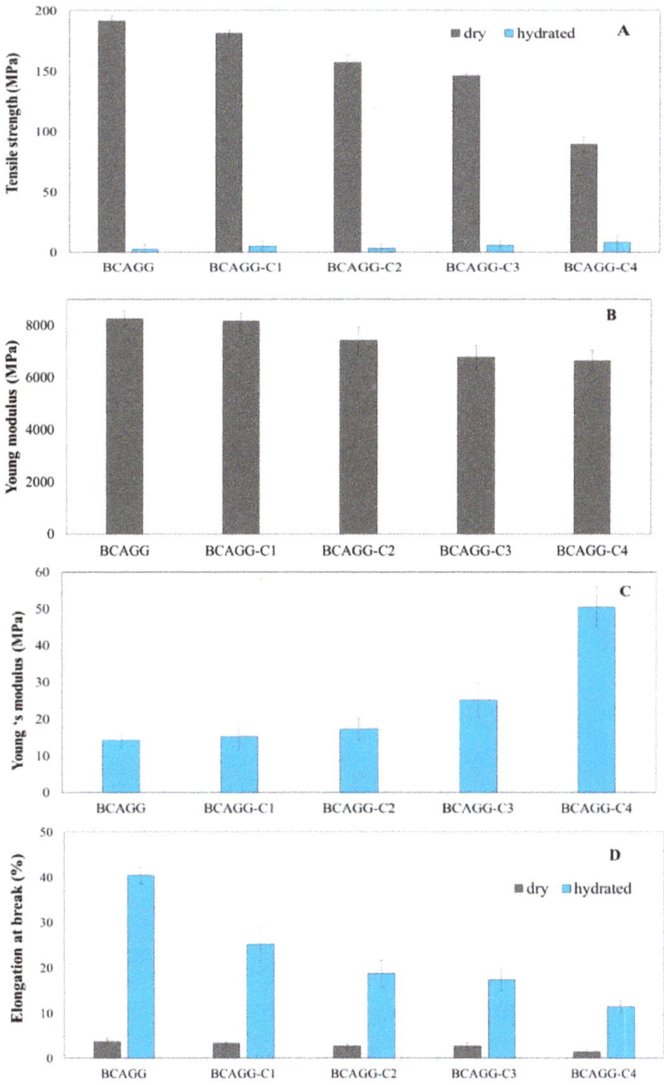

Figure 5. Mechanical properties of films: (**A**) tensile strength of dry and hydrated films; (**B**) Young's modulus of dry films; (**C**) Young's modulus of hydrated films; and (**D**) elongation at break (%) of dry and hydrated films.

2.8. In Vitro Cell Studies

The direct culture method was adopted to evaluate in vitro cytotoxicity of the fabricated films. Cell viability after cultivation for 48 h is shown in Figure 6. All samples were nontoxic to normal healthy cells (HaCaT and GF). According to ISO10993-5, the materials are classified as a noncytotoxic material if cell viability is greater than 70%. An increase in cell number on BCAGG, BCAGG-C1 and BCAGG-C2 was observed at 48 h incubation time. Along with curcumin loading, CAL-27 cell viability on BCAGG-C2 and BCAGG-C3 decreased by almost half in comparison to the control, which appears to be due to the anticancer property of curcumin. Morphologies of HaCaT, GF and CAL-27 are shown in Figure 7. HaCaT and GF cells tended to grow in flat monolayer with cells oriented in a parallel manner on coverslips. HaCaT and GF cells spread well on the fabricated films and retained their structural and morphological characteristic of cells. HaCaT exhibited and maintained the polygonal cell shape, which is typical of HaCaT cell morphology. GF cells tended to be randomly oriented, elongated and growing into multiple layers. Increasing curcumin content, to some extent, induced some cellular morphology changes. As shown in Figure 7, some GF cells exhibited a short spindle-shaped morphology on BCAGG-C3 film. Some round cells of HaCaT were also observed on BCAGG-C3. Meanwhile, CAL-27 cells displayed distinct morphological changes, including cell shrinkage, rounding up, irregularity in shapes, ruffled cell membrane surface and partial detachment on BCAGG-C films. CAL-27 cells were isolated, with the presence of cellular debris.

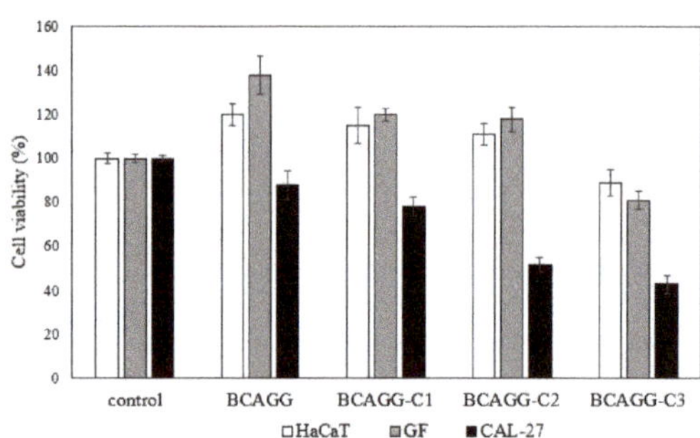

Figure 6. Cell viability (%) of HaCaT, GF and CAL-27 at 48 h after seeding.

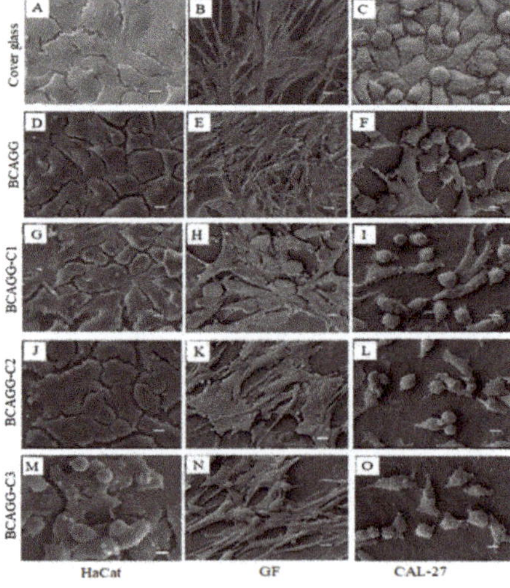

Figure 7. SEM images of HaCaT, GF and CAL-27 on film samples for 48 h (scale bar = 10 μm).

2.9. Antibacterial Tests

The results of antibacterial tests are shown in Figure 8A,B. BCAGG-C1, BCAGG-C2 and BCAGG-C3 films as well as commercially available antibacterial wound dressing, Bactrigra® were tested against *E. coli* and *S. aureus*. From our preliminary study, BCAGG had no antibacterial activities. By loading curcumin into the biocomposite matrix, the antibacterial activities against *E. coli* and *S. aureus* were enhanced with the increase in curcumin concentration. The bacterial counts of *E. coli* and *S. aureus* cells from 48 h incubation on surface of BCAGG-C3 were comparable to those of the commercial wound dressing. It was noticed that the composite films had greater inhibition effect against Gram-positive bacteria (*S. aureus*) than Gram-negative bacteria (*E. coli*).

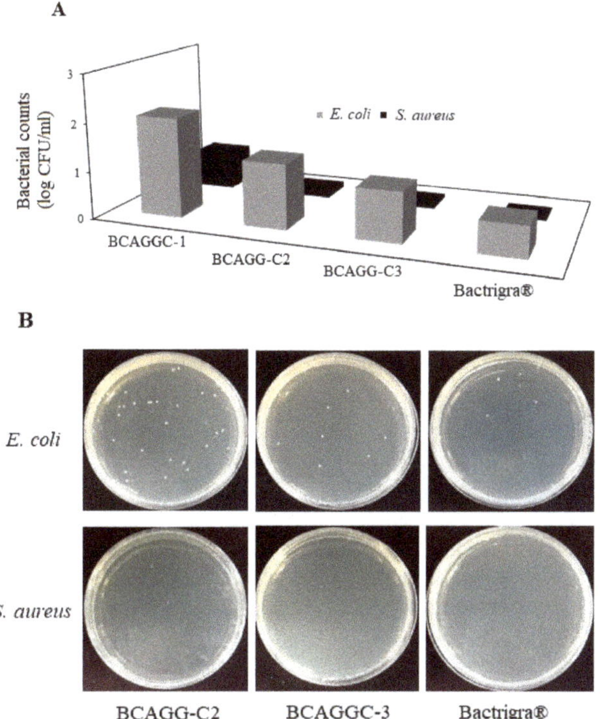

Figure 8. (**A**) Number of viable bacteria cells on the sample surface; and (**B**), agar plates of *E. coli* and *S. aureus* colonies at 48 h incubation on BCAGG-C2, BCAGG-C3 and Bactrigra®.

3. Discussion

The fibrillated BC appeared highly entangled and formed a web-like network structure. The high aspect ratio fibers were suitable for applying as a reinforcing agent in composite materials [23]. BCAGG films demonstrated a uniform structure. According to SEM analysis, there was no significant effect of curcumin loading on the surface morphology. A good distribution of curcumin without agglomerations was obtained. This suggests reasonably good compatibility between curcumin and other polymers in the matrix.

The FTIR results of all curcumin-loaded films show characteristic absorption bands of curcumin, suggesting that the films should retain the functions of curcumin. It was found that there were no additional peaks, excluding the specific bands of BC, alginate, gelatin and curcumin. However, typical peaks of curcumin-incorporated film were shifted to some extent as compared to BCAGG film.

The results are in accordance with the previous report [24]. Similar observations were reported in gelatin/curcumin composite films [25], in which the FT-IR and NMR results indicate the hydrogen bonding between curcumin and gelatin. According to the previous work of BC–alginate matrix, the slight shift and intensity changes were observed due to intermolecular hydrogen bonding formation between anionic chains of BC and alginate [6]. The opposite charge of polyanionic polysaccharide chains and positively charged gelatin could electrostatically attract each other, leading to the formation of stable polyelectrolyte complexes [18]. These findings were consistent with the previous study on complex formation between oppositely charged macromolecules [26]. The amino groups of gelatin and carboxyl groups of alginate had electrostatic interactions with each other. In addition, it was reported that cellulose could form hydrogen bonding with alginate and gelatin within the network and enriched a dense and rigid network structure [27]. Curcumin-loaded BCAGG depicted similar characteristic peaks of all parental molecules. The results show that there was no new peak generated or disappeared after the addition of curcumin. However, small shifts and intensity changes were observed, which might be attributed to local modifications leading to small variations of rotation and vibration frequencies via hydrogen bonds. During the drying step, a dense and stable structure of films was formed and curcumin was entrapped inside the cross-linked BCAGG network. In addition, it was found that the peaks of –OH stretching in the spectrum of curcumin-loading composites were obviously shifted and intensity changed in the ranges of 3200–3500 and 1650–1616 cm^{-1} as compared to that of BCAGG, which indicated the intermolecular interaction between phenolic groups of curcumin and BCAGG matrix. Therefore, curcumin likely interacted via hydrogen bonding in the polymer matrix. The negatively charged groups could interact with –OH groups of curcumin through hydrogen bonding [28].

The fluid uptake ability of the films, which decreased with the increase in curcumin loading, can be attributed to hydrophobic property of curcumin. It was also found that curcumin did not diffuse out of the films throughout 48 h immersion in PBS and artificial saliva. It was previously reported that cellulose crystals acted as a diffusion barrier that retarded the release of curcumin from the film samples [12]. Hydrogen bonding interactions within the film matrix restrict the relaxation and motion of polymeric chains. Such physical crosslinks would impede water molecules diffusion and retard curcumin release. At the beginning, the fluid uptake ability in artificial saliva solution was slightly higher than that of PBS buffer. The fluid of artificial saliva, which contains carboxymethyl cellulose (CMC) as a viscosity modifier to mimic human saliva, had better wetting properties on the material surface. CMC is commonly known as humectant, which is a substance that actually bonds with water molecules to increase the water content to itself. The fluid uptakes increased rapidly and reached the maximum after 24 h of incubation. Alginate does not swell in acidic environments but possesses hydrophilic and water retaining properties in neutral or mildly alkaline environments [29]. Gelatin/polyglycerol sebacate copolymer hydrogels were also found to swell less under acidic pH conditions [30]. Similar behavior was also reported for calcium alginate/chitosan bi-polymeric beads in acidic stimulating intestinal fluid [31]. Under acidic conditions, the amino groups were protonated and the electrostatic interactions between carboxyl groups (COO-) within the polymeric network and protonated amine groups (-NH$_3$+) were strengthened, resulting in a denser structure with lower water uptake. Consequently, the films without or with low curcumin loading (BCAGG and BCAGG-C1) exhibited higher PBS absorption than that of artificial saliva after the 24 h immersion in the fluids.

Typically, wound pH values vary from 5.5 to 7.8. After prolonged periods, it will settle at pH 7.4–7.0 [32]. Wound dressing absorption capacity is important for wound healing. According to the study of absorption capacity of commercial super-absorbent dressings [33], the fluid absorption capacity was found in a range of 50–220%. The curcumin-containing films developed in this study were classified as thin film. It was demonstrated that the films could adhere and stick directly to skin (without adhesive) with high fluid uptake ability in the range of 100–700%.

The composite films became more hydrophobic with the addition of curcumin. Generally, surfaces with a water contact angle of less than 90° are considered hydrophilic, while those with a water contact angle greater than 90° are considered hydrophobic. To create sufficient film adhesion, liquid has to both wet the contact surface and flow easily over the entire surface. Topical drug carrier films should be capable of adhering to surfaces in order to bring drugs in contact with the target site for a sufficient period of time. Surface wettability also influences protein adherence, as well as cell attachment and cell proliferation, and it is one of the critical factors in determining cell behavior. Based on the hydrophilic properties, all hydrated films in this study provided perfect coverage and were able to neatly adhere to the skin. The composite films were also able to cope with curves and uneven textures.

Diffusion of water vapor through films affects the barrier properties of those films. Topical films in biomedical applications should be occlusive to bacteria and viruses; however, water vapor, oxygen and carbon dioxide should be easily exchanged. Films are supposed to allow moisture to evaporate and to maintain normal skin function under the films. High WVTR films lead to quick moisture loss, and they cause wound dryness, while low WVTR might cause exudate storing, which may retard the wound healing process [34]. Intact human skin WVTR ranges 240–1920 $g/m^2/24$ h, whereas the WVTR of an uncovered wound is around 4800 $g/m^2/24$ h [35]. WVTR values of existing commercial film and hydrocolloid dressings range between 436 and 900 $g/m^2/24$ h [36,37]. Injured skin (first degree to third degree burns) can range from 278.4 to 4274 $g/m^2/24$ h [38]. The WVTR characteristic of the fabricated films in this study was found to be comparable to human skin and commercial wound dressings.

Adhesion of film to the surface is an important key factor for retaining the film on the target site. It was demonstrated that the composite films presented good adhesion properties and maintained their integrity on the target sites. For the further application as oral wound dressing, the films were submerged into artificial saliva. It was found that hydrated films adhered well to the mucosa surface, and the lack of air between surfaces was attributed to good surface wettability. A vacuum system is formed and the external air pressure keeps the material and surface together, allowing the films to maintain intimate contact. The functional groups of hydrophilic polymers such as hydroxyl, carboxyl and amine groups are able to form hydrogen binding between the material and superficial surfaces [39]. In addition, positive charges of gelatin molecules were able to form electrostatic bonds with the negative charge of mucosal surfaces. Therefore, the developed films can be applied as adhering films for the oral cavity. Once films are place in the mouth, they adhere strongly, maintaining moisture without causing drying. Previously, the in vitro adhesion residence time of alginate-based mucoadhesive bi-layered buccal patches for antimigraine treatment reportedly varied from 20 to 25 min [40]. Another study on mucoadhesive patches containing a miconazole drug reported that the adhesion residence time was 2 h [41]. It has been shown that the fabricated film in this work has advantages in terms of flexibility and long bioadhesion residence time.

The mechanical properties of the films decreased with the increase in curcumin content. This was possibly due to the disruption of hydrogen bonding between BC fibrils by the addition of an interfering compound and a decrease in intermolecular interactions between polymeric backbone chains and the presence of additives. The dispersed curcumin particles restrict the motions of polymeric chains, thus reducing their strength and elongation tendencies. Similar patterns of changes in mechanical properties have been observed in other studies. For example, mechanical properties were reduced by the integration of turmeric extract into chitosan films [19]; tensile strength, Young's modulus and elongation at break all tended to decrease as a function of increasing curcumin loading in BC film [16]; and phenolic–protein inter-chain interaction resulted in a decline in flexibility of gelatin films [42].

Mechanical behavior after hydration is also important in predicting the in vivo clinical performance of the films. All samples in hydrated condition were found to be softer and more flexible compared to samples tested in the dry condition. Those results suggest that hydration of samples before use could provide the samples with more flexibility. Under hydrated condition, clusters of water molecules could fill intervening voids in the matrix. Water behaved as a plasticizer by creating bonds with hydrophilic groups of the polymers resulting in the increase of ductility. It was observed that mechanical strength of hydrated curcumin-loaded films increased in comparison to non-curcumin loaded ones. Curcumin solid content increased rigidity of the polymeric matrix. Curcumin molecules partially limit the diffusion of water molecule into the matrix. This is possibly due to interlocking effects, such as micro-pins within the polymeric matrix [43]. Curcumin restricts the segmental motions of the polymer chains, leading to the increase in matrix stiffness. Under wet environment, water molecules penetrated the composite and diffused along fiber–matrix interfaces causing debonding between fiber and matrix [44]. Mechanical properties showed that the fabricated films possessed the appreciable strength and were conformable enough for practical use. This good balance between stiffness and elasticity needs to be achieved for the clinical approach. Film should be stiff enough to bear forces and maintain its integrity during clinical uses. Meanwhile, a very stiff film does not allow for good clinical utilization. The design and adoption of leave-on skin applications should be a good match with natural skin in terms of mechanical strength and flexibility. The oral mucosa possessed a wide range of possible elastic modulus from 0.06 to 8.89 MPa [45]. Typically values of 1–5 MPa were adopted for simulations [46]. Tensile strength and Young's modulus of commercial wound dressings are 0.9–28.3 and 0.08–8.85 MPa, respectively [47,48]. Based on the mechanical characteristics exhibited by the commercial benchmark, BCAGG, BCAGG-C1, BCAGG-C2 and BCAGG-C3 have appropriate mechanical characteristics for application as wound dressings.

Overall, cell growth of HaCaT and GF exhibited consistent trends without great differences among all samples. Both HaCat and GF cells tended to orient themselves and presented numerous cellular extensions and 3D-like structure that more closely resemble in vivo tissue. The collagen fibers in the wound tissue will organize along the stress lines of injury. The collagen fibers in scar tissues are arranged in parallel to the skin surface, as opposed to a non-parallel conformation in healthy skin [49]. Matrix stiffness plays an important role in the cell signaling that regulates cell behavior. Substrate rigidity regulated cancer cell invasiveness [50]. Breast cancer cells exhibited increased proliferation and migration through stiffer surface [51]. The migration speed of U-474 MG human glioblastoma cancer cells increased as substrate stiffness increased [52]. In most models of cell spread, cells must generate traction forces at adhesion sites so that cells can be stretched. Traction forces are not able to form without adequate external resistance from a soft substrate. In this study, CAL-27 cell viability gradually decreased on non-curcumin loaded BCAGG. Possibly, stiffness-related effects prevailed over hydrophilic and surface properties. The number of CAL-27 cells significantly dropped in curcumin-loaded BCAGG. This was attributed to the synergistic effect of stiffness and addition of curcumin. It was postulated that functional adaptation was acquired by the cancer cells to survive and thrive in the altered environment. Curcumin has anticancer activity against oral squamous cell carcinoma (OSCC) via both autophagy and apoptosis [53]. It has been suggested that, during cellular stress, to survive starvation or environmental toxicity, cells utilize autophagy to adapt to the microenvironment. Autophagy due to excessive stress could lead to cell death. According to our previous report, the release of curcumin from curcumin-loaded BC films showed strong cytotoxic effect against A375 human melanoma cancer cells, without significant cytotoxic effect against human keratinocytes and human dermal fibroblasts [16]. In this study, although no significant release of curcumin was observed in PBS and artificial saliva, curcumin-loaded BCAGG films (BCAGG-C2 and BCAGG-C3) showed cytotoxicity against oral cancer cells (CAL-27), but non-cytotoxicity to HaCaT and GF cells. Since curcumin was not released, the regulation of this response should be from the contact and interaction of curcumin to some kinds of cell surface molecules, which then modulated the cell behaviors. Recent studies demonstrated that curcumin can bind to cell membrane-bound toll-like receptors (TLRs) [54,55]. Curcumin demonstrates an antagonist effect on the TLR2 and TLR4 downstream signaling pathway resulting in an anti-inflammatory effect or induction of cell apoptosis [54]. TLR2 and TLR4 are found in several cell types including oral epithelial cell, gingival fibroblast and with overexpression by oral cancer cells [56–58]. Curcumin effectively modulates the TLR response and thereby exerts its potent therapeutic effects against a range of diseases such as inflammation, infection and autoimmune diseases including cancer [59]. Curcumin also has abilities to induce cell apoptosis, suppress the expression of multiple oncogenes and transcription factors inducing epithelial-to-mesenchymal transition (EMT) in cancer cells through interfering with the binding of Wnt to its cell surface receptor in the Wnt/β-catenin signaling pathway, suggesting curcumin as a potential target for anticancer therapies [60]. A further study should be performed find the actual mechanism on cancer viability of curcumin-loaded film.

The curcumin-loaded films had substantial antibacterial activity against *E. coli* and *S. aureus*. It was noticed that the composite films had greater inhibition effect against Gram-positive bacteria (*S. aureus*) than Gram-negative bacteria (*E. coli*) Several reports also indicate that curcumin is more effective on Gram-positive than Gram-negative bacteria. Bacteriostatic of gelatin/curcumin composite films showed stronger activity against Gram-positive than Gram-negative ones. Curcumin-loaded BCAGG films may possibly be identified as self-disinfecting, bacterial resistant composites as they did not accumulate bacteria growth over their surfaces. They could also be categorized as a passive antibacterial material as they did not release antimicrobial agents to the environment yet were able to inhibit bacterial growth. The alteration in antibacterial effect might be due to the difference in bacteria cell membrane structures. Gram-positive bacteria were found to be more susceptible to many natural antibacterial products [61,62]. Gram-positive bacteria have a thicker layer of peptidoglycan. On the other hand, Gram-negative bacteria have a cell wall composed of a thin layer of peptidoglycan surrounded by an outer membrane. The outer membrane of Gram-negative bacteria contains lipopolysaccharide, in addition to proteins and phospholipids [63]. The distinctive structure of the Gram-negative bacteria outer membrane could prevent some antimicrobial agents from entering to the cell. The results in this study demonstrate that curcumin exhibits some antibacterial activities despite being trapped in a biocomposite matrix of BCAGG. Auto-oxidation, one of the bactericidal properties of curcumin, could produce reactive oxygen species (ROS) such as hydrogen peroxide and superoxide, thereby disrupting integrity of the bacterial cell wall leading to cell death. Only the contact between the curcumin and bacterial cells is required for this mechanism as these free radicals have a short half-life [64]. Previously, it was shown that binding of curcumin, an amphipathic and lipophilic molecule, to the bacterial cell wall increases the cell membrane permeabilization of *S. aureus* and *E. coli* causing cell membrane damage and leading to cell lysis. This mode of antibacterial activity of curcumin on Gram-positive bacteria is more effective than that on Gram-negative bacteria according to the composition of cell wall [65]. However, further studies should be clarified for the definite mechanisms of action by the curcumin loaded film.

4. Materials and Methods

4.1. Preparation of Bacterial Cellulose/Alginate/Gelatin Films Incorporated with Curcumin

The preparation of BC/alginate/gelatin composite (BCAGG) was performed according to our previous work [23]. Films of BCAGG were fabricated from a blend of the BC, alginate and gelatin at a weight ratio of 60:20:20, added with glycerol at 2 g/10 g of gelatin solution. To prepare the curcumin-loaded BC/alginate/gelatin (BCAGG-C), solutions of curcumin (>95% purity, Sigma-Aldrich, St. Louis, MO, USA) at concentrations of 2, 4, 6 and 8 mg/mL were prepared by dispersing curcumin in absolute ethanol. Curcumin solutions of 1 mL were added into 100 g of BC/alginate/gelatin blends, and the mixtures were added with glycerol at 2 g/10 g of gelatin solution. All mixtures were thoroughly stirred together for 4 h until homogeneous mixtures were obtained. The mixtures were placed in a 14.5-cm diameter sterile Petri-dish (40 g/plate) and air-dried at room temperature for 24 h. After that, all dried films were cross-linked in a 1% (*w/v*) solution of calcium chloride for 1 h. The films were then rinsed with DI water and dried at room temperature (30 °C). The films were then placed in airtight containers at room temperature. BCAGG-C1, BCAGG-C2, BCAGG-C3 and BCAGG-C4 represented BCAGG-C prepared with 2, 4, 6 and 8 mg/mL curcumin solution, respectively. The actual curcumin content in each film was also determined. Each sample (2 cm × 2 cm) was immersed in 10 mL absolute ethanol and stirred at 100 rpm for 6 h and 1 mL of supernatant was then collected. The actual content of curcumin was quantified by UV-visible spectrophotometry (Shimadzu UV-2550, Tokyo, Japan) at the wavelength of 420 nm.

4.2. Characterization of Composite Films

Scanning Electron Microscope (SEM) was used to observe morphology of the samples. Films were cut and sputter-coated with gold in a Balzers-SCD-040 sputter coater (Balzers, Liechtenstein). Surface feature analysis was visualized using a JSM-5410LV scanning electron microscope (JOEL, Japan) operated at an acceleration voltage of 15 kV.

The information on chemical structural of the composite films was collected by FT-IR analysis using Spectrum One FT-IR spectrometer (PerkinElmer, Waltham, MA, USA). The dried samples were mixed with KBr and compressed into semitransparent KBR pellets before the measurement. The FT-IRs of samples were analyzed within the range of 4000–800 cm^{-1} at a scanning resolution of 2 cm^{-1}.

Fluid uptake ability of the samples was measure by immersing the weighed samples in PBS (pH 7.4) and artificial saliva (pH 6.2) at 37 °C for 6 and 24 h. Excess fluid on the sample surface was removed by blotting out with tissue papers. The weights of wet samples were then recorded. All tests were done in triplicate. Fluid absorption capacity was calculated using the following equation:

$$\%WAC = \frac{(Ws - Wd)}{Wd} \times 100$$

where Wd and Ws are the weights of dry sample and wet sample, respectively.

Surface wettability of the films was evaluated as static water contact angle using OCA 20 contact angle analyzer (Dataphysics, Filderstadt, Germany). Each of samples was placed on the platform of the contact angle analyzer just below the needle, from which a 2 µL water droplet was dispensed. The contact angle was measured in degrees using the SCA20 software. Measurements were performed at least 6 different locations for each sample.

WVTRs of samples were analyzed by water vapor permeability analyzer, PERMETRAN-W® model 398 (Mocon, Minneapolis, MN, USA). The test condition followed ASTM E398-03 standard. The determination of WVTR was carried out at 38 °C and 98% relative humidity. The test specimen was sealed to the open mount of test dish containing a desiccant and the assembly placed in a controlled atmosphere. Periodic weighting was performed to determine the rate of water vapor movement through the specimen into the desiccant.

In vitro mucoadhesion time of samples was evaluated by assessing the time for samples to detach from the porcine buccal mucosa. The tests were based on modified method of the previous work [40]. The porcine buccal mucosa was bought from a local butcher house and fixed on the internal side of a well-stirred beaker containing 50 mL of artificial saliva maintained at 37 °C. The samples were wetted with artificial saliva and adhered to buccal mucosa. The time for detachment of the samples from buccal mucosa was recorded and taken as in vitro mucoadhesion time.

The tensile strength, Young's modulus and elongation at break of dry and hydrated films were measured by an INSTRON 5567 Universal Testing Machine (INSTRON, Norwood, MA, USA) equipped with a 1.0 kN load cell. Hydrated samples were prepared by immersing dry films in DI water for 24 h prior to testing. The samples were cut into strip-shaped specimens with a width of 10 mm and a length of 10 cm. The test conditions in this assay followed ASTM D882. The tensile strength, Young's modulus and elongation at break values were reported as average values determined from at least six specimens.

Human keratinocyte (HaCaT), human oral cancer (CAL-27) cell lines and primary human gingival fibroblasts (GF) were used to evaluate in vitro cytocompatibility as a direct contact test. Human primary gingival fibroblasts (GF, passage 2) were kindly provided by the Department of Anatomy, Faculty of Dentistry, Chulalongkorn University, Thailand. All cell types were cultured in DMEM supplemented with 10% *v/v* FBS and 1% of antibiotics in a humidified 37 °C incubator containing 5% (*v/v*) CO$_2$. The cell cultures of 5×10^4 cells/well were seeded on each film placed in 24-well plates. Cells were allowed to proliferate for 48 h. Cells cultured on glass coverslips were served as control groups. The relative cell viability was evaluated by MTT assay. The optical densities were measured by a Multiskan EX microplate reader (Thermo Scientific, West Columbia, SC, USA) at wavelength of 570 nm. To observe cell morphology on samples, films containing cells were washed with PBS and fixed in 2.5 vol % fixative solution. The samples were dehydrated in serial dilutions of ethanol afterwards and dried in an automated Leica EM-CPD300 critical point dryer (Leica Microsystems, Wien, Austria). Samples were then sputter-coated with gold and observed under a JSM-5410 Scanning Electron Microscope (JEOL, Tokyo, Japan).

Antibacterial testing was carried out according to Japanese International Standard, JISZ2801 [56]. The antibacterial activity analysis of all composite films was tested against Gram-positive *S. aureus* (ATCC 29213) and Gram-negative *E. coli* (ATCC 25922). Briefly, bacteria were suspended in nutrient broth medium. After the cultivation, the bacterial suspensions were diluted with some amount of sterilized water to obtain a suspension with the range of bacteria concentration of $1.6–3.1 \times 10^6$ cfu/mL. The bacteria suspension of 0.040 mL was dropped on the surface of each sample, which was horizontally placed on a sterilized petri dish. Each sample size was 5×5 cm^2. The samples were subsequently placed in an incubator and were incubated for 48 h at 37 °C and relative humidity of 90%. After 48 h of incubation, the surfaces of samples were rinsed to harvest bacteria exposed to the samples. The numbers of surviving bacteria on the surface of samples were counted after 48 of incubation by the standard agar plate colony counting method. The images of the agar plates were captured. The above experiments were done in triplicate.

All results were reported as mean ± standard deviation. Statistical analysis was performed using Minitab 17 software (Minitab Inc., State College, PA, USA). The level of statistical significance for all tests was set at p-value ≤ 0.05.

5. Conclusions

In this study, to develop a multifunctional biopolymer composite film, curcumin was integrated into BCAGG composite film. SEM images indicated a uniform structure of the composites. Mechanical properties showed that BCAGG-C films possessed the appreciable strength and flexibility for practical use as wound dressings. Hydrated films could adhere firmly onto the skin. In vitro mucoadhesion time was found in the range of 0.5–6 h with porcine mucosa as model membrane under an artificial saliva medium. The release of curcumin was not observed in the testing media during 48 h immersion in PBS and artificial saliva, where the fluids uptakes were in the range of 100–700%. Despite being trapped within the biopolymer matrix composite, curcumin could possess useful biological activities. The curcumin-loaded films demonstrated antibacterial activities against *E. coli* and *S. aureus* infection. The films showed anticancer activity against oral cancer cells (CAL-27), but non-cytotoxicity to HaCaT and GF cells. Therefore, BCAGG-C films have potential future applications as wound care dressings and oral mucoadhesive patches for periodontitis or oral cancer treatment.

Author Contributions: Conceptualization, M.P. and N.C.; methodology, M.P. and N.C.; validation, T.S., N.S. and M.P.; formal analysis, N.C.; investigation, M.P. and N.C.; resources, M.P. and N.S.; data curation, N.C.; writing—original draft preparation, N.C.; writing—review and editing, M.P. and N.S.; supervision, M.P. and N.S.; project administration, M.P.; and funding acquisition, M.P. and N.S. All authors have read and agreed to the published version of the manuscript.

Funding: This research was supported by the 100th Anniversary Chulalongkorn University Fund for Doctoral Scholarship, The 90th Anniversary of Chulalongkorn University Fund (Ratchadaphiseksomphot Endowment Fund) and The Graduate School, Chulalongkorn University. The authors also acknowledge the support from The Thailand Research Fund, TRF (RGU62).

Conflicts of Interest: The authors have no conflict of interest to declare.

References

1. AshaRani, P.V.; Low Kah Mun, G.; Hande, M.P.; Valiyaveettil, S. Cytotoxicity and Genotoxicity of Silver Nanoparticles in Human Cells. *ACS Nano* **2008**, *3*, 279–290. [CrossRef] [PubMed]
2. Mukherjee, S.G.; O'Claonadh, N.; Casey, A.; Chambers, G. Comparative *in vitro* cytotoxicity study of silver nanoparticle on two mammalian cell lines. *Toxicol. In Vitro* **2012**, *26*, 238–251. [CrossRef] [PubMed]

3. Zhang, T.; Wang, L.; Chen, Q.; Chen, C. Cytotoxic Potential of Silver Nanoparticles. *Yonsei Med. J.* **2014**, *55*, 283. [CrossRef] [PubMed]

4. Vieira, L.F.; Lins, M.P.; Viana, I.M.; Santos, J.E.; Smaniotto, S.; Reis, M.D. Metallic nanoparticles reduce the migration of human fibroblasts in vitro. *Nanoscale Res. Lett.* **2017**, *12*, 200. [CrossRef] [PubMed]

5. Tayeb, A.H.; Amini, E.; Ghasemi, S.; Tajvidi, M. Cellulose nanomaterials-binding properties and applications: A review. *Molecules* **2018**, *23*, 2684. [CrossRef]

6. Chiaoprakobkij, N.; Sanchavanakit, N.; Subbalekha, K.; Pavasant, P.; Phisalaphong, M. Characterization and biocompatibility of bacterial cellulose/alginate composite sponges with human keratinocytes and gingival fibroblasts. *Carbohydr. Polym.* **2011**, *85*, 548–553. [CrossRef]

7. Basnet, P.; Skalko-Basnet, N. Curcumin: An anti-inflammatory molecule from a curry spice on the path to cancer treatment. *Molecules* **2011**, *16*, 4567–4598. [CrossRef]

8. Prasad, S.; Tyagi, A.K.; Aggarwal, B.B. Recent developments in delivery, bioavailability, absorption and metabolism of curcumin: The golden pigment from golden spice. *Cancer Res. Treat.* **2014**, *46*, 2–18. [CrossRef]

9. Pulido-Moran, M.; Moreno-Fernandez, J.; Ramirez-Tortosa, C.; Ramirez-Tortosa, M. Curcumin and health. *Molecules* **2016**, *21*, 264. [CrossRef]

10. Luo, N.; Varaprasad, K.; Reddy, G.V.S.; Rajulu, A.V.; Zhang, J. Preparation and characterization of cellulose/curcumin composite films. *RSC Adv.* **2012**, *2*, 8483. [CrossRef]

11. Raghavendra, G.M.; Jayaramudu, T.; Varaprasad, K.; Ramesh, S.; Raju, K.M. Microbial resistant nanocurcumin-gelatin-cellulose fibers for advanced medical applications. *RSC Adv.* **2014**, *4*, 3494–3501. [CrossRef]

12. Bajpai, S.K.; Chand, N.; Ahuja, S. Investigation of curcumin release from chitosan/cellulose micro crystals (CMC) antimicrobial films. *Int. J. Biol. Macromol.* **2015**, *79*, 440–448. [CrossRef] [PubMed]

13. Tong, W.Y.; bin Abdullah, A.Y.K.; binti Rozman, N.A.S.; bin Wahid, M.I.A.; Hossain, M.S.; Ring, L.C.; Tan, W.-N. Antimicrobial wound dressing film utilizing cellulose nanocrystal as drug delivery system for curcumin. *Cellulose* **2017**, *25*, 631–638. [CrossRef]

14. Udeni Gunathilake, T.M.S.; Ching, Y.C.; Chuah, C.H. Enhancement of curcumin bioavailability using nanocellulose reinforced chitosan hydrogel. *Polymers* **2017**, *9*, 64. [CrossRef]

15. Ching, Y.C.; Gunathilake, T.M.S.U.; Chuah, C.H. Curcumin/Tween 20-incorporated cellulose nanoparticles with enhanced curcumin solubility for nano-drug delivery: Characterization and in vitro evaluation. *Cellulose* **2019**, *26*, 5467–5481. [CrossRef]

16. Subtaweesin, C.; Woraharn, W.; Taokaew, S.; Chiaoprakobkij, N.; Sereemaspun, A.; Phisalaphong, M. Characteristics of curcumin-loaded bacterial cellulose films and anticancer properties against malignant melanoma skin cancer cells. *Appl. Sci.* **2018**, *8*, 1188. [CrossRef]

17. Sharma, A.; Gescher, J.; Steward, P. Curcumin: The story so far. *Eur. J. Cancer* **2005**, *41*, 1955–1968. [CrossRef]

18. Chiaoprakobkij, N.; Seetabhawang, S.; Sanchavanakit, N.; Phisalaphong, M. Fabrication and characterization of novel bacterial cellulose/alginate/gelatin biocomposite film. *J. Biomater. Sci. Polym. Ed.* **2019**, *30*, 961–982. [CrossRef]

19. Kalaycıoğlu, Z.; Torlak, E.; Akın-Evingür, G.; Özen, İ.; Erim, F.B. Antimicrobial and physical properties of chitosan films incorporated with turmeric extract. *Int. J. Biol. Macromol.* **2017**, *101*, 882–888. [CrossRef]

20. Liu, Y.; Cai, Y.; Jiang, X.; Wu, J.; Le, X. Molecular interactions, characterization and antimicrobial activity of curcumin–chitosan blend films. *Food Hydrocoll.* **2016**, *52*, 564–572. [CrossRef]

21. Mamidi, N.; Romo, I.L.; Barrera, E.V.; Elías-Zúñiga, A. High throughput fabrication of curcumin embedded gelatin-polylactic acid forcespun fiber-aligned scaffolds for the controlled release of curcumin. *MRS Commun.* **2018**, *8*, 1395–1403. [CrossRef]

22. Musso, Y.S.; Salgado, P.R.; Mauri, A.N. Smart edible films based on gelatin and curcumin. *Food Hydrocoll.* **2017**, *66*, 8–15. [CrossRef]

23. Phomrak, S.; Phisalaphong, M. Reinforcement of natural rubber with bacterial cellulose via a latex aqueous microdispersion process. *J. Nanomater.* **2017**, *4*, 1–9. [CrossRef]

24. Dai, X.; Liu, J.; Zheng, H. Nano-formulated curcumin accelerates acute wound healing through Dkk-1-mediated fibroblast mobilization and MCP-1-mediated anti-inflammation. *NPG Asia Mater.* **2017**, *9*, e368. [CrossRef]

25. Roy, S.; Rhim, J.W. Preparation of antimicrobial and antioxidant gelatin/curcumin composite films for active food packaging application. *Colloids Surf. B* **2020**, *188*, 110761. [CrossRef]

26. Jeong, S.; Kim, B.; Lau, H.C.; Kim, A. Gelatin-Alginate Complexes for EGF Encapsulation: Effects of H-Bonding and Electrostatic Interactions. *Pharmaceutics* **2019**, *11*, 530. [CrossRef]
27. Shan, Y.; Li, C.; Wu, Y.; Li, Q.; Liao, J. Hybrid cellulose nanocrystal/alginate/gelatin scaffold with improved mechanical properties and guided wound healing. *RSC Adv.* **2019**, *9*, 22966–22979. [CrossRef]
28. Sarika, P.; James, N. Polyelectrolyte complex nanoparticles from cationised gelatin and sodium alginate for curcumin delivery. *Carbohydr Polym.* **2016**, *148*, 354–361. [CrossRef]
29. Berger, F.M.; Ludwig, B.J.; Wielich, K.H. The hydrophilic and acid binding properties of alginates. *J. Dig. Dis.* **1953**, *20*, 39–42. [CrossRef]
30. Yoon, S.; Chen, B. Elastomeric and pH-responsive hydrogels based on direct crosslinking of the poly (glycerol sebacate) pre-polymer and gelatin. *Polym. Chem.* **2018**, *9*, 3727–3740. [CrossRef]
31. Bajpai, S.K.; Tankhiwale, R. Investigation of water uptake behavior and stability of calcium alginate/chitosan bi-polymeric beads: Part–1. *React. Funct Polym.* **2006**, *66*, 645–658. [CrossRef]
32. Schneider, L.A.; Korber, A.; Grabbe, S.; Dissemond, J. Influence of pH on wound-healing: A new perspective for wound-therapy. *Arch. Dermatol Res.* **2006**, *29*, 413–420. [CrossRef] [PubMed]
33. Cutting, K.F.; Westgate, S.J. Super-absorbent dressings: How do they perform in vitro? *Br. J. Nurs.* **2012**, *21*, 14–19. [CrossRef] [PubMed]
34. Xu, R.; Xia, H.; He, W.; Li, Z.; Zhao, J.; Liu, B.; Wang, Y.; Lei, Q.; Kong, Y.; Bai, Y.; et al. Controlled water vapor transmission rate promotes wound-healing via wound re-epithelialization and contraction enhancement. *Sci. Rep.* **2016**, *6*, 24596. [CrossRef] [PubMed]
35. Yusof, N.L.B.M.; Wee, A.; Lim, L.Y.; Khor, E. Flexible chitin films as potential wound-dressing materials: Wound model studies. *J. BioMed Mater. Res. A* **2003**, *66*, 224–232. [CrossRef] [PubMed]
36. Thomas, S.; Loveless, P.; Hay, N.P. Comparative review of the properties of six semipermeable film dressings. *Pharm. J.* **1988**, *240*, 785–789.
37. Wu, P.; Fisher, A.C.; Foo, P.P.; Queen, D.; Gaylor, J.D.S. In vitro assessment of water vapour transmission of synthetic wound dressings. *Biomaterials* **1995**, *16*, 171–175. [CrossRef]
38. Ngadaonye, J.I.; Geever, L.M.; McEvoy, K.E.; Killion, J.; Brady, D.B.; Higginbotham, C.L. Evaluation of Novel Antibiotic-Eluting Thermoresponsive Chitosan-PDEAAm Based Wound Dressings. *Int. J. Polym. Mater.* **2014**, *63*, 873–883. [CrossRef]
39. Russo, D.; Ollivier, J.; Teixeira, J. Water hydrogen bond analysis on hydrophilic and hydrophobic biomolecule sites. *Phys. Chem. Chem. Phys.* **2008**, *10*, 4968–4974. [CrossRef]
40. Asthana, A.; Shilakari, S.; Asthana, A. Formulation and evaluation of alginate-based mucoadhesive buccal patch for delivery of antimigraine drug. *Asian J. Pharm. Clin. Res.* **2018**, *11*, 185–191. [CrossRef]
41. Nafee, N.A.; Boraie, M.A.; Ismail, F.A.; Mortada, L.M. Design and characterization of mucoadhesive buccal patches containing cetylpyridinium chloride. *Acta Pharm.* **2003**, *53*, 199–212. [PubMed]
42. Zhang, X.; Ma, L.; Yu, Y.; Zhou, H.; Guo, T.; Dai, H.; Zhang, Y. Physico-mechanical and antioxidant properties of gelatin film from rabbit skin incorporated with rosemary acid. *Food Packag. Shelf Life* **2019**, *19*, 121–130. [CrossRef]
43. Chan, M.; Lau, K.; Wong, T.; Ho, M.; Hui, D. Mechanism of reinforcement in a nanoclay/polymer composite. *Compos. B Eng.* **2011**, *42*, 1708–1712. [CrossRef]
44. Alamri, H.; Low, I.M. Effect of water absorption on the mechanical properties of nanoclay filled recycled cellulose fibre reinforced epoxy hybrid nanocomposites. *Compos. Part A Appl. Sci. Manuf.* **2013**, *44*, 23–31. [CrossRef]
45. Chen, J.; Ahmad, R.; Li, W.; Swain, M.; Li, Q. Biomechanics of oral mucosa. *J. R. Soc. Interface* **2015**, *12*, 20150325. [CrossRef]
46. Lima, J.B.G.; Orsi, I.A.; Borie, E.; Lima, J.H.F.; Noritomi, P.Y. Analysis of stress on mucosa and basal bone underlying complete dentures with different reliner material thicknesses: A three-dimensional finite element study. *J. Oral Rehabil.* **2013**, *40*, 767–773. [CrossRef]
47. Elsner, J.J.; Zilberman, M. Novel antibiotic-eluting wound dressings: An in vitro study and engineering aspects in the dressing's design. *J. Tissue Viability* **2010**, *19*, 54–66. [CrossRef]
48. Yu, B.; Kang, Y.; Akthakul, A.; Ramadurai, N.; Pilkenton, M.; Patel, A.; Nashat, A.; Anderson, D.; Sakamoto, F.; Gilchrest, B.; et al. An elastic second skin. *Nat. Mater.* **2016**, *15*, 911–918. [CrossRef]
49. Mulholland, E.J. Electrospun Biomaterials in the Treatment and Prevention of Scars in Skin Wound Healing. *Front. Bioeng. Biotechnol.* **2020**, *8*, 481. [CrossRef]

50. Lo, C.M.; Wang, H.B.; Dembo, M.; Wang, Y.L. Cell movement is guided by the rigidity of the substrate. *Biophys. J.* **2000**, *79*, 144–152. [CrossRef]
51. Kostic, A.; Lynch, C.D.; Sheetz, M.P. Differential Matrix Rigidity Response in Breast Cancer Cell Lines Correlates with the Tissue Tropism. *PLoS ONE* **2009**, *4*, e6361. [CrossRef] [PubMed]
52. Pathak, A.; Kumar, S. Independent regulation of tumor cell migration by matrix stiffness and confinement. *Proc. Natl. Acad. Sci. USA* **2012**, *109*, 10334–10339. [CrossRef] [PubMed]
53. Kim, J.Y.; Cho, T.J.; Woo, B.H.; Choi, K.U.; Lee, C.H.; Ryu, M.H.; Park, H.R. Curcumin-induced autophagy contributes to the decreased survival of oral cancer cells. *Arch. Oral Biol.* **2012**, *57*, 1018–1025. [CrossRef] [PubMed]
54. Boozari, M.; Butler, A.E.; Sahebkar, A. Impact of curcumin on toll-like receptors. *J. Cell Physiol.* **2019**, *234*, 12471–12482. [CrossRef] [PubMed]
55. Gao, Y.; Zhuang, Z.; Lu, Y.; Tao, T.; Zhou, Y.; Liu, G.; Wang, H.; Zhang, D.; Wu, L.; Dai, H.; et al. Curcumin mitigates neuro-inflammation by modulating microglia polarization through inhibiting TLR4 axis signaling pathway following experimental subarachnoid hemorrhage. *Front. Neurosci.* **2019**, *13*, 1223. [CrossRef]
56. Andreani, V.; Gatti, G.; Simonella, L.; Rivero, V.; Maccioni, M. Activation of toll-like receptor 4 on tumor cells in vitro inhibits subsequent tumor growth in vivo. *Cancer Res.* **2007**, *67*, 10519–10527. [CrossRef]
57. Rich, A.M.; Hussaini, H.M.; Parachuru, V.P.; Seymour, G.J. Toll-like receptors and cancer, particularly oral squamous cell carcinoma. *Front. Immunol.* **2014**, *24*, 464. [CrossRef]
58. Pisani, L.P.; Estadekka, D.; Ribeiro, D.A. The role of TLRs in oral carcinogenesis. *Anticancer Res.* **2017**, *37*, 5389–5394.
59. Mirzaei, H.; Naseri, G.; Rezaee, R.; Mohammadi, M.; Banikazemi, Z.; Mirzaei, H.R.; Salehi, H.; Peyvandi, M.; Pawelek, J.M.; Sahebkar, A. Curcumin: A new candidate for melanoma therapy? *Int. J. Cancer* **2016**, *139*, 1683–1695. [CrossRef]
60. Wang, M.; Jiang, S.; Zhou, L.; Yu, F.; Ding, H.; Li, P.; Zhou, M.; Wang, K. Potential mechanisms of action of curcumin for cancer prevention: Focus on cellular signaling pathways and miRNAs. *Int. J. Biol. Sci.* **2019**, *15*, 1200–1214. [CrossRef]
61. Oliveira, D.A.; Salvador, A.A.; Smânia, A.; Smânia, E.F.; Maraschin, M.; Ferreira, S.R. Antimicrobial activity and composition profile of grape (Vitis vinifera) pomace extracts obtained by supercritical fluids. *J. Biotechnol.* **2013**, *164*, 423–432. [CrossRef] [PubMed]
62. Kozlowska, M.; Laudy, A.E.; Przybyl, J.; Ziarno, M.; Majewska, E. Chemical composition and antibacterial activity of some medicinal plants from Lamiaceae family. *Acta Pol. Pharm.* **2015**, *72*, 757–767. [PubMed]
63. Breijyeh, Z.; Jubeh, B.; Karaman, R. Resistance of Gram-Negative Bacteria to Current Antibacterial Agents and Approaches to Resolve It. *Molecules* **2020**, *25*, 1340. [CrossRef] [PubMed]
64. Zheng, D.; Huang, C.; Huang, H.; Zhao, Y.; Khan, M.R.U.; Zhao, H.; Huang, L. Antibacterial Mechanism of Curcumin: A Review. *Chem. Biodivers.* **2020**, *17*, e2000171. [CrossRef]
65. Tyagi, P.; Singh, M.; Kumari, H.; Kumari, A.; Mukhopadhyay, K. Bactericidal activity of curcumin I is associated with damaging of bacterial membrane. *PLoS ONE* **2015**, *26*, e0121313. [CrossRef]

Sample Availability: Samples of the compounds are not available from the authors.

 *molecules*

Article

Photoinduced Polymerization of Eugenol-Derived Methacrylates

Samantha Molina-Gutiérrez [1,2], **Sara Dalle Vacche** [2], **Alessandra Vitale** [2], **Vincent Ladmiral** [1], **Sylvain Caillol** [1], **Roberta Bongiovanni** [2,*] and **Patrick Lacroix-Desmazes** [1,*]

1   Institut Charles Gerhardt Montpellier (ICGM), University of Montpellier, CNRS, ENSCM, 34095 Montpellier, France; samantha.molina-gutierrez@enscm.fr (S.M.-G.); vincent.ladmiral@enscm.fr (V.L.); sylvain.caillol@enscm.fr (S.C.)
2   Department of Applied Science and Technology, Politecnico di Torino, Corso Duca degli Abruzzi 24, 10129 Torino, Italy; sara.dallevacche@polito.it (S.D.V.); alessandra.vitale@polito.it (A.V.)
*   Correspondence: roberta.bongiovanni@polito.it (R.B.); patrick.lacroix-desmazes@enscm.fr (P.L.-D.)

Academic Editor: Dimitrios Bikiaris
Received: 23 June 2020; Accepted: 24 July 2020; Published: 29 July 2020

Abstract: Biobased monomers have been used to replace their petroleum counterparts in the synthesis of polymers that are aimed at different applications. However, environmentally friendly polymerization processes are also essential to guarantee greener materials. Thus, photoinduced polymerization, which is low-energy consuming and solvent-free, rises as a suitable option. In this work, eugenol-, isoeugenol-, and dihydroeugenol-derived methacrylates are employed in radical photopolymerization to produce biobased polymers. The polymerization is monitored in the absence and presence of a photoinitiator and under air or protected from air, using Real-Time Fourier Transform Infrared Spectroscopy. The polymerization rate of the methacrylate double bonds was affected by the presence and reactivity of the allyl and propenyl groups in the eugenol- and isoeugenol-derived methacrylates, respectively. These groups are involved in radical addition, degradative chain transfer, and termination reactions, yielding crosslinked polymers. The materials, in the form of films, are characterized by differential scanning calorimetry, thermogravimetric, and contact angle analyses.

Keywords: biobased monomer; photoinduced-polymerization; eugenol-derived methacrylate

---

1. Introduction

The need for more environmentally friendly materials and processes has led to the development of suitable biobased building blocks to produce polymers [1]. However, the use of energy-efficient polymerization techniques is also paramount. Photoinduced polymerization is a suitable option, as it allows fast processes, low energy consumption, room temperature reactions, and solvent-free conditions with the concomitant reduction or elimination of volatile organic compounds (VOCs) [2]. Thanks to these advantages, it has found wide application in industrial processes. It is an established technique in the fields of coatings, inks, adhesives, and wood finishing [3]. Products from photopolymerization are present in everyday life, such as contact lenses [4], filling for dental cavities [5], and credit cards [6].

Photopolymerization processes are also characterized by spatial and temporal control, which means that they only occur in the irradiated area and they are stop-and-go reactions, i.e., they start and stop simply by switching on and off the light [7]. Therefore, they are key reactions for the emerging additive manufacturing technologies [8–10]. For most applications, conventional radical reactions (proceeding via propagation of macromolecular radicals after initiation triggered by irradiation) are employed. Most monomers do not produce any radical initiating species with sufficiently high yields under light irradiation; thus, a photoinitiator (PI) may be required [3]. Most PIs form radicals either by homolytic cleavage or by hydrogen abstraction. Then, the radicals initiate the polymerization by reacting with the monomers. The most commonly used monomers in radical photopolymerization are acrylates and methacrylates. In search of sustainability, it is crucial to replace oil-based monomers with bio-based ones produced from renewable sources. Among the available biobased building blocks, some natural molecules can undergo autooxidation reactions, cyclization, isomerization, dimerization, and oligomerization in the presence of light [11,12]. However, most of these require being suitably functionalized prior to photoinduced polymerization processes. The introduction of polymerizable functions on biobased building blocks is thus a crucial step.

Recently, the use of naturally occurring phenols, such as eugenol and eugenol-derivatives, has gained attention for producing biobased monomers, as they can be obtained by lignin depolymerization [13–15]. In addition, eugenol-derived monomers are attractive because they possess antioxidant, antiseptic, and antibacterial properties [16,17], which could be exploited in photopolymers for dentistry or food packaging [18]. In particular, isoeugenol has a higher antibacterial activity than eugenol and is not genotoxic [19]. However, as many other biobased building blocks, eugenol and its derivatives do not possess functional groups that react readily through photoinduced polymerization. In addition, phenols scavenge free radicals and inhibit polymerization [20,21]. Thus, suitable functional groups must be inserted to avoid this inhibition and promote polymerization.

Besides the functionalization of eugenol with epoxy groups for cationic photopolymerization [22–24], methacrylate functional groups have also been introduced in the molecule. By reacting to the allylic double bond with 3-mercaptopropionic acid and thiomalic acid (*via* thiol-ene chemistry) and then reacting to the resulting carboxylic acid product and phenol group of eugenol with glycidyl methacrylate, eugenol methacrylic derivatives were obtained. These monomers were then used in photoinduced copolymerization with AESO (acrylated epoxidized soybean oil) to produce biobased coatings [25]. Moreover, allyl-etherified eugenol-derivatives were copolymerized through thiol-ene reactions with pentaerythritol-based primary and secondary tetrathiol and with isocyanurate-based secondary trithiol, to prepare crosslinked polymers [26]. Similarly, allyl-etherified eugenol and linalool were copolymerized with trimethylolpropane tris(3-mercaptopropionate) to form crosslinked networks endowed with antioxidant and antibacterial properties [18]. Later, a trifunctional allyl compound, tris(4-allyl-2 methoxyphenolyl) phosphate, was synthesized and reacted with thiols with two to four functionalities via thiol-ene chemistry and the influence of crosslink density on the different materials was studied [27]. Thiol-ene chemistry was also employed to covalently attach eugenol through its allylic double bond to a limonene-derived polymer network and prepare antibacterial coatings [28].

An easy two-step synthetic alternative to produce methacrylated eugenol derivatives was described in a previous work [29]. A methacrylate group was introduced in eugenol structures taking advantage of its phenol group. Ethoxy eugenyl methacrylate (EEMA), ethoxy isoeugenyl methacrylate (EIMA) and ethoxy dihydroeugenyl methacrylate (EDMA) (Figure 1) were synthesized and successfully employed in solution and emulsion polymerization [29,30].

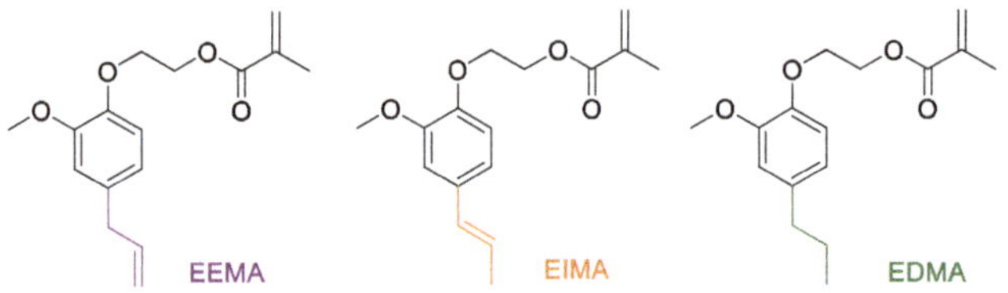

Figure 1. Eugenol-derived methacrylates.

Biobased polymers obtained in the form of homogeneous and transparent films are potentially interesting for industrial development and could find application in coatings, food packaging or dentistry. Therefore, in the present article, we investigate the photopolymerization of films of these eugenol methacrylates under irradiation in different conditions: with or without the radical photoinitiator and in the presence or in the absence of air. Moreover, the conversion of the methacrylic double bond of the three monomers as well as the conversion of the allylic (EEMA) or propenyl (EIMA) double bonds are monitored, and the properties of the polymers are tested.

2. Results and Discussion

2.1. Kinetic Monitoring of Photoinduced Polymerization of Eugenol-Derived Methacrylates

The photoinduced polymerization of ethoxy eugenyl methacrylate (EEMA), ethoxy isoeugenyl methacrylate (EIMA) and ethoxy dihydroeugenyl methacrylate (EDMA) was conducted by irradiating the monomers spread on a solid substrate in the form of films. Different experimental conditions were investigated. At first, the reactions were attempted in the absence of any photoinitiator (PI). Avoiding the use of PI is a crucial step in the development of new products for many real life applications (e.g., inks for food packaging, dental materials) as photoinitiators decompose into harmful species which can uncontrollably migrate [31]. Then, reactions were done in the presence of two different Norrish Type I photoinitiators. Azo-initiators, largely used in radical polymerization, can also be used as photoinitiators. However, they have been reported to have low efficiency compared to acyl photoinitiators [32]. Thus, Darocur 1173 and Irgacure 819 have been selected. As the reaction proceeds via a radical mechanism, the effect of oxygen was studied by irradiating the monomers either in the presence or absence of air. Experiments in the absence of air were carried out by covering the monomer films with a polypropylene (PP) film. This is a common strategy to protect polymerization samples from oxygen and reduce inhibition [33].

2.1.1. Photopolymerization Without Photoinitiator

The kinetics of the reactions of the eugenol-derivatives EDMA, EEMA, and EIMA were monitored by Real-Time FT-IR in transmission mode while they were exposed to a UV-light source (L9566-02A, 240 to 400 nm, 260 mW cm^{-2}) [34] in the presence of air or in the absence of air. The band corresponding to the methacrylate double bond at 1638 cm^{-1} (C=C stretching vibration) [35], and the aromatic band at 1514 cm^{-1} (C-H aromatic in-plane bending) [36] as reference, were monitored over the irradiation time. The conversion of the methacrylate double bonds (MDB) for EDMA, EEMA, and EIMA are presented in Figure 2. Simultaneously, the conversion of the allylic double bonds (ADB) from EEMA and propenyl double bonds (PDB) from EIMA were monitored using the bands at 995 cm^{-1} and 960 cm^{-1} respectively. The results comparing allylic and propenyl double bond conversion with regards to the presence or absence of air are plotted in Figure 3.

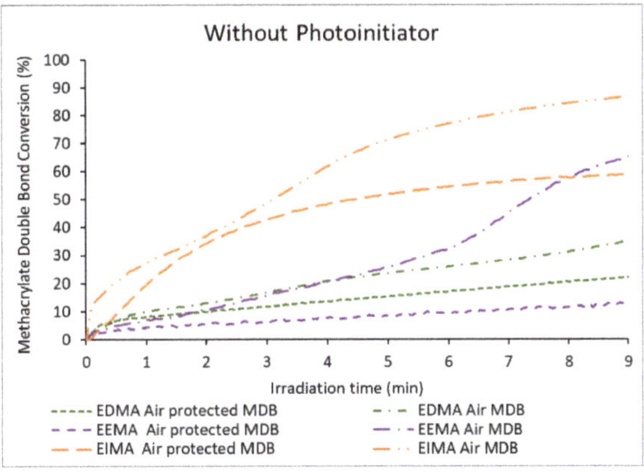

Figure 2. Methacrylate double bond (MDB) conversion of eugenol-derived monomers versus irradiation time in the absence of photoinitiator.

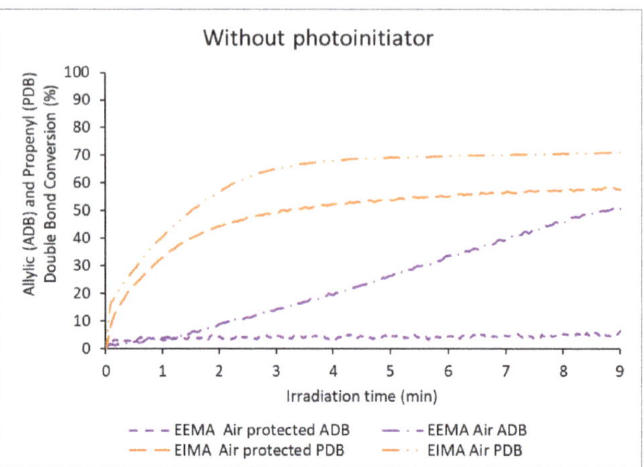

Figure 3. Allylic (ADB) and Propenyl double bond (PDB) conversion of eugenol-derived monomers versus irradiation time in the absence of photoinitiator.

Figure 2 shows that the methacrylic double bonds (MDB) of all monomers can react upon light exposure even in the absence of any photoinitiator. This is not surprising, as it has been previously reported that (meth)acrylates could undergo photopolymerization without a photoinitiator due to self-initiation [37–40]. Interestingly, the final conversion of MDB and the conversion rate (i.e., the slope of the conversion versus time curve) were different for the three monomers. The reactivity trend is as follows: EIMA >> EDMA > EEMA. EIMA was the most reactive monomer (higher slope, final conversion of 59%), while the conversion of EDMA and EEMA remained low reaching only 22% and 12% respectively at the end of the irradiation (Table 1). Conversion rates of EDMA and EEMA were much lower than that of EIMA. The different reactivities of the MDBs may be explained by the difference in the UV absorption spectra of the monomers (Figure S4). The monomers UV absorption curves overlap with the emission spectrum of the Hg lamp used as irradiation source [34]. The absorption of EIMA is significantly higher than that of EDMA and EEMA. EIMA is thus more likely to undergo faster self-initiation. To confirm that the monomers self-initiate due to UV absorption [37], further polymerization experiments protected from air were performed under UV irradiation but using a filter to stop wavelengths below 365 nm. As expected, since the monomers absorb below 320 nm, no reaction was observed for any of the monomers. These experiments confirmed the hypothesis of self-initiation responsible for the polymerization occurring in the absence of PI.

Besides the methacrylic group, the reactivity of the allyl and propenyl groups, present in EEMA and EIMA respectively, were studied during photopolymerization processes protected from air (Figure 3). Indeed, allylic and propenyl double bonds can experience secondary reactions such as (degradative) chain transfer reactions (allylic hydrogen abstraction produce poorly-reactive highly-stabilized radicals) and radical addition (cross-propagation) (Figure 4) [29,41–43]. The radicals formed from these secondary reactions can undergo further propagation or termination yielding branched and even crosslinked polymers (in the case of termination by combination). In the absence of air, the propenyl double bond (PDB) of EIMA was quite reactive and reached nearly the same conversion as the methacrylic double bonds (58%). On the other hand, the allylic double bond (ADB) of EEMA displayed a very low conversion (6%). For this monomer, a lower reaction rate and MDB conversion were obtained (Figure 2). EEMA allylic hydrogens can be abstracted and form highly stabilized radicals (main secondary reaction). This affects the propagation rate as the corresponding radicals become less reactive (degradative chain transfer). This effect was not seen for EIMA, implying that PDB reacts mainly though cross-propagation reactions between propenyl and methacrylic groups.

Table 1. Methacrylate (MDB), allylic (ADB) and propenyl (PDB) double bond conversions under different conditions of irradiation.

Monomer	Condition	Conversion (%)			
		220–600 nm		365 nm	
		Without PI	**Darocur 1173**	**Darocur 1173**	**Irgacure 819**
EDMA MDB		22	100	94	96
EEMA MDB	Air protected	12	66	74	76
EIMA MDB		59	100	65	78
EEMA ADB		6	7	6	3
EIMA PDB		58	56	40	12
EDMA MDB		35	61	8	8
EEMA MDB	Under air	66	81	0	7
EIMA MDB		86	92	39	40
EEMA ADB		49	64	2	9
EIMA PDB		68	76	58	30

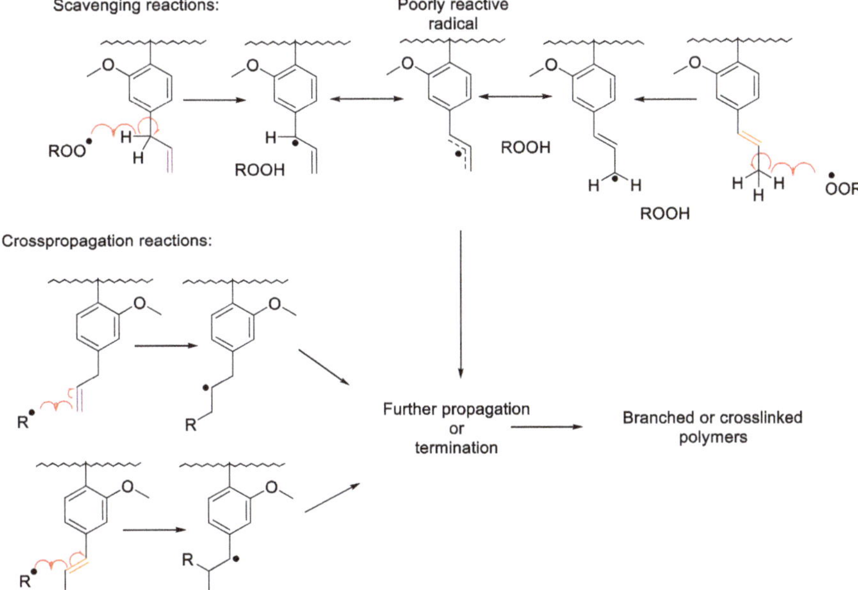

Figure 4. Scavenging and cross-propagation reactions on allylic and propenyl double bonds of eugenol-derived methacrylates.

Figure 2 shows that the methacrylate double bond conversions under air have different profiles compared to those of the polymerization protected from air. Changes in the curve slope are visible, signaling a noticeable variation of the speed of the reaction during the irradiation. This behavior is particularly clear for EIMA and to a lower extent for EEMA (Figure 2), but it is negligible for EDMA. The sigmoidal profile appearing in the curves is caused by the occurrence of two polymerization regimes. These are due to the formation of hydroperoxides in the presence of oxygen, as reported in literature [44]. Indeed, radicals produced by irradiation of the monomers can react with oxygen according to Equations (1)–(3).

$$R^{\cdot} + O_2 \rightarrow RO_2^{\cdot} \tag{1}$$

$$RO_2^{\cdot} + RH \rightarrow RO_2H + R^{\cdot} \tag{2}$$

$$RO_2H \xrightarrow{h\nu} RO^{\cdot} + HO^{\cdot} \tag{3}$$

Once peroxyl radicals (RO_2) are formed (Equation (1)), hydroperoxides (RO_2H) are generated by hydrogen abstraction (Equation (2)) [33]. The three monomers possess abstractable hydrogen atoms: bis-allylic hydrogens in EEMA (Ph- CH_2- CH= CH_2), propenylic hydrogens in EIMA (Ph- CH= CH- CH_3), and benzylic hydrogens in EDMA (Ph- CH_2- CH_2- CH_3). Hydroperoxides react slowly, therefore oxygen is often described as an inhibitor of radical polymerization. The phenomenon is particularly severe in photopolymerization when monomers are irradiated in films. In such a situation, a large area is exposed to oxygen, and oxygen can be continuously replaced by diffusion at the surface of the reacting formulation. However, hydroperoxides can decompose, through continued irradiation, to produce new radicals (Equation (3)) that are able to trigger additional initiation and a second polymerization regime [44]. Herein, IR analyses confirmed the hydroperoxides formation during the photopolymerization reactions carried out in the presence of air (Figures S12–S14).

Contrary to what was expected, all the monomers showed a higher MDB conversion in the presence of air than in the absence of air. During the first minute of irradiation, EEMA and EDMA displayed very similar MDB conversion under air or protected from air. However, conversion increased significantly at higher irradiation time in the presence of air. Specifically, for EEMA, MDB final conversion reached 66% under air (and only 12% when protected from air over the same irradiation time). For EDMA, the final conversion was 35% in the presence of air and 22% when protected from air (Table 1). Finally, for EIMA, the MDB conversion and conversion rates under air were always higher than in the absence of air from the onset of irradiation. Similarly to EEMA, after the first polymerization regime, the conversion rate of EIMA MDB increased (producing a second polymerization regime) and a final conversion of 86% was reached (59% in the absence of air).

Figure 3 shows that EEMA ADB are consumed up to 49% in the presence of air, while they are almost non-reactive in the absence of air. This can explain the high reactivity of the EEMA MDBs under air. Allylic double bonds undergo hydrogen abstraction leading to radicals that can react with oxygen to form peroxy radicals which scavenge oxygen and thus prevent the oxygen inhibition of MDB polymerization. The peroxy radicals can then form hydroperoxides that decompose to provide additional radicals for further MDB polymerization. In the case of EIMA, the conversion of PDB reached relatively high values both under air (68%) and while air-protected (58%) and was always higher under air, independent of the irradiation time (Figure 3 and Table 1). PDB can be consumed not only by the formation of peroxy and hydroperoxy radicals when oxygen is present but also by cross-propagation reactions (Figure 4). Noteworthy is that only a slightly higher consumption of PDB was observed in the polymerizations carried out under air compared to the polymerizations protected from it. This suggests that although hydroperoxides are formed under air, cross-propagation is the main secondary reaction.

In addition, higher conversion of the monomers under air can also be related to the formation of ozone induced by the UV irradiation at 242 nm and its subsequent photolysis into singlet oxygen ($^1O_2$) [45,46]. Singlet oxygen can react with the ADB and PDB of both EEMA and EIMA, again forming peroxy radicals and hydroperoxides which dissociate into other radicals [47–50].

Styrene can polymerize in the absence of a photoinitiator, due its capacity to form charge-transfer complexes with oxygen [51,52]. These complexes lead to the production of peroxides eventually leading to the production of radicals. Recently, Krueger et al. [53] concluded that in photoinitiator-free styrene polymerizations, oxygen reacts photochemically with styrene at the beginning of their polymerization reactions but that peroxides are not the sole source of radical formation. The photochemical radical generation via photo electron transfer (PET) requires a donor-acceptor pair. In the absence of oxygen, the PET between styrene-polystyrene leads to the generation of radical ions continuing the polymerization in the absence of PI. A similar process could occur in the case of isoeugenol. Moreover, the triplet state of isoeugenol derivatives has been suspected to produce singlet oxygen able to react with the double bond to form dioxetane, which can cleave to produce aldehydes [52,54]. Nonetheless, no increment or change was noticed in the bands at 2827 cm^{-1} and 2725 cm^{-1}, corresponding to the Fermi resonance characteristic of aldehydes. Hence, the consumption of PDB does not follow this pathway here.

To avoid the absorption of light by the monomers and the possible formation of ozone, further polymerization experiments under air were performed using a 355–375 nm bandpass filter. Unsurprisingly, no reaction was observed for any of the monomers, since neither monomer homolytic cleavage nor ozone production (both leading to radicals) occur at this longer irradiation wavelength.

In conclusion, the eugenol-derived methacrylates can photopolymerize in the absence of the photoinitiator both in the presence or absence of air as long as the irradiation wavelengths are short (from 220 to 355 nm). The presence of oxygen (while irradiating at short wavelengths <365 nm) leads to higher conversion of the methacrylic, allylic, and propenyl double bonds of the eugenol-derived methacrylates as a consequence of the production of hydroperoxides and their decomposition. The presence of ADB and PDB causes secondary reactions such as allylic hydrogen abstraction and cross-propagation which could lead to branched or crosslinked structures.

2.1.2. Photopolymerization with Photoinitiator

Experiments proceeded with the use of common photoinitiators. Darocur 1173 was added to the monomers at 2% wbm (weight based on monomer). It is a Norrish Type I photoinitiator that undergoes homolytic cleavage to produce two carbon-centered radicals (Figure 5 and Figure S5).

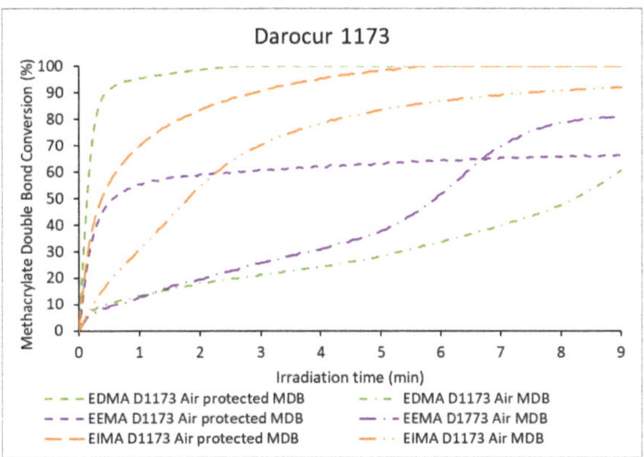

Figure 5. Homolytic cleavage under light of Darocur 1173.

The evolutions with irradiation time of the MDB conversions of the three monomers for the photopolymerizations carried out in the presence and in the absence of air using Darocur 1173 are shown in Figure 6. The evolutions with time of the conversions of ADB and PDB in the same conditions are displayed in Figure 7. The comparison of these data with those observed for polymerizations carried out in the absence of the photoinitiator demonstrates, as expected, that the PI accelerates the polymerization.

Figure 6. Methacrylate double bond (MDB) conversion of eugenol-derived monomers versus irradiation time in the presence of Darocur 1173.

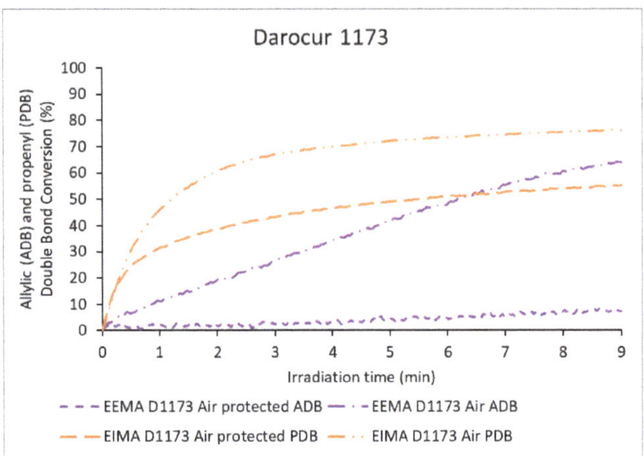

Figure 7. Allylic (ADB) and propenyl double bond (PDB) conversion of eugenol-derived monomers versus irradiation time in the presence of Darocur 1173.

In the absence of air, the conversion of EDMA MDB was fast and reached 100%. The polymerization rates of the difunctional methacrylates, EEMA and EIMA, were slower. EEMA MDB conversion reached 66%, whereas that of EIMA MDB reached 100% although at a lower rate than EDMA (Table 1). In the case of EEMA, radicals were presumed to be consumed by the allyl groups (degradative chain transfer), even to a small extent, to form highly stabilized radicals that resulted in a lower polymerization rate and ultimately in termination reactions limiting the conversion. No increment or appearance of the band at 960 cm^{-1} corresponding to the propenyl double bonds was observed, suggesting that the isomerization of EEMA into EIMA does not occur under these experimental conditions. In addition, the conversion of allylic double bonds to propenyl double bonds would lead to the decrease of the 995 cm^{-1} peak area (corresponding to allylic double bond), which did not occur, as conversion was <10%. In the case of EIMA, the PDBs were consumed up to 56% (Table 1) most likely via cross-propagation (vide supra). This cross-propagation slightly slows down the polymerization but does not prevent the quantitative conversion of EIMA MDB. Moreover, as discussed above, EIMA has a higher absorption than EEMA and forms propagating species by itself, thus enhancing the conversion.

In the presence of air, the polymerization rates were lower than those observed for polymerizations carried out in the absence of air. This decrease of the polymerization rate was likely caused by oxygen inhibition. EDMA was strongly inhibited by air and presented the lowest MDB conversion (61%). EIMA and EEMA were less affected and reached high conversions: 92% and 81%, respectively (Table 1). As previously discussed, reactions with oxygen can lead to the formation of hydroperoxides, which decompose, causing a second polymerization regime. The corresponding sigmoidal curve is observed quite clearly for EEMA, but not for EIMA. In addition, the conversion of PDB was higher than that of ADB (76% and 64% respectively, Figure 7). PDBs were also highly consumed when protected from air (56%), while ADBs were not (7%). This may be because cross-propagation is the dominant reaction in the consumption of PDBs (only a fraction might be consumed by hydrogen abstraction or hydroperoxide formation). In contrast, in the case of EEMA, for which bis-allylic H-abstraction and radical termination dominate, the absence of air limits the overall polymerization. However, in the presence of air, hydroperoxide dissociation provides the necessary radicals to continue EEMA MDB polymerization.

The effect of the monomer light absorption and the possible formation of ozone on the kinetics of polymerization remained to be investigated. Thus, a bandpass filter centered at 365 nm, preventing monomer light absorption and ozone formation (<242 nm) was again used to irradiate the formulations. Experiments with Darocur 1173 were performed and results are shown in Figures 8 and 9.

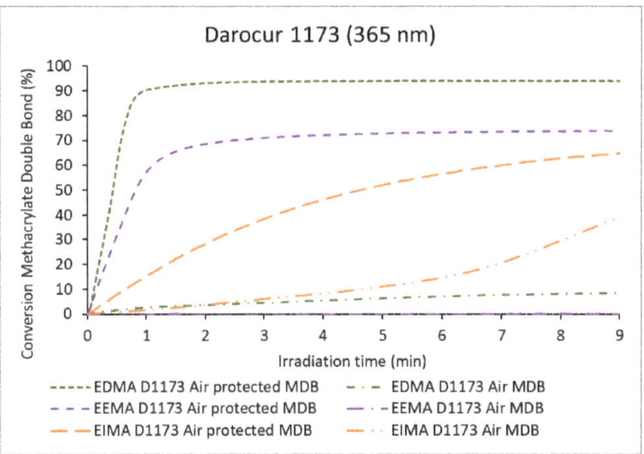

Figure 8. Methacrylate double bond (MDB) conversion of eugenol-derived monomers with irradiation time in the presence of Darocur 1173, irradiation under λ = 365 nm.

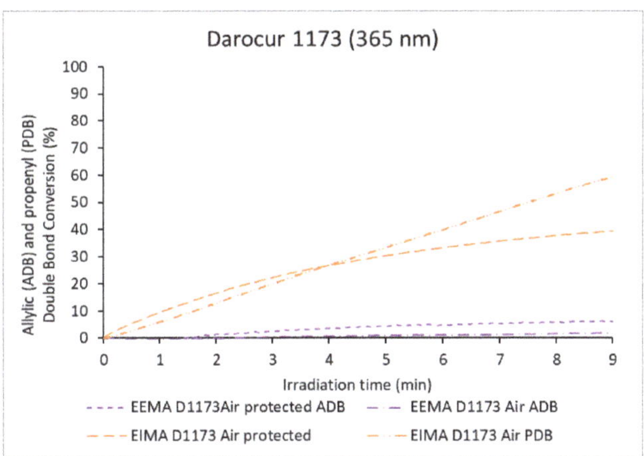

Figure 9. Allylic (ADB) and propenyl double bond (PDB) conversion of eugenol-derived monomers with irradiation time in the presence of Darocur 1173, irradiation under λ = 365 nm.

In the presence of air, almost no conversion of MDB could be measured for EEMA and EDMA (conversion <10%, Figure 8 and Table 1). Similar results were observed for EEMA ADB with a conversion close to zero (2%, Figure 9 and Table 1). In the presence of the filter, the production of radicals by cleavage of the photoinitiator was strongly diminished and the scarce quantity of radicals could quickly be quenched by oxygen, while no peroxides nor hydroperoxides could be generated [33]. Nevertheless, the considerable consumption of both EIMA MDB and PDB (39% and 58% respectively, Table 1 was observed in spite of the presence of oxygen. This may be explained by the formation of charge-transfer complexes of EIMA with oxygen (as reported for styrene [51,52]) which lead to the production of radicals and allow propagation.

In the absence of air, both the polymerization rate and the final conversion increased significantly. MDB conversion followed the trend: EDMA (94%) > EEMA (74%) > EIMA (65%) (Table 1). The polymerization rate was lower than that observed for the reaction carried out using light, including shorter irradiation wavelengths (i.e., without filter). This may be explained by a lower radical production both from Darocur 1173 (which absorbs weakly at 365 nm) and from the monomers (which do not absorb at 365, see Figure S4). Moreover, the irradiance decreases because of the filter (78 mW.cm^{-2} UVA). However, contrary to the experiments carried out without filter, EIMA showed lower MDB conversion than EEMA. This means that EIMA UV light absorption and cleavage (responsible for the reaction in the absence of PI) contribute to the formation of reactive species. Moreover, the consumption of the PDBs reached 40% (Table 1), while ADB consumption remained low (6%).

The comparison of the results of the polymerizations irradiated with and without filter using Darocur 1173 (Norrish Type I photoinitiator) suggests that the monomer absorptions play an important role in their reactivity, especially for EIMA. Moreover, in the presence of air, reactions of peroxides and hydroperoxides, ozone formation and photolysis to singlet oxygen, contribute to the polymerization mechanism.

In a final study, a passband filter centered at 365 nm and another Norrish type I PI with high absorption at longer wavelengths (absorption in the UVA region), Irgacure 819 [55] were used. The results obtained in the different conditions (with and without air) are gathered in the SI (Figures S16 and S17). In this case, the potential cleavage of the methacrylates was prevented by the pass band filter and oxygen inhibition or the production of hydroperoxides was avoided by protecting the samples from air. In addition, the flux of radicals had been raised by using Irgacure 819, which has a higher molar extinction coefficient and quantum yield than Darocur 1173 in the UVA region [55]. A behavior similar to that of Darocur 1173 was observed.

The experiments executed in different conditions (with or without initiator, in the presence and absence of air, with or without filter) revealed that EDMA polymerization was always strongly inhibited in the presence of air. On the contrary, the presence of the pending allylic (EEMA) or propenyl (EIMA) double bonds could produce a second polymerization regime due to dissociation of hydroperoxides (formed in-situ in the presence of air under shorter (<320 nm) wavelength irradiation). It was also shown that the dominant reaction mechanism for PDB is cross-propagation rather than hydrogen abstraction or hydroperoxide formation, as they were consumed to a high extent even in the absence of air. The polymerization of EIMA was the least affected by air.

2.2. Polymers Characterization

Properties of the polymers prepared by photoinduced polymerization in the presence of Darocur 1173, both in the presence and absence of air, were measured (Table 2). The polymerization conditions (i.e., use of a UV irradiation spectrum from 320 to 390 nm and of Darocur 1173 as PI) were selected to guarantee high conversions. Polymers obtained from EDMA had a linear structure and were soluble (gel content $\approx$ 0%). In contrast, polymers from EEMA and EIMA were crosslinked and completely insoluble (gel content = 100%), suggesting that the unreacted functional groups potentially present (when the conversion was not quantitative, as reported in Table S2) were dangling from the network and that no free oligomer or monomer were present. The glass transition temperature did not vary much between the samples irradiated in the presence or in the absence of air except for poly(EDMA) prepared by irradiation under air. In this case, the presence of oligomers or unreacted monomer plasticized the resulting polymer and reduced its T_g. The obtained T_gs were in agreement with the results previously obtained with polymers prepared by emulsion polymerization [30].

Table 2. Thermal properties, gel content, and contact angle of homopolymers produced with Darocur 1173.

Monomer	Polymerization Condition	Gel Content (%)	T_g (°C)	$T_{d5\%}$ (°C)	Contact Angle DI Water (°)	Contact Angle Hexadecane (°)
EDMA	with air	2	8	236	92	24
	no air	3	23	269	84	30
EEMA	with air	100	35	298	89	33
	no air	98	34	294	85	34
EIMA	with air	100	56	246	85	24
	no air	100	58	258	82	25

The TGA results showed that the starting degradation temperatures of the polymers were always higher than 230 °C (Figures 10 and 11). Polymerization carried out in the presence of air led to crosslinked poly(EEMA) with higher decomposition temperatures, due to a higher consumption of ADBs. A slightly lower degradation temperature was registered for poly(EIMA) prepared in the presence of air but both polymers (produced under air or in the absence of air) exhibited complex profiles, indicating complex polymeric architectures. Their glass transition temperatures (ranging from 8 °C and 58 °C) as well as their degradation temperatures (above 230 °C) make these materials suitable for application in coatings.

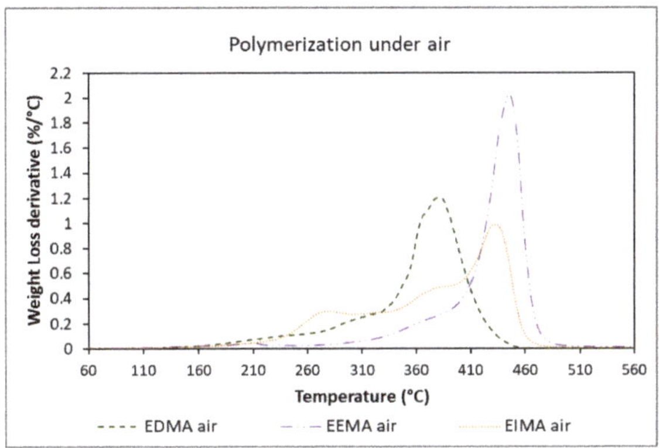

Figure 10. Thermogravimetric analysis of the different eugenol-derived methacrylates polymers from polymerization with Darocur 1173 under air (1° derivative).

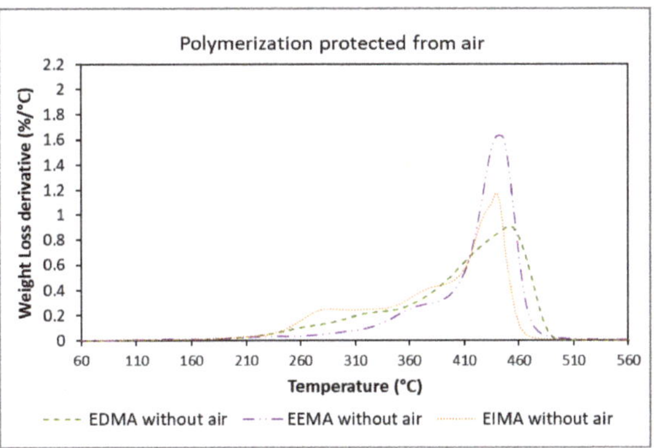

Figure 11. Thermogravimetric analysis of the different eugenol-derived methacrylates polymers from polymerization with Darocur 1173 protected from air (1° derivative).

The water and hexadecane contact angles (Table 2) indicated that the wettability of all the polymers were independent of the structure. The polymers were almost hydrophobic and displayed moderate oleophilicity.

3. Materials and Methods

3.1. Materials

Ethoxy eugenyl methacrylate (EEMA), ethoxy isoeugenyl methacrylate (EIMA), ethoxy dihydroeugenyl methacrylate (EDMA) monomers (Figure 1) were synthesized as described in a previous article from our group [29]. Toluene (>98%) was purchased from Sigma-Aldrich (Milano, Italy). The radical photoinitiators 2-hydroxy-2-methylpropiophenone (tradename Darocur 1173) and phenylbis(2,4,6-trimethylbenzoyl)phosphine oxide (tradename Irgacure 819), kindly given by Badische Anilin und Soda Fabrik (BASF) (Ludwigshafen, Germany) were used as received. Darocur 1173 and Irgacure 819 structures are presented in Figure 12.

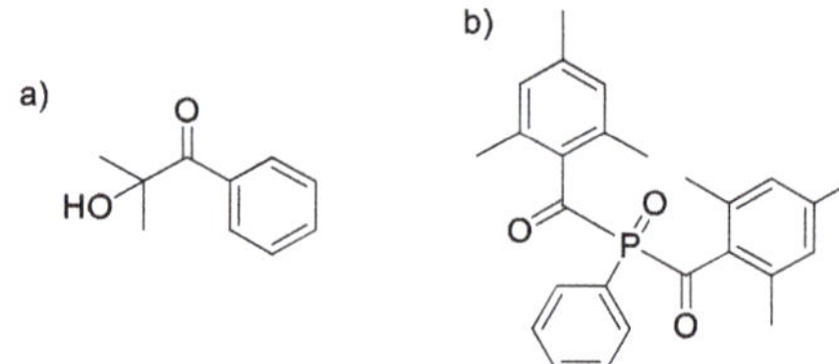

Figure 12. Photoinitiators (**a**) Darocur 1173 (**b**) Irgacure 819.

3.2. Photoinduced Polymerization of Eugenol Derived Methacrylates

3.2.1. Samples Preparation

To monitor the photopolymerization kinetics of each monomer, a mixture of monomer and PI at 2% wbm (weight based on monomer) was spread over a silicon wafer using a rod coater, forming a film with a thickness of 10 µm. Samples were irradiated up to 9 min either under air or protected from air with a 30 µm-thick polypropylene (PP) film.

3.2.2. Kinetics Monitoring

Photopolymerization was monitored using Real-Time Fourier Transform Infrared (FT-IR) spectroscopy on a Nicolet iS50 spectrometer (Thermo Fisher Scientific Inc., Waltham, MA, USA). The spectra were acquired in transmission mode, in the 650–4000 cm^{-1} range, with 1 scan per spectrum and a resolution of 4 cm^{-1}.

A high-pressure mercury-xenon lamp Lightning Cure LC8 from Hamamatsu equipped with a flexible light guide was used as UV-light source (L9566-02A, 220 to 600 nm) [34] and an EIT Powerpuck® II radiometer (EIT, LLC., Leesburg, VA, USA) was used to measure the UV irradiance. The samples were irradiated with 260 mW cm^{-2} (sum of UVA, UVB, UVC, UVV). In some experiments, light was filtered using a A9616-07 filter (Hamamatsu, Shizuoka, Japan) with a transmittance wavelength of 355–375 nm (centered at 365 nm) The filtered light had and intensity of 78 mW cm^{-2} (UVA).

3.2.3. Conversion Determination

The methacrylate double bonds (MDB) conversion was monitored using the band at 1638 cm^{-1}, the allylic double bonds (ADB) conversion was monitored using the band at 995 cm^{-1}, and the propenyl double bonds (PDB) conversion was determined using the 960 cm^{-1} band (Figure S1) [26,36,56,57]. Each conversion was calculated using the following Equation (4):

$$Conversion\%_{t=x} = 100 \times \left(1 - \frac{\frac{A_{t=x}}{Ref.A_{t=x}}}{\frac{A_{t=0}}{Ref.A_{t=0}}}\right) \tag{4}$$

Where A is the absorbance of the IR band of the functional group monitored during irradiation; Ref A is the absorbance of the band of the aromatic ring (C-C stretching) taken as a reference (1540 cm^{-1} to 1490 cm^{-1}).

Absorbances were estimated as the area of the vibrational bands under examination. Data were processed using OMNIC software (Madison, WI, USA). All curves were smoothed using the Savitzky–Golay method with 20 points window and second polynomial order. For the determination of EEMA methacrylate double bonds conversion, an approximation was made as peaks corresponding to the allylic and methacrylate double bond superimposed at circa 1638 cm^{-1}. The calculation details are explained in SI (Figures S2 and S3 and Table S1).

3.3. Characterization Methods

Samples for thermogravimetric analysis (TGA), differential scanning calorimetry (DSC), and gel content were prepared by coating a glass slide with 200 µm films and irradiating it for 10 min using a DYMAX 5000 EC UV flood lamp (Dymax Corporation, Torrington, CT, USA) in the range of 320 to 390 nm with an intensity on the sample of 156 mW cm^{-2} (UVA and UVV).

IR spectra to determine the conversion were acquired on a Thermo Scientific Nicolet 6700 FTIR apparatus (Madison, WI, USA) in the 650–4000 cm^{-1} range, with 32 scan per spectrum and a resolution of 4 cm^{-1} (using attenuated total reflectance technique, ATR) on both faces of the film: the one exposed to the atmosphere and the one in contact with the glass slide.

Ultraviolet and visible spectroscopy (Figures S4 and S5): UV-Vis spectra were recorded on a Varian Cary 50 Scan spectrophotometer (Victoria, Australia) from 200 to 800 nm with a scan rate of 4800 nm min^{-1}.

Thermogravimetric Analysis: TGA was performed on 5–10 mg samples on a TGA Q50 apparatus from TA Instruments (New Castle, DE, USA) from 20 °C to 580 °C, in an aluminum pan, at a heating rate of 20 °C min^{-1}, under nitrogen.

Differential Scanning Calorimetry: (Figures S6–S11) DSC measurements were performed on 10–15 mg samples, under nitrogen atmosphere, with a Netzsch DSC 200 F3 instrument (Selb, Germany) using the following heating/cooling cycle: first cooling ramp from room temperature (ca. 20 °C) to −40 °C at 20 °C min^{-1}, isothermal plateau at −40 °C for 10 min, first heating ramp from −40 to 150 °C at 20 °C min^{-1}, second cooling stage from 150 to −40 °C at 20 °C min^{-1}, isothermal plateau at −40 °C for 10 min, second heating ramp from −40 to 150 °C at 20 °C min^{-1}, third cooling stage from 150 to −40 °C at 20 °C min^{-1}, isothermal plateau at −40 °C for 10 min, third heating ramp from −40 to 150 °C at 20 °C min^{-1} and last cooling stage from 150 °C to room temperature (ca. 20 °C). T_g values are given from the evaluation of the third heating ramp. Calibration of the instrument was performed with noble metals and checked with an indium sample.

Gel content measurements: The gel content of the polymers was measured by placing approximately 30–50 mg of polymer in a Teflon pocket which was subsequently immersed in 10 mL of toluene for 24 h, then dried in a ventilated oven at 50 °C for 4 h. The gel content was calculated based on the initial (W_i) and final (W_f) polymer mass according to Equation (5).

$$Gel\ content(\%) = \frac{W_f \times 100}{W_i} \tag{5}$$

4. Conclusions

Three eugenol-derived methacrylates (EDMA, EEMA, EIMA) were polymerized via photopolymerization without a photoinitiator and with two Norrish Type I photoinitiators (Darocur 1173 and Irgacure 819), under air or without air. Their polymerization behavior under the different conditions was described. The monomers were shown to polymerize in the absence of a photoinitiator, especially in the presence of oxygen, due to self-initiation and oxidation reactions. In the presence of air, EIMA showed the highest conversions in any of the conditions studied. The second polymerization regimes, due to the formation and photolysis of hydroperoxides, were observed upon irradiation at a wavelength shorter than 365 nm in air with or without PI. This effect was clearly visible for EIMA and EEMA. Moreover, both allylic and propenyl double bonds were reactive in the presence of air. In addition, PDBs were shown to be predominantly polymerized via cross-propagation reactions while ADBs were mainly consumed under air via hydrogen abstraction and hydroperoxides formation. In the absence of air and using PI, EDMA reached the highest conversions. To eliminate the self-initiation of the monomers as well as the formation of hydroperoxides, a 365 nm passband filter and air-protected conditions were used. Under these conditions, the polymerization rate followed the order EDMA > EEMA > EIMA. EEMA displayed a significant reduction of the propagation rate, due to the formation of highly stabilized bis-allylic radicals. EIMA exhibited a lower MDB conversion, due to cross-propagation with the PDB. The polymers properties indicated that their use in applications in coatings and in dentistry could be envisaged.

Supplementary Materials: The following are available online. Figure S1. IR spectra in transmission of the different eugenol derived methacrylates: (A) EDMA, (B) EEMA and (C) EIMA, Figure S2. Eugenol, ethoxy eugenol, and ethoxy eugenyl methacrylate, Table S1. Normalization of the allylic band area for Eugenol and EE, Figure S3. ATR Spectra of eugenol, ethoxyeugenol, and ethoxy eugenyl methacrylate, Figure S4. UV absorption spectra of the eugenol-derived methacrylates, in acetonitrile 0.002 wt%, Figure S5. UV absorption of photoinitiators, Figure S6. DSC measurement of poly(EDMA) obtained from photoinduced polymerization with Darocur 1173 (2% wbm) under nitrogen, Figure S7. DSC measurement of poly(EDMA) obtained from photoinduced polymerization with Darocur 1173 (2% wbm) under air, Figure S8. DSC measurement of poly(EEMA) obtained from photoinduced polymerization with Darocur 1173 (2% wbm) under nitrogen, Figure S9. DSC measurement of poly(EEMA) obtained from photoinduced polymerization with Darocur 1173 (2% wbm) under air, Figure S10. DSC measurement of poly(EIMA) obtained from photoinduced polymerization with Darocur 1173 (2% wbm) under nitrogen, Figure S11. DSC measurement of poly(EIMA) obtained from photoinduced polymerization with Darocur 1173 (2% wbm) under air, Figure S12. FT-IR spectra at time = 0 and time = 9 min denoting the formation of hydroperoxides during photopolymerization of EEMA in the absence of PI under air, Figure S13. FT-IR spectra at time = 0 and time = 9 min denoting the formation of hydroperoxides during photopolymerization of EIMA in the absence of PI under air, Figure S14. FT-IR spectra at time = 0 and time = 9 min denoting the formation of hydroperoxides during photopolymerization of EDMA in the absence of PI under air, Figure S15. Irgacure 819, its homolytic cleavage under light, Figure S16. Methacrylate double bond conversion for eugenol-derived monomer versus irradiation time with Irgacure 819 and filter (dotted lines = irradiation under air; solid lines = irradiation with air protection), Figure S17. Allylic and propenyl double bond conversion for eugenol-derived monomers versus irradiation time with Irgacure 819 and filter (dotted lines = irradiation under air; solid lines = irradiation with air protection), Table S2. Monomer conversion of films used for polymer characterization (by ATR-FTIR).

Author Contributions: Conceptualization, S.M.-G., R.B., and S.D.V.; methodology, S.M.-G. and R.B., formal analysis, S.M.-G., R.B., A.V., and S.D.V.; investigation, S.M.-G. and S.D.V.; resources, P.L.-D. and R.B.; data curation, S.M.-G., R.B., and P.L.-D.; writing—original draft preparation, S.M.-G. and R.B.; writing—review and editing, S.M.-G., R.B., P.L.-D., S.D.V., A.V., S.C., and V.L.; supervision, P.L.-D., R.B., S.C., and V.L.; project administration, P.L.-D.; funding acquisition, P.L.-D. All authors have read and agreed to the published version of the manuscript.

Funding: This work was funded through a SINCHEM Grant for the doctoral studies of S.M.G. SINCHEM is a Joint Doctorate program selected under the Erasmus Mundus Action 1 Program (FPA 2013-0037). S.D.V. acknowledges financial support from EU through the project ComBIOsites which has received funding from the European Union's Horizon 2020 Research and Innovation program under the Marie Sklodowska-Curie grant agreement no. 789454. The authors also thank Synthomer (UK) Ltd. for financial support of their research on biobased polymers as well as for fruitful discussions (Peter Shaw and Renaud Perrin).

Acknowledgments: Samantha Molina-Gutiérrez would like to thank Parnian Kianfar, Gustavo González, and Giuseppe Trusiano for their support during the development of this work. This manuscript is in honor of the 50 year anniversary of the French Polymer Group (Groupe Français des Polymères–GFP).

Conflicts of Interest: The authors declare no conflict of interest. The funders had no role in the design of the study; in the collection, analyses, or interpretation of data; in the writing of the manuscript, or in the decision to publish the results.

References

1. Gandini, A.; Lacerda, T.M.; Carvalho, A.J.F.; Trovatti, E. Progress of polymers from renewable resources: Furans, vegetable oils, and polysaccharides. *Chem. Rev.* **2015**, *116*, 1637–1669. [CrossRef] [PubMed]
2. Fouassier, J.; Allonas, X.; Burget, D. Photopolymerization reactions under visible lights: Principle, mechanisms and examples of applications. *Prog. Org. Coat.* **2003**, *47*, 16–36. [CrossRef]
3. Crivello, J.V.; Reichmanis, E. Photopolymer materials and processes for advanced technologies. *Chem. Mater.* **2013**, *26*, 533–548. [CrossRef]
4. Childs, A.; Li, H.; Lewittes, D.M.; Dong, B.; Liu, W.; Shu, X.; Sun, C.; Zhang, H.F. Fabricating customized hydrogel contact lens. *Sci. Rep.* **2016**, *6*, 34905. [CrossRef] [PubMed]
5. Stansbuty, J.W. Curing dental resins and composites by photopolymerization. *J. Esthet. Restor. Dent.* **2000**, *12*, 300–308. [CrossRef] [PubMed]
6. Lloret, T.; Navarro-Fuster, V.; Ramírez, M.G.; Ortuño, M.; Neipp, C.; Beléndez, A.; Pascual, I. Holographic lenses in an environment-friendly photopolymer. *Polymers* **2018**, *10*, 302. [CrossRef]
7. Corrigan, N.A.; Yeow, J.; Judzewitsch, P.; Xu, J.; Boyer, C. Seeing the light: Advancing materials chemistry through photopolymerization. *Angew. Chem. Int. Ed.* **2019**, *58*, 5170–5189. [CrossRef]
8. Vitale, A.; Hennessy, M.G.; Matar, O.K.; Cabral, J.T. A unified approach for patterning via frontal photopolymerization. *Adv. Mater.* **2015**, *27*, 6118–6124. [CrossRef]
9. Ligon, S.C.; Liska, R.; Stampfl, J.; Gurr, M.; Mülhaupt, R. Polymers for 3D printing and customized additive manufacturing. *Chem. Rev.* **2017**, *117*, 10212–10290. [CrossRef]
10. Zhang, J.; Xiao, P. 3D printing of photopolymers. *Polym. Chem.* **2018**, *9*, 1530–1540. [CrossRef]
11. Fertier, L.; Koleilat, H.; Stemmelen, M.; Giani, O.; Joly-Duhamel, C.; Lapinte, V.; Robin, J.-J. The use of renewable feedstock in UV-curable materials—A new age for polymers and green chemistry. *Prog. Polym. Sci.* **2013**, *38*, 932–962. [CrossRef]
12. Goldberg, M.Z.; Burke, L.A.; Samokhvalov, A. Selective activation of C=C bond in sustainable phenolic compounds from ligninvia photooxidation: Experiment and density functional theory calculations. *Photochem. Photobiol.* **2015**, *91*, 1332–1339. [CrossRef]
13. Khalil, A.A.; Rahman, U.U.; Khan, M.R.; Sahar, A.; Mehmood, T.; Khan, M. Essential oil eugenol: Sources, extraction techniques and nutraceutical perspectives. *RSC Adv.* **2017**, *7*, 32669–32681. [CrossRef]
14. Sakai, K.; Takeuti, H.; Mun, S.-P.; Imamura, H. Formation of isoeugenol and eugenol during the cleavage of β-aryl ethers in lignin by alcohol-bisulfite treatment. *J. Wood Chem. Technol.* **1988**, *8*, 29–41. [CrossRef]
15. Llevot, A.; Grau, E.; Carlotti, S.; Grelier, S.; Cramail, H. From lignin-derived aromatic compounds to novel biobased polymers. *Macromol. Rapid Commun.* **2015**, *37*, 9–28. [CrossRef]
16. Kong, X.; Liu, X.; Li, J.; Yang, Y. Advances in pharmacological research of eugenol. *Curr. Opin. Complement. Altern. Med.* **2014**, *1*, 8–11.
17. Da Silva, F.F.M.; Monte, F.J.; Lemos, T.L.G.; Nascimento, P.G.G.D.; Costa, A.K.D.M.; De Paiva, L.M.M. Eugenol derivatives: Synthesis, characterization, and evaluation of antibacterial and antioxidant activities. *Chem. Central J.* **2018**, *12*, 34. [CrossRef]
18. Modjinou, T.; Versace, D.-L.; Abbad-Andallousi, S.; Bousserrhine, N.; Dubot, P.; Langlois, V.; Renard, E. Antibacterial and antioxidant bio-based networks derived from eugenol using photo-activated thiol-ene reaction. *React. Funct. Polym.* **2016**, *101*, 47–53. [CrossRef]
19. Munerato, M.C.; Sinigaglia, M.; Reguly, M.L.; De Andrade, H.H.R. Genotoxic effects of eugenol, isoeugenol and safrole in the wing spot test of Drosophila melanogaster. *Mutat. Res. Toxicol. Environ. Mutagen.* **2005**, *582*, 87–94. [CrossRef]
20. Lartigue-Peyrou, F. The use of phenolic compounds as free-radical polymerization inhibitors. In *The Roots of Organic Development*; Elsevier BV: Amsterdam, The Netherlands, 1996; Volume 8, pp. 489–505.
21. Fujisawa, S.; Kadoma, Y. Action of eugenol as a retarder against polymerization of methyl methacrylate by benzoyl peroxide. *Biomaterials* **1997**, *18*, 701–703. [CrossRef]

22. Guzmán, D.; Ramis, X.; Fernández-Francos, X.; De La Flor, S.; Serra, A. New bio-based materials obtained by thiol-ene/thiol-epoxy dual curing click procedures from eugenol derivates. *Eur. Polym. J.* **2017**, *93*, 530–544. [CrossRef]

23. Guzmán, D.; Ramis, X.; Fernández-Francos, X.; De La Flor, S.; Serra, A. Preparation of new biobased coatings from a triglycidyl eugenol derivative through thiol-epoxy click reaction. *Prog. Org. Coat.* **2018**, *114*, 259–267. [CrossRef]

24. Modjinou, T.; Versace, D.-L.; Abbad-Andaloussi, S.; Langlois, V.; Renard, E. Enhancement of biological properties of photoinduced biobased networks by post-functionalization with antibacterial molecule. *ACS Sustain. Chem. Eng.* **2018**, *7*, 2500–2507. [CrossRef]

25. Dai, J.; Jiang, Y.; Liu, X.; Wang, J.; Zhu, J. Synthesis of eugenol-based multifunctional monomers via a thiol–ene reaction and preparation of UV curable resins together with soybean oil derivatives. *RSC Adv.* **2016**, *6*, 17857–17866. [CrossRef]

26. Yoshimura, T.; Shimasaki, T.; Teramoto, N.; Shibata, M. Bio-based polymer networks by thiol–ene photopolymerizations of allyl-etherified eugenol derivatives. *Eur. Polym. J.* **2015**, *67*, 397–408. [CrossRef]

27. Miao, J.-T.; Yuan, L.; Guan, Q.; Liang, G.; Gu, A. Water-phase synthesis of a biobased allyl compound for building uv-curable flexible thiol–ene polymer networks with high mechanical strength and transparency. *ACS Sustain. Chem. Eng.* **2018**, *6*, 7902–7909. [CrossRef]

28. Breloy, L.; Ouarabi, C.A.; Brosseau, A.; Dubot, P.; Brezova, V.; Abbad-Andaloussi, S.; Malval, J.-P.; Versace, D.-L. β-Carotene/limonene derivatives/eugenol: Green synthesis of antibacterial coatings under visible-light exposure. *ACS Sustain. Chem. Eng.* **2019**, *7*, 19591–19604. [CrossRef]

29. Molina-Gutiérrez, S.; Manseri, A.; Ladmiral, V.; Bongiovanni, R.; Caillol, S.; Lacroix-Desmazes, P. Eugenol: A promising building block for synthesis of radically polymerizable monomers. *Macromol. Chem. Phys.* **2019**, *220*. [CrossRef]

30. Molina-Gutiérrez, S.; Ladmiral, V.; Bongiovanni, R.M.; Caillol, S.; Lacroix-Desmazes, P. Emulsion polymerization of dihydroeugenol-, eugenol-, and isoeugenol-derived methacrylates. *Ind. Eng. Chem. Res.* **2019**, *58*, 21155–21164. [CrossRef]

31. Lalevée, J.; Fouassier, J.P.; Graff, B.; Zhang, J.; Xiao, P. *Chapter 6. How to Design Novel Photoinitiators for Blue Light*; The Royal Society of Chemistry (RSC): London, UK, 2018; Volume 48, pp. 179–199.

32. Detrembleur, C.; Versace, D.-L.; Piette, Y.; Hurtgen, M.; Jérôme, C.; Lalevée, J.; Debuigne, A. Synthetic and mechanistic inputs of photochemistry into the bis-acetylacetonatocobalt-mediated radical polymerization of n-butyl acrylate and vinyl acetate. *Polym. Chem.* **2012**, *3*, 1856–1866. [CrossRef]

33. Ligon, S.C.; Husár, B.; Wutzel, H.; Holman, R.; Liska, R. Strategies to reduce oxygen inhibition in photoinduced polymerization. *Chem. Rev.* **2013**, *114*, 557–589. [CrossRef]

34. Hamamatsu Spotlight Sources Lightningcure Series, March 2019. Available online: https://www.hamamatsu.com/resources/pdf/etd/LC8_TLSZ1008E.pdf (accessed on 19 May 2020).

35. Nyquist, R.A.; Fiedler, S.; Streck, R. Infrared study of vinyl acetate, methyl acrylate and methyl methacrylate in various solvents. *Vib. Spectrosc.* **1994**, *6*, 285–291. [CrossRef]

36. Chowdhry, B.Z.; Ryall, J.P.; Dines, T.J.; Mendham, A.P. Infrared and raman spectroscopy of eugenol, isoeugenol, and methyl eugenol: Conformational analysis and vibrational assignments from density functional theory calculations of the anharmonic fundamentals. *J. Phys. Chem. A* **2015**, *119*, 11280–11292. [CrossRef] [PubMed]

37. Wang, H.; Brown, H. Self-initiated photopolymerization and photografting of acrylic monomers. *Macromol. Rapid Commun.* **2004**, *25*, 1095–1099. [CrossRef]

38. Huang, L.; Li, Y.; Yang, J.; Zeng, Z.; Chen, Y. Self-initiated photopolymerization of hyperbranched acrylates. *Polymer* **2009**, *50*, 4325–4333. [CrossRef]

39. Hoijemberg, P.A.; Chemtob, A.; Croutxé-Barghorn, C. Two routes towards photoinitiator-free photopolymerization in miniemulsion: Acrylate self-initiation and photoactive surfactant. *Macromol. Chem. Phys.* **2011**, *212*, 2417–2422. [CrossRef]

40. Furutani, M.; Ide, T.; Kinoshita, S.; Horiguchi, R.; Mori, I.; Sakai, K.; Arimitsu, K. Initiator-free photopolymerization of common acrylate monomers with 254 nm light. *Polym. Int.* **2018**, *68*, 79–82. [CrossRef]

41. Rojo, L.; Vazquez-Lasa, B.; Parra, J.; Bravo, A.L.; Deb, S.; Roman, J.S. From natural products to polymeric derivatives of "eugenol": A new approach for preparation of dental composites and orthopedic bone cements. *Biomacromolecules* **2006**, *7*, 2751–2761. [CrossRef]

42. Zhang, Y.; Li, Y.; Wang, L.; Gao, Z.; Kessler, M.R. Synthesis and characterization of methacrylated eugenol as a sustainable reactive diluent for a maleinated acrylated epoxidized soybean oil resin. *ACS Sustain. Chem. Eng.* **2017**, *5*, 8876–8883. [CrossRef]

43. Gazzotti, S.; Hakkarainen, M.; Adolfsson, K.H.; Ortenzi, M.A.; Farina, H.; Lesma, G.; Silvani, A. One-pot synthesis of sustainable high-performance thermoset by exploiting eugenol functionalized 1,3-dioxolan-4-one. *ACS Sustain. Chem. Eng.* **2018**, *6*, 15201–15211. [CrossRef]

44. Pynaert, R.; Buguet, J.; Croutxé-Barghorn, C.; Moireau, P.; Allonas, X. Effect of reactive oxygen species on the kinetics of free radical photopolymerization. *Polym. Chem.* **2013**, *4*, 2475. [CrossRef]

45. Schiff, H.I. Kinetics of ozone photochemistry. *Pure Appl. Geophys. PAGEOPH* **1973**, *106*, 1464–1467. [CrossRef]

46. Matsumi, Y.; Kawasaki, M. Photolysis of atmospheric ozone in the ultraviolet region. *Chem. Rev.* **2003**, *103*, 4767–4782. [CrossRef] [PubMed]

47. Frimer, A.A. The reaction of singlet oxygen with olefins: The question of mechanism. *Chem. Rev.* **1979**, *79*, 359–387. [CrossRef]

48. Singleton, D.A.; Hang, C.; Szymanski, M.J.; Meyer, M.P.; Leach, A.G.; Kuwata, K.T.; Chen, J.S.; Greer, A.; Foote, C.S.; Houk, K.N. Mechanism of ene reactions of singlet oxygen. a two-step no-intermediate mechanism. *J. Am. Chem. Soc.* **2003**, *125*, 1319–1328. [CrossRef] [PubMed]

49. Ghogare, A.A.; Greer, A. Using singlet oxygen to synthesize natural products and drugs. *Chem. Rev.* **2016**, *116*, 9994–10034. [CrossRef]

50. Parker, D.H. Laser photochemistry of molecular oxygen. *Accounts Chem. Res.* **2000**, *33*, 563–571. [CrossRef]

51. Kodaira, T.; Hayashi, K.; Ohnishi, T. Photopolymerization of styrene in the presence of oxygen. role of the charge-transfer complex. *Polym. J.* **1973**, *4*, 1–9. [CrossRef]

52. Kodaira, T.; Hashimoto, K.; Sakanaka, Y.; Tanihata, S.; Ikeda, K. The role of the charge-transfer complex in the photocopolymerization of oxygen with styrene and α-methylstyrene. *Bull. Chem. Soc. Jpn.* **1978**, *51*, 1487–1489. [CrossRef]

53. Krüger, K.; Tauer, K.; Yagci, Y.; Moszner, N. Photoinitiated bulk and emulsion polymerization of styrene–evidence for photo-controlled radical polymerization. *Macromolecules* **2011**, *44*, 9539–9549. [CrossRef]

54. Gellerstedt, G.; Pettersson, E.-L.; Weeks, O.B. Light-induced oxidation of lignin. The behaviour of structural units containing a ring-conjugated double bond. *Acta Chem. Scand.* **1975**, *29*, 1005–1010. [CrossRef]

55. Eibel, A.; Fast, D.E.; Gescheidt, G. Choosing the ideal photoinitiator for free radical photopolymerizations: Predictions based on simulations using established data. *Polym. Chem.* **2018**, *9*, 5107–5115. [CrossRef]

56. Silverstein, R.M.; Webster, F.X.; Kiemle, D.J. *Spectrometric Identification of Organic Compounds*, 7th ed.; John Wiley & Sons: Hoboken, NJ, USA, 2005; ISBN 0-471-39362-2.

57. Wang, L.-H.; Sung, W.-C. Rapid evaluation and quantitative analysis of eugenol derivatives in essential oils and cosmetic formulations on human skin using attenuated total reflectance–infrared spectroscopy. *Spectroscopy* **2011**, *26*, 43–52. [CrossRef]

Sample Availability: Samples of the compounds EDMA, EEMA and EIMA are available from the authors.

Article

Adhesion and Stability of Nanocellulose Coatings on Flat Polymer Films and Textiles

Raha Saremi [1,2], Nikolay Borodinov [3], Amine Mohamed Laradji [1], Suraj Sharma [1,2], Igor Luzinov [3] and Sergiy Minko [1,2,*]

1   Nanostructured Materials Laboratory, University of Georgia, Athens, GA 30602, USA; raha@uga.edu (R.S.); alaradji@wustl.edu (A.M.L.); ssharma@uga.edu (S.S.)
2   Department of Textiles, Merchandising and Interiors, the University of Georgia, Athens, GA 30602, USA
3   Department of Materials Science and Engineering, Clemson University, Clemson, SC 29634, USA; nikolab@g.clemson.edu (N.B.); luzinov@clemson.edu (I.L.)
*   Correspondence: sminko@uga.edu

Academic Editor: Sylvain Caillol
Received: 21 June 2020; Accepted: 13 July 2020; Published: 16 July 2020

Abstract: Renewable nanocellulose materials received increased attention owing to their small dimensions, high specific surface area, high mechanical characteristics, biocompatibility, and compostability. Nanocellulose coatings are among many interesting applications of these materials to functionalize different by composition and structure surfaces, including plastics, polymer coatings, and textiles with broader applications from food packaging to smart textiles. Variations in porosity and thickness of nanocellulose coatings are used to adjust a load of functional molecules and particles into the coatings, their permeability, and filtration properties. Mechanical stability of nanocellulose coatings in a wet and dry state are critical characteristics for many applications. In this work, nanofibrillated and nanocrystalline cellulose coatings deposited on the surface of polymer films and textiles made of cellulose, polyester, and nylon are studied using atomic force microscopy, ellipsometry, and T-peel adhesion tests. Methods to improve coatings' adhesion and stability using physical and chemical cross-linking with added polymers and polycarboxylic acids are analyzed in this study. The paper reports on the effect of the substrate structure and ability of nanocellulose particles to intercalate into the substrate on the coating adhesion.

Keywords: nanocellulose; polymer; coating; textile; adhesion

---

1. Introduction

Cellulose is the most abundant [1,2], renewable, biodegradable, and environmentally friendly organic material found in nature with great potential for the development of new applications with minimal health, environmental, or safety concerns [3]. Plant cell walls are composed of assembled cellulosic fibrils that are stabilized by intra- and interchain hydrogen bonds and van der Waals forces [4]. The fibrils are semicrystalline cellulose with 50–75% crystalline regions [5]. The fibrils can be separated by mechanical [2,6–10], chemical [11–13], or a combination of both treatments [14] to make cellulose nanoparticles in the form of nanofibers (nanofibrillated cellulose, NFC) or whiskers (nanocrystalline cellulose, NCC) [15,16] forming hydrogels in water. These nanoparticles, owing to their dimensions, shape, and high mechanical characteristics, attracted great interest in the engineering of nanostructured materials [14,17–21]. Nanocellulose-based materials, including coatings, were explored for many applications, such as packaging films [22–26], engineered composites [27], adsorbents [28], materials for health care [29], cosmetics [30], thermal insulation [31], paper [32], and filtration [33,34].

For time immemorial, one of the traditional application of the cellulosic material is the textile industry. Cotton, linen, hemp, and many other plant fibers are the major feedstock for the most demanded and comfortable cloth. In recent decades, there is a clear disposition for the shift from conventional clothing to smart textiles that integrate emerging technologies, such as communication devices, flexible electronics, and sensors [35]. Importantly, strong environmental and societal concerns demand a shift to the development of renewable and compostable materials along with sustainable technologies with minimal negative environmental impact [36]. A combination of wearable clothing systems and sustainability is advancing innovation in the traditional areas of textile manufacturing by endowing textiles with functional properties to address current health, safety, and environmental concerns associated with the textile industry.

Nanoscale dimensions and large specific surface of NFC and NCC allow them to intercalate into hierarchically organized fibrous structures, such as woven, knits, nonwoven, and composite textiles. Hence, nanocellulose materials can deliver functional molecules or particles bearing functionality covalently bonded or physically entrapped (caged) into a nanocellulose particle network. The functionalized NFC and NCC network is subsequently anchored to the textile surfaces via hydrogen/covalent bonds and physical interlocking. This method of functionalization of natural and synthetic fibers and fabrics is an environmentally sound approach without compromising the compostability and biocompatibility of the compostable textiles. For example, we have recently demonstrated that dyeing of textiles using NFC particles conjugated with commonly used reactive dyes can decrease the use of water, salts, and alkali by one order of magnitude with no change of the textile performance such as colorfastness referred to as conventional textile dyeing technology [37–39].

The key aspects of NFC and NCC textile coatings are adhesion and mechanical stability during dry (wearing due to abrasion) and wet (laundry) conditions. Cellulose has a natural self-adhesive characteristic, which relies primarily on interchain hydrogen bonding between hydroxyl groups of the adjacent cellulose chains and between cellulose chains and polymers of the textile, entanglements (specifically for NFC), and interlocking through the entanglement and intercalation into the fabric structures [40,41]. Many pretreatment methods such as plasma, ozone, and exposure to alkali solutions were successfully used for the materials incapable of the formation of hydrogen bonds with nanocellulose. For example, the treatment of polypropylene with ozone resulted in an improved adhesion to NCC owing to the hydrogen bonds with the oxidized surface of polypropylene bearing hydroxyl, carboxylic, and other oxygen-containing functional groups [42]. Nevertheless, strongly hydrogen-bonded nanocellulose materials swell in an aqueous environment [43,44]. The swelling can cause film degradation and loss of functional properties bound to the nanocellulose coatings. Therefore, it is important to understand the effect of nanocellulose swelling on the coating stability and develop methods for mitigating this problem.

This paper reports on a systematic study of adhesion and adhesion resistance to swelling (in water) of NFC and NCC thin film coatings on the surface of polymer films and fabrics made of cellulose (CL), cotton, poly(ethylene terephthalate) (PET), and nylon 6,6 (PA 6,6). We studied several methods for improvement of the adhesion and coating's stability, such as the use of a cationic polyelectrolyte poly(ethylene imine) (PEI), functional copolymers, and covalent cross-linking to elucidate major mechanisms for the improvement of the stability of nanocellulose coatings via combinations of adhesive and cohesive properties of the coatings. PEI is added to enforce the physical network of nanocellulose particles and nanocellulose-polymer substrate (films and textiles) interfaces via strong hydrogen bonds between primary and secondary amino groups of PEI and hydroxyl, amide, and ester groups of nanocellulose and textile materials (PET and nylon). A functional polymer—a copolymer of glycidyl methacrylate (GMA) and oligo(ethylene glycol) methacrylate (OEGMA) (P(GMA-OEGMA))—was selected and synthesized based on the compatibility with nanocellulose hydrogels, the ability to form hydrogen bonds with cellulose, and functional epoxy groups to cross-link the polymer and form a network for reinforcing of the coating. Alternatively, a commonly used cellulose-crosslinking method with polycarboxylic acids was applied to probe the effect of cross-linking on the coating stability.

2. Results and Discussion

2.1. Fabrication of Uniform Nanocellulose Coatings on the Surface of Polymeric Materials

The formation of nanocellulose coatings on the surface of fabrics is affected by the infiltration of nanocellulose hydrogels into a complex structure of the fabric. The permeation dynamics of hydrogels depends on the fabric density, structure of the yarn, interfacial tension, and rheological properties of the hydrogel. Many of these complications can be eluded using single filament fibers for coating, where the film formation is only limited by the wetting thermodynamics and rheology of the hydrogel. For low nanocellulose concentrations (<1%), when the hydrogel viscosity is low, it spreads over the fiber surface and forms an enclosed nanocellulose coating upon evaporation of water, as can be observed from the scanning electron microscopy (SEM) images of the coated polyester, cotton, and nylon single fibers (Figure 1). The image of a nylon fiber (Figure 1c) exhibits a peeled off NFC film at the edge of the cut fiber surface visualizing the coating film morphology and thickness. The peeled fraction of the coating corroborates a uniform layer of NFC around the fiber surface.

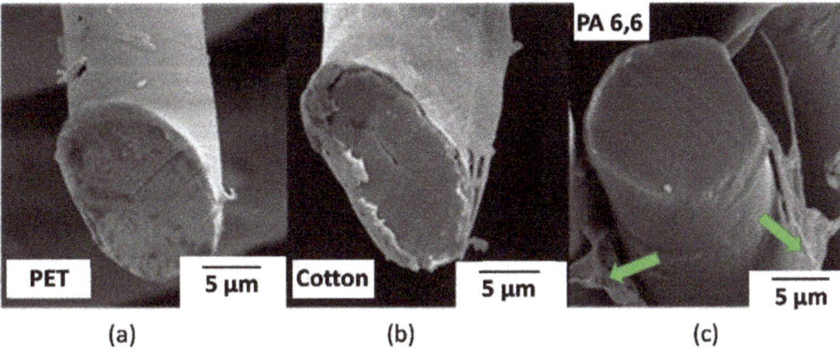

Figure 1. Scanning electron microscopy (SEM) images of (**a**) poly(ethylene terephthalate) (PET), (**b**) cotton, and (**c**) nylon 6,6 (PA 6,6) single fibers coated with nanofibrillated cellulose (NFC). Arrows point to the peeled fraction of the NFC coating.

The results of the experiments with single filament fibers show the formation of uniform smooth coatings over the fiber surface (Figure 1a,b). This uniform coating of NFC justifies the use of model flat substrates (e.g., polymer films) to probe morphology, adhesive behavior, and stability of nanocellulose coatings on the surfaces of different polymeric materials to monitor the coating structure and changes upon different treatment methods. Figure 2 exhibits differences in the surface morphology of NFC and NCC coatings on the Si-wafers. NFC coatings show a higher roughness owing to the higher particle size polydispersity in contrast to smoother and more uniform NCC coatings.

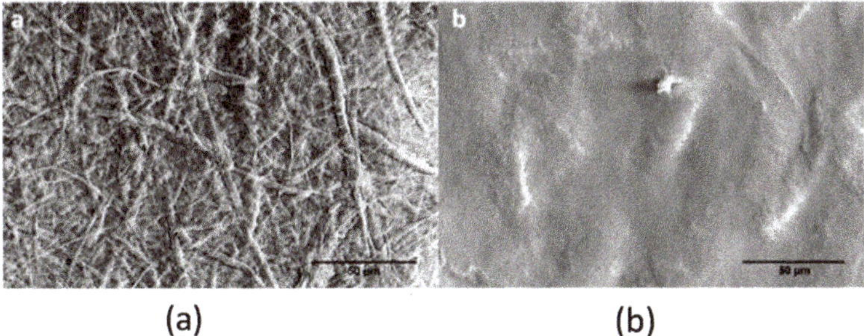

(a) (b)

Figure 2. SEM images of the morphology of (**a**) NFC and (**b**) nanocrystalline cellulose (NCC) spin-coated on the Si-wafers.

In this work, we used plane polymer films made of CL, PET, and PA 6,6 polymer solutions deposited on the surface of polished Si-wafers to minimize possible effects of the surface roughness of the substrate on the film formation. We prevented possible instabilities that could originate from a poor polymer-Si-wafer adhesion via the strengthening of the polymer-Si-wafer interactions by pretreatment of the Si-wafers with PEI polycations prior to deposition of the CL, PET, and PA 6,6 polymer films. Thickness and surface roughness of the polymer films were estimated with ellipsometry and atomic force microscopy (AFM) (Figure 3). The root mean square (RMS) roughness of the films did not exceed 7 nm, with the highest roughness observed for the cellulose films.

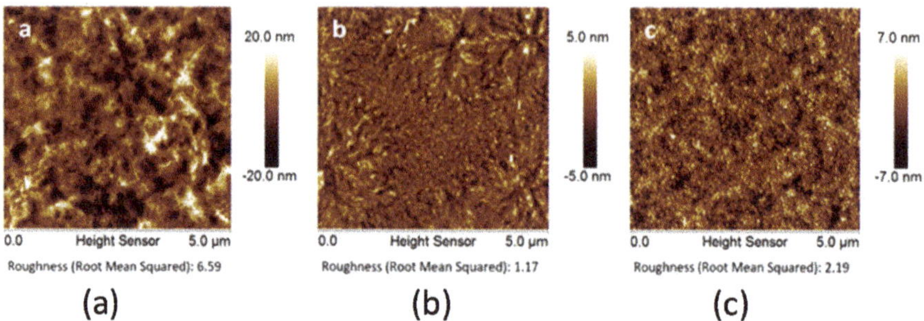

Figure 3. Representative atomic force microscopy (AFM) images of about 50–180 nm thick spin-coated films of (**a**) cellulose (CL), root mean square (RMS) roughness is 6.59 nm; (**b**) nylon, RMS roughness is 1.17 nm; and (**c**) PET, RMS roughness is 2.19 nm on the surface of the Si-wafers.

NFC and NCC films were deposited (spin-coated) on the surface of the polymer-coated Si-wafers using several different protocols (Figure S1). (i) Protocol 1: NFC and NCC aqueous dispersions were deposited on the CL, PET, and PA 6,6 coated Si-wafers; (ii) Protocol 2: NFC and NCC aqueous dispersions were deposited on the PEI pretreated CL, PET, and PA 6,6 coated Si-wafers; (iii) Protocol 3: NFC and NCC aqueous dispersions were mixed with PEI, and spin-coated on the CL, PET and PA 6,6 coated Si-wafers; and (iv) Protocol 4: NFC and NCC aqueous dispersions were mixed with P(GMA-OEGMA) copolymer, spin-coated on the CL, PET, and PA 6,6 coated Si-wafers. In all cases, the nanocellulose coatings were annealed after the deposition at 120 °C for 1 h.

The NFC and NCC coatings were prepared first using Protocol 1. We discovered a poor coverage of the PET and PA 6,6 surface with the nanocellulose materials. Then, we applied PEI pretreatment of all polymer substrates in Protocol 2 to improve wetting and coverage with nanocellulose. NFC- and NCC-coated samples from Protocol 2 are labeled as PEI-NFC and PEI-NCC, respectively.

Alternatively, we applied two different protocols to improve the adhesive and cohesive properties of nanocellulose coatings. According to Protocol 3, we mixed NFC and NCC hydrogels with PEI in solutions prior to the deposition on the polymer surfaces. Obtained by Protocol 3, NFC- and NCC-coated samples are labeled as NFC+PEI and NCC+PEI, respectively. Protocol 4 was used to mix NFC and NCC hydrogels with P(GMA-OEGMA) copolymer solutions; the samples were labeled as NFC+CP and NCC+CP, respectively.

To summarize the nanocellulose coatings preparation, the resulting films are multilayered structures constituted of the Si-wafer substrate, a native SiO_2 layer (0.5–1 nm), PEI adsorbed layer (typically 0.2–0.5 nm thick), polymer coating (CL, PET, or PA 6,6, typically 50–180 nm thick), and a nanocellulose (NFC or NCC) top layer with or without mixing with PEI or the copolymer. In some samples, the polymer layers are pre-coated with PEI (Protocol 2) before applying NC coatings. Representative 3D-plots for the layered structures obtained with imaging ellipsometry demonstrate uniform layered structure across the multicomponent coatings (Figure 4).

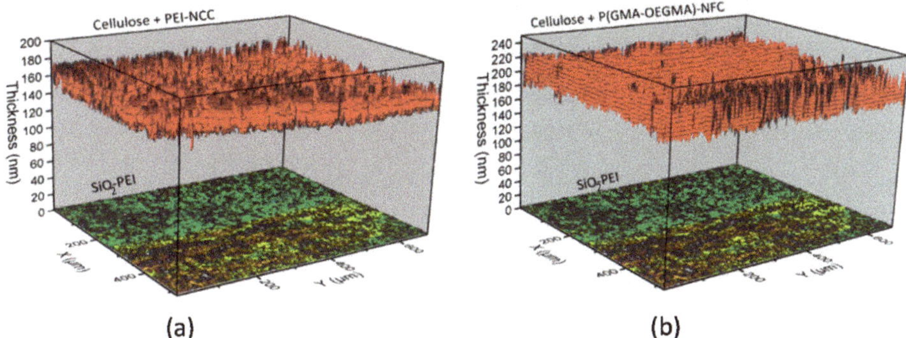

Figure 4. Representative examples of 3D-plots of layer-by-layer ellipsometric mapping of nanocellulose films on the polymer-coated Si-wafers constituted of the layers: (**a**) native SiO_2 (1 nm), poly(ethylene imine) (PEI) (1 nm), cellulose film (150 nm), PEI (1 nm), and NCC (20 nm); (**b**) native SiO_2 (1 nm), PEI (1 nm), CL (180 nm), PEI (1 nm), and NFC-copolymer of glycidyl methacrylate (GMA) and oligo(ethylene glycol) methacrylate (OEGMA) (P(GMA-OEGMA)) mixture (40 nm).

2.2. Mechanisms of the Nanocellulose Coating Degradation in a Wet State

In aqueous solution, nanocellulose coatings become swollen owing to the strong hydrogen bonding of water molecules and cellulose [44]. The developed osmotic pressure, in combination with share forces, can cause complete defoliation of the coatings from the substrate surface or partial delamination. The prevalence of one of the two mechanisms of degradation is defined by the balance of adhesive and cohesive interactions in the film. The complete or very large depletion of the film materials is likely associated with adhesive failure, while fractional losses or partial delamination of the coating film are caused by cohesive failure.

The stability of the deposited NFC and NCC films in an aqueous environment was estimated with a simple test. The coated samples were exposed to 50 °C aqueous solution at stirring for 1 h. Comparing the AFM images of the film before and after exposure to the aqueous medium in most cases did not reveal changes in the film morphology (Figure 5). For these nanocellulose coatings, we monitored changes in average film thickness. Only in the case of very poor adhesion, as for untreated PET and some PA 6,6 substrates, the AFM images demonstrate a low surface coverage by NFC and NCC, respectively (Figures 6 and 7). For these coatings, we monitored the surface coverage using an AFM "flooding analysis"—a statistical method where the polymer coating layer is set as a threshold and the surface areas of all other structures above the threshold are added to estimate the overall coverage with the nanocellulose material.

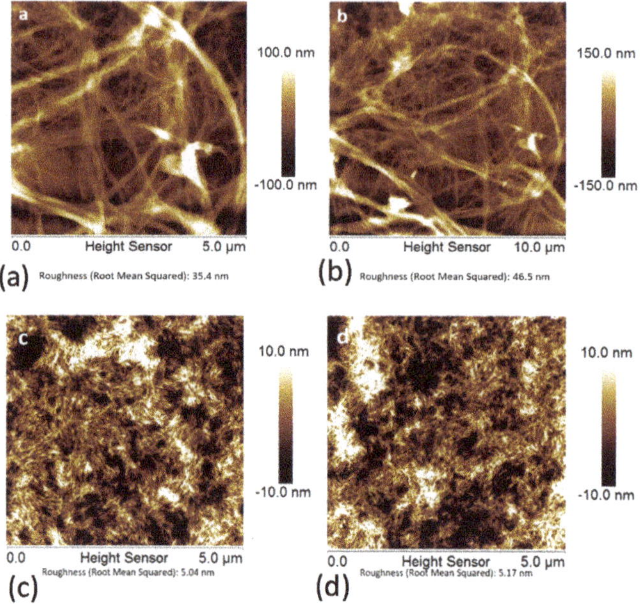

Figure 5. AFM topography images of nanocellulose coatings (**a**,**b**) PEI-NFC and (**c**,**d**) NCC+PEI on (**a**,**c**) PA 6,6 as-deposited and (**b**,**d**) after rinsing in water.

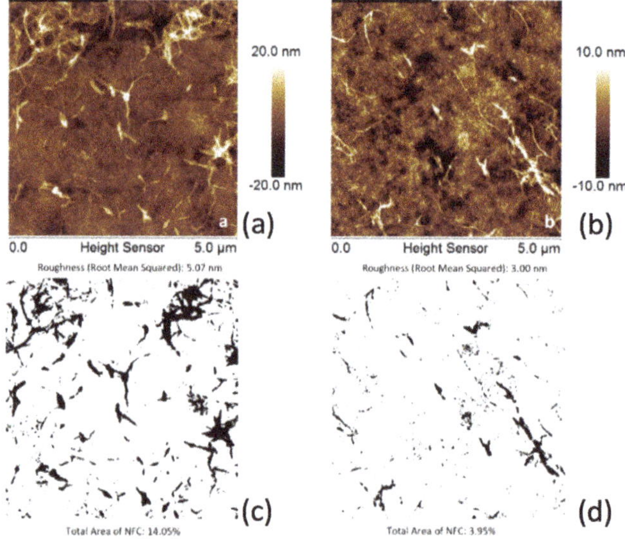

Figure 6. AFM images of NFC coatings on PET (**a**,**c**) as-deposited and (**b**,**d**) after rinsing in water: (**a**,**b**) topography images and (**c**,**d**) flooding analysis images showing the surface coverage by NFC particles (dark areas).

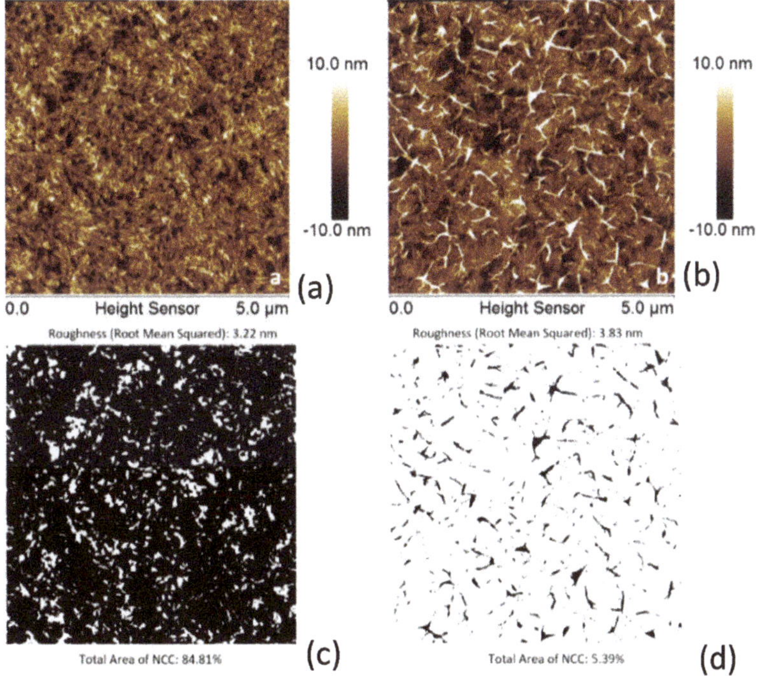

Figure 7. AFM images NCC coatings on PET as-deposited (**a,c**) and after rinsing in water (**b,d**): topography images (**a,b**) and flooding analysis images showing the surface coverage by NCC particles (**c,d**).

Changes in film thickness of the NFC and NCC coatings in all cases, with exceptions of untreated PET substrates, after rinsing in water, are reported in Table 1 as a percent (%) of the detached coating materials. For untreated PET substrates, the results report changes of the surface coverage using the flooding method. The coating thickness prepared by a spin-coating method depends on the rheological characteristics of NFC and NCC hydrogels and wetting of the polymer films. The rheological properties of the hydrogels depend on the concentration and presence of additives. Consequently, we analyze the relative changes in film thickness prepared using different modification methods.

Table 1. Changes in thickness (surface coverage) of nanofibrillated cellulose (NFC) and nanocrystalline cellulose (NCC) coatings on the surfaces of cellulose (CL), poly(ethylene terephthalate) (PET), and nylon 6,6 (PA 6,6) after rinsing in water. PEI, poly(ethylene imine); P(GMA-OEGMA), copolymer of glycidyl methacrylate (GMA) and oligo(ethylene glycol) methacrylate (OEGMA).

Sample	Film Thickness, H, and A fraction of Washed-Out Coating from Different Substrates, F (10% Error)					
	CL		PET		PA 6,6	
	H, nm	F, %	H, nm	F, %	H, nm	F, %
NFC	50	68	non-uniform	71	non-uniform	40
PEI-NFC	46	22	42	18	-	-
NFC+PEI	20	70	77	81	53	37
NFC+P(GMA-OEGMA)	33	77	-	-	401	25
NCC	21	5	non-uniform	93	non-uniform	93
PEI-NCC	89	53	113	60	-	-
NCC+PEI	40	25	77	12	65	77
NCC+P(GMA-OEGMA)	8	11	80	1	133	2

The analysis of the experimental data shows that the most common mechanism of coating degradation is partial delamination. Only for the PET substrate, we observed almost a complete adhesive detachment of the nanocellulose. The result shows that the nanocellulose coating has the lowest adhesion to the PET surface and the strongest interaction with the PA 6,6 surface among the synthetic polymers. NCC coatings demonstrate a higher adhesion to different CL substrates than NFC coatings. It is likely owing to the denser packed NCC particles in the coating in contrast to NFC fibers, and hence a lower swelling of the coating. The addition of PEI and P(GMA-OEGMA) improves the stability of the coatings. The latter effect is likely owing to the switching from the adhesive defoliation mechanism to the partial delamination of the film.

Notably, the film is much more stable on PET and PA 6,6 substrates when the coating is mixed with P(GMA-OEGMA). We may speculate that the major strengthening contribution of the copolymer is in the improvement of the cohesive properties of the film. The much greater thickness of the mixed films supports this conclusion.

We may speculate about the following mechanism of the improvement of the stability of the coating in an aqueous environment. Cationic anchoring polymers bearing amino-functional groups are used to treat different substrates for improvement in their interaction with cellulose [45–53]. Cationic polyelectrolytes interact with cellulose coatings through an electrostatic, donor-acceptor type of interactions, and hydrogen bond formation [45]. Nanocellulose, cotton, and polyester fibers are negatively charged in an aqueous environment, whereas nylon possesses amphoteric properties. In all cases, swollen in water, nanocellulose materials and polymers experience repulsive electrostatic forces. These repulsive interactions could be compensated by surface modification of the interacting materials with polycations such as PEI [52]. However, an excess of PEI will result in an overcharge of the surfaces, and repulsion between negatively charged materials will be replaced by repulsion among positively charged materials in water. PEI and other polyelectrolytes may also enhance swelling of the coatings in water. The results show no benefits of the use of PEI for the treatment of cotton and nylon fabrics; however, the interaction with PET is slightly improved. The latter is explained by poor hydrogen bonding between PET and nanocelluloses, which can be improved owing to the PEI-PET interactions.

P(GMA-OEGMA) copolymer bears ethylene oxide and epoxy functional groups. These two types of functional groups provide a combination of strong hydrogen bonds and covalent cross-linking. The covalent cross-linking mechanism involves the opening of epoxy rings and the formation of covalent bonds between epoxy groups of P(GMA-OEGMA) [54]. Reactivity of cellulosic -OH with epoxy-groups is not high enough to provide substantial effect for the cross-linking involving nanocellulose [55]. However, surface carboxylic functional groups that may be present because of oxidative degradation in the process of the production of nanocellulose could interact with epoxy groups and form covalent cross-links. The nanocellulose and P(GMA-OEGMA) copolymer blends upon drying, and thermal annealing will form an interpenetrated network owing to the covalently cross-linked polymer and physical cross-links via hydrogen bonds of nanocellulose materials. These two interpenetrating networks are also co-cross-linked via some fraction of carboxylic groups on the surface of cellulose. Epoxy groups secure good adhesion to various polar substrates. The experiments show that the presence of the copolymer improves adhesion to PET and nylon surfaces. In all experiments, we observe the obvious improvement of the nanocellulose coating stability in the presence of the copolymer.

2.3. Adhesive Behavior of Nanocellulose Coatings

Mechanical stability of the nanocellulose coating in the dry state is another important property for practical applications. Upon mechanical forces, the coating could be peeled off the polymer surface (adhesive failure) or partially delimited (cohesive failure). Hence, the coating performance can be analyzed using similar concepts of the degradation mechanisms, as was discussed for the aqueous environment. For the experiments, we used T-peel tests. Two identical materials were adhered using NFC or NCC hydrogels sandwiched between the materials, followed by drying and annealing the samples. These tests were performed in two series of experiments. In the first series, NFC and NCC were used to bind two identical samples of cellophane, PET, and PA 6,6 polymer films. In the second series of experiments, NFC and NCC were used to bind two identical samples of cotton, PET, and PA 6,6 fabrics. The NFC and NCC coatings on the polymer films and fabrics were prepared using the same Protocols 1–4 as in the tests of the stability of the coating in water.

The results of the peel tests with the polymer films show similar tendency as for the experiments on the coating stability in water; that is, NCC demonstrates a higher strength as compared with NFC, and the interaction with PET is the lowest among other polymers (Figure 8). However, in contrast with the experiments in water, nanocellulose interactions with the cellophane film are much stronger than with nylon. This difference in adhesion between nanocellulose materials and cellulose substrates in the dry state and in water provides evidence that the nanocellulose coating degradation in water is affected by swelling of the coating and weakening of the interfacial hydrogen bonds, while in the dry state, the intermolecular interactions remain strong.

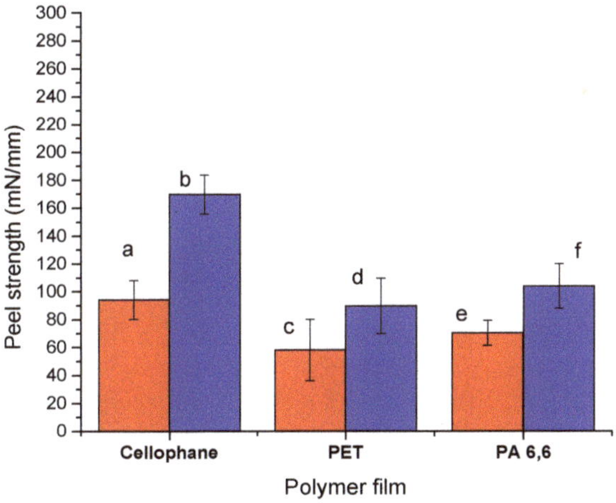

Figure 8. Peel strength for polymer films made of cellophane, PET, and PA 6,6 adhered with (**a,c,e**) NFC and (**b,d,f**) NCC.

The outcomes changed for the peel tests using fabrics instead of films made of the same polymers. NCC binding is stronger than NFC for all cases (compare Figures 9 and 10). We observe that the peel strength increases in the order cotton < PET < PA 6,6 for NFC and NCC. For both types of nanocelluloses, the pretreatment with PEI and mixing with the copolymer improves the peel strength. Similar conclusions about peel strength are applied to a blended (50:50) cotton-PET fabric (Figure 11). The most surprising result is the lowest peel strength for cotton textiles. This result was in conflict with the experiment in water (Table 1) and with the peel test for polymeric films (Figure 8) when NFC and NCC coatings showed the strongest stability and adhesion to the cellulose substrates. We hypothesized an additional factor that may impact the peel strength is the structure of the fabrics or the ability of nanocellulosic materials to infiltrate the fabric structure. The latter will result in a greater contact area between the fabric and nanocellulose and the formation of mechanical interlocks between intercalated fibrillary structures.

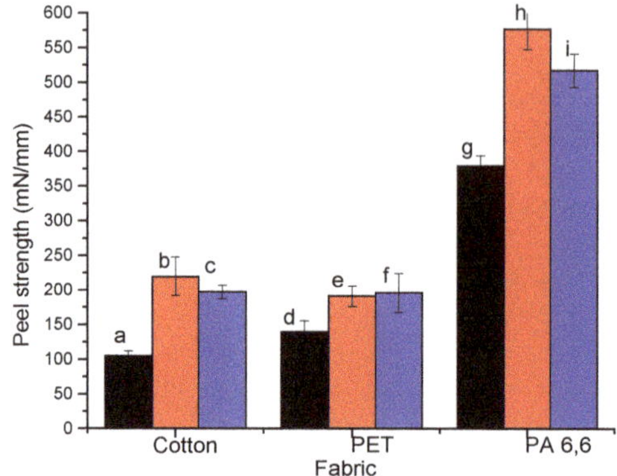

Figure 9. Peel strength for cotton, PET, and PA 6,6 fabrics adhered using NFC: (**a**,**d**,**g**) NFC with no additives, (**b**,**e**,**h**) NFC+PEI, and (**c**,**f**,**i**) NFC+P(GMA-OEGMA) mixtures.

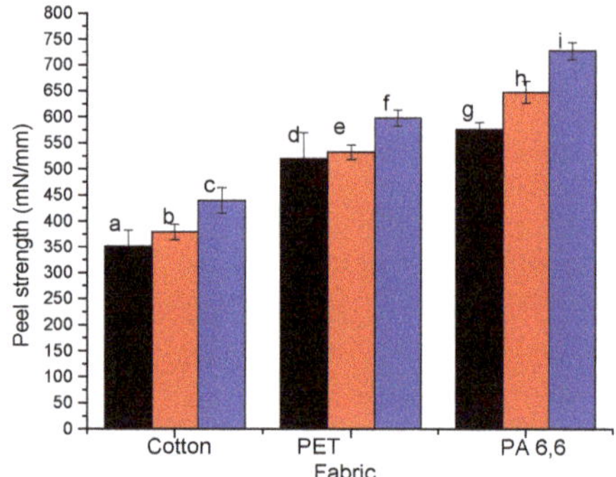

Figure 10. Peel strength for cotton, PET, and PA 6,6 fabrics adhered using NCC: (**a**,**d**,**g**) NCC with no additives, (**b**,**e**,**h**) NCC+PEI, and (**c**,**f**,**i**) NCC+P(GMA-OEGMA) mixtures.

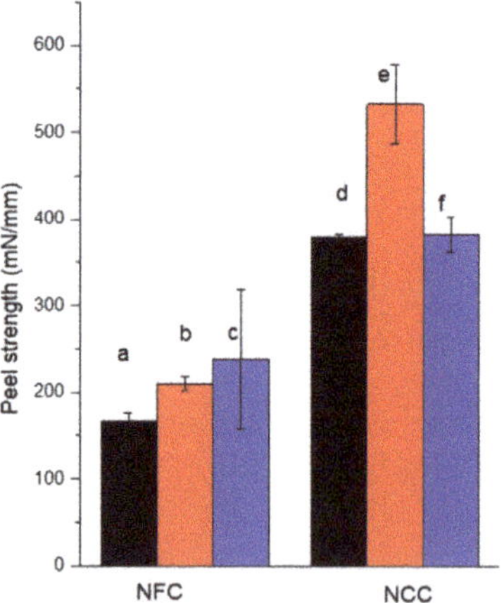

Figure 11. Peel strength for cotton-PET (50:50) fabric blend adhered with (**a**,**d**,**c**) NFC and (**d**–**f**) NFC: (**a**,**d**) with no additives, (**b**) NFC+PEI, (**e**) NCC+PEI, (**c**) NFC+P(GMA-OEGMA), and (**f**) NCC+P(GMA-OEGMA) mixtures.

This hypothesis was verified with the analysis of the structure and porosity of the fabrics. The warp and weft density and mean flow pore diameter (MFPD) of the fabrics are presented in Table 2. The result indicates that warp and weft density for cotton fabrics are substantially higher than those for other samples. While warp densities between polyester, nylon, and cotton/polyester are similar, the weft density of nylon is the lowest among the samples. Fabrics with higher weft density are less permeable for functional additives [56]. This structural property explains the highest peel strength for nylon and the lowest for the cotton fabrics.

Table 2. Structural characteristics of the fabrics. MFPD, mean flow pore diameter.

Samples	Warp Density, Yarns/cm	Weft Density, Yarns/cm	MFPD, μm
Cotton	33	29	67
Polyester	22	17	18
Nylon	20	11	64
Cotton/Polyester (50%/50%)	21	19	40

The mean flow pore diameter of the cotton and nylon fabrics also shows the highest values with broader pore size distribution in nylon fabric. This is another factor correlating with the uptake of NFC and NCC into the fabrics. Polyester has the lowest MFPD of 18 μm, but the weft density of 17 yarns/cm makes the fabric less dense for the infiltration of nanocellulose hydrogels into the fabrics, contributing to the higher peel strength as compared with the cotton fabric.

For the cotton fabric, the pore size ranges from 10 to 128 μm, while for the nylon fabric, the range is very broad, and the pore size reaches ~200 μm. The cotton/polyester blend pore size distribution shows that the majority of the pores are less than ~100 μm, and for the PET fabric, the pore size range is between ~5 and ~50 μm. The pore size distributions of the cotton, polyester, nylon, and cotton/polyester blended fabrics are shown in Figures S3–S6. The peel test results correlate with pore size distribution as the highest peel strength is observed for the nylon fabrics with a skew distribution towards large size pores.

2.4. Covalent Cross-Linking of Nanocellulose Coatings

An alternative approach to stabilize nanocellulose coatings is the cross-linking of NFC and NCC particles. Cross-linking of cotton fabrics is a widely used method to fabricate wrinkle resistance cotton products. Polycarboxylic acids for cross-linking of cotton cellulose were first introduced in the 1960s. The cross-linking can be catalyzed with sodium hypophosphite NaH_2PO_2 [57–62]. Polycarboxylic acids, for example, maleic acid (MA), in the presence of sodium hypophosphite, form ester bonds with cellulose hydroxyls at 160–180 °C [63,64]. The formation of the cross-links is confirmed by the appearance of the ester carbonyl band at 1720 and 1718 cm^{-1} (Figure S7).

The results of the peel tests for the cross-linked NFC and NCC coatings on the cotton fabric are shown in Figure 12. For both nanocellulose materials, the cross-linked system is stronger as compared with the reference non-cross-linked materials. However, the results are comparable to those obtained with P(GMA-OEGMA) copolymer. The results with cross-linked nanocellulose materials on the cotton fabric are compared to those on the cellophane film in Figure 12. This comparison reveals the synergistic effect of cross-linking and infiltration of nanocellulosic materials into the fabric structure. Smaller NCC particles infiltrate into the dense cotton fabric structure more efficiently as compared with NFC. This infiltration results in an increased adhesive interface. At the same time, the infiltration underlines the contribution of the mechanical interlocking enforced by the cross-linking of the cellulosic materials.

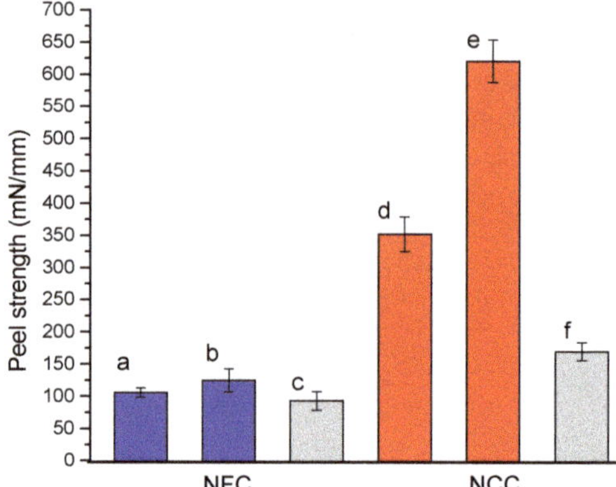

Figure 12. Peel strength for cotton textiles adhered with (**a,b**) NFC and (**d,e**) NCC with (**a,d**) no cross-linking and (**b,e**) with covalent cross-linking using MA compared with (**c**) NFC and (**f**) NCC adhered cellophane films.

3. Materials and Methods

3.1. Materials

Nanofibrillated cellulose hydrogel (2%) was prepared as previously reported using a mechanical homogenization method [37,38]. Nanocrystalline cellulose hydrogel (11.9%) was purchased from the Process Development Center, University of Maine. m-Cresol, lithium chloride, N,N-dimethylacetamide, chlorobenzene, phenol, polyethylenimine (PEI) (number average molecular mass Mn = 60 kg/mol), maleic acid (MA), sodium hypophosphite, glycidyl methacrylate (GMA, 97%), azoisobutyronitrile (AIBN), oligo (ethylene glycol) methyl ether methacrylate (OEGMA, Mn = 950 g/mol, stabilized with inhibitors), and inhibitor removers (kits for removing hydroquinone and monomethyl ether hydroquinone (MEHQ) and tert-butylcatechol (BHT)) were purchased from Sigma-Aldrich (St. Louis, MO, USA).

Silicon wafers (100 crystal plane) with a native oxide layer were purchased from University Wafer (South Boston, MA, USA). A polyethylene terephthalate (PET) film (0.50 mm thick) and a nylon 6,6 (PA 6,6) film (0.5 mm thick) were purchased from GoodFellow, Coraopolis, PA, USA. A cellophane film (regenerated cellulose) 0.03 mm thick was provided by Thermetrics. Cotton (100%, plain weave), nylon 6,6 (100%, spun, plain weave), cotton/polyester, PET (50%/50%, plain weave), polyester, and PET (100%, spun, plain weave) fabrics were purchased from Testafabrics, Inc., West Pittston, PA, USA.

3.2. Synthesis of P(GMA-OEGMA) Copolymer

P(GMA-OEGMA), a random copolymer of glycidyl methacrylate (GMA) and oligo(ethylene glycol) methacrylate (OEGMA), M_w = 2000 kg/mol, was synthesized by solution free-radical polymerization [65,66]. The inhibitor MEHQ and BHT removers were added to purify GMA and OEGMA for 45 min each. After filtration and purging with nitrogen for 45 min, the monomer solution (0.5 M) at GMA/OEGMA + 1:4 in MEK was used for polymerization initiated by 0.01 M AIBN at 50 °C for 1.5 h. The copolymer was extracted and purified by repetitive precipitation (three times) in diethyl ether. The copolymer was stored in a MEK solution in the absence of light. According to an NMR analysis (Bruker AVANCE-300, Billerica, MA, USA), the copolymer composition is 66 mol% (93 wt%) OEGMA and 34 mol% GMA. The copolymer is characterized by a glass transition temperature of −50 °C and a melting temperature of 35 °C (DSC2920, TA Instruments, New Castle, DE, USA). The copolymer is soluble in water.

3.3. Preparation of Polymer Substrates for Nanocellulose Deposition

Si-wafers were cut into square pieces (1 cm × 1 cm) and then cleaned in a solution of 28% NH_4OH/30% H_2O_2/H_2O (1:1:1) at 65 °C for 1 h. They were rinsed with deionized water (DI) water and dried under a flux of argon gas. The cleaned wafers were dipped into a solution of PEI (1%) for 15 min and rinsed with DI water and dried with argon gas. The resulting modified silicon wafers were stored at room temperature in a clean desiccator.

PET was dissolved in a solution of phenol-chlorobenzene (1:1) in a boiling water bath (100 °C). Once fully dissolved, 20 µL of the solution was spin-coated (3000 rpm for 20 s) on the Si-wafers. The substrates were transferred to an oven heated at 180 °C and annealed for 24 h to remove the residual solvent.

PA 6,6 was dissolved in m-cresol at 100 °C by stirring for several hours. The nylon films were prepared by spin-coating (3000 rpm for 20 s) 20 µL of the solution on the Si-wafers. The substrates were immediately transferred to an oven and dried at 180 °C for 24 h.

Cellulose films were prepared by heating of a cellulose powder in a solution of lithium chloride (LiCl, 1–3%) and N,N-dimethylacetamide (DMAc, 3–9%) to 150 °C, and then allowed the solution to cool slowly to room temperature [67,68]. Afterward, 20 µL of the cellulose solution was spin-coated (3000 rpm for 10 s) onto the cleaned Si-wafers. The samples were submerged in DI water for 20 min to remove the remaining LiCl; they were dried with argon gas and heated for 20 min at 180 °C to allow for the evaporation of any residual solvent. The samples were stored at room temperature.

3.4. Nanocellulose Coatings on Si-Wafers and Polymer-Coated Si-Wafers

NFC and NCC films were prepared by spin-coating (3000 rpm for 30 s) 20 µL of the diluted NFC and NCC hydrogels (1%) onto the substrates (Si-wafers and cellulose, PET, and PA 6,6 coated Si-wafers) and annealed at 120 °C for 24 h.

Nanocellulose coatings were prepared on the surface of CL, PET, and PA 6,6 coated Si-wafers after pretreatment of the substrates with PEI or P(GMA-OEGMA). These samples were labeled as PEI-NFC and PEI-NCC, respectively. These samples were prepared by submerging of the polymer-coated Si-wafers into a 1% PEI solution for 30 min. Then, the samples were rinsed with DI water and dried with argon gas. NFC and NCC solutions were spin coated (3000 rpm 30 s). The samples were annealed at 120 °C for 1 h.

Alternatively, NFC (0.1%) and NCC (1%) were mixed with a 1% PEI or P(GMA-OEGMA) solution (20:1 by volume) and stirred for 1 h, for nanocellulose-PEI and nanocellulose- P(GMA-OEGMA) blends, respectively. Then, 20 µL of the PEI-modified or P(GMA-OEGMA)-modified hydrogels was spin-coated onto the polymer-coated Si-wafers and dried at 120 °C for 1 h. These samples were labeled PEI+NFC, PEI+NCC, CP+NFC, and CP+NCC, for NFC and NCC mixed coatings, respectively.

3.5. Characterization of Coatings

Scanning electron microscopy (SEM) imaging was carried out using an FEI Teneo (FEI Co., Hillsboro, OR, USA), a field emission scanning electron microscope. Atomic force microscopy (AFM) images were obtained using a Bruker Multimode Nanoscope instrument (Bruker, Billerica, MA, USA) with the ScanAssyst-Air probe (Bruker) spring constant 0.4 N/n, a silicon oxide tip). All the measurements were performed under ambient conditions at room temperature and at a relative humidity (RH) of 50–55%. All AFM data analysis and data processing were done with the NanoScope Analysis software version 1.40 (Bruker).

The thickness of the films at three different locations for each sample after the deposition of each layer on the substrates was measured by a single wavelength imaging ellipsometer ep4sw (Accurion, Göttingen, Germany) with a fixed angle of incidence of 70°. Ellipsometry thickness maps were generated using the Accurion software package, DataStudio, for selected samples to verify the uniformity of the coatings and justify measurements of the samples series using the three-location-approach. An attenuated total reflection fourier transform infrared spectroscopy (ATR-FTIR, ThermoElectron Nicolet 6700) was used to collect the infrared spectra. AFM was used as an alternative method for coating thickness by scratching the coating with a steel needle and measuring the profile of the scratch (Figure S8).

The spectra were presented using absorbance mode ($-\log R/R_o$). The resolution for all the infrared spectra was 4 cm^{-1}, 120 scans for each spectrum.

3.6. Structural Characterization

A porometer (Porous Materials Inc., Ithaca, NY, USA) was used to measure the fabric porosity. A capillary flow porometer (CFP) was used to evaluate an average pore diameter, pore size distribution, and mean flow pore size by assessing the relationship between pressure and gas flow rate [69]. The fabrics were cut into circular pieces of 25 mm in the diameter, soaked into Galwick wetting liquid with the surface tension of 16 Dyne/cm, placed, and sealed into the sample holder for the measurements.

3.7. Cross-Linking of Nanocellulose Coatings

The cross-linking was conducted as published elsewhere [63,64], after drying the nanocellulose coatings were treated with 6%MA and 4% NaH_2PO_2 solutions and annealed at 185 °C for 2 min.

3.8. T-Peel Tests

The tests were performed according to the method described in the ASTM D1876-08 standard with five tested samples for each material. Samples of the fabrics were cut into 50 mm × 152 mm stripes along the warp direction. The test panel (Figure S9) consisted of two fabrics stripes bonded together with 5 g of 2% NFC or NCC hydrogels along 127 mm of their length. A 2 kg load was applied to the top of the test panels while drying at 85 °C. Then, the specimens were placed in a conditioning chamber (Caron®) for 7 days at a relative humidity of 50 ± 2% at 23 ± 1 °C. Finally, we cut the bonded panels into 25 mm wide test specimens with a sharp cutter. During the tests, the peeling force was recorded at a constant head speed of 254 mm/min. The same method for the sample preparation was used for PEI and P(GMA-OEGMA) blended NFC and NCC hydrogels, respectively, as well as for the samples with cross-linked nanocellulose coatings. In the latter case, the samples were annealed at 185 °C for 2 min. Using ICPeel software [70], we estimated that the bending energy contributions is 15%, which is comparable to the peel test experimental error. Variations in nanocellulose coating thickness in a range of 1–5 μm have a very low effect on the contribution of the bending energy.

4. Conclusions

Wet and dry tests of nanocellulose coatings on the surface of cellulose, PET, and nylon (PA 6,6) polymer films revealed that the coatings have the highest adhesion to the nylon, cellulose, and cellophane surfaces, while adhesion is the lowest for the coatings on PET. The coatings stability is improved using treatment with a polycation polymer PEI and a reactive P(GMA-OEGMA) copolymer capable of forming a cross-linked network. In the latter case, the highest coating adhesion and stability were observed. Alternatively, the coating is reinforced by the cross-linking of nanocellulose with polycarboxylic acids. NCC coatings demonstrate higher adhesion to all substrates than NFC coatings. The experiments with cotton, PET, and PA 6,6 fabrics revealed that the fabric structure is an additional important factor for the stability and adhesion of the nanocellulose coatings. The lower density of the textile and higher porosity is beneficial for stronger adhesion of the coatings. Hydrogen bonding, swelling in water, physical, covalent cross-linking, overall contact area, and porosity of the substrate, which provide intercalation of the nanocellulose particles into the fabric structure, are all characteristics that contribute to the nanocellulose coating adhesion and stability.

Supplementary Materials: The following are available online, Figure S1: Schematics for the multilayered samples of nanocellulose coatings on the surface of polymer materials according to Protocols 1–4, Figure S2: Structure of the P(GMA-OEGMA) copolymer, Figure S3: Pore size distribution of cotton fabrics, Figure S4: Pore size distribution of nylon fabrics, Figure S5: Pore size distribution of cotton/polyester fabrics, Figure S6: Pore size distribution of polyester fabrics, Figure S7: FT-IR spectra of (a) cotton-NCC and cotton-NCC treated with MA and (b) cotton-NFC and cotton- NFC treated with MA, Figure S8: AFM topography image of the cellulose coating on the Si-wafer after the scratch with a steel needle, Figure S9: Schematic of a sample for the T-peel test.

Author Contributions: Conceptualization, S.M., S.S., and I.L.; methodology, S.M., S.S., and I.L.; investigation, R.S., N.B., and A.M.L.; writing—original draft preparation, R.S.; writing—review and editing, S.M. All authors have read and agreed to the published version of the manuscript.

Funding: This research was partially funded by NATO Science for Peace and Security Program under grant G5330, the National Science Foundation via EPSCoR OIA-1655740, and Clemson University Research Foundation.

Conflicts of Interest: The authors declare no conflict of interest.

References

1. Henriksson, M.; Berglund, L.A. Structure and properties of cellulose nanocomposite films containing melamine formaldehyde. *J. Appl. Polym. Sci.* **2007**, *106*, 2817–2824. [CrossRef]
2. Iwamoto, S.; Nakagaito, A.; Yano, H. Nano-fibrillation of pulp fibers for the processing of transparent nanocomposites. *Appl. Phys. A Mater. Sci. Process.* **2007**, *89*, 461–466. [CrossRef]
3. Azzam, F.; Moreau, C.L.; Cousin, F.; Menelle, A.; Bizot, H.; Cathala, B. Cellulose nanofibril-based multilayered thin films: Effect of ionic strength on porosity, swelling, and optical properties. *Langmuir* **2014**, *30*, 8091–8100. [CrossRef] [PubMed]
4. Kalia, S.; Dufresne, A.; Cherian, B.M.; Kaith, B.; Avérous, L.; Njuguna, J.; Nassiopoulos, E. Cellulose-based bio-and nanocomposites: A review. *Int. J. Polym. Sci.* **2011**, *2011*, 1–35. [CrossRef]
5. Daicho, K.; Saito, T.; Fujisawa, S.; Isogai, A. The Crystallinity of Nanocellulose: Dispersion-Induced Disordering of the Grain Boundary in Biologically Structured Cellulose. *ACS Appl. Nano Mater.* **2018**, *1*, 5774–5785. [CrossRef]
6. Chakraborty, A.; Sain, M.; Kortschot, M. Cellulose microfibrils: A novel method of preparation using high shear refining and cryocrushing. *Holzforschung* **2005**, *59*, 102–107. [CrossRef]
7. Johnson, R.K.; Zink-Sharp, A.; Renneckar, S.H.; Glasser, W.G. A new bio-based nanocomposite: Fibrillated TEMPO-oxidized celluloses in hydroxypropylcellulose matrix. *Cellulose* **2009**, *16*, 227–238. [CrossRef]
8. Lavoine, N.; Desloges, I.; Dufresne, A.; Bras, J. Microfibrillated cellulose—Its barrier properties and applications in cellulosic materials: A review. *Carbohydr. Polym.* **2012**, *90*, 735–764. [CrossRef]
9. Nishiyama, Y. Structure and properties of the cellulose microfibril. *J. Wood Sci.* **2009**, *55*, 241–249. [CrossRef]
10. Wang, S.; Cheng, Q. A novel process to isolate fibrils from cellulose fibers by high-intensity ultrasonication, Part 1: Process optimization. *J. Appl. Polym. Sci.* **2009**, *113*, 1270–1275. [CrossRef]
11. Missoum, K.; Belgacem, M.N.; Bras, J. Nanofibrillated cellulose surface modification: A review. *Materials* **2013**, *6*, 1745–1766. [CrossRef] [PubMed]
12. Saito, T.; Kimura, S.; Nishiyama, Y.; Isogai, A. Cellulose nanofibers prepared by TEMPO-mediated oxidation of native cellulose. *Biomacromolecules* **2007**, *8*, 2485–2491. [CrossRef]
13. Saito, T.; Nishiyama, Y.; Putaux, J.-L.; Vignon, M.; Isogai, A. Homogeneous suspensions of individualized microfibrils from TEMPO-catalyzed oxidation of native cellulose. *Biomacromolecules* **2006**, *7*, 1687–1691. [CrossRef] [PubMed]
14. Moon, R.J.; Martini, A.; Nairn, J.; Simonsen, J.; Youngblood, J. Cellulose nanomaterials review: Structure, properties and nanocomposites. *Chem. Soc. Rev.* **2011**, *40*, 3941–3994. [CrossRef] [PubMed]
15. Klemm, D.; Kramer, F.; Moritz, S.; Lindström, T.; Ankerfors, M.; Gray, D.; Dorris, A. Nanocelluloses: A new family of nature-based materials. *Angew. Chem. Int. Ed.* **2011**, *50*, 5438–5466. [CrossRef] [PubMed]
16. Atalla, R.H.; Brady, J.W.; Matthews, J.F.; Ding, S.Y.; Himmel, M.E. Structures of plant cell wall celluloses. In *Biomass Recalcitrance: Deconstructing the Plant Cell Wall for Bioenergy*; Wiley-Blackwell: Chichester, UK, 2008; pp. 188–212.
17. Salas, C.; Nypelö, T.; Rodriguez-Abreu, C.; Carrillo, C.; Rojas, O.J. Nanocellulose properties and applications in colloids and interfaces. *Curr. Opin. Colloid Interface Sci.* **2014**, *19*, 383–396. [CrossRef]

18. Habibi, Y.; Lucia, L.A.; Rojas, O.J. Cellulose nanocrystals: Chemistry, self-assembly, and applications. *Chem. Rev.* **2010**, *110*, 3479–3500. [CrossRef]

19. Johar, N.; Ahmad, I.; Dufresne, A. Extraction, preparation and characterization of cellulose fibres and nanocrystals from rice husk. *Ind. Crops Prod.* **2012**, *37*, 93–99. [CrossRef]

20. Usov, I.; Nyström, G.; Adamcik, J.; Handschin, S.; Schütz, C.; Fall, A.; Bergström, L.; Mezzenga, R. Understanding nanocellulose chirality and structure-properties relationship at the single fibril level. *Nat. Commun.* **2015**, *6*, 7564. [CrossRef]

21. Jiang, F.; Hsieh, Y.L. Chemically and mechanically isolated nanocellulose and their self-assembled structures. *Carbohydr. Polym.* **2013**, *95*, 32–40. [CrossRef]

22. Zhao, Y.; Simonsen, J.; Cavender, G.; Jung, J.; Fuchigami, L.H. Nano-Cellulose Coatings to Prevent Damage in Foodstuffs. U.S. Patent 10,334,863, 2 July 2019.

23. Vilarinho, F.; Sanches-Silva, A.; Vaz, M.F.; Farinha, J.P. Nanocellulose: A benefit for green food packaging. *Crit. Rev. Food Sci. Nutr.* **2016**, *58*, 1526–1537. [CrossRef] [PubMed]

24. Rampazzo, R.; Mascheroni, E.; Fasano, F.; Mari, M.; Piergiovanni, L. Strategies for implementing nano-cellulose coatings in flexible packaging. *Ital. J. Food Sci.* **2015**, 13–17.

25. Li, F.; Biagioni, P.; Bollani, M.; Maccagnan, A.; Piergiovanni, L. Multi-functional coating of cellulose nanocrystals for flexible packaging applications. *Cellulose* **2013**, *20*, 2491–2504. [CrossRef]

26. Aulin, C.; Gallstedt, M.; Lindstrom, T. Oxygen and oil barrier properties of microfibrillated cellulose films and coatings. *Cellulose* **2010**, *17*, 559–574. [CrossRef]

27. Jabbar, A.; Militký, J.; Wiener, J.; Kale, B.M.; Ali, U.; Rwawiire, S. Nanocellulose coated woven jute/green epoxy composites: Characterization of mechanical and dynamic mechanical behavior. *Compos. Struct.* **2017**, *161*, 340–349. [CrossRef]

28. Jiang, F.; Hsieh, Y.L. Amphiphilic superabsorbent cellulose nanofibril aerogels. *J. Mater. Chem. A* **2014**, *2*, 6337–6342. [CrossRef]

29. Wang, C.; Li, Y.L.; Hong, F.; Tang, S.J.; Wang, Y.Y. Nano-cellulose coating small-caliber artificial blood vessel. *Adv. Mater. Res.* **2011**, *332*, 1794–1798. [CrossRef]

30. Ullah, H.; Santos, H.A.; Khan, T. Applications of bacterial cellulose in food, cosmetics and drug delivery. *Cellulose* **2016**, *23*, 2291–2314. [CrossRef]

31. Zhou, J.; Hsieh, Y.L. Nanocellulose aerogel-based porous coaxial fibers for thermal insulation. *Nano Energy* **2020**, *68*, 9. [CrossRef]

32. Osong, S.H.; Norgren, S.; Engstrand, P. Processing of wood-based microfibrillated cellulose and nanofibrillated cellulose, and applications relating to papermaking: A review. *Cellulose* **2016**, *23*, 93–123. [CrossRef]

33. Carpenter, A.W.; de Lannoy, C.F.; Wiesner, M.R. Cellulose nanomaterials in water treatment technologies. *Environ. Sci. Technol.* **2015**, *49*, 5277–5287. [CrossRef]

34. Nemoto, J.; Saito, T.; Isogai, A. Simple freeze-drying procedure for producing nanocellulose aerogel-containing, high-performance air filters. *ACS Appl. Mater. Interfaces* **2015**, *7*, 19809–19815. [CrossRef] [PubMed]

35. Tao, X. *Handbook of Smart Textiles*; Springer: Singapore, 2015.

36. Song, J.H.; Murphy, R.J.; Narayan, R.; Davies, G.B.H. Biodegradable and Compostable Alternatives to Conventional Plastics. *Philos. Trans. R. Soc. B-Biol. Sci.* **2009**, *364*, 2127–2139. [CrossRef] [PubMed]

37. Minko, S.; Sharma, S.; Hardin, I.; Luzinov, I.; Daubenmire, S.W.; Zakharchenko, A.; Saremi, R.; Kim, Y.S. Textile Dyeing Using Nanocellulosic Fibers. U.S. Patent US20160010275A1, 29 November 2016.

38. Kim, Y.; McCoy, L.T.; Lee, E.; Lee, H.; Saremi, R.; Feit, C.; Hardin, I.R.; Sharma, S.; Mani, S.; Minko, S. Environmentally sound textile dyeing technology with nanofibrillated cellulose. *Green Chem.* **2017**, *19*, 4031–4035. [CrossRef]

39. Liyanapathiranage, A.; Peña, M.J.; Sharma, S.; Minko, S. Nanocellulose-Based Sustainable Dyeing of Cotton Textiles with Minimized Water Pollution. *ACS Omega* **2020**, *5*, 9196–9203. [CrossRef] [PubMed]

40. Gardner, D.J.; Oporto, G.S.; Mills, R.; Samir, M.A.S.A. Adhesion and surface issues in cellulose and nanocellulose. *J. Adhes. Sci. Technol.* **2008**, *22*, 545–567. [CrossRef]

41. Sinko, R.; Qin, X.; Keten, S. Interfacial mechanics of cellulose nanocrystals. *MRS Bull.* **2015**, *40*, 340–348. [CrossRef]

42. d'Eon, J.; Zhang, W.; Chen, L.; Berry, R.M.; Zhao, B.X. Coating cellulose nanocrystals on polypropylene and its film adhesion and mechanical properties. *Cellulose* **2017**, *24*, 1877–1888. [CrossRef]

43. Hossain, L.; Raghuwanshi, V.S.; Tanner, J.; Wu, C.M.; Kleinerman, O.; Cohen, Y.; Garnier, G. Structure and swelling of cross-linked nanocellulose foams. *J. Colloid Interface Sci.* **2020**, *568*, 234–244. [CrossRef]

44. Aulin, C.; Ahola, S.; Josefsson, P.; Nishino, T.; Hirose, Y.; Osterberg, M.; Wagberg, L. Nanoscale Cellulose Films with Different Crystallinities and Mesostructures-Their Surface Properties and Interaction with Water. *Langmuir* **2009**, *25*, 7675–7685. [CrossRef]

45. Ahola, S.; Salmi, J.; Johansson, L.-S.; Laine, J.; Österberg, M. Model films from native cellulose nanofibrils. Preparation, swelling, and surface interactions. *Biomacromolecules* **2008**, *9*, 1273–1282. [CrossRef] [PubMed]

46. Eriksson, M.; Notley, S.M.; Wågberg, L. Cellulose thin films: Degree of cellulose ordering and its influence on adhesion. *Biomacromolecules* **2007**, *8*, 912–919. [CrossRef] [PubMed]

47. Fält, S.; Wågberg, L.; Vesterlind, E.-L.; Larsson, P.T. Model films of cellulose ID–improved preparation method and characterization of the cellulose film. *Cellulose* **2004**, *11*, 151–162. [CrossRef]

48. Gunnars, S.; Wågberg, L.; Stuart, M.C. Model films of cellulose: I. Method development and initial results. *Cellulose* **2002**, *9*, 239–249. [CrossRef]

49. Hoeger, I.; Rojas, O.J.; Efimenko, K.; Velev, O.D.; Kelley, S.S. Ultrathin film coatings of aligned cellulose nanocrystals from a convective-shear assembly system and their surface mechanical properties. *Soft Matter* **2011**, *7*, 1957–1967. [CrossRef]

50. Notley, S.M.; Eriksson, M.; Wågberg, L.; Beck, S.; Gray, D.G. Surface forces measurements of spin-coated cellulose thin films with different crystallinity. *Langmuir* **2006**, *22*, 3154–3160. [CrossRef] [PubMed]

51. Sczech, R.; Riegler, H. Molecularly smooth cellulose surfaces for adhesion studies. *J. Colloid Interface Sci.* **2006**, *301*, 376–385. [CrossRef]

52. Wågberg, L.; Decher, G.; Norgren, M.; Lindström, T.; Ankerfors, M.; Axnäs, K. The build-up of polyelectrolyte multilayers of microfibrillated cellulose and cationic polyelectrolytes. *Langmuir* **2008**, *24*, 784–795. [CrossRef]

53. Yokota, S.; Kitaoka, T.; Wariishi, H. Surface morphology of cellulose films prepared by spin coating on silicon oxide substrates pretreated with cationic polyelectrolyte. *Appl. Surf. Sci.* **2007**, *253*, 4208–4214. [CrossRef]

54. Savchak, M.; Borodinov, N.; Burtovyy, R.; Anayee, M.; Hu, K.; Ma, R.; Grant, A.; Li, H.; Cutshall, D.B.; Wen, Y.; et al. Highly conductive and transparent reduced graphene oxide nanoscale films via thermal conversion of polymer-encapsulated graphene oxide sheets. *ACS Appl. Mater. Interfaces* **2018**, *10*, 3975–3985. [CrossRef]

55. Doszlop, S.; Vargha, V.; Horkay, F. Reactions of epoxy with other functional groups and the arising sec-hydroxyl groups. *Periodica Polytechn. Chem. Eng.* **1978**, *22*, 253–275.

56. Cay, A.; Atrav, R.; Duran, K. Effects of warp-weft density variation and fabric porosity of the cotton fabrics on their colour in reactive dyeing. *Fibres Text. East. Eur.* **2007**, *60*, 91–94.

57. Allen, T. Non-aqueous ester cross-linking of cotton cellulose. *Text. Res. J.* **1964**, *34*, 331–336. [CrossRef]

58. Chen, D.; Yang, C.Q.; Qiu, X. Aqueous polymerization of maleic acid and cross-linking of cotton cellulose by poly (maleic acid). *Ind. Eng. Chem. Res.* **2005**, *44*, 7921–7927. [CrossRef]

59. Gagliardi, D.; Shippee, F. Crosslinking of cellulose with polycarboxylic acids. *Am. Dyest. Rep.* **1963**, *52*, 74–77.

60. Rowland, S.P.; Welch, C.M.; Brannan, M.A.F.; Gallagher, D.M. Introduction of ester cross links into cotton cellulose by a rapid curing process. *Text. Res. J.* **1967**, *37*, 933–941. [CrossRef]

61. Welch, C. *Formaldehyde Free Durable Press Finishing in Surface Characterization of Fibres and Textiles*; Marcel Dekker: New York, NY, USA, 2000.

62. Welch, C.M. Tetracarboxylic acids as formaldehyde-free durable press finishing agents: Part i: Catalyst, additive, and durability studies. *Text. Res. J.* **1988**, *58*, 480–486. [CrossRef]

63. Yang, C.Q.; Chen, D.; Guan, J.; He, Q. Cross-linking cotton cellulose by the combination of maleic acid and sodium hypophosphite. 1. Fabric wrinkle resistance. *Ind. Eng. Chem. Res.* **2010**, *49*, 8325–8332. [CrossRef]

64. Yang, C.Q.; Wang, X. Formation of cyclic anhydride intermediates and esterification of cotton cellulose by multifunctional carboxylic acids: An infrared spectroscopy study. *Text. Res. J.* **1996**, *66*, 595–603. [CrossRef]

65. Yadavalli, N.S.; Borodinov, N.; Choudhury, C.K.; Quiñones-Ruiz, T.; Laradji, A.M.; Tu, S.; Lednev, I.K.; Kuksenok, O.; Luzinov, I.; Minko, S. Thermal Stabilization of Enzymes with Molecular Brushes. *ACS Catalysis* **2017**, *7*, 8675–8684. [CrossRef]

66. Borodinov, N.; Gil, D.; Savchak, M.; Gross, C.E.; Yadavalli, N.S.; Ma, R.; Tsukruk, V.V.; Minko, S.; Vertegel, A.; Luzinov, I. En route to practicality of the polymer grafting technology: One-step interfacial modification with amphiphilic molecular brushes. *ACS Appl. Mater. Interfaces* **2018**, *10*, 13941–13952. [CrossRef] [PubMed]

67. McCormick, C.L. Novel Cellulose Solutions. U.S. Patent 4,278,790, 14 July 1981.

68. McCormick, C.L.; Callais, P.A.; Hutchinson, B.H., Jr. Solution studies of cellulose in lithium chloride and N, N-dimethylacetamide. *Macromolecules* **1985**, *18*, 2394–2401. [CrossRef]
69. Mayer, E. Porometry Characterization of Filtration Media. *Filtn News* **2002**, *20*, 1–7.
70. Kawashita, L.F.; Moore, D.R. ICPeel Digitised Stress-Strain. Available online: http://www.imperial.ac. uk/media/imperial-college/research-centres-and-groups/adhesion-and-adhesives-group/17285696.XLS (accessed on 8 February 2017).

Sample Availability: Samples of the nanocellulose coatings labeled with reactive dyes are available from the authors for a limited period.

Article

Tuning Lignin Characteristics by Fractionation: A Versatile Approach Based on Solvent Extraction and Membrane-Assisted Ultrafiltration

Chiara Allegretti [1], Oussama Boumezgane [1], Letizia Rossato [1,2], Alberto Strini [3], Julien Troquet [4], Stefano Turri [1], Gianmarco Griffini [1,*] and Paola D'Arrigo [1,5,*]

1 Department of Chemistry, Materials and Chemical Engineering "Giulio Natta", Politecnico di Milano, p.zza L. da Vinci 32, 20133 Milano, Italy; chiara.allegretti@polimi.it (C.A.); oussama.boumezgane@polimi.it (O.B.); l.rossato1@campus.unimib.it (L.R.); stefano.turri@polimi.it (S.T.)
2 Dipartimento di Biotecnologie e Bioscienze, Università di Milano-Bicocca, Piazza della Scienza 2, 20126 Milano, Italy
3 Istituto per le Tecnologie della Costruzione, Consiglio Nazionale delle Ricerche (ITC-CNR), via Lombardia 49, 20098 San Giuliano Milanese, Italy; alberto.strini@itc.cnr.it
4 Biobasic Environnement, Biopôle Clermont-Limagne, 63360 Saint-Beauzire, France; jtroquet@biobasicenvironnement.com
5 Istituto di Scienze e Tecnologie Chimiche "Giulio Natta", Consiglio Nazionale delle Ricerche (SCITEC-CNR), via Mario Bianco 9, 20131 Milano, Italy
* Correspondence: gianmarco.griffini@polimi.it (G.G.); paola.darrigo@polimi.it (P.D.)

Academic Editor: Sylvain Caillol
Received: 22 May 2020; Accepted: 21 June 2020; Published: 23 June 2020

Abstract: Technical lignins, typically obtained from the biorefining of lignocellulosic raw materials, represent a highly abundant natural aromatic feedstock with high potential in a sustainable economy scenario, especially considering the huge primary production volumes and the inherently renewable nature of this resource. One of the main drawbacks in their full exploitation is their high variability and heterogeneity in terms of chemical composition and molecular weight distribution. Within this context, the availability of effective and robust fractionation processes represents a key requirement for the effective valorization of lignin. In the present work, a multistep fractionation of two different well known technical lignins obtained from two distinct delignification processes (soda vs. kraft pulping) was described. A comprehensive approach combining solvent extraction in organic or aqueous medium with membrane-assisted ultrafiltration was developed in order to maximize the process versatility. The obtained lignin fractions were thoroughly characterized in terms of their chemical, physical, thermal, and structural properties, highlighting the ability of the proposed approach to deliver consistent and reproducible fractions of well-controlled and predictable characteristics, irrespective of their biomass origin. The results of this study demonstrate the versatility and the reliability of this integrated multistep fractionation method, which can be easily adapted to different solvent media using the same ultrafiltration membrane set up, thereby enhancing the potential applicability of this approach in an industrial scale-up perspective for a large variety of starting raw lignins.

Keywords: lignin; fractionation; biobased polymers; solvent extraction; membrane-assisted ultrafiltration

---

1. Introduction

Lignin is a biobased aromatic amorphous polymer composed of a complex methoxylated phenylpropane skeleton, which represents on average about 24% of the total components of plants [1]

and, consequently, a substantial amount of the organic carbon in the biosphere [2]. It acts as a natural glue, providing the plants' structural integrity and resistance against environmental and biological degradation [3]. Technical lignins are typically obtained from the biorefining of vegetal feedstocks for the valorization of cellulose and hemicelluloses, the so-called sugar-based platform [4–6]. Another major source of technical lignins is the pulp and paper manufacturing sector, where it is produced at an approximate rate of about 50 million tons per year [7]. Within these contexts, lignin is usually considered a waste byproduct with low residual value. To date, only a small part of the technical lignins produced worldwide is actually used for material manufacturing and chemical synthesis, as the majority is being burnt for energy recovery at the processing plant itself [8]. The main existing pathways for lignin valorization aim at its direct utilization (e.g., as micro or nanostructured filler in blend with polymers [9,10] or as a water-reducing agent for concrete [11]) with limited processing after the initial cellulose separation from the original biomasses. Lignin represents, however, a highly abundant natural aromatic feedstock with potentially very interesting applications in a sustainable economy scenario, especially considering the high primary production volumes and the inherently renewable nature of this resource [12]. Several studies are actually under way for the development of valuable lignin derivatives, such as raw chemicals [6], polymers [13–18], high-performance concrete admixtures [19,20], and hydrogels [21].

Different factors lead to the current marginal use of lignin as renewable feedstock, including its very complex and bulky macromolecular structure and its chemical recalcitrance in relation to the difficulties associated with the development of controlled depolymerization processes [12,22,23]. One of the main drawbacks in the full exploitation of the lignin value chain potentials is, however, its high variability. Unlike sugars, lignins are delivered as very complex mixtures and their structure and physico-chemical properties, especially in terms of composition, heterogeneity, and molecular size distribution, are strongly dependent on their natural origin (plant species and growing conditions) and on the specific process employed for their separation from the original biomass [1]. High-value lignin applications, such as the production of polymers and fine chemicals, typically require strict specifications of the starting material, particularly in terms of the chemical composition and molecular mass distribution. Variations in these characteristics due to different lignin origins and processing/extraction methods can cause significant alterations in the performance of the final product [24]. A critical dependence on lignin properties was found for several high-value applications [25–28], such as the production of polymers [16], antioxidants [29], aromatic compounds [30], the synthesis of vanillin [31], and the development of high-performance concrete water-reducing agents [32]. The availability of effective fractionation processes for obtaining lignins with tailored physico-chemical properties and a defined molecular weight distribution is therefore a key requirement to effectively cope with both the upstream material variability and the downstream application requirements.

Several innovative techniques are under study for the extraction of natural product from biomasses, including, among others, supercritical fluids, pulsed electric fields, and acoustic cavitation [33]. These are interesting approaches that are, however, still under development and not yet ready for a widespread implementation. In this work, the combined use of two well-known methods was instead addressed, namely solvent extraction and membrane ultrafiltration, in order to develop an enhanced lignin fractionation process using established technologies.

Solvent extraction is a widely adopted strategy to isolate technical lignin fractions with controlled properties [34–37]. This approach is typically implemented with the treatment of raw lignin using a sequence of different solvents with increasing solubilization capability in order to obtain selected fractions with defined characteristics. The main drawbacks of this method are the costs and the environmental issues related to the large amount of complex solvent systems involved in its implementation. Moreover, the characteristics of final fractions are strictly dependent on biomass–solvent interactions and variations in the lignin source can lead to the need of a whole process readjustment.

Membrane ultrafiltration is another interesting method for technical lignin fractionation [38,39]. Specific advantages are the ease of implementation, the low environmental impact, the high separation efficiency, and the consistency of results with different solutes and solvents. Membrane ultrafiltration is therefore a recognized approach for the development of a green separation process [40]. The main difficulty related to the application of this technology to lignin fractionation resides in the poor solubility of most technical lignins, with consequent membrane fouling and process performance degradation [41].

The combination of solvent extraction and membrane ultrafiltration appears as a promising strategy for the development of a viable lignin fractionation process. An application of this method was described for the delignification of non-woody lignocellulosic biomasses [42]. In previous works from our groups, an integrated approach was described that combines a first solubilization phase followed by membrane ultrafiltration in order to obtain a versatile cascade process for the fractionation of a technical lignin obtained from the soda process. This strategy was implemented with both aqueous [43] and organic [44] solvents. In the first case (hereafter referred to as the aqueous process), a solubilization in water/ethanol (60/40 *v/v*) mixture was performed, followed by a microfiltration step in order to obtain a clear lignin solution. In the second one (hereafter referred to as the organic process), the solubilization was carried out by Soxhlet extraction with the selected solvent. Both approaches implemented a two-section cascade membrane ultrafiltration downstream in order to obtain three different lignin fractions.

The combined strategy allows some drawbacks of each method to be overcome. The upstream solubilization phase allows the selection of the best conditions needed for each specific biomass. This ensures the delivery of a neat particle-free process fluid to the ultrafiltration system, with consequent mitigation of fouling-related problems. The downstream fine fractionation by membrane ultrafiltration allows in turn the use of a single solvent instead of a complex series needed for the implementation of the whole fractionation sequence with a multistage solvent extraction, providing great benefits in terms of decreased process costs, reduced complexity, and enhanced sustainability. Moreover, the integrated approach combines the versatility of the first solvent fractionation (inherent to the solubilization step) with the reliability and performance consistency of membrane ultrafiltration in the delivery of the final products.

In the present work, this approach was extended to two well-known technical lignins obtained from two distinct delignification processes (soda vs. kraft pulping) and originated from a non-woody (wheat straw/Sarkanda-grass) vs. a woody (softwood) biomass, respectively. In particular, a direct comparison between these two lignin feedstocks in terms of process yield and the chemical, physical, thermal, and structural characteristics of the resulting fractions was performed using both the aqueous and the organic processes. While previous studies focused on the investigation of single (either aqueous or organic) fractionation approaches, a comprehensive strategy was purposely developed here in order to maximize the control over the characteristics of the obtained lignins and to take advantage of the versatility of the proposed approach. Particularly, for both process implementations, the same type of membrane was adopted, allowing the use of a single ultrafiltration setup and ultimately optimizing the consistency of the two process outcomes.

The approach presented in this work clearly demonstrates the versatility and the reliability of this integrated fractionation method, using two very different solvent systems and lignins of different origin and characteristics. Furthermore, the switchability between different solvents using the same ultrafiltration membrane systems enhances the potential of this fractionation method for a large variety of starting raw lignins, while the recyclability of the organic solvents used in the process in a close-circuit recirculation plant considerably reduces possible environmental concerns.

2. Results and Discussion

This work demonstrates an integrated approach for lignin fractionation based on a cascade multistep process that combines a first solubilization step followed by a membrane ultrafiltration sequence. The process was implemented using both aqueous and organic solvents, and two different lignin feedstocks were investigated and thoroughly compared. In the aqueous process, the insoluble fraction was separated by microfiltration, while in the organic process, the solubilization was performed by Soxhlet extraction. Both processes included a cascade membrane ultrafiltration for the separation of fractions with tailored molecular weight distribution. The overall fractionation process is depicted in Figure 1.

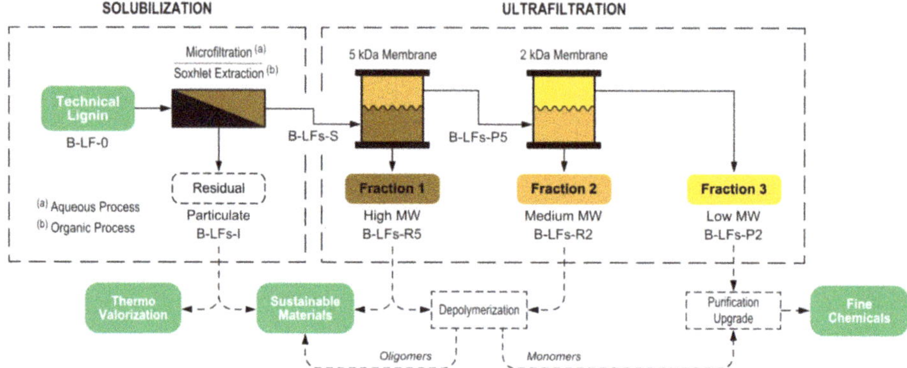

Figure 1. Overview of the overall fractionation process. Possible applications of obtained fractions and related downstream processes are indicated in the bottom line (dashed arrows).

Because of the relevant number of different fractions handled, in Figure 1 and throughout the present work a labelling scheme for the working lignin fractions was adopted, allowing a quick identification of the related biomass, solvent, and process step (Figure 2).

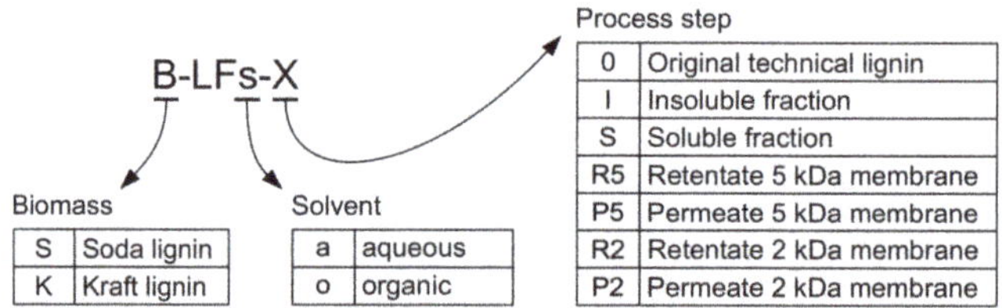

Figure 2. Labelling scheme for lignin fractions.

A key design feature of this study was the implementation of the same ultrafiltration membrane setup for both the aqueous and the organic processes. This was obtained by the use of stabilized cellulose membranes, thanks to their very wide solvent compatibility that spans from water-based mixtures to a broad range of organic solvents. The possibility of using the same ultrafiltration membrane system with different process solvents is a very important result from a preparative perspective, allowing the selection of the most appropriate solubilization conditions for each specific biomass while maintaining high consistency in the delivered fractions. In addition, solvent switching costs are reduced, also as a result of the easy membrane cleaning (e.g., washing with 1 M NaOH solution at 40 °C and then conditioning with organic solvent for the aqueous to organic process switch). The simple process switching allows optimization of the fractionation conditions, mitigating the major drawbacks of each technology (i.e., membrane fouling and final solvent extraction in the case of the aqueous process, and the need of several different solvents in the case of the organic process).

The two processes were studied using two distinct model technical lignin systems, namely a wheat straw-Sarkanda grass lignin (Protobind 1000) obtained from the soda pulping process (S-LF-0) and a softwood lignin (Indulin AT) obtained from the kraft pulping process (K-LF-0). The soda lignin is a herbaceous sulphur-free lignin, whose chemical composition is close to that of natural lignin, given the relatively mild delignification conditions used to extract it. The choice of this commercially available lignin material for this study reflects the great impact of non-woody, agricultural, and crop-derived biomasses on the global biobased economy both in terms of economic turnaround and mass availability [13]. The kraft lignin is a byproduct of the sulphate cooking process and constitutes around 85% of the total world lignin production. During the kraft process, lignin is dissolved in severe conditions in an aqueous solution of sodium hydroxide and sodium sulphite and degraded into fragments. Kraft lignin contains a significant amount of sulphur as sulphonic groups, strongly affecting its solubility and ultimate chemical-physical characteristics.

2.1. Solubilization and Solvent Selection

The solubilization phase is the key step in order to allow the processing of different lignin sources. Particularly, the selection of the best solubilization conditions, governed in turn by the selection of the solvent process mixture, is of fundamental importance for the success of the fractionation methodology.

The aqueous process was performed by using a water-based solvent mixture that could be finely tuned by co-solvent selection and pH adjustment. In the present study, an ethanol/basic water mixture (60/40 *v/v*, adjusted to pH 10 with NaOH) was selected. The final pH after solubilization was around 8. This allowed a 75 and 95 g/L operating concentration for Soda (S-LFa-S) and Kraft lignin (K-LFa-S), respectively, to be reached. These conditions permitted working with a much higher lignin solution concentration with respect to the maximum value of 15 g/L obtained in a previous work for the same soda lignin [43] using pure water/ethanol (60/40 *v/v*). In the aqueous process, it was very important to implement a microfiltration step before the subsequent ultrafiltration in order to separate the insoluble fraction and avoid possible membrane clogging.

The organic process solubilization was performed with a Soxhlet extraction, which allowed the direct separation of the insoluble fraction and the recovery of a clean solute stream. The solvent selection is critical in order to obtain a viable process, with both biomass solubility and relevant solvent properties (e.g., easy evaporability because of high vapor pressure, low viscosity of the final solutions, recyclability, and uncritical handling) to be carefully considered. In order to determine the solubility characteristics of the two investigated technical lignins, a solubilization study was performed with six different organic solvents, namely *tert*-butyl methyl ether (MTBE), *n*-butyl acetate (BuOAc), tetrahydrofuran (THF), ethyl acetate (EtOAc), 2-butanone (MEK), and methanol (MeOH). The resulting extraction yields (as *w/w* percent of solubilized fraction vs. total fraction) are reported in Figure 3 (with aqueous solvent solubility as the comparison).

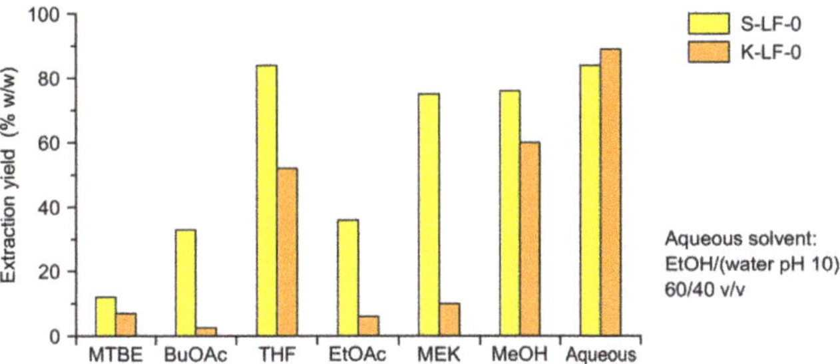

Figure 3. Extraction yields as the percent (*w/w*) of solubilized fraction vs. total fraction for S-LF-0 (yellow) and K-LF-0 (orange) for different organic solvents and for the (basic) aqueous system.

As expected, marked differences were demonstrated for the solubility of the two technical lignins in the various organic solvents. The highest extraction yields (75~80%) were obtained in THF, MeOH, and MEK for S-LF-0, and in THF and MeOH (50~60%) for K-LF-0. Among these solvents, MEK presented several advantages in laboratory handling, including easy evaporation, good recyclability, uncritical handling, low toxicity, and low viscosity of the extracts (that significantly limited ultrafiltration membrane clogging issues). Because of these benefits, and considering also its extensive use at the industrial scale, MEK was selected in this study as the working organic solvent for both the soda and kraft lignin fractionation. This also led to a comparison of the process performances in both high solubilization conditions (with S-LF-0) and low solubilization conditions (with K-LF-0). The obtained operative biomass concentrations in the extracted MEK solutions were 95 g/L for soda (S-LFo-S) and 15 g/L for kraft (K-LFo-S) lignin.

2.2. Membrane Ultrafiltration

The membrane ultrafiltration phase was pivotal for the quality and consistency of the lignin fractions delivered by the process. As the key design requirement was the implementation of the same membrane set for both aqueous and organic processes, two modified cellulose ultrafiltration membranes (with a nominal cut-off at 5 and 2 kDa) were selected for their wide solvent and pH compatibility. This selection allowed the limited solvent compatibility of the polyether sulfone ultrafiltration membranes used in a previous work to be overcome [43]: These membranes in fact were found to work quite well in aqueous solution but suffered from limited compatibility with organic solvents. In this study, process parameters, such as the flow rate, pressure, temperature, and recirculation cycle, were thoroughly optimized according to the requirement of each solvent. As will be discussed in the following sections, the limited dependence of the molecular weight distribution of delivered fractions on the solvent used confirmed a good consistency of the biomass fractionation products.

2.3. Lignin Fractionation

According to the results discussed in Section 2.1 and reported in Figure 3, the aqueous process was designed for a starting solution of lignin (both kraft and soda) in ethanol/water (60/40) in basic conditions. Preparation of the lignin starting solution was optimized by adding sodium hydroxide to facilitate the solubilization process. The final pH of the solution was around 8. The sample recovery from the aqueous process fractions included four main steps: Evaporation under vacuum of the ethanol present in the solutions; acidification of the aqueous solutions with hydrochloric acid (or sulfuric acid) at a pH close to 1; extraction in organic solvent, such as ethyl acetate; and evaporation of the organic phase and recovery of the solid fractions. On the other hand, the fractions coming from the organic process, after recovery, were evaporated under reduced pressure without further processing.

The best suitable concentrations for S-LF-0 and K-LF-0 in the aqueous process were found to be around 75 and 95 g/L, respectively, whereas the best concentrations in the organic process were 95 and 15 g/L due to the great chemical differences between the two lignins.

Aliquots of the permeates and retentates obtained from S-LF-0 and K-LF-0 for the two fractionation processes (aqueous and organic) were withdrawn after successive steps and fully characterized in terms of the chemical composition, molecular weight, and physical properties. It should be pointed out that the determination and the possible variations in the molecular weight and the composition of the lignin fractions are fundamental for their successful valorization as precursors of chemicals and polymeric materials. Moreover, the quantification and characterization of the functional groups present in the samples constitute essential steps for the effective assessment of a lignin valorization setup. Therefore, one of the goals of this work was the extraction and the full characterization of each fraction in order to recover homogeneous fractions with specific properties.

2.4. Gel Permeation Chromatography (GPC)

GPC analyses were performed on all fractions in order to evaluate the effectiveness of the aqueous and organic processes on the molecular weights and molecular weight distributions of the technical lignins considered in this work.

The characteristic values of the number-average molecular weight (Mn), weight-average molecular weight (Mw), and polydispersity index ($Đ$) are listed in Table 1 for S-LF-0 and Table 2 for K-LF-0. Analyses were performed using mono-dispersed polystyrene standards as the reference (see the materials and methods section for details).

Table 1. GPC analysis (molecular weights Mn and Mw, polydispersity index $Đ$ of all examined soluble lignin fractions from S-LF-0 obtained with the two fractionation methods).

Sample [a]	% of Starting Material (*w/w*)	Mn (Da)	Mw (Da)	$Đ$
S-LF-0	-	1390	4660	3.3
S-LFa-R5	49.3	1650	3090	1.9
S-LFa-R2	7.6	860	1100	1.3
S-LFa-P2	6.0	590	720	1.2
S-LFo-R5	66.3	1150	1950	1.7
S-LFo-R2	7.0	790	1190	1.5
S-LFo-P2	1.6	570	700	1.2

[a] Samples were eluted after acetylation. The reported values for molecular weights are relative to polystyrene standards.

Table 2. GPC analysis (molecular weights Mn and Mw, polydispersity index $Đ$ of all delivered fractions of aqueous and organic processes fractionation of K-LF-0.

Sample [a]	% of Starting Material (*w/w*)	Mn (Da)	Mw (Da)	$Đ$
K-LF-0	-	1600	4260	2.7
K-LFa-R5 [b]	72	n.a.	n.a.	n.a.
K-LFa-R2	4.8	1060	2000	1.9
K-LFa-P2	2.5	575	840	1.5
K-LFo-R5	7.9	740	1320	1.8
K-LFo-R2	1.7	530	860	1.6
K-LFo-P2	1.1	480	740	1.5

[a] Samples were eluted after acetylation. The reported values for molecular weights are relative to polystyrene standards. [b] This fraction was found to be poorly soluble in the eluting solvent (THF) even after acetylation, preventing the possibility of performing GPC analysis.

Both in the aqueous and in the organic process, the soda and kraft lignins exhibited a similarly noticeable decrease in their molecular weight after permeation through membranes of increasingly finer mesh. In particular, the filtration through a 2-kDa cut-off membrane was shown in all cases to provide lignin fractions with Mn in the range of ~500 g/mol, both for soda and kraft lignin. Such a reduction in the molecular weight was accompanied by a significant narrowing of the molecular weight distribution, as evidenced by the lower value of $Đ$ obtained in the lighter fractions ($Đ = 1.2$ for S-LFa-P2 and S-LFo-P2, $Đ = 1.5$ for K-LFa-P2 and K-LFo-P2) with respect to the parent lignin material ($Đ = 3.3$ and 2.7 for S-LF-0 and K-LF-0, respectively). It is also evident how retentate fractions in general exhibited higher molecular weights and larger $Đ$ values when compared with permeate fractions, likely due to a broader distribution of longer lignin macromolecules in the former. These trends are in line with previously reported results on both kraft and soda lignin fractionation processes selectively undertaken in organic solvent or the aqueous phase [39,43–46]. Interestingly, fraction K-LFa-R5 was found to be poorly soluble in the solvent used as eluting medium for the GPC analysis (THF), leaving a large amount of solid residue as precipitate even after acetylation. This result suggests a significant enrichment in high molecular weight chains in the retentate material of kraft lignin after the first aqueous-based ultrafiltration step (5 kDa cut-off).

Finally, it is worth noting that the same trends on the obtained molecular weights (decreasing) and molecular weight distributions (narrower) for lignin fractions recovered after filtration through increasingly finer membranes were observed in both aqueous and organic processes (Figure 4). In particular, these trends were found to be comparable for soda and kraft lignins. This evidence highlights the key role played by the ultrafiltration process in providing consistent and reproducible fractions of narrowly controlled molecular characteristics. In addition, it suggests that the performance of the ultrafiltration membranes is not significantly affected by the solvent used, indicating a factual independence between the extraction operation (controlled by the solvent) and the downstream fractionation (controlled by the ultrafiltration membranes). This aspect is a very relevant feature for an industrial scale-up perspective because it represents a clear advantage in terms of process versatility and straightforward optimization when tackling different types of biomass as input feedstock.

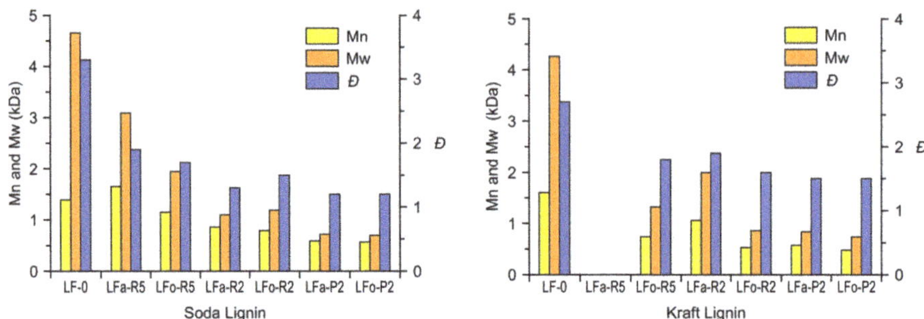

Figure 4. Comparison of the results of the aqueous and organic processes for S-LF-0 and K-LF-0 fractionation as Mn, Mw, and Đ by GPC analysis of resulting delivered fractions.

2.5. Gel Filtration Chromatography (GFC)

GFC analyses are reported in Figure 5 below, showing the performance of the two fractionation processes for S-LF-0 and K-LF-0, respectively. A new procedure using a basic water solution as eluent was developed in this work. Moving through the fractionation processes, the GFC analysis showed very well the shift of the peaks towards lower retention times, proving the effectiveness of the proposed approach in a straightforward manner. As it is clear in Figure 5, moving from the pristine lignin (in black), to the retentate at 5 kDa (in blue) and more clearly to the chromatogram of the fractions exiting the membrane with the finest cut-off (permeate B-LFo/a-P2, 2 kDa membranes, yellow lines), in all cases the enrichment in small molecules represented by the peaks at a higher retention time was striking. This analysis demonstrated that the two fractionation processes were very effective even when applied to two different biomasses.

2.6. Gas-Chromatography/Mass Spectrometry (GC/MS)

The identification and quantification of low molecular weight compounds present in each fraction was performed by using GC/MS. This analytical tool allowed the analysis of the distribution of the small compounds present in each fraction, divided into three main classes—namely, benzaldehydes and acetophenone derivatives (ArCHO, ArCOR), benzoic and coumaric acids (ArCOOH, ArCHCHCOOH), and aliphatic long-chain carboxylic acids (RCOOH).

The analyzed samples were prepared with a small-scale chromatography on silica gel in order to eliminate all the polymeric/oligomeric fractions and to recover only the suitable fraction for GC/MS analysis (% of mass recovered after chromatography is reported in Tables 3 and 4). The distribution trends confirmed clearly the efficacy of the fractionation procedures for the two considered lignins, as the percentages of total monomers in each sample increase when proceeding through the two processes. As reported in Table 3 for S-LF-0 fractionation, the percentage of monomers recovered through the aqueous process, which were analyzed in GC/MS, moved from 8% (*w/w*) of monomers in the higher molecular weight fraction to a maximum of 81% (*w/w*) in the lower molecular weight one whereas in the organic process the percentage of monomers shifted from 7% to 63%.

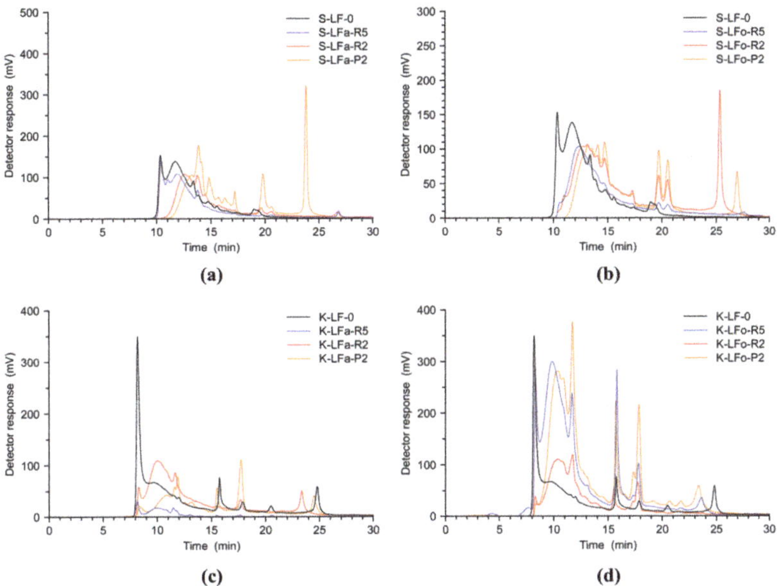

Figure 5. GFC chromatograms of selected soda (**a,b**) and kraft (**c,d**) lignin fractions recovered from the aqueous process (**a,c**) and the organic process (**b,d**).

Table 3. Summary of the GC/MS results of fractions coming from aqueous and organic processes applied on S-LF-0 illustrating the estimated amount of volatile compounds, divided in three main groups. Data reported as the total mass of compounds/mass of initial fraction (%). Values were rounded up to the nearest two significant figures with an estimated relative error of ± 0.01.

Compounds	S-LFa-R5	S-LFa-R2	S-LFa-P2	S-LFo-R5	S-LFo-R2	S-LFo-P2
-	(8.0%)[#]	(15.9%)[#]	(81%)[#]	(7%)[#]	(36%)[#]	(63%)[#]
ArCHO + ArCOR + ArOH + Ar(CH₂)₃OH	-	1.1	22	14	13	31
Ar-COOH + ArCHCHCOOH	-	-	5.3	-	5.6	16
R-COOH + R-COOR′	2.6	2.1	4.1	-	2.2	-
Total monomers	2.6	3.2	31	14	21	47

[#] Fraction analyzed; Ar: aromatic residue; R: aliphatic chain.

Table 4. Summary of the GC/MS results of fractions coming from aqueous and organic processes applied on K-LF-0 illustrating the estimated amount of volatile compounds, divided in three main groups. Data reported as the total mass of compounds/mass of initial fraction (%). Values were rounded up to the nearest two significant figures with an estimated relative error of ± 0.01.

Compounds	K-LFa-R5	K-LFa-R2	K-LFa-P2	K-LFo-R5	K-LFo-R2	K-LFo-P2
-	(21.5%)[#]	(17%)[#]	(66%)[#]	(21%)[#]	(39%)[#]	(77%)[#]
ArCHO + ArCOR + ArOH + Ar(CH₂)₃OH	16	-	17	7.5	13	20
Ar-COOH + ArCHCHCOOH	2.8	-	4.6	-	-	1.3
R-COOH + R-COOR′	22	33	22	4.4	4.3	-
Total monomers	41	33	44	12	17	21

[#] Fraction analyzed; Ar: aromatic residue; R: aliphatic chain.

For K-LF-0 fractionation, the results of the GC/MS analysis are reported in Table 4. Additionally, in this case, the efficacy of the fractionation procedure, expressed as percentages of total monomers in each sample, increased when proceeding through the processes: K-LFa-R5 and K-LFo-R5, constituting the higher molecular weight retentates, contain only 21% of small molecules, which could be inferred from GC-MS; on the contrary, K-LFa-P2 and K-LFo-P2, constituting the fractions permeated from the finest membrane (2 kDa cutoff), were considerably enriched in small molecules (around 70%).

The chromatogram profiles of the different fractions and the structures of the most abundant identified compounds with their relative abundance are reported in Table 3 and Figure 6 for S-LF-0 and Table 4 and Figure 7 for K-LF-0. As discussed above (Tables 3 and 4), the enrichment in smaller molecular weight products clearly appeared when moving throughout the fractionation processes for S-LF-0 from fractions S-LFa/o-R5 to S-LFa/o-P2 (as shown in Figure 6), whereas for K-LF-0 the number of peaks in the chromatograms becomes much more relevant in K-LFa/o-P2 (in Figure 7). The peaks from 7 to 12 min are mainly due to the silylating agent used for the derivatization of the analyzed samples. The peak at 6 min is attributed to benzaldehyde, which was used as the internal standard.

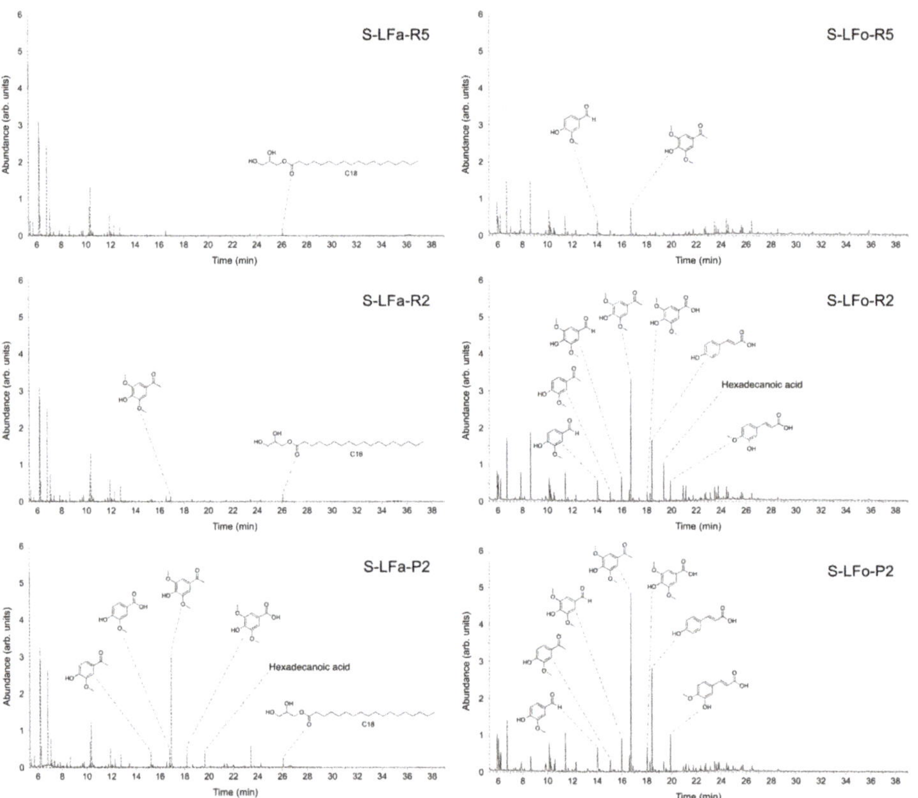

Figure 6. GC/MS total ion chromatograms of delivered fractions from aqueous and organic processes using S-LFa/o-0. Selected component identification by MS library matching is reported for each fraction.

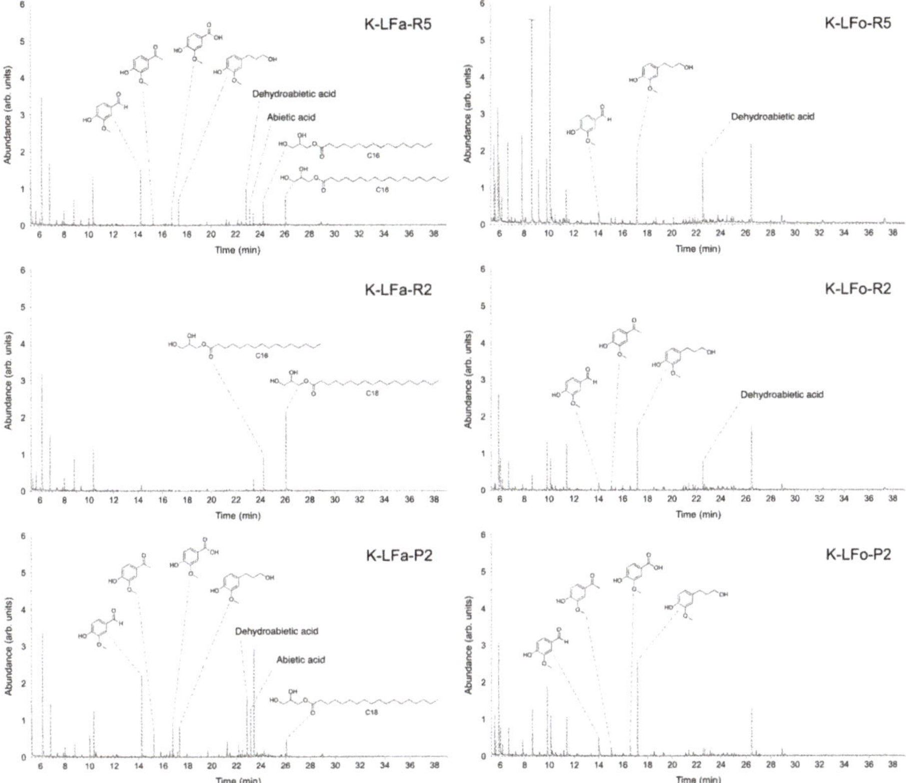

Figure 7. GC/MS total ion chromatograms of delivered fractions from aqueous and organic processes using K-LFa/o-0. Selected component identification by MS library matching is reported for each fraction.

2.7. Determination of the Hydroxylation Levels in the Recovered Fractions

The determination of hydroxylation levels in lignins constitutes an essential step for the effective valorization of the fractions obtained from the fractionation processes. The classical methods are the Folin–Ciocalteu titration method (FC), which allows the quantification of phenolic groups, and ^{31}P-NMR analysis, which leads to a quantitative determination of aliphatic and aromatic hydroxyl groups as well as of carboxyl moieties.

2.7.1. Phenolic Hydroxyl Group Determination

The FC assay was carried out to determine the total phenolic content in the original lignin and all process fractions. It is based on the reaction of phenolic hydroxyl groups with a specific redox reagent (FC reagent), which leads to the formation of a blue chromophore, which is, however, sensitive and unstable in strong bases. Therefore, based on an analytical protocol previously demonstrated by our group [41], DMSO was used as solvent for the samples in order to obtain their complete solubilization in neutral conditions.

The phenolic content results are reported in Tables 5 and 6 as vanillin equivalents (mmol/g of dry lignin). In the case of soda lignin, S-LFa/o-R2 and S-LFa/o-P2 appeared to be the fractions with a higher content in phenolic hydroxyl groups, in agreement with the trends observed with ^{31}P-NMR, as will be discussed later in the text.

Table 5. Results of the determination of phenolic hydroxyl groups expressed as vanillin equivalents in the lignin fractions recovered from aqueous and organic processes applied on S-LF-0 (expressed in mmol/g) in the dry lignin. Estimated standard errors ±0.2 mmol/g vanillin equivalent (1 σ, from calibration data).

Fraction	Vanillin Equivalent Content (mmol/g)
S-LF-0	3.06
S-LFa-R5	2.91
S-LFa-R2	3.48
S-LFa-P2	3.35
S-LFo-R5	4.38
S-LFo-R2	5.91
S-LFo-P2	6.35

Table 6. Results of the determination of phenolic hydroxyl groups expressed as vanillin equivalents in the lignin fractions recovered from aqueous and organic processes applied on K-LF-0 (expressed in mmol/g) in the dry lignin. Estimated standard errors ±0.2 mmol/g vanillin equivalent (1 σ, from calibration data).

Fraction	Vanillin Equivalent Content (mmol/g)
K-LF-0	4.79
K-LFa-R5	2.28
K-LFa-R2	2.99
K-LFa-P2	3.87
K-LFo-R5	5.90
K-LFo-R2	5.80
K-LFo-P2	4.16

If we compare the two starting lignins, K-LF-0 appears to contain an increased amount of phenolic hydroxyl groups compared to S-LF-0, probably due to the extensive cleavage of β-aryl group bonds during the cooking process. In S-LF-0 fractionation, a very high enrichment in phenolic groups appears in the fractions R2 and P2 as reported in Table 5, in excellent agreement with the GC/MS analysis reported in Figure 6. For K-LF-0 fractionation, as reported in Table 6, the highest value of phenolic content appears to be associated with K-LFa-P2 for the aqueous process, whereas in the organic one, the R5 fraction appears to be enriched in phenolic groups.

2.7.2. Total Hydroxyl Groups' Quantification with ^{31}P-NMR

The different types of hydroxyl groups present in the starting lignins and fractionated samples were deeply investigated by ^{31}P-NMR. The ^{31}P-NMR spectra of the phosphitylated lignin fractions are shown with chemical shift assignments in Figures 8 and 9, where the parent materials S-LF-0 and K-LF-0 are also shown for easy comparison. In detail, the following spectral regions were integrated to acquire the information about the chemical nature of the different hydroxyl groups: 150–147 ppm for the signals associated with aliphatic hydroxyl groups, and 145–138 ppm for the signals associated with aromatic hydroxyl groups. The signals of carboxylic acids groups were centered at 136 ppm. The peak integration in the three main portions of the spectrum led to the quantification of the total hydroxyl groups expressed in mmol of functional groups per gram of dry compound, as reported in Tables 7 and 8 for S-LF-0 and K-LF-0, respectively.

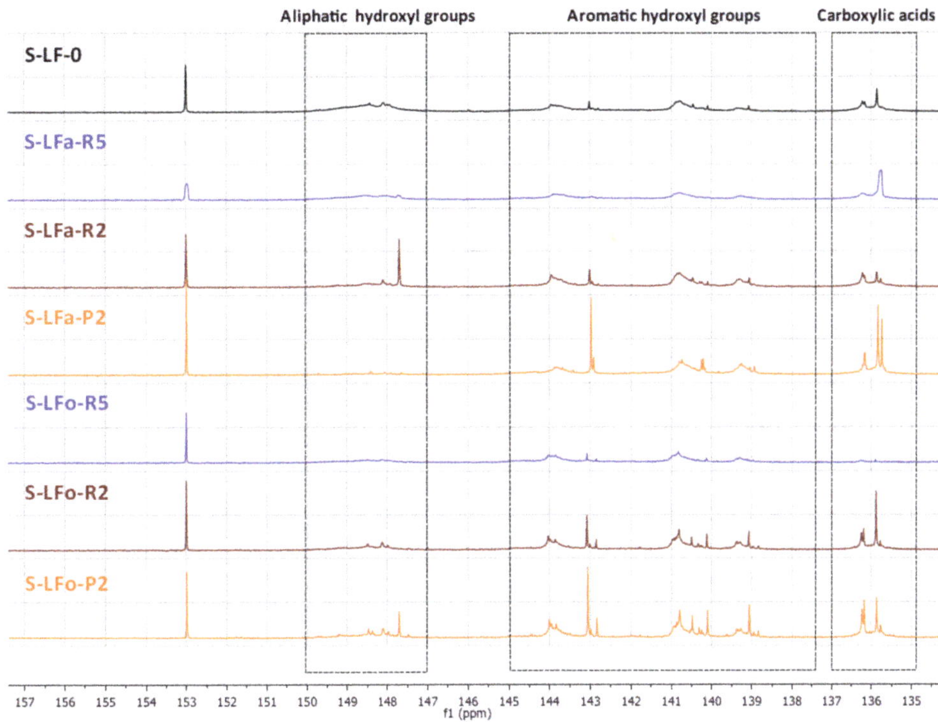

Figure 8. ^{31}P-NMR spectra of lignin fractions recovered from the aqueous and the organic process applied on S-LF-0.

Table 7. Detailed hydroxyl/carboxyl quantification by ^{31}P-NMR for the fractions recovered from aqueous and organic processes applied on S-LF-0.

Fraction	-OH Aliphatic (mmol/g)	-OH Phenolic (mmol/g)	-COOH (mmol/g)
S-LF-0	1.83	3.54	0.92
S-LFa-R5	1.32	2.63	1.33
S-LFa-R2	1.11	3.61	0.9
S-LFa-P2	0.18	3.33	1.26
S-LFo-R5	1.24	4.19	0.09
S-LFo-R2	1.27	4.52	1.29
S-LFo-P2	1.48	5.2	1.52

Table 8. Detailed hydroxyl/carboxyl quantification by ^{31}P-NMR for the fractions recovered from aqueous and organic processes applied on K-LF-0.

Fraction	-OH Aliphatic (mmol/g)	-OH Phenolic (mmol/g)	-COOH (mmol/g)
K-LF-0	3.31	5.32	0.57
K-LFa-R5	*	*	*
K-LFa-R2	1.39	2.98	0.50
K-LFa-P2	1.11	4.32	1.51
K-LFo-R5	1.64	5.65	0.48
K-LFo-R2	1.10 **	3.50 **	0.45 **
K-LFo-P2	2.52	4.44	0.55

* ^{31}P-NMR measurements could not be performed for this fraction due to extensive precipitation of the sample in the conditions used for the analysis; ** possible underestimation of these values due to the limited solubility of the sample in the solvent used for the analysis.

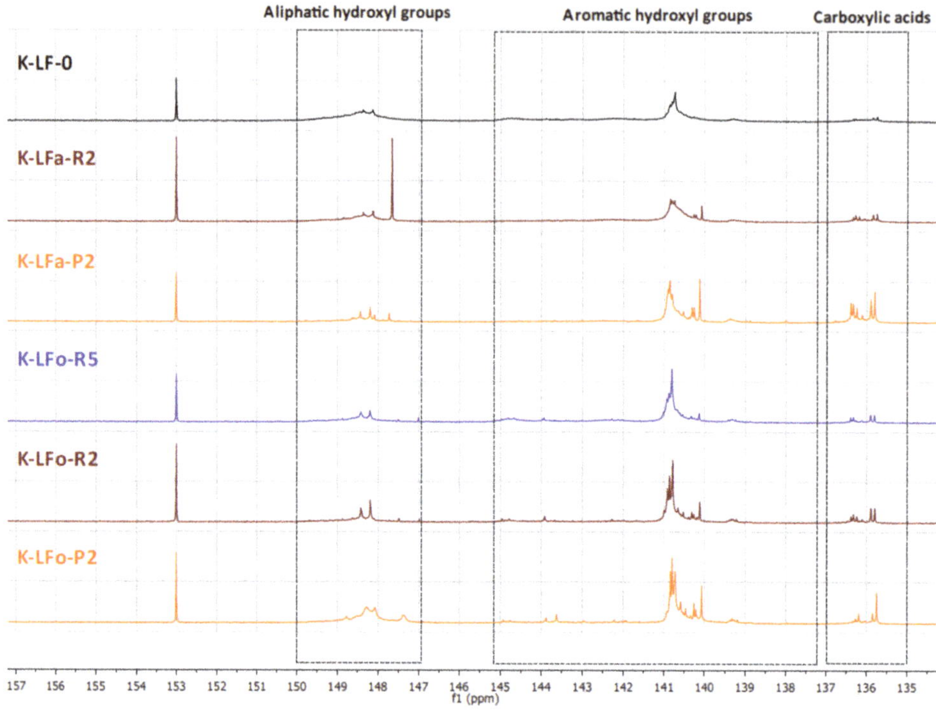

Figure 9. ^{31}P-NMR spectra of lignin fractions recovered from the aqueous and the organic process applied on K-LF-0.

Both FC assay and ^{31}P-NMR data indicated a slight increase in phenolic hydroxyl groups in the lower molecular weight fractions B-LFa/o-R2 and B-LFa/o-P2. A similar trend was also reported from ^{31}P-NMR analysis on aliphatic hydroxyl and carboxyl groups, in agreement with recent works on analogous technical lignins [43–45]. These results have great interest in terms of further selection of an appropriate fraction for a specific application, where the relative abundance of the functional groups is essential for the further processing.

2.8. Fourier-Transform Infrared Spectroscopy (FTIR)

FTIR spectroscopy was employed to analyze the chemical characteristics of the lignin fractions recovered after solvent extraction and subsequent membrane-assisted ultrafiltration. FTIR spectra of both the soda and kraft lignin fraction systems are presented in Figure 10, along with the FTIR spectrum of each parent lignin also shown for comparisons. All fractions presented a broad absorption band in the 3600–3100 cm^{-1} range that could be attributed to stretching vibrations of hydrogen-bonded phenolic and aliphatic O–H groups present in lignin, with a peak at 3390 cm^{-1} and a shoulder at 3200 cm^{-1}, respectively [47]. Signals observed in the 3050–2800 cm^{-1} region were attributed to the symmetrical and asymmetrical CH stretching of the methyl and methylene groups. At around 1700 cm^{-1}, a clear peak was observed in all spectra with the exception of K-LF-0, where more of a shoulder was observed, attributable to the stretching vibration of C-O bonds in conjugated aldehydes and carboxylic acids. In particular, it was observed that after solvent extraction (in both kraft and soda lignin streams), an increase in this signal was registered compared with the pristine samples (especially in low molecular weight fractions S-LFa/o-P2 and K-LFa/o-P2), likely indicating a higher concentration of carbonyl and carboxylic groups. The presence of the intense peak at 1515 cm^{-1} was associated to the pure aromatic skeletal vibrations in lignin. In the 1400−1000 cm^{-1} spectral region, signals of variable intensity were observed, which could be attributed as follows: Bending vibrations of phenolic O−H and aliphatic C−H in methyl groups (1370 cm^{-1}); C−O, C−C, and C−O stretching vibrations (1270 and 1210 cm^{-1}); C−H in plane deformations (1125 cm^{-1}); and C−O deformations in primary (1030 cm^{-1}) alcohols.

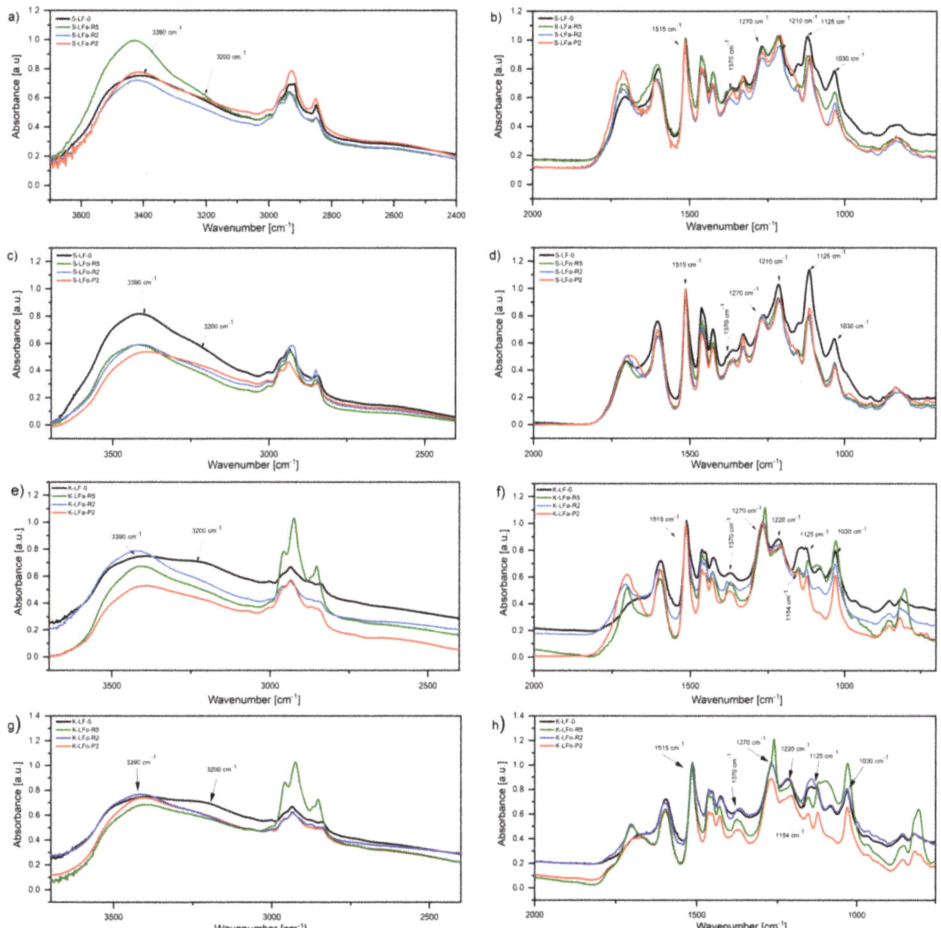

Figure 10. FT-IR spectra of soda (**a,b,c,d**) and kraft (**e,f,g,h**) lignin fractions recovered from the aqueous process (**a,b,e,f**) and the organic process (**c,d,g,h**). Different regions of the infrared spectrum are reported for clarity.

In the kraft fractions, both in aqueous and organic solvent, a clear peak at 1154 cm^{-1} was observed, which could be assigned to C-O deformations in conjugated ester groups present in guaiacyl (G), syringyl (S), and *p*-hydroxyphenyl (H) groups. This signal indicated a higher abundance of such moieties, not present in soda fractions.

2.9. Differential Scanning Calorimetry (DSC)

The thermal transitions in all lignin fractions, in particular the glass transition temperature (Tg), were evaluated by means of DSC analysis and compared with the pristine samples S-LF-0 and K-LF-0. DSC traces of both soda and kraft lignin fraction systems are presented in Figure 11. As it can be seen in the plots, in the case of organic solvent fractionation, the solvent extraction step led to a decrease in Tg of all the recovered materials, with samples S-LFo-P2 (Tg = 58 °C) and K-LFo-P2 (Tg = 44 °C) showing the lowest values, in accordance with the trends observed on molecular weights in the GPC results and in line with what was recently reported on analogous systems [43,44]. In the case of aqueous fractionation, a similar trend was observed on S-LFa-P2 and K-LFa-P2, with comparable values of Tg (Tg = 50 °C and 46 °C for S-LFa-P2 and K-LFa-P2, respectively). Sample S-LFa-R5 showed a slightly higher Tg than the parent sample S-LF-0, likely due to the slightly higher value of Mn observed for this fraction in GPC measurements. Similarly, the Tg of sample K-LFa-R5 was not easily detectable from the DSC trace but was higher (Tg = 180 °C) than the corresponding parent material K-LF-0 (Tg = 155 °C), in line with the high molecular weight inferred for this fraction from GPC analysis (Table 2).

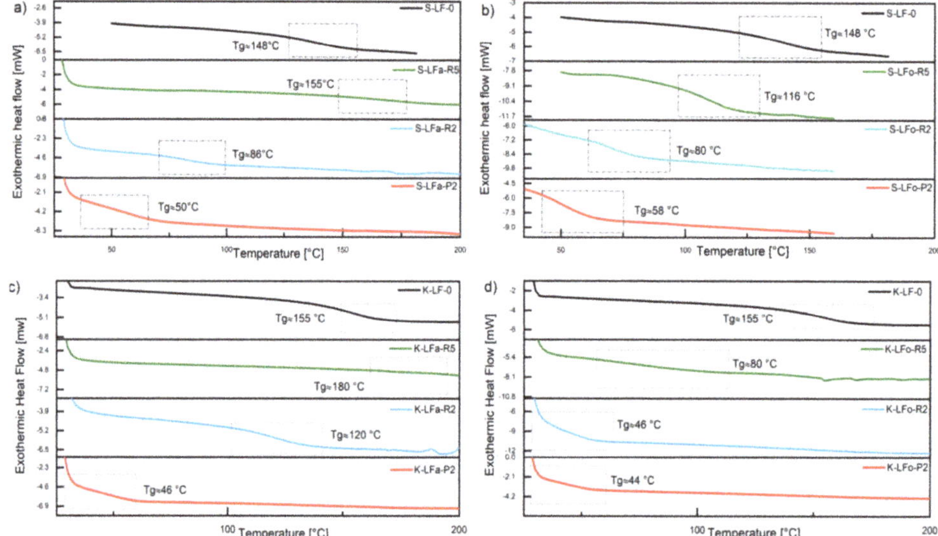

Figure 11. DSC traces of soda (**a,b**) and kraft (**c,d**) lignin fractions recovered from the aqueous process (**a,c**) and the organic process (**b,d**).

The similar trends observed on the resulting Tg of the two different lignin systems (soda vs. kraft) upon fractionation in both aqueous and organic phase further demonstrated the versatility of the proposed process, which did not appear to be significantly affected by the solvent used for the extraction operation. Moreover, they provided additional proof of the ability of this approach to deliver consistent and reproducible fractions of well-controlled and predictable thermal characteristics irrespective of the biomass origin, with this aspect being particularly useful for potential use in the development of lignin-based macromolecular materials.

3. Materials and Methods

3.1. Materials

All chemicals and analytical grade solvents, such as tetrahydrofuran (THF), methanol (MeOH), ethyl acetate (EtOAc), n-butyl acetate (BuOAc), and *tert*-butyl methyl ether (MTBE), were purchased from Sigma-Aldrich. 2-Butanone (methyl ethyl ketone, MEK) was provided by BCD Chemie GmbH (Hamburg, Germany). Lignin S-LF-0 (Protobind 1000, a mixed wheat straw/Sarkanda grass lignin from soda pulping of non-woody biomass) was provided by Tanovis (Alpnach, Switzerland). Lignin K-LF-0 (softwood Kraft lignin, Indulin AT) was provided by Meadwestvaco (Charleston, SC, USA). Lignins were used as received. All solutions were prepared in Milli-Q water (Elix Millipore Purification System, Molsheim, France). All analyses were carried out at least in duplicate unless otherwise stated.

3.2. Lignin Solubility in Organic Solvents

Lignin solubilities in organic solvent were determined by treating 10 g of starting lignin with 100 mL of the different solvents stirring at 400 rpm. Each test was carried out overnight at room temperature. The suspensions were then filtered, and the solvents were evaporated at reduced pressure and the final residues were dried until a constant weight was achieved prior to quantification.

3.3. Fractionation via Aqueous Process

The aqueous process fractionation comprised a microfiltration step followed by a two-membrane cascade ultrafiltration (shown in Figure 1 as part of an idealized lignin valorization scheme) and a final fraction recovery step, as described below (see also the discussion section).

3.3.1. Microfiltration

The lignin solution in ethanol/water was prepared by dissolution of 150 g of lignin per L of a mixture of ethanol and NaOH aqueous solution at pH 10. The insoluble material was eliminated by a microfiltration on a 0.7-μm fiberglass filter under vacuum.

3.3.2. Membrane-Assisted Fractionation

Lignin fractionation was performed by means of an ultrafiltration apparatus (Sartoflow Advanced filtration module purchased from Sartorius Stedim, Aubagne, France) equipped with flowmeters and pressure sensors to control the permeate flow, trans-membrane pressure, and filtration time. The membranes (Hydrosart membranes, Sartorius Stedim) in stabilized cellulose had nominal molecular weight cutoffs of 5 and 2 kDa and a filtration area of 0.1 m^2 each. The membrane regeneration and storage were performed at 40 °C using 1 M and 0.1 M NaOH solution, respectively.

These membranes were compatible with a high range of different solvents, allowing to switch easily from an aqueous solvent to an organic one. The procedure required the cleaning of the membrane with a solution of 1M NaOH at 40 °C, and then the washing and conditioning of the membrane in the appropriate solvents for the successive fractionation sequence. The membranes could be also stored without any problems in 0.1 M NaOH.

3.3.3. Fraction Extraction Procedure

The recovered fractions from the aqueous process were evaporated under reduced pressure to remove ethanol. The aqueous solutions were acidified to pH close to 1 with 37% HCl, and extracted three times with ethyl acetate. The combined organic phases were dried over sodium sulfate and evaporated under reduced pressure. The final solid residues were dried until a constant weight was achieved prior to analysis and quantification.

3.4. Fractionation via Organic Process

The organic process fractionation comprised a Soxhlet extraction followed by a two-membrane cascade ultrafiltration (shown in Figure 1 as part of an idealized lignin valorization scheme) and a final fraction recovery step, as described below (see also the results and discussion section).

3.4.1. Soxhlet Extraction

Soxhlet extractions were performed using a standard glass apparatus (Buchi extraction system B-811, Villebon sur Yvette, France), which allowed 4 extractions to be achieved at the same time. Each extraction unit was composed of a 150-mL working volume bottom solvent flask, a 330-mL-capacity thimble holder and a water-cooled condenser. For each extraction process, 150 mL of MEK were placed in the solvent flask and about 15 g of lignin were inserted inside a 41-mm diameter and 123-mm height cellulose paper thimble (FiltraTECH, Saran, France). The solvent reflux was kept for 8 h, adjusting the heating power in order to have 4 extraction cycles/h.

3.4.2. Membrane Ultrafiltration

Lignin fractionation was performed in MEK directly on the Soxhlet extraction solution by means of the same apparatus described in Section 3.3.2 with the membranes conditioned and used only in MEK.

3.4.3. Fraction Recovery Procedure

The recovered fractions from the organic process were evaporated under reduced pressure without further processing. The final solid residues were dried until a constant weight was achieved prior to analysis and quantification.

3.5. Gel Permeation Chromatography (GPC)

A Waters 510 HPLC system equipped with a refractive index detector was used for GPC analyses. Tetrahydrofuran (THF) was used as eluent. The analyzed lignin sample (volume 200 µL, concentration 1 mg/mL in THF) was injected into a system of three columns connected in series (Ultrastyragel HR, Waters–dimensions 7.8 mm × 300 mm) and the analysis was performed at 30 °C at a flow rate of 0.5 mL/min. The GPC system was calibrated against polystyrene standards in the 10^2–10^4 g/mol molecular weight range. To allow complete solubility in the THF eluent, before the analysis, the parent lignin and the fractions were acetylated following a standard literature procedure. The estimation of the number-average and weight-average molecular weights of the obtained lignin fractions was performed excluding the signals related to the solvent (THF) and the solvent stabilizer (butylated hydroxytoluene), visible at long elution times (>29.5 min).

3.6. Gel Filtration Chromatography (GFC)

GFC analyses were performed using a Merck-Hitachi L4000 apparatus (Tokyo, Japan) equipped with a UV detector set at λ of 254 nm. A hydroxylated acrylic polymer GFC column (Polysep-GFC-P 2000, 300 × 7.8 mm, Phenomenex, Aschaffenburg, Germany) was used with a water-based eluent (sodium borate pH 10 buffer 10 mM plus 300 mM NaCl) at 0.5 mL/min flow rate.

The samples (5 mg) were completely dissolved in a small amount of NaOH. The pH of the solution was adjusted to pH 10 with HCl 6 M, and diluted to the final concentration of 2 mg/mL with sodium borate buffer 10 mM pH 10. The sample was then centrifugated, filtered, and analyzed.

3.7. Gas-Chromatography/Mass Spectrometry (GC/MS)

The GC/MS apparatus used was an Agilent GC System 7890A, with an inert MSD with Triple-Axis Detector 7975C (Cernusco sul Naviglio, Italy). The separation was performed on a DB-5MS column (30 m × 250 µm × 0.25 µm, Phenomenex) with a helium flow rate of 1.18 mL/min, a temperature program of 50 °C (1 min) to 280 °C at 10 °C/min, 280 °C at 15 min (total run time 39 min, temperature of the injector 250 °C, injection volume 1.00 µL, injection mode split, split ratio 5:1). A solvent delay of 4 min was selected. The samples were prepared by derivatization and dissolved in methanol or acetone in a concentration around 0.5-1 mg/mL as previously described [44]. Compound identification was performed by means of an NIST 2008 mass spectral library search.

3.8. Folin–Ciocalteu Assay (FC Assay)

Total phenolic contents of the different fractions were determined by the classical Folin–Ciocalteu method with some modifications in the sample preparation step. The samples were prepared by dissolving lignin in dimethylsulfoxide (DMSO) with a final concentration of 1 mg/mL. DMSO was chosen because, being completely miscible in water, it allowed a complete lignin solubilization and did not interfere with the FC assay.

For each determination, 5 µL of the working solution (or the standard solution) were then mixed with 120 µL of deionized water, 125 µL of FC reagent (Sigma 47641), and kept for 6 min at r.t. after 30 s of vortex stirring. Then, after the addition of 1.25 mL of 5% sodium carbonate and mixing, the vial was incubated on a thermoshaker at 40 °C for 30 min. The reaction mixture absorbance was measured using a UV/Vis spectrophotometer (Jasco V-560) equipped with a temperature-controlled cuvette holder and a thermostatic water bath (Haake K10, Karlsruhe, Germany). All spectrophotometric measurements were carried out at 760 nm, 25 °C, using a 1-cm optical path cuvette. Vanillin was chosen as the reference standard. The calibration curve was constructed with nine different vanillin solutions in DMSO with the concentration in the range 0–800 µg/mL. Each FC assay determination was carried out in triplicate.

3.9. ^{31}P-NMR Analysis

^{31}P-NMR spectroscopic analyses were recorded on a Bruker Instrument AVANCE400 spectrometer (Milano, Italy). Acquisition and data treatment were performed with Bruker TopSpin 3.2 software (Milano, Italy). The spectra were collected at 29 °C with a 4-s acquisition time, 5-s relaxation delay, and 256 scans. Prior to analysis, samples were dried for 24 h under vacuum and then derivatized according to the following procedure.

A lignin sample (40 mg) was completely dissolved in 300 µL of *N,N*-dimethylformamide. To this solution, the following components were added: 200 µL of dry pyridine, 100 µL of solution of internal standard (10 mg of Endo-*N*-hydroxy-5-norbornene-2,3-dicarboximide (Sigma 226378) dissolved in 0.5 mL of a mixture of pyridine and $CDCl_3$ 1.6:1 *v/v*), 50 µL of solution of relaxation agent (5.7 mg of chromium (III) acetylacetonate (Sigma 574082) dissolved in 0.5 mL of a mixture of pyridine and $CDCl_3$ 1.6:1 *v/v*), 100 µL of 2-chloro-4,4,5,5-tetramethyl-1,3,2-dioxaphospholane (Sigma 447536), and at the end 200 µL $CDCl_3$. The solution was centrifuged if necessary. All chemical shifts reported were related to the reaction product of the phosphorylating agent with water, which gave a signal at 132.2 ppm.

3.10. Fourier-Transform Infrared Spectroscopy (FT-IR)

FT-IR spectra of all lignin fractions were recorded in transmission mode on lignin-containing KBr pellets (amount of lignin was approximately 5 mg). The analysis was performed by means of a Nicolet 760-FTIR spectrophotometer (Thermo Fisher Scientific, Rodano, Italy) at room temperature in air in the $4000–700\ cm^{-1}$ wavenumber range with 64 accumulated scans and a resolution of $2\ cm^{-1}$.

3.11. Differential Scanning Calorimetry (DSC)

DSC was performed on solid-state samples (about 10–15 mg) by means of a Mettler-Toledo DSC/823e instrument (Milano, Italy) at a scan rate of 20 °C/min under nitrogen flux.

4. Conclusions

Lignin constitutes a byproduct accumulated worldwide in significant quantities in the pulp and paper industry, despite possessing a great potential for generating novel biobased platform chemicals and polymeric materials alternative to those currently derived from petrochemical routes. One of the most important tasks when targeting the establishment of any lignin valorisation process is the ability to set up robust and reliable fractionation processes enabling the recovery of lignin streams with controllable and predictable composition and properties. Indeed, the great variability in its chemical composition, structural characteristics, molecular weight, and reactivity, intimately linked to the vast geographical and seasonal distribution of lignocellulosic biomass feedstocks, makes lignin processing and valorization particularly challenging.

In an attempt to address this issue, in this work, a straightforward fractionation method was demonstrated combining solvent extraction and membrane-assisted ultrafiltration. This approach was applied to two commercially available technical lignins with different origins (non-woody vs. woody biomass) obtained from two distinct delignification routes (soda vs. kraft pulping), starting from an extraction process using two different solvent (aqueous vs. organic) systems. Based on this strategy, a comprehensive method was demonstrated for the tunable and predictive control over the characteristics of the recovered lignin fractions. In particular, the key role played by the ultrafiltration process in providing consistent and reproducible fractions of narrowly controlled chemical, structural, molecular, and thermal characteristics was shown. In addition, the performance of the ultrafiltration membranes was found to be relatively unaffected by the solvent used (either aqueous or organic), indicating a relative independence between the extraction operation (controlled by the solvent) and the downstream fractionation (controlled by the ultrafiltration membranes). This aspect is a very relevant feature from an industrial scale-up perspective, because it represents a clear advantage in terms of process tunability and optimization when tackling different types of biomass as input feedstock.

This versatile, robust, and easy to implement fractionation approach paves the path for the delivery of consistent and reproducible lignin fractions of well-controlled and predictable properties irrespective of the biomass origin, with this aspect being of particular interest in view of their potential use in the development of lignin-based macromolecular materials.

Author Contributions: Conceptualization, P.D., G.G., S.T., J.T. and A.S.; investigation, C.A., O.B., L.R. and J.T.; writing, C.A., P.D., G.G., A.S. and J.T.; supervision, P.D. and G.G.; funding acquisition, P.D. and G.G. All authors have read and agreed to the published version of the manuscript.

Funding: This research received funding from Regione Lombardia and Fondazione Cariplo, through grant number 2018-1739 (project POLISTE).

Acknowledgments: The authors kindly acknowledge Maria Elisabetta Brenna for instrumental support in GC-MS analyses and Francesco Gilberto Gatti for help in NMR analyses optimization.

Conflicts of Interest: The authors declare no conflict of interest. The funders had no role in the design of the study; in the collection, analyses, or interpretation of data; in the writing of the manuscript, or in the decision to publish the results.

References

1. Calvo-Flores, F.G.; Dobado, J.A.; Isac-García, J.; Martín-Martínez, F.J. *Lignin and Lignans as Renewable Raw Materials*; John Wiley & Sons Ltd.: Chichester, UK, 2015.
2. Boerjan, W.; Ralph, J.; Baucher, M. Lignin biosynthesis. *Annu. Rev. Plant Biol.* **2003**, *54*, 519–546. [CrossRef] [PubMed]
3. Liu, Q.Q.; Luo, L.; Zheng, L.Q. Lignins: Biosynthesis and Biological Functions in Plants. *Int. J. Mol. Sci.* **2018**, *19*, 335. [CrossRef] [PubMed]
4. Li, T.; Takkellapati, S. The current and emerging sources of technical lignins and their applications. *Biofuels Bioprod. Biorefin.* **2018**, *12*, 756–787. [CrossRef] [PubMed]
5. Ragauskas, A.J.; Beckham, G.T.; Biddy, M.J.; Chandra, R.; Chen, F.; Davis, M.F.; Davison, B.H.; Dixon, R.A.; Gilna, P.; Keller, M.; et al. Lignin Valorization: Improving Lignin Processing in the Biorefinery. *Science* **2014**, *344*. [CrossRef] [PubMed]
6. Schutyser, W.; Renders, T.; Van den Bosch, S.; Koelewijn, S.F.; Beckham, G.T.; Sels, B.F. Chemicals from lignin: An interplay of lignocellulose fractionation, depolymerisation, and upgrading. *Chem. Soc. Rev.* **2018**, *47*, 852–908. [CrossRef]
7. Tribot, A.; Amer, G.; Alio, M.A.; de Baynast, H.; Delattre, C.; Pons, A.; Mathias, J.D.; Callois, J.M.; Vial, C.; Michaud, P.; et al. Wood-lignin: Supply, extraction processes and use as bio-based material. *Eur. Polym. J.* **2019**, *112*, 228–240. [CrossRef]
8. Vishtal, A.; Kraslawski, A. Challenges in Industrial Applications of Technical Lignins. *Bioresources* **2011**, *6*, 3547–3568.
9. Garcia Gonzalez, M.N.; Levi, M.; Turri, S.; Griffini, G. Lignin nanoparticles by ultrasonication and their incorporation in waterborne polymer nanocomposites. *J. Appl. Polym. Sci.* **2017**, *134*, 45318. [CrossRef]

10. Kun, D.; Pukanszky, B. Polymer/lignin blends: Interactions, properties, applications. *Eur. Polym. J.* **2017**, *93*, 618–641. [CrossRef]

11. Edmeades, R.M.; Hewlett, P.C. Cement admixtures. In *Lea's Chemistry of Cement and Concrete*, 4th ed.; Hewlett, P.C., Ed.; Butterworth-Heinemann: Oxford, UK, 2003. [CrossRef]

12. Strassberger, Z.; Tanase, S.; Rothenberg, G. The pros and cons of lignin valorisation in an integrated biorefinery. *RSC Adv.* **2014**, *4*, 25310–25318. [CrossRef]

13. de Haro, J.C.; Magagnin, L.; Turri, S.; Griffini, G. Lignin-Based Anticorrosion Coatings for the Protection of Aluminum Surfaces. *ACS Sustain. Chem. Eng.* **2019**, *7*, 6213–6222. [CrossRef]

14. Scarica, C.; Suriano, R.; Levi, M.; Turri, S.; Griffini, G. Lignin Functionalized with Succinic Anhydride as Building Block for Biobased Thermosetting Polyester Coatings. *ACS Sustain. Chem. Eng.* **2018**, *6*, 3392–3401. [CrossRef]

15. Duval, A.; Lawoko, M. A review on lignin-based polymeric, micro- and nano-structured materials. *React. Funct. Polym.* **2014**, *85*, 78–96. [CrossRef]

16. Upton, B.M.; Kasko, A.M. Strategies for the Conversion of Lignin to High-Value Polymeric Materials: Review and Perspective. *Chem. Rev.* **2016**, *116*, 2275–2306. [CrossRef] [PubMed]

17. Grossman, A.; Vermerris, W. Lignin-based polymers and nanomaterials. *Curr. Opin. Biotechnol.* **2019**, *56*, 112–120. [CrossRef]

18. Collins, M.N.; Nechifor, M.; Tanasa, F.; Zanoaga, M.; McLoughlin, A.; Strozyk, M.A.; Culebras, M.; Teaca, C.A. Valorization of lignin in polymer and composite systems for advanced engineering applications—A review. *Int. J. Biol. Macromol.* **2019**, *131*, 828–849. [CrossRef] [PubMed]

19. Kalliola, A.; Vehmas, T.; Liitia, T.; Tamminen, T. Alkali-O_2 oxidized lignin—A bio-based concrete plasticizer. *Ind. Crop. Prod.* **2015**, *74*, 150–157. [CrossRef]

20. Gupta, C.; Nadelman, E.; Washburn, N.R.; Kurtis, K.E. Lignopolymer Superplasticizers for Low-CO_2 Cements. *ACS Sustain. Chem. Eng.* **2017**, *5*, 4041–4049. [CrossRef]

21. Thakur, V.K.; Thakur, M.K. Recent advances in green hydrogels from lignin: A review. *Int. J. Biol. Macromol.* **2015**, *72*, 834–847. [CrossRef]

22. Sun, Z.H.; Fridrich, B.; de Santi, A.; Elangovan, S.; Barta, K. Bright Side of Lignin Depolymerization: Toward New Platform Chemicals. *Chem. Rev.* **2018**, *118*, 614–678. [CrossRef]

23. Wendisch, V.F.; Kim, Y.; Lee, J.H. Chemicals from lignin: Recent depolymerization techniques and upgrading extended pathways. *Curr. Opin. Green Sustain. Chem.* **2018**, *14*, 33–39. [CrossRef]

24. de Haro, J.C.; Allegretti, C.; Smit, A.T.; Turri, S.; D'Arrigo, P.; Griffini, G. Biobased Polyurethane Coatings with High Biomass Content: Tailored Properties by Lignin Selection. *ACS Sustain. Chem. Eng.* **2019**, *7*, 11700–11711. [CrossRef]

25. Gordobil, O.; Olaizola, P.; Banales, J.M.; Labidi, J. Lignins from Agroindustrial by-Products as Natural Ingredients for Cosmetics: Chemical Structure and In Vitro Sunscreen and Cytotoxic Activities. *Molecules* **2020**, *25*, 1131. [CrossRef] [PubMed]

26. Ortiz, P.; Vendamme, R.; Eevers, W. Fully Biobased Epoxy Resins from Fatty Acids and Lignin. *Molecules* **2020**, *25*, 1158. [CrossRef]

27. Tayeb, A.H.; Tajvidi, M.; Bousfield, D. Paper-Based Oil Barrier Packaging using Lignin-Containing Cellulose Nanofibrils. *Molecules* **2020**, *25*, 1344. [CrossRef]

28. Gouveia, J.R.; Garcia, G.E.S.; Antonino, L.D.; Tavares, L.B.; dos Santos, D.J. Epoxidation of Kraft Lignin as a Tool for Improving the Mechanical Properties of Epoxy Adhesive. *Molecules* **2020**, *25*, 2513. [CrossRef]

29. Li, M.F.; Sun, S.N.; Xu, F.; Sun, R.C. Sequential solvent fractionation of heterogeneous bamboo organosolv lignin for value-added application. *Sep. Purif. Technol.* **2012**, *101*, 18–25. [CrossRef]

30. Kim, J.Y.; Park, S.Y.; Lee, J.H.; Choi, I.G.; Choi, J.W. Sequential solvent fractionation of lignin for selective production of monoaromatics by Ru catalyzed ethanolysis. *RSC Adv.* **2017**, *7*, 53117–53125. [CrossRef]

31. Fache, M.; Boutevin, B.; Caillol, S. Vanillin Production from Lignin and Its Use as a Renewable Chemical. *ACS Sustain. Chem. Eng.* **2016**, *4*, 35–46. [CrossRef]

32. Li, S.Y.; Li, Z.Q.; Zhang, Y.D.; Liu, C.; Yu, G.; Li, B.; Mu, X.D.; Peng, H. Preparation of Concrete Water Reducer via Fractionation and Modification of Lignin Extracted from Pine Wood by Formic Acid. *ACS Sustain. Chem. Eng.* **2017**, *5*, 4214–4222. [CrossRef]

33. Chemat, F.; Abert Vian, M.; Fabiano-Tixier, A.-S.; Nutrizio, M.; Režek Jambrak, A.; Munekata, P.E.S.; Lorenzo, J.M.; Barba, F.J.; Binello, A.; Cravotto, G. A review of sustainable and intensified techniques for extraction of food and natural products. *Green Chem.* **2020**, *22*, 2325–2353. [CrossRef]

34. Jia, Z.A.; Li, M.F.; Wan, G.C.; Luo, B.; Guo, C.Y.; Wang, S.F.; Min, D.Y. Improving the homogeneity of sugarcane bagasse kraft lignin through sequential solvents. *RSC Adv.* **2018**, *8*, 42269–42279. [CrossRef]

35. Ramakoti, B.; Dhanagopal, H.; Deepa, K.; Rajesh, M.; Ramaswamy, S.; Tamilarasan, K. Solvent fractionation of organosolv lignin to improve lignin homogeneity: Structural characterization. *Bioresour. Technol. Rep.* **2019**, *7*. [CrossRef]

36. Kumar, N.; Vijayshankar, S.; Pasupathi, P.; Kumar, S.N.; Elangovan, P.; Rajesh, M.; Tamilarasan, K. Optimal extraction, sequential fractionation and structural characterization of soda lignin. *Res. Chem. Intermed.* **2018**, *44*, 5403–5417. [CrossRef]

37. Griffini, G.; Passoni, V.; Suriano, R.; Levi, M.; Turri, S. Polyurethane Coatings Based on Chemically Unmodified Fractionated Lignin. *ACS Sustain. Chem. Eng.* **2015**, *3*, 1145–1154. [CrossRef]

38. Sevastyanova, O.; Helander, M.; Chowdhury, S.; Lange, H.; Wedin, H.; Zhang, L.; Ek, M.; Kadla John, F.; Crestini, C.; Lindström Mikael, E. Tailoring the molecular and thermo–mechanical properties of kraft lignin by ultrafiltration. *J. Appl. Polym. Sci.* **2014**, *131*. [CrossRef]

39. Toledano, A.; Garcia, A.; Mondragon, I.; Labidi, J. Lignin separation and fractionation by ultrafiltration. *Sep. Purif. Technol.* **2010**, *71*, 38–43. [CrossRef]

40. Ziwei, D.; Jindui, J.; Amirhossein, B.; Ji, L.; Yue, S. Mechanism and Control of Spalling for Friable Sandstone Pillars in a Room and Pillar Mine. *J. Eng. Sci. Technol. Rev.* **2018**, *11*, 197–205. [CrossRef]

41. Weis, A.; Bird, M.R.; Nystrom, M. The chemical cleaning of polymeric UF membranes fouled with spent sulphite liquor over multiple operational cycles. *J. Membr. Sci.* **2003**, *216*, 67–79. [CrossRef]

42. Alriols, M.G.; Garcia, A.; Llano-ponte, R.; Labidi, J. Combined organosolv and ultrafiltration lignocellulosic biorefinery process. *Chem. Eng. J.* **2010**, *157*, 113–120. [CrossRef]

43. Allegretti, C.; Fontanay, S.; Krauke, Y.; Luebbert, M.; Strini, A.; Troquet, J.; Turri, S.; Griffini, G.; D'Arrigo, P. Fractionation of Soda Pulp Lignin in Aqueous Solvent through Membrane-Assisted Ultrafiltration. *ACS Sustain. Chem. Eng.* **2018**, *6*, 9056–9064. [CrossRef]

44. Allegretti, C.; Fontanay, S.; Rischka, K.; Strini, A.; Troquet, J.; Turri, S.; Griffini, G.; D'Arrigo, P. Two-Step Fractionation of a Model Technical Lignin by Combined Organic Solvent Extraction and Membrane Ultrafiltration. *ACS Omega* **2019**, *4*, 4615–4626. [CrossRef]

45. Passoni, V.; Scarica, C.; Levi, M.; Turri, S.; Griffini, G. Fractionation of Industrial Softwood Kraft Lignin: Solvent Selection as a Tool for Tailored Material Properties. *ACS Sustain. Chem. Eng.* **2016**, *4*, 2232–2242. [CrossRef]

46. Jääskeläinen, A.S.; Liitia, T.; Mikkelson, T.; Tamminen, T. Aqueous organic solvent fractionation as means to improve lignin homogeneity and purity. *Ind. Crop. Prod.* **2017**, *103*, 51–58. [CrossRef]

47. Bock, P.; Gierlinger, N. Infrared and Raman spectra of lignin substructures: Coniferyl alcohol, abietin, and coniferyl aldehyde. *J. Raman Spectrosc.* **2019**, *50*, 778–792. [CrossRef] [PubMed]

Sample Availability: Samples of the compounds are not available from the authors.

Article

Multiscale Structure of Starches Grafted with Hydrophobic Groups: A New Analytical Strategy

Chloé Volant [1], Alexandre Gilet [2], Fatima Beddiaf [3], Marion Collinet-Fressancourt [4,5], Xavier Falourd [3,6], Nicolas Descamps [7], Vincent Wiatz [7], Hervé Bricout [2], Sébastien Tilloy [2], Eric Monflier [2,*], Claude Quettier [7], Ahmed Mazzah [1] and Agnès Rolland-Sabaté [3,8,*]

[1] University Lille, CNRS, USR3290—MSAP—Miniaturisation pour la Synthèse, l'Analyse et la Protéomique, F-59000 Lille, France; chloe.volant@univ-ubs.fr (C.V.); ahmed.mazzah@univ-lille.fr (A.M.)

[2] University Artois, CNRS, Centrale Lille, University Lille, UMR 8181—UCCS—Unité de Catalyse et Chimie du Solide, F-62300 Lens, France; alexandre.r.gilet@gmail.com (A.G.); herve.bricout@univ-artois.fr (H.B.); sebastien.tilloy@univ-artois.fr (S.T.)

[3] INRAE, UR BIA, F-44316 Nantes, France; fb.beddiaf@gmail.com (F.B.); xavier.falourd@inrae.fr (X.F.)

[4] CIRAD, UPR Recyclage et Risque, F-97743 Saint-Denis, Réunion, France; marion.collinet@cirad.fr

[5] University Montpellier, Recyclage et Risque, CIRAD, 34398 Montpellier, France

[6] INRAE, BIBS Facility, F-44316 Nantes, France

[7] ROQUETTE Frères, Rue de la Haute Loge, 62136 Lestrem, France; nicolas.descamps@roquette.com (N.D.); vincent.wiatz@roquette.com (V.W.); claude.quettier@roquette.com (C.Q.)

[8] INRAE, Université d'Avignon, UMR SQPOV, F-84914 Avignon, France

[*] Correspondence: eric.monflier@univ-artois.fr (E.M.); agnes.rolland-sabate@inrae.fr (A.R.-S.); Tel.: +33-(0)3-2179-1772 (E.M.); +33-(0)4-3272-2522 (A.R.-S.)

Academic Editor: Sylvain Caillol

Received: 28 May 2020; Accepted: 17 June 2020; Published: 18 June 2020

Abstract: Starch, an abundant and low-cost plant-based glucopolymer, has great potential to replace carbon-based polymers in various materials. In order to optimize its functional properties for bioplastics applications chemical groups need to be introduced on the free hydroxyl groups in a controlled manner, so an understanding of the resulting structure-properties relationships is therefore essential. The purpose of this work was to study the multiscale structure of highly-acetylated (degree of substitution, $0.4 < DS \leq 3$) and etherified starches by using an original combination of experimental strategies and methodologies. The molecular structure and substituents repartition were investigated by developing new sample preparation strategies for specific analysis including Asymmetrical Flow Field Flow Fractionation associated with Multiangle Laser Light Scattering, Nuclear Magnetic Resonance (NMR), Raman and Time of Flight Secondary Ion Mass spectroscopies. Molar mass decrease and specific ways of chain breakage due to modification were pointed out and are correlated to the amylose content. The amorphous structuration was revealed by solid-state NMR. This original broad analytical approach allowed for the first time a large characterization of highly-acetylated starches insoluble in aqueous solvents. This strategy, then applied to characterize etherified starches, opens the way to correlate the structure to the properties of such insoluble starch-based materials.

Keywords: acetylated starch; etherified starch; chemical composition; macromolecular characteristics; surface characterization

---

1. Introduction

Starch is, after cellulose and hemicelluloses, one of the most abundant carbohydrates in plants. Its semi-crystalline granules are mainly composed of two glucopolysaccharides: amylose, a quasi-linear chain of D-glucosyl units linked in α-1,4, and amylopectin which is constituted by a complex arborescent arrangement of linear chains of D-glucosyl units linked in α-1,4 with about 5–6% of branching points in

α-1,6 [1]. Starch is abundant, biodegradable and low cost, and it can easily be processed by extrusion or thermomoulding to obtain biomaterials [2,3]. Unfortunately starch hydrophilicity and poor mechanical properties make it not suitable for bioplastic application [3]. A way to improve its physicochemical properties is introducing functional groups on the free hydroxyl groups by esterification or etherification with a maximum theoretical degree of substitution (DS) of 3 [4]. The properties of such chemically modified starches depend on the starch structure, the DS, the type of grafted groups and their repartition on the macromolecules [5] but these relations are not well known because of structure analysis limitations.

Indeed, because of the physicochemical impacts of starch modifications, the characterization of low DS and high DS modified starch-based products requires different strategies. The macromolecular structure of low DS high molar mass starches has been investigated by Size-Exclusion Chromatography (SEC) [6–8] and Asymmetrical Flow Field Flow Fractionation (AF4) coupled with Multiangle Laser Light Scattering (MALLS) [9–11]. The localization of acetyl groups was approached by liquid-state Nuclear Magnetic Resonance (NMR) [12] and by combining specific enzymatic hydrolyses and Matrix Assisted Laser Desorption Ionization-Time of Flight (MALDI-TOF) Mass Spectroscopy [13] or SEC [6,14–16]. Yet, when the DS increases, water solubility becomes lower, as a result the macromolecular analyses of highly modified starch become very difficult as no adequate investigation systems can be run in adapted organic solvents. In addition to the poor solubility, the low accessibility to the enzymes caused by the chemical groups grafted onto the starch macromolecules prevents the use of specific enzymes [17]. Because of these difficulties, so far, only very few molecular characterizations performed on highly acetylated starch (DS > 2) with high molar mass amylopectins have been reported in literature except with liquid-state NMR and SEC-MALLS [18,19], this latter being inappropriate for amylopectin analysis [20,21].

To design new chemically modified starch-based products, a better understanding of their structure-properties relationships is mandatory. Thereby, accessing structural knowledge at different scales, and in particular at the molecular scale, is essential. In order to answer the characterization needs of starch-based materials developers, the aim of this work was then to develop new tools combination to investigate the molecular structure, localization of substituents and supramolecular structure of high molar mass hydrophobically modified starches with DS from 0.8 up to 3 using original analytical approaches.

The molecular structure was investigated by combining new sample preparation strategies with chromatographic techniques including AF4-MALLS, solid and liquid-state NMR, Raman and Time of Flight Secondary Ion Mass (TOF-SIMS) spectroscopies and Atomic Force Microscopy (AFM). The first step consisted in selecting and implementing the various techniques then, the analytical strategy was developed to characterize the supramolecular organization and the localization of substituents for two types of modified starches: highly acetylated starches and etherified starches with various substituents (Figure 1). The amorphous structuration was revealed by solid-state NMR which also emerged as the best way to obtain the DS for acetylated starches. Molar mass decrease and specific ways of chain breakage due to modification were highlighted and correlated with the amylose content, while changes in supramolecular structure were associated with the DS.

2. Results and Discussion

2.1. Development of the Experimental Strategy

In order to develop a new general strategy for the multiscale structural characterization of chemically modified starches insoluble in water and with high molar mass, we focused first on one acetylated sample obtained from waxy maize starch (WMS), i.e., without amylose.

2.1.1. Determination of the Degree of Substitution

The degree of substitution (DS) of the acetylated waxy maize starch (AWMS) (average number of acetyls grafted per anhydroglucose unit, AGU) was determined by three different methods: titration as previously described [22], liquid-state ^{1}H-NMR in dimethylsulfoxide-d$_6$ (DMSO-d$_6$) and solid-state ^{13}C-NMR. The DS values obtained in liquid-state (titration and ^{1}H-NMR) were 1.8 ± 0.2 whereas solid-state ^{13}C-NMR allowed to obtain the expected value of 2.6 (± 0.01). This indicated a bias for liquid-state methods which could result from the poor solubility of AWMS. This limitation of the liquid-state NMR had already been reported for acetylated starches with DS greater than 2 [23]. In fact, with uncomplete solubilization, the detected signals did not represent the entire sample, hence a characterization of the soluble fraction only and therefore a partial view of the material. Solid-state NMR seemed then to be the most adapted method for the DS determination of acetylated starches with DS > 2.

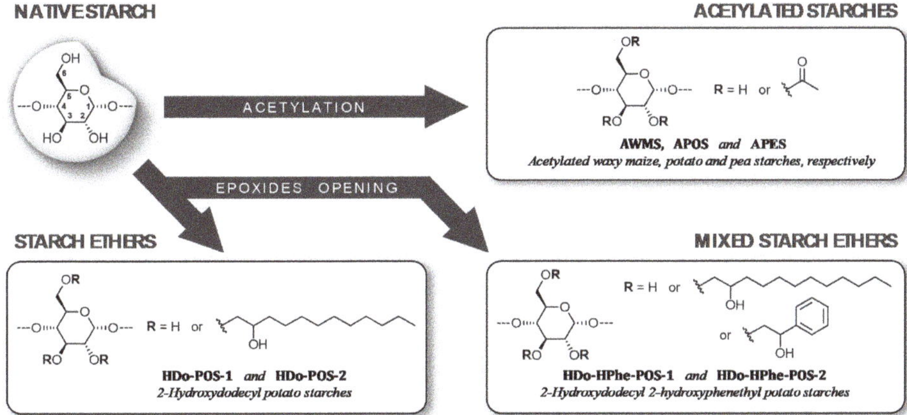

Figure 1. Synthesis, representation and codification of modified starches.

2.1.2. Macromolecular and Molecular Structure

First AWMS solubility was evaluated. It was soluble in pure dimethylsulfoxide (DMSO) but insoluble in water (Supplementary Materials S1, Table S1a) as well as in DMSO/water mixtures. High Performance Size-Exclusion Chromatography (HPSEC) analysis of high molar mass non-modified starches [20] had shown to be inappropriate in DMSO and we confirmed that for AWMS (results not shown). In addition, perform AF4 analysis, a more suitable technique for high molar mass starches, was not possible in DMSO (even with a system compatible with organic solvents). For these reasons, AWMS was analyzed after a deacetylation process in order to increase its water solubility and allow its analysis by aqueous AF4-MALLS. The deacetylation process was developed specifically for AWMS in order to avoid starch chains depolymerization by KOH (see the Materials and Methods section), and the complete deacetylation was checked by Fourier Transformed Infrared Spectroscopy FTIR (Supplementary Materials S2). The solubilization recoveries (in water for native and in 0.1 M KOH for acetylated starch) were good (> 93%) and elution recoveries were 93% for native starch and lower for AWMS (75%) which meant that the analysis was representative. The lower elution recovery obtained for AWMS was probably due to the loss of few small sugars produced by the acetylation process through the membrane during the fractionation procedure.

The AF4 elugram of WMS exhibited one peak at 17–21 mL (Figure 2a), corresponding to amylopectin as expected [21,24]. The elugram of AWMS showed in addition a peak shoulder at 17 mL which corresponded to the elution of smaller macromolecules (as elution volume is proportional to molecular size), produced by a breakage of amylopectin during the acetylation process. Molar mass distributions and the weight average molar mass $\overline{M}_w$) also showed slightly lower values for AWMS (Figure 2a, Table 1), which confirmed the slight amylopectin degradation. Yet the dispersity ($\overline{M}_w/\overline{M}_n$) and the z-average radius of gyration ($\overline{R}_G$) were stable for WMS after acetylation (Figure 2b), in line with small molar mass variations.

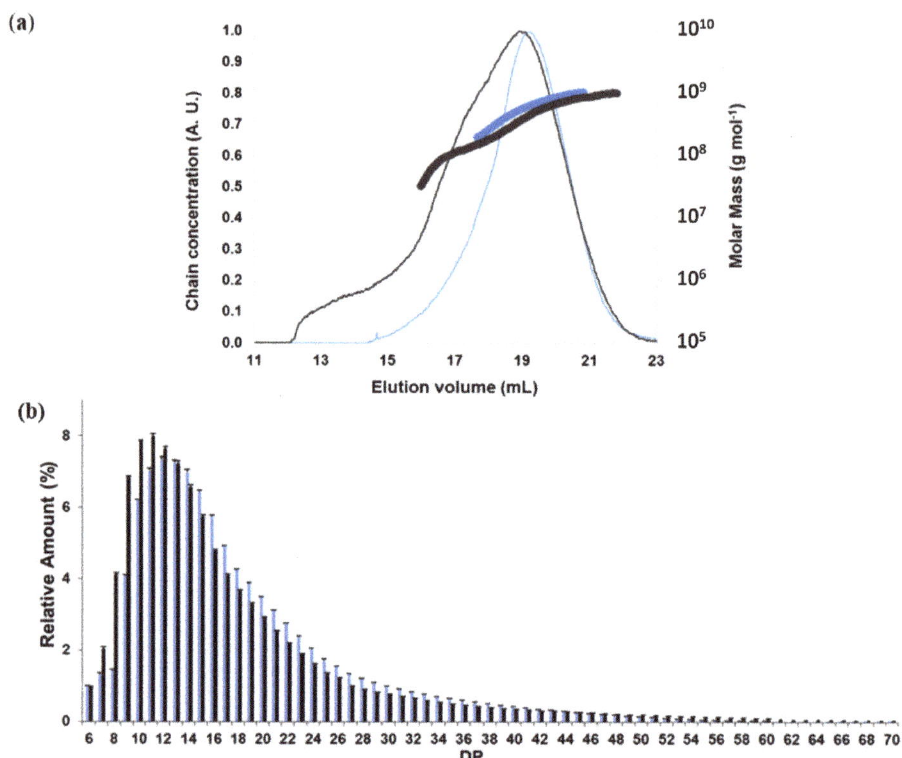

Figure 2. Macromolecular structure of native and acetylated waxy maize starch. (**a**) Elugrams (thin lines) and molar mass distributions (thick lines) of native WMS (blue) and acetylated WMS (AWMS) after deacetylation (black) obtained by AF4-MALLS; (**b**) Chain length distributions of WMS (Blue) and AWMS after deacetylation (Black). DS before deacetylation: 2.6.

Table 1. Weight-average molar mass ($\overline{M}_w$), z-average radius of gyration ($\overline{R}_G$) and dispersity index ($\overline{M}_w/\overline{M}_n$) determined by AF4-MALLS for WMS and AWMS after deacetylation.

Reference	$\overline{M}_w \times 10^{-7}$ (g mol⁻¹)	$\overline{R}_G$ (nm)	$\overline{M}_w/\overline{M}_n$
WMS	60.0	282	1.12
AWMS	52.0	283	1.16

$\overline{M}_w$, $\overline{R}_G$ and $\overline{M}_w/\overline{M}_n$ values were taken over the whole peak. The experimental uncertainties were 5%.

Debranching enzymes do not work with more than 20% DMSO in the solution and acetyl groups inhibit their action. In order to determine the chain length distribution of AWMS, it was first deacetylated to allow its solubilization in aqueous solvent and to allow the enzymes to act on it reliably. The distributions of debranched chains for deacetylated AWMS and WMS displayed a major peak corresponding to a degree of polymerization (DP) 11–13 with a large tail (Figure 2b and Supplementary Materials S3), as expected for waxy maize starch [25]. Larger proportions (37.5%) of short A chains (DP 6–12) in the amylopectin cluster [1] and smaller proportions (55.9%) of intermediate chains (DP 13-36, defined as B1a and B1b chains in amylopectin [1]) were observed for AWMS compared to WMS which showed 28.4% and 65.6% for the short A and intermediate B chains respectively (Figure 2b and Table S3). This meant that even if the molar mass of WMS was only slightly affected by the acetylation process, specific chain breakings occurred and the amylopectin fine structure was modified. B1a and B1b chains in the clusters (DP 13-36 [1]), i.e., chains with intermediate length that bear at least one chain, seemed to be cut more specifically and short A chains liberated (DP 6-12).

2.1.3. Composition, localization of Acetyl Groups and Supramolecular Structure

The composition and supramolecular structure of AWMS and its native counterpart were studied by solid-state ^{13}C-NMR on powdered samples. Spectral chemical shifts of native WMS were in accordance with literature [26,27]. AWMS spectrum was characteristic of acetylated species, with signals at 20.2 and 170.6 ppm assigned to the acetyl carbons (methyl and carbonyl respectively) (Figure 3). AWMS exhibited a C-6 shift from 61 ppm (for WMS, results not shown) to 65 ppm due to acetylation on C-6, and a C-1 shift from 100 ppm to 96 ppm due to acetylation on C-2, as was previously assigned by liquid-state ^{13}C-NMR in D_2O [12]. A slight shift from 73 to 70 ppm was also observed for the C-3 signal which signed for acetylation of this carbon as well, even if it was in slighter proportions. This shift included probably the C-2 shift. Indeed, the poor resolution of this spectral region did not allow us to clearly discriminate the signal of C-2 from that of C-3.

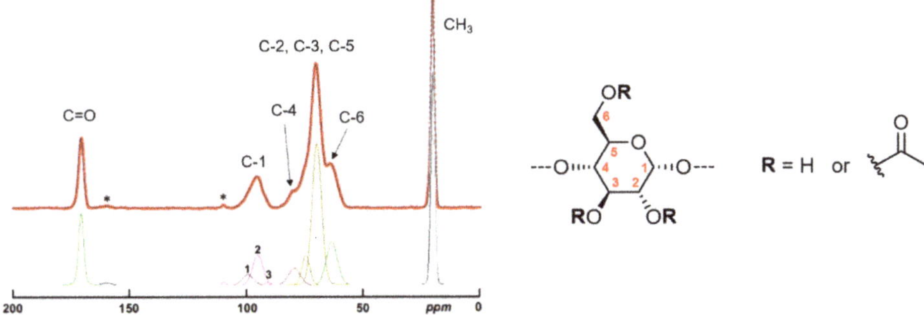

Figure 3. Solid-state ^{13}C-NMR spectrum obtained for acetylated waxy maize starch (DS 2.6) and its spectral decomposition. * Spinning side band.

Moreover, the decomposition of C-1 signal (90–105 ppm) with Lorentz-Gauss function was performed to identify the local short-range organization of starches [28]. Native WMS showed three peaks for C-1 (results not shown), standing for A-type double helices, in line with the A-type crystallinity known for native maize starch [27]. For AWMS, the decomposition of C-1 peak (90–105 ppm) gave three resonances that have already been identified on amorphous starches [27–29] (Figure 3): 95.4 ppm (73% of total signal) typical for amorphous very unfavorable and/or constrained conformations for α-1,4 and α-1,6 linkages, 99.7 ppm (26% of total signal) typical for B-type double helices or paracrystalline bundles in amorphous samples (isolated double helices or double helices embryos, on a scale small enough to not give any long-range ordering visible with wide-angle X-Ray scattering) and 91.6 ppm (2% of total signal) corresponding to reducing extremities.

C-1 and C-6 signals showed typical amorphous starch behavior, as expected for such gelatinized starches and in agreement with the absence of crystallinity observed by X-Ray diffraction [30], with a minor proportion of residual double helices and a major proportion of constrained conformations for α-1,4 and α-1,6 linkages that could be attributed to the acetylation points on C6 but also on C2 and C3.

2.1.4. Surface Characterization of Materials

The surface characterization was made by Raman spectroscopy on powder and film (Figure 4a). The most suitable analysis conditions were determined on native powdered starch and allowed to attribute the chemical shifts of the starch backbone linkages. In native WMS, intense bands relative to organized amylopectin (CCC around 480 cm^{-1}) as well as glycoside ring vibrations (COC) were detected [31]. In AWMS the introduction of acetyl groups was visible with the vibration around 1740 cm^{-1} attributed to the deformation of the C=O bond [18,32]. The disappearance of CCC vibration around 480 cm^{-1} confirmed the loss of crystalline structure of WMS with acetylation process as was evidenced by solid-state ^{13}C-NMR. Nevertheless, Raman spectroscopy did not allow characterizing the repartition of hydrophobic groups nor the DS.

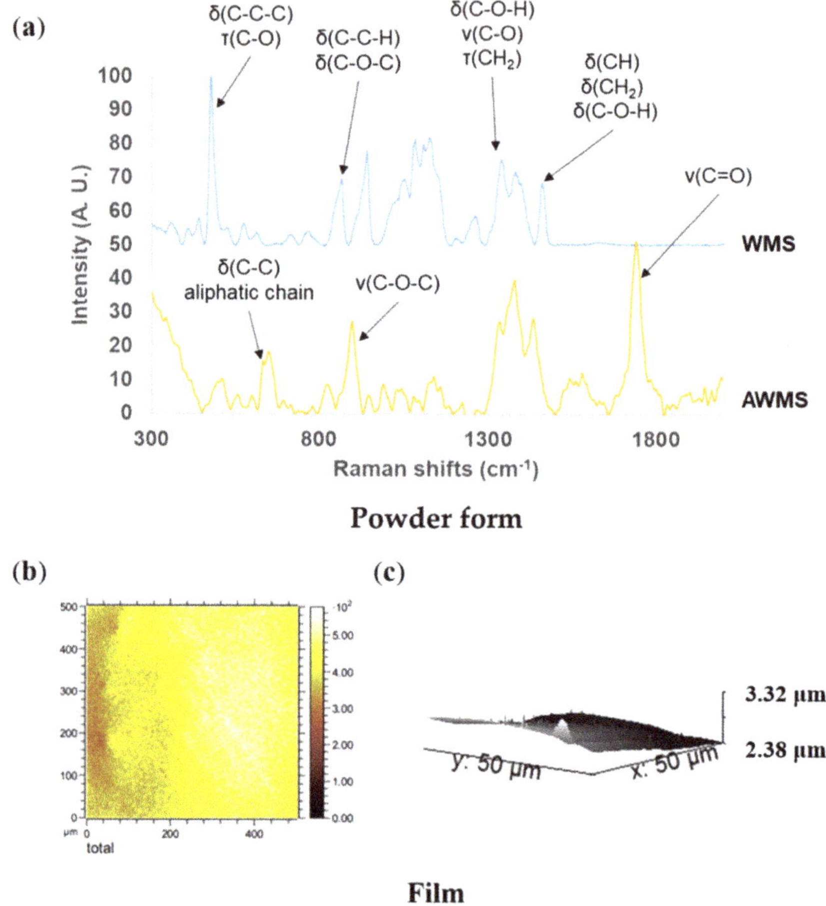

Figure 4. Surface characterization of acetylated waxy maize starch (DS 2.6). (**a**) Raman spectra of WMS and AWMS, (**b**) TOF-SIMS 2D mapping of the acetate fragment in AWMS, (**c**) AFM topography image of AWMS.

The surface composition of AWMS film was approached at the µm scale by TOF-SIMS. Identification of ionized fragments from substituents followed by 2D mapping revealed a uniform distribution of acetyl fragments on film surface on the 500 µm^2 observed area (Figure 4b and Supplementary Materials

S4). No remaining granules were observed which was in agreement with the absence of Maltese cross under polarized light (results not shown) and confirm that all starch granules have been destroyed during the acetylation process. AFM observations, which did not reveal any particular topography (Figure 4c), confirmed the homogeneity and smoothness of AWMS film surface.

2.2. Characterization of Acetylated Starches with Various DS and Amylose Content

Acetylated starches with DS of 2.4 and 0.7 (± 0.01) determined by solid-state ^{13}C-NMR, and obtained from starches with amylose contents of 22 and 35% (potato and pea starches, respectively) were studied by comparison of AWMS (0% amylose) and their native counterparts using the strategy established in the previous section.

2.2.1. Macromolecular Characteristics

Macromolecular characteristics obtained by aqueous AF4-MALLS for acetylated potato starch (APOS) and acetylated pea starch (APES) after deacetylation, and their native counterparts are reported in Figure 5 and Table 2. Solubilization recoveries were higher than 89% for all the starches whereas elution recoveries were 40 and 85% for APOS and APES respectively (Table 2) which indicated a higher loss of small sugars produced by the acetylation process during the analysis procedure for APOS.

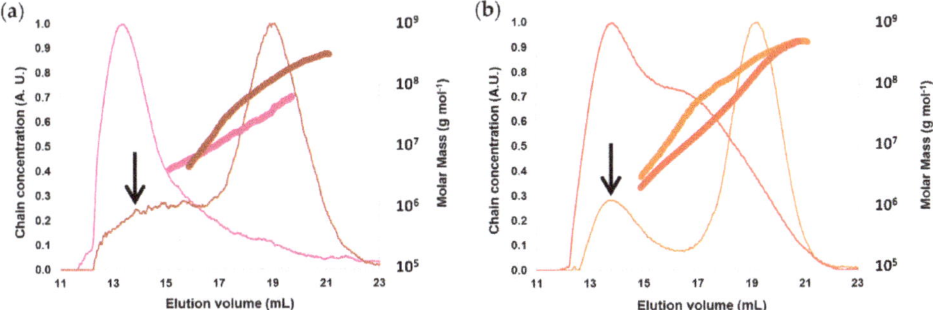

Figure 5. Macromolecular characteristics of native and acetylated potato and pea starches obtained by AF4-MALLS. (**a**) POS (dark red) and APOS after deacetylation (pink); (**b**) PES (orange) and APES after deacetylation (red). Elugram (thin lines): chain concentration versus elution volume; molar mass distributions (thick lines); the arrows indicate the amylose peak. Native potato (POS), pea (PES) starches and corresponding acetylated starches: APOS and APES, respectively. The DS of APOS and APES before deacetylation were 2.4 and 0.7, respectively. $\overline{M}_w$, $\overline{R}_G$ and $\overline{M}_w/\overline{M}_n$.

Table 2. Solubilization and elution recoveries, weight-average molar mass ($\overline{M}_w$), z-average radius of gyration ($\overline{R}_G$) and dispersity index ($\overline{M}_w/\overline{M}_n$) determined by integrating the whole elugrams.

Reference	Solubilization Recovery (%)	Elution Recovery (%)	$\overline{M}_w \times 10^{-7}$ (g mol^{-1})	$\overline{R}_G$ (nm)	$\overline{M}_w/\overline{M}_n$
POS	100	70	10.0	179	4.79
PES	89	100	20.4	219	3.49
APOS	100	40	1.5	104	1.35
APES	100	85	5.1	181	5.71

$\overline{M}_w$, $\overline{R}_G$ and $\overline{M}_w/\overline{M}_n$ values were taken over the whole peak. The experimental uncertainties were 5%.

AF4 elugrams exhibited two peaks for native potato (POS) and pea (PES) starches: a main peak corresponding to amylopectin (17–21 mL), and a minor peak (13–15 mL) corresponding to amylose (Figure 5), in agreement with previously reported distributions for native potato and maize starches [21,24]. Acetylated starches elugrams showed a peak at 12–17 mL which signed the presence of smaller macromolecules resulting from the scission of starch macromolecules during the acetylation process. This phenomenon was accompanied by a shift of the amylopectin population to lower macromolecular sizes, which indicated amylopectin degradation. Molar mass distributions also showed lower values for acetylated starches (Figure 5), which confirmed the amylopectin breakage during the acetylation process. This macromolecular degradation was particularly important for APOS (Figure 5a), for which the elugram did not show any amylopectin peak.

Molar mass decrease of 75% and 85% were observed for PES and POS respectively (Table 2) whereas only 13% of the molar mass was loss during the acetylation of WMS (Table 1). The dispersities of PES and POS were higher than WMS as expected because of the presence of amylose. Moreover, when it was stable between WMS and AWMS, it increased a lot for PES after acetylation (from 3.49 to 5.71) because of the high production of short macromolecular chains. This meant that amylopectin but also amylose molecules were degraded during the acetylation process. On the contrary, the dispersity of POS decreased because of a dramatic degradation of the largest macromolecules in APOS. The low elution recovery of APOS was in line with the high amount of small macromolecules in this sample: a large part of these molecules were too small compared to the cut off of the AF4 membrane and were not analyzed because they were eliminated during the fractionation process. This result was also confirmed on non-deacetylated APOS by the higher proportion of reducing units observed by solid-state ^{13}C-NMR for this sample (see next section). Starch macromolecules breakage have been reported to depend on acetylation process [18], thus the low molar mass and macromolecular size of APOS is probably due to the particularly drastic process applied to this starch to obtain a DS of 2.4.

2.2.2. Chain-Length Distributions of Acetylated Starches

The distributions of debranched chains of APOS and APES after deacetylation and corresponding native starches displayed two peaks (Figure 6a,b and Table S3): a major peak corresponding to DP 11–13 and a minor peak corresponding to DP 44–49. Such distributions were expected for native starches [25,33–36]. These chain length (CL) distributions showed similar patterns with some differences depending on the amylose content and the DS. APES and APOS had more chains of DP >37 (>13.3%) than AWMS (6.5%) because of the presence of amylose, whereas AWMS and APOS, which had a DS ≥ 2.4, had more very short A chains (Figure 2b, Figure 6a,b and Table S3).

APOS had larger proportions (75.0%) of short A chains (DP 6–12) and intermediate B1a chains (DP 13–24) and lower proportions (14.4%) of long chains (DP > 37) compared to POS which showed 69.3% and 18.3% of DP 6-24 chains and DP >37 chains respectively (Figure 6c and Table S3). This decrease of long chains was in line with specific chain breakings on the long B2 and B3 chains of potato amylopectin [1] during the acetylation process and agreed well with the drastic molar mass decrease observed by AF4-MALLS that corresponded to the liberation of low molar mass dextrins. The same chain breaking trend was observed for PES, exhibiting the highest amylose content, yet in very slight proportions (Figure 6b and Table S3) compared to POS. This was due to a less drastic acetylation process in link with its lower DS (0.7), by analogy to the observations made on acetylated barley starches by Bello-Pérez et al. [19] which reported a decrease of long chains and an increase of short chains with an increasing effect when the DS increase. Moreover these slight variations of CL distribution of amylopectin despite a decrease of 75% of the molar mass of PES caused by the acetylation process could also be related to a preferential hydrolysis of amylose chains and extra-long amylopectin chains in starches containing amylose. Whereas for starches without amylose, such as WMS, a preferential hydrolysis of chains with intermediate length was observed in the amylopectin during the acetylation process (Figure 6c and Table S3).

2.2.3. Localization of Acetyl Groups, Supramolecular Structure and Surface Characteristics of Acetylated Starches

Solid-state ^{13}C-NMR spectra obtained on powdered samples showed the typical acetyl carbons signals at 20.2–20.4 and 169.4–170.60 ppm for both APOS and APES, and exhibited a C-6 shift from 61 ppm for the native starches (results not shown) to 65 ppm due to acetylation (Figure 7). Only APOS and

AWMS spectra showed a shift for the C-1 resonance from 100 ppm to 96 ppm (Figure 2a,b) indicating they were acetylated on C-2, in line with the results presented in the previous section, and a shift of C-3 resonance (from 72 ppm to 70 ppm).

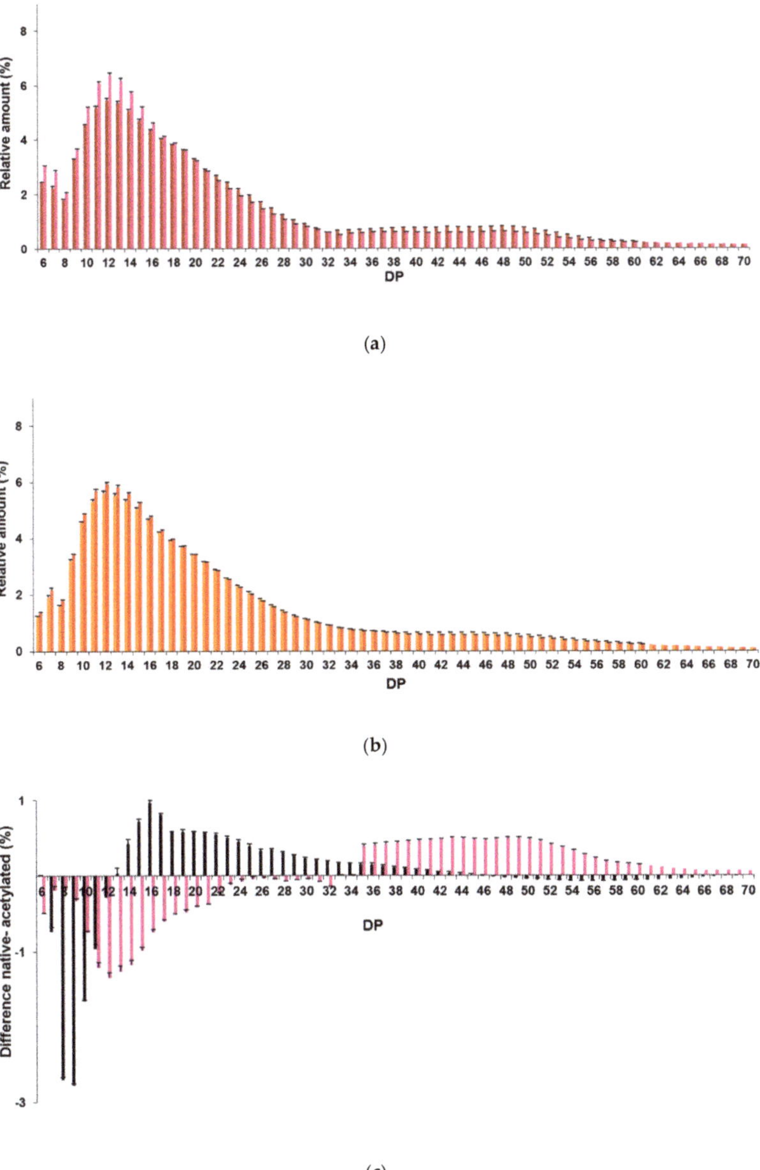

(a)

(b)

(c)

Figure 6. Chain length distributions of native and acetylated starches. (**a**) POS (dark red) and APOS (pink), (**b**) PES (Orange) and APES (Red), (**c**) difference of chain length between native and acetylated starches for AWMS (Black) and APOS (pink). Native waxy maize (WMS), potato (POS), pea (PES) starches and corresponding acetylated starches: AWMS, APOS and APES, respectively. The DS of AWMS, APOS and APES before deacetylation were 2.6, 2.4 and 0.7, respectively.

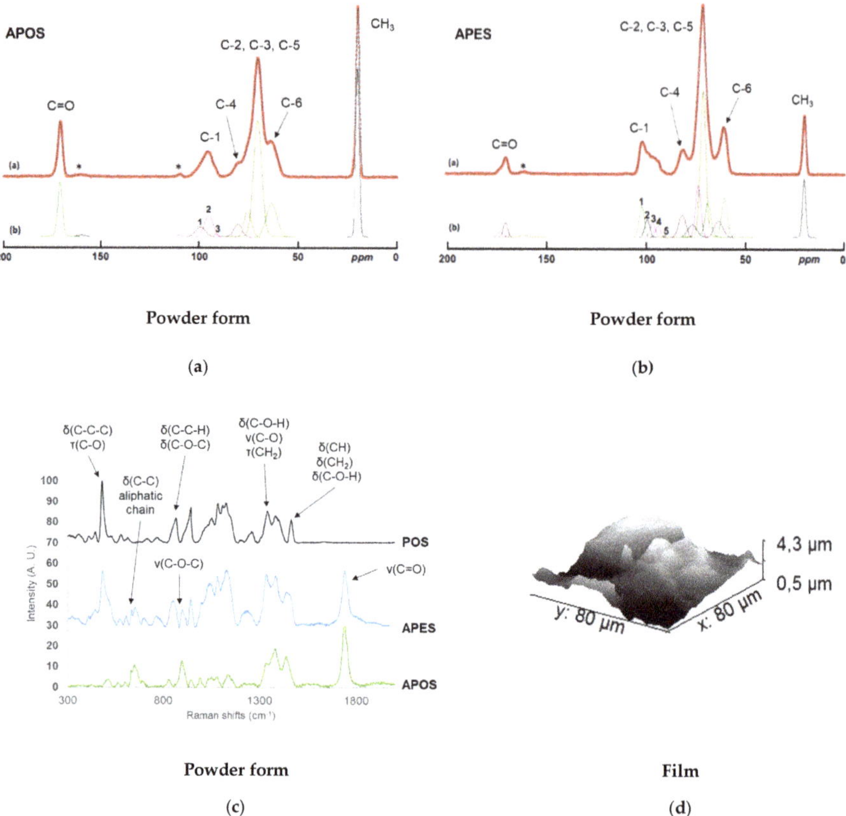

Powder form

(a)

Powder form

(b)

Powder form

(c)

Film

(d)

Figure 7. Bulk and surface characterization of APES and APOS Solid-state ^{13}C-NMR spectra obtained for APOS (**a**) and APES (**b**) and their spectral decomposition, Raman spectra of POS, APOS and APES (**c**), AFM topography image of APES (**d**). The DS of APOS and APES before deacetylation were 2.4 and 0.7, respectively. *Spinning side band.

Thus, AWMS and APOS were also acetylated in C-2 and C-3 whereas APES, which had a lower DS, was acetylated only in C-6. This privileged substitution on C-6 had already been identified for acetylated rice starches with DS < 1 [22] and was in agreement with the reactivity of hydroxyl groups: the primary hydroxyl group on C6 was the most reactive and accessible, followed by the secondary hydroxyl groups on C2 and C3 [37].

The decomposition of C-1 signal allowed the identification of local short-range organization of starches. PES showed three peaks for C-1, like WMS, standing for A-type double helices, and POS two peaks (results not shown), indicative of B-type double helices, as expected [27,29]. For APOS the decomposition of C-1 peak gave the same three resonances typical for amorphous starches found for AWMS but with different proportions (Figure 6a): 64% of total signal corresponded to typical for amorphous material very unfavorable and/or constrained conformations linked to the acetylation points on C6, C2 and C3 (95.4 ppm), 33% of total signal to typical for B-type double helices (99.6 ppm) and 3% of total signal to reducing extremities (91.6 ppm). This confirmed the absence of crystallinity in APOS like in AWMS. Moreover the higher amount of amorphous material very unfavorable and/or constrained conformations found in AWMS and the lower B-type double helices could be explained by the absence of amylose in AWMS, and by the high amount of very short chains found in AWMS (as chains with DP < 10 could not be involved in double helices), respectively. Besides APES showed the most defined C-1 signal which was deconvoluted onto five main resonances (Figure 7b): 102.5 ppm typical for V-type single helices in amorphous starches and branching points for amylopectin (41%), 99.6 ppm corresponding to B-type double helices or paracrystallinity (24%), 96.9 ppm typical for constrained α-1,6 linkages (19%), 94.2 ppm typical for constrained α-1,4 linkages (15%) and 91.2 ppm corresponding to reducing extremities (1%). APES had then a higher local order than the two other acetylated starches which could be related to its lower macromolecular degradation (Figure 6; Figure 7 and also attested by a lower amount of reducing extremities) and a lower acetylation. Increasing the DS would then produce not only the destruction of crystallinity but of local order in amorphous starch as well. Though C-1 signals showed higher local organization, C-6 signal for APES was typical of amorphous starch, indicating that the substitution was preferentially on amorphous regions, near branching points, confirming the hypothesis made by Shogren [38] for low substituted corn starch.

The surface characterization of APOS and APES was made by Raman spectroscopy (Figure 7c). The introduction of acetyl groups was visible (1740 cm^{-1}) and the disappearance of CCC vibration (~480 cm^{-1}) observed for APOS, as well as for AWMS, but not for APES, which allowed to follow the crystalline structure loss when DS increased. APES spectrum only contained C-O-C bonds and had a band around 1260 cm^{-1} attributed to V-helices [31]. Raman spectroscopy confirmed on the surface area the supramolecular information obtained in bulk by solid-state ^{13}C-NMR and X-ray diffraction (results not shown). Moreover, Raman spectroscopy imaging showed a homogeneous repartition of acetyl groups on the surface whatever the DS (results not shown).

The surface composition of starch films at the µm scale, approached by TOF-SIMS, revealed a uniform distribution of acetyl groups for APOS and APES (Supplementary Materials S4) and confirmed that all starch granules have been destroyed during the acetylation process independently of the DS and the amylose content. Under AFM, APES film surface showed large nodules (approximately 2 µm, Figure 7d) and surface roughness, consistent with the observations made by Hong et al. [39] on cassava acetylated starches with low DS. These authors showed that higher roughness can prove a good compatibility despite a heterogeneous distribution of hydrophobic groups on starch matrix. On the opposite, AWMS film surface did not reveal any particular topography (Figure 4c). According to the previous hypothesis on low-acetylated starch surface, this could indicate a more reliable compatibility when higher substitution in spite of a probable covering of starch matrix by hydrophobic groups.

2.3. Characterization of Etherified Starches

The four starch ethers were studied according to the analytical strategy used for starch acetates.

2.3.1. Degree of Substitution

The DS of starch ethers were determined by liquid (liq) ^{1}H-NMR (DS$_{liqNMR}$), for the modified starches soluble in NMR solvents and for the others by elemental analysis (EA) (Table 3).

Table 3. Degrees of substitution of starch ethers determined by different techniques.

Starch Ethers	DS$_{HDo.liqNMR}$ [a] (DMSO-d$_6$)	DS$_{HDo.EA(C)}$ [b]	DS$_{HDo.EA(H)}$ [c]	DS$_{HDo.liqNMR}$ [d] (CDCl$_3$ and THF-d$_8$)	DS$_{HPhe.liqNMR}$ [e] (CDCl$_3$ and THF-d$_8$)
HDo-POS-1	0.40 ± 0.05	0.39 ± 0.05	0.41 ± 0.06	ND [f]	ND [f]
HDo-POS-2	ND [g]	1.60 ± 0.09	1.58 ± 0.11	ND [f]	ND [f]
HDo-HPhe-POS-1	ND [g]	-	-	1.45 ± 0.07	0.10 ± 0.02
HDo-HPhe-POS-2	ND [g]	-	-	1.67 ± 0.04	0.19 ± 0.03

[a] Number of 2-hydroxydodecyl groups (HDo) per AGU determined by ^{1}H-NMR in DMSO-d6, [b] by elemental analysis from wt% of C, [c] by elemental analysis from wt% of H, [d] by ^{1}H-NMR in CDCl$_3$ and THF-d8. [e] Number of 2-hydroxyphenethyl groups (HPhe) per AGU determined by ^{1}H-NMR in CDCl$_3$ and THF-d8. [f] Products insoluble in chloroform and in tetrahydrofuran. [g] Products insoluble in dimethylsulfoxide. ND: Not determined.

A DS$_{HDo.liqNMR}$ of 0.40 (average number of 2-hydroxydodecyl grafts per AGU) was determined for HDo-POS-1 by ^{1}H-NMR analysis of the product dissolved in DMSO-d6 using the formulas reported in Supplementary Materials S6. This DS of 0.40 was confirmed for HDo-POS-1 by EA, using the formulas demonstrated in Supplementary Materials S5 from %C (DS_(HDo.EA(C)) = 0.39) and from %H (DS_(HDo.EA(H)) = 0.41). A DS of 1.6 was determined for HDo-POS-2 (insoluble in NMR solvents) by EA with the same formulas (DS = 1.60 from %C and 1.58 from %H).

By analyzing the ^{1}H-NMR spectra of the two mixed starch ethers in $CDCl_3$ and in THF-d8, the DS corresponding to 2-hydroxydodecyl/2-hydroxyphenethyl groups (calculated for the two types of grafts as demonstrated in Supplementary Materials S7) were found to be equal to 1.45/0.10 for HDo-HPhe-POS-1, and to 1.67/0.19 for HDo-HPhe-POS-2.

2.3.2. Macromolecular Characteristics

Etherified starches were insoluble in water and soluble or partially soluble in DMSO and THF (Supplementary Materials Table S1b). Because of these poor solubilities, we were able to analyze the macromolecular characteristics of HDo-POS-1 only without removing the grafted groups. After a first solubilization in pure DMSO, the DMSO content in the solution was reduced to 5% with water addition (final concentration of HDo-POS-1: 0.5 mg mL^{-1}) in order to be able to carry on the macromolecular characteristics analysis of HDo-POS-1 by aqueous AF4-MALLS with good elution recovery (91%). HDo-POS-1 exhibited a minor peak at 20 mL and a major peak at 14 mL (Figure 8). This latter corresponded to a large amount of small macromolecules, produced by the hydrolysis of the amylopectin chains during the etherification process. On the other hand, the residual fraction of amylopectin in HDo-POS-1 showed a higher molar mass distribution (Figure 8) and a higher $\overline{R}_G$ (241 nm) than the POS amylopectin (Figure 8, $\overline{R}_G$ of 179 nm), in line with its later elution (20 mL instead of 19 mL for POS amylopectin) which signed for a higher hydrodynamic radius. Thus, even if the chemical modification induced differences in the dn/dc which could considerably affect the results obtained with MALLS, the increase of size of the amylopectin fraction after chemical modification was obvious. Moreover, the residual amylopectin fraction seemed to be densified, as for the same elution volume, i.e., the same macromolecular size, it exhibited a higher molar mass. This increase of size and densification could not be solely due to an increase of molar mass caused by grafting on the amylopectin chains; more likely it could be explained by a supramolecular structuration of the prototypes in the aqueous medium thanks to hydrophobic interactions between the newly grafted chains. The prototype macromolecules could indeed aggregate by forming denser structures with a hydrophobic core with the fatty chains and a hydrophilic ring corresponding to the amylopectin chains.

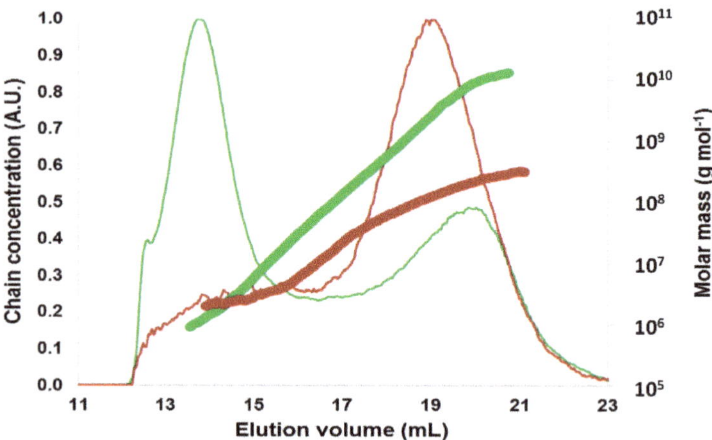

Figure 8. Macromolecular characteristics of native and etherified potato starches obtained by AF4-MALLS. Elugrams, i.e., chain concentration versus elution volume (thin lines), and molar mass distributions (thick lines) for POS (dark red), and HDo-POS-1 (green). Native potato (POS) starch and corresponding etherified starch with DS 0.40: HDo-POS-1.

2.3.3. Composition, Localization of Ether Groups and Supramolecular Structure

Figure 9a shows the solid-state ^{13}C-NMR spectra of starch ethers. Even if some typical signals for fatty chain (12.5–32.5 ppm) were observed for both HDo and HDo-HPhe grafted starches, the signal at 127.5 ppm corresponding to the phenyl was only visible for HDo-HPhe-POS-1 (DS 1.45/0.10), which did not allow determining the DS from these spectra. Moreover, contrary to ester starches with carbonyl function that influenced carbon shifts depending on the location, ether did not have characteristic shifts thus it was not possible to localize functionality.

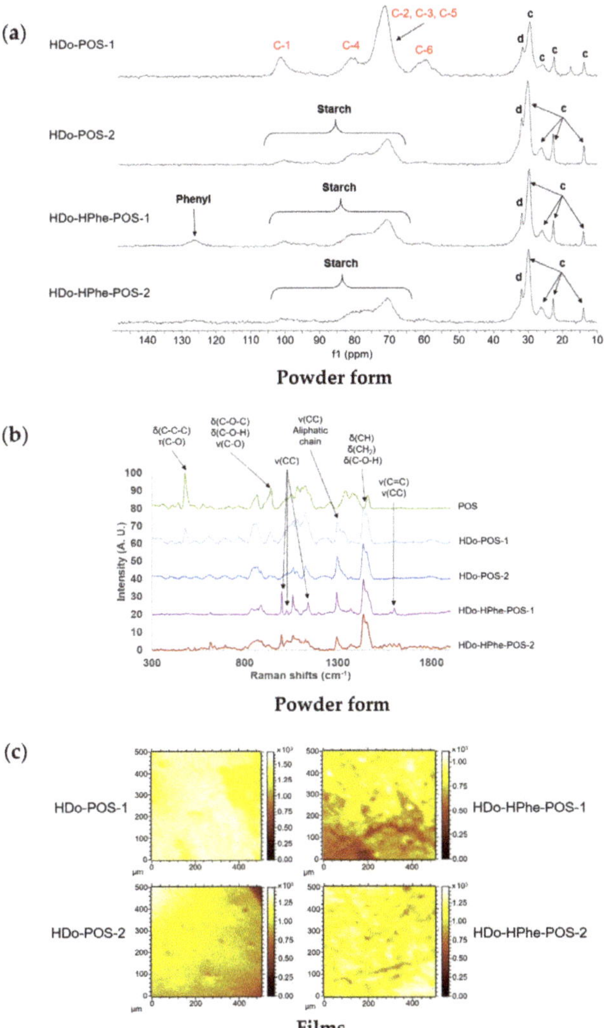

Figure 9. Bulk and surface characterization of ethers produced from potato starch. (**a**) Solid-state ^{13}C-NMR spectra obtained for HDo-POS-1, HDo-POS-2, HDo-HPhe-POS-1 and HDo-HPhe-POS-2 and their spectral decomposition, (**b**) Raman spectra of potato starch (POS) and corresponding ether starches (HDo-POS-1, HDo-POS-2, HDo-HPhe-POS-1 and HDo-HPhe-POS-2), (**c**) TOF-SIMS 2D mapping of POS ethers films. The DS of HDo-POS-1, HDo-POS-2, HDo-HPhe-POS-1 and HDo-HPhe-POS-2 were 0.40, 1.60, 1.45/0.10, and 1.67/0.19, respectively.

As the signal resolution of starch on solid-state ^{13}C-NMR spectrum was low, we could not deconvolute the spectra and detect the characteristic resonances of semi-crystalline or amorphous starch, except for HDo-POS-1, which was less substituted. C-1 signal deconvolution of HDo-POS-1 showed the major presence of V-type single helices (102.5 ppm) which could be the result of the complexation of the linear chains of D-glucosyl units linked in α-1,4 with the 2-hydroxydodecyl grafted groups. In HDo-POS-1, C-1 and C-6 signals showed typical amorphous starch pattern, as well as C-4 signal and C-2, C-3, C-5 peak [27] which signed for crystalline structure loss during chemical modification as was also observed for acetylated starches. This was in agreement with the absence of crystallinity observed by X-Ray diffraction for all these etherified starches [40].

On Raman spectra (Figure 9b), we identified characteristic vibrations from aliphatic chains (1080–1125, 1300, 1450 cm^{-1}) and supplementary aromatic linkages for HDo-HPhe-POS-1 and HDo-HPhe-POS-2 (999, 1025 and 1601 cm^{-1}). HDo-POS-1 only exhibited the organized-polysaccharide C-C-C vibration at 480 cm^{-1}, this was consistent with well-resolved starch signals in solid-state ^{13}C-NMR and signed for a higher structuration of the amorphous starch in this sample. In addition, a typical vibration at 1260 cm^{-1} corresponding to V-type helices was detected in both HDo-POS samples, with a lower signal in HDo-POS-2, but was absent in HDo-HPhe-POS samples. The amorphous structuration of chemically modified samples at the surface would then decrease with DS by analogy with acetylated samples, and probably decrease as well with the size of the grafted group.

Films surface composition determined by TOF-SIMS (Figure 9c and Supplementary Materials S4) for HDo-POS-1 and HDo-POS-2 exhibited uniform distribution of polysaccharide and alkyl fragments without phase separation indicating that its repartition did not depend on composition and reagents ratio. TOF-SIMS spectra and 2D mapping of HDo-HPhe-POS-1 and HDo-HPhe-POS-2 showed also uniformly distributed fragments from alkyl chains (upon tetramer) (Figure 9c and Supplementary Materials S4) and phenyl derivatives even in low intensity, probably because of a low substitution on surface (Supplementary Materials S4). The smaller and more intense zones of about 20 μm observed in HDo-HPhe-POS-1 maybe caused by a difference in the height of the surface.

3. Materials and Methods

3.1. Materials

Native waxy maize (WMS), potato (POS) and pea (PES) starches containing 0, 22 and 35% amylose respectively were obtained from Roquette Frères (Lestrem, France). Three corresponding acetylated starches, named AWMS, APOS and APES respectively, were produced by Roquette Frères in homogeneous media, as previously described by Quettier [41]. APOS was obtained with a basic catalyst whereas the others were produced with an acid catalyst. Isoamylase (EC 3.2.1.68) was from Sigma-Aldrich (St. Louis, MO, USA). Four starch ethers were obtained by reaction of POS with epoxides in basic aqueous media by classical protocols [42–44]. Among them, two 2-hydroxydodecyl POS (HDo-POS-1 and 2) were synthesized by reaction of POS with different amounts of 1,2-epoxydodecane (1,2-EDD), and the two others are mixed ethers of 2-hydroxydodecyl 2-hydroxyphenethyl POS named HDo-HPhe-POS-1 and 2 which were obtained by reactions of POS with two different mixtures of 1,2-EDD and styrene oxide (SO) (Figure 1). The structural and physical characterizations of materials were made on 200 μm thick films elaborated from raw powder by compression molding at 200 °C. For pea acetate starch, 40% wet basis of water was added to the raw powder prior to compression molding.

3.2. Methods

The degree of substitution (DS) of modified starches were determined for acetylated starches by titration as previously described [22] and NMR, and for etherified starches by elemental analysis (Supplementary Materials S6) and NMR.

3.2.1. Nuclear Magnetic Resonance Measurements

The DS were determined by liquid-state ^{1}H-NMR using a 300 MHz or 400 MHz spectrometer (Bruker, Wissembourg, France) after solubilization of the samples in deuterated dimethylsulfoxide (DMSO-d_6) at 80 °C for acetylated starches as previously described [18] and for etherified starches (Supplementary Materials S6); after solubilization in tetrahydrofuran (THF-d_8, ambient temperature) or in chloroform (CDCl$_3$, ambient temperature) (Supplementary Materials S7). Solid-state ^{13}C-NMR experiments were carried on a Bruker Avance III 400 WB NMR spectrometer operating at 100.62 MHz for ^{13}C (B$_0$ = 9.4 T), equipped with a double-resonance H/X CP-MAS 4-mm probe for Cross-Polarization Magic Angle Spinning (CP-MAS) solid-state experiments. The samples were packed in 4 mm ZrO2 rotors. The magic-angle-spinning (MAS) rate was fixed at 9 kHz and each acquisition was recorded at room temperature (293 °K). Chemical shifts were calibrated with external glycin, assigning the carbonyl at 176.03 ppm. The chemical shift, half-width and area of peaks were determined using a least-squares fitting method using the Peakfit® software (Systat Software Inc., San Jose, CA, USA). The DS of acetylated starches were determined with solid-state ^{13}C-NMR based on the methods developed in liquid-state ^{1}H-NMR [18].

3.2.2. Determination of size, Molar Mass and Chain Length Distributions

- Pretreatment of the samples

Acetylated starches (10 mg) were first deacetylated in KOH 1 M (1 mL) for 1 day at room temperature under mild stirring. This procedure also allowed their solubilization without degradation of the molar mass of starch polysaccharides. Actually, this procedure was carefully adjusted in order to avoid macromolecular degradation of the chains by β-elimination mechanism due to the pH of this solvent. In particular, stirring, temperature (4 °C and 25 °C) and time of solubilization (1 h, 3 h, 24 h, 2 days, 5 days and 7 days) were tested (results not shown). Samples solutions were then diluted 10 times with water. Native starches were first solubilized in DMSO, precipitated with ethanol, dried and then solubilized in water by microwave heating under pressure, as previously described [24]. Etherified starches were first solubilized in DMSO at 10 mg mL^{-1}, and then diluted 20 times with water, as they were soluble in DMSO/water mixture. All the starches solutions were filtered on 5 μm Durapore^{TM} membranes (Waters, Bedford, MA, USA) before analysis.

- Determination of macromolecular characteristics

Molar mass and radius of gyration distributions were determined by AF4-MALLS. The AF4 equipment was an Eclipse system (Wyatt Technology Corporation (WTC), Santa Barbara, CA, USA) with a AF4 long channel (tip–to–tip length of 291 mm) placed in a ThermosPRO oven regulated at 25 °C and an ISO-3000SD pump (Thermo Scientific, Waltham, MA, USA). A 350 μm polyester spacer and a regenerated cellulose membrane with a cutoff of 10 KDa (Merck Millipore, Darmstadt, Germany) were used. The carrier (0.02% NaN$_3$ in Millipore water) was filtered through a 0.1 μm Durapore membrane (Millipore), and degassed. All samples were introduced into the channel using an autosampler WPS-3000SL (Thermo Scientific) and eluted with a flow method adapted from Rolland-Sabaté et al. [21]. The cross flow (F$_c$) was set at 0.84 mL min^{-1} for the sample introduction (injection at 0.20 mL min^{-1} for 300 s) and the focusing/relaxation period (300 s). F$_{out}$ was then set at 0.84 mL min^{-1} and F$_c$ decreased in 480 s from 0.4 to 0.05 mL min^{-1}, maintained 600 s at 0.05 mL min^{-1}, and finally 300 s at 0 mL min^{-1}. Two on-line detectors were used: a MALLS instrument (Dawn® HELEOS™, WTC, Santa Barbara, CA, USA) fitted with a K5 flow cell and a GaAs laser (λ = 658 nm), and an refractometer (Optilab, WTC, Santa Barbara, CA, USA) operating at the same wavelength (WTC). Solubilization and elution recovery rates, and acquisition and data processing using ASTRA® software from WTC were established as previously described [21,24].

- Determination of chain length (CL) distributions

Deacetylated starches solubilized at 1 mg mL^{-1} in 0.1 M KOH were neutralized with 0.1 M HCl and debranched using isoamylase (591 U g^{-1} of dry starch) at 40 °C for 24 h in 50 mM acetate buffer pH 3.6 with 0.02% NaN$_3$ and 0.1% BSA. The CL distributions of debranched deacetylated starches were examined by high-performance anion-exchange chromatography coupled with pulsed amperometric detection (HPAEC-PAD) (ICS-500+, ThermoFisher), using a Carbopac PA200 (4 mm × 250 mm) column. The elution process involved 500 mM NaOH (eluent A), 1 M NaOAc (eluent B) and water eluted at 0.4 mL min^{-1}. The elution gradient was composed of 30% eluent A and (i) 0–30 min with a linear gradient from10% to 25% eluent B and 60% to 45% water, (ii) 30–52.5 min with a second linear gradient from 25% to 34% eluent B and 45% to 36% water, (iii) 52.5–82.5 min with a third linear gradient from 34% to 40% eluent B and 36% to 30% water, (iv) 82.5–85 min with a fourth linear gradient from 40% to 50% eluent B and 30% to 20% water. The concentration of each chain was determined by using the linear relationship between the detector response per mole of α(1,4) chains and CL. The linear curve coefficients were determined from maltooligosaccharide standards of DP1 to 7 and were used for longer compound quantification. It should be noticed that as in HPAEC-PAD response coefficients decrease with increasing DP, the % area of long chains is not representative of the exact weight fraction of each DP.

3.2.3. Topology and Surface Chemical Composition

Raman spectroscopy analyzes were carried out on powders and films by a LabRAM Visible Raman micro-spectrometer (Horiba Jobin-Yvon, Kyoto, Japan), with a He-Ne laser (λ = 632.8 nm) and a network of 600 t mm^{-1}. The device was equipped with a confocal microscope (Olympus, Tokyo, Japan) with a ×100 magnification lens. Spectral resolution was 1.1 cm^{-1} per pixel. The data were acquired and processed with NGSLabSpec (version 5.45.09).

Time of flight secondary ion mass spectroscopy (TOF-SIMS) spectra were obtained on films using a TOF SIMS5 (IONTOF GmbH, Münster, Germany) comprising a pulsed ion source (Bi^{3+}) with a current of 0.35 pA and with charge compensation. Both positive and negative secondary ion spectra were collected for each sample with a range of m/z = 0–200 Da and accumulated from 50 scans. Three spectra were recorded for each sample on 3 different areas of 500 μm × 500 μm with 128 × 128 pixels each.

Atomic force microscopy (AFM) analyzes were carried out on films by a NTEGRA atomic force microscope (NT-MDT, Moscow, Russia), under air, with a semi-contact mode (lever and tetrahedral silicone tip 14 to 16 μm high and doped with antimony). The resonance frequency was 320 kHz.

4. Conclusions

Macromolecular and molecular characterizations of highly substituted starches by conventional methods are a real challenge because of their low solubility in aqueous solvents and their resistance to enzyme action. We developed new analytical strategies (Figure 10) allowing, for the first time, the determination of macromolecular parameters, chemical composition and topography for highly acetylated and etherified starches with various substituents showing the almost universal character of this analytical approach which can thus help the design of eco-friendly materials.

In particular, despite limitations linked to the solubility of the samples, we showed that the NMR technique, used in its liquid and solid aspects, allowed obtaining the DS of low to highly modified starches and localizing grafted groups. Moreover, using solid-state ^{13}C-NMR we showed that the loss of crystallinity as well as the modification of the amorphous structure depended on the process, the DS and the nature of the grafted groups. Finally, thanks to AF4-MALLS, which permitted the size and molar mass distributions analysis of starches with high molar mass, we showed for the first time that the chemical modification process caused reduction of molar mass depending on the process, the DS, the nature of the grafted groups and the amylose content for high molar mass starches with DS from 0.4 to 3.

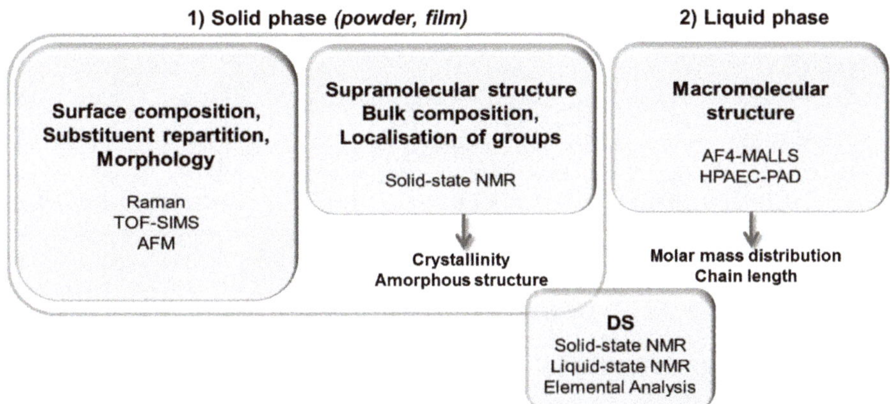

Figure 10. Designed strategy for multi-scale and multi-state characterization of high molar mass chemically-modified starches with 0.4 < DS ≤ 3.

Supplementary Materials: The following are available online, Table S1a: Solubility of acetylated starches, Table S1b: Solubility of etherified starches; Figure S2: FTIR spectra of AWMS and deacetylated AWMS, Table S3: Chain length distribution of debranched native and acetylated starches obtained from HPAEC-PAD; S4: Surface composition of acetylated and etherified starches studied by TOF-SIMS; S5: Determination of the degree of substitution (DS) of the 2-hydroxydodecyl potato starches (HDo-POS-1 and HDo-POS-2) by elemental analysis (EA); S6: Determination of the degree of substitution (DS) of HDo-POS-1 by 1H-NMR analysis; S7: Determination of the degree of substitution (DS) of the 2-hydroxydodecyl 2-hydroxyphenethyl potato starches (HDo-HPhe-POS-1 and HDo-HPhe-POS-2) by 1H-NMR analysis.

Author Contributions: Conceptualization, A.R.-S., A.M., S.T., E.M., V.W.; methodology, C.V., A.R.-S., F.B., M.C.-F., A.M.; validation, A.R.-S., A.M., M.C.-F., H.B., S.T., E.M.; formal analysis, C.V., A.G., F.B., M.C.-F., X.F.; investigation, C.V., A.G., F.B., X.F., C.Q.; resources, C.Q., N.D., V.W., A.G., H.B., S.T., E.M.; data curation, C.V., A.G., F.B., M.C.-F.; writing—original draft preparation, C.V.; writing—review and editing, A.R.-S., E.M., A.M., M.C.-F., S.T., X.F.; visualization, A.R.-S, C.V.; supervision, A.R.-S., A.M., S.T., E.M.; project administration, A.R.-S.; funding acquisition, A.R.-S. All authors have read and agreed to the published version of the manuscript.

Funding: This research was funded by Agence Nationale de la Recherche (ANR) under the "Investissements d'Avenir" Program, grant number: ANR-10-IEED-0004-01.

Acknowledgments: The authors acknowledge Sophie Guilois for excellent technical assistance and Philippe Looten for helpful discussions. Solid state NMR experiments were performed at the BIBS facility (UR1268 BIA, IBiSA, Phenome-Emphasis-FR (grant number ANR-11-INBS-0012)).

Conflicts of Interest: The authors declare no conflict of interest. The funders had no role in the design of the study; in the collection, analyses, or interpretation of data; in the writing of the manuscript, or in the decision to publish the results.

References

1. Pérez, S.; Bertoft, E. The molecular structures of starch components and their contribution to the architecture of starch granules: A comprehensive review. *Starch Stärke* **2010**, *62*, 389–420. [CrossRef]
2. Avérous, L.; Halley, P. Biocomposites based on plasticized starch. *Biofuels Bioprod. Biorefining* **2009**, *3*, 329–343. [CrossRef]
3. Xie, F.; Luckman, P.; Milne, J.; McDonald, L.; Young, C.; Tu, C.Y.; Di Pasquale, T.; Faveere, R.; Halley, P.; Xie, F. Thermoplastic Starch. *J. Renew. Mater.* **2014**, *2*, 95–106. [CrossRef]
4. Tomasik, P.; Schilling, C.H. CHEMICAL MODIFICATION OF STARCH. In *Advances in Carbohydrate Chemistry and Biochemistry*; Elsevier BV: Amsterdam, The Netherlands, 2004; Volume 59, pp. 175–403.
5. Ashogbon, A.O.; Akintayo, E.T. Recent trend in the physical and chemical modification of starches from different botanical sources: A review. *Starch Stärke* **2013**, *66*, 41–57. [CrossRef]

6. Miao, M.; Li, R.; Jiang, B.; Cui, S.W.; Zhang, T.; Jin, Z. Structure and physicochemical properties of octenyl succinic esters of sugary maize soluble starch and waxy maize starch. *Food Chem.* **2014**, *151*, 154–160. [CrossRef] [PubMed]

7. Simsek, S.; Ovando-Martínez, M.; Whitney, K.; Bello-Pérez, L.A. Effect of acetylation, oxidation and annealing on physicochemical properties of bean starch. *Food Chem.* **2012**, *134*, 1796–1803. [CrossRef]

8. Sun, S.; Ganwei, Z.; Ma, C. Preparation, physicochemical characterization and application of acetylated lotus rhizome starches. *Carbohydr. Polym.* **2016**, *135*, 10–17. [CrossRef]

9. Lee, S.; Kim, S.T.; Pant, B.R.; Kwen, H.D.; Song, H.H.; Lee, S.K.; Nehete, S.V. Carboxymethylation of corn starch and characterization using asymmetrical flow field-flow fractionation coupled with multiangle light scattering. *J. Chromatogr. A* **2010**, *1217*, 4623–4628. [CrossRef]

10. Modig, G.; Nilsson, L.; Bergenståhl, B.; Wahlund, K.-G. Homogenization-induced degradation of hydrophobically modified starch determined by asymmetrical flow field-flow fractionation and multi-angle light scattering. *Food Hydrocoll.* **2006**, *20*, 1087–1095. [CrossRef]

11. Nilsson, L.; Leeman, M.; Wahlund, K.-G.; Bergenståhl, B. Mechanical Degradation and Changes in Conformation of Hydrophobically Modified Starch. *Biomacromolecules* **2006**, *7*, 2671–2679. [CrossRef]

12. Heins, D.; Kulicke, W.-M.; Käuper, P.; Thielking, H. Characterization of Acetyl Starch by Means of NMR Spectroscopy and SEC/MALLS in Comparison with Hydroxyethyl Starch. *Starch Stärke* **1998**, *50*, 431–437. [CrossRef]

13. Chen, Z.; A Schols, H.; Voragen, A.G. Differently sized granules from acetylated potato and sweet potato starches differ in the acetyl substitution pattern of their amylose populations. *Carbohydr. Polym.* **2004**, *56*, 219–226. [CrossRef]

14. Huang, J.; Schols, H.A.; Klaver, R.; Jin, Z.; Voragen, A.G. Acetyl substitution patterns of amylose and amylopectin populations in cowpea starch modified with acetic anhydride and vinyl acetate. *Carbohydr. Polym.* **2007**, *67*, 542–550. [CrossRef]

15. Richardson, S.; Gorton, L. Characterisation of the substituent distribution in starch and cellulose derivatives. *Anal. Chim. Acta* **2003**, *497*, 27–65. [CrossRef]

16. Wang, Y.-J.; Wang, L. Characterization of Acetylated Waxy Maize Starches Prepared under Catalysis by Different Alkali and Alkaline-Earth Hydroxides. *Starch Stärke* **2002**, *54*, 25–30. [CrossRef]

17. Chen, Z.; Huang, J.; Suurs, P.; Schols, H.A.; Voragen, A.G. Granule size affects the acetyl substitution on amylopectin populations in potato and sweet potato starches. *Carbohydr. Polym.* **2005**, *62*, 333–337. [CrossRef]

18. Volkert, B.; Lehmann, A.; Greco, T.; Nejad, M.H. A comparison of different synthesis routes for starch acetates and the resulting mechanical properties. *Carbohydr. Polym.* **2010**, *79*, 571–577. [CrossRef]

19. Bello-Pérez, L.A.; Agama-Acevedo, E.; Zamudio-Flores, P.B.; Mendez-Montealvo, G.; Rodriguez-Ambriz, S.L. Effect of low and high acetylation degree in the morphological, physicochemical and structural characteristics of barley starch. *LWT* **2010**, *43*, 1434–1440. [CrossRef]

20. Cave, R.A.; Seabrook, S.A.; Gidley, M.J.; Gilbert, R.G. Characterization of Starch by Size-Exclusion Chromatography: The Limitations Imposed by Shear Scission. *Biomacromolecules* **2009**, *10*, 2245–2253. [CrossRef] [PubMed]

21. Rolland-Sabaté, A.; Guilois, S.; Jaillais, B.; Colonna, P. Molecular size and mass distributions of native starches using complementary separation methods: Asymmetrical Flow Field Flow Fractionation (A4F) and Hydrodynamic and Size Exclusion Chromatography (HDC-SEC). *Anal. Bioanal. Chem.* **2010**, *399*, 1493–1505. [CrossRef]

22. Colussi, R.; Pinto, V.Z.; El Halal, S.L.M.; Vanier, N.L.; Villanova, F.A.; Silva, R.; Zavareze, E.D.R.; Dias, A.R.G. Structural, morphological, and physicochemical properties of acetylated high-, medium-, and low-amylose rice starches. *Carbohydr. Polym.* **2014**, *103*, 405–413. [CrossRef] [PubMed]

23. Elomaa, M. Determination of the degree of substitution of acetylated starch by hydrolysis, 1H NMR and TGA/IR. *Carbohydr. Polym.* **2004**, *57*, 261–267. [CrossRef]

24. Rolland-Sabaté, A.; Colonna, P.; Mendez-Montealvo, M.G.; Planchot, V. Branching Features of Amylopectins and Glycogen Determined by Asymmetrical Flow Field Flow Fractionation Coupled with Multiangle Laser Light Scattering. *Biomacromolecules* **2007**, *8*, 2520–2532. [CrossRef]

25. Rolland-Sabaté, A.; Sánchez, T.; Buleon, A.; Colonna, P.; Jaillais, B.; Ceballos, H.; Dufour, D. Structural characterization of novel cassava starches with low and high-amylose contents in comparison with other commercial sources. *Food Hydrocoll.* **2012**, *27*, 161–174. [CrossRef]

26. Gidley, M.J.; Bociek, S.M. Molecular organization in starches: A carbon 13 CP/MAS NMR study. *J. Am. Chem. Soc.* **1985**, *107*, 7040–7044. [CrossRef]

27. Paris, M.; Bizot, H.; Emery, J.; Buzaré, J.; Buleon, A. Crystallinity and structuring role of water in native and recrystallized starches by 13C CP-MAS NMR spectroscopy. *Carbohydr. Polym.* **1999**, *39*, 327–339. [CrossRef]

28. Chevigny, C.; Foucat, L.; Rolland-Sabaté, A.; Buleon, A.; Lourdin, D. Shape-memory effect in amorphous potato starch: The influence of local orders and paracrystallinity. *Carbohydr. Polym.* **2016**, *146*, 411–419. [CrossRef]

29. Paris, M.; Bizot, H.; Emery, J.; Buzaré, J.; Buléon, A. NMR local range investigations in amorphous starchy substrates I. Structural heterogeneity probed by (13)C CP-MAS NMR. *Int. J. Boil. Macromol.* **2001**, *29*, 127–136. [CrossRef]

30. David, A. Etude de dérivés d'amidon: Relation entre la structure et le comportement thermomécanique. Ph.D. Thesis, Université de Lille, Lille, France, 2017.

31. Kizil, R.; Irudayaraj, J.; Seetharaman, K. Characterization of Irradiated Starches by Using FT-Raman and FTIR Spectroscopy. *J. Agric. Food Chem.* **2002**, *50*, 3912–3918. [CrossRef]

32. Phillips, D.L.; Liu, H.; Pan, D.; Corke, H. General Application of Raman Spectroscopy for the Determination of Level of Acetylation in Modified Starches. *Cereal Chem. J.* **1999**, *76*, 439–443. [CrossRef]

33. Chung, H.-J.; Liu, Q. Impact of molecular structure of amylopectin and amylose on amylose chain association during cooling. *Carbohydr. Polym.* **2009**, *77*, 807–815. [CrossRef]

34. Jane, J.; Chen, Y.Y.; Lee, L.F.; McPherson, A.E.; Wong, K.S.; Radosavljević, M.; Kasemsuwan, T. Effects of Amylopectin Branch Chain Length and Amylose Content on the Gelatinization and Pasting Properties of Starch. *Cereal Chem. J.* **1999**, *76*, 629–637. [CrossRef]

35. Ratnayake, W.; Hoover, R.; Shahidi, F.; Perera, C.; Jane, J. Composition, molecular structure, and physicochemical properties of starches from four field pea (Pisum sativum L.) cultivars. *Food Chem.* **2001**, *74*, 189–202. [CrossRef]

36. McPherson, A. Comparison of waxy potato with other root and tuber starches. *Carbohydr. Polym.* **1999**, *40*, 57–70. [CrossRef]

37. Xu, Y.; Miladinov, V.; Hanna, M. Synthesis and Characterization of Starch Acetates with High Substitution. *Cereal Chem. J.* **2004**, *81*, 735–740. [CrossRef]

38. Shogren, R.L. Scandium triflate catalyzed acetylation of starch at low to moderate temperatures. *Carbohydr. Polym.* **2008**, *72*, 439–443. [CrossRef]

39. Hong, J.; Zeng, X.; Buckow, R.; Han, Z.; Wang, M.-S. Nanostructure, morphology and functionality of cassava starch after pulsed electric fields assisted acetylation. *Food Hydrocoll.* **2016**, *54*, 139–150. [CrossRef]

40. Gilet, A. Développement de résines thermoplastiques pour fonctionnalisation de matières premières amyglacées. Ph.D. Thesis, Université d'Artois, Arras, France, 2016.

41. Quettier, C. Procédé de préparation de dérivés acétylés de matière amylacée. International Patent Application No. PCT/FR2010/052147, 11 October 2010.

42. Bien, F.; Wiege, B.; Warwel, S. Hydrophobic Modification of Starch by Alkali-Catalyzed Addition of 1,2-Epoxyalkanes. *Starch Stärke* **2001**, *53*, 555–559. [CrossRef]

43. Funke, U.; Lindhauer, M.G. Effect of Reaction Conditions and Alkyl Chain Lengths on the Properties of Hydroxyalkyl Starch Ethers. *Starch Stärke* **2001**, *53*, 547. [CrossRef]

44. Gilet, A.; Quettier, C.; Wiatz, V.; Bricout, H.; Ferreira, M.; Rousseau, C.; Monflier, E.; Tilloy, S. Synthesis of 2-Hydroxydodecyl Starch Ethers: Importance of the Purification Process. *Ind. Eng. Chem. Res.* **2018**, *58*, 2437–2444. [CrossRef]

Sample Availability: Samples of the compounds are not available from the authors.

Article

Influence of Drainage on Peat Organic Matter: Implications for Development, Stability, and Transformation

Lech W. Szajdak [1,*], Adam Jezierski [2], Kazimiera Wegner [3], Teresa Meysner [1] and Marek Szczepański [1]

[1] Institute for Agricultural and Forest Environment, Polish Academy of Sciences, 60-809 Poznań, Poland; teresa.meysner@isrl.poznan.pl (T.M.); morfeush@tlen.pl (M.S.)
[2] Faculty of Chemistry, University of Wrocław, 50-383 Wrocław, Poland; adam.jezierski@chem.uni.wroc.pl
[3] Faculty of Agriculture and Biotechnology, University of Science and Technology, 85-029 Bydgoszcz, Poland; kwegner@utp.edu.pl
* Correspondence: lech.szajdak@isrl.poznan.pl; Tel.: +48-61-8475-601

Received: 20 April 2020; Accepted: 25 May 2020; Published: 2 June 2020

Abstract: The agricultural use of peatlands, the stabilization of the substrate for building or road construction or for increasing the capacity of soil to support heavy machinery for industrial activities (peat and petroleum extraction), harvesting to provide peat for energy, and the growing media and initiation of chemical processes must be preceded by drainage. As a consequence of drainage, peat underwent an irreversible conversion into moorsh (secondary transformation of the peat). The object of the study was to investigate comparatively the organic matter composition and molecular structure of humic acids (HAs) in the raised bog, fen, and peat-moorsh soils developed in various compositions of botanical cover, peat-forming species, and oxic and anoxic conditions as a result of the oscillation of ground water during drainage as well as to evaluate the vulnerability of soil organic matter (SOM) to decomposition. Drainage was shown to be the principal factor causing the various chemical compositions and physicochemical properties of HAs. Large and significant differences in chemical composition of peat and the properties of HAs were found to be related to the degree of decomposition. The HAs from drained peatlands were less chemically mature. In contrast, the HAs from fen and raised bog were found to be more mature than that of the corresponding drained peatlands. The above findings showed the distinguishable structure of HAs within the soil profile created by the plant residue biodegradation and formed in both oxic and anoxic conditions. The analytical methods of thermal analysis together with the optical densities and paramagnetic behaviour are suitable and effective tools for studying structure–property relationships characterizing the origin and formation process of HAs in various environmental conditions.

Keywords: drained and undrained peatlands; peats; humic acids; thermal; paramagnetic and optical properties

---

1. Introduction

There are peatlands in every climatic zone. Total Earth's area of these elements of the landscape is more than 3 million km^2 (about 3% of the Earth's land surface). Peatlands accumulate some 70% of natural freshwater. Each year, they absorb 0.37 giga- tonnes of carbon dioxide (CO_2), which corresponds to a greater capacity for carbon storage than that attributed to all other vegetative organisms in the world [1].

The conversion of vegetable matter into peat structures changes depending on the vegetation type, ability of tissues to be decompose, deepness of roots and water table, pH, ionic strength, balance between humification and degradation processes and also depends on access to organic and inorganic compounds [2–10].

Over 20 forms of peat bogs, 19 forms of peat bogs and about 6 forms of swamps could be treated as peatlands [11]. Ombrotrophic bogs are mainly rain fed and have nutrient deficiency. Fens supplied by surface and ground water are richer in nutrients then bogs [12,13]. Bogs and fens grow in conditions of full water saturation what significantly slows down a degradation of organic matter.

The peatlands must undergo drainage for their application in agriculture as a soil substrate and for their stabilization as substrate in building and road construction as well as a support for heavy machinery for industry used e.g. in peat and petroleum extraction for energy producing [14–16].

In wetlands located in the cool climate based on an average storage rate of 200 kg C ha^{-1} and assuming an area of about 350×10^6 ha. The annual accumulation not disturbed wetlands has been calculated as 0.06–0.08 Pg C y^{-1} [17]. The total drained area in the period 1795–1980 was 8.219×10^6 ha; 5.5×10^6 ha; 9.4×10^6 ha for crops, pastures, and forests, respectively.

Up to 35×10^6 ha of wetlands has been drained [18]. However, in the following countries the rate of subsidence of drained organic soils ranged from 1 to 8 cm y^{-1} (The Netherlands: 1.75 cm y^{-1}, Quebec in Canada: 2.07 cm y^{-1}, Everglades in the United States: 3 cm y^{-1}, San Joaquin Delta in the United States: 7.6 cm y^{-1}, and Hula Valley in Israel cm y^{-1}). Drying shrinkage, loss of the buoyant force of groundwater, compaction, wind erosion, burning, and microbial oxidation belong to the main reasons for subsidence of histosols.

As reported by Terry [19] approximately 73% of the loss of surface elevation in Everglades histosols is caused by microbial oxidation. In addition, an assumed C release from drained wetlands by oxidation of the organic material of 10 t C ha^{-1} y^{-1} a global annual C release is 0.05 to 0.35 Pg C. In gleysols the long-term drainage of 106 ha impacts on an extra release of 0.01 Pg C y^{-1}. Total release from histosols and glaysols ranges from 0.03 to 0.37 Pg C y^{-1}.

In the period 1795–1980, about 4% of the wetland was drained in the tropics. In cool regions, the annual shift (loss of sink strength and gain of source strength) in the global C balance is 0.063–0.085 Pg C due to drainage of histosols. However, including tropical histosols, the global shift would be 0.15–0.184 Pg C y^{-1} [18]. Under cultivation the potential to increase C levels in soils is largely promoted to upland soils. Restoring C sinks in wetland soils drained artificially is improbable unless they are transferred from agricultural production to natural wetlands [20].

At present time, a total amount of decomposed worldwide peatlands is equal to 65 million ha. The drainage is the key driver of the degradation of peat soils [21]. From 14 to 20% of peatlands in the world and 14% in Europe are applied for agriculture. The meadows and pastures are the great majority of peatlands used for agriculture in Europe. The percentage use of European peatlands for cultivation is as follows: Hungary (98%), Greece (90%), The Netherlands (85%), Germany (85%), and Poland (70%).

In countries of other continents e.g. Canada the national resources of peatlands were mostly undrained and forested; only 15% were drained and used for agriculture. In USA, more than 230,000 hectares of fen in the Florida Everglades are mainly used for cultivation of sugar cane and rice. 20% of Indonesia's peatlands were drained and utilized for agriculture. In recent years due to increasing nature protection and for economic cause the total area of agricultural peatlands has reduced [12,22].

Drainage enhances the emission of CO_2 and N_2O and decreases emission of methane from the peat. The evolution from 15 to 17 Mg CO_2 from grassland and of 41.1 Mg CO_2 from ploughed fens was determined. In addition, drained peatlands emit of nitrous oxide with fluxes varying from 2 to 56 kg N_2O–N ha^{-1} year^{-1}, while CH_4 fluxes ranged from −4.9 to 9.1 kg ha^{-1} year^{-1} [23].

Drainage, besides the impacts on the mineralization of organic matter and the evolution of gasses reduces also the diversity of peatland vegetation and favours forest plant species. The first species to decrease and vanish are those that thrive on wet lawn and flark levels. The plants related to drier hummocks should adapt to changing conditions and initially even benefit from drainage. Later, the growth of tree stands and increased shade will limit their opportunities to thrive. More mature tree stands will also lose more water through evapotranspiration, increasing the drying-out effect, and accelerating the mineralization of organic matter [21,24]. The rate of the changes caused by drainage will depend on factors including: the concentration of mineral and organic nutrients, availability of moisture, the efficiency of the drainage ditches, and intensity of tree growth [24,25]. Typical members of the plant kingdom are resistant to drying stress [26] whereas wetter and more nutrient-rich peatlands are changed more dramatically [27,28].

Maslov et al. [29] published the basic information on eighty peatland experiments which were established at various times within the territory of the Russian Federation. The aims of these field experiments were to study the peatland ecosystem as well as site transformation under the influence of drainage, forestry, and agricultural use. Materials obtained on the experimental sites include multidiscipline investigations and characterize wetlands in general, peat soil properties, drainage system arrangements, etc.

Annual raising of the Carex peat in eastern Europe is from 0.5 to 1 mm per year, while in northern mires, the Carex peat layer had been growing slowly and thus almost two meters of peat has accumulated over 9800 years (~0.2 mm per year) [30]. The drainage of peatlands impacts on the lowering of this element of the landscape. In New Zealand and in Norway the lowering is 3.4 cm year^{-1}, and 2.5 cm year^{-1}, respectively [31,32]. Thus, deposits that have accumulated over many millennia can disappear over a time scale that is very relevant to human activity.

Last advances in peatland restoration methods have led to the establishment of *Sphagnum* moss on the remnant cutover peat surface following peat extraction; however, evaluating restoration success remains a key issue. The study of Lucchese et al. [33] showed an increase of organic matter accumulation from 2.3 ± 1.7 cm 4 years post-restoration to 13.6 ± 6.5 cm 8 years post-restoration. For comparison, at an adjacent non-restored section of the peatland organic matter, accumulation was significantly lower ($p < 0.001$ for all years), with mean thicknesses of 0.2 ± 0.6 and 0.8 ± 1.2 cm for 24 and 28 years post-extraction, respectively.

The changes of oxic conditions in peatlands as a result of the oscillation of ground water during melioration activates the irreversible conversion of peat into moorsh (synonyms: peat-moorsh soils, secondary transformed peat, and mucks) which is usually treated as the secondary transformation of peat [34]. This conversion is responsible for the changes in the structure of organic mass constituting these soils, causing modification of the properties of high molecular weight substances from hydrophilic to hydrophobic, the disappearance of peatland, and a decrease of anisotropy of peat deposit [35–38]. As the result of the above processes the soil's abilities to swell again, to disperse, and hence to re-soak are lost as well. The moorsh formed from peat seems to be fine-grained, more colloidal, and degraded due to particle size and a higher percentage of mineral matter [39,40].

The irreversible loss of wettability due to drying is responsible for damage of colloidal behavior in peat. The secondary transformation of peat showed the disintegration of the thermodynamic equilibrium in peat. The decline in peat soil moisture content resulting from drainage implicates shrinkage of the peat structures. Volume change from shrinkage is generated by several forces acting at the microscale, whereby its mechanism and magnitude differ from those in mineral (clay) soils. Drying and wetting of peat soils giving soil volume changes is manifested in soil vertical movement and bulk density changes [41–43]. In addition, biotic and abiotic conversions and degradation of peat organic matter is observed as effect of drainage [44–47]. Säurich et al. [15] indicated that bog peat samples tend to be more sensitive to anthropogenic disturbance than fen peat samples.

Drainage is the main direct cause of fen habitat degradation, either due to reclamation of fen or the changes in the water flow within fen systems. Reducing of the water level in peatlands activates anaerobic conditions into aerobic ones and accelerates the peat mineralization. In the first period after drainage, this causes usually an increase of nutrient availability, especially nitrogen and phosphorus, which are released during mineralization [48]. However, the raised fertility is usually only a short-term effect [49], and therefore, additional fertilization is needed to sustain economically prospective agricultural production on drained fens, which has a further negative impact on biodiversity. Apart from increasing nutrient availability, drainage also lowers the water storage capacity of peat soils, making them more susceptible to water-table fluctuations and droughts. A further aspect of fen habitat degradation is acidification. This may be related to drainage, which results in partial replacement of groundwater by rainwater [50], and to the increased atmospheric deposition of nitrogen and sulphur compounds [51].

In general, drainage of peat leads to the progressive differentiation of the hydrophobic peptides and total amino acid content in organic matter. In proteins of peats, hydrophobic contacts exist between hydrophobic and hydrophilic structural elements (between the side chains of the radicals of phenylalanine, leucine, isoleucine, valine, proline, methionine, and tryptophan). Hydrophobic forces stabilize the tertiary structure of proteins and determine the properties of lipids and biological membranes. The presence of amino acids, hydrocarbon chains, and other nonpolar fragments in their composition are related to hydrophobic properties of humic substances [37,52].

Since organic matter is a major component of the soil phase of peat and moorsh soil causing soil water repellency, it is important to study the effect of chemical soil properties on their wettability. It was observed that the significant changes of chemical properties of transformed organic matter in peat have a significant influence on the sequential modifications in physical and hydraulic features created by lowering of water table for agriculture. Van Dijk [53] postulated relationship between a high increase of shrinkage, changes in the number of many chemical and physical properties and humic components in peats [53].

The retention of water by peat can be considered in terms of reaction of water molecules with the surfaces of peat particles. The phenomenon is therefore amenable to the analytical methods of colloid and surface chemistry. Considerable attention should be paid to the chemical composition of peat, identifying the molecular structure of substances and aggregates most likely to hold water strongly. Knowledge of this structure can provide a rationale for treatments intended to remove or render less water-retentive the most hydrophilic fraction.

The substances of greatest immediate interest are humic acids (HAs). HAs are created in peats by degradation, poly-condensation, polymerization, and poly-addition of organic substances as a result of habitat and anthropogenic processes, including the degradation of plants and animal residues which are characterized by a complex macromolecular structure with aromatic and aliphatic units; peptide chain; and nitrogen in aliphatic, cyclic, and aromatic forms. The HAs represent macromolecular polydisperse biphyllic systems, including both hydrophobic domains (saturated hydrocarbon chains and aromatic structural units) and hydrophilic functional groups, i.e., having an amphiphilic character. The hydrophilicity of peat surfaces is generally attributed to the availability of organic functional groups capable of hydrogen-bonding. Such groups, well known to organic chemists, include carboxyls as well as phenolic and alcoholic hydroxyls. HAs have been regarded as peat component principally responsible for water retention in peat. The hydrophilicity of HAs depend not only on the numbers of hydroxyl and other polar units but on their ability to form hydrogen-bonding with water as there are hydroxyl groups inaccessible to water.

These substances are formed of similar but not identical substrates; therefore, no two HAs are identical. The HAs from various types of peats and organic and inorganic components of HAs matrices are altered to different extents, significantly differing both in composition and properties. The quantity and quality of HAs in soils organic matter depend on the balance between primary productivity and the rate of decomposition [38]. Chemical properties and structural characteristics of humic substances were shown to be better predictors of soil organic matter turnover rate in vertisols than soil organic matter content.

This suggests the possibility of using humic substances as indicators of soil organic matter turnover because they are sources of intermediates and energy for many chemical and biochemical pathways in the soil [39,40].

The study outlined here was conducted to propose tools and analytical methods for the quantitative and qualitative evaluation of the turnover processes occurring in the oxic and anoxic conditions of developed peat deposits.

The aim of this study is to analyse comparatively the organic matter composition and molecular structure of HAs in the raised bog, fen and peat-moorsh soils developed in various composition of botanical cover, peat-forming species, and oxic and anoxic conditions as a result of the oscillation of ground water during drainage as well as to evaluate the vulnerability of soil organic matter (SOM) to decomposition.

2. Methods

2.1. Study Sites

Six sites from 3 peatlands were found to vary according to their macrofossil analysis, their state of decomposition, and their GPS parameters, and an outline of their location has been given (Figures 1 and 2, Table 1).

Figure 1. Location of the study sites.

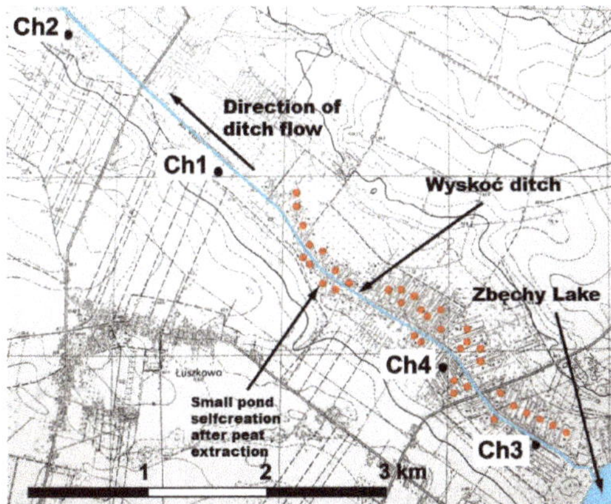

Figure 2. Location of the sites Ch1, Ch2, Ch3, and Ch4 (peat-moorsh soils—Turew).

Table 1. The classification of peat (WRB 2015), type of peat, degree of decomposition, and chemical properties of peat soils.

Sampling Site		GPS	Classification of Peat (WRB 2015)	Depth (cm)	Type of Peat Based on Macrofossil Analysis	Degree of Decomposition (Von Post)	TOC (g kg^{-1})	C_{WHE} (g kg^{-1})	N-total (g kg^{-1})	C/N
Baltic-type raised bog	Kusowo	53° 48′ 07.83″ N 16° 32′ 42.03″ E	Ombric Hemic Fibric Histosols (Dystric)	0–25	Sphagnum	H2	570.05 ± 24.83	15.42 ± 1.59	12.69 ± 1.53	44.92 ± 5.37
				25–50	Cotton grass-Sphagnum	H3/H4	575.66 ± 31.34	12.51 ± 1.31	10.83 ± 0.94	53.15 ± 7.30
				50–75	Sphagnum, cotton grass-Sphagnum	H3	590.44 ± 36.83	12.43 ± 0.38	9.80 ± 0.75	60.25 ± 8.86
				75–100	Cotton grass-Sphagnum	H4/H5	596.56 ± 23.60	11.50 ± 1.03	9.42 ± 0.61	63.33 ± 7.51
Fen	Stążka	53° 36′ 17.58″ N 17° 57′ 20.38″ E	Ombric Hemic Fibric Histosols (Dystric Lignic)	0–25	Sedge-Hypnum	H3	501.94 ± 37.17	12.49 ± 2.29	20.01 ± 1.26	25.08 ± 4.30
				50–75	Sedge-Hypnum	H4	557.51 ± 47.54	5.39 ± 0.72	18.75 ± 2.23	29.73 ± 5.78
				75–100	Sedge	H5	581.50 ± 42.23	5.41 ± 0.47	18.11 ± 2.51	32.11 ± 6.57
Peat-moorsh soils, Turew	Ch1	52° 01′ 35.45″ N 16° 52′ 34.80″ E	Rheic Murshic Sapric Histosols (Limnic Dystric)	0–25	Moorsh soil	H8	294.04 ± 31.96	8.81 ± 0.64	19.19 ± 3.73	15.32 ± 3.19
				25–50	Sedge	H8	445.85 ± 35.80	7.80 ± 0.48	24.38 ± 1.71	18.29 ± 2.51
				50–75	Sedge	H7	498.20 ± 33.03	6.09 ± 0.64	26.66 ± 1.49	18.69 ± 2.16
				75–100	Sedge	H7	496.60 ± 43.24	5.79 ± 0.44	25.62 ± 1.45	19.38 ± 2.25
	Ch2	52° 02′ 21.70″ N 16° 51′ 09.50″ E	Rheic Murshic Sapric Histosols (Lignic)	0–25	Moorsh soil	H8	395.39 ± 35.77	15.89 ± 1.25	31.03 ± 1.65	12.74 ± 2.72
				25–50	Alder swamp	H8	455.81 ± 45.72	10.15 ± 0.89	27.66 ± 2.84	16.48 ± 2.40
				50–75	Sedge	H8	497.46 ± 31.90	7.27 ± 0.70	24.86 ± 2.18	20.01 ± 2.67
				75–100	Sedge	H8	496.11 ± 47.08	5.06 ± 0.53	21.69 ± 2.39	22.87 ± 2.16
	Ch3	52° 00′ 57.50″ N 16° 53′ 49.75″ E	Rheic Sapric Dystric Histosols (Calcic Limnic)	0–25	Moorsh soil	H8	246.63 ± 21.40	7.62 ± 1.45	19.51 ± 2.23	12.64 ± 2.10
				50–75	Alder swamp	H8	466.20 ± 32.60	7.48 ± 0.97	24.50 ± 2.30	19.03 ± 3.55
				75–100	Sedge with wooden	H8	484.90 ± 32.20	5.74 ± 0.68	23.97 ± 1.52	20.23 ± 3.13
	Ch4	52° 01′ 12.61″ N 16° 53′ 23.38″ E	Rheic Sapric Dystric Histosols (Limnic Lignic)	0–25	Moorsh soil	H7	370.90 ± 33.70	11.58 ± 0.91	24.81 ± 3.86	14.95 ± 2.76
				50–75	Sedge	H8	471.21 ± 31.04	8.75 ± 0.82	25.83 ± 2.01	18.24 ± 2.53
				75–100	Sedge	H8	488.46 ± 31.70	7.28 ± 0.69	25.02 ± 1.14	19.52 ± 2.03

TOC—total organic carbon; C_{WHE}—hot water extractable organic carbon.

2.2. Collection of Smples (WRB classification 2015)

Peat samples were collected from the following:

a)   Baltic-type raised bog (Kusowo)
b)   fen (Stążka)
c)   peat-moorsh soils: Ch1, Ch2, Ch3, and Ch4 (Turew)

Peat samples were collected in triplicate from the field using a 5.0-cm diameter Instorf peat auger and various depths from 0 to 100 cm in the stratigraphic profile of each peat deposit, transported to the laboratory at ca. 4 °C and stored at −20 °C. The samples were dried at 20 °C and homogenized in a grinder after removal of any visible live plant material, after which they were passed through a 1-mm sieve to remove rock fragments and large organic debris. The botanical composition of peat was analyzed by the microscopic method and subsequently classified according to the Polish standards (PN-76/G-02501 1977). Peat samples were used for the description, classification, and physicochemical analysis [54] (Table 1).

The Kusowo Bog is situated in the West Pomeranian Voivodship. This is likely the best-preserved Polish Baltic-type raised bog. The reserve has an area of 326.56 ha and is entirely included into the Szczecineckie Lake Natura 2000. The mean peat thickness may exceed 12 m. The mean annual air temperature is 7.2 °C. The mean annual precipitation is 760.1 mm. (Table 2) (Figure 1). The age of peatland is 710 years (Table 3). Moisture contents of peat samples ranged from 89.69% to 92.31% (Table 5). Among raised bogs, that of the Baltic-type is distinguished, one peculiar to a humid climate, with a high rainfall. The bog, formerly a lake, developed on a moraine with kames in an extensive depression. The bog is clearly divided into two parts—northern and southern—separated by three mineral mounds.

Table 2. Annual precipitation from 2009 to 2019 (mm).

Sampling Site	Year											Average
	2009	2010	2011	2012	2013	2014	2015	2016	2017	2018	2019	
Peat-moorsh soils Ch1-4 Turew	576.0	698.1	460.6	534.5	552.3	410.2	485.7	793.3	565.9	425.8	429.9	539.3
Baltic-type raised bog Kusowo	754.2	938.9	673.8	758.4	643.0	661.9	630.5	837.9	1092.2	647.1	723.4	760.1
Fen Stążka	556.0	835.6	487.4	652.9	510.0	472.0	485.4	701.0	885.1	456.0	546.0	598.9

The highest average annual precipitation was measured for Baltic-type raised bog (Kusowo) at 760.1 mm, and the lowest was measured for peat-moorsh soils at 539.3 mm.

Table 3. The radiocarbon ^{14}C dates for Baltic-type raised bog Kusowo.

Depth [cm]	Age ^{14}C Date	Calibrated Range 95.4%	BC/AD
9–10	106.87 ± 0.33 pMC	1694–1919	AD
16–17	111.4 ± 0.36 pMC	1692–1919	AD
33–34	80 ± 30 BP	1690–1926	AD
53–54	40 ± 40 BP	1690–1925	AD
65–66	310 ± 30 BP	1485–1650	AD
76–77	375 ± 30 BP	1446–1633	AD
90–91	550 ± 30 BP	1310–1435	AD
98–99	555 ± 30 BP	1310–1431	AD

The northern part is better preserved, with a gently sloping "living" dome, about 3–4 m high, with some peat ponds. The dome is mostly composed of *Sphagnum* hummocks and hollows of the plant association *Sphagnetummagellanici*. In places with high water content, e.g., near peat ponds, the *Rhynchosporetumalbae*, *Eriophoroangustifolii-Sphagnetumrecurvi*, and *Caricetumlimosae* are located. Nearly half of the northern part consists of wooded habitats, primarily pine and birch bog forests (*Vacciniouliginosi-Pinetum* and *Vacciniouliginosi-Betuletumpubescentis*) [55,56] (Figure 1)

The Stążka fen is a part of the "Bagna nad Stążką"—a mire complex located in Northern Poland in the region of the Tuchola Forest. The fen is part of the Nature Reserve, where the whole complex of natural peatlands is under protection. This peatland measures some 478.45 ha. The maximum thickness of peat deposits is 1.4 m. This region was characterized by a mean annual air temperature of 7.2 °C and by mean annual precipitation 598.9 mm (Table 2) [26]. The age of peatland is 1400 years (Table 4). Moisture contents of peat samples ranged from 93.15% to 93.92% (Table 5).

Table 4. The radiocarbon ^{14}C dates for fen Stążka.

Depth (cm)	Age ^{14}C Date	Calibrated Range 95.4%	BC/AD
12–13	120.43 ± 0.4 pMC	1689–1928	AD
25–26	195 ± 30 BP	1648–1955	AD
40–41	170 ± 30 BP	1659–1954	AD
54–55	155 ± 30 BP	1666–1953	AD
67–68	1005 ± 30 BP	977–1153	AD
83–84	1125 ± 30 BP	783–991	AD
90–91	1295 ± 70 BP	620–890	AD
105–106	1295 ± 35 BP	655–779	AD

BC—before Christ; AD—anno Domini.

Table 5. Moisture content of sampling sites.

Sampling Site		Depth (cm)	Moisture (%)
Baltic-type raised bog (Kusowo)		0–25	91.57 ± 1.92
		25–50	92.31 ± 3.92
		50–75	89.69 ± 1.21
		75–100	91.82 ± 1.44
Fen (Stążka)		0–25	93.92 ± 1.84
		50–75	93.15 ± 3.98
		75–100	93.17 ± 0.86
Peat-moorsh soils (Turew)	Ch1	0–25	66.03 ± 7.55
		25–50	77.62 ± 4.27
		50–75	83.21 ± 2.59
		75–100	83.55 ± 2.60
	Ch2	0–25	74.44 ± 5.88
		25–50	81.18 ± 4.25
		50–75	82.15 ± 1.44
		75–100	79.24 ± 1.73
	Ch3	0–25	61.31 ± 4.49
		50–75	79.80 ± 2.67
		75–100	82.33 ± 1.90
	CH4	0–25	70.29 ± 5.38
		50–75	86.82 ± 1.97
		75–100	86.04 ± 1.47

Drainage significantly impacted the concentration of the moisture in peats. The moisture contents in undrained peats (90–94%) were higher than in drained peats (61–83%) (Table 5).

Peat-moorsh samples were taken from four chosen sites marked as Ch1, Ch2, Ch3, and Ch4 on the 4.5-km long transect of peatland located in the Chłapowski Agro-ecological Landscape Park of the West Polish Lowland, about 40 km southwest of Poznań. These are the sites investigated along the Wyskoć Canal. The thickness of peat deposit ranges from 1.5 to 2.75 m. The mean annual air temperature is 8.4 °C, the mean annual precipitation is 539.3 mm (Table 2), and the growing season is about 200 days (Figures 1 and 2). Moisture content of peat samples ranged from 61% to 86.82% (Table 5).

All examined samples were classified as raised bog, fen, and peat-moorsh soils according to World Reference Base Soil Resources (2015) [57].

2.3. Organic Carbon, Nitrogen, C/N, and Degree of Decomposition

Total carbon (TC) and inorganic carbon (IC) concentrations in dried peat samples were analyzed by means of a Total Organic Carbon Analyzer (TOC 5050A) with a Solid Sample Module SSM-5000A, Shimadzu, Japan. Total Organic Carbon (TOC) was analyzed by placing about 50 mg of a soil sample in Total Organic Carbon Analyzer (TOC 5050A) with Solid Sample Module (SSM-5000A) produced by Shimadzu (Kyoto, Japan). Total organic carbon (TOC) was calculated as the difference between total and inorganic carbon.

Air-dried peat samples for the measurements of hot water extractable organic carbon (C_{HWE}) were mixed with deionized water and heated at 100 °C for two hours under a reflux condenser. Extracts were filtered through 0.45-μm pore-size filters and analyzed on TOC 5050A facilities, Shimadzu, Japan [58].

Total nitrogen was evaluated by the Kjeldahl methods, using Vapodest 10s analyser (Gerhardt, Germany). N-total was evaluated by semimicro-Kjeldahl method; 0.5 g of soil sample was set in Kjeldahl digestion tube. Five mL of twice distilled water was added. After 30 min, 2.5 mL of 95% H_2SO_4 and powder zinc were added and heated 20 min. Fifteen mL of 95% H_2SO_4, 5 g of K_2SO_4, and a small amount of selenium mixture were added and heated at 350 °C. A Kjeldahl digestion tube was cooled to room temperature. Mineralized sample was set in a 100-mL volumetric flask and filled to the line by twice distilled water. A mineralized sample was centrifuged (MLW K 23 D, Germany) in 3000 rpm by 15 min. Twenty-five mL of the sample was set in Parnas–Wagner apparatus. Fifty mL of 30% NaOH was added. Fifty mL of distillate was collected to a 100-mL Erlenmeyer flask filled by 20 mL of 4% H_3BO_3 and indicator (bromocresol green and methyl red mixture). The blue distillate was titrated to red color by 0.02 N HCl.

The atomic ratios of C/N were calculated after determining the concentrations of carbon and nitrogen.

Peat decomposition was determined at the sampled depths according to the field squeezing method by means of the von Post classification scale [59]. This method identifies ten classes of decomposition, with H1 being undecomposed peat and H10 being completely decomposed peat. Peat type was determined based on plant macrofossil analysis [60].

2.4. Extraction and Purification of HAs

Isolation of HAs was performed according to the recommendations of the International Humic Substances Society procedure under relatively gentle conditions [61,62]. The humic substances were extracted from air-dried peat soils using the grain size fraction smaller than one millimeter with 0.1 M NaOH using an extractant/peat ratio of *v/v* 5:1) at pH 7.00 under a N_2 gas atmosphere and shaken for four hours and, after that, were stored overnight for the coagulation of HA fractions which were separated by centrifugation. The purification of the HAs was performed by the following method. The suspension was centrifuged at 4000× *g* at 24 °C for 1 h. The solution was acidified with 6 *M* HCl to pH 1.3 to precipitate the HAs and was allowed to stand overnight. Then, the solution was centrifuged to eliminate the supernatant. The procedure was repeated three times. Finally, precipitated HAs were freeze dried and stored in a vacuum desiccator over P_2O_5.

Moisture analyzer MAX series (Radwag, Poland) was used to determine hygroscopic humidity content in peats. The content of hygroscopic humidity was taken into account in the analysis of peats.

^{14}C dates were provided by the Poznan Radiocarbon Laboratory (Poland).

2.5. Elemental Composition

The content of C, N, H, and S in every HAs fraction was analysed using the Vario Micro Cube Elemental Analyser. The samples were burned in an oxygen atmosphere at 1150 °C. The gases (CO_2, N_2, H_2O, and SO_2) were separated chromatographically and measured by a thermal conductivity detector (TCD). Oxygen content was obtained by subtracting the sum of other elemental contents from 100%, and the results were expressed as percentages calculated on the basis of total element amounts (Table 6).

Table 6. Elemental analysis (wt.%) and atomic ratios of humic acids (HAs).

Sampling Site		Depth (cm)	C	H	N	O	S	H/C	C/N*	O/C
Baltic-type raised bog (Kusowo)	Kusowo	0–25	43.41	4.66	3.26	47.85	0.82	1.28	15.53	0.83
		25–50	41.79	4.24	2.56	50.78	0.63	1.21	19.04	0.91
		50–75	46.90	4.78	2.65	45.20	0.47	1.21	20.64	0.72
		75–100	47.04	4.74	2.74	45.10	0.38	1.20	20.02	0.72
Fen	Stążka	0–25	40.76	4.24	2.85	51.23	0.92	1.24	16.68	0.94
		50–75	44.01	4.14	2.64	48.55	0.66	1.24	19.44	0.83
		75–100	44.38	3.74	2.08	49.25	0.55	1.12	24.88	0.83
Peat-moorsh soils	Ch1	0–25	42.84	4.50	3.64	48.08	0.94	1.00	13.73	0.84
		25–50	41.62	3.98	2.86	50.66	0.88	1.25	16.97	0.91
		50–75	46.03	4.31	2.69	45.86	1.11	1.14	19.96	0.75
		75–100	44.63	4.27	2.86	46.90	1.34	1.12	18.20	0.79
	Ch2	0–25	36.09	3.73	2.86	56.36	0.96	1.14	14.72	1.17
		25–50	42.66	4.13	3.17	48.90	1.14	1.23	15.69	0.86
		50–75	42.61	4.07	2.93	49.00	1.39	1.15	16.96	0.86
		75–100	43.19	4.08	2.90	48.15	1.68	1.14	17.37	0.84
	Ch3	0–25	41.43	4.27	3.45	49.60	1.25	1.13	14.00	0.90
		50–75	42.13	4.10	3.03	48.80	1.94	1.23	16.21	0.87
		75–100	43.41	4.03	2.73	47.58	2.25	1.24	18.54	0.82
	Ch4	0–25	40.57	4.07	3.00	51.15	1.21	1.16	15.77	0.95
		50–75	43.21	4.11	3.07	48.21	1.40	1.11	16.41	0.84
		75–100	43.21	4.03	3.11	48.00	1.65	1.19	16.20	0.83

C/N* ratio in HAs.

2.6. VIS-Spectroscopy of HAs

E_4/E_6 ratios were determined by dissolving 3 mg of HAs in 10 mL of 0.05 M $NaHCO_3$ (pH = 9.0) and by measuring optical densities at λ = 465 nm (E_4) and λ = 665 nm (E_6) spectrophotometrically on SHIMADZU UVmini-1240 (Japan) with 1 cm thickness (Table 7) [63].

Table 7. Parameters of VIS-spectroscopy and electron paramagnetic resonance of HAs.

Sampling Site		Depth (cm)	E_4/E_6	g-Value	*Spin Concentration $\times 10^{17}$
BalticRaised bog	Kusowo	0–25	6.36	2.0035	1.29
		25–50	4.88	2.0029	3.13
		50–75	4.00	2.0021	3.92
		75–100	3.98	2.0023	5.64
Fen	Stążka	0–25	5.67	2.0035	1.04
		50–75	5.57	2.0036	1.38
		75–100	4.46	2.0036	3.41

Table 7. *Cont.*

Peat-moorsh soils	Ch1	0–25	6.78	2.0036	0.77
		25–50	6.12	2.0036	1.71
		50–75	5.94	2.0036	2.22
		75–100	6.19	2.0036	2.77
	Ch2	0–25	6.95	2.0035	2.20
		25–50	6.73	2.0036	1.23
		50–75	6.49	2.0036	1.43
		75–100	5.59	2.0036	1.12
	Ch3	0–25	6.65	2.0036	0.95
		50–75	6.14	2.0036	2.27
		75–100	5.18	2.0036	3.82
	Ch4	0–25	6.37	2.0035	1.48
		50–75	6.19	2.0035	3.34
		75–100	5.70	2.0035	4.30

g-value, the spin concentrations are given in spins/g units, e.g., the last value is 3.41×10^{17} spins/g organic matter.

2.7. Electron Paramagnetic Resonance of HAs

Electron paramagnetic resonance (EPR) X-band (9.8 GHz) was recorded at room temperature on a Bruker ElexSys E500 instrument equipped with an NMR teslameter ER 036TM and E 41 FC frequency counter. The concentration of spins was measured using a double integral procedure applying the Bruker WinEPR program. Leonardite was used as a spin concentration standard (International Humic Substance Society) [64].

2.8. Differential Thermal Analysis of HAs

Thermal properties of HAs were evaluated by means of an OD-103 derivatograph (MOM-Paulik-Paulik-Erdey, Hungary) [65]. The curves of differential thermal analysis (DTA), thermogravimetry (TGA), and differential thermogravimetry (DTG) were recorded simultaneously. Weight losses at different steps of thermal decomposition were calculated from the TGA curves. The Z index was calculated, and aliphatic character of HAs was pointed out. All chemical analyses were run in triplicate, and the results were averaged.

2.9. Statistical Analysis

The confidence intervals were calculated using the following formula: $\bar{x} \pm t_{\alpha\,(n-1)}$ SE, where: $\bar{x}$ is the mean; $t_{\alpha\,(n-1)}$ is the value of the Student test for $\alpha = 0.05$; $n-1$ is the degree of freedom; and SE is standard error. Linear correlations between the values were calculated. Normal distribution of the results and homogenous variances were checked before statistical analysis.

Principal component analysis (PCA) using Statistica version 9.1 was performed to determine the correlations between Baltic-type raised bog, fen, and peat-moorsh soils in the peat deposit properties and physicochemical properties of soil organic matter. The number of factors extracted from the variables was determined by a scree test according to Kaiser's rule. With this criterion, the first two principal components with an eigenvalue greater than a third were retained. Principal component analysis (PCA) using Statistica version 9.1 was performed to determine the correlations between raised bog, fen, and peat-moorsh soils and of peat deposits properties and physicochemical properties of soil organic matter.

3. Results and Discussion

3.1. Characteristics of Peats

3.1.1. Decomposition Degree, Carbon, Nitrogen, and C/N Ratio of Peat Soils

Decomposition Degree

Degree of decomposition is an important property of the organic matter in soils and other deposits which contain fossil carbon. It describes the intensity of transformation or the humification degree of the original living organic matter. Our knowledge of the degree of decomposition and chemical characteristics of peat soil may be translated into peat physicochemical properties, therefore, both the degree of decomposition and type of peats described above [7,66].

All peats were decomposed, however, in varying degrees (Table 1). A gradual increase of the degree of decomposition has been observed in the order from the Baltic-type raised bog (H2–H5) throughout fen (H3–H5) to peat-moorsh soils (H7–H8). A drop of the water level in the peat-moorsh soils increased the oxygen content and accelerated the mineralization of organic matter, leading to an increase of the degree of decomposition (H8). Therefore, a decrease in the degree of peat decomposition is treated as an indicator of a rise of the ground water table [67].

During decomposition of organic matter, labile compounds such as peptides and polysaccharides are preferentially decomposed while refractory aromatic or aliphatic compounds become residually enriched [68].

Almost all nitrogen in organic matter is in organic form (0–90%). Nevertheless, the chemical composition of nitrogen in organic soil fraction is not completely understood and little is known of the factors affecting the distribution of organic nitrogen forms in soils.

Humus is composed from 20% to 60% of HAs. The nitrogen (20–40%) in HAs consists of amino acids or peptides, the main unit of proteins, and is connected to the central core by hydrogen bonds [69]. Therefore, amino acids and proteins are the main fraction of total nitrogen in organic soils.

Soil amino acids are components of protein conglomerates. They occur in stable form. Soil proteins included in organic colloids are hydrophilic colloids. These colloids are water-related. Drainage causes the denaturation of colloids and a change in the properties of proteins from hydrophilic to hydrophobic. During the drainage process, a progressive increase in the hydrophobic amino acid content was observed. In proteins of peats, hydrophobic contacts exist between hydrophobic and hydrophilic structural elements (between the side chains of the radicals of phenylalanine, leucine, isoleucine, valine, proline, methionine, and tryptophan). The presence of amino acids, hydrocarbon chains, and other nonpolar fragments in their composition are related to hydrophobic properties of humic substances [37,52].

TOC

Significantly increasing concentrations of TOC exhibited a trend with a gradually increasing down-core for all investigated sites (Table 1). A lower TOC was determined in peat-moorsh soils in comparison with Baltic-type raised bog and fen. This relationship was found to be related to the degree of decomposition. These changes observed for the degree of decomposition, and TOC could have been caused by a change in any one of a number of the following factors, e.g., water level and the balance between accumulation/decomposition, indicating most likely that they are attributable to the combination of such factors.

The lower TOC content in the upper layer may be related to the increased rate of decomposition rather than accumulation. In deeper layers of peat profile under saturated conditions, lower decomposition rates of organic matter are observed.

Our results are in line with the results of Benavides [70], who showed that peat and carbon accumulation rates were lower in drained sites, indicating either greater decomposition rates of the upper peat column or lower production by the changed plant communities. The ecological services offered by peatlands to agrarian communities downstream are important. In addition, he observed that species composition was much affected by drainage, which resulted in a reduction in cover of *Sphagnum* and other peat-forming species, and by the encroachment of sedges and Juncus effusus. The ability of peat to store water and carbon was also reduced in drained peatlands. Vegetation records show a shift towards sedge-Juncus communities around 50 years ago when agricultural use of water increased.

Batjes [71], Chambers et al. [72], Fontaine et al. [73], and Wang et al. [74] suggested that a lack of a supply of fresh carbon may prevent the decomposition of organic carbon pool in deeper layers of peat deposits.

C_{HWE}

C_{HWE} is responsible for the microbiological activity in depth profile to be exuded from plant roots, acting as a significant source of carbon fueling microbial metabolism [75]. This fraction is a potential source of carbon and energy for heterotrophic organisms and contributes significantly to stream ecosystem metabolism. The flux of C_{HWE} from an ecosystem can be a significant component of carbon (C) budgets, especially in watersheds containing wetlands. In watersheds containing organic wetland soils or peatlands, the flux from the watershed can be 4–8% of annual net primary production, a significant fraction that should be addressed when performing a carbon mass balance [76].

A significant decline of C_{HWE} with increasing depth in all sites was observed (Table 1). The highest decrease of C_{HWE} (68.2%) with increasing depth was measured in Ch2, representing peat-moorsh soil, and can be attributable to the most degraded peat-moorsh soil.

In contrast, the lowest decrease of C_{HWE} (25.4%) with increasing depth was determined in the lowest decomposed peats of Baltic-type raised bog as compared to that of peat-moorsh soils. Thus, C_{HWE} can be said to increase gradually with the increase of peat during decomposition (Table 1).

The decreases in C_{HWE} concentrations were generally accompanied by the increases of the TOC in all sites of sampling. We found a consistent pattern in parallel changes as per decreasing/increasing C_{HWE} and TOC concentrations among all sites.

The values of the ratios C_{HWE}/TOC in undrained peat (raised bog and fen) ranged from 0.9 to 2.7, while in drained peat, it ranged from 1.2 to 4.1. The highest value of the ratio, 4.1, was measured in the Ch2, which showed the highest value of the degree of decomposition.

Monitoring the properties of C_{HWE} in the peat profile is frequently used to evaluate changes in peat quality and to explain shifts in peatland ecosystem function [77].

The purification of ground water by the transect consisting of peat-moorsh soils of 4.5 km length was observed (Figure 2). Peatland plays a positive function as a biogeochemical barrier, which reduces the content of chemical compounds moving in the groundwater throughout the peatland. In groundwater between Ch3 and Ch2, the concentrations of nitrates (38.5%), N-organic (10%), N-total (24.5%), ammonium (38.7%), dissolved total carbon (33.1%), dissolved total inorganic carbon (10%), and dissolved organic carbon (57.5%) were significantly decreased [78].

C/N

Another effective and common indicator of peat decomposition is the use of C/N ratios [79]. This indicator is based on the observed residual enrichment of N relative to C during mineralization of organic matter.

The concentrations of N-total in Baltic-type raised bog were significantly lower compared to the N-total in fen and in peat-moorsh soils (Table 1). Baltic-type raised bog receives water and nutrients primarily from atmospheric deposition.

Significant differences of C/N ratios between less decomposed Baltic-type raised bog, fen, and highly decomposed peat-moorsh soils were observed (Table 1). The C/N values in undrained peats were significantly higher as compared to that of corresponding drained peats. The highest C/N ratio and the lowest degree of decomposition in the Baltic-type raised bog was measured.

The high C/N ratios in the deeper layer of peat indicates more intensity in the accumulation of organic matter with the formation of mature HAs as compared to that in the upper layer. In contrast, the low values of the C/N in the surface layer of all peat-moorsh soils correspond to drier conditions and the intensity of organic nitrogen mineralization, leading to the formation of gaseous substances such as ammonia and the further emission of N_2O and N_2 into the atmosphere [80–82].

Baltic raised bog and fen represent two types peats with similar values of the degree of decomposition H2–H5 (Table 1). Baltic raised bog as nitrogen deficiency system was shown to have lower concentrations of N total and higher C/N ratios compared with that of the corresponding fen. Drainage was shown to be the principal factor causing the increase of the degree of decomposition and the decrease of C/N ratios. Peat-moorsh soils were created as a result of secondary transformation of fen. Therefore, in peat-moorsh soils, higher degrees of decomposition H7–H8, higher C/N ratios, and lower concentrations of TOC compared with that of the corresponding in fen were measured. For all peat-moorsh soils, the values of decomposition degree were shown to be in line with the results calculated for C/N ratios (Table 1).

Our results are in line with the results of Malmer and Holm [83] and of Kuhry and Vitt [84]. They pointed out that lower C/N ratios characterize more decomposed peat material.

According to Broder et al. [85], the lowest humification and high C/N ratios in *Sphagnum* bogs justified the high polyphenol content. On the other hand, Rice and MacCarthy [86] and Anderson [81] pointed out that the C/N ratio does not always reflect soil organic matter mineralization and, moreover, that the percentage share of C_{WHE} would seem to be a better indicator of organic matter mineralization as compared to the C/N ratio (expanding the trend in the C/N ratios with depth), which is bound to hydrological changes of nitrogen concentration.

3.2. Analytical Data of HAs

3.2.1. Elemental Composition of HAs

One of the most fundamental characteristics of HAs is its elemental composition expressed in the weight and atomic percentage structure of particular elements [86,87].

The C content ranged from 41.79% to 47.04%, from 40.76% to 44.38%, and from 36.09% to 46.03% in Baltic-type raised bog, fen, and peat-moorsh soils, respectively (Table 6). The C concentration increased with an increase in the depth profile for highly decomposed peat, while the opposite was shown for oxygen content, which decreased in all peat profiles with increasing depth.

The H and N amounts here demonstrated higher variability.

The S concentration decreased with increasing depth in Baltic-type raised bog and fen. In contrast, the concentrations of S increased with increasing depth in peat-moorsh soil compared to the increasing depth in Baltic-type raised bog and fen.

The ratio of H/C, C/N* (in HAs), and O/C bears much more diagnostic information than the elemental composition of HAs. The magnitude of the H/C ratio has been used to indicate the degree of aromaticity (a small value) or aliphaticity (a large value) of a substance [52]. The values of H/C ratio in studied HA samples are in the range from 1.20 to 1.28 in Baltic-type raised bog, from 1.12 to 1.24 in fen, and from 1.00 to 1.25 in peat-moorsh soils (Table 6). However, these H/C and C/N* values among peat soils are not significantly different.

The decrease of the H/C ratio in HAs with increasing depth in undrained peats (Baltic-type raised bog and fen) is related to the peat accumulation rate and development of the aromatic structures, where more labile structures are destroyed or transformed into the appearance of more stable aromatic and polyaromatic structures.

On the other hand, in drained peat (peat-moorsh soil) with increasing depth, the increase of the H/C in HAs increased, indicating less amounts of the C in HAs as compared to that of undrained peats (Baltic-type raised bog and fen).

The C/N* and O/C ratios increased with increasing depth. The high values of C/N* in the deeper layer are related to the presence of proteinaceous materials of living organic matter, indicating a higher intensity of humification than mineralization process with the formation of mature HAs (Table 6).

The significant negative correlations between the O/C vs. C/N* (r = −0.448) and H/C vs C/N* ratio (r = −0.632) indicate that, with an increase in depth, the decarboxylation processes were in line with the increase of the N concentration in HAs (Table 9).

According to Van Krevelen [88], the H/C atomic ratios from 0.7 to 1.5 correspond to aromatic systems coupled with aliphatic chains and contain up to ten carbon atoms. The O/C ratio, for its part, is considered as an indicator of carbohydrate and carboxylic group contents and can be directly related to the aromatization of the peat-forming organic matter. DiDonato et al. [87] suggested that more of the carboxyl-containing aliphatic molecules are sourced from lignin.

3.2.2. VIS spectra of HAs

The E_4/E_6 ratio is a valid and informative index for the characterization of aromatic condensation and poly-conjugation in the humic molecule [35,89,90].

Table 7 shows that the E_4/E_6 ratio of HAs fraction in peat-moorsh soils ranged from 5.18 to 6.95; in fens, from 4.46 to 6.2; and in Baltic-type raised bog, from 3.98 to 6.36. In peat-moorsh soil profiles, the variability of this value was generally low. As expected, the E_4/E_6 ratio of HAs decreased with increasing depth profile in all investigated sites related to progressive humification and condensation of aromatic constituents. These findings are in line with our data of elemental analysis of HAs, indicating more chemically mature HAs in the bottom rather than in the upper layers.

Moreover, the E_4/E_6 trend was found to be higher in drained peats compared to E_4/E_6 from undrained peats. This reflects a lower degree of aromatic condensation and poly-conjugation and a lower degree of humification in the molecules of HAs from drained rather than undrained peats [88,91,92].

In this context, Kļaviņš and Sire [93] showed strong negative correlations between total acidity values and the E_4/E_6 ratios in peat bog profiles. They pointed out that an increase of the acidic groups in the HAs samples resulted in a reduced E_4/E_6 ratio.

3.2.3. Electron paramagnetic resonance of HAs

Our study shows the impact of peat type and the degree of decomposition on the EPR spectra, which consists of one signal with ΔH_{pp} at about 4 G typical for radical species (Figure 3) (Table 7).

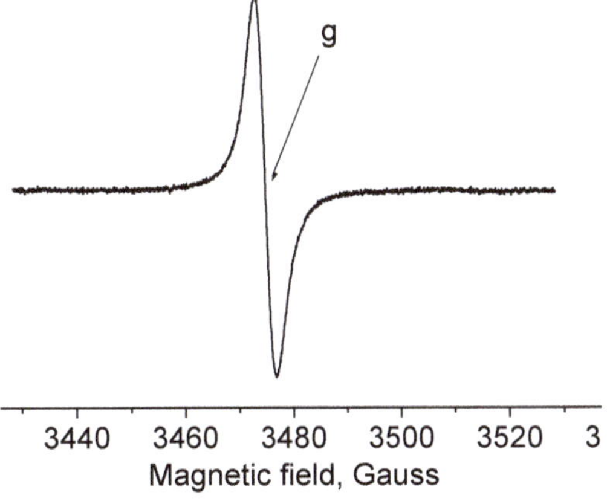

Figure 3. The electron paramagnetic resonance (EPR) spectra of the studied samples of HAs consists of one signal with ΔH_{pp} at about 4 G typical for the radical species.

A strong aromatization in the deeper layer is distinctly observed for the Baltic-type raised bog. The drop of oxidative conditions in the depth has led to an increase in EPR signal intensity for HAs (Table 7) and decline of the g-value from 2.0035 to 2.0021.

The maximum of spin concentrations determined in the deepest layers are in line with the VIS-spectra of HAs and reflects more chemically mature HAs from the Baltic-type raised bog compared with the fen and peat-moorsh soils.

Our results are in line with Czechowski and Jezierski [94], who observed a steady decrease in the g parameter towards the g = 2.0023 value of free electrons for free radicals in bituminous carbon components in parallel with an increase in the aromaticity of the free radicals. In addition, the effect of g parameter reduction (determined by EPR studies) of semiquinone radicals naturally occurring in HAs of increasing aromaticity (Knüpling at al. [95]) was confirmed and explained theoretically by Witwicki and Jezierska [96] on the basis of quantum mechanical calculations of g parameters for model semiquinone radicals.

Inspection of EPR spectra of HAs from fen and peat-moorsh soils (Ch1 and Ch3) reveals g-value characteristics for HAs (2.0036) [64]. The deeper layers of these deposits include an increasing concentration of semiquinone radicals (Table 7) connected with the formation of oxygen containing polymers without growing aromaticity, as the g-values are practically unchanged. In addition, the same g-values of the HAs in all deeper layers are related to similar redox conditions.

The sample Ch2 (peat-moorsh soil) reveals a rather unexpected change of the spin concentration that appeared to be highest in the upper layer. The oxidative conditions of the upper layer impacted the g-value and the spin concentration in this HAs. The g-value is smaller for the upper layer (2.0035) as compared to the g-values (the difference exceeds the experimental error) for lower layers (2.0036), showing a slightly greater aromaticity. This phenomenon is related to the high degree of decomposition of the upper layer; the most transformed peat soil from all studied peat-moorsh soils; the highest O/C ratio in molecule of HAs; and the highest TOC, C_{HWE}, and N-total (Table 1). In the context of CH2, due to a high in situ, the oxidative properties in the decomposition rate of the organic matter are faster than its accumulation.

3.2.4. Thermal Properties of HAs

All shapes of DTA and DTG curves are compatible. Thus, each thermal effect on the DTA curve is related to a weight loss measured in TGA of thermal degradation of HAs exhibiting endothermic and exothermic peaks associated with these in all samples (Figure 4; Table 8) [97–99]. The DTA curves of all HAs showed one endothermic effect and three exothermic effects.

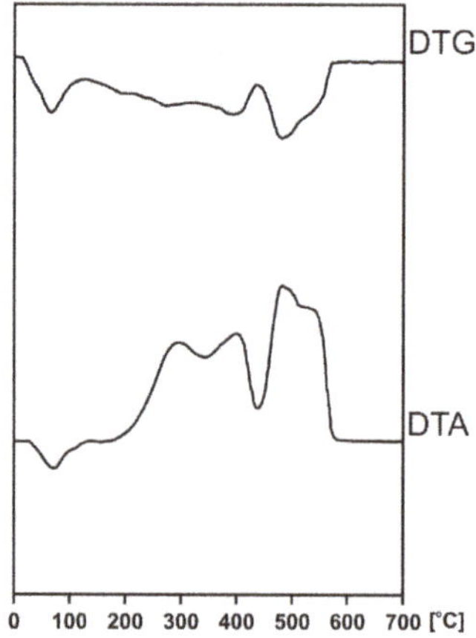

Figure 4. Thermogram of HAs isolated from Kusowo Bog at the depth 25–50 cm.

Table 8. Parameters of thermal decomposition of HAs.

Sampling Site		Depth (cm)	Temperature of Effects on DTA Curves (°C)				Mass Loss Corresponding with the Effects on DTG Curves (%)				The Ratios of Area under DTA and DTG Curves for Exothermal Effects			Z $\frac{endo+exo1}{exo2+3}$
			endo	exo 1	exo 2	exo 3	Endo	exo 1	exo 2	exo 3	$\frac{DTA_1}{DTG_1}$	$\frac{DTA_{2+3}}{DTG_{2+3}}$	$\frac{\sum DTA}{\sum DTG}$	
Raised bog	Kusowo	0–25	74	297	403	489	13.6	27.4	24.0	35.0	3.32	4.08	3.84	0.70
		25–50	75	294	427	485	15.4	28.8	24.1	31.7	3.41	6.10	5.18	0.79
		50–75	76	295	400	473	13.2	29.4	24.6	32.8	4.07	7.83	6.55	0.74
		75–100	74	294	418	490	15.7	28.8	24.7	30.8	4.75	8.02	6.90	0.80
Fen	Stążka	0–25	76	294	390	476	16.0	28.1	22.2	33.7	2.60	2.92	2.81	0.79
		50–75	74	294	399	490	15.7	28.8	24.7	30.8	2.48	3.92	3.29	0.67
		75–100	77	304	419	486	13.5	28.1	28.5	29.9	2.53	3.38	3.44	0.71
Peat-moorsh soils	Ch1	0–25	79	287	365	467	12.7	31.6	25.4	30.4	3.07	4.82	3.88	0.79
		25–50	79	290	372	461	12.7	29.5	28.4	29.4	3.80	5.46	4.50	0.76
		50–75	78	320	421	483	12.6	27.8	32.2	26.3	4.62	6.75	5.31	0.69
		75–100	76	314	426	496	14.2	28.8	32.4	24.6	2.76	5.65	4.68	0.76
	Ch2	0–25	76	277	348	437	21.7	29.1	9.1	40.1	2.15	5.99	4.64	1.03
		25–50	75	281	350	439	17.1	30.0	10.9	41.4	2.99	6.78	5.38	0.92
		50–75	75	294	370	450	18.3	28.0	14.4	39.3	2.24	6.06	4.76	0.86
		75–100	77	292	367	450	18.1	28.8	16.0	37.1	3.08	7.47	5.94	0.88
	Ch3	0–25	80	291	377	476	17.3	30.2	19.8	33.1	2.50	3.58	4.59	1.08
		50–75	84	296	394	481	13.9	28.3	21.3	36.6	3.23	3.70	4.87	0.95
		75–100	81	293	391	480	15.9	31.2	19.3	32.1	4.58	5.17	6.23	0.91
	Ch4	0–25	79	289	367	458	16.4	30.7	16.2	36.7	2.32	4.41	3.64	0.89
		50–75	78	302	394	479	13.7	29.9	23.2	33.2	2.60	4.03	3.38	0.76
		75–100	79	300	399	482	12.1	30.4	28.1	29.4	4.22	7.57	6.44	0.71

DTA—differential thermal analysis; DTG—differential thermogravimetry; Z—the loss ratio of HAs in the low range to those in the high temperature range; endo—endotermic; exo—exotermic effect.

The endothermic effect ranges from 74 to 84 °C, with weight loss from 12% to 20% and corresponding to the evaporation of absorbed water and dehydration reaction. This effect does not demonstrate any significant changes *downsizing into* the peat profile, and the intensity of weight losses is related to weakly bounded water by hygroscopic components.

Our results are in line with Schnitzer et al. [100], Kodama and Schnitzer [101], Shurygina et al. [102], Leinweber and Schulten [103], and Gołębiowska et al. [65], who pointed out that, at low temperature (about 100 °C), the evaporation of hydroscopic moisture is observed. In addition, Kļaviņš and Sire [59] showed that the amount of hygroscopically bound water in HAs from peat is higher compared to peat from the corresponding layer and does not differ in peat profile.

The thermal analysis of HAs in all peat samples showed an increasing combustion temperature with a downsizing depth of the profile for three exothermic effects (Table 8), which are related to more thermally labile fractions in HAs of the upper layer compared to bottom layers.

The first exothermic reaction ranged from 277 to 320 °C with a mass loss from 27.4% to 31.6% related to the thermal combustion of carbohydrates, peptides, lignin, external functional groups of HAs, decarboxylation of acidic groups in aliphatic compounds, and dehydration of hydroxylate aliphatic structures.

The second exothermic reaction varied from 348 to 427 °C with a mass loss from 9.1% to 32.4%, corresponding to the combustion of less mature components of HAs.

The third exothermic reaction ranged from 437 to 496 °C with a mass loss from 24.6% to 41.4%, which is related to the combustion of bitumens, to highly mature HAs of increasing thermal stability, and to the cleavage of C–C bonds [104,105].

In contrast, in Ch1 and Ch2 (peat-moorsh soils with the highest degree of decomposition), the mass loss of the second exothermic effect increased with an increase in depth.

The lowest mass loss was measured for the second and the highest mass losses for the third exothermic effect in Ch2 with the highest degree of decomposition.

Schnitzer and Levesque. [66] detected two exothermic peaks between 200 and 500 °C for humic substances of podzolic soil. The authors hypothesized that the first peak was associated with dehydratation and decarboxylation and that the second was associated with carbon oxidation.

The thermal analysis provides some important information on HAs structure based on Z parameter values, which expresses the ratio between thermally labile and stable parts in HAs (mainly aliphatic and aromatic) [97].

Inspection of Z parameters revealed that, with increasing depth, HAs contain more structures resistant to oxidation in high temperatures. The HAs from Ch2 and Ch3 (peat-moorsh soils) revealed a more aliphatic character in contrast to the CH1 and Ch4 Baltic-type raised bog and fen, which were found to be more aromatic (Table 8).

The significant negative correlation between the heat of the combustion (DTA_1/DTG_1) and E_4/E_6 (r = −0.528) (Table 9) may be attributable to more condensed stable aromatic structures of HAs in the deeper layers of peat deposits.

The thermal analysis of HAs in all peat samples showed an increasing combustion temperature with a downsizing depth of the profile for three exothermic effects (Table 8), which are related to more thermally labile fractions in HAs of the upper layer compared to bottom layers.

The first exothermic reaction ranged from 277 to 320 °C with a mass loss from 27.4% to 31.6% related to the thermal combustion of carbohydrates, peptides, lignin, external functional groups of HAs, decarboxylation of acidic groups in aliphatic compounds, and dehydration of hydroxylate aliphatic structures.

The second exothermic reaction varied from 348 to 427 °C with a mass loss from 9.1% to 32.4%, corresponding to the combustion of less mature components of HAs.

The third exothermic reaction ranged from 437 to 496 °C with a mass loss from 24.6% to 41.4%, which is related to the combustion of bitumens, to highly mature HAs of increasing thermal stability, and to the cleavage of C–C bonds [104,105].

Table 9. Correlation coefficients between chemical properties of peat deposits, VIS-spectroscopy, electron paramagnetic resonance, and thermal decomposition of HAs.

	Peat Deposit						HAs						
Parameter	TOC	C_{WHE}	N-Total	C/N	E_4/E_6	g-Value	Spin Concentr.	$\frac{DTA_1}{DTG_1}$	$\frac{DTA_{2+3}}{DTG_{2+3}}$	$\frac{\sum DTA}{\sum DTG}$	Z	H/C	C/N*
C_{WHE}	0.065	-											
N-total	−0.481 *	−0.301	-										
C/N	0.736 *	0.384	−0.900 *	-									
E_4/E_6	−0.726 *	0.039	0.697 *	−0.806 *	-								
g-value	−0.480 *	−0.420	0.729 *	−0.865 *	0.744 *	-							
Spin conc.	0.552 *	−0.059	−0.313	0.548 *	−0.756 *	−0.620 *	-						
$\frac{DTA_1}{DTG_1}$	0.358	−0.103	−0.305	0.426	−0.528 *	−0.454 *	0.612 *	-					
$\frac{DTA_{2+3}}{DTG_{2+3}}$	0.283	0.134	−0.106	0.374	−0.313	−0.584 *	0.552 *	0.607 *	-				
$\frac{\sum DTA}{\sum DTG}$	0.228	−0.008	−0.179	0.376	−0.425	−0.569 *	0.669 *	0.759 *	0.850 *	-			
Z	−0.645 *	0.186	0.347	−0.406	0.429	0.159	−0.239	−0.313	−0.126	0.104	-		
H/C	−0.303	0.772 *	−0.326	0.203	0.266	−0.285	−0.339	−0.095	−0.005	−0.051	0.295	-	
C/N *	0.743 *	−0.322	−0.382	0.530 *	−0.815 *	−0.369	0.553 *	0.296	0.152	0.163	−0.532 *	−0.632 *	-
O/C	−0.454 *	0.432	0.458 *	−0.462 *	0.536 *	0.379	−0.421	−0.590 *	−0.307	−0.384	0.570 *	0.279	−0.488 *

* significant correlation coefficient α = 0.05; abbreviations as in Tables 1–8.

In contrast, in Ch1 and Ch2 (peat-moorsh soils with the highest degree of decomposition), the mass loss of the second exothermic effect increased with an increase in depth.

The lowest mass loss was measured for the second and the highest mass loss for the third exothermic effect in Ch2 with the highest degree of decomposition.

Schnitzer et al. [100] detected two exothermic peaks between 200 and 500 °C for humic substances of podzolic soil. The authors hypothesized that the first peak was associated with dehydratation and decarboxylation and that the second was associated with carbon oxidation.

The thermal analysis provides some important information on HAs structure based on Z parameter values, which expresses the ratio between thermally labile and stable parts in HAs (mainly aliphatic and aromatic) [97].

Inspection of Z parameters revealed that, with increasing depth, HAs contain more structures resistant to oxidation in high temperatures. The HAs from Ch2 and Ch3 (peat-moors soils) revealed a more aliphatic character in contrast to the CH1 and Ch4 Baltic-type raised bog and fen, which were found to be more aromatic (Table 8).

The significant negative correlation between the heat of the combustion (DTA_1/DTG_1) and E_4/E_6 (r = −0.528) (Table 9) may be attributable to more condensed stable aromatic structures of HAs in the deeper layers of peat deposits.

The findings illustrated in Table 9 show a significant negative correlation between TOC concentrations and Z values (r = −0.645), which is related to the increasing contents of aromatic structures in line with the downsizing of peat profile. The above data agrees with E_4/E_6 and C/N values (Table 9), where the decrease of both ratios with an increase of depth corresponded to a high degree of aromatic condensation and poly-conjugation of aromatic structures and to the presence of relatively low proportions of aliphatic structures.

As shown in the study of thermal organic matter stability [106], these properties are a function of the chemical composition, degree of humification, and mineral association. However, Kļaviņš and Sire [93] showed that the decay of more labile structure in HAs decreases as the degree of peat decomposition increases, though more condensed aromatic structures were positively correlated with the peat degree of decomposition.

Francioso et al. [107] showed in peat HAs various biochemical fractions of plants preserved during peat formation characteristic of thermal decomposition and more labile structures in the origin and formation process of HAs. This preservation therefore might be the result of anoxic environmental conditions occurring due to peat accumulation.

3.2.5. PCA Analysis

A variety of multiple causes are responsible for the composition of peat and the properties of HAs from undrained and drained peats. Therefore, a comparative study was carried out of the chemical composition of peats and the properties of HAs from various undrained and drained types of peats developed in various compositions of botanical cover, peat-forming species, and oxic and anoxic conditions as a result of the oscillation of ground water during drainage.

The PCA analysis was conducted to determine the impact of various types of peat and decomposition degree on the properties of peat deposits and HAs. The results reveal that the properties of organic matter and HAs vary in different environments (Figure 5).

The results of the PCA explain 80.26% of the total variability of organic matter properties. The first two axes were statistically significant ($\alpha < 0.05$): the first axis (PC1) explains 45.98%, and the second (PC2) explains 19.84% of the total variability, while the third axis (PC3) explained 14.43% (Table 10). The PC1 was negatively associated with the TOC, C/N, spin concentration DTA_1/DTG_1, and C/N* and positively coordinates with E_4/E_6, g-value, and carbon in HAs on this axis in peat soils. The PC2 showed a close negative association with C_{HWE} and H/C.

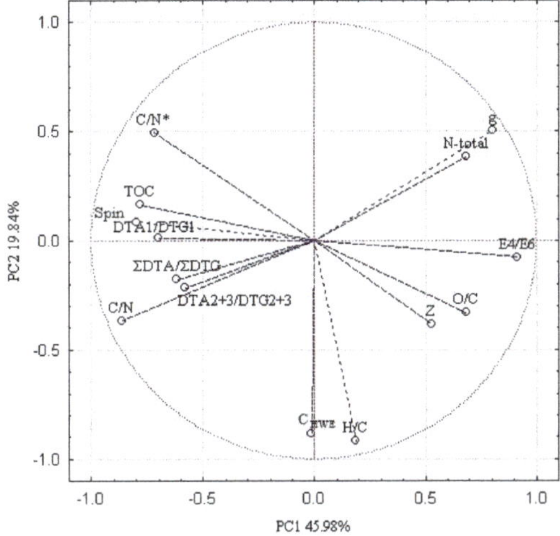

Figure 5. Eigenvectors of soil chemical variables: Abbreviations as in the Tables 1–9.

Table 10. Analysis of principal components for Baltic-type raised bog, fen, and peat-moorsh soils.

Principal Components	Eigenvalues	% of total Variance	Cumulative Eigenvalues	Cumulative % of Variance
PC1	6.44	45.98	6.44	45.98
PC2	2.78	19.84	9.22	65.83
PC3	2.02	14.43	11.24	80.26

The results of the PCA analysis showed on PC1, with 45.98% variability, a significantly positive correlation between TOC, spin, and DTA_1/DTG_1 (Figure 5). The data suggest significantly an increased TOC, spin, and DTA_1/DTG_1 with the depth, indicating a decrease in the content of thermolabile structural units (carbohydrates, free, and bound functional groups) and an increase in thermostable skeleton part of the HAs molecules. This reflects that the transformation of the organic matter is strongly connected with the humification process and the molecular structure of HAs. Moreover, there were significant negatively associate of TOC, spin, and DTA_1/DTG_1 with E_4/E_6. The observed changes can be explained by a higher degree of aromatic condensation and poly-conjugation in the molecules of HAs, in the deeper layers (Table 10).

Additionally, on the second axis PC2 with 19.84% variability, the C_{HWE} vs. H/C was significantly positively associated (Figure 4). The decrease of C_{HWE} content and H/C values with the depth suggests the decline of aliphatic bridges between aromatic structural units HAs. On this basis, it can be stated dominantly the accumulation processes rather than ones of decomposition with the depth of peat profile.

The third axis showed a significant positive correlation with ΣDTA/ΣDTG (Table 11). These variables suggest that the first principal component describes the dependence of VIS-spectra, electron paramagnetic resonance, and thermal properties of HAs on the organic matter. Therefore, a higher g-value and E_4/E_6 in peat-moorsh soils as well as fens compared to those from Baltic-type raised bogs affects a lower degree of aromatic condensation and humification of HAs.

Table 11. Factor loadings and explained variance of three principal components in PCA.

Variable	PC1	PC2	PC3
TOC	−0.78	0.17	−0.33
C_{WHE}	−0.01	−0.89	−0.20
N-total	0.68	0.38	0.46
C/N	−0.86	−0.37	−0.32
E_4/E_6	0.92	−0.08	0.18
g-value	0.80	0.50	0.01
Spin concentration	−0.80	0.09	0.30
$\frac{DTA_1}{DTG_1}$	−0.70	0.01	0.47
$\frac{DTA_{2+3}}{DTG_{2+3}}$	−0.58	−0.22	0.63
$\frac{\sum DTA}{\sum DTG}$	−0.62	−0.18	0.74
Z	0.52	−0.38	0.41
H/C	0.19	−0.92	−0.12
C/N*	−0.72	0.49	−0.27
O/C	0.68	−0.33	−0.03

C/N*—atomic ratios of the HAs; abbreviations as in Tables 1–9.

4. Conclusions

This study has examined the vulnerability of organic matter of Baltic-type raised bog, fen, and peat-moorsh soils to decomposition by determining the chemical composition and physicochemical properties of HAs.

Drainage was shown to be the principal factor causing the chemical composition and physicochemical properties of HAs. The latter properties of HAs were found to be in line with their chemical composition of peats.

Conversion of undrained to drained peatlands (modifications in *oxic–anoxic properties*) has led to an increase of the degree of decomposition, relating to the mass loss of organic matter, particularly labile organic fractions, and was confirmed by a correspondingly lower content of TOC, lower C/N values, and higher E_4/E_6 values. In line with an increase of the degree of organic matter decomposition, the N-total; E_4/E_6, g-values; mass loss for exo1, exo 2, and exo 3; Z values; and spin concentrations all increased while C/N decreased.

In light of these results, the HAs from undrained peatlands were found to be richer in aromatic structures and can be regarded therefore as more humified and mature (therefore, more stable) than that of peat-moorsh soils. Overall, the most important indicators for the propensity of organic matter to decomposition as a consequence of drainage identified in the present study were more aliphatic structures in the HAs. There would appear to be therefore a positive feedback loop of drainage and the consequent vulnerability of soil organic matter to decomposition.

In this context, the analytical methods of thermal analysis together with optical densities and paramagnetic behaviour can be said to be suitable and effective tools for characterizing the origin and formation process of HAs in various environmental properties.

Author Contributions: L.W.S.: conceptualization, methodology, writing—original draft preparation, writing—review & editing, supervision, funding acquisition; A.J.: EPR investigations; K.W.: differential thermal analysis; T.M.: formal analysis, investigation; M.S.: formal analysis, investigation, software. All authors have read and agreed to the published version of the manuscript.

Funding: This work was supported by the National Science Centre Poland (grant number 2013/09/B/NZ9/03169).

Acknowledgments: We wish to express our sincere gratitude to Wioletta Gaca for her valuable discussions.

Conflicts of Interest: No potential conflict of interest was reported by the authors.

References

1. Lappalainen, E. *Global Peat Resources*; International Peat Society–Geological Survey of Finland, Saarijärven Offset Oy: Saarijärvi, Finland, 1996.
2. Okruszko, H. The principles of the identification and classification of hydrogenic soils according to the need of reclamation. *Bibl. Wiad. IMUZ.* **1976**, *52*, 7–53.
3. Gorham, E. Northern Peatlands: Role in the Carbon Cycle and Probable Responses to Climatic Warming. *Ecol. Appl.* **1991**, *1*, 182–195. [CrossRef] [PubMed]
4. Johnson, L.; Damman, A.W. Decay and its regulation in Sphagnum peatlands. *Adv. Bryolog.* **1993**, *5*, 249–296.
5. Belyea, L.R. Separating the effects of litter quality and microenvironment on decomposition rates in a patterned peatland. *Oikos* **1996**, *77*, 529. [CrossRef]
6. Clymo, R.S.; Turunen, J.; Tolonen, K. Carbon Accumulation in Peatland. *Oikos* **1998**, *81*, 368. [CrossRef]
7. Bambalov, N. Regularities of peat soils anthropic evolution. *Acta Agroph.* **2000**, *26*, 179–203.
8. Bridgham, S.D.; Richardson, C.J. Endogenous versus exogenous nutrient control over decomposition and mineralization in North Carolina peatlands. *Biogeochemistry* **2003**, *65*, 151–178. [CrossRef]
9. Bragazza, L.; Siffi, C.; Iacumin, P.; Gerdol, R. Mass loss and nutrient release during litter decay in peatland: The role of microbial adaptability to litter chemistry. *Soil Biol. Biochem.* **2007**, *39*, 257–267. [CrossRef]
10. Gierlach-Hładoń, T.; Szajdak, L. Physicochemical properties of humic acids isolated from an Eriophorum Sphagnum raised bog. In *Mires and Peat*; Kļaviņš, M., Ed.; University of Latvia Press: Riga, Latvia, 2010; pp. 143–157.
11. Lindsay, R. Peatland Classification. In *The Wetland Book. I: Structure and Function, Management, and Methods*; Finlayson, M.C., Everard, M., Irvine, K., McInnes, R.J., Middleton, B.A., van Dam, A.A., Davidson, N.C., Eds.; Springer: Dordrecht, The Netherlands, 2018; pp. 1515–1528.
12. Joosten, H.; Clarke, D. *Wise Use of Mires and Peatlands–Background and Principles Including a Framework for Decision-Making*; International Mire Conservation Group and International Peat Society: Saarijärvi, Finland, 2002.
13. Szajdak, L.W.; Szatylowicz, J.; Kõlli, R. Peat and peatlands, physical properties. In *Encyclopedia of Agrophysics, Encyclopedia of Earth Sciences*; Gliński, J., Horabik, J., Lipiec, J., Eds.; Springer: Berlin, Germany, 2011; pp. 551–555.
14. Joosten, H.; Tanneberger, F.; Moen, A. *Mires and Peatlands of Europe: Status, Distribution and Conservation*; Schweizerbart Science Publishers: Stuttgart, Germany, 2017.
15. Säurich, A.; Tiemayer, B.; Don, A.; Fiedler, S.; Bechtold, M.; Wulf Amelung, W.; Freibauer, A. Drained organic soils under agriculture—The more degraded the soil the higher the specific basal respiration. *Geoderma* **2019**, *355*, 11391. [CrossRef]
16. Clarke, D.; Rieley, J. *Strategy for Responsible Peatlands Management*; International Peat Society: Jyväskylä, Finland, 2019.
17. Armentado, T.V.; Menges, E.S. Patterns of change in the carbon balance of organic soil-wetlands of the temperate zone. *J. Ecol.* **1986**, *74*, 755–774. [CrossRef]
18. Bouwman, A.F. Exchange of greenhouse gases between terrestrial ecosystems and the atmosphere. In *Soils and the Greenhouse Effect*; John Willey and Sons: Chichester, UK, 1990; pp. 61–129.
19. Terry, R.E. Nitrogen transformations in Histosols. In *The Role of Organic Matter in Modern Agriculture*; Chen, Y., Avnimelech, Y., Eds.; Martinus Nijhoff Publishers: Leiden, The Netherlands, 1986; pp. 55–70.
20. Nieder, R.; Benbi, D.K.; Isermann, K. Soil organic matter dynamics. In *Handbook of Processes and Modeling in the Soil-Plant System*; Benbi, D.K., Nieder, R., Eds.; Food Products Press: New York, NY, USA, 2003; pp. 345–408.
21. Joosten, H. *Peatlands, Climate Change Mitigation and Biodiversity Conservation*; Nordic Council of Minister: Copenhagen, Denmark, 2015.
22. Ilnicki, P. *Peatlands and Peat*; Wydawnictwo AR: Poznań, Poland, 2002.
23. Oleszczuk, R.; Regina, K.; Szajdak, L.; Maryganova, V. Impacts of Agricultural Utilization of Peat—Soil on the Greenhouse Gas Balance. In *Peatlands and Climate Change*; Strack, M., Ed.; International Peat Society, Saarijärven Offset Oy: Saarijärvi, Finland, 2008; pp. 70–97.
24. Laine, J.; Vanha-Majamaa, I. Vegetation ecology along a trophic gradient on drained pine mires in southern Finland. *Ann. Bot. Fen.* **1992**, *29*, 213–233.

25. Laine, J.; Vasander, H.; Laiho, R. Long-term effects of water level drawdown on the vegetation of drained pine mires in southern Finland. *J. App. Ecol.* **1995**, *32*, 785–802.

26. Vasander, H. Effect of forest amelioration on diversity in an ombrotrophic bog. *Ann. Bot. Fen.* **1984**, *21*, 7–15.

27. Mälson, K.; Backéus, I.; Rydin, H. Long-term effects of drainage and initial effects of hydrological restoration on rich fen vegetation. *App. Veg. Sci.* **2008**, *11*, 99–106. [CrossRef]

28. Similä, M.; Aapala, K.; Penttinen, J. *Ecological Restoration in Drained Peatlands—Best Practices from Finland*; Natural Heritage Services: Vantaa, Finland, 2014.

29. Maslov, B.S.; Konstantinov, V.K.; Babikov, B.V.; Ahti, E. *Permanent Experiments on Drained Peatlands in Russia*; Finish Forest Research Institute: Vantaa, Finland, 2006.

30. Mäkilä, M. The sufficiency of peat for energy use on the basis of carbon accumulation. In *Geoscience for Society 125th Anniversary Volume*; Geological Survey of Finland: Espoo, Finland, 2011; Volume 49, pp. 163–170.

31. Schipper, L.A.; McLeod, M. Subsidence rates and carbon loss in peat soils following conversion to pasture in the Waikato region, New Zealand. *Soil Use Manag.* **2002**, *18*, 91–93. [CrossRef]

32. Minkkinen, K.; Laine, J. Long-term effect of forest drainage on the peat carbon stores of pine mires in Finland. *Can. For. Res.* **1998**, *28*, 1267–1275. [CrossRef]

33. Lucchese, M.; Waddington, J.M.; Poulin, M.; Pouliot, R.; Rochefort, L.; Strack, M. Organic matter accumulation in a restored peatland: Evaluating restoration success. *Ecol. Eng.* **2010**, *36*, 482–488. [CrossRef]

34. Gawlik, J. Division of differently silted peat formation into classes according to their state of secondary transformations. *Acta Agroph.* **2000**, *26*, 17–24.

35. Chason, D.B.; Siegel, D.I. Hydraulic conductivity and related properties of peat, Lost River Peatland, northern Minnesota. *Soil Sci.* **1986**, *142*, 91–99. [CrossRef]

36. Beckwith, C.W.; Baird, A.J.; Heathwaite, A.L. Anisotropy and depth-related heterogeneity of hydraulic conductivity in a bog peat. I: Laboratory measurements. *Hydrol. Process.* **2003**, *17*, 89–101. [CrossRef]

37. Sokołowska, Z.; Szajdak, L.; Matyka-Sarzyńska, D. Impact of the degree of secondary transformation on acid–base properties of organic compounds in mucks. *Geoderma* **2005**, *127*, 80–90. [CrossRef]

38. Szajdak, L.; Szatyłowicz, J. Impact of drainage on hydrophobicity of fen peat-peat-morsh soils. In *Mires and Peat*; Kļaviņš, M., Ed.; University of Latvia Press: Riga, Latvia, 2010; pp. 158–174.

39. Okruszko, H.; Ilnicki, P. The moorsh horizons as quality indicators of reclaimed organic soils. In *Organic Soils and Peat Materials for Sustainable Agriculture*; Parent, L.-E., Ilnicki, P., Eds.; CRC Press: Boca Raton, FL, USA, 2003; pp. 12–25.

40. Ilnicki, P.; Szajdak, L.W. *Peatland Disappearance*; Ilnicki, P., Szajdak, L.W., Eds.; PTPN: Poznań, Poland, 2016.

41. Schnitzer, M. Water retention by humic substances. In *Peat and Water. Aspects of Water Retention and Dewatering in Peat*; Fuchsman, C.H., Ed.; Elsevier Applied Science Publishers Ltd.: Amsterdam, The Netherlands, 1986; pp. 159–176.

42. Brandyk, T.; Szatyłowicz, J.; Oleszczuk, R.; Gnatowski, T. Water-related physical attributes of organic soils. In *Organic Soils and Peat Materials for Sustainable Agriculture*; Parent, L.-E., Ilnicki, P., Eds.; CRC Press: Boca Raton, FL, USA, 2002; pp. 33–66.

43. Brandyk, T.; Oleszczuk, T.; Szatyłowicz, J. Investigation of soil water dynamics in a fen peat-moorsh soil profile. *Int. Peat J.* **2001**, *11*, 15–24.

44. Grootjans, A.P.; Schipper, P.C.; van der Windt, H.J. Influence of drainage on N-mineralization and vegetation response in wet meadows. II. Cirsio-Molinietum stands. *Oecology Plant* **1986**, *7*, 3–14.

45. Kwak, J.C.; Ayub, A.L.; Shepard, J.D. The role of colloid science in peat dewatering: Principles and dewatering studies. In *Peat and Water*; Fuchsman, C.H., Ed.; Elsevier Applied Science Publishers: London, UK, 1986; pp. 95–118.

46. Lüttig, G. Plants to peat. In *Peat and Water*; Fuchsman, C.H., Ed.; Elsevier Applied Science Publishers: London, UK, 1996; pp. 9–19.

47. Inisheva, L.I.; Dementieva, T.V. Mineralization rate of organic matter in peats. *Pochvovedenie* **2000**, *2*, 196–203.

48. Koeselman, W.; Verhoeven, J.T.A. Eutrophication of fen ecosystems: External and internal nutrient sources and restoration strategies. In *Restoration of Temperate Wetlands*; Wheeler, B.D., Show, S.C., Fojt, W.J., Robertson, R.A., Eds.; Willey: Chichester, UK, 1995; pp. 91–112.

49. Kajak, A.; Okruszko, H. Grasslands on drained peats in Poland. In *Ecosystems of the world 17A: Managed Grasslands*; Breymeyer, A.I., Ed.; Elsevier Sc. Publ.: Amsterdam, The Netherlands, 1990; pp. 213–253.

50. Van Diggelen, R.; Molenaar, W.J.; Kooijman, A.M. Vegetation succession in a floating mire in reaction to management and hydrology. *J. Veg. Sci.* **1996**, *7*, 809–820.

51. Kotowski, W.; Fen Communities. Ecological Mechanisms and Conservation Strategies. Ph.D. Thesis, University of Groningen, Groningen, The Netherlands, 2002.

52. Szajdak, L.; Matuszewska, T.; Gawlik, J. Effect of secondary transformation state of peat-muck soils on total amino acid content. *Inter. Peat J.* **1998**, *8*, 76–80.

53. Van Dijk, H. Colloid chemical properties of humic matter. In *Soil Biochemistry*; McLaren, A.D., Skujins, J., Eds.; Marcel Dekker: New York, NY, USA, 1971; p. 21.

54. Lamentowicz, M.; Tobolski, K.; Mitchell, E.A.D. Palaeoecological evidence for anthropogenic acidification of a kettle-hole peatland in northern Poland. *Holocene* **2007**, *17*, 1185–1196. [CrossRef]

55. Ellenberg, H. *Vegetation Ecology of Central Europe*, 4th ed.; Cambridge University Press: Cambridge, UK, 1988.

56. Cedro, A.; Sotek, Z. Natural and Anthropogenic Transformations of A Baltic Raised Bog (Bagno Kusowo, North West Poland) in the light of dendrochronological analysis of *Pinus sylvestris* L. *Forests* **2016**, *7*, 202–216. [CrossRef]

57. IUSS Working Group WRB. *World Reference Base for Soil Resources 2014, Update 2015 International Soil Classification System for Naming Soils and Creating Legends for Soil Maps*; World Soil Resources Reports No. 106: Rome, Italy, 2015.

58. Szajdak, L.W.; Gaca, W. Nitrate reductase activity in soil under shelterbelt and an adjoining cultivated field. *Chem. Ecol.* **2010**, *26*, 123–134. [CrossRef]

59. von Post, L. Sveriges Geologiska Undersöknings torvinventering och några av dess hitills vunna resultat. *Sven. Mosskulturföreningens Tidsk.* **1922**, *36*, 1–27.

60. Gałka, M.; Tobolski, K.; Górska, A.; Milecka, K.; Fiałkiewicz-Kozieł, B.; Lamentowicz, M. Disentangling the drivers for the development of a Baltic bog during the Little Ice Age in northern Poland. *Quatern. Int.* **2014**, *329*, 323–327. [CrossRef]

61. Swift, R.S. Organic matter characterization. In *Methods of Soil Analysis, Part 3: Chemical Methods*; Sparks, D.L., Ed.; American Society of Agronomy, Inc.: Madison, WI, USA, 1996; pp. 1011–1106.

62. Österberg, R.; Szajdak, L.; Mortensen, K. Temperature-dependent restructuring of fractal humic acids: A proton-dependent process. *Environ. Int.* **1994**, *20*, 77–80. [CrossRef]

63. Chen, Y.; Senesi, N.; Schnitzer, M. Information Provided on Humic Substances by E4/E6 Ratios. *Soil Sci. Soc. Am. J.* **1977**, *41*, 352. [CrossRef]

64. Jezierski, A.; Czechowski, F.; Jerzykiewicz, M.; Drozd, J. EPR investigations of structure of humic acids from compost, soil, peat and soft brown coal upon oxidation and metal uptake. *Appl. Magn. Reason* **2000**, *18*, 127–136. [CrossRef]

65. Gołębiowska, D.; Ptak, W.; Wegner, K. Correlation between derivatographic and chemiluminescence analysis data in relation to elemental composition of humic acids. *Environ. Int.* **1996**, *22*, 495–500. [CrossRef]

66. Schnitzer, M.; Levesque, M. Electron spin resonance as a guide to the degree of humification of peats. *Soil Sci.* **1979**, *127*, 140–145.

67. Boelter, D.H. Physical properties of peats as related to degree of decomposition. *Soil Sci Soc. Am. Proc.* **1969**, *33*, 606–609. [CrossRef]

68. Klavins, M.; Sire, J.; Purmalis, O.; Melecis, V. Approaches to estimating humification indicators for peat. *Mires Peat* **2008**, *3*, 1–15.

69. Harworth, R.D. The chemical nature of humic acid. *Soil Sci.* **1971**, *106*, 188–192.

70. Benavides, J.C. The effect of drainage on organic matter accumulation and plant communities of high-altitude peatlands in the Colombian tropical Andes. *Mires Peat* **2014**, *15*, 1–15.

71. Batjes, N.H. Total carbon and nitrogen in the soils of the world. *Euro Soil Sci.* **2014**, *65*, 10–21. [CrossRef]

72. Chambers, F.M.; Beilman, D.W.; Yu, Z. Methods for determining peat humification and for quantifying peat bulk density, organic matter and carbon content for palaeostudies of climate and peatland carbon dynamics. *Mires Peat* **2010**, *7*, 1–10.

73. Fontaine, S.; Barot, S.; Barré, P.; Bdioui, N.; Mary, B.; Rumpel, C. Stability of organic carbon in deep soil layers controlled by fresh carbon supply. *Nature* **2007**, *450*, 277–280. [CrossRef]

74. Wang, M.; Moore, T.R.; Talbot, J.; Riley, J.L. The stoichiometry of carbon and nutrients in peat formation. *Global Biogeochem. Cy.* **2015**, *29*, 113–121. [CrossRef]

75. Smolander, A.; Kitunen, V. Soil microbial activities and characteristics of dissolved organic C and N in relation to tree species. *Soil Biol. Biochem.* **2002**, *34*, 651–660. [CrossRef]

76. Kolka, R.; Weishampel, P.; Fröberg, M. Measurement and importance of dissolved organic carbon. In *Field Measurements for Forest Carbon Monitoring*; A Landscape-Scale Approach; Hoover, C.M., Ed.; Springer: Dordrecht, The Netherlands, 2008; pp. 171–176.

77. Freeman, C.; Fenner, N.; Ostle, N.J.; Kang, H.D.; Dowrick, J.; Reynolds, B.M.; Lock, M.A.; Sleep, D.; Hughes, S.; Hudson, J. Export of dissolved organic carbon from peatlands under elevated carbon dioxide levels. *Nature* **2004**, *430*, 195–198. [CrossRef]

78. Szajdak, L.; Szczepański, M.; Bogacz, A. Impact of secondary transformation of peat-moorsh soils on the decrease of nitrogen and carbon compounds in ground water. *Agr. Res.* **2007**, *5*, 189–200.

79. Biester, H.; Knorr, K.-H.; Schellekens, J.; Basler, J.; Hermanns, Y.-M. Comparison of different methods to determine the degree of peat decomposition in peat bogs. *Biogeosciences* **2014**, *11*, 2691–2707. [CrossRef]

80. Szajdak, L. Chemical properties of peat. In *Peatlands and Peat*; Ilnicki, P., Ed.; Wydawnictwo AR: Poznań, Poland, 2002; pp. 432–450.

81. Anderson, D.E. Carbon accumulation and C/N ratios of peat bogs in North-West Scotland. *Scot. Geogr. J.* **2002**, *118*, 323–341.

82. Rezanezhad, F.; William, J.S.P.; Quinton, L.; Lennartz, B.; Milojevic, T.; Van Cappellen, P. Structure of peat soils and implications for water storage, flow and solute transport: A review update for geochemists. *Chem. Geol.* **2016**, *429*, 75–84. [CrossRef]

83. Malmer, N.; Holm, E. Variation in the C/N-quotient of peat in relation to decomposition rate and age determination with 210 pb. *Oikos* **1984**, *43*, 171–182. [CrossRef]

84. Kuhry, P.; Vitt, D.H. Fossil carbon/nitrogen ratios as a measure of peat decomposition. *Ecology.* **1996**, *77*, 271–275. [CrossRef]

85. Broder, T.; Blodau, C.; Biester, H.; Knorr, K.H. Peat decomposition records in three pristine ombrotrophic bogs in southern Patagonia. *Biogeosciences* **2012**, *9*, 1479–1491. [CrossRef]

86. Rice, J.A.; MacCarthy, P. Statistical evaluation of the elemental composition of humic substances. *Org. Geochem.* **1991**, *7*, 635–648. [CrossRef]

87. DiDonato, N.; Chen, H.; Waggoner, D.; Hatcher, P.G. Potential origin and formation for molecular components of humic acids in soils. *Geochim. Cosmochim. Acta* **2016**, *178*, 210–222. [CrossRef]

88. Van Krevelen, D. Graphical statistical method for the study of structure and reaction processes of coal. *Fuel* **1950**, *29*, 269–284.

89. Schnitzer, M. Humic Substances: Chemistry and Reactions. In *Soil Organic Matter*; Schnitzer, M., Khan, S.U., Eds.; Elsevier: Amsterdam, The Netherlands, 1978; pp. 1–64.

90. Stevenson, F.J. *Humus Chemistry: Genesis, Composition, Reactions*, 2nd ed.; Wiley & Sons: New York, NY, USA, 1994; p. 512.

91. Kalbitz, K.; Geyer, W.; Geyer, S. Spectroscopic properties of dissolved humic substances a reflection of land use history in a fen area. *Biogeochemistry* **1999**, *47*, 219–238. [CrossRef]

92. Zaccone, C.; Miano, T.M.; Shotyk, W. Qualitative comparison between raw peat and related humic acids in an ombrotrophic bog profile. *Org. Geochem.* **2007**, *38*, 151–160. [CrossRef]

93. Kļaviņš, M.; Sire, J. Variations of humic acid properties within peat profiles. In *Mires and Peat*; Kļaviņš, M., Ed.; University of Latvia Press: Riga, Latvia, 2010; pp. 175–197.

94. Czechowski, F.; Jezierski, A. EPR studies on petrographic constituents of bituminous coals chars of brown coals group components, and humic acids 600 °C char upon oxygen and solvent action. *Energy Fuels* **1997**, *11*, 951–964. [CrossRef]

95. Knüpling, M.; Tiörring, J.T.; Un, S. The relationship between the molecular structure of semiquinone radicals and their g-values. *Chem. Phys.* **1997**, *219*, 291–304. [CrossRef]

96. Witwicki, M.; Jezierska, J. Protonated o-semiquinone radical as a mimetic of the humic acids native radicals: A DFT approach to the molecular structure and EPR properties. *Geochim. Cosmochim. Acta* **2012**, *86*, 384–391. [CrossRef]

97. Gonet, S.S.; Cieślewicz, J. Differential thermal analysis of sedimentary humic acids in the light of their origin. *Environ. Int.* **1988**, *24*, 629–636. [CrossRef]

98. Dell'Abate, M.T.; Benedetti, A.; Brookes, P.C. Hyphenated techniques of thermal analysis for characterisation of soil humic substances. *J. Seph. Sci.* **2003**, *26*, 433–440. [CrossRef]

99. Purmalis, O.; Porsnovs, D.; Klavins, M. Differential Thermal Analysis of Peat and Peat Humic Acids. *Mat. Sci. Appl. Chem.* **2011**, *24*, 89–94. [CrossRef]
100. Schnitzer, M.; Turner, R.C.; Hoffman, I.A. Thermogravimetric study of organic matter of representative canadian podzol soils. *Can. J. Soil Sci.* **1964**, *44*, 7–13. [CrossRef]
101. Kodama, H.; Schnitzer, M. Kinetics and mechanism of the thermal decomposition of fulvic acids. *Soil Sci.* **1970**, *109*, 265–271. [CrossRef]
102. Shurygina, E.A.; Larina, N.K.; Chubarova, M.A.; Kononova, M.M. Differential thermal analysis (DTA) and thermogravimetry (TG) of soil humus substances. *Geoderma* **1971**, *6*, 169–177. [CrossRef]
103. Leinweber, P.; Schulten, H.-R. Differential thermal analysis, thermogravimetry and in-source pyrolysis-mass spectrometry studies on the formation of soil organic matter. *Thermochim. Acta* **1992**, *200*, 151–167. [CrossRef]
104. Naucke, W. Die Untersuchung des Naturstiffe Torf und seiner Inhaltstoffe. *Chemi. App.* **1968**, 261–280.
105. Naucke, W. *Möglichkeiten zur Analyse von Torfinhaltstoffen mit Physikalisch-Chemischen Methoden.* 4; Torf-Kolloquium DDR-VR Polen: Rostock, Germany, 1968.
106. Plante, A.F.; Fernández, J.M.; Leifeld, J. Application of thermal analysis techniques in soil science. *Geoderma* **2009**, *153*, 1–10. [CrossRef]
107. Francioso, O.; Montecchio, D.; Gioacchini, P.; Cavani, L.; Ciavatta, C.; Trubetskoj, O.; Trybetskaja, O. Thermal analysis (TG–DTA) and isotopic characterization (13C–15N) of humic acids from different origins. *Appl. Geochem.* **2005**, *20*, 537–544. [CrossRef]

Sample Availability: Samples of compound **1** are available from the authors.

Article

Cellulose Nanofibrils Filled Poly(Lactic Acid) Biocomposite Filament for FDM 3D Printing

Qianqian Wang [1], Chencheng Ji [1], Lushan Sun [2,*], Jianzhong Sun [1,*] and Jun Liu [1]

[1] Biofuels Institute, School of the Environment and Safety Engineering, Jiangsu University, Zhenjiang 212013, China; wqq@ujs.edu.cn (Q.W.); cecilji@amecnsh.com (C.J.); junliu115142@ujs.edu.cn (J.L.)
[2] Institute of Textiles and Clothing, The Hong Kong Polytechnic University, Hong Kong, China
* Correspondence: sarina.sun@polyu.edu.hk (L.S.); jzsun1002@ujs.edu.cn (J.S.)

Academic Editor: Sylvain Caillol
Received: 30 April 2020; Accepted: 13 May 2020; Published: 15 May 2020

Abstract: As direct digital manufacturing, 3D printing (3DP) technology provides new development directions and opportunities for the high-value utilization of a wide range of biological materials. Cellulose nanofibrils (CNF) and polylactic acid (PLA) biocomposite filaments for fused deposition modeling (FDM) 3DP were developed in this study. Firstly, CNF was isolated by enzymatic hydrolysis combined with high-pressure homogenization. CNF/PLA filaments were then prepared by melt-extrusion of PLA as the matrix and CNF as the filler. Thermal stability, mechanical performance, and water absorption property of biocomposite filaments and 3D-printed objects were analyzed. Findings showed that CNF increased the thermal stability of the PLA/PEG600/CNF composite. Compared to unfilled PLA FDM filaments, the CNF filled PLA biocomposite filament showed an increase of 33% in tensile strength and 19% in elongation at break, suggesting better compatibility for desktop FDM 3DP. This study provided a new potential for the high-value utilization of CNF in 3DP in consumer product applications.

Keywords: melt extrusion; 3D printing; cellulose nanofibrils; biocomposite filaments; physical property

---

1. Introduction

From aerospace engineering to the maker communities, 3D printing (3DP), as a form of direct digital manufacturing and additive manufacturing, has gained exponential growth in a variety of fields during the last few decades [1–5]. Many have found promising potential in its mass adoption resulting in products such as furniture and architectural design that use various 3DP technologies. Among them, the apparel and textile fields have already been using 3DP in footwear and accessory products that require less flexibility in the 3DP materials applied [6]. Considering the unique advantages of design efficiency and customization in 3DP, some industry brands and professional designers and engineers have been exploring its integration into more wearable textile and apparel products [7,8]. Through unique 3D computer-aided design (CAD) techniques and approaches [9,10], garments developed with 3DP components are now much more versatile and even resilient towards the human body's movements and fitting. However, few examples are found for the use of a sustainable material base in 3DP-integrated textiles and apparel. Currently, the textile and apparel industry has recognized the negative environmental effects of fiber and fabric production processes, all of which can often involve extensive toxic or biologically modifying reagents [6]. Further, there is a challenge with regards to efficiency and cost in recycling and upcycling postconsumer textile and apparel products [11,12]. Therefore, it is essential that new fabrication technology, such as 3DP, adopts more environmentally sustainable approaches and materials.

Although researchers have begun to explore recycled 3DP materials [13], biocomposite materials used in desktop fused deposition modeling (FDM) 3DP are expected to also become highly impactful in the development of bio-based materials for consumer products [2]. The biomaterials are often derived from wood or agricultural waste. Polylactic acid (PLA), commonly made using corn starch and sugar cane has been one of the FDM thermoplastic options. However, as the 3DP technology rapidly advances to integrate into fields like textiles and apparel, conventional virgin PLA filament can no longer meet the need for diverse products. To improve the physicochemical properties of PLA for FDM 3DP, some researchers have begun to introduce biomass raw materials into the PLA matrix. The addition of microcrystalline cellulose (MCC) as a reinforcing material to the PLA matrix has improved the crystallinity and the strength of the composite material as compared to that of virgin PLA [14,15].

Further, nanocellulose has also been studied as a key additive to strengthen PLA. It is produced from natural cellulose. Its low density, ultra-fine structure, and high strength properties are attractive in the field of reinforcing materials [16]. In a small proportion, nanocellulose can improve the PLA matrix in crystallinity, mechanical properties, degradability, etc. [17]. Scholars have further recognized that 3D printing with nanomaterials could be used in leveraging greater control in fundamental material properties, providing multifunctionality and custom geometry to product needs [18]. Wang et al. [19] prepared micro-nanocellulose using a colloid mill and combined it into the PLA matrix at a relatively high proportion. The developed composite filament had a higher tensile strength.

Additionally, the preparation of nanocellulose is critical in composite development and mainly includes chemical, mechanical, and biological methods [20–22]. Nanocellulose prepared by a concentrated sulfuric acid hydrolysis is poor in thermal stability due to the introduction of sulfate half ester groups [23]. Nanocellulose can also be prepared by high-pressure homogenization with long fibers as raw materials. Mechanical homogenization requires high energy consumption. Moreover, the high-pressure homogenizer was occasionally blocked by the long fiber, which led to the failure to discharge. This limitation still exists in the PLA/cellulose nanofibrils (CNF) composite technique when applied to 3D printing processing. Particularly, the uniform dispersion of nanocellulose into the PLA matrix is a real challenge that will significantly affect the printability of the fabricated composite filament [24] and the physical properties of 3D-printed architectures. In addition, PEG, as a plasticizer, commonly used in the melting process for polymers, is another important factor involved in biocomposite fabrication.

In this study, the enzymatic hydrolysis of MCC was used to a pretreatment step for high-pressure homogenization. The obtained CNF was freeze-dried. The PLA/CNF composite filaments for 3DP were prepared by melt-extrusion as shown in Figure 1. The properties of the composite filament were characterized and its printability with an FDM 3D printer was verified. The kinetic thermal behavior of the CNF filled PLA filaments were systematically investigated in our previous study [25]. The purpose of this study is to develop CNF filled PLA biocomposite filaments for 3D FDM printing with CNF as the filler, polyethylene glycol 600 (PEG600) as the smoothing agent, and virgin PLA granules as the host matrix. The isolated CNF was tested through morphological studies. The PLA/CNF composite specimens were evaluated in thermal, mechanical, and water absorption tests. The best-performing PLA/CNF composite specimen was extruded into 3DP filament and then used in 3D printing for prototype objects.

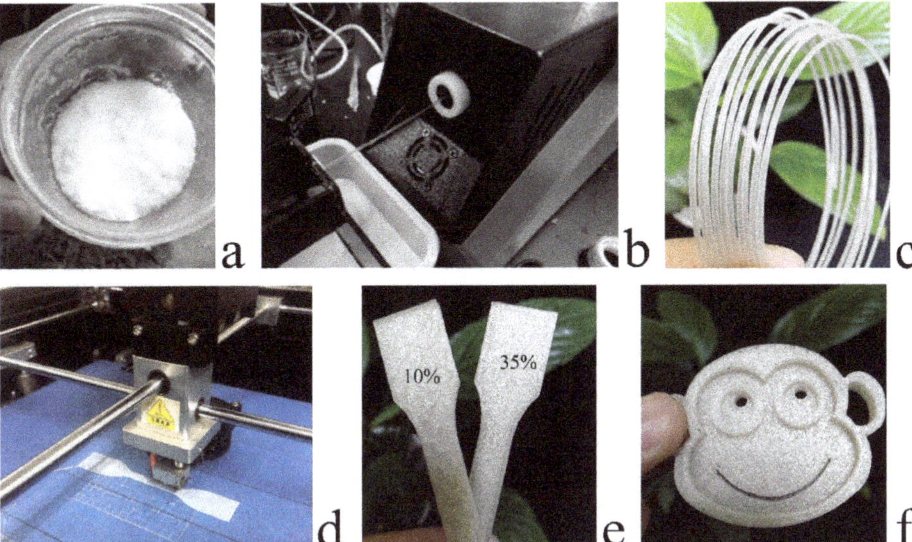

Figure 1. Cellulose nanofibrils (CNF) powder: (**a**) composite filament extruded on Wellzoom desktop extruder; (**b**) composite filament (1.75 mm diameter); (**c**) specimen in printing on M3036 FDM desktop 3D printer; (**d**) 3D-printed specimen with 10% and 35% infill; (**e**) 3D-printed specimen monkey face with 40 mm/s printing speed and 100% infill (**f**).

2. Results

2.1. CNF Morphological Observation

As can be seen from the TEM images (Figure 2c), the CNF was successfully isolated with 10 passes using enzymatic hydrolyzed MCC. The length of the CNF was in the range of a few hundred nanometers and 1 micron and the diameter was less than 50 nm. This indicated that CNF can be obtained via the combined enzymatic hydrolysis and high-pressure homogenization techniques. The CNF has a large number of hydroxyl groups and a high surface area, thus, a great tendency to form networks during freeze-drying, as shown in Figure 2d. The mechanically milled CNF exhibits a similar network structure (Figure 2e).

The CNF prepared by enzymatic hydrolysis and high-pressure homogenization showed good homogeneity in length and diameter as determined by the statistical analysis of the TEM images. A large CNF fibrous network structure can be seen in the SEM and TEM images. Dried CNF forms complex three-dimensional network structures. The high-pressure homogenization process destroys part of the crystalline structure, resulting in a decrease in the crystallinity of the developed CNF [20]. FTIR spectra (Figure 2a) show typical IR bands for the cellulose structure. The peak at 3450 cm^{-1} was attributed to the O-H group stretching vibration. The peak at 2900 cm^{-1} was derived from the C-H stretching, while the peak at 1060 cm^{-1} was attributed to the vibration of C-O-C in the glucose ring [26].

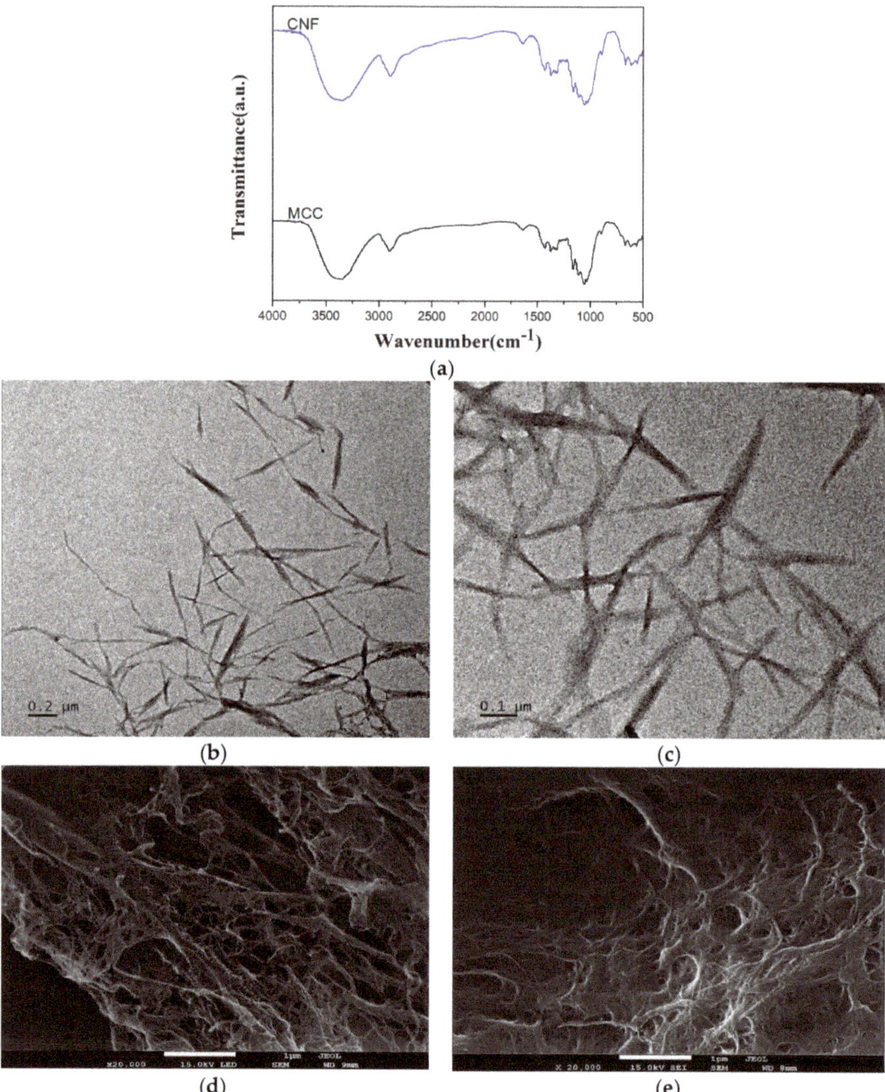

Figure 2. FTIR Spectrum of microcrystalline cellulose (MCC) and CNF. (**a**) TEM images of enzymatic hydrolysis of cellulose with high-pressure homogenization under different scales. (**b**) and (**c**) after 10 passes through the homogenizer. FE-SEM micrographs of CNF at a magnification of 20,000 (**d**) after freeze-drying and (**e**) after mechanical dispersion (following freeze drying).

2.2. PLA/CNF Thermal Stability Analysis

In our study, PEG600 is not only working as a plasticizer but also as an adhesive in the filament preparation procedure. The CNF and PLA particles with PEG600 can be homogeneously mixed at a temperature above the PEG600 melting point (17–22 °C). The CNF/PEG600 mixture stuck to the PLA particles when the temperature dropped below the melting point of the PEG600. Without PEG600, the dispersion of CNF in the PLA particles is a real challenge for the single screw extruder. High-quality filaments cannot be prepared without the addition of the PEG600 using the single screw extruder, and only the filaments with high quality are suitable for further analysis.

In Table 1 the weight loss temperature at 5% wt. of mass loss (T5%) was used as the initial decomposition temperature because we decided that is more reliable than T_{onset}. In the TG and DTG curves of the MCC and CNF the peak at 100 °C due to the adsorbed water can be observed (Figure 3a). The values of $T_{5\%}$, $T_{10\%}$, $T_{50\%}$, and T_{vmax} for the CNF are lower than that of the MCC, suggesting that the thermal stability of the CNF is lower than that of the MCC (Table 1).

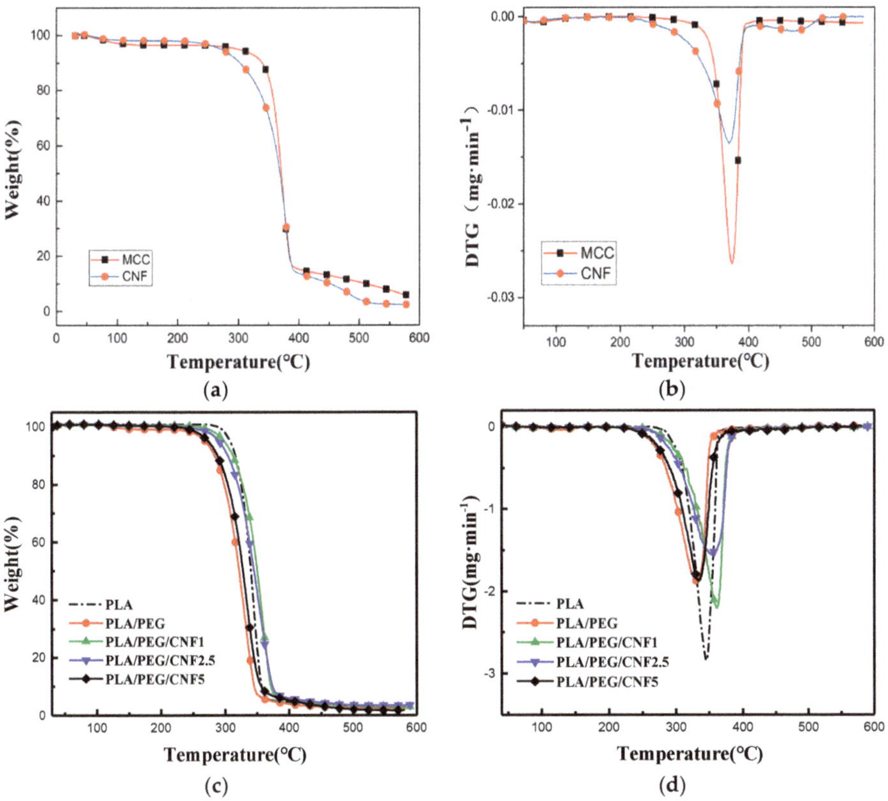

Figure 3. TGA analysis of MCC and CNF: (**a**) TG curves, (**b**) DTG curves; TGA analysis of virgin polylactic acid (PLA) and composites with different wt.% CNF: (**c**) TG curves and (**d**) DTG curves.

Table 1. Weight loss temperature of MCC, CNF, PLA, PLA/PEG, and PLA/PEG/CNF composites.

Specimen *	$T_{onset}(°C)$	$T_{5\%}(°C)$	$T_{10\%}(°C)$	$T_{50\%}(°C)$	$T_{vmax}(°C)$
MCC	353	301	340	371	372
CNF	321	257	292	362	369
PLA	311	308	316	340	346
PLA/PEG	295	267	284	321	330
PLA/PEG/CNF 1 wt.%	320	298	312	350	361
PLA/PEG/CNF 2.5 wt.%	311	290	304	344	352
PLA/PEG/CNF 5 wt.%	302	273	288	328	335

*: PLA/PEG/CNF 1 wt.%, PLA/PEG/CNF 2.5 wt.%, PLA/PEG/CNF 5 wt.% means CNF loading were 1, 2.5 and 5 wt.%, respectively.

Compared to composites with and without PEG600 (Figure 3c–d), the thermal stability of the PLA with PEG600 is lower than the virgin PLA. This suggests that the small molecule PEG600 can alter the thermal behavior of the PLA [27], making the intermolecular heat transfer faster and accelerating the break of the molecular chain. Table 1 shows that the virgin PEG600 composite began to decompose at 267 °C, which contrasts with the PLA/PEG600/CNF composite's initial decomposition temperature that is higher, up to 298 °C at 1 wt.% CNF. Also, the composite's $T_{5\%}$ appears to improve when the CNF is at 1 wt.% that decreases to 273 °C than when the CNF is at 5 wt.%, which is still higher than the virgin PLA and PEG600 composite. The thermal stability of the composite shows the tendency of rising at a low CNF addition and declining as the CNF wt.% increases that suggests the CNF proportion increase results in the gradual change of the composite's thermal degradation behavior.

The DTG curves (Figure 3d) show that the PLA/PEG600/CNF composite's maximum rate of weight loss starts at 361 °C and tends to decrease as the CNF wt.% increases. The best thermal stability was found for composites with 1 wt.% of CNF addition.

2.3. PLA/PEG600/CNF Composites Mechanical Performance Analysis

Figure 4a shows the tensile strength of the PLA/PEG600/CNF filaments. The addition of PEG600 increased the elongation at break and reduced the tensile strength of the PLA filament. This is due to the plasticizing effect of PEG600. The mechanical properties of the composite material first increased and then decreased with the increase of the CNF content. The maximum mechanical stability was observed at 2.5 wt.% loading. The CNF can be well-dispersed in the PLA matrix thanks to the compatibilizing effect of the PEG600, and the hydrogen bonds could be formed among CNF, PLA, and PEG600, leading to the increase in the tensile strength and elongation at break. When the CNF content is low, the cross-linked structure of the CNF is loose in the composite system, which has little effect on the mechanical properties. When 5 wt.% of CNF is added, the CNF can be partially agglomerated in the PLA matrix, and this becomes a defect point in the composite material. The specimen's cross section was analyzed by FE-SEM (Figure 4b–e). Compared to the high, flat surface of the virgin PLA, the cross-section of the various CNF composites appeared increasingly uneven with fibrous structures.

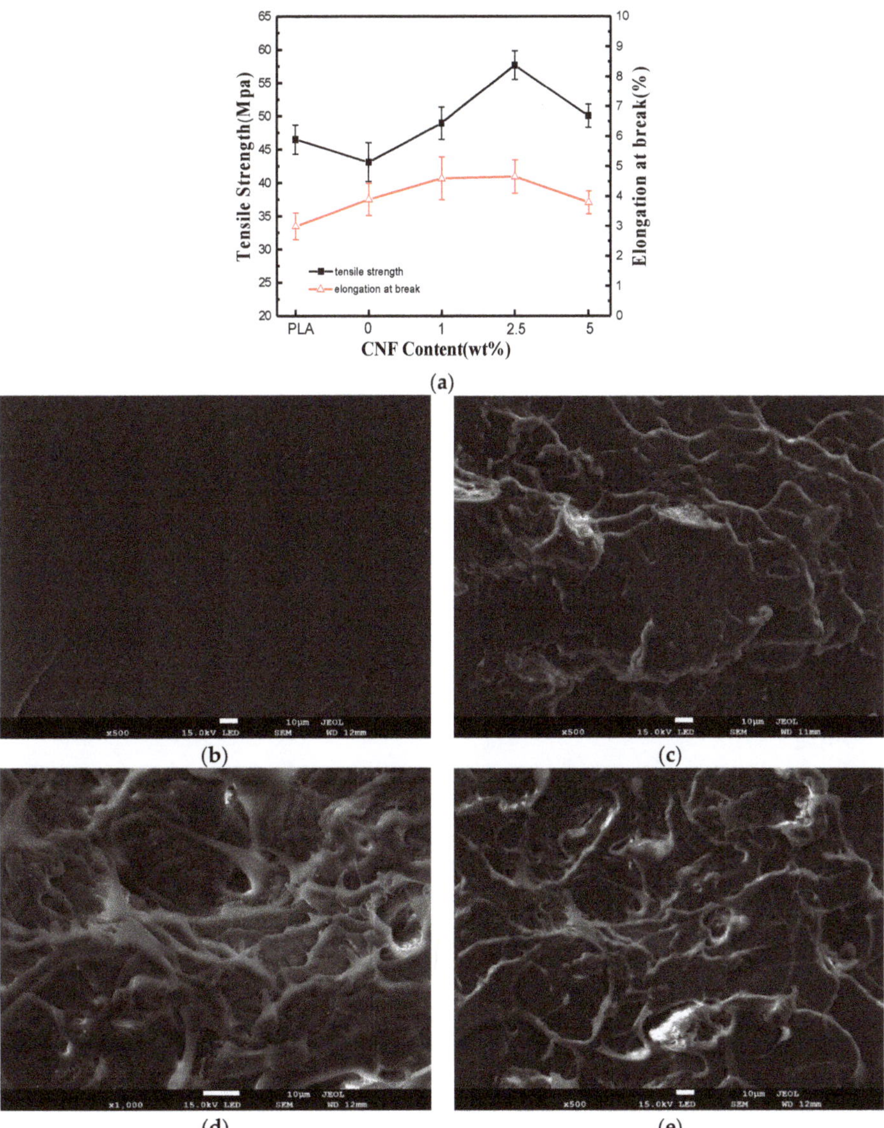

Figure 4. (**a**) Tensile properties of PLA/PEG/CNF composites with different CNF loading; (**b**) FE-SEM micrographs of cross-section of virgin PLA; (**c**) PLA/PEG/CNF 1 wt.%; (**d**) PLA/PEG/ CNF 2.5 wt.%; (**e**) PLA/PEG/CNF 5 wt.%. Scale bar = 10 μm.

2.4. PLA/PEG600/CNF Composites Water Absorption

Although water absorption may harm the structural stability of the specimen, it may also have a positive effect on the biodegradability. Since CNF is hydrophilic, it is vital to determine the water absorption rate. In water absorption results, the specimens showed a slight weight increase during the 96 h of the experiment (Figure 5). Overall, compared to the virgin PLA, the composites with CNF at various wt.% showed a similar trend, and absorption of water was the fastest in the first 24 h and stabilized afterward around hour 48 of the experiment. The higher the wt.% of CNF, the higher the water absorption rate of the specimen. Furthermore, the virgin PLA/PEG600 composite reflects a drastic increase in water absorption rate as compared to neat PLA. The water absorption rate increased gradually with the increase of CNF content because the cellulose surface contains a large number of hydrophilic hydroxyl groups. The terminal hydroxyl groups in PEG600 may also affect water absorption rate. Because PEG600 can alter the formation of crystallites of PLA, it allowed more water molecules to bond with hydroxyl groups in the composites' structure, thus increasing the water absorption rate. PLA degradation is water-based. Thus, the improvement in the water absorption rate may enhance the PLA degradation rate.

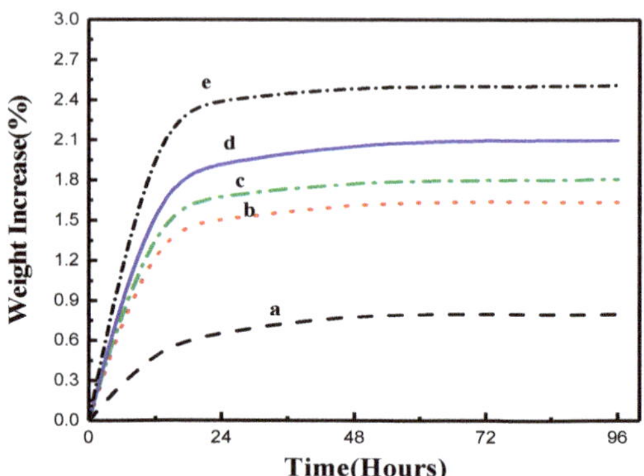

Figure 5. (a) Water absorption rate of virgin PLA; (b) PLA/PEG; (c) PLA/PEG/CNF 1 wt.%; (d) PLA/PEG/CNF 2.5 wt.%; (e) PLA/PEG/CNF 5 wt.%.

3. Discussion

3.1. Filament Extrusion and 3D Printing

The PLA/PEG/CNF 2.5 wt.% filament was tested in 3D-printed prototypes (Figure 1b,c). Overall, the resulting filaments revealed a smooth surface with improved properties compared to the conventional PLA filament. It suggests the compatibility of this kind of material in desktop FDM 3D printing. In evaluating the surface quality of the composite filament, the developed prototypes (Figure 1b,c) are compared in a simple strip design with 10% and 35% fill rates. The outcome with the higher fill rate (35%) showed higher compactness than the sparse texture in the lower fill rate (10%) prototype, which can be suitable for application needs. In the more complex structure (Figure 1f), the outcome surface remained smooth, and no clogging occurred in the printer nozzle during the printing process.

In terms of the tensile strength and elongation at break for the final filament, the results show that the maximum tensile strength and elongation at break of the PLA/PEG600/CNF composite filaments range from 19.5% to 33.8%, which is higher compared to the pure PLA filament [28]. Two aspects are vital in developing the CNF filled PLA filaments: (1) PEG600 served as a plasticizer in improving the composite's compatibility. In the post-enzymatic reaction, the higher-pressure homogenized CNF leads to a larger surface area of CNF. Therefore, mixing of the PEG600 with the CNF allowed the additives to blend with the PLA granules more efficiently. It is important to note that PEG600 also provides lubrication in the final filament extrusion. (2) When the CNF wt.% is low (2.5 wt.%), the composites resulted in a looser crosslinked structure that provided minimal impact on the composite's final tensile strength. On the contrary, when the CNF is high (5 wt.%), the CNF is not able to evenly disperse in the PLA matrix, which leads to agglomeration and weakness of tensile strength. Compared to the pure or conventional PLA, the final PLA/PEG600/CNF composite filaments show the maximum tensile strength and elongation at break of 33.8% and 19.5%, respectively.

3.2. Preparation CNF and PLA/PEG600/CNF Composites

To develop high-quality PLA-based composites, several factors are critical.

1) The 2-stage preparation procedure for CNF is prerequisite for the later effective composite mixing. The combined use of enzymatic hydrolysis and high-pressure homogenization enabled efficient CNF modification. Further, previous research has resulted in visible granules in the specimens with high CNF concentration. This may become one of the major limiting factors in the PLA composite's development [29]. Therefore, the mechanical pretreatment for freeze-dried composite is critical to increase the material dispersion and reduce agglomeration in the PLA composite.

2) The unique use of PEG600 is very important in achieving high performance in the final filament extrusion and 3D printing. The CNF developed already shows higher water absorption and thermal stability. After adding PEG600, these qualities were further enhanced.

3) The CNF used in composite development is often customized but limited in the preparation techniques or recipes in former studies. Previous research [30] employed fine wood powders to mix into the PLA that resulted in a rough and uneven texture. Here, the mixing of the PEG600 is much simpler and cheaper, and largely reduces the need for industrial processing to achieve an even composite mixture. In addition, the properties of the CNF produced by different techniques vary greatly, so it is critical to ensure the stability and quality of the CNF for 3D-printing-related applications.

3.3. Textile and Apparel Product Applications

In the textiles and apparel fields, ready-to-wear products and experimental explorations using FDM 3DP are currently limited to thermoplastic materials that are not biodegradable. The PLA/PEG600/CNF biocomposite filament developed in this study indicates great potential for more renewable material in future 3DP-integrated wearables in several ways: (1) Based on the mechanical performance measures, the increased tensile strength and elongation at break provides a foundation for potential 3D-printed structures that can be used in textiles and apparel [7,8]. Many of the currently available filament options for FDM printing have limitations in the material's overall resilience and durability for comfort considerations. For complex or detailed textile or apparel 3DP structures, it is also important to test the mechanical responses in filament form and outcomes from different printing directions [30]. (2) The hydrophilic nature of the CNF helps to enhance the quality of the developed biocomposite and, more importantly, allows the later coloring or dying post-processing to be more accessible for varying the 3D-printed parts in textile and apparel products.

At a larger scale, (4) the PLA/PEG600/CNF biocomposite filament has the potential of providing a naturally derived material for consumer products and allows higher efficiency in post-consumer recycling and upcycling. (5) It also promotes compatibility in combining with other natural fiber-based traditional fabrics in other applications and design possibilities. However, a limitation still exists in achieving the market-ready quality filament for wearable products. In previous filament explorations, scholars have pointed out the limitations in processing, cost, consistency, volume reliability, and high lead time in nanocomposite production [18]. Further investigation must be conducted to perfect the techniques for higher production yield, as well as its economic performance for various applications.

4. Material and Methods

CNF was produced from microcrystalline cellulose (MCC) from the Qu Fu Tian Li Medical Supplement Corporation (Qufu, China). Cellic CTec2 (cellulase cocktail, enzyme blend) was obtained from Novozymes (Suzhou, China), while PEG600 and virgin PLA (4032D) were purchased from Sinopharm Chemical Reagent Limited Company (Shanhai, China) and Natureworks (Minnetonka, MN, USA), respectively. The density of PLA was 1.24 g.cm^{-3} and it had a melting point in range of 155 to 170 °C [31].

4.1. CNF Isolation

A 2-stage process was utilized in the CNF preparation. First, the MCC was mixed with acetic acid-sodium acetate buffer solution (pH = 4.6) in 10 wt.% consistency, and 0.5 mL cellic CTec2 (cellulase cocktail, 128 FPU/mL) per gram MCC was added. The enzymatic hydrolysis was kept in a 50 °C water bath for 2 h and inactivated in an 80 °C water bath for 30 min. The resulting cellulose suspension was vacuum filtrated and washed with deionized water. The filtered cellulose suspension was diluted into 2% solid content and homogenized 10 times at 80 Mpa (AH-BASIC, ATS Engineering Co. Ltd., Jiangsu, China). The homogenized cellulose suspension was finally pre-frozen, freeze-dried, and milled to pass through a 300 mesh (Herb Grinder 2, Yongkang Boou hardware products Co. Ltd., Zhejiang, China). The morphology of the resulting CNF was tested with a Tecnai F30 TEM and a Hitachi 7800F FE-SEM. The FTIR spectra of the isolated CNF were determined with a Fourier-transform infrared instrument (Nexus 470, Nicolet Instruments, Offenbach, Germany) in transmission mode. Freeze-dried CNF samples were ground and pelletized using KBr and the spectra were recorded in the range from 400 to 4000 cm^{-1}.

4.2. PLA/CNF Composite Development and Testing

4.2.1. Preparing the PLA/CNF Composite

The lab for filament extrusion is not air conditioned. The temperature of the lab was around 0 °C in winter. PEG600, which is in a solid state at 0 °C, was heated to liquid form and added into the PLA granules at 4% loading to provide a more even and well-blended consistency. CNF powder was then added at three different wt.% (1, 2.5, and 5) to the final granule mixture for refrigeration (Figure 1a). The PLA/PEG/CNF mixture was melt-extruded using a Wellzoom desktop single screw extruder specially designed for manufacturing 3D printing filaments by Shenzhen Si Da Technology Limited (175 °C at mixing region and 170 °C at extrusion region). The speed of the single screw extruder was set to 35 rpm. A 1.8 mm round hole is arranged at the front of the cooling water tank to control the diameter of the filament and the speed of the extractor was tuned manually according to the speed of the extruder. The resulting filament was maintained with a diameter of 1.75 ± 0.05 mm and was air-cooled and water-cooled. Total of five specimens (virgin PLA, PLA/PEG, PLA/PEG/CNF 1 wt.%, PLA/PEG/CNF 2.5 wt.%, PLA/PEG/CNF 5 wt.%) were used in the following tests.

4.2.2. Thermal Stability

The specimen (~5 mg) was analyzed in an alumina crucible by thermal gravimetric analysis (TGA) for its thermal stability using the temperature range of 30–600 °C at a heating rate of 10 °C/min under nitrogen atmosphere, and the gas flow rate of 20 mL/min.

4.2.3. Mechanical Performance

Tensile strength was measured with a JF-9003 tensile tester by Dongguan Jianfeng Instrument Limited with a gauge of 50 mm, and a speed of 5 mm/min. Filaments with various formulations were tested at least 5 times each and the average value was reported.

4.2.4. Water Absorption

The specimens were first oven-dried at 60 °C for 48 h and weighted, and then submerged in PBS buffer (Phosphate-buffered saline) with pH 7.4 for 24 h at room temperature (20 °C). The specimen was then surface-cleaned with absorbent paper and weighted with an analytical balance (Sartorius TE214S with an accuracy of 0.1 mg, Goettingen, Germany). The water absorption rate was calculated using the following equation [32]:

$$\text{water absorption rate} = [(M_2 - M_1)/M_1] \times 100\% \tag{1}$$

Here, the M_1 represents pre-absorbent mass, and M_2 is the post-absorbent mass. The results were reported as an average of two measurements.

4.2.5. 3D Printing and Evaluation

The fabricated filaments were used in printing prototypes for evaluation. The M3036 FDM desktop 3D printer by Shenzhen Technology Limited Co. and Cura slicing software was used to process the CAD models. As for the 3D printing processing, a 0.4 mm nozzle was used. The extruder temperature and printing speed were set as 210 °C and 40 mm/s, respectively. A simple model sample was compared at different fill rates (10%, 35%), and a complex model sample was printed for printability evaluation.

5. Conclusions

In this paper, PLA/PEG600/CNF biocomposite filaments were developed and tested for desktop FDM 3D printing. The CNF was prepared using a 2-stage technique involving enzymatic hydrolysis with high-pressure homogenization. PEG600 was added to the composite to improve material performance. The final PLA/PEG600/CNF composite material was extruded to 3DP filament and printed into prototypes. With a CNF content of 2.5 wt.%, the test results showed that the composite filament can meet the requirements of desktop FDM 3D printing. The water absorption rates of CNF composites may indicate good structural stability in a humid environment. The study further pinpoints some valuable properties for applications in the highly prospected textiles and apparel industry. The potential advantage in using the biodegradable 3DP composite in relevant products could lead to higher energy-efficient processing in product recycling and upcycling. Traditional fabric customization by 3D printing may be possible with less labor-intensive processing and could be explored in a future study. Further studies should evaluate the functional and comfort properties and potential of the 3D-printed parts, particularly in different textile and apparel product integrations. Additionally, the effects of infill structures in the 3D-printed parts should be considered for different functional applications. Future studies can also expand to the fields of architecture, interiors, and packaging applications. More efforts in structural design, 3D printing process optimization, and post-processing are needed to increase the mechanical performance in nanosized natural fiber-reinforced PLA filament and 3D-printed products.

Author Contributions: Conceptualization, Q.W., J.S., and L.S.; methodology, C.J.; data curation, C.J.; writing, review, and editing, L.S., Q.W., J.S., and J.L.; resource and supporting, J.S. All authors have read and agreed to the published version of the manuscript.

Funding: This research was funded by the National Key R&D Program of China (2018YFE0107100), the Academic Program Development of Jiangsu Higher Education Institutions (4013000019), Jiangsu Provincial Key Laboratory of Biomass Energy, and Material Open Fund (JSBEM202012).

Conflicts of Interest: The authors declare no conflict of interest.

References

1. Wang, Q.Q.; Sun, J.Z.; Yao, Q.; Ji, C.C.; Liu, J.; Zhu, Q.Q. 3D printing with cellulose materials. *Cellulose* **2018**, *25*, 4275–4301. [CrossRef]
2. Liu, J.; Sun, L.; Xu, W.; Wang, Q.; Yu, S.; Sun, J. Current advances and future perspectives of 3D printing natural-derived biopolymers. *Carbohydr. Polym.* **2019**, *207*, 297–316. [CrossRef] [PubMed]
3. Wong, K.V.; Hernandez, A. A review of additive manufacturing. *Int. Sch. Res. Not.* **2012**, *2012*, 208760. [CrossRef]
4. Horn, T.J.; Harrysson, O.L. Overview of current additive manufacturing technologies and selected applications. *Sci. Prog.* **2012**, *95*, 255–282. [CrossRef] [PubMed]
5. Bose, S.; Vahabzadeh, S.; Bandyopadhyay, A. Bone tissue engineering using 3D printing. *Mater. Today* **2013**, *16*, 496–504. [CrossRef]
6. Sun, L.; Zhao, L. Envisioning the era of 3D printing: A conceptual model for the fashion industry. *Fash. Text.* **2017**, *4*, 25. [CrossRef]

7. Robinson, F. Chanel's Couture Progression: 3D Printing Materiality. Available online: http://www.disruptivemagazine.com/opinion/chanel%E2%80%99s-couture-progression-3d-printing-materiality (accessed on 1 April 2018).
8. Danit, P. How I 3D Printed a 5 Piece Fashion Collection at Home. Available online: http://danitpeleg.com/3d-printing-fashion-process/ (accessed on 1 April 2018).
9. Melnikova, R.; Ehrmann, A.; Finsterbusch, K. *IOP Conference Series: Materials Science and Engineering, Proceedings of the 3D Printing of Textile-Based Structures by Fused Deposition Modelling (FDM) with Different Polymer Materials, Ningbo, China, 27–29 May 2014*; IOP Publishing: Bristol, UK, 2014; p. 012018.
10. Lussenburg, K.; Van Der Velden, N.; Doubrovski, Z.; Karana, E. Designing (with) 3D Printed Textiles. Master's Thesis, Delft University of Technology, Delft, The Netherlands, 2014.
11. Tanney, D.; Meisel, N.A.; Moore, J. Investigating Material Degradation through the Recycling of PLA in Additively Manufactured Parts. In Proceedings of the 28th Annual International Solid Freeform Fabrication Symposium, Austin, TX, USA, 7–9 August 2017; pp. 519–531.
12. Lanzotti, A.; Martorelli, M.; Maietta, S.; Gerbino, S.; Penta, F.; Gloria, A. A comparison between mechanical properties of specimens 3D printed with virgin and recycled PLA. *Procedia Cirp* **2019**, *79*, 143–146. [CrossRef]
13. Anderson, I. Mechanical properties of specimens 3D printed with virgin and recycled polylactic acid. *3d Print. Addit. Manuf.* **2017**, *4*, 110–115. [CrossRef]
14. Murphy, C.A.; Collins, M.N. Microcrystalline cellulose reinforced polylactic acid biocomposite filaments for 3D printing. *Polym. Compos.* **2018**, *39*, 1311–1320. [CrossRef]
15. Mathew, A.P.; Oksman, K.; Sain, M. Mechanical properties of biodegradable composites from poly lactic acid (PLA) and microcrystalline cellulose (MCC). *J. Appl. Polym. Sci.* **2005**, *97*, 2014–2025. [CrossRef]
16. Lendvai, L.; Karger-Kocsis, J.; Kmetty, Á.; Drakopoulos, S.X. Production and characterization of microfibrillated cellulose-reinforced thermoplastic starch composites. *J. Appl. Polym. Sci.* **2016**, *133*, 42397. [CrossRef]
17. Suryanegara, L.; Nakagaito, A.N.; Yano, H. The effect of crystallization of PLA on the thermal and mechanical properties of microfibrillated cellulose-reinforced PLA composites. *Compos. Sci. Technol.* **2009**, *69*, 1187–1192. [CrossRef]
18. Campbell, T.A.; Ivanova, O.S. 3D printing of multifunctional nanocomposites. *Nano Today* **2013**, *8*, 119–120. [CrossRef]
19. Wang, Z.; Xu, J.; Lu, Y.; Hu, L.; Fan, Y.; Ma, J.; Zhou, X. Preparation of 3D printable micro/nanocellulose-polylactic acid (MNC/PLA) composite wire rods with high MNC constitution. *Ind. Crop. Prod.* **2017**, *109*, 889–896. [CrossRef]
20. Wang, Q.; Wei, W.; Chang, F.; Sun, J.; Xie, S.; Zhu, Q. Controlling the Size and Film Strength of Individualized Cellulose Nanofibrils Prepared by Combined Enzymatic Pretreatment and High Pressure Microfluidization. *BioResources* **2016**, *11*, 2536–2547. [CrossRef]
21. Wang, Q.; Zhao, X.; Zhu, J.Y. Kinetics of Strong Acid Hydrolysis of a Bleached Kraft Pulp for Producing Cellulose Nanocrystals (CNCs). *Ind. Eng. Chem. Res.* **2014**, *53*, 11007–11014. [CrossRef]
22. Wang, Q.Q.; Zhu, J.Y.; Gleisner, R.; Kuster, T.A.; Baxa, U.; McNeil, S.E. Morphological development of cellulose fibrils of a bleached eucalyptus pulp by mechanical fibrillation. *Cellulose* **2012**, *19*, 1631–1643. [CrossRef]
23. Roman, M.; Winter, W.T. Effect of sulfate groups from sulfuric acid hydrolysis on the thermal degradation behavior of bacterial cellulose. *Biomacromolecules* **2004**, *5*, 1671–1677. [CrossRef]
24. Mokhena, T.; Sefadi, J.; Sadiku, E.; John, M.; Mochane, M.; Mtibe, A. Thermoplastic processing of PLA/cellulose nanomaterials composites. *Polymers* **2018**, *10*, 1363. [CrossRef]
25. Wang, Q.; Ji, C.; Sun, J.; Yao, Q.; Liu, J.; Saeed, R.M.Y.; Zhu, Q. Kinetic thermal behavior of nanocellulose filled polylactic acid filament for fused filament fabrication 3D printing. *J. Appl. Polym. Sci.* **2019**, *137*, 48374. [CrossRef]
26. Wulandari, W.T.; Rochliadi, A.; Arcana, I.M. Nanocellulose prepared by acid hydrolysis of isolated cellulose from sugarcane bagasse. *Iop Conf. Ser. Mater. Sci. Eng.* **2016**, *107*, 012045. [CrossRef]
27. Jia, S.; Yu, D.; Zhu, Y.; Wang, Z.; Chen, L.; Fu, L. Morphology, crystallization and thermal behaviors of PLA-based composites: Wonderful effects of hybrid GO/PEG via dynamic impregnating. *Polymers* **2017**, *9*, 528. [CrossRef] [PubMed]

28. Tymrak, B.; Kreiger, M.; Pearce, J.M. Mechanical properties of components fabricated with open-source 3-D printers under realistic environmental conditions. *Mater. Des.* **2014**, *58*, 242–246. [CrossRef]

29. Tao, Y.; Wang, H.; Li, Z.; Li, P.; Shi, S.Q. Development and Application of Wood Flour-Filled Polylactic Acid Composite Filament for 3D Printing. *Materials* **2017**, *10*, 339. [CrossRef] [PubMed]

30. Song, Y.; Li, Y.; Song, W.; Yee, K.; Lee, K.-Y.; Tagarielli, V.L. Measurements of the mechanical response of unidirectional 3D-printed PLA. *Mater. Des.* **2017**, *123*, 154–164. [CrossRef]

31. Ingeo™ Biopolymer 4032D Technical Data Sheet. Available online: https://www.natureworksllc.com/~{}/media/Technical_Resources/Technical_Data_Sheets/TechnicalDataSheet_4032D_films_pdf.pdf (accessed on 8 May 2019).

32. Wenjuan, Z.; Zhihua, S. The Preparation of Nano Cellulose Whiskers/Polylactic Acid Composites. In Proceedings of the 2011 International Conference on Future Computer Science and Education, Xi'an, China, 20–21 August 2011; IEEE: Piscataway, NJ, USA, 2011; pp. 123–125.

Sample Availability: Samples of the used materials are available from the authors.

 *molecules* 

Article

Bio-Based Thermo-Reversible Aliphatic Polycarbonate Network

Pierre-Luc Durand, Etienne Grau and Henri Cramail *

CNRS, University Bordeaux, Bordeaux INP, LCPO, UMR 5629, F-33600 Pessac, France;
durand.pierreluc@gmail.com (P.-L.D.); egrau@enscbp.fr (E.G.)
* Correspondence: cramail@enscbp.fr; Tel.: +33-5568-46184

Academic Editor: Sylvain Caillol
Received: 27 November 2019; Accepted: 12 December 2019; Published: 24 December 2019

Abstract: Aliphatic polycarbonates represent an important class of materials with notable applications in the biomedical field. In this work, low Tg furan-functionalized bio-based aliphatic polycarbonates were cross-linked thanks to the Diels–Alder (DA) reaction with a bis-maleimide as the cross-linking agent. The thermo-reversible DA reaction allowed for the preparation of reversible cross-linked polycarbonate materials with tuneable properties as a function of the pendent furan content that was grafted on the polycarbonate backbone. The possibility to decrosslink the network around 70 °C could be an advantage for biomedical applications, despite the rather poor thermal stability of the furan-functionalized cross-linked polycarbonates.

Keywords: polycarbonates; furan-maleimide; Diels-Alder; bio-based; fatty acids

---

1. Introduction

Aliphatic polycarbonates (APC) raise considerable interest in the last decades for their specific characteristics i.e., biocompatibility and biodegradability [1]. Different routes can achive their synthesis, but these polymers are mainly synthesized by ring-opening polymerization (ROP) of cyclic carbonate monomers allowing for a good control over the polymer's microstructure [2]. The advantage of designing functional polycarbonates when compared to traditional polytrimethylenecarbonate (PTMC) lies in the modulation of their physico-chemical properties for specific needs, thus broadening and improving their performance characteristics. Functional polycarbonates can be synthesized by direct polymerization of a functional monomer or by post-polymerization chemical modification [3]. Besides, the use of APC from renewable resources, such as vegetable oils, is very interesting due to environmental concerns and also to seek new functionalities [4].

In addition, polycarbonate networks with elastomeric properties are desirable for a large number of biomedical applications, particularly in the emerging field of soft tissue engineering or drug delivery [5–10]. Thus, several research groups have studied various cross-linking methods for obtaining polycarbonate materials [11–16]. However, the main disadvantage of these cross-linked materials is their inability to be reformed or recycled. As a result, the development of self-healing polymers has been a very active area of research over the last decade [17–21]. In 2013, the World Economic Forum even named self-healing materials among the top 10 emerging technologies and could be of interest in many areas, such as protective coatings, biomedical applications, piping, and electronics [18,22]. Various stimuli can be employed to activate the reversibility of the transformation. Nevertheless, temperature and light are the two stimuli, which are most commonly used in the synthesis of "stimuli-responsive" self-healing materials.

A thermal process is convenient and effective for the treatment of materials and it can be easily implemented to any object. As a result, heat-induced reversible reactions are attractive for building up thermal-responsive self-healing materials. The most studied thermally reversible reactions are based on the Diels–Alder (DA) chemistry. The latter has been widely used in macromolecular chemistry and mostly based on the furan/maleimide chemistry [23–32].

Therefore, the purpose of this investigation is to synthesize new, thermally reversible, polycarbonate networks via Diels–Alder and retro-Diels–Alder reactions. The general method is based on the synthesis of low Tg polymers containing dangling furan groups, followed by their Diels–Alder cross-linking with an appropriate bis-maleimide cross-linking agent.

First, the grafting of pendent furan moieties will be detailed on a fatty acid based polycarbonate. Subsequently, the thermal reversibility of the cross-linking will be demonstrated. Finally, the influence of the cross-linking density on the polycarbonate properties will be studied.

2. Results and Discussion

2.1. Grafting of Furan Moieties on the Polymer

Aliphatic polycarbonate P(**NH-Und-6CC**) was used as the starting materials, since its synthesis by ROP of **NH-Und-6CC** is well-controlled and it can be performed on a multi-gram scale according o already published protocols (see Supplementary Materials for the full protocols) [16,21].

Therefore, the first task was the preparation of a furan-functionalized polycarbonate. Thiol-ene click chemistry appears to be straightforward to graft a desired functionality onto the polymer backbone, thanks to the available terminal unsaturation on each monomer unit.

The thiol-ene reaction between P(**NH-Und-6CC**) and the commercially available furfuryl mercaptan was performed at room temperature in ethanol at 2 mol L^{-1} for 8 h under UV irradiation (365 nm) (Scheme 1). DMPA was used as a radical initiator. The mixture was irradiated through an optical fiber (365 nm) to improve the conversion and decrease the reaction time. Additionally, three equivalents of furfuryl mercaptan with respect to C=C double bond were added for the same reasons. The polymer was recovered by precipitation in cold methanol.

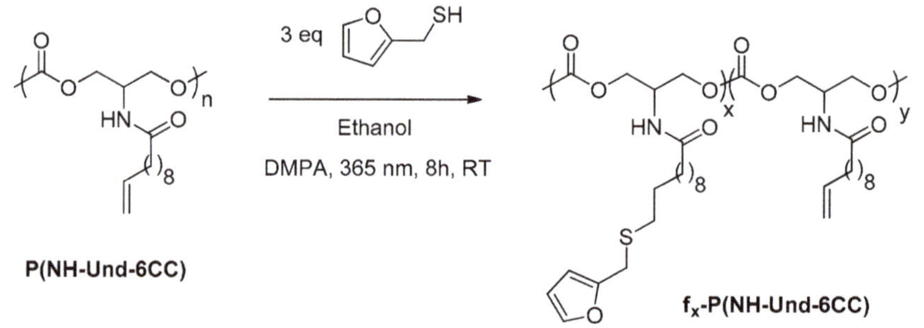

Scheme 1. Experimental procedure for the preparation of furan-functionalized P(NH-Und-6CC).

^{1}H NMR spectroscopy followed the thiol-ene reaction for 8 h (Figure S1). The disappearance of the characteristic signals of the terminal double bond (5.0 and 5.8 ppm) and the appearance at 2.50 ppm of the proton in α position of the sulfur atom (towards the polymer chain) enable the monitoring of the reaction. The signal at 2.25 ppm (integrated for 2) of the protons in α position of carbonyl from the amide group was taken as an internal reference to determine the furan content in the copolymer, thanks to the ratio of the peak at 2.25 and 2.50 ppm.

The evolution of the furan and the double bond content in the polycarbonate was plotted as a function of time (see Figure 1). The polymer displayed a moderate reactivity towards furfuryl mercaptan, since only 85 mol% of furan was grafted on the polycarbonate backbone after 8 h of reaction. However, the amount of furan that is grafted on the polycarbonate backbone can easily be adjusted by varying the reaction time. In addition, one can assume that the higher the furan content, the higher the cross-linking density. Thus, this control of the furan content is interesting to investigate the influence of the cross-linking density on the network properties.

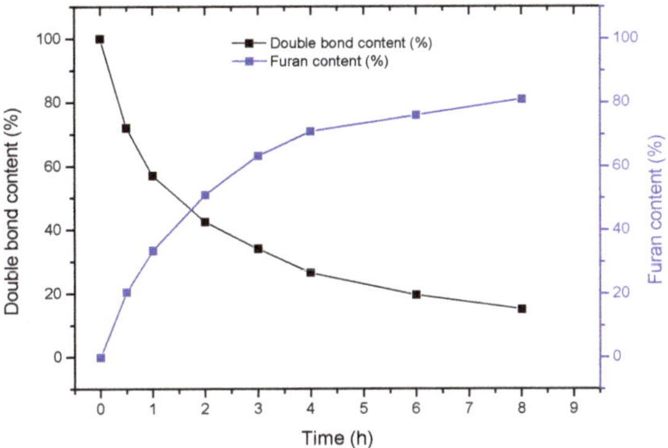

Figure 1. Kinetic profile of the P(NH-Und-6CC) functionalization by furan moieties.

^{1}H NMR characterized the resulting 85 mol% furan-functionalized polycarbonate, f_{85}-P(**NH-Und-6CC**) (see Figure 2). All of the peaks were assigned confirming the structure of the furan-functionalized polymer.

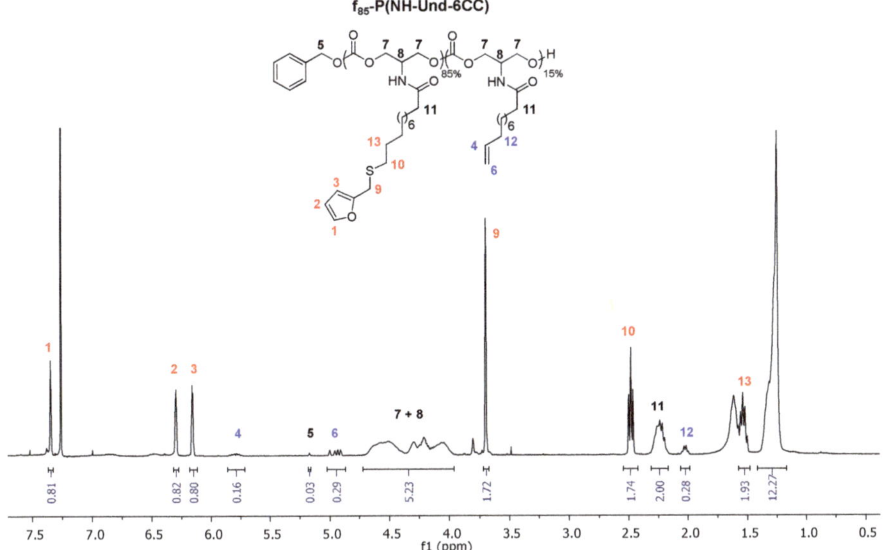

Figure 2. ^{1}H NMR of f_{85}-P(NH-Und-6CC) in CDCl$_3$.

Various polycarbonates were synthesized with different targeted furan contents while using the kinetic master curve. Table 1 summarizes the results.

First, it is noteworthy that the furan content is in a good agreement with respect to the kinetic profile (1). T_gs of these polycarbonates (Figure 1) range from 23.1 to −59.6 °C as a function of the furan content increase. Such a T_g decrease can be explained by the flexibility that the sulfur atoms introduced to the polymer chains. In addition, the SEC traces (Figure S2) indicated that the polymer molar masses increase is proportional to the furan content and testified no apparent degradation of the polymer during the grafting reaction.

Table 1. Characteristics of f_x-P(NH-Und-6CC) with pendent furan moieties.

Polymer	Reaction Time (h)	Furan Content [a] (mol%)	M_n [b] (g mol^{-1})	[Đ] [b]	T_g (°C) [c]
P(NH-Und-6CC)	0	0	5700	1.07	23.1
f_{31}-P(NH-Und-6CC)	1	31	6200	1.15	−6.6
f_{51}-P(NH-Und-6CC)	2	51	6650	1.12	−13.5
f_{69}-P(NH-Und-6CC)	4	69	7300	1.19	−49.0
f_{88}-P(NH-Und-6CC)	12	88	7850	1.22	−59.6

[a]: Determined by ^{1}H NMR; [b]: Determined by SEC in THF (PS Std); [c]: Determined by DSC at 10 °C min^{-1} from the second cycle (see Figure S3).

Therefore, several pendent furan-containing polycarbonates have been successfully synthesized. The reversible cross-linking will now be investigated thanks to the introduction of a bis-maleimide as cross-linking agent.

2.2. Reversible Cross-Linking and Reprocessability

The thermal reversibility of the cross-linking reaction was first demonstrated while taking f_{85}-P(**NH-Und-6CC**) as starting furan-containing polymer. Scheme 2 summarizes the mechanism that is involved in the cross-linking reaction.

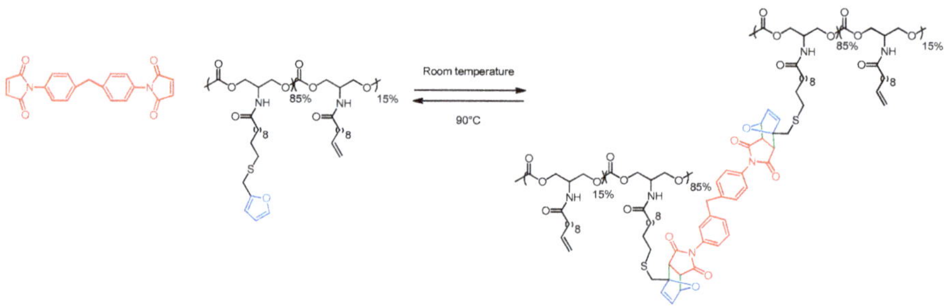

Scheme 2. Cross-linking reaction between f_{85}-P(**NH-Und-6CC**) and bis-maleimide.

The cross-linked polycarbonate material was obtained, as follows. The polymer f_{85}-P(**NH-Und-6CC**) was dissolved in chloroform (CHCl$_3$) at 1 g mL^{-1} in a vial. The bis-maleimide cross-linking agent was then added to the previous solution (0.5 equiv., with respect to furan groups) and the mixture was homogenized by vortex stirring until a clear yellow solution appears (the color is due to the bis-maleimide molecule). The vial was closed tightly and then heated up to 60 °C for 10 min.

The warm solution was poured in a Teflon mold and the chloroform was allowed to gently evaporate overnight at room temperature. A cross-linked film was obtained (Diels–Alder reaction) and was dried under reduced pressure for several hours to remove the traces of chloroform.

Placed in chloroform, this cross-linked polycarbonate should be back in solution upon heating at 90 °C for 10 min. (retro-Diels–Alder reaction). Figure 3 shows the solubility of the polymer after heating at 90 °C and the formation of an insoluble network when the solution is cooled down to room temperature. These transformations can be done several times in a row without any decomposition of the polymer.

FT-IR analysis was performed to demonstrate the reversibility of the cross-linking reaction (Figure 4). First, de-cross-linking of the material was followed by increasing the temperature (2 °C/min) (Figure 4a). The decreasing of the characteristic DA cyclo-adduct peak at 1190 cm^{-1} confirms the de-cross-linking reaction. Once thede-cross-linking of the material was achieved, the temperature was allowed to slowly decrease (2 °C/min). The re-cross-linking reaction was followed by transmission FT-IR (Figure 4b). This time, a significant increase of the DA cyclo-adduct peak was observed at 1190 cm^{-1}, highlighting the re-cross-linking reaction.

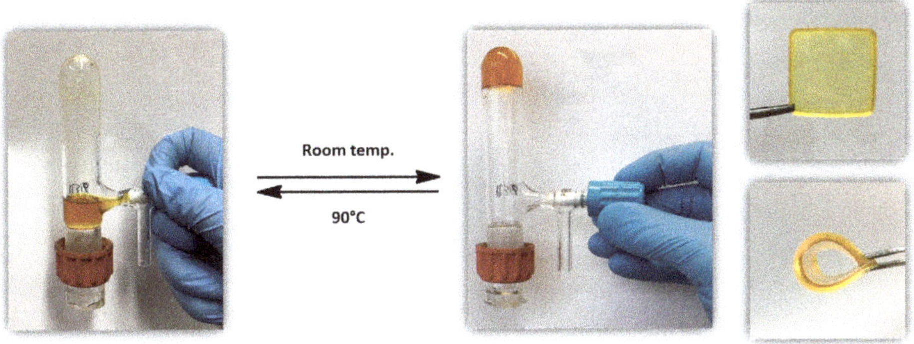

Figure 3. Thermo-reversible bio-based polycarbonate networks: from a soluble polymer solution (**left** picture) to an insoluble network (**right** pictures).

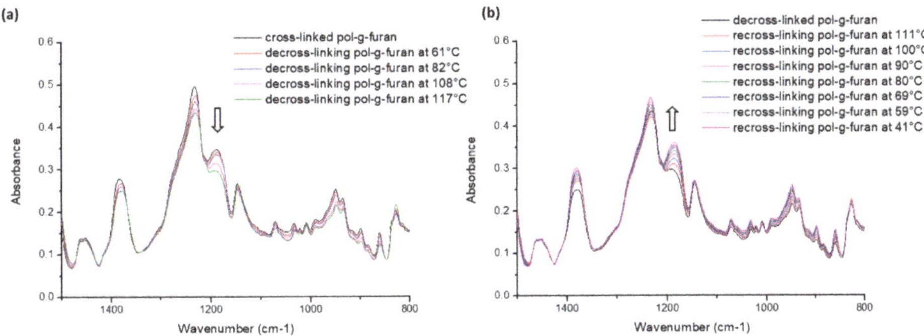

Figure 4. FT-IR traces following (**a**) the de-cross-linking reaction upon heating and (**b**) the cross-linking reaction upon cooling.

Figure S4 shows the stacked transmission FT-IR spectra of DA (de/re)-cross-linked furan-containing polycarbonate. Some characteristic furan peaks, such as ν(COC) at 1013 cm^{-1} and ν(C-H for mono-substituted furan) at 710 cm^{-1}, decrease upon cross-linking and increase after de-cross-linking. The ν(COC) peak, for example, is clearly visible in the spectrum of f$_{85}$-P(**NH-Und-6CC**). Its magnitude decreases upon the addition of the bis-maleimide cross-linking agent (cross-linked f$_{85}$-P(**NH-Und-6CC**)). Directly after heating the sample at 120 °C, the magnitude of this peak increases again as the furan groups are decoupled (de-cross-linked f$_{85}$-P(**NH-Und-6CC**)). The peak disappears again after the sample has been cooled down at room temperature (re-cross-linked f$_{85}$-P(**NH-Und-6CC**)). The reverse phenomenon is observed with the cyclo-adduct peak ν(DA cycloadduct) at 1190 cm^{-1}.

Both of the observations prove cross-linking and decross-linking take place via a reversible DA reaction between the grafted furan groups and the added bis-maleimide cross-linking agents.

The film obtained with f$_{51}$-P(**NH-Und-6CC**) was cut in several pieces then re-mold in order to demonstrate further the reprocessability of the cross-linked material, as illustrated in Figure 5.

Subsequently, the mechanical properties of the reprocessed materials were compared to the pristine ones to gauge self-healing behaviors.

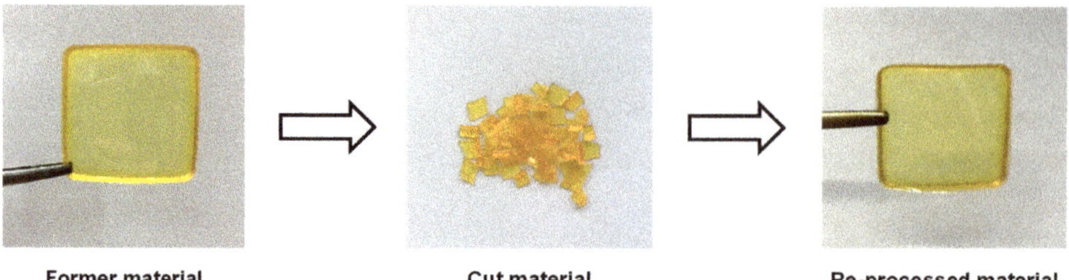

Figure 5. Reprocessability of the cross-linked polycarbonate material.

Dynamic Mechanical Analysis (DMA) measured the temperature response of the material's mechanical behavior (Figure 6a). The moduli of the samples were determined while heating and cooling at a controlled rate (4 °C/min.) with an oscillation frequency of 1 Hz and a strain of 0.04%. The tensile tests at room temperature were also performed with a strain rate of 50 mm/min on three samples (Figure 6b).

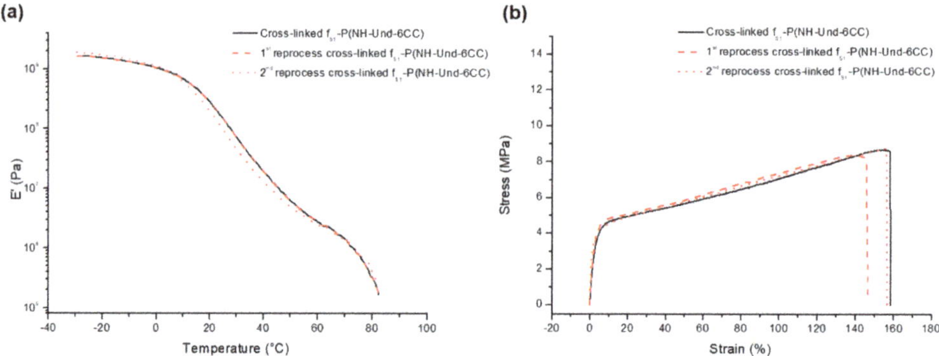

Figure 6. Reprocessability testing: (**a**) dynamic mechanical analyses (DMA) traces and (**b**) Strain-Stress traces of f_{51}-P(NH-Und-6CC).

The DMA traces show that DA cross-linked polycarbonate, before and after two reprocessing, display exactly the same mechanical behavior upon heating (Figure 6a). Indeed, the storage modulus (E′), which largely determines the elasticity of the rubber, follows the same trend before and after reprocessing. The cross-linked samples show high moduli (>1000 MPa) before 0 °C (glassy state), with only a modest decrease upon heating. A broad glass transition is then observed from 0 °C to 50 °C associated with a significant decrease of E′ upon heating. A drastic loose of mechanical resistance appears after 70 °C, although a very narrow elastic plateau (E′ ≈ 3 MPa) from 50 °C to 70 °C (rubbery state) is detected where the decrease of E′ is less marked, which testifies the start of the retro-DA reaction (de-cross-linking of the material).

Following the same trend, the tensile tests prove that the reprocessing of the polycarbonate materials does not affect its mechanical properties, such as young modulus, elongation at break, and strain at break (Figure 6b).

The next part discusses the influence of the cross-linking density on the mechanical properties of these materials, aving demonstrated the thermo-reversibility of the DA cross-linking and the reprocessing ability of the so-formed cross-linked polycarbonates.

2.3. Tuneable Network Properties

In this part, we aim at investigating the influence of several parameters on the thermo-mechanical properties of the networks. For this purpose, the cross-linking procedure exposed in Scheme 2 was applied to all the polycarbonates with different furan contents. The thermo-mechanical properties of the DA cross-linked polycarbonates are summarized in Table 2.

Table 2. Thermo-mechanical properties of Diels–Alder (DA) cross-linked polycarbonate materials.

			Performed at RT	Performed at 65 °C				
Furan Content [a]	$T_{d(5\%)}$ f_x-Network (°C) [b]	T_g f_x-Network (°C) [c]	Young Modulus (MPa) [d]	Young Modulus (MPa) [e]	Max Stress (MPa) [e]	Elongation at Break (%) [e]	Gel Content (%)	Swelling Ratio (%)
88%	108	4.6	31 ± 2	5.3	1.7	70	96.5	238
69%	107	6.8	43 ± 2	3.9	1.5	79	93.7	296
51%	110	13.4	90 ± 15	2.7	1.3	98	96	318
31%	112	10.2	230 ± 70	0.85	0.8	170	71	765

[a] Calculated with ^{1}H NMR. [b] Measured by TGA analysis. [c] Measured by DSC analysis (see Figure S5). [d] Calculated with tensile tests at RT. [e] Calculated with tensile tests at 65 °C.

Table 2 shows that the DA cross-linked polycarbonates are not very thermally stable. Indeed, the decomposition of the networks starts at around 110 °C, which is just after the retro-DA reaction temperature. Such a feature can be a limitation for these materials. Table 2 also indicates that the T_g of the different cross-linked polycarbonates range from 4.6 °C to 13.4 °C. These values are not following a trend with respect to the furan content grafted on the P(**NH-Und-6CC**). Indeed, the highest furan content network (i.e., 88 mol%) that has potentially the highest cross-linking density, exhibits a T_g of 4.6 °C, while the network that was prepared from the linear precursor only containing 31 mol% of furan moieties displays a higher T_g of 10.2 °C. At a first glance, this result seems to be not logical, but while taking into account that the T_g of the polymer f_{88}-P(**NH-Und-6CC**) is −59.8 °C and those of f_{31}-P(**NH-Und-6CC**), −6.6 °C, the difference between the T_g of the network and the T_g of the linear precursor is much more significant for high furan content samples.

Nevertheless, the mechanical properties of the networks evaluated at room temperature will be influenced by the state of the material (rubbery or glassy) and not only by the cross-linking density, due to these different T_g values.

Thus, the mechanical properties were tested at 65 °C, where all of the cross-linked polymers are in a rubbery state and before the r-DA. Tensile tests were performed on the various cross-linked samples (Figure 7a). The tensile strength, the Young's modulus and the elongation at break were determined from these tensile tests and averaged over four measurements (Figure 7b). These values are also reported in Table 2. Figure 7 shows that the high furan-containing materials exhibit significantly higher tensile modulus and lower elongation at break values when compared to those containing lower furan moieties. This is typical, as high tensile moduli and low elongation values are indicative of material with high cross-linking densities. Consequently, the cross-linking density increases as a function of the dangling furan content in the former polycarbonate, as expected.

In addition, DMA measured the mechanical behaviour with temperature of the cross-linked polycarbonates (Figure 8). The test was performed under the same conditions than mentioned previously.

Figure 8 confirms the results that were obtained by DSC i.e., the T_α slightly decreases with the increase of cross-linking density. Additionally, the cross-linking density also influences the retro DA reaction. Indeed, the DMA traces show that the higher the cross-linking density, the later the retro-DA reaction. The slope of the curve corresponding to the rDA is also less steep, meaning that the rDA reaction is slower.

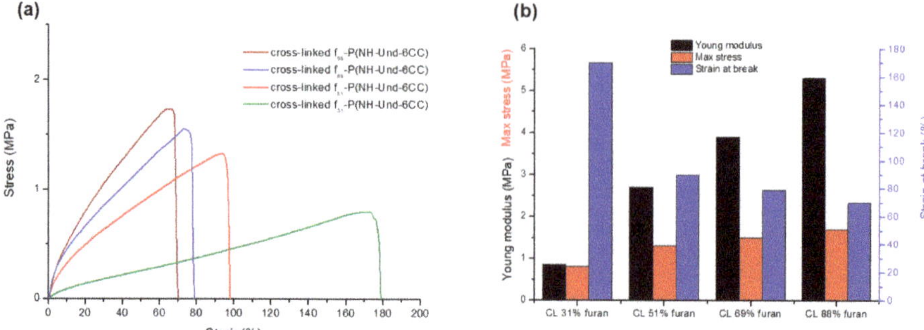

Figure 7. (**a**) Tensile tests performed at 65 °C for various polycarbonate materials and (**b**) the corresponding Young's modulus, max stress, and strain at break.

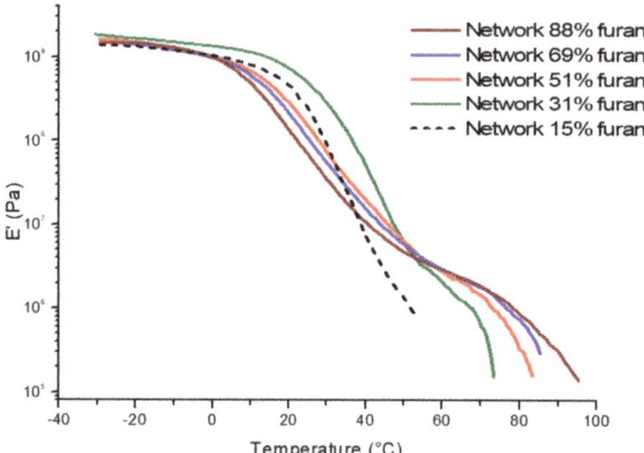

Figure 8. DMA traces for cross-linked polycarbonates samples.

Finally, gel content and swelling ratio of the cross-linked materials provide information about the cross-linking density. A high gel content combined with a low swelling ratio indicates that the material is highly cross-linked. The gel fraction [33] was calculated according to

$$Gel\ fraction(\%) = \frac{m_d}{m_0} \times 100 \qquad (1)$$

where m_d is the mass of dried (extracted) samples and m_0 is the mass of the specimens before swelling. Equation (1): Formula used to calculate the gel fraction.

The swelling ratio was calculated according to Equation (2), where m_t is the mass of the swollen samples in DCM.

$$Swelling\ ratio\ (\%) = \frac{m_t - m_d}{m_d} \times 100 \qquad (2)$$

Equation (2): Formula used to calculate the swelling ratio.

Table 2 provides the gel contents and the equilibrium swelling ratios values of the networks. The effect of the pendent furan content on the polymer can be clearly seen. The resulting polycarbonate networks had greatly increased gel contents and decreased swelling ratios with increasing pendent furan rings content. Indeed, the gel content of the network increases from 71% to 96.5% for 31 mol% and 88 mol% of pendent furan content, respectively. Simultaneously, the swelling ratio of the network decreased from 765% to 238% with the above-mentioned pendent furan content. As expected, the cross-linking density is higher when high furan content is grafted on P(**NH-Und-6CC**).

3. Materials and Methods

3.1. Materials

All of the products and solvents (reagent grade) were used as received, except otherwise mentioned. The solvents were of reagent grade quality and they were purified whenever necessary according to the methods reported in the literature. The polycarbonate, P(NH-Und-6CC) has been synthesized, as previously described [21].

3.2. Characterization

^{1}H and ^{13}C-NMR spectra were recorded on Bruker Avance 400 spectrometer (400.20 MHz or 400.33 MHz and 100.63 MHz for ^{1}H and ^{13}C, respectively, Bruker, Wissembourg, France) by using CDCl$_3$ as a solvent at room temperature, except otherwise mentioned. Two-dimensional analyses, such as ^{1}H-^{1}H COSY (COrrelation SpectroscopY), and ^{1}H-^{13}C HSQC (Heteronuclear Single Quantum Spectroscopy) were also performed.

Size Exclusion Chromatography (SEC) analyses were performed in THF (25 °C) on a PL GPC50 with four TSK columns: HXL-L (guard column), G4000HXL (particles of 5 mm, pore size of 200A, and exclusion limit of 400,000 g/mol), G3000HXL (particles of 5 mm, pore size of 75 A, and exclusion limit of 60,000 g/mol), G2000HXL (particles of 5 mm, pore size of 20 A, and exclusion limit of 10,000 g/mol), at an elution rate of 1 mL/min. The elution times of the filtered samples were monitored while using UV and RI detectors and SEC were calibrated using polystyrene standards.

The differential scanning calorimetry (DSC) thermograms were measured using a DSC Q100 apparatus from TA instruments (New Castle, DE, USA). For each sample, two cycles from −80 to 100 °C at 10 °C min^{-1} were performed and then the glass transition and melting temperatures were calculated from the second heating run.

Thermogravimetric (TGA) analyses were performed on TGA-Q50 system from TA instruments (New Castle, DE, USA) at a heating rate of 10 °C min^{-1} under nitrogen atmosphere from room temperature to 600 °C.

The dynamic mechanical analyses (DMA) were performed on RSA 3 (TA instrument, New Castle, DE, USA). The sample temperature was modulated from −50 °C to 150 °C, depending on the sample at a heating rate of 4 °C min^{-1}.

Tensile tests were performed on a MTS Qtest 25 Elite controller (Cary, NC, USA) at room temperature. The initial grip separation was set at 20 mm and the crosshead speed at 50 mm min^{-1}. The results were obtained from at least three replicates for each sample. Tensile test at 65 °C was also performed on one sample while using the same procedure.

Strip-shaped specimens (0.6 mm thick, 0.5 mm width, 3 mm length) were placed in 10 mL of DCM for 48 h to determine the equilibrium swelling ratios and gel contents. DCM was replaced once after 24 h. We assume that this procedure ensured complete removal of the sol fraction. Subsequently, the swollen gels were dried to constant weight at room temperature in vacuum and weighed.

3.3. General Procedure for the Synthesis of Furan-Functionalized fx-P(NH-Und-6CC)

In a round-bottom flask equipped with magnetic stirrer, P(**NH-Und-6CC**) (1 equiv.), furfuryl mercaptan (3 equiv.) and DMPA (1 mol%) were dissolved in ethanol (2 mol/L^{-1}). The reaction was allowed to proceed at room temperature during the appropriate period of time under UV irradiation through an optical fiber (365 nm). The solvent was removed to produce f$_x$-P(**NH-Und-6CC**), which was purified by precipitation in cold methanol (−63 °C, mixture liquid nitrogen/chloroform). The functional polymer was obtained as waxy solid. Yield = 80–90%, depending on the degree of functionalization.

3.4. General Procedure for Polycarbonates Cross-Linking by Diels-Alder Reaction

The polymer f$_x$-P(**NH-Und-6CC**) was dissolved in chloroform (CHCl$_3$) at 1 g mL^{-1} in a vial. The bis-maleimide cross-linking agent was then added to the previous solution (0.5 equiv. with respect to furan groups) and the mixture was homogenized by vortex stirring until a clear yellow solution appeared. The vial was closed tightly and then heated up to 60 °C for 10 min.

The warm solution was then poured in a Teflon mold, and the chloroform was allowed to gently evaporate overnight at room temperature. A cross-linked film CL-f_x-P(**NH-Und-6CC**) was obtained and then dried under reduced pressure for several hours to remove traces of chloroform. DSC, TGA, DMA, and tensile tests characterized the obtained cross-linked film.

3.5. General Procedure for Polycarbonates Decross-Linking by Retro-Diels–Alder Reaction

Placed in chloroform, the cross-linked polycarbonate CL-f_x-P(**NH-Und-6CC**) was heated at 90 °C for 10 min. The retro-DA reaction occurs and the polymer is dissolved in the chloroform. These transformations can be done several times in a row without any decomposition of the polymer.

4. Conclusions

To conclude, several furan-containing polycarbonates were synthesized via the thiol-ene reaction between P(**NH-Und-6CC**) and furfuryl mercaptan. The content of dangling furan rings that were grafted on the polymer backbone can be tuned by monitoring the reaction time. Such furan-functionalized polycarbonates were then cross-linked thanks to the Diels–Alder reaction with a bis-maleimide as cross-linking agent. The reversibility of the cross-linking reaction was proven by FT-IR analyses. Moreover, cross-linked polycarbonate materials have been reprocessed twice without altering their mechanical properties. Finally, the thermo-reversible DA reaction affords the preparation of reversible cross-linked polycarbonate materials with tuneable properties as a function of the pendent furan content that was grafted on the polycarbonate backbone.

While, the poor thermal stability of the furan-functionalized cross-linked polycarbonates represent a hindrance for other applications, the possibility to decrosslink the network around 70 °C could be an advantage in biomedical applications.

Supplementary Materials: The following are available online at http://www.mdpi.com/1420-3049/25/1/74/s1, Experimental procedure to synthesize P(NH-Und-6CC), Figure S1: Stacked ^{1}H NMR monitoring the reaction between furfuryl mercaptan and P(NH-Und-6CC); Figure S2: SEC traces in THF of fx-P(NH-Und-6CC) as a function of furan content; Figure S3: DSC traces of furan-functionalized P(NH-Und-6CC); Figure S4: FT-IR absorption spectra of furan-containing polycarbonate and DA (de/re)-cross-linked furan-containing polycarbonate; Figure S5: DSC traces of cross-linked furan-functionalized P(NH-Und-6CC); Figure S6 TGA traces of cross-linked furan-functionalized P(NH-Und-6CC).

Author Contributions: P.-L.D. performed all the experiments in the article. E.G. and H.C. managed the project, designed the experiments and write the article. All authors have read and agreed to the published version of the manuscript.

Funding: This work was funded by P.I.V.E.R.T., as part of the Investments for the Future, by the French Government under the reference ANR-001-01.

Acknowledgments: This work was performed, in partnership with the SAS PIVERT, within the frame of the French Institute for the Energy Transition (Institut pour la Transition Energétique-ITE) P.I.V.E.R.T. (www.institut-pivert.com) selected as an Investment for the Future ("Investissements d'Avenir"). This work was supported, as part of the Investments for the Future, by the French Government under the reference ANR-001-01. The authors thank Equipex Xyloforest ANR-10-EQPX-16. The financial support from the CPER CAMPUSB project funded by the French state and the Région Nouvelle Aquitaine is gratefully acknowledged.

Conflicts of Interest: The authors declare no conflict of interest.

References

1. Artham, T.; Doble, M. Biodegradation of aliphatic and aromatic polycarbonates. *Macromol. Biosci.* **2008**, *8*, 14–24. [CrossRef]
2. Rokicki, G.; Parzuchowski, P.G. ROP of Cyclic Carbonates and ROP of Macrocycles. In *Polymer Science: A Comprehensive Reference*; Matyjaszewski, K., Möller, M., Eds.; Elsevier BV: Amsterdam, The Netherlands, 2012; Volume 4.

3. Tempelaar, S.; Mespouille, L.; Coulembier, O.; Dubois, P.; Dove, A.P. Synthesis and post-polymerisation modifications of aliphatic poly(carbonate)s prepared by ring-opening polymerisation. *Chem. Soc. Rev.* **2013**, *42*, 1312–1336. [CrossRef] [PubMed]

4. Montero De Espinosa, L.; Meier, M.A.R.; De Espinosa, L.; Meier, M.A.R. Plant oils: The perfect renewable resource for polymer science? *Eur. Polym. J.* **2011**, *47*, 837–852. [CrossRef]

5. Martina, M.; Hutmacher, D.W. Biodegradable polymers applied in tissue engineering research: A review. *Polym. Int.* **2007**, *56*, 145–157. [CrossRef]

6. Amsden, B. Curable, biodegradable elastomers: Emerging biomaterials for drug delivery and tissue engineering. *Soft Matter* **2007**, *3*, 1335. [CrossRef]

7. Pêgo, A.P.; Poot, A.A.; Grijpma, D.W.; Feijen, J. Biodegradable elastomeric scaffolds for soft tissue engineering. *J. Control. Release* **2003**, *87*, 69–79. [CrossRef]

8. Place, E.S.; George, J.H.; Williams, C.K.; Stevens, M.M. Synthetic polymer scaffolds for tissue engineering. *Chem. Soc. Rev.* **2009**, *38*, 1139–1151. [CrossRef] [PubMed]

9. Fukushima, K. Poly(trimethylene carbonate)-based polymers engineered for biodegradable functional biomaterials. *Biomater. Sci.* **2016**, *4*, 9–24. [CrossRef]

10. Feng, J.; Zhuo, R.X.; Zhang, X.Z. Construction of functional aliphatic polycarbonates for biomedical applications. *Prog. Polym. Sci.* **2012**, *37*, 211–236. [CrossRef]

11. Pascual, A.; Tan, J.P.K.; Yuen, A.; Chan, J.M.W.; Coady, D.J.; Mecerreyes, D.; Hedrick, J.L.; Yang, Y.Y.; Sardon, H. Broad-Spectrum Antimicrobial Polycarbonate Hydrogels with Fast Degradability Biomacromolecules. *Biomacromolecules* **2015**, *16*, 1169–1178. [CrossRef]

12. Yuen, A.Y.; Lopez-Martinez, E.; Gomez-Bengoa, E.; Cortajarena, A.L.; Aguirresarobe, R.H.; Bossion, A.; Mecerreyes, D.; Hedrick, J.L.; Yang, Y.Y.; Sardon, H. Preparation of Biodegradable Cationic Polycarbonates and Hydrogels through the Direct Polymerization of Quaternized Cyclic Carbonates. *ACS Biomater. Sci. Eng.* **2017**, *3*, 1567–1575. [CrossRef]

13. Martín, C.; Kleij, A.W. Terpolymers Derived from Limonene Oxide and Carbon Dioxide: Access to Cross-Linked Polycarbonates with Improved Thermal Properties. *Macromolecules* **2016**, *49*, 6285–6295. [CrossRef]

14. Stevens, D.M.; Tempelaar, S.; Dove, A.P.; Harth, E. Nanosponge Formation from Organocatalytically Synthesized Poly(carbonate) Copolymers. *ACS Macro Lett.* **2012**, *1*, 915–918. [CrossRef] [PubMed]

15. Schüller-Ravoo, S.; Feijen, J.; Grijpma, D.W. Preparation of Flexible and Elastic Poly(trimethylene carbonate) Structures by Stereolithography. *Macromol. Biosci.* **2011**, *11*, 1662–1671. [CrossRef] [PubMed]

16. Durand, P.-L.; Chollet, G.; Grau, E.; Cramail, H. Versatile cross-linked fatty acid-based polycarbonate networks obtained by thiol–ene coupling reaction. *RSC Adv.* **2019**, *9*, 145–150. [CrossRef]

17. Guimard, N.K.; Oehlenschlaeger, K.K.; Zhou, J.; Hilf, S.; Schmidt, F.G.; Barner-Kowollik, C. Current Trends in the Field of Self-Healing Materials. *Macromol. Chem. Phys.* **2012**, *213*, 131–143. [CrossRef]

18. Binder, W.H. *Self-Healing Polymers: From Principles to Applications*; Wiley-VCH: Weinheim, Germany, 2013.

19. Billiet, S.; Hillewaere, X.K.D.; Teixeira, R.F.A.; Du Prez, F.E. Chemistry of crosslinking processes for self-healing polymers. *Macromol. Rapid Commun.* **2013**, *34*, 290–309. [CrossRef]

20. Bekas, D.G.; Tsirka, K.; Baltzis, D.; Paipetis, A.S. Self-healing materials: A review of advances in materials, evaluation, characterization and monitoring techniques. *Compos. Part B Eng.* **2016**, *87*, 92–119. [CrossRef]

21. Durand, P.-L.; Chollet, G.; Grau, E.; Cramail, H. Simple and Efficient Approach toward Photosensitive Biobased Aliphatic Polycarbonate Materials. *ACS Macro. Lett.* **2018**, *7*, 250–254. [CrossRef]

22. The Top 10 Emerging Technologies for 2013. Available online: https://www.weforum.org/agenda/2013/02/top-10-emerging-technologies-for-2013/ (accessed on 19 November 2019).

23. Liu, Y.-L.; Chuo, T.-W. Self-healing polymers based on thermally reversible Diels–Alder chemistry. *Polym. Chem.* **2013**, *4*, 2194–2205. [CrossRef]

24. Gandini, A. The furan/maleimide Diels–Alder reaction: A versatile click–unclick tool in macromolecular synthesis. *Polym. Sci.* **2013**, *38*, 1–29. [CrossRef]

25. Bai, N.; Saito, K.; Simon, G.P. Synthesis of a diamine cross-linker containing Diels–Alder adducts to produce self-healing thermosetting epoxy polymer from a widely used epoxy monomer. *Polym. Chem.* **2013**, *4*, 724–730. [CrossRef]

26. Vilela, C.; Silvestre, A.J.D.; Gandini, A. Thermoreversible nonlinear Diels-Alder polymerization of furan/plant oil monomers. *J. Polym. Sci. Part A Polym. Chem.* **2013**, *51*, 2260–2270. [CrossRef]

27. Zhang, J.; Niu, Y.; Huang, C.; Xiao, L.; Chen, Z.; Yang, K.; Wang, Y. Self-healable and recyclable triple-shape PPDO–PTMEG co-network constructed through thermoreversible Diels–Alder reaction. *Polym. Chem.* **2012**, *3*, 1390–1393. [CrossRef]

28. Tasdelen, M.A. Diels–Alder "click" reactions: Recent applications in polymer and material science. *Polym. Chem.* **2011**, *2*, 2133–2145. [CrossRef]

29. Gheneim, R.; Perez-Berumen, C.; Gandini, A. Diels–Alder Reactions with Novel Polymeric Dienes and Dienophiles: Synthesis of Reversibly Cross-Linked Elastomers. *Macromolecules* **2002**, *35*, 7246–7253. [CrossRef]

30. Diaz, M.M.; Van Assche, G.; Maurer, F.H.J.; Van Mele, B. Thermophysical characterization of a reversible dynamic polymer network based on kinetics and equilibrium of an amorphous furan-maleimide Diels-Alder cycloaddition. *Polymer* **2017**, *120*, 176–188. [CrossRef]

31. Polgar, L.M.; Van Duin, M.; Broekhuis, A.A.; Picchioni, F. Use of Diels–Alder Chemistry for Thermoreversible Cross-Linking of Rubbers: The Next Step toward Recycling of Rubber Products? *Macromolecules* **2015**, *48*, 7096–7105. [CrossRef]

32. Defize, T.; Riva, R.; Thomassin, J.-M.; Jérôme, C.; Alexandre, M. Thermo-Reversible Reactions for the Preparation of Smart Materials: Recyclable Covalently-Crosslinked Shape Memory Polymers. *Macromol. Symp.* **2011**, *309–310*, 154–161. [CrossRef]

33. Roper, T.M.; Guymon, C.A.; Jönsson, E.S.; Hoyle, C.E. Influence of the alkene structure on the mechanism and kinetics of thiol–alkene photopolymerizations with real-time infrared spectroscopy. *J. Polym. Sci. Part A Polym. Chem.* **2004**, *42*, 6283–6298. [CrossRef]

 *molecules*

Article

New Insight on the Study of the Kinetic of Biobased Polyurethanes Synthesis Based on Oleo-Chemistry

Julien Peyrton [1], Clémence Chambaretaud [1,2] and Luc Avérous [1,*]

[1] BioTeam/ICPEES-ECPM, UMR CNRS 7515, Université de Strasbourg, 25 rue Becquerel, CEDEX 2,
    67087 Strasbourg, France; julien.peyrton@etu.unistra.fr (J.P.); clemence.chambaretaud@uha.fr (C.C.)
[2] Soprema, 14 rue de Saint-Nazaire, CEDEX 1, 67025 Strasbourg, France
*   Correspondence: luc.averous@unistra.fr; Tel.: +33-3-6885-2784

Academic Editor: Sylvain Caillol PhD
Received: 11 November 2019; Accepted: 24 November 2019; Published: 27 November 2019

Abstract: Nowadays, polyols are basic chemicals for the synthesis of a large range of polymers, such as polyurethane foams (PUF), which are produced with several other compounds, such as polyisocyanates. During the last decades, the oleo-chemistry has developed several routes from glycerides to polyols for the polyurethanes (PU) industry to replace mainly conventional fossil-based polyols. A large range of biobased polyols can be now obtained by epoxidation of the double bonds and ring-opening (RO) of the subsequent epoxides with different chemical moieties. In preliminary studies, the RO kinetics of an epoxidized model molecule (methyl oleate) with ethanol and acetic acid were investigated. Subsequently, polyols that were derived from unsaturated triglycerides were explored in the frame of e.g., PUF formulations. Different associations were studied with different mono-alcohols derived from epoxidized and ring-opened methyl oleate while using several ring-openers to model such systems and for comparison purposes. Kinetic studies were realized with the pseudo-first-order principle, meaning that hydroxyls are in large excess when compared to the isocyanate groups. The rate of isocyanate consumption was found to be dependent on the moiety located in β-position of the reactive hydroxyl, following this specific order: tertiary amine >> ether > ester. The tertiary amine in β-position of the hydroxyl tremendously increases the reactivity toward isocyanate. Consequently, a biobased reactive polyurethane catalyst was synthesized from unsaturated glycerides. These approaches offer new insights regarding the replacement of current catalysts often harmful, pungent, and volatile used in PU and PUF industry, in order to revisit this chemistry.

Keywords: kinetics; epoxide; ring-opening; biobased; polyurethane foam; catalyst

---

1. Introduction

The polyurethane (PU) is a very versatile family of polymer that is mainly obtained by polyaddition between polyols and polyisocyanates [1]. PUs can be used in various forms to fulfill different applications for a worldwide market of $50 Billion in 2016 due to the multiplicity of their structures. With more than 60%, foams are the largest part of this market, with segments including the furniture, bedding, insulation, building, or construction materials. Foams are elaborated through a complex formulation that is based on polyols, polyisocyanates, blowing agent, and several other additives [2,3]. Commercial foams are mainly formulated with fossil-based components. However, increasing foams are obtained from renewable resources nowadays.

The abundance and versatility of vegetable oils are the key points for replacing petrochemical products in polymer synthesis and developing very promising renewable compounds while using a well-established oleo-chemistry. During the last decades, starting from unsaturated triglycerides, extensive research [4–6] has been performed to obtain new macromolecular architectures. Multiple strategies were developed to obtain polyols from unsaturated glycerides for the PU industry [7] (i) the hydroformylation and ozonolysis, followed by a catalytic reduction is developed at an industrial scale, (ii) the transesterification, (iii) the introduction of hydroxyl groups via double-bonds with microorganisms is promising [8], and (iv) the epoxidation of the double bonds and subsequent the ring-opening (RO) of the epoxides. The last way keeps the initial glyceride structure, and the opportunity to synthetized different polyols structures, even at an industrial level.

In the case of vegetable oils and fats, the epoxide is mainly di-substituted and, consequently, less reactive than a terminal one. Nevertheless, several different types of reagents can be considered for the RO, such as amines [9–11], alcohols [12–14], carboxylic acids [15,16], or hydrogen halides [17,18]. The epoxidation procedure is carried out while using a short carboxylic acid [19]. Although it is not fully new, for instance, few publications are focused on the study of the RO kinetics of epoxide by acetic or formic acid [20,21].

The presently studied way to obtain PUs from poly-unsaturated triglycerides contains three steps: 1. Epoxidation of double bonds, 2. RO reaction, and 3. Polymerization with polyisocyanate. In our study, the polyunsaturated triglycerides were modeled by a fatty ester only containing one double-bond: methyl oleate. The double bond was chemically converted into epoxide by a peracetic acid that was formed in situ. The objective of this preliminary study was to understand the acid-catalyzed RO of disubstituted epoxide. To do so, a new kinetic method that was based on Nuclear magnetic resonance (NMR) was developed to monitor the epoxide RO reaction. Subsequently, it was applied to the kinetic study of acid-catalyzed RO reactions of epoxidized methyl oleate. In the second part of this paper, the reactivity of different alcohols (models) that were obtained from RO of epoxidized fatty esters with various conditions was compared in the frame of PU synthesis. The model-alcohols were synthesized by the RO of the epoxide, with acetic acid, ethanol, hydrogen halide, or diethylamine.

2. Materials and Methods

2.1. Materials

Fatty Acid Methyl Ester of Very High Oleic Sunflower Oil (FAMEVHOSO) with 3.32 mmol double bond/g was kindly supplied by the ITERG group (Canéjan, France). The FAMEVHOSO is composed of 83% of oleic acid. Table S1 presents the distribution of fatty methyl esters of unsaturated FAMEVHOSO. Glacial acetic acid (AA), toluene (99%), H_2O_2 30%, ethyl acetate (99%), and ethanol (99.9%) were obtained from Fisher Scientific (Illkirch-Graffenstaden, France). Amberlyst® 15H (strongly acidic cation exchanger dry), Amberlite® IR120H (strongly acidic hydrogen form), $CDCl_3$, phenylisocyanate (98%), dibutylamine (DBA) (99.5%), HBr (48% in water), HCl (37% in water), and diethylamine (99%) (DEA) were provided by Sigma-Aldrich (Saint-Quentin-Fallavier, France). Ethanol absolute (EtOH) was purchased from VWR (Briare, France). All of the chemicals were used without any purification.

2.2. Epoxidation of FAMEVHOSO

According to a previously described protocol [22], 200 g of FAMEVHOSO (0.66 mol, 1 eq), 50 g of Amberlite® IR 120H (25 wt% of FAMEVHOSO) were introduced in a 1 L three-neck flask that was equipped with a reflux condenser, a magnetic stirrer, and a dropping funnel. 20 mL of acetic acid (0.35 mol, 0.5 eq) and 200 mL of toluene were added. The mixture was heated to 70 °C under vigorous magnetic stirring. Afterwards, 90 mL of H_2O_2 30% (1.15 mmol, 1.7 eq) was added dropwise by the dropping funnel for 30 min. to prevent overheating and epoxide RO. The mixture was heated at 70 °C for 7 h additional hours. At the end, the mixture was recovered in 500 mL of ethyl acetate. The Amberlite® IR 120H was filtered off. The organic phase was washed with saturated $NaHCO_3$ solution until neutral pH. Afterwards, it was washed with brine solution, dried with anhydrous sodium sulfate, and then filtered. The solvent was evaporated under reduced pressure. The epoxidized FAMEVHSOSO (EVHOSO) was dried overnight in a vacuum oven at 40 °C. The yield was 90 mol%.

2.3. Ring-Opening of EVHOSO with Acetic Acid

The reaction was carried out in a round bottom flask that was equipped with a reflux condenser and a magnetic stirrer. The flask was filled with 50 g of EVHOSO 3.05 mmol epoxide/g, (0.24 mol, 1 eq) and 100 mL of acetic acid (2.8 mol, 11.5 eq). The mixture was stirred at 90 °C for 7 h. At the end, the mixture was recovered in 300 mL of ethyl acetate. The organic phase was washed with saturated $NaHCO_3$ solution until neutral pH. Subsequently, it was washed with brine solution, dried with anhydrous sodium sulfate, and then filtered. The solvent was evaporated under reduced pressure. The ring-opened EVHOSO with acetic acid (EVHOSO-AA) was dried overnight in a vacuum oven at 40 °C. The yield was 82 mol%.

2.4. Ring-Opening of EVHOSO with Ethanol

The protocol was adapted from a previously published work [23]. The reaction was carried out in a round bottom flask that was equipped with a reflux condenser and a magnetic stirrer. The flask was filled with 50 g of EVHOSO 3.05 mmol epoxide/g, (0.15 mol, 1 eq) and 2 g of Amberlyst® 15 H. 100 mL of ethanol (1.7 mol, 11.5 eq). The mixture was stirred at 70 °C for 9 h. At the end the mixture was recovered in 300 mL of ethyl acetate. The organic phase was washed with saturated $NaHCO_3$ solution until neutral pH. Subsequently, it was washed with brine solution, dried with anhydrous sodium sulfate, and then filtered. The solvent was evaporated under reduced pressure. The ring-opened EVHOSO with ethanol (EVHOSO-EtOH) was dried overnight in a vacuum oven at 40 °C. The yield was 92 mol%.

2.5. Ring-Opening of EVHOSO with Diethylamine

The protocol was adapted from a previously published work [24]. The reaction was carried out in a round bottom flask that was equipped with a reflux condenser and a magnetic stirrer. The flask was filled with 5 g of anhydrous $ZnCl_2$ (0.5 eq). 25 g of EVHOSO 3.05 mmol epoxide/g (76 mmol, 1 eq) was dissolved in 20 mL of diethylamine (193 mmol, 2.5 eq). The solution was added in the flask. The mixture was stirred at reflux for 15 h. At the end, the mixture was recovered in 300 mL of ethyl acetate and deionized water. The organic phase was washed with saturated $NaHCO_3$ solution until neutral pH. Afterwards, it was washed with brine solution, dried with anhydrous sodium sulfate, and then filtered. The solvent was evaporated under reduced pressure. The ring-opened EVHOSO with diethylamine (EVHOSO-DEA) was dried overnight in a vacuum oven at 40 °C. The yield was 90 mol%.

2.6. Ring-Opening of EVHOSO with Different Hydrogen Halides

The reaction was carried out in a round bottom flask that was equipped with a reflux condenser and a magnetic stirrer. The flask was filled with 25 g of EVHOSO 3.05 mmol epoxide/g (76 mmol, 1 eq) dissolved in 15 mL of acetone. The halogen halide (1.5 eq) was added dropwise for 15 min. to avoid overheating. The mixture was stirred at room temperature for 30 min. At the end, the mixture was recovered in 300 mL of ethyl acetate and deionized water. The organic phase was washed with saturated $NaHCO_3$ solution until neutral pH. Subsequently, it was washed with brine solution, dried over anhydrous sodium sulfate, and then filtered. The solvent was evaporated under reduced pressure. The ring-opened EVHOSO with hydrochloric acid (EVHOSO-HCl) or hydrobromic acid (EVHOSO-HBr) was dried overnight in a vacuum oven at 40 °C. The yield was 99 mol%.

2.7. Kinetic Study of Epoxide Ring Opening with Ethanol

In a typical procedure, 0.5 g of EVHOSO (1.7 mmol, 1 eq) and between 4 to 20 wt% (depending on the experiment) of Amberlyst® 15H in a 50 mL round bottom flask was heated to the desired temperature while using a hot plate that was equipped with magnetic stirring. When the temperature was attained, 3.8 mL of absolute ethanol was added (65 mmol, 38 eq). A few drops of the reaction mixture were taken at different reaction times and recovered in ethyl acetate and washed two times with water to remove any trace of acid. The solvent was evaporated on a rotary evaporator. The drops of oil were recovered with $CDCl_3$ and analyzed by NMR.

2.8. Kinetic Study of Urethane Formation

In a typical procedure, 3 g of VHOSO-AA (7.2 mmol, 11.5 eq) was introduced in a round bottom flask that was equipped with a magnetic stirrer. The flask was heated to the desired temperature. The flask was put under vacuum for 15 min. and then flushed with argon. 1 mL of a 0.62 mol/L solution of phenyl isocyanate in toluene was introduced. The conditions were set to avoid any contact with humidity or another nucleophile. The isocyanate content ([NCO]) over time was determined by taking aliquots of 0.4 mL quenched by a DBA solution. [NCO] was determined by potential titration of the excess of DBA by an acid.

2.9. Kinetic Model of Ring-Opening

2.9.1. Mechanism of Ring-Opening

The RO of the epoxide is catalyzed by acid [25], as shown in Scheme 1. The nucleophilic attack is favored by the enhancement of the electrophilic character of epoxide group carbons. The nucleophilic species can be alcohol, carboxylic acid, water, hydrogen halide, etc.

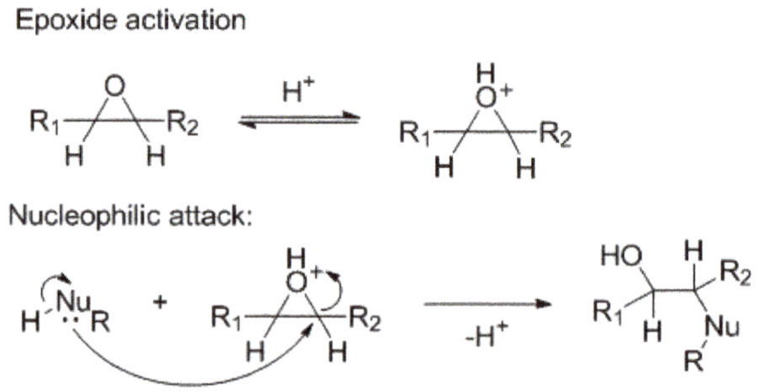

Scheme 1. General mechanism of an acidic catalyzed nucleophile ring-opening (RO).

The reaction forms a secondary hydroxyl and another group, depending on the nature of the nucleophile. Reactions with acid groups, such as carboxylic acid or halogen halide, are self-catalyzed.

2.9.2. Kinetic Equations

The activation step that is presented in Scheme 1 is considered to be fast. Subsequently, the nucleophilic attack is the rate-determining step. The acid catalyzed RO rate is described by Equation (1):

$$r = k * [Ep]^\alpha * [B]^\beta * [Cat]^\delta \tag{1}$$

where k is the reaction rate constant, $[Ep]$ is the epoxide concentration, $[B]$ the RO reagent concentration, $[Cat]$ the catalyst concentration, and α, β and δ are the respective partial orders. The consummation of epoxide over time, depending on the reaction rate, is written as Equation (2):

$$-\frac{d[Ep]}{dt} = k * [Ep]^\alpha * [B]^\beta * [Cat]^\delta \tag{2}$$

The kinetic parameters were determined while using a pseudo-first order assumption [26]. The nucleophile (B) was in large excess as compared to the epoxide groups (Ep). Subsequently, the concentration of the nucleophile ([B]) was considered to be constant during the reaction. Furthermore, the catalyst is regenerated along the reaction, [Cat] is a constant over time. When considering these hypotheses, the rate of reaction that is expressed in Equation (2) is transformed in Equation (3):

$$-d\frac{[Ep]}{dt} = k_{app} * [Ep]^{\alpha} \tag{3}$$

where k_{app} is a constant, as expressed in Equation (4):

$$k_{app} = k * [B]^{\beta} * [Cat]^{\delta} \tag{4}$$

2.10. NMR

The NMR analyses were realized on a 400 MHz Bruker spectrometer. The ^{1}H number of scans was set to 32. Each spectrum was calibrated with the CDCl$_3$ signals, being set at 7.26 ppm.

2.11. NCO Concentration Measurement

The isocyanate content [NCO] was determined by the adaptation of ISO 14896:2009. Aliquots were dissolved in a 20 mL solution of 5×10^{-3} mol/L dibutylamine in toluene. The resulting mixture was stirred for 20 min. Afterwards, 20 mL of acetone was added to avoid a dephasing of the solution and the excess of dibutylamine was titrated with an automatic titrator by a 4.6×10^{-3} mol/L molar solution of HCl. The equivalence was determined by a potential leap. [NCO] was calculated with Equation (5):

$$[NCO] = \frac{((V_{BI} - V_{eq}) * [HCl]}{V_{Aliquot}} \tag{5}$$

with V_{BI} the equivalence volume of 20 mL of dibutylamine solution, V_{eq} the equivalence volume of the aliquot, $[HCl]$ the chlorhydric acid concentration, and $V_{aliquot}$ the volume of solution taken from the solution for each kinetic measurement.

3. Results

3.1. Synthesis of EVHOSO

Table S1 presents FAMEVHOSO data. The average double bond per molecule is 1 by calculation. It is well known that unsaturated fatty acids are sensitive to UV oxidation [27]. An NMR measurement was undertaken to control the double bond quantity before the epoxidation. 0.93 double bonds per molecule were calculated by the integration of the proton of the double bond on the NMR spectrum (Figure S1). The FAMEVHOSO was in-situ epoxidized with peracid in a biphasic system. The reaction converts 90 % of the double bonds in epoxides (Figure S2). Side reactions, such as RO by acetic acid, limit the reaction yield. The number of epoxides per molecule is only 0.83.

3.2. Kinetics of Epoxide Ring-Opening by Ethanol

The NMR method that is described in Appendix A.1 on the RO with acetic acid was applied to the RO with ethanol.

3.2.1. Determination of the Epoxide Partial Order

The epoxide partial order was determined at three temperatures, with all other parameters remaining equal. The pseudo-first order was applied by introducing a large excess (11 eq) of ethanol and constant catalyst content. Integrating Equation (3) with $\alpha = 1$ gives Equation (6):

$$\ln\left(\frac{[Ep]_0}{[Ep]}\right) = \ln\left(\frac{1}{1-\chi}\right) = k_{app\_EtOH} * t \tag{6}$$

where χ is the yield of the reaction, t the time in minute, and k_{app_EtOH} the pseudo-reaction rate constant. $\ln[Ep]_0/[Ep]$ was presented as a function of time in Figure 1.

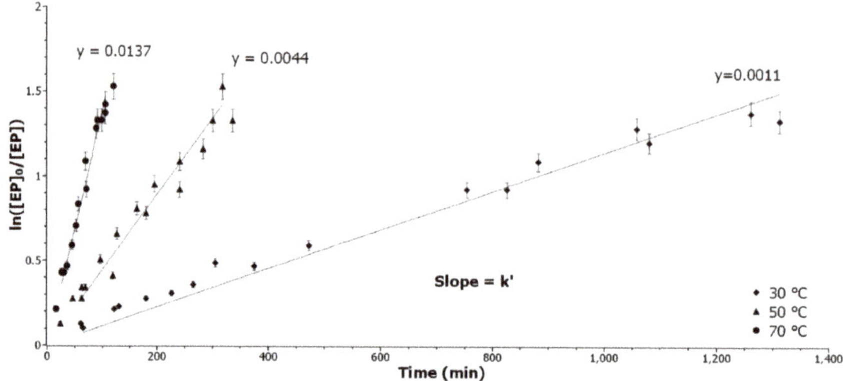

Figure 1. Determination of the epoxide partial order. Representation of ln([Ep]₀/[Ep]) as a function of time at 30 °C (♦), 50 °C (▲), and 70 °C (●).

The model of a partial order of 1 for the epoxide is well confirmed by experiments at 30, 50, and 70 °C, respectively. The range of validity of the method is between 20–95 % of conversion due to the exponential character of the conversion against time, as demonstrated with the acetic acid kinetic experiment (Appendix A.2).

3.2.2. Determination of the Catalyst Partial Order

Experiments with catalyst content variations from 4 to 20 wt% were performed at 30, 50, and 70 °C to determine the partial order of [Cat]. k_{app_EtOH} were determined by the linear regression of ln([Ep]₀/[Ep]) as a function of time, and Table S2 presents the results. The catalyst is an acidic resin of divinylbenzene and styrene sulfonated [28]. The correlation between $[H^+]$ and the mass of catalyst was determined by the pH measurement of 49 mg to 1 g of resin in 20 mL of water (Figure S3). Equation (7) expresses the application of Equation (4) in the ethanol RO:

$$k_{app\_EtOH} = k_{EtOH} * [EtOH]^\beta * \left[H^+\right]^\delta \tag{7}$$

where $[EtOH]$ and $[H^+]$ are the concentration of ethanol and acid, respectively. The partial order of catalyst was determined by the slope of the linear regression of ln(k_{app_EtOH}) as a function of ln([Cat]), as presented in Figure 2.

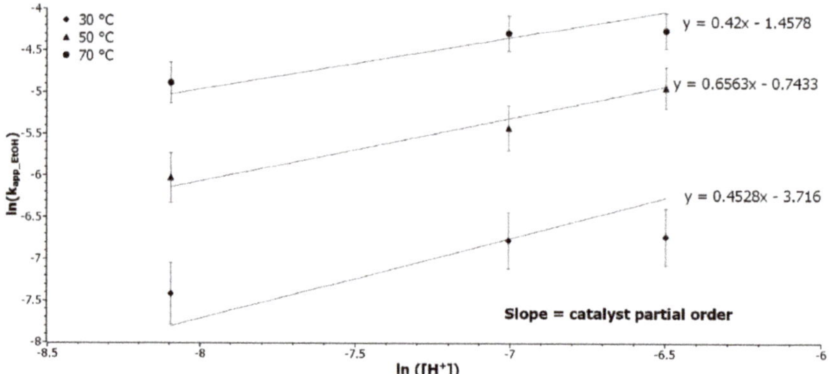

Figure 2. Determination of the catalyst partial order. Representation of ln(k_{app_EtOH}) as a function of ln(H⁺) at 30 (♦), 50 (▲), and 70 °C (●).

The coefficients of determination are between 0.93 and 0.98 and the three slopes are tending toward 0.5, which represent the catalyst partial order. It was expected that the catalyst would have a higher influence on the reaction rate. The catalyst is decreasing the activation energy in the thermodynamic side of the reaction, but it has less influence on the kinetics. The right amount of catalyst can be selected, depending on the thermal sensibility of the epoxidized studied molecule.

3.2.3. Determination of the Ethanol Partial Order

Experiments with [EtOH] variations from 15 to 4.5 mol/L were performed at 70 °C to determine the partial order of [EtOH]. The catalyst loading was kept constant then, with the volume variation, $[H^+]$ was varying along experiments. k_{app_EtOH}, as expressed in Equation (7), can be expressed as a function of the total volume of solution, giving Equation (8):

$$k_{app\_EtOH} = k_{EtOH} * \left(\frac{V_{EtOH} * d_{EtOH}}{V_t} \right)^{\beta} * \left(\frac{m_{cat} * IA_{eq}}{V_t} \right)^{\delta} \tag{8}$$

where V_{EtOH} is the volume of ethanol introduced in the solution, m_{cat}, the mass of catalyst, and IA_{eq} is the acid equivalent of the resin determined by Figure S3. By logarithmic transformation and rearrangement, $\ln(k_{app\_EtOH})$ is expressed in Equation (9) as a linear function of $\ln(1/V_t)$, with the sum of partial order as the slope.

$$\ln(k_{app\_EtOH}) = K + (\beta + \delta) * \ln\left(\frac{1}{V_t} \right) \tag{9}$$

where K is a constant, including V_{EtOH}, d_{EtOH}, m_{cat}, IA_{Eq}, and k_{EtOH}. Figure 3 represents the experimental results.

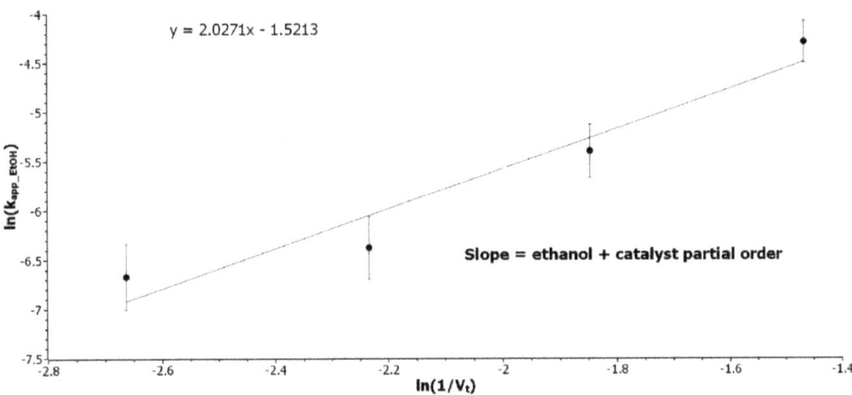

Figure 3. Determination of the sum of the ethanol and catalyst partial order. Representation of $\ln(k_{app\_EtOH})$ as a function of $\ln(1/V_t)$ at 70 °C with 12 wt% catalyst loading.

The sum of the partial order of $[H^+]$ and $[EtOH]$ is 2, as the partial order of $[H^+]$ was determined to be 0.5, the partial order with respect to $[EtOH]$ is 1.5. These results were confirmed by an experiment design to compensate for the dilution of catalyst that was induced by the change of concentration. Equation (10) describes the final equation rate:

$$r = k_{EtOH} * [Ep]^1 * [EtOH]^{1.5} [H^+]^{0.5} \tag{10}$$

The major factor impacting the reaction rate is the $[EtOH]$, followed by the $[Ep]$, and then the $[H^+]$. This order is only accurate if the catalyst is present; otherwise, the reaction is slower. In this case, there is no activation step when compared to the mechanism presented in Scheme 1.

3.2.4. Thermodynamics Data

Once the reaction rate determined, the thermodynamic constants were calculated and are summarized in Table 1. The sum of the catalyst partial order and the ethanol is the same as the partial order of $[AA]$. It can also confirm hypotheses that are given in the literature [20,21] regarding the autocatalysis of the acetic acid due to its acidic properties. The separation between the catalysis and the reagent action in the acetic acid case would be possible by the isolation of the reaction intermediate. Overall, the order of the reaction is the same for both reactions: third order.

Table 1. Thermodynamics data of the epoxidized Fatty Acid Methyl Ester of Very High Oleic Sunflower Oil (EVHOSO) ring-opening reaction with acetic acid and ethanol.

Thermodynamics Data	Acetic Acid [19] [1]	Acetic Acid	Ethanol
Equation rate	$r = k_{AA} * [Ep]^1 * [AA]^2$		$r = k_{EtOH} * [Ep]^1 * [EtOH]^{1.5} * [H^+]^{0.5}$
k (70 °C) ($L^2 \cdot mol^{-2} \cdot min^{-1}$)	3.3×10^{-5}	$(2 \pm 0.2) \times 10^{-5}$	$(7.8 \pm 0.4) \times 10^{-3}$
Frequency factor (min)	2.31×10^7	$(8.8 \pm 0.8) \times 10^4$	$(1.28 \pm 0.06) \times 10^6$
E_a (kJ/mol)	66	63 ± 6	54 ± 3
$\Delta H^{\ddagger}$ (70 °C) (kJ/mol)	63	60 ± 5	51 ± 3
$\Delta S^{\ddagger}$ (70 °C) (J/mol/K)	-182	-160 ± 10	-138 ± 7
$\Delta G^{\ddagger}$ (70 °C) (kJ/mol)	126	120 ± 10	98 ± 5

[1] RO kinetic study made on epoxidized soybean oil by measuring the epoxide function concentration.

The rate constant at 70 °C, k (70 °C), is an indication of the rate of a chemical reaction at this specified temperature [29]. The corresponding unit depends on the reaction order. From a kinetic point of view, the ethanol RO is faster when compared to the case with acetic acid due to the two decades difference in terms of k (70 °C). The reaction constant is a function of the temperature following an Arrhenius law, as described by Equation (11):

$$k(T) = A * \exp\left(\frac{E_A}{R * T}\right) \tag{11}$$

where A is the frequency factor, E_a the activation energy of the reaction, R the gas constant, and T the temperature in Kelvin. E_a is determined by the slope of the linear regression of $\ln(k)$ as a function of the inverse of the temperature (Figure S4). It represents the energy barrier to overcome to form the intermediate state. E_a can be decreased while using a catalyst, but the catalyst loading does not affect it (Figure S4). The EVHOSO-EtOH intermediate state requires less energy to be formed than EVHOSO-AA. The enthalpy of activation ($\Delta H^{\ddagger}$) is the energy difference between the reagents and the intermediate state [30]. It was calculated according to Equation (12)

$$\Delta H^{\neq} = E_A - RT \tag{12}$$

An exothermic reaction has a negative $\Delta H^{\ddagger}$ and both intermediate RO reactions are endothermic. They absorb energy from the environment to reach an intermediate state. The entropy of activation variation ($\Delta S^{\ddagger}$) is a measurement of the disorders of the reaction. Negative $\Delta S^{\ddagger}$ represents the loss of freedom or an order increase [30]. Kinetics data $\Delta S^{\ddagger}$ is calculated via Equation (13):

$$\Delta S^{\neq} = R * \left[\ln\left(\frac{h * k_b}{k * T}\right) + \frac{\Delta H^{\neq}}{RT}\right] \tag{13}$$

where h is the Planck constant, k the Boltzman constant, k_b the rate constant, T the temperature, and R the gas constant. In RO, by acetic acid or ethanol, the intermediate states have fewer degrees of freedom than the reactants. The entropy variation is more important in a previous publication [20], because the model molecule studied is based on epoxidized soybean oil with more than one epoxide. The loss of freedom degree is then higher and can explain the gaps in $\Delta S^{\ddagger}$ and E_a.

The sign of the free enthalpy of activation $\Delta G^{\ddagger}$ is the combination of the entropic and enthalpic contribution, representing the spontaneity of the reaction (Equation (14)).

$$\Delta G^{\neq} = \Delta H^{\neq} - T\Delta S^{\neq} \tag{14}$$

A negative $\Delta G^{\ddagger}$ indicates a spontaneous reaction [30]. From the thermodynamic point of view, both RO intermediates are not spontaneously formed. Thermodynamics is then not favorable to reach the intermediate of the RO of epoxide with acetic acid or ethanol, but the temperature and the kinetic are making the reaction possible in less than 10 h.

3.3. Kinetic Study of Urethane Formation from Fatty Acid

The kinetics of the reaction between an isocyanate and an alcohol derived from a fatty acid was investigated. The phenyl isocyanate was chosen to model the methylene diphenyl diisocyanate (MDI) that was used in foaming processes. EVHOSO-AA was the alcohol used for this study. The measurements were made while using the pseudo first order principle [30] by putting the hydroxyl moieties in large excess and measuring the concentration of isocyanate [NCO] over time. The same model as for the RO kinetic can be applied. The reaction proceeds without catalyst, and Equations (15) and (16) are then expressed.

$$-d\frac{[NCO]}{dt} = k_{app\_u} * [NCO]^{\alpha} \tag{15}$$

where α is the partial order of the isocyanate concentration ([NCO]) and k_{app_u} is:

$$k_{app\_u} = k * [OH]^{\beta} \tag{16}$$

The determination of isocyanate content method was tested with known phenylisocyanate contents before starting the kinetics measurement (Figure S5). The precision of the method was around 96% and the studied concentration range was from 7×10^{-3} to 0.6 mol/L.

3.3.1. Determination of the Isocyanate Partial Order

The isocyanate partial order was determined by variation of the temperature, with all other parameters being equal. The pseudo first order was applied by introducing a large excess (11 eq) of VHOSO-AA. While considering Equation (15), the logarithm of $[NCO]_0/[NCO]$ was represented as a function of time on Figure 4.

The partial order of 1 for the isocyanate is well confirmed by experiments at different temperatures, which are 42, 53, 62, and 74 °C. For the conversion above 99 %, the limit of the titration techniques was attained. The back titration of the [NCO] is determined by Equation (5) and a 1% difference between V_{eq} and V_{bl} is the limit of detection of the titration method.

3.3.2. Determination of the Hydroxyl Partial Order

The determination of the hydroxyl partial order is made by varying the concentration of hydroxyl, while maintaining the great excess as compared to the isocyanate group. The variation of concentration was made by diluting the solution with toluene. The determination of k_{app_U} with 0.51, 0.72, 1.24, 1.51, 1.73, and 1.98 mol/L of hydroxyl are, respectively, presented in Figure S6. The slope of the linear regression of $\ln(k_{app\_U})$ in function of $\ln([OH])$ that is represented in Figure 5 is able to establish the partial order of hydroxyl.

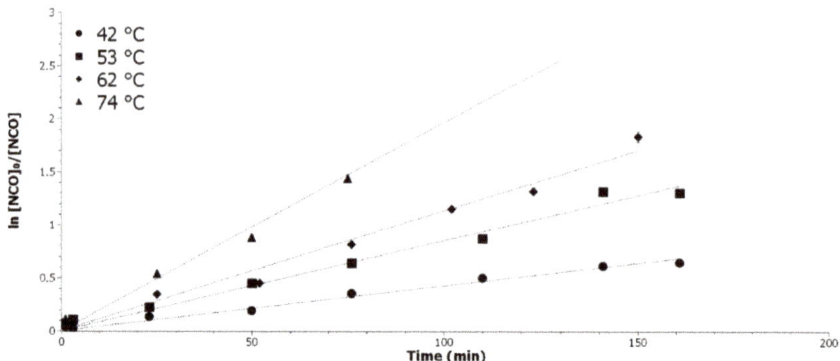

Figure 4. Determination of the [NCO] partial order. Variation of $\ln([NCO]_0/[NCO])$ in function of time at 42 °C (●), 53 °C (■), 62 °C (◆), and 74 °C (▲).

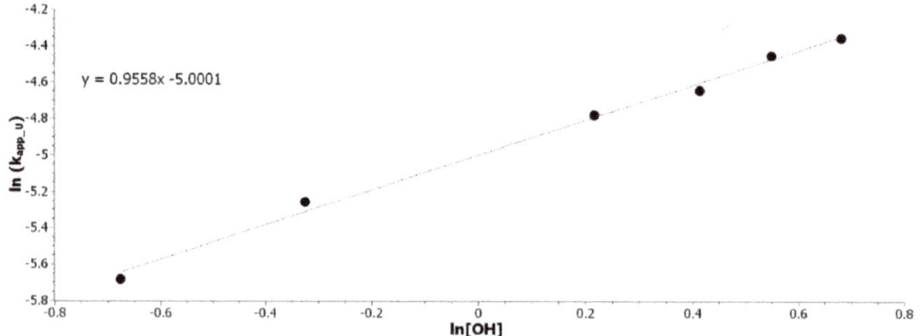

Figure 5. Determination of hydroxyl partial order. Linear regression of $\ln(k_{app\_U})$ as a function of $\ln([OH])$.

The coefficient of determination is 0.995. The partial order of [OH] can be round up to 1. Table 2 presents the final rate equation with the thermodynamic data of the intermediate state. The partial order of hydroxyl and isocyanate were both equal to 1. This result is in perfect agreement with the results that were found in the literature [31,32].

The energy of activation (44 kJ/mol) is in good agreement with the results from the literature. The energy of activation reported for a reaction between phenyl isocyanate and 2-butanol is 41 kJ/mol [32] and 52 kJ/mol [33] for a reaction with or without xylene as solvent, respectively. The reaction rate constant that is presented in Table 2 (1.0×10^{-3} L mol^{-1} min^{-1}) is six times smaller than data from the literature [32] for the reaction between secondary alcohol and phenylisocyanate. The 2-butanol is less sterically hindered than the mono-alcohols that were derived from methyl oleate. For instance, the reaction rate constant ratio between 2-hexanol and 3-hexanol was recently proved to be 1.5 for a reaction with an isocyanate [34].

From the value of the activation energy and the reaction constant of 6.0×10^{-3} L mol^{-1} min^{-1} enthalpy, entropy, and free energy were calculated with Equations (12)–(14), respectively. The large negative entropy of activation is an indication of the dissociation of the charged centers in the activation complex [33]. The difference of free energy of activation can be explained by better stabilization of the intermediate complex by toluene when compared to xylene [35].

The rate constant is similar to the one that was found for EVHOSO-EtOH with 20 °C of difference. By applying the Arrhenius Equation (11), the reaction rate at 70 °C for urethane formation is 1×10^{-2} L mol^{-1} min^{-1}. The reaction between hydroxyl and aromatic isocyanate is faster than the RO by ethanol of a di-substituted epoxide. The urethane formation is not catalyzed. Catalysts, such as tertiary amine or tin salt, substantially reduce the E_a, which increases the reaction rate by several orders of magnitude. The frequency factor makes the rate of the RO with ethanol faster with temperature, despite the difference of activation energies (Figure S7). The $\Delta S^{\ddagger}$ of the urethane formation is higher than the one of RO with ethanol. The transition state has fewer degrees of freedom in urethane formation.

Table 2. Thermodynamics data of urethane formation by the reaction between phenyl isocyanate and EVHOSO with acetic acid (EVHOSO-AA).

Thermodynamics Data	Ref [32] [1]	PIC [2]/EVHOSO-AA
Equation rate	$r = k_{AA} * [OH]^1 * [NCO]^1$	
k (25 °C) (L mol^{-1} min^{-1})	6.0×10^{-3}	$(1.0 \pm 0.05) \times 10^{-3}$
Frequency factor (min)		$(5.3 \pm 0.3) \times 10^4$
E_a (kJ/mol)	41	44 ± 2
$\Delta H^{\ddagger}$ (50 °C) (kJ/mol)	39	41 ± 2
$\Delta S^{\ddagger}$ (50 °C) (J/mol/K)	-198	-160 ± 8
$\Delta G^{\ddagger}$ (50 °C) (kJ/mol)	103	92 ± 5

[1] Reaction between 2-propanol and phenylisocyanate in xylene, [2] Phenylisocyanate.

The kinetics and thermodynamics parameters of the reaction between a fatty ester alcohol and an aromatic isocyanate were determined. These parameters are specific to the studied reaction and they cannot be generalized to all isocyanates and hydroxyl substrates.

3.4. Study of the Reactivity with Different Isocyanates Structures

This section aims to compare the reactivity of different isocyanate and fatty ester alcohol, since different chemical architectures are available for polyols and/or polyisocyanates to produce a broad range of materials for the polyurethane industry. The reactivity of the system is an important factor, especially for foam, where the polyaddition must be fast.

Different isocyanate structures were compared in terms of reactivity with our model EVHOSO-AA: aromatic, aliphatic, and cycloaliphatic. The results are presented in Figure 6 once again demonstrated the higher reactivity of the aromatic isocyanate on agreement with previous results [36,37].

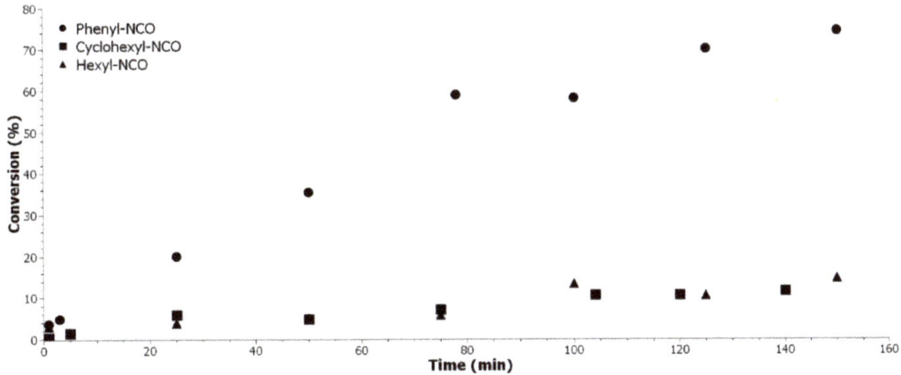

Figure 6. Reactivity comparison in urethane formation with EVHOSO-AA of different isocyanate structure: aromatic (●), aliphatic (▲), and cyclo-aliphatic (■).

The reactions from Figure 6 were performed in the same conditions with an excess of hydroxyl moiety. Both aliphatic isocyanates have the same reactivity toward hydroxyl. The steric hindrance due to the cycle (cyclo-aliphatic chemical) does not influence the beginning of the reaction (less than 15%

of conversion) with a secondary alcohol. The higher reactivity of aromatic isocyanate is due to the tautomer conformation that is presented in Figure 7.

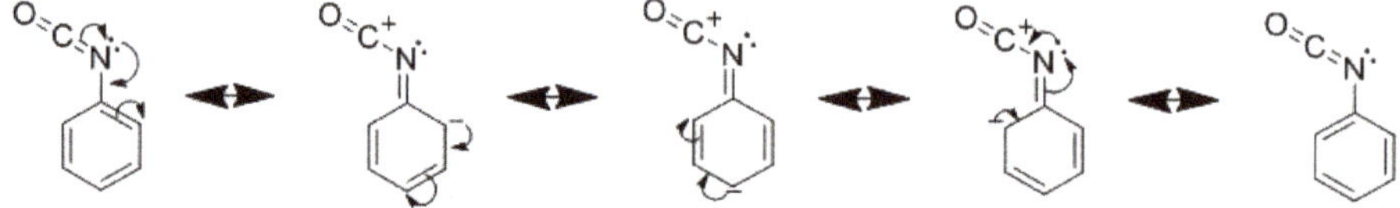

Figure 7. Tautomer conformation of phenyl isocyanate.

The delocalization of the lone pair of electrons of the nitrogen atom on the aromatic cycle is increasing the electrophile character of the carbon [36,37]. It can be easily attacked by nucleophile like hydroxyl moieties. Our model that is based on fatty ester hydroxyl confirms this trend. For foams, the necessary fast polyaddition can be reached by the use of p-MDI, due to its aromatic character, which also leads to higher mechanical properties. Our model reaction confirms that, despite the steric hindrance of the aromatic moieties and the secondary character of the corresponding hydroxyl, aromatic isocyanates are the most suitable chemical for foam formulation with hydroxyl that is derived from fatty esters.

3.5. Synthesis of a Mono-Alcohol Model from Epoxidized Fatty Ester

RO can be conducted with alcohol, carboxylic acid, hydrogen halide, and secondary amine leading, respectively, to ether, ester, halide, and tertiary amine groups plus a secondary hydroxyl group. A series of different alcohols that were derived from fatty ester were synthesized by epoxide RO, with ethanol, acetic acid, HCl, HBr, and DEA (Figure 8).

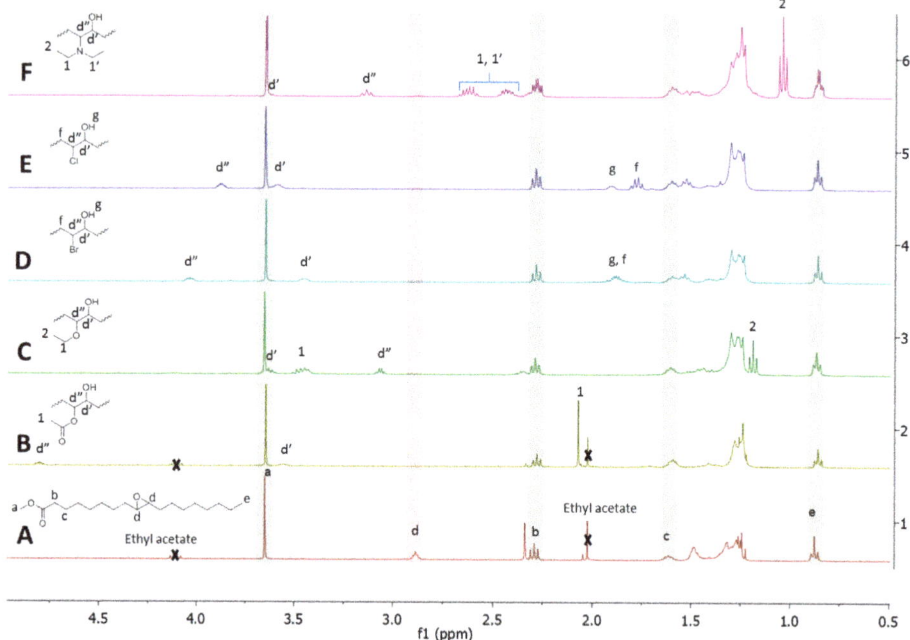

Figure 8. NMR Spectrum of the initial epoxidized fatty methyl ester (**A**) and after the RO with acetic acid (**B**), ethanol (**C**), hydrobromic acid (**D**), hydrochloric acid (**E**), and diethylamine (**F**).

There is a clear disappearance of the epoxide signal (d) located at 2.8 ppm on the NMR spectrum for all the RO reagents. Signals that are characteristic of the proton located in alpha position to the newly created hydroxyl (d') are located in the 3.2–3.7 ppm zone. The signal located at 3.7 ppm, which is characteristic of the methyl close to the fatty ester bond (a), is constant in all products indicating non-significant ester breaking by transesterification, hydrolysis, or amidation. The RO protocols that were developed with the EVHOSO model molecule are efficient and they cause no significant ester bond breaking. These characteristics are important for the RO of more complex oils, such as triglycerides bearing several epoxide groups.

3.6. Evaluation of the Potential from Alcohols Derived from Fatty Acid in Polyurethane Application

The reactivity toward isocyanate of different models based on hydroxyl groups was investigated. The potential of the previously synthesized model alcohol in a urethane material was investigated through the scope of the reactivity. The reactions were carried out with a constant concentration of hydroxyl and isocyanate. The phenyl isocyanate concentration was followed over time by taking aliquots of the reaction. The results that are presented in Figure 9 show a clear tendency of reactivity with this evolution from the lowest to the highest, EVHOSO-HCl/HBr < EVHOSO-AA < EVHOSO-EtOH << EVHOSO-DEA.

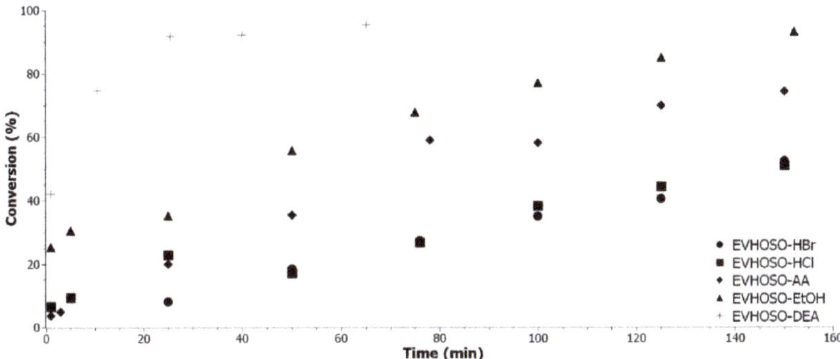

Figure 9. Reactivity comparison in urethane formation with phenyl isocyanate with different RO FAMEVHOSO: EVHOSO-HBr (●), EVHOSO-HCl (■), EVHOSO-AA (♦), EVHOSO-EtOH (▲), and EVHOSO-DEA (+).

As clearly demonstrated, the aromatic isocyanates are very reactive toward nucleophilic attack. The surrounding of the active hydrogen group impacts the reactivity. The electron releasing groups increase the electron density of the hydroxyl by mesomeric or inductive effect. The studied groups have an inductive withdrawing effect. Except for tertiary amine, they are all electron releasing groups by mesomeric effect [38]. In general, the mesomeric effect is predominant. Furthermore, the inductive effect is decreasing with distance [39]. In our case, the two carbons distance between the inductive electron withdrawing group and the hydroxyl decreased the effect. The reactivity is then explained by the difference in strength of electron releasing groups according to the evolution that is presented in Figure 10. The EVHOSO-EtOH has a richer electron density around the hydroxyl group when compared to EVHOSO-AA and EVHOSO-Cl/Br.

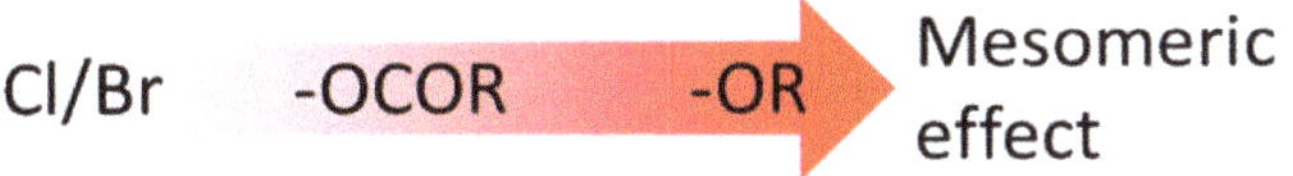

Figure 10. Electron releasing effect by mesomeric effect.

When considering the tertiary amine, the inductive effect reduced by the two carbons distance is counterbalanced by a catalytic effect. In the polyurethane industry, the catalytic activity of the amine by complexation with the isocyanate or by alcohol deprotonation is well established [40–42]. Afterwards, the experiments confirmed the catalytic activity of the EVHOSO-DEA with aromatic or aliphatic isocyanate (Figure S8). The proximity of the reactive hydroxyl and tertiary amine must increase the corresponding catalytic activity.

4. Conclusions

This paper leads to different main and new results while considering the literature. An innovative NMR method was developed and successfully applied to determine the complete kinetic parameters of the EVHOSO RO reaction with ethanol. The calculated activation energy was determined, with 54 and 63 kJ/mol for the RO reaction with ethanol and acetic acid, respectively. This study led to a better understanding of the acid catalyzed RO reaction of epoxides.

Model-alcohols were successfully synthetized by epoxide RO reaction with acetic acid, ethanol, hydrogen halide, and diethylamine. The reactivity comparison with the phenyl isocyanate shows a clear tendency, with a clear evolution from the lowest to the highest, EVHOSO-HCl/HBr < EVHOSO-AA < EVHOSO-EtOH << EVHOSO-DEA.

This work can be considered as a model to better understand the oleo-chemistry approaches leading to biobased polyols. In this frame, the RO of multi-epoxidized oil with hydrogen halide, ethanol and acetic acid, or diethylamine is leading to biobased reactive additives, polyols or catalysts, respectively. The transition from model to practical application is often complex due to the limitations of the model. However, these approaches offer new insights on the replacement of current catalysts, additives, and polyols often fossil-based, used in the PUF industry, in order to revisit this chemistry. In the future, the accuracy of the kinetic model for the RO reaction needs to be tested on more complex oils. The reactivity of future triglyceride based polyols could be adjusted, depending on the RO reagent used. The potential of each RO reaction to provide a biobased substitute for the actual additives or polyols in the PUF industry could be investigated.

Supplementary Materials: The following are available online at http://www.mdpi.com/1420-3049/24/23/4332/s1. Table S1: Lipid profile. Fatty acids distribution in unsaturated FAMEVHOSO. Table S2: k_{app_EtOH} (min^{-1}) with 30, 50 and 70 °C and 4, 12 and 20 wt % of amberlyst. Figure S1: NMR Spectra of the FAMEVHOSO unsaturated. Figure S2: NMR spectra of epoxidized FAMEVHOSO. Figure S3: Correlation between the mass of Amberlyst® 15H and the proton quantity. Figure S4: Arrhenius law for the RO of FAMEVHOSO with ethanol at 4 (♦), 12 (▲) and 20 (●) wt% of catalyst loading. Figure S5: Representation of the gap between theoretical and experimental dosage of the NCO concentration. Figure S6: Determination of the k_{aap_U}. Variation of ln([NCO]$_0$/[NCO]) in function of time at 1.24 (●), 1.51 (■), 1.73 (♦), 1.98 (▲), 0.51 (▼) and 0.72 (+)mol/L. Figure S7: Application of the Arrhenius equation to determine the temperature dependency of reaction rate for the urethane formation (●) and the RO with ethanol (■). Figure S8: Conversion as a function of time for the reaction between phenyl isocyanate and VHOSO-AA (●), Hexyl isocyanate and VHOSO-AA (■) and Hexyl isocyanate and VHOSO-DEA (♦).

Author Contributions: The following statements should be used "conceptualization, L.A. and J.P.; methodology, J.P.; validation, L.A. and J.P.; formal analysis, J.P. and C.C.; investigation, J.P. and C.C.; resources, L.A.; writing—original draft preparation, J.P.; writing—review and editing, Pr L.A., J.P. and C.C.; visualization, J.P.; supervision, L.A. and J.P.; project administration, L.A.; funding acquisition, L.A.

Funding: This study was funded by the Programme d'investissements d'avenir (PIA) of Bpifrance.

Conflicts of Interest: The authors declare no conflict of interest.

Appendix A Method Development: Kinetics of Epoxide Ring-Opening by Acetic Acid

Appendix A.1 NMR Method for the Yield Determination

In the literature, the acetic acid rate law was determined by chemical dosage [20,21], and never by NMR. A new NMR method was developed to determine the kinetic of RO reactions. The reaction between EVHOSO and an excess of acetic acid was carried out in bulk. Aliquots were washed to remove the excess of acetic acid and then analyzed by NMR spectroscopy. The chemical shifts between the EVHOSO and the ring-opened are defined and detectable (Figure A1).

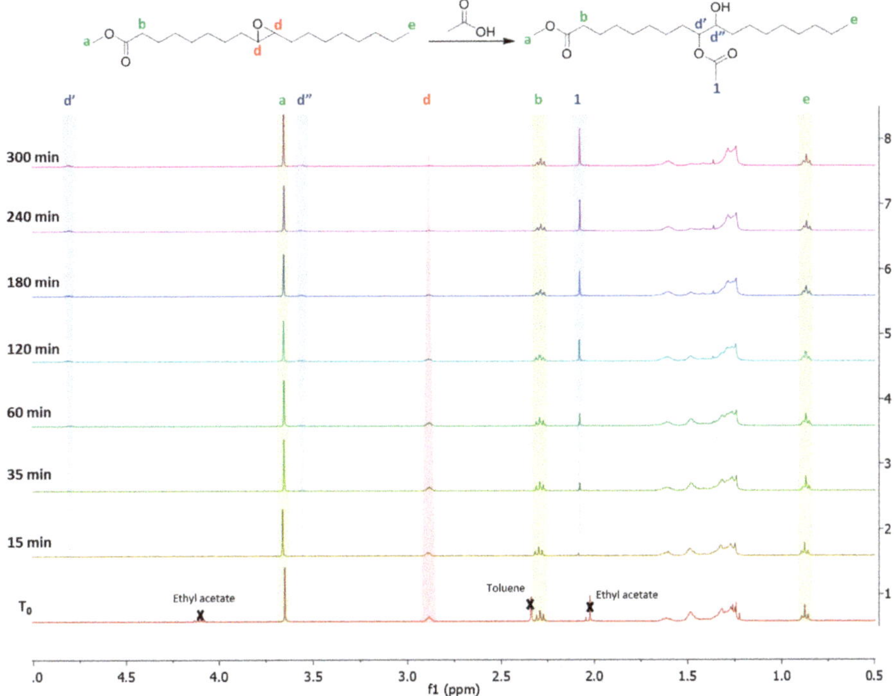

Figure A1. NMR analysis of epoxidized FAMEVHOSO (T_0) and the oil after **15, 35, 60, 120, 180, 240** and 300 min. of reaction at 90 °C with acetic acid.

The absence of singlet at 2.10 ppm indicates the efficiency of the aliquots washing. The signals of the backbone's protons of the methyl fatty ester were constant while the integration of the protons d adjacent to the epoxide was decreased over time. Signals of protons characteristic of the VHOSO-AA d', d" and 1 were increased over time indicating a progression of the reaction until completion. Among the three signals indicating the progress of the reaction, the signal 1 is not specific to the RO reaction because it can be the result of transesterification. The signals of free acetic and ethyl acetate are close. The signal d" is too close to the signal of the methyl ester to be quantitative. The signal d' was chosen to calculate the yield by comparison with the d signal, according to Equation (A1):

$$Yield(\%) = \chi = \frac{I_{d'}}{\frac{1.66}{2}} \tag{A1}$$

where $I_{d'}$ is the integration of the d' signal, 1.66 is the integration of the d signal on the EVHOSO. Based on this calculation the determination of the different partial orders and the activation energy were accomplished and compared to the literature for a validation method.

Appendix A.2 Determination of the Epoxide Partial Order

The reaction is done without catalyst, the Equations (3) and (4) applied to the acetic acid become Equation (A2):

$$-d\frac{[Ep]}{dt} = k_{app\_AA} * [Ep]^\alpha \tag{A2}$$

where [Ep] is the concentration of epoxide over time, α the partial order of [Ep]. k_{app_AA} is detailed by Equation (A3):

$$k_{app\_AA} = k_{AA} * [AA]^\beta \qquad (A3)$$

where k_{AA} is the rate constant of the reaction, [AA] the concentration of acetic acid and β the partial order of [AA].

The epoxide partial order was determined by variation of the temperature with an excess of acetic acid. The pseudo first order was applied by introducing a large excess (11 eq) of acetic acid.

The integration of the Equation (A2) with $\alpha = 1$ gives Equation (A4):

$$\ln\left(\frac{[Ep]_0}{[Ep]}\right) = \ln\left(\frac{1}{1-\chi}\right) = k_{app\_AA} * t \qquad (A4)$$

where χ is the yield of the reaction and t the time in minute. The logarithm of $[Ep]_0/[Ep]$ was represented as a function of time in Figure A2. The linear regression fit the experimental data with a correlation coefficient of 0.85 for 110 °C, 0.99 for 90 and 70 °C, respectively. The lack of correlation at high temperature is the result of the fast rates of reaction and so an increased error and the side reactions.

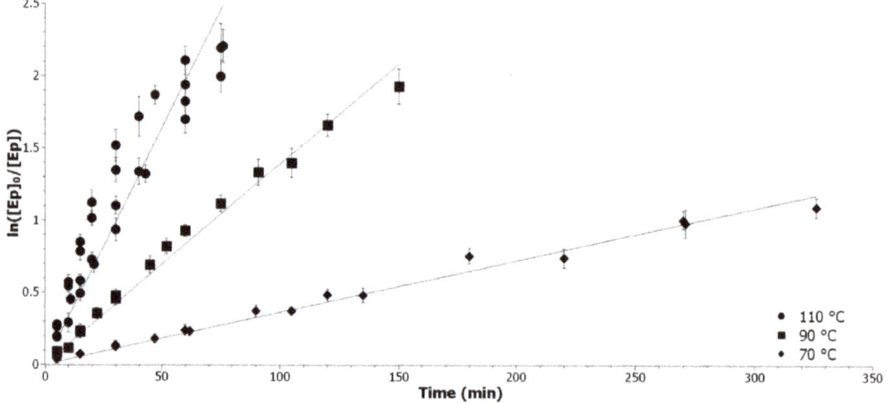

Figure A2. Determination of the epoxide partial order. Representation of ln([Ep]/[Ep]$_0$) as a function of time at 70 °C (♦), 90 °C (▲) and 110 °C (●).

The model of a partial order of 1 for the epoxide is well confirmed by experiments at different temperatures. The precision diminishes when the conversion is less than 20% due to the noise on the NMR spectra. The average NMR precision on the conversion is around 5% on the 20–95% interval. The precision is similar to the one obtained in previous studies [20,21].

Appendix A.3 Determination of the Acetic Acid Partial Order

The determination of the acetic acid partial order is performed by varying the concentration of acetic acid while maintaining the great excess compared to epoxide. The acetic acid concentration was set at 8.6, 12.8 or 16.6 mol/L, by diluting with ethyl acetate. To determined k_{app_AA}, the linear regression of ln([Ep]$_0$/[Ep]) in function of time was plotted (Figure A3).

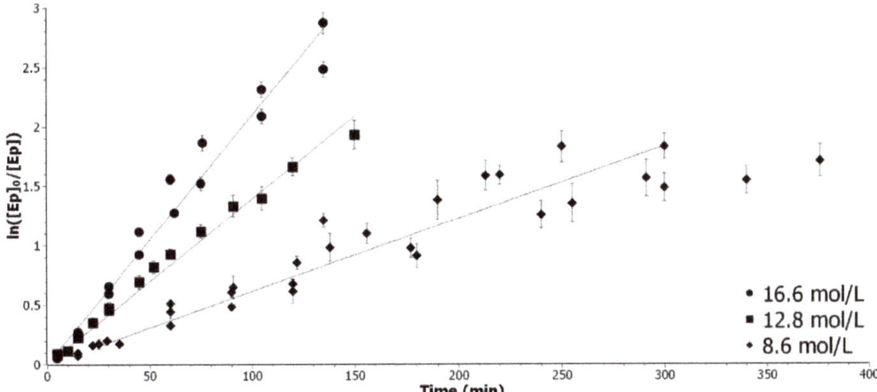

Figure A3. Determination of the k_{app_AA} with 8.6 mol/L (♦), 12.8 mol/L (▲) and 16.6 mol/L (●) of acetic acid.

For the highest concentration, the coefficient of correlation is superior to 0.97. For the smallest concentration, the data points are sparser, due to the logarithm character of the model. The rate of reaction decreases at the end and reaches a plateau. In Equation A2, k_{app_AA} depends only on $[AA]$, the other factors are constant. The logarithm of Equation A2 gives the Equation (A5).

$$\ln(k_{app\_AA}) = \ln(k_{AA}) + \beta * \ln([AA]), \tag{A5}$$

The combination of Equation A5 and the k_{app_AA} obtained at different $[AA]$, linear regression of the logarithm of k_{app_AA} as a function of the $[AA]$ was presented in Figure A4, in order to determine β.

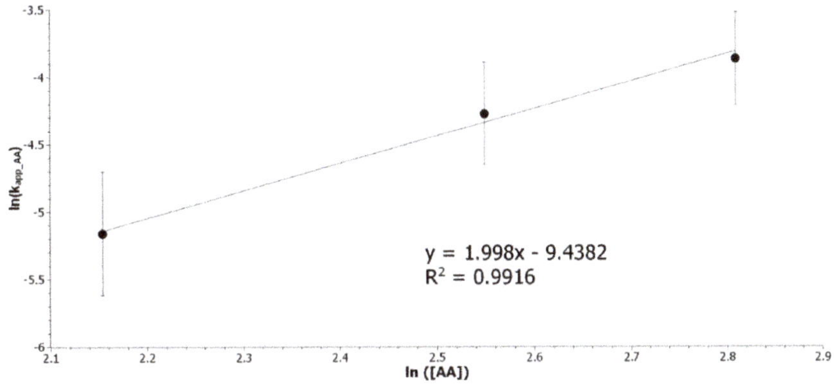

Figure A4. Determination of the acetic acid partial order. Representation of $\ln(k_{app\_AA})$ as a function of $\ln([AA])$.

The partial order for [AA] was determined to be 2. The overall kinetic rate is expressed as Equation (A6):

$$r = k * [Ep] * [AA]^2, \tag{A6}$$

which is in agreement with the previous results found in the literature [20,21]. The rate constant k can be calculated from all the k_{app_AA} by dividing them by $[AA]^2$. The average k was calculated at 70, 90 and 110 °C. The temperature dependence of k was modeled with an Arrhenius law described by

Equation (11). The linear regression of $\ln(k_{AA})$ in function of $1/T$ (Figure A5) was used to determined E_{A_AA}.

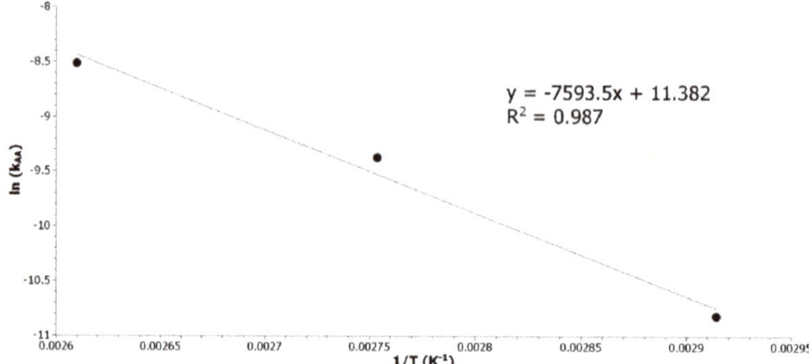

Figure A5. Determination of E_{A_AA} by plotting the $\ln(k_{AA})$ as a function of the inverse of the temperature.

The energy of activation (63 kJ/mol) is in good agreement with results from the literature. The energies of activation reported are 66 kJ/mol [20] and 73 kJ/mol [21] for the RO with acetic acid of epoxidized soybean oil or epoxidized methyl ester of palm oil, respectively. The RO of disubstituted epoxide by acetic acid was studied. Despite the difference in systems and methods, the kinetic rate was found to be the same as the one in the literature and the calculated activation energy is close to the one matching our system. This is validating our NMR method for the determination of the reaction rate. Thus, it can be applied to the study of the RO of ethanol with acid catalysis.

References

1. Bayer, O. Das Di-Isocyanat-Polyadditionsverfahren (Polyurethane). *Angew. Chem.* **1947**, *59*, 257–272. [CrossRef]
2. Furtwengler, P.; Boumbimba, R.M.; Avérous, L. Elaboration and Characterization of Advanced Biobased Polyurethane Foams Presenting Anisotropic Behavior. *Macromol. Mater. Eng.* **2018**, *303*, 1700501. [CrossRef]
3. Obi, B.E. Fundamentals of Polymeric Foams and Classification of Foam Types. In *Polymeric Foams Structure-Property-Performance*; Elsevier: Amsterdam, The Netherlands, 2018; pp. 93–129. ISBN 978-1-4557-7755-6.
4. Zhang, C.; Garrison, T.F.; Madbouly, S.A.; Kessler, M.R. Recent advances in vegetable oil-based polymers and their composites. *Prog. Polym. Sci.* **2017**, *71*, 91–143. [CrossRef]
5. Petrovic, Z. Polyurethanes from Vegetable Oils. *Polym. Rev.* **2008**, *48*, 109–155. [CrossRef]
6. Maisonneuve, L.; Chollet, G.; Grau, E.; Cramail, H. Vegetable oils: A source of polyols for polyurethane materials. *OCL* **2016**, *23*, D508. [CrossRef]
7. Pfister, D.P.; Xia, Y.; Larock, R.C. Recent Advances in Vegetable Oil-Based Polyurethanes. *ChemSusChem* **2011**, *4*, 703–717. [CrossRef]
8. Takeuchi, M.; Kishino, S.; Tanabe, K.; Hirata, A.; Park, S.-B.; Shimizu, S.; Ogawa, J. Hydroxy fatty acid production by *Pediococcus* sp. *Eur. J. Lipid Sci. Technol.* **2013**, *115*, 386–393. [CrossRef]
9. Durán Pachón, L.; Gamez, P.; van Brussel, J.J.M.; Reedijk, J. Zinc-catalyzed aminolysis of epoxides. *Tetrahedron Lett.* **2003**, *44*, 6025–6027. [CrossRef]
10. Biswas, A.; Adhvaryu, A.; Gordon, S.H.; Erhan, S.Z.; Willett, J.L. Synthesis of Diethylamine-Functionalized Soybean Oil. *J. Agric. Food Chem.* **2005**, *53*, 9485–9490. [CrossRef]
11. Yang, L.-T.; Zhao, C.-S.; Dai, C.-L.; Fu, L.-Y.; Lin, S.-Q. Thermal and Mechanical Properties of Polyurethane Rigid Foam Based on Epoxidized Soybean Oil. *J. Polym. Environ.* **2012**, *20*, 230–236. [CrossRef]
12. Dai, H.; Yang, L.; Lin, B.; Wang, C.; Shi, G. Synthesis and Characterization of the Different Soy-Based Polyols by Ring Opening of Epoxidized Soybean Oil with Methanol, 1,2-Ethanediol and 1,2-Propanediol. *J. Am. Oil Chem. Soc.* **2009**, *86*, 261–267. [CrossRef]

13. Chen, R.; Zhang, C.; Kessler, M.R. Polyols and polyurethanes prepared from epoxidized soybean oil ring-opened by polyhydroxy fatty acids with varying OH numbers. *J. Appl. Polym. Sci.* **2015**, *132*. [CrossRef]

14. Turco, R.; Tesser, R.; Vitiello, R.; Russo, V.; Andini, S.; Serio, M.D. Synthesis of Biolubricant Basestocks from Epoxidized Soybean Oil. *Catalysts* **2017**, *7*, 309. [CrossRef]

15. Shuo, X.; Ligong, C.; Lan, X.; Liang, L.; Xin, Y.; Liye, Z. Diester Derivatives from Chemically Modified Waste Cooking Oil as Substitute for Petroleum Based Lubricating Oils. *China Pet. Process. Petrochem. Technol.* **2015**, *17*, 76–83.

16. Schuster, H.; Rios, L.A.; Weckes, P.P.; Hoelderich, W.F. Heterogeneous catalysts for the production of new lubricants with unique properties. *Appl. Catal. A Gen.* **2008**, *348*, 266–270. [CrossRef]

17. Durbetaki, A.J. Direct Titration of Oxirane Oxygen with Hydrogen Bromide in Acetic Acid. *Anal. Chem.* **1956**, *28*, 2000–2001. [CrossRef]

18. Guo, A.; Cho, Y.; Petrović, Z.S. Structure and properties of halogenated and nonhalogenated soy-based polyols. *J. Polym. Sci. Part A Polym. Chem.* **2000**, *38*, 3900–3910. [CrossRef]

19. Petrović, Z.S.; Zlatanić, A.; Lava, C.C.; Sinadinović-Fišer, S. Epoxidation of soybean oil in toluene with peroxoacetic and peroxoformic acids—kinetics and side reactions. *Eur. J. Lipid Sci. Technol.* **2002**, *104*, 293–299. [CrossRef]

20. Zaher, F.A.; El-Mallah, M.H.; El-Hefnawy, M.M. Kinetics of oxirane cleavage in epoxidized soybean oil. *J. Am. Oil Chem. Soc.* **1989**, *66*, 698–700. [CrossRef]

21. Gan, L.H.; Goh, S.H.; Ooi, K.S. Kinetic studies of epoxidation and oxirane cleavage of palm olein methyl esters. *J. Am. Oil Chem. Soc.* **1992**, *69*, 347–351. [CrossRef]

22. Arbenz, A.; Perrin, R.; Avérous, L. Elaboration and Properties of Innovative Biobased PUIR Foams from Microalgae. *J. Polym. Environ.* **2017**, *26*, 254–262. [CrossRef]

23. Palaskar, D.V.; Boyer, A.; Cloutet, E.; Le Meins, J.-F.; Gadenne, B.; Alfos, C.; Farcet, C.; Cramail, H. Original diols from sunflower and ricin oils: Synthesis, characterization, and use as polyurethane building blocks. *J. Polym. Sci. Part A Polym. Chem.* **2012**, *50*, 1766–1782. [CrossRef]

24. Harry-O'kuru, R.E.; Tisserat, B.; Gordon, S.H.; Gravett, A. Osage Orange (*Maclura pomifera* L.) Seed Oil Poly(α-hydroxydibutylamine) Triglycerides: Synthesis and Characterization. *J. Agric. Food Chem.* **2015**, *63*, 6588–6595. [CrossRef]

25. Parker, R.E.; Isaacs, N.S. Mechanisms of Epoxide Reactions. *Chem. Rev.* **1959**, *59*, 737–799. [CrossRef]

26. Espenson, J.H. *Chemical Kinetics and Reaction Mechanisms*, 2nd ed.; McGraw Hill series in advanced chemistry; McGraw Hill: New York, NY, USA, 1995; ISBN 978-0-07-020260-3.

27. Kenaston, C.B.; Wilbur, K.M.; Ottolenghi, A.; Bernheim, F. Comparison of methods for determining fatty acid oxidation produced by ultraviolet irradiation. *J. Am. Oil Chem. Soc.* **1955**, *32*, 33–35. [CrossRef]

28. Rios, L.A. *Heterogeneously Catalyzed Reactions with Vegetable Oils: Epoxidation and Nucleophilic Epoxide Ring-Opening with Alcohols*; Technology RWTH-Aachen: Aachen, Germany, 2003.

29. Menzinger, M.; Wolfgang, R. The Meaning and Use of the Arrhenius Activation Energy. *Angew. Chem. Int. Ed. Eng.* **1969**, *8*, 438–444. [CrossRef]

30. Avery, H.E. *Basic Reaction Kinetics and Mechanisms*; Macmillan: London, UK, 1974; ISBN 978-0-333-15381-9.

31. Burkus, J.; Eckert, C.F. The Kinetics of the Triethylamine-catalyzed Reaction of Diisocyanates with 1-Butanol in Toluene. *J. Am. Chem. Soc.* **1958**, *80*, 5948–5950. [CrossRef]

32. Dyer, E.; Taylor, H.A.; Mason, S.J.; Samson, J. The Rates of Reaction of Isocyanates with Alcohols. I. Phenyl Isocyanate with 1- and 2-Butanol. *J. Am. Chem. Soc.* **1949**, *71*, 4106–4109. [CrossRef]

33. Lovering, E.G.; Laidler, K.J. Kinetic Studies of Some Alcohol-Isocyanate Reactions. *Can. J. Chem.* **1962**, *40*, 31–36. [CrossRef]

34. Nagy, L.; Nagy, T.; Kuki, Á.; Purgel, M.; Zsuga, M.; Kéki, S. Kinetics of Uncatalyzed Reactions of 2,4'- and 4,4'-Diphenylmethane-Diisocyanate with Primary and Secondary Alcohols. *Int. J. Chem. Kinet.* **2017**, *49*, 643–655. [CrossRef]

35. SATO, M. The Rate of the Reaction of Isocyanates with Alcohols. II. *J. Org. Chem.* **1962**, *27*, 819–825. [CrossRef]

36. Baker, J.W.; Holdsworth, J.B. 135. The mechanism of aromatic side-chain reactions with special reference to the polar effects of substituents. Part XIII. Kinetic examination of the reaction of aryl isocyanates with methyl alcohol. *J. Chem. Soc. (Resumed)* **1947**, 713. [CrossRef]

37. Sato, M. The Rates of Reaction of 1-Alkenyl Isocyanates with Methanol. *J. Am. Chem. Soc.* **1960**, *82*, 3893–3897. [CrossRef]
38. Taft, R.W.; Ehrenson, S.; Lewis, I.C.; Glick, R.E. Evaluation of Resonance Effects on Reactivity by Application of the Linear Inductive Energy Relationship.1,2 VI. Concerning the Effects of Polarization and Conjugation on the Mesomeric Order. *J. Am. Chem. Soc.* **1959**, *81*, 5352–5361. [CrossRef]
39. Stock, L.M. The origin of the inductive effect. *J. Chem. Educ.* **1972**, *49*, 400. [CrossRef]
40. Ionescu, M. *Chemistry and Technology of Polyols for Polyurethanes*; Rapra Technology: Shawbury, Shropshire, UK, 2005; ISBN 978-1-60119-664-4.
41. Van Maris, R.; Tamano, Y.; Yoshimura, H.; Gay, K.M. Polyurethane Catalysis by Tertiary Amines. *J. Cell. Plast.* **2005**, *41*, 305–322. [CrossRef]
42. Silva, A.L.; Bordado, J.C. Recent Developments in Polyurethane Catalysis: Catalytic Mechanisms Review. *Catal. Rev.* **2004**, *46*, 31–51. [CrossRef]

Sample Availability: Not available.

 *molecules*

Article

Synthesis of Resins Using Epoxies and Humins as Building Blocks: A Mechanistic Study Based on In-Situ FT-IR and NMR Spectroscopies

Xavier Montané, Roxana Dinu and Alice Mija *

Institut de Chimie de Nice, Université Côte d'Azur, Université Nice-Sophia Antipolis, UMR CNRS 7272, CEDEX 02, 06108 Nice, France; Xavier.Montane@unice.fr (X.M.); Roxana.CONDRUZ@unice.fr (R.D.)
* Correspondence: alice.mija@unice.fr; Tel.: +33489150174

Academic Editor: Sylvain Caillol
Received: 10 October 2019; Accepted: 12 November 2019; Published: 14 November 2019

Abstract: The combination of eco-respectful epoxy compounds with the humins, a by-product of biomass chemical conversion technologies, allow the obtention of materials with high added value. In this work, we propose a chemical connection study of humins with two aliphatic bis-epoxides through copolymerization reactions to synthesize sustainable, bio-based thermosets. The mechanism insights for the crosslinking between the epoxides and humins was proposed considering the different functionalities of the humins structure. Fourier Transform InfraRed (FT-IR), one dimensional (1D) and two-dimensional (2D) Nuclear Magnetic Resonance (NMR) spectroscopy techniques were used to build the proposed mechanism. By these techniques, the principal chain connections and the reactivity of all the components were highlighted in the synthesized networks.

Keywords: biomass; green chemistry; mechanims; humins; epoxy resins; thermosets

1. Introduction

Nowadays, there is an increasing demand to produce renewable resource-based materials due to the rapid exhaustion of fossil fuels [1–4]. The design and synthesis of polymeric materials developed from renewable resources like vegetable oils or lignocellulosic biomass have successfully shown their potential as alternatives to the petroleum-based polymers [5,6]. A class of polymers that are widely used due to their variety of excellent properties including adhesion, mechanical performance, thermal resistance and chemical environmental stability are the epoxy resins. Nowadays, epoxy monomers can be obtained from plant oils or from other bio-based raw materials [7–9]. One of these examples is the glycerol diglycidyl ether (GDE), which is obtained from glycerol. Glycerol is a triol component of triglyceride vegetable oils. Moreover, glycerol can be obtained as a by-product in the synthesis of biodiesel via the transesterification of vegetable oils [10]. Poly(ethylene glycol) diglycidyl ether (PEGDE) is a further bis-epoxide that is used in a wide variety of applications, from industrial applications to medicine. PEGDE is obtained from ethylene glycol, a green raw material that can be obtained from starch [5].

On the other hand, the valorization of industrial waste has started to receive more attention owing to the required transition from a linear economy to a circular one. This transition is nowadays attracting a great deal of interest due to the high price and future depletion of fossil fuel stocks. One of the industrial wastes that has been used for the synthesis of polymers are humins [11,12]. Humins are carbonaceous, heterogeneous and polydisperse by-products obtained during the acid-catalyzed dehydration of sugars (ACD). In these processes, the formation of humins is still unavoidable. Humins chemical structure is based on a network of furanic rings linked via aliphatic chains bearing reactive oxygen-based functional groups (hydroxyls, acids, ketones, aldehydes, esters, …) [13–15]. The study of the formation and characterization of these side products are of great interest. In fact, different routes have been proposed for humins formation [16–20].

Humins demonstrate high potential in different applications during the recent years. They can be used as a source of sustainable H_2 and as a model feedstock for gasification [21]. Besides, they can be used in the preparation of nanocomposites, sorbents and electrode materials [22–24]. Likewise, other polymeric materials were synthesized by using humins as raw materials. As an example, our group reported also the synthesis of humins-derived porous materials [25,26].

The combination of eco-friendly epoxy monomers with humins allows for the development of materials with high added value: The introduction of furanic structures into the polymeric networks will provide an improvement of the rigidity and give a high thermal stability to the resulting epoxy thermosets.

In this work, we propose the investigation of chemical connection of the humins as macromonomers with epoxide structures through copolymerization reactions to synthesize fully sustainable thermosets. Humins were combined with two epoxy monomers using different weight ratios. To promote their covalent interactions, a tertiary amine (*N,N*-Benzyldimethylamine, BDMA) was used as initiator. For the synthesis of the thermosets, neither chemical modifications nor solvent were needed, which agrees with the green chemistry principles. The mechanism insights for the crosslinking between the epoxy molecules and the humins was proposed considering the different functionalities of the compounds. The main starting hypothesis of this work is that the carboxylic acid groups and the hydroxyl groups of humins could open the epoxy rings of both glycerol digylcidyl ether (GDE) and poly(ethylene glycol) diglycidyl ether (PEGDE) [27–30]. However, different secondary reactions could theoretically occur: The catalyzed homopolymerization of epoxides, the formation of Diels–Alder adducts between different furan rings of humins, the nucleophilic attack of alkoxide anions to the carbonyl groups in humins, or the autopolycondensation of humins.

Recently, we reported on the thermomechanical properties of the humins/PEGDE, humins/GDE and humins/PEGDE/GDE copolymers [31]. The prepared humins copolymers have a ductile and elastomeric character, with a tensile strain at break reaching ≈ 60 % that is an excellent result for a polyfuranic network. Continuing the interesting results of our precedent study, this work focusses on the elucidation of the chemical pathway and reaction scenarios conducting at the synthesis of these new copolymers. The purpose is to determine the functionalities that lead to the chemical connections between humins and the epoxides and to highlight the principal reactivities by in situ Fourier Transform InfraRed (FT-IR) spectroscopy, 1D and 2D Nuclear Magnetic Resonance (NMR) spectroscopy.

2. Results

Synthesis of Copolymers

Three different formulations were prepared. The composition of each formulation is summarized in Table 1. The composition of the three different formulations were studied to evaluate the effect of the two aliphatic epoxy monomers: GDE and PEGDE, when they were mixed with humins. The three proposed formulations contain a high amount of humins, the main objective being the valorization of this by-product. The reactivity of each bis-epoxide and the crosslinking reactions were studied. In addition, a third formulation that contains the same ratio of both bis-epoxides was prepared and analyzed.

Table 1. Composition of the formulations and conditions of the two-step curing process.

Formulation	Initial *w/w* Ratios: Humins/Diepoxy Monomer/Initiator
HG40B5	Humins (55)/GDE (40)/BDMA (5)
HP40B5	Humins (55)/PEGDE (40)/BDMA (5)
HP20G20B5	Humins (55)/PEGDE (20)/GDE (20)/BDMA (5)

3. Discussion

3.1. Plausible Polymerization Mechanism

The complexity of humins was studied by different authors. In fact, the combination of different techniques such FT-IR, pyrolysis GC-MS and liquid and solid-state NMR suggested a furan-rich structure containing several oxygen functionalities (carboxylic acids, aldehydes, ketones, hydroxyl groups, etc.), in which the furan rings are connected directly or via aliphatic linkages [18].

The presence of different functionalities in humins could lead to different reactivities in the presence of bis-epoxide comonomers. Nevertheless, one could expect that the predominant reaction in these systems is the ring-opening polymerization of diepoxy monomers. The initiation of the ring-opening polymerization of an epoxy compound using a tertiary amine as initiator (as in the case of BDMA in our study) requires the presence of a proton donor or electrophilic agent such as a carboxylic acid or hydroxyl groups [32].

The large number of hydroxyl groups that are present in humins combined with the carboxylic acid groups could initiate epoxy ring polymerization through a nucleophilic attack to the less-hindered carbon, as shown in Scheme 1. In the case of humins, the tertiary amine initiator helps to transfer the carboxylic acid and the hydroxyl to carboxylate and alkoxide anions respectively, which are the true nucleophiles that opens the epoxy rings (initiation step a). Furthermore, the hydroxyl and carboxylic acid groups can form a complex with the oxygen in the epoxy ring and the tertiary amine. This interaction increases the positive charge in the oxirane ring methylene carbons, which then undergoes via nucleophilic attack by the lone pair of electrons of the tertiary amine. In this manner, a carboxylate, an alkoxide and a mobile quaternary ammonium ion are formed between the oxirane and the amine (initiation step b). Then, the alkoxide formed is active to opening another epoxy ring and the reaction further proceeds until completion (propagation step) [33]. In the terpolymerization (formulation HP20G20B5), a mixture of poly(ethylene glycol) diglycidyl ether and glycerol diglycidyl ether in the same weight ratio were mixed with humins. Due to the presence of both epoxide structures, a random combination between both PEGDE and GDE epoxy monomers in the growing polymer chains are expected.

Furthermore, humins present other functionalities that can proceed via other kind of reactions during the thermocuring process. Indeed, the carbonyl functionalities of humins can undergo via aldol addition/condensation with other aldehydes or ketones from humins [34]. Patil et al. proposed the formation of humins via aldol addition and condensation during the acid-catalyzed conversion of 5-hydroxymethylfurfural (HMF) [35]. In fact, the possible intramolecular aldol addition/condensation or between different molecules of humins can be considered an auto-crosslinking of the preformed humins molecules, apart from the reaction that induces humins formation. This reaction can take place between carbonyl groups located in the same molecule of humins (intramolecular reaction) or between carbonyl groups from different molecules (intermolecular reaction). The proposed mechanism is shown in Scheme 2. In humins systems, the relatively acidic hydrogen in the α-position to aldehydes or ketones allow the formation of an enolate in a basic media generated by BDMA. After that, the enolate reacts as a carbon nucleophile and attacks a carbonyl carbon of another aldehyde or ketone function. If possible, this reaction continues via dehydration to give α,β-unsaturated aldehyde or ketone. The overall process is then referred as an aldol condensation. This reaction has also been described at room temperature between small molecules that contains aldehyde groups and ketones using tertiary amines like DBU or diisopropylethylamine (DIPEA) as initiators [36,37]. So, one can anticipate that the aldol addition/condensation could occur in our systems between aldehydes and ketones from the humins simultaneously with the ring-opening copolymerization of the epoxides.

Initiation

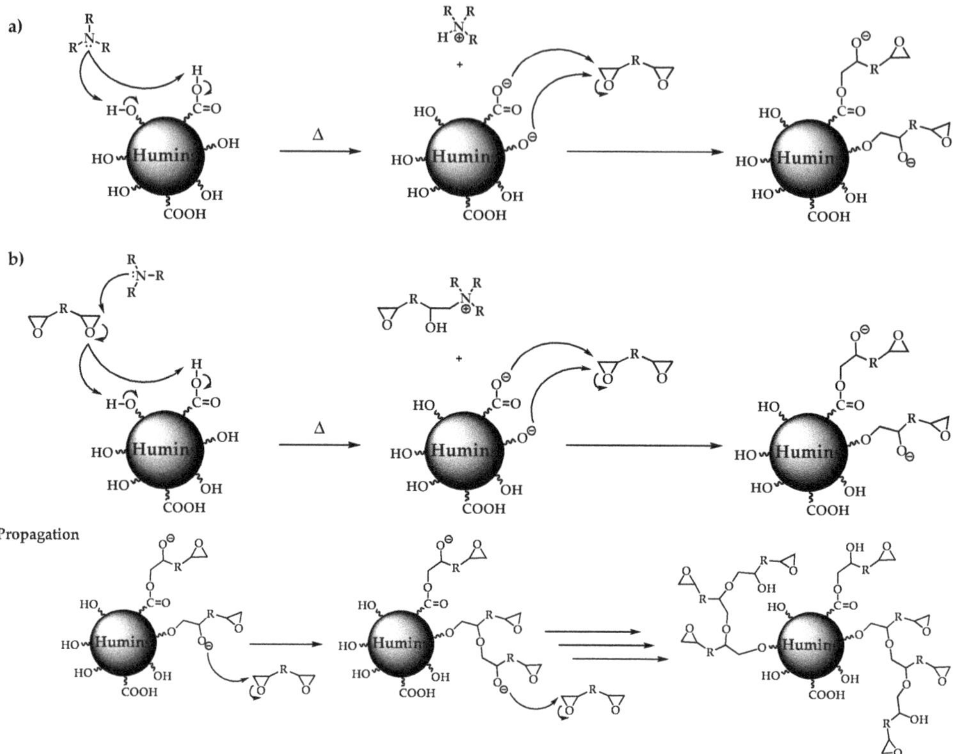

Scheme 1. Proposed mechanism for the initiation and propagation of the ring-opening polymerization of the diepoxy monomers with humins using benzyldimethylamine (BDMA) as initiator.

Scheme 2. Proposed mechanism of the aldol addition between carbonyl groups of humins and their subsequent dehydration (condensation).

3.2. FT-IR Spectroscopy

3.2.1. Raw Materials

The FT-IR spectra of the raw humins, PEGDE, GDE and BDMA are shown in Figure S1. These spectra were used as a reference to elucidate the changes occurred on the functional groups present in humins and in the epoxy compounds during the copolymerization. The assignment of the major bands of humins, in agreement with literature is also summarized in Table S1. The complexity of humins is evidenced by the presence of different functional groups in its structure: Hydroxyl groups, ethers, aldehydes, ketones, carboxylic acids, esters and other carbonyl derivatives linked to furan rings.

The characteristic bands associated to both diepoxy monomers used and to BDMA are assigned in Table S2. The spectrum of GDE contains the characteristic bands associated to the epoxy groups: The C–H stretching vibrations in the oxirane rings appear at 3056 cm^{-1} and 3000 cm^{-1}, while the C–O–C stretching vibration is observed at 1253 cm^{-1}, the C–O–C asymmetric stretching vibration appears at 906 cm^{-1} and the bending deformation of the epoxy group at 837 cm^{-1} (Figure S1). Furthermore, a band at 3482 cm^{-1} associated to the O–H stretching vibration is attributed to the hydroxyl group. In the case of PEGDE, the same bands associated to the oxirane rings commented for GDE were observed except the O–H stretching vibration.

In the IR spectrum of BDMA, we observe the characteristic C–N stretching vibration related to a tertiary amine present at 1024 cm^{-1}. Furthermore, the bands related to the C–H stretching vibrations in the aromatic ring are observed between 3100 and 3000 cm^{-1}, whereas the C–H stretching vibrations between 3000 and 2700 cm^{-1} correspond to the C–H stretching vibration of the aliphatic methylene and methyl groups of the amine.

3.2.2. Copolymers Based on Humins–Epoxides Thermosets

The evolution of IR bands during the polymerization of the system HG40B5 is shown in Figure 1 and Figure S2. When the humins were copolymerized with the GDE using BDMA as initiator, the intensity of the bands associated to the epoxide groups decreases progressively until their total disappearance at a temperature of reaction ~130 °C. This observation supports our assumption that the ring-opening of the epoxy groups is the predominant reaction in the humins–epoxy reaction mixtures. Nevertheless, other changes in the spectra are also observed. On the one hand, only one band between 3600 and 3300 cm^{-1} is observed during the curing. This band, that appears at the same wavenumber for the three systems, could be related to the O–H stretching vibration of the hydrogen bonds. Compared with the other two systems, the intensity of this band increases due to the presence of more reactive alkoxide anions generated from GDE during the propagation of the ring-opening polymerization.

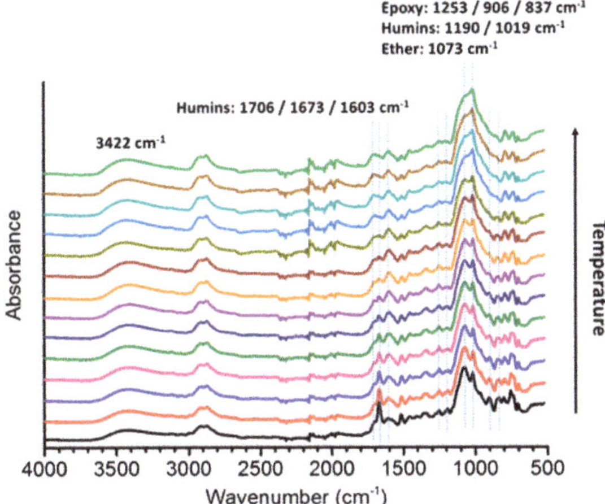

Figure 1. FT-IR spectra evolution during the polymerization of HG40B5 formulation.

In the region between 1800 and 1400 cm^{-1}, four main bands are observed: One at 1673 cm^{-1}, related to the C=O stretching vibration of the aldehyde groups from humins structure [25,38]. During the curing process, the intensity of this band gradually decreases until their complete disappearance. This observation reinforces the proposed hypothesis that the aldehyde groups of humins would react via aldol additions. In fact, some studies demonstrate that the humins itself are obtained via the aldol addition/condensation reactions between 5-Hydroxymethylfurfural, 5-Methyoxymetylfurfural, levulinic acid and other small organic compounds [17,39]. Another band that supports this hypothesis appears at 1603 cm^{-1}. This broad signal is related to the C=C stretching vibration of double bonds conjugated with carbonyl groups. As commented before, the obtained product of an aldol condensation is an α, β-conjugated system. Finally, a band at 1518 cm^{-1} is assigned to the C=C stretching vibration in the furan rings of humins linked to aldehyde groups. The decrease of the intensity of this band during the polymerization reaction means that the chemical environment of these furan moieties of humins is modified. On the other hand, a band at 1508 cm^{-1} associated to the same C=C stretching vibration in other furan rings not linked to aldehydes keeps constant during the polymerization. In fact, this band is clearly observable in the final FT-IR spectra of each system.

Finally, in the region between 1300 and 700 cm^{-1}, different bands associated to the humins, the epoxy monomers and the resulting copolymers are detected. Three bands at 768, 804 and 1190 cm^{-1} are assigned to humins (Table S1). While the intensity of the bands at 768 and 1190 cm^{-1} gradually decreases, the band at 804 cm^{-1} associated to the C–H out-of-plane deformation in the furan rings moves to 798 cm^{-1}. This variation to a small wavenumber is consistent with the furan substitution in the C$_5$ position. Two other bands at 1073 and 1019 cm^{-1} were noted, which were assigned to the C–O–C asymmetric stretching vibration of aliphatic ethers and to the C–O stretching vibration of the furan rings, respectively. As observed in Figure S2, at the end of copolymerization the contribution of the band assigned to the C–O stretching in the furan rings is higher than the contribution of the band attributed to the C–O stretching of the aliphatic ethers, in contrast with the proportion of these peaks at the beginning of reaction. The presence of the hydroxyl groups in GDE (AB$_2$ monomer) increments the number of reactive positions during the propagation of the ring-opening polymerization in the HG40B5 system. Hence, the formation of more alkoxide anions increases the degree of branching of the final polymer leading to structures with shorter chains between the crosslinks in which the contribution of humins is higher compared with the epoxide polyether chain [40]. Consequently, once the alkoxides cannot continue the ring-opening propagation due to steric impediments and topological restrictions, the aldol addition/condensation could be favoured. The faster decrease of the band attributed to the aldehyde's functionalities in humins (1673 cm^{-1}) in HG40B5 system confirms this hypothesis. Besides, the band associated to the C=C stretching vibration at around 1600 cm^{-1} shows a higher contribution in the formulation HG40B5. On the other hand, the contribution of the band at~ 1079 cm^{-1} is higher than the band associated to the C-O stretching vibration of the furan rings when more PEGDE was added (Figure 2). PEGDE contains only the oxirane functionalities and the concentration of the propagation species is smaller than in the systems containing the GDE. So, the ring-opening polymerization is slower when increasing the amount of PEGDE and at the same time, the aldol additions are less favoured. Moreover, less hyperbranched polymeric structures are obtained, which results in a higher contribution of the polyether chain in the final polymeric networks.

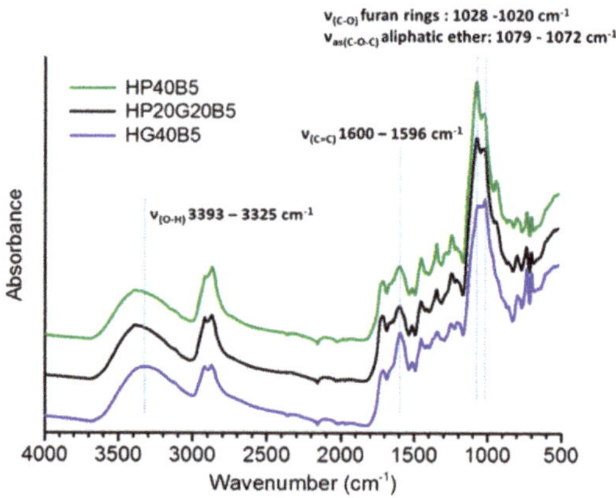

Figure 2. FT-IR spectra of the three thermoset copolymers at the end of the reaction.

Two well-defined peaks appear at 1714 and 1732 cm^{-1} when PEGDE is a comonomer (Figure S3). As described in literature [41], these two peaks are associated to the carbonyl groups present in the aliphatic regions of humins, the chemical environment of which has been modified during the curing. The formation of these two peaks at 1715 and 1733 cm^{-1} are also clearly observed for the formulation HP20G20B5, in which a mixture of GDE and PEGDE was used (Figure S4). The observation of these two bands supports the ester formation during the curing process from the carboxylic acid groups of the starting humins.

For the formulation HP20G20B5, the evolution of the IR spectra during the curing is very similar with that of the system HP40B5. Surprisingly, all the peaks that appear in the region between 1650 and 1400 cm^{-1} (assigned to the C–C stretching in aromatic rings, furan rings, C–H asymmetric and symmetric bending in methyl and methylene groups) became thin and well defined during the copolymerization. This observation is related to the different contribution of each monomer in the final composition and proves that all of them are chemically grafted. In other words, different chemical environments were detected by FT-IR during the curing of this formulation [42].

When the FT-IR spectra of the three final thermoset copolymers were compared, the O–H stretching band shows a different value for the formulation HG40B5 (3325 cm^{-1}) in contrast with the other two formulations that contains PEGDE (3393 cm^{-1}). The higher reactivity expected of GDE vs. PEGDE could help to explain this difference. When both comonomers were combined, the observable O–H stretching vibration is the one attributed to PEGDE contribution. This fact means that PEGDE units are mainly in the external part of the resulting polymers.

One possible side reaction that we proposed was the Diels–Alder reaction between different furan rings in humins. Nevertheless, the typical bands associated to the Diels–Alder adducts at around 1720, 1690 and 868 cm^{-1} are not observed during the reaction evolution neither in the final thermosets. The absence of a clear molecule that can act as a dienophile and the prove that the Diels–Alder cycloaddition between furan rings is not energetically favoured was reported in literature [43].

3.3. NMR Spectroscopy

The NMR investigations are presented in two parts. In the first part, we tackle the characterization of humins, the epoxy monomers, BDMA and the initial mixtures. The aim of the second part is to illustrate the covalent linkages that are developed between the epoxides and the humins network. For this purpose, comparative studies between each prepared formulation at t = 0 h and at the end of the curing process (t = 6 h) have been performed.

3.3.1. Raw Materials and Initial Mixtures

The ^{1}H and ^{13}C NMR spectra of GDE, BDMA, humins and the formulation HG40B5 at t = 0 h and t = 6 h are shown in Figure 3a,b, respectively. The assignment of the peaks in ^{1}H and ^{13}C NMR spectra of humins is in agreement with the one described in literature by van Zandvoort et al. [18] Different methylene and methyne units that are linked to the furan rings appear in the region between 2.5 and 4.0 ppm. The signal assigned as F (^{1}H δ = 3.30 ppm) is attributed to the end methyl groups of methoxy units linked to the furan rings by a methylene unit (furan-CH$_2$-O-$\underline{CH_3}$). A broad peak centred at 3.40 ppm is assigned to the hydroxyl protons. Other broad peaks with smaller intensity can be seen between 4.5 and 5.6 ppm, which demonstrate the complexity of the humins structure. Two singlet signals at 4.45 (H$_E$) and 4.50 ppm (H$_D$) are associated with the methylene groups that link the furan rings with a methoxy and hydroxyl units. The carbon signals for this methylene groups appear between ^{13}C δ = 54 and 58 ppm respectively. In the aromatic region, the protons linked to the C$_3$ and C$_4$ position of the furan rings appears at 6.60 and 7.50 ppm. The carbon signals attributed to these C$_3$ and C$_4$ in the furan rings appear in the region between 109 and 125 ppm. The free C$_2$ or C$_5$ in furan rings, which are expected around 142 ppm are not observed. [18] Furthermore, we can confirm that humins have three aldehyde groups with a different chemical environment, i.e., three different signals between 9.50–9.60 ppm in the ^{1}H NMR spectrum correlated with three signals between 178 and

179 ppm in the ^{13}C NMR spectrum. In the ^{13}C NMR spectrum, a peak at 174 ppm corresponds to the carboxylic acid functionalities present in humins (Signal 3* in Figure 3b and Figure S8).

The assignment of the signals of GDE and BDMA are shown in the same figures. For GDE, the methylene group of the epoxy ring gives rise to two different proton signals centred at ^{1}H δ = 2.55 (H$_a$) and 2.70 ppm (H$_b$), while the methyne group gives a signal centred at 3.10 ppm (H$_c$). The methylene protons in the α-position of the oxirane ring give two signals at ^{1}H δ = 3.30 (H$_d$) and 3.70 ppm (H$_e$), respectively. The methyne linked to the hydroxyl group appears at 3.70 ppm (H$_h$). The resonances of the carbon atoms from the oxirane rings are found at 43 ppm (C$_1$) and 50 ppm (C$_2$), whilst the other carbon signals in the ether chain appear between 70 and 78 ppm. In the case of PEGDE, the methylene protons in the α-position of the oxirane ring give two signals at ^{1}H δ = 3.25 (H$_d$) and 3.70 ppm (H$_e$), while this carbon appear at 72 ppm. The polyether chain of PEGDE shows a broad peak between 3.50 and 3.60 ppm in the ^{1}H NMR spectrum and two signals in the ^{13}C NMR spectrum around 70 ppm (Figures S7 and S8, respectively). Related to BDMA, two singlet signals at ^{1}H δ = 2.10 ppm and ^{1}H δ = 3.35 ppm are assigned to the methyl and methylene units respectively in its ^{1}H NMR spectrum. The protons of the aromatic ring appear between 7.20 and 7.35 ppm. In the ^{13}C NMR spectrum, the methyl groups appear at 45 ppm, the methyne group at 63 ppm and the carbons of the aromatic ring are seen between 127 and 139 ppm.

3.3.2. Humins–Epoxy Thermosets

The chemical structures that result from the reaction of humins with diepoxy monomers were investigated by 1D and 2D NMR spectroscopy. Unlike in the FT-IR analyses, the polymerization was carried out in a solution to allow a good homogeneity and chemical interaction of the reagents using DMSO-d_6 as solvent. To perform the NMR, an aliquot of the homogeneous polymerized system in an oven at 130 °C was diluted in DMSO-d_6 then analysed by liquid-state NMR at 25 °C, as described in Section 4.2. As observed in ^{1}H NMR spectra acquired at t = 0 h of the three formulations prepared (Figure 3a, Figures S7 and S18), new peaks between 7.50 and 7.60 ppm appeared in the spectra, which are accompanied by a decrease of the aromatic signals of the starting BDMA between 7.20 and 7.35 ppm. Moreover, these new peaks are broader than the ones of BDMA. These quick changes in the chemical shifts of the aromatic protons assigned to BDMA support that this one is covalently attached onto the polymer as a quaternary ammonium salt. Moreover, this fact confirms that the hydroxyl groups from humins act as a proton donor. This observation proves the effectiveness of the selected initiator. Furthermore, the higher intensity of these new peaks in the HG40B5 system also suggests that the hydroxyl groups from GDE can act as a proton donor and start the ring-opening polymerization of the epoxide groups. Moreover, these hydroxyl groups can lead to intra- as well as intermolecular transfer reactions during the propagation, generating new alkoxide anions that further propagates, resulting in structures with a higher degree of cross-linking when GDE was present in the mixtures [44,45].

a) ^{1}H NMR

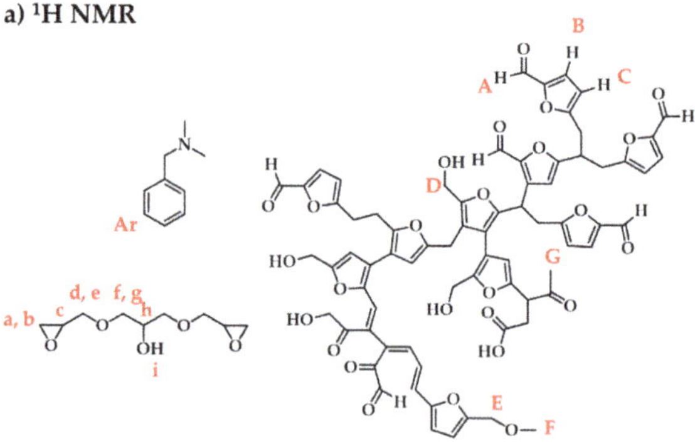

Figure 3. *Cont.*

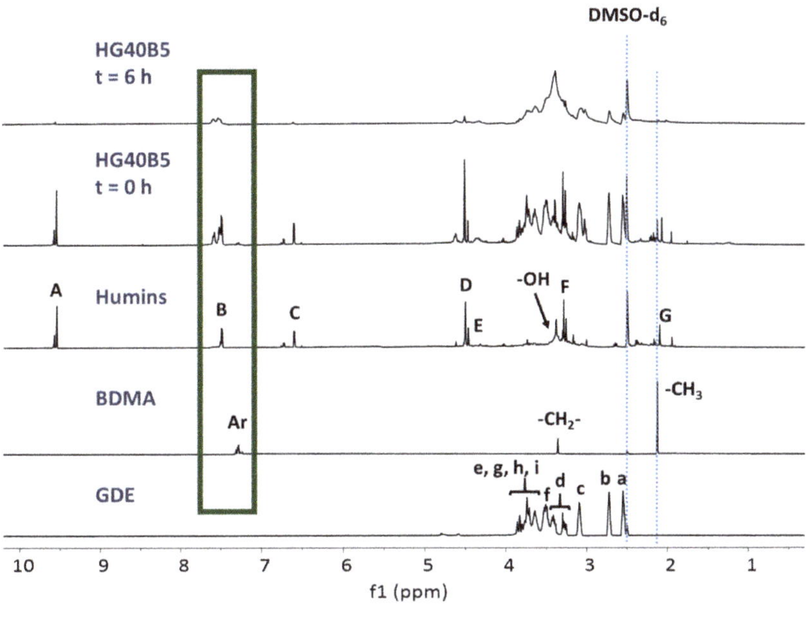

b) ^{13}C NMR

Figure 3. *Cont.*

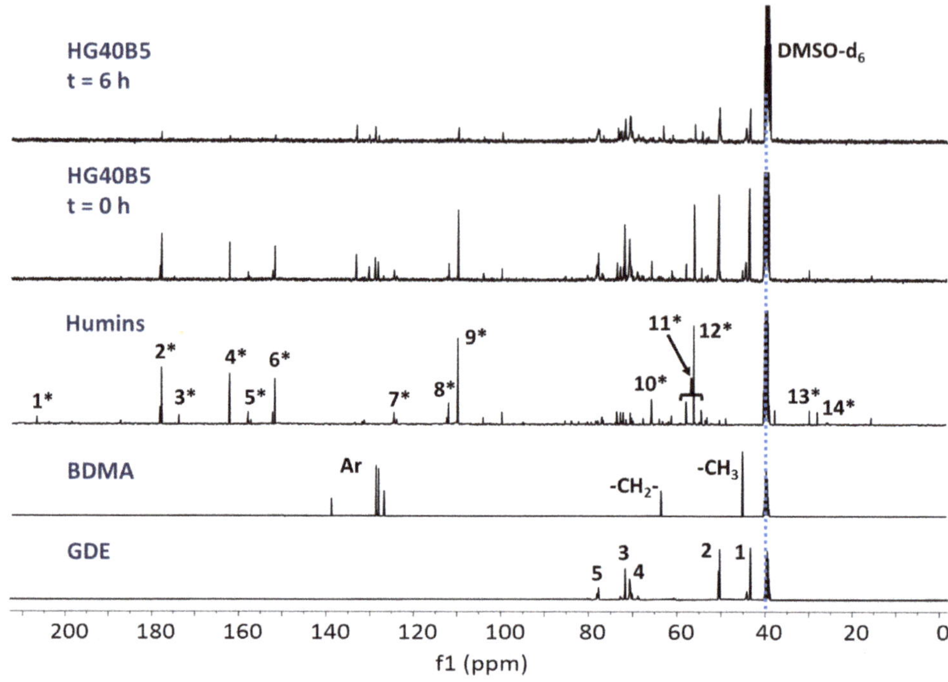

Figure 3. (a) ^{1}H NMR spectra and (b) ^{13}C NMR spectra of GDE, BDMA, humins, HG40B5 at t = 0 h and HG40B5 at t = 6 h.

For the formulation HG40B5, the comparison of the ^{1}H and ^{13}C NMR spectra of the starting and the final mixture is shown in Figure 3. A clear broadening of the peaks at the end of the curing in both ^{1}H NMR and ^{13}C NMR spectra was observed, correlated with the increase of the molecular weight during the polymerization. The presence of the peaks related to the oxirane rings at the end of the reaction means that in presence of DMSO and at the 130 °C polymerization temperature, not all the epoxy groups were consumed. A zoom of the ^{1}H NMR in the regions of the methylene units linked to furan rings and of the alkoxide units that start the ring opening polymerization is shown in Figure S6; this highlights the new signal at 4.40 ppm that appears in the ^{1}H NMR as a result of the crosslinking between humins and bis-epoxide monomers. Furthermore, the decrease of the intensity of the signal at 56.1 ppm related to these methylene units between the furan rings and the hydroxyl groups in humins is also noted in the ^{13}C NMR spectra (Figure S6), which corroborates the hypothesis of the initiation step of the ring-opening polymerization by the humins.

The broad peaks observed in the ^{1}H NMR spectra of HG40B5 at the end of the reaction and the low intensity of the new peaks that appeared in the ^{13}C NMR make difficult the assignment of the new signals related to the crosslinking between humins and diepoxy monomers by 1 D NMR. Thus, different multidimensional NMR experiments were used to confirm the expected crosslinks. First of all, ^{1}J$_{CH}$ correlations were investigated by Heteronuclear Single Quantum Correlation (HSQC) NMR spectroscopy. The HSQC spectra of the formulation HG40B5 at t = 0 h and t = 6 h are shown in Figure S9 and Figure 4, respectively. A cross-peak centred at ~ ^{1}H δ=4.40 ppm and at ^{13}C δ=65.3 ppm that appears during the curing process corresponds to the methylene group linked to the furan rings in humins chemically bonded to an opened epoxy ring of glycerol diglycidyl ether by an ether linkage: furan-CH$_2$-O-CH$_2$-CH(R′)-O-GDE (Signal a,1 in Figure 4).

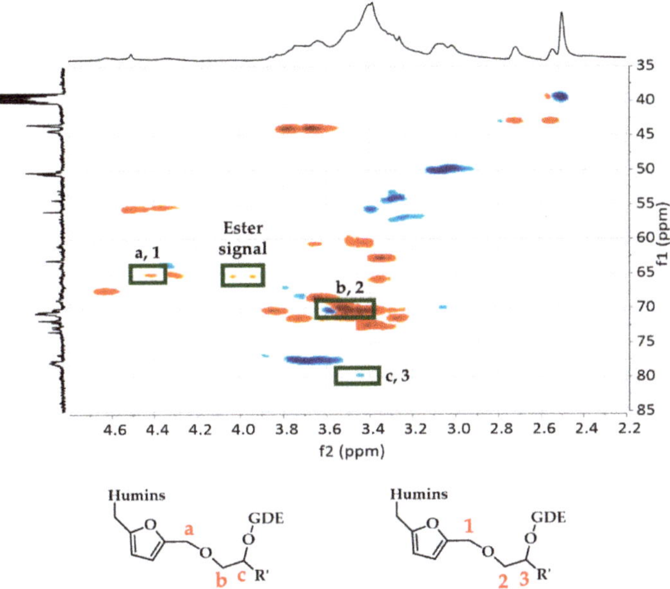

Figure 4. HSQC NMR spectra of HG40B5 at t = 6 h, which shows the cross-peaks attributed to the humins–GDE covalent linkages (green boxes). In the spectra, the blue signals correspond to –CH– and –CH$_3$ signals, while the red ones correspond to –CH$_2$– signals.

Furthermore, a methylene signal at 70.3 ppm is correlated to two signals at 3.40 and 3.60 ppm in the HSQC spectra (Signals b,2; Table 2). These protons and carbon signals are related to the methylene group of the opened oxirane ring linked to the humins (signal b,2 in Figure 4). The correlations observed in the Heteronuclear Multiple Bond Correlation (HMBC) spectra at the end of the polymerization: ^{1}H δ = 3.40 ppm and ^{1}H δ = 3.60 ppm with ^{13}C δ = 79.8 ppm (Correlation H$_b$-C$_3$ in Figure 5) aided us to confirm that this carbon signal could be attributed to the methyne group of the opened oxirane ring. This chemical shift is in agreement with the assignment presented by Vandenberg et al., in which they study the obtention of polyether from glycidol. [44] The cross-peak observed at ^{1}H δ = 3.45 ppm and ^{13}C δ = 79.8 ppm in the HSQC spectra (signal c, 3 in Figure 4) confirms the presence of this methyne signal in the final structure.

Table 2. Assignment of the protons and carbons involved in the resulting covalent ether linkage between humins and GDE in the formulation HG40B5.

Signals	^{1}H NMR Chemical Shift (ppm)	^{13}C NMR Chemical Shift (ppm)
a, 1	4.40	65.3
b, 2	3.40–3.60	70.3
c, 3	3.45	79.8

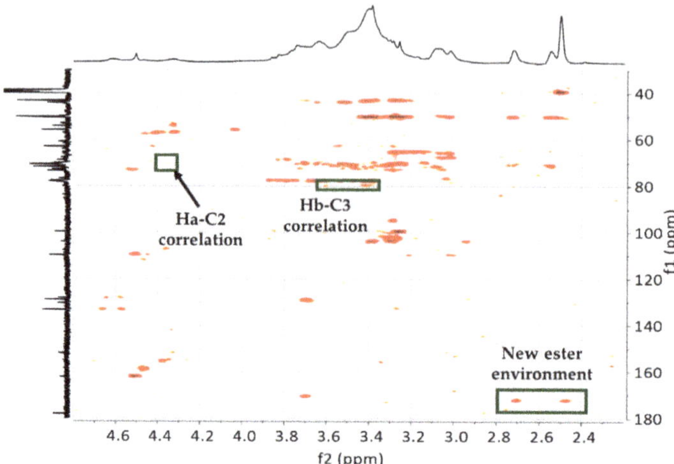

Figure 5. HMBC NMR spectra of HG40B5 at t = 6 h, which shows the additional cross-peaks attributed to the new covalent linkages between humins and epoxides (green boxes).

As previously noticed, the carboxylic acid groups found in humins structure could also react with the epoxy groups forming ester linkages. The two new cross-peaks at ^{1}H δ = 3.90 ppm and ^{13}C δ = 65.5 ppm and ^{1}H δ = 4.00 ppm and ^{13}C δ = 65.5 ppm that appeared during the polymerization of HG40B5 in the HSQC spectra confirms a new ester environment formed after the ring-opening polymerization of the epoxy rings by the carboxylic acid groups. Nevertheless, the low intensity of these correlations indicates that humins contain a small amount of carboxylic acid groups.

Additionally, the signal that appears at 4.40 ppm during the polymerization attributed to the H_a protons (Figure 4) shows new cross-peaks in the HMBC spectrum at t = 6 h (Figure 5), which confirms the covalent connections between the humins and the GDE by an ether linkage. The HMBC spectrum of the formulation HG40B5 at initial time of the polymerization is shown in Figure S10. In the new ether linkage between humins and diepoxy monomers, a ^{2}J connection between the H_a protons (Figure 5) δ = 4.40 ppm and ^{13}C δ = 155.5 ppm (C_5 in furan rings) appears at the end of the curing. Another cross-peak attributed to a ^{3}J connection at ^{1}H δ = 4.40 ppm and ^{13}C δ = 107.5 ppm (C_4 in the furan rings) confirms the chemical modification of the hydroxyl groups of humins. At the same time, the observation of a weak ^{3}J connection at ^{1}H δ = 4.40 ppm and ^{13}C δ = 70.3 ppm could be related to the ring-opening of the epoxy by the primary hydroxyl groups of the humins. This ^{13}C carbon signal is related to the methylene group of the opened oxirane ring (Signal b,2 in Figure 4), while the ^{1}H signal corresponds to the methylene group linked to the furan rings in humins (Signal a,1 in Figure 4). In the final ^{13}C NMR spectrum, this peak appears to overlap with the signal assigned as 4 in the ^{13}C NMR spectra of GDE (Figure 3b). Nevertheless, the appearance of a small broad peak in the ^{13}C NMR spectrum of the cured HP40B5 confirms the chemical shift of the expected methylene group of the opened epoxide linked to the humins (Figure S11).

Besides, the displacement of two cross-peaks from ^{1}H δ = 2.35 ppm and ^{13}C δ = 174.0 ppm (^{2}J) and ^{1}H δ = 2.65 ppm and ^{13}C δ = 174.0 ppm (^{3}J) to ^{1}H δ = 2.50 ppm and ^{13}C δ = 172.4 ppm (^{2}J) and ^{1}H δ = 2.70 ppm and ^{13}C δ = 172.4 ppm (^{3}J) are related to the new esters formed by the ring-opening polymerization of the carboxylic acid groups of humins with the diepoxy monomers.

Regarding the occurrence of GDE homopolymerization, ^{1}H NMR spectra of GDE monomer and of GDE homopolymer are shown in Figure S12. Both spectra are practically identical. The main difference between them is that the signals of GDE homopolymer are broader than in the GDE monomer related to the increase of the molecular weight during the formation of GDE homopolymer. Furthermore, the correlations observed in the HSQC spectrum of GDE homopolymer after the curing (t = 6 h, Figure S14) are the same with ones observed in the spectrum at t = 0 h (Figure S13). So, the main cross-peaks at ^{1}H δ = 3.50 ppm and ^{13}C δ = 70.5 ppm and at ^{1}H δ = 3.85 ppm and ^{13}C δ = 70.5 ppm suggested that the protons of the methylene signal attributed to the C$_2$ in the cross-linked structure between humins and GDE appear overlapped with the methylene signals of the polyether chain. Again, the observation of the cross-peaks at ^{1}H δ = 2.55 and 2.70 ppm and ^{13}C δ = 43 ppm and ^{1}H δ = 3.10 ppm and ^{13}C δ = 50 ppm at the end of the polymerization confirms that not all the epoxy groups were consumed in the conditions of samples preparation for the NMR experiments. This evidence helped us to identify the possible structure of our systems. Hence, the formation of hyperbranched polymers in which humins act as a core from which polyether chains grow is the most probable structure [46–48].

The homopolymerization of humins with BDMA using the same conditions was also studied. In this case, only the two peaks from the methylene and methyl groups of BDMA changes their chemical displacement (from 2.25 to 2.35 ppm in the case of the methyl groups and from 3.58 to 3.72 ppm in the case of the methylene group, as shown in Figure S15), which corroborates the protonation of the tertiary amine. Apart from this evidence, no other changes were detected by NMR for this system. Moreover, no evidences of the aldol/addition condensation reaction of humins were found by NMR spectroscopy when their evolution was studied in a solution of DMSO-d_6. The use of DMSO-d_6 can affect this reaction because a gradually decrease of the aldehyde band (1673 cm^{-1}) combined with the increase of the broad signal at 1603 cm^{-1} (related to the C=C bond formed in the aldol addition) were detected by FT-IR when the polymerizations were carried in bulk (Figures S2–S4) [49,50]. To confirm this hypothesis, the final products obtained in solution were also analysed by FT-IR. As observed in Figure S5, the intensity of the band attributed to the C=C bond around 1600 cm^{-1}, which is generated during the aldol condensation, is clearly lower compared with the thermoset obtained in bulk for the formulation HG40B5. Furthermore, a C=O band related to the remainder aldehyde groups of the humins is still observed in the mixture prepared in solution at the end of the polymerization.

On the other hand, a new correlation between two proton signals at 3.40 ppm and 3.60 ppm appeared in the homonuclear correlation spectroscopy (COSY) NMR spectra of the system at t = 6 h (Figure 6). This new correlation, which was not observed at t = 0 h (Figure S16), confirms that these two signals are close in the chemical environment of the molecule. In fact, both protons belong to the methylene units of the opened oxirane ring linked to the humins (H$_b$ in Figure 4).

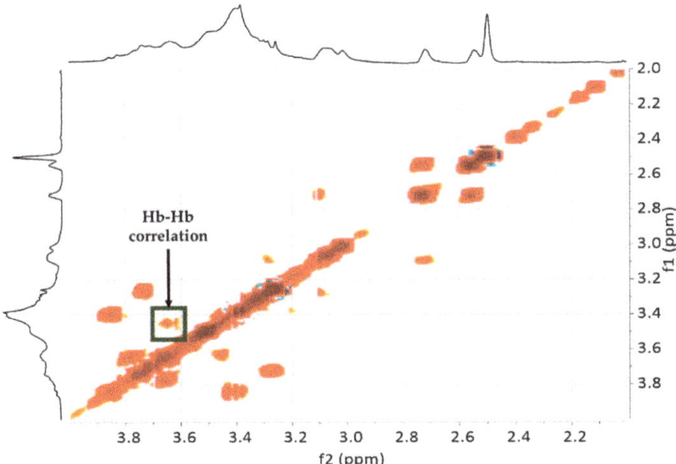

Figure 6. COSY NMR spectra of HG40B5 at t = 6 h, which shows the additional cross-peak attributed to the new covalent linkages between humins and epoxides (green).

All these predictions are in agreement with an article published by Fréchet group, in which they explore the preparation of hyperbranched polyether from epoxy monomers [51]. To sum up, the assignment of the ^{1}H NMR and ^{13}C NMR peaks involved in the ether linkage between humins and GDE is shown in Table 2.

The presence of very similar chemical shifts in NMR suggest that the other two formulations follow the same mechanism previously proposed. The evolution of the ^{1}H NMR during the curing for the system HP20G20B5 is shown in Figure S18. When both diepoxy where mixed with humins, it seems that the alkoxides have preference to open the epoxy rings of GDE monomer first. In this case, PEGDE act as a reactive diluent to GDE. A reactive diluent is a compound that improves the final conversion by enhancing the kinetics of the propagation reaction [52]. In fact, it will be expected that the addition of the linear flexible PEGDE to a GDE formulation reduces the steric and topological restrictions in the network formation improving the conversion of GDE. The appearance of different broad peaks in the region between 66 and 76 ppm in the ^{13}C NMR spectrum of the final formulation, confirms that both epoxy monomers were combined in the resulting polymeric chains (Figure S19). In the same way as in the other two systems, a final conversion of the epoxy groups was not achieved with the conditions described in the experimental section.

According to the information deduced from FT-IR and NMR studies, the proposed structure of the obtained networks is depicted in Figure 7. In our systems, it will be expected different humins cores are crosslinked by polyether chains, which have grown from the polymerization of bis-epoxide molecules. In the presented structures, humins are the cores of hyperbranched polymers that have hydroxyl and epoxy groups as chain-end. For the formulations HP40G5 and HP20G20B5, the hydroxyl and epoxy end groups are attributed to PEGDE, while in the case of HG40B5, the chain-end groups are attributed to GDE.

4. Materials and Methods

4.1. Materials

Glycerol diglycidyl ether (GDE, purity: Technical grade), with and epoxy equivalent weight (EEW) of 142 g·equivalent^{-1}, and poly(ethylene glycol) diglycidyl ether (PEGDE, average Mn = 500), with an EEW of 275 g·equivalent^{-1}, were used as monomers (The epoxy content was calculated by ^{1}H NMR). *N,N*-Dimethylbenzylamine (BDMA, purity: ≥ 99 %) was used as polymerization initiator. Both monomers and the initiator were purchased from Sigma-Aldrich and were used as received. Humins are directly produced by Avantium Chemicals at their Pilot Plant in Geleen (The Netherlands) by ACD process of carbohydrates into methoxymethylfurfural (MMF). Humins have the appearance of a very viscous, shiny, black bitumen.

4.2. Bulk Preparation of Resins

Thermoset bulk preparation was performed following the procedure described below. The components were mixed in the following order: Humins, epoxy monomers and finally BDMA was added. The *w/w* ratio material was chosen for the preparation of the mixtures. The complete homogeneity of the mixture was confirmed before the addition of each component. Once all the components were added, the formulations were heated in an oven at 80 °C for 4 h and then at 130 °C for 1.5 h more to ensure the completion of the cure. Selected times and temperatures for the curing reaction were selected according to the data presented in a previous work recently published, in which we reported on the thermomechanical properties of the humins/PEGDE, humins/GDE and humins/PEGDE/GDE copolymers. [31]

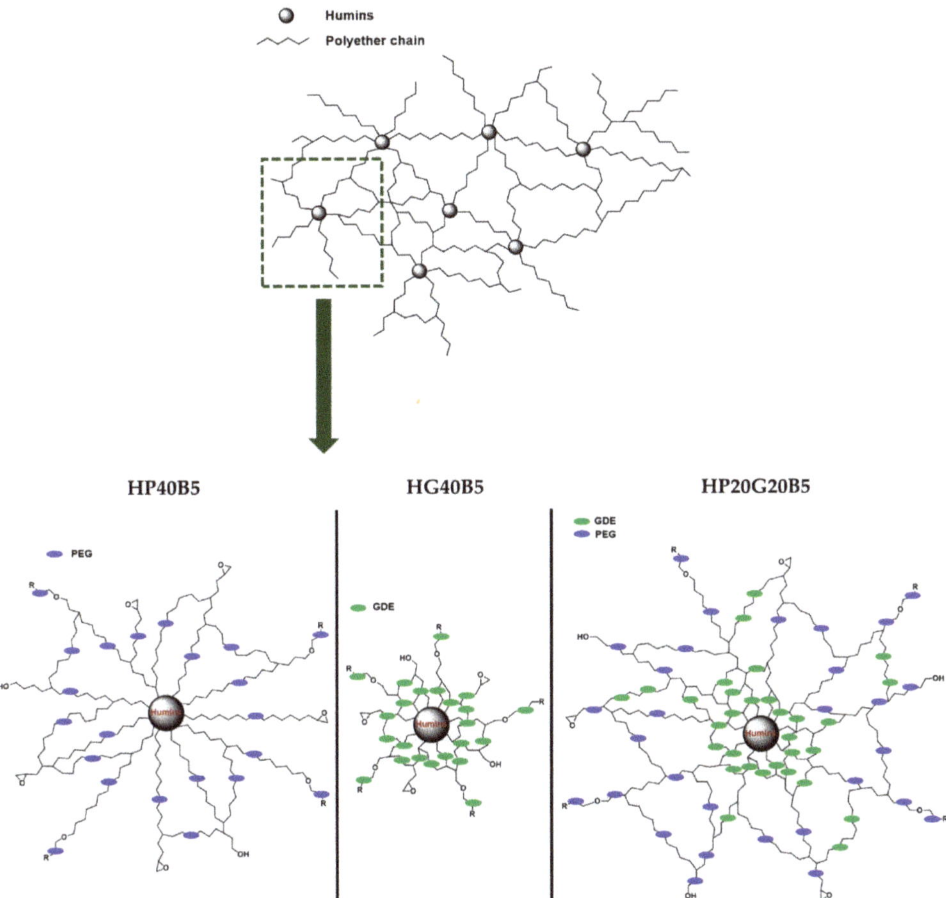

Figure 7. Proposed structures of the obtained polymer networks.

4.3. FT-IR Spectroscopy

The same procedure described for the preparation of bulk resins was applied. Thus, the formulations were heated at 80 °C for 4 h and then at 130 °C for 2 h more. Samples were taken every 30 min and directly analyzed by FT-IR. A NICOLET iS50 FT-IR spectrophotometer with a GladiATR single diamond was used. A spectrum of air was recorded as the background. A total of 32 scans with a resolution of 4 cm^{-1} were accumulated for the background as well as for each sample in the range 4000–500 cm^{-1}.

4.4. NMR Spectroscopy

1D and 2D NMR spectra (^{1}H, ^{13}C, COSY, HSQC, HMBC) were recorded on a Bruker Nano Bay AVANCE III HD 400 spectrometer with a direct BBF-H-D-05-Z-probe that operated at 400.17 MHz for ^{1}H and 100.32 MHz for ^{13}C, respectively. The NMR acquisitions were done at a temperature of 25 °C using DMSO-d_6 as deuterated solvent. The solvent signal of DMSO-d_6 at 2.50 ppm was used as a standard reference.

All NMR experiments were carried out using pulse sequences supplied by the spectrometer manufacturer (BRUKER – TOPSPIN 3.6) and processed via MestreNOVA software (version 14.1.0). ^{1}H spectrum was acquired using: 8 KHz spectral width (SW), 64K complex data point, acquisition time (aq) of 4.08s, relaxation delay (D1) of 1s, number of scan (ns) of 32 and a 30° flip angle pulse width. $^{13}C\{^1H\}$ spectra were acquired using: 29.7 KHz for SW, 64 K complex data point, 0.9s for aq, D1 of 2s, ns of 4000 and 30° flip angle pulse width. ^1H decoupling was achieved using WALTZ 16 pulse sequence. Prior to Fourier transformation, the fids were multiplied by an exponential line broadening function of 1 Hz. Gradient Selective COSY (gs-COSY) spectra were obtained with a spectral width of 5.19 KHz in both dimension, 2 K complex data point in F2, 256 t1 increments (8 scans by increment) in F1, 0.19s for aq, D1 of 2s. Prior to Fourier transformation, the data were zero filled in F1. gs-HSQC phase sensitive (echo–antiecho mode) was obtained with a spectral width of 5.19 KHz and 1K complex data point in F2 and a spectral width of 22.1KHz and 256 t1 increment (32 scans by increment) in F1. Others main parameters are: 0.98s for aq, 1.5s for D1. Prior to Fourier transformation, a QSINE window function (SSB =2) was applied in both dimensions and the data were zero filled and linear predicted (NC=32) to 1K data points in F1. gs-HMBC was acquired with a spectral width of 5.19KHz and 1K complex data point in F2 and a spectral width of 29.7KHz and 256 t1 increments (64 scans by increment) in F1. Others main parameters are: 0.39s for aq, 1.5s for D1, 8Hz for J(X-H) long range coupling and 145 Hz for 1j (X-H). Prior to Fourier transformation, a SINE window function (SSB =0) was applied in both dimensions and the data were zero filled and linear predicted (NC=32) to 1K data points in F1.

To do the NMR analyses, the samples were prepared by adding a small quantity of DMSO-d_6 (1g mixture/ 1mL solvent) to the fresh mixture. Then the polymerization was conducted using the same program protocol as in bulk: 4 h at 80 °C followed by 1.5 h at 130 °C. At the end of the polymerization, a viscous mixture was obtained. Thereafter, to perform the NMR tests, an aliquot (or small amount) of the homogeneous polymerized system was diluted in more DMSO-d_6. These diluted samples were analysed by liquid-state NMR at 25 °C.

5. Conclusions

The synthesis of completely bio-based thermosets derived from diepoxy monomers and humins were achieved. By means of FT-IR and NMR spectroscopy, the covalent connections between the hydroxyl and the carboxylic acid groups of humins and the oxirane rings of diepoxy monomers were found. We determined that the hydroxyl and the carboxylic acid groups of humins play a key role in the initiation step of the ring-opening polymerization of the diepoxy monomers. It was proved that the presence of BDMA promotes the generation of the alkoxide and carboxylate anions from these hydroxyl and carboxylic acid groups, that subsequently attack the epoxy ring by a nucleophilic addition. Furthermore, we demonstrate that the hydroxyl groups of GDE can act as a proton donor during the propagation of the ring-opening polymerization. On the other hand, the aldol addition/condensation taking place between the carbonyl functionalities of humins once the epoxy ring-opening polymerization has commenced cannot continue due to steric impediments. The obtention of hyperbranched polymeric networks consisting on a core of humins surrounded by an aliphatic polyether in the periphery is envisioned.

Supplementary Materials: The following are available online: FT-IR and NMR spectra of the raw materials used, the formulations HG40B5, HP40B5, HG20G20B5, GDE and humins homopolymerziation are in the supplementary data for this paper.

Author Contributions: X.M. and A.M. contributed to the conception, design and critical revisions of the experimental part and of the manuscript. X.M. and R.D. collected the information and performed the experiments. X.M. analyzed the data and drafted the article. A.M. corrected and approved the final version of the manuscript.

Funding: This research was funded by the European Union's Horizon 2020 Research and Innovation program under Grant Agreement 723268 for project KaRMA2020.

Conflicts of Interest: The authors declare no conflicts of interest.

References

1. Gandini, A.; Lacerda, T.M. From monomers to polymers from renewable resources: Recent advances. *Prog. Polym. Sci.* **2015**, *48*, 1–39. [CrossRef]
2. Sousa, A.F.; Vilela, C.; Fonseca, A.C.; Matos, M.; Freire, C.S.R.; Gruter, G.J.M.; Coelho, J.F.J.; Silvestre, A.J.D. Biobased Polyesters and Other Polymers from 2,5-Furandicarboxylic Acid: A Tribute to Furan Excellency. *Polym. Chem.* **2015**, *6*, 5961–5983. [CrossRef]
3. Llevot, A.; Grau, E.; Carlotti, S.; Grelier, S.; Cramail, H. From Lignin-Derived Aromatic Compounds to Novel Biobased Polymers. *Macromol. Rapid Commun.* **2016**, *37*, 9–28. [CrossRef] [PubMed]
4. Imre, B.; García, L.; Puglia, D.; Vilaplana, F. Reactive Compatibilization of Plant Polysaccharides and Biobased Polymers: Review on Current Strategies, Expectations and Reality. *Carbohydr. Polym.* **2019**, *209*, 20–37. [CrossRef] [PubMed]
5. Zhu, Y.; Romain, C.; Williams, C.K. Sustainable Polymers from Renewable Resources. *Nature* **2016**, *540*, 354–362. [CrossRef]
6. Gandini, A.; Lacerda, T.M.; Carvalho, A.J.F.; Trovatti, E. Progress of Polymers from Renewable Resources: Furans, Vegetable Oils, and Polysaccharides. *Chem. Rev.* **2016**, *116*, 1637–1669. [CrossRef]
7. Mashouf Roudsari, G.; Mohanty, A.K.; Misra, M. Green Approaches to Engineer Tough Biobased Epoxies: A Review. *ACS Sustain. Chem. Eng.* **2017**, *5*, 9528–9541. [CrossRef]
8. Kumar, S.; Krishnan, S.; Mohanty, S.; Nayak, S.K. Synthesis and Characterization of Petroleum and Biobased Epoxy Resins: A Review. *Polym. Int.* **2018**, *67*, 815–839. [CrossRef]
9. Xu, J.; Li, Z.; Wang, B.; Liu, F.; Liu, Y.; Liu, F. Recyclable Biobased Materials Based on Diels-Alder Cycloaddition. *J. Appl. Polym. Sci.* **2019**, *136*, 1–10. [CrossRef]
10. Haas, M.J.; McAloon, A.J.; Yee, W.C.; Foglia, T.A. A Process Model to Estimate Biodiesel Production Costs. *Bioresour. Technol.* **2006**, *97*, 671–678. [CrossRef]
11. Schweizer, A. Caramel and Humin. A Contribution to the Knowledge of the Decomposition Products of Sugars. *Recl. Des Trav. Chim. Des Pays Bas* **1938**, *57*, 345–382. [CrossRef]
12. Schweizer, A. The Composition of the Humins Produced by the Action of Sulphuric Acid on Some Organic Substances. *Recl. Des Trav. Chim. Des Pays Bas* **1940**, *59*, 781–784. [CrossRef]
13. Cheng, Z.; Everhart, J.L.; Tsilomelekis, G.; Nikolakis, V.; Saha, B.; Vlachos, D.G. Structural Analysis of Humins Formed in the Brönsted Acid Catalyzed Dehydration of Fructose. *Green Chem.* **2018**, *20*, 997–1006. [CrossRef]
14. Van Zandvoort, I.; Koers, E.J.; Weingarth, M.; Bruijnincx, P.C.A.; Baldus, M.; Weckhuysen, B.M. Structural Characterization of 13C-Enriched Humins and Alkali-Treated 13C Humins by 2D Solid-State NMR. *Green Chem.* **2015**, *17*, 4383–4392. [CrossRef]
15. Herzfeld, J.; Rand, D.; Matsuki, Y.; Daviso, E.; Mak-Jurkauskas, M.; Mamajanov, I. Molecular Structure of Humin and Melanoidin via Solid State NMR. *J. Phys. Chem. B* **2011**, *115*, 5741–5745. [CrossRef] [PubMed]
16. Sumerskii, I.V.; Krutov, S.M.; Zarubin, M.Y. Humin-like Substances Formed under the Conditions of Industrial Hydrolysis of Wood. *Russ. J. Appl. Chem.* **2010**, *83*, 320–327. [CrossRef]
17. Patil, S.K.R.; Heltzel, J.; Lund, C.R.F. Comparison of Structural Features of Humins Formed Catalytically from Glucose, Fructose, and 5-Hydroxylmethylfurfuraldehyde. *Energy Fuels* **2012**, *26*, 5281–5293. [CrossRef]
18. Van Zandvoort, I.; Wang, Y.; Rasrendra, C.B.; Van Eck, E.R.H.; Bruijnincx, P.C.A.; Heeres, H.J.; Weckhuysen, B.M. Formation, Molecular Structure, and Morphology of Humins in Biomass Conversion: Influence of Feedstock and Processsing Conditions. *ChemSusChem* **2013**, *6*, 1745–1758. [CrossRef]
19. Tsilomelekis, G.; Orella, M.J.; Lin, Z.; Cheng, Z.; Zheng, W.; Nikolakis, V.; Vlachos, D.G. Molecular Structure, Morphology and Growth Mechanisms and Rates of 5-Hydroxymethyl Furfural (HMF) Derived Humins. *Green Chem.* **2016**, *18*, 1983–1993. [CrossRef]
20. Shi, N.; Liu, Q.; Ju, R.; He, X.; Zhang, Y.; Tang, S.; Ma, L. Condensation of α-Carbonyl Aldehydes Leads to the Formation of Solid Humins during the Hydrothermal Degradation of Carbohydrates. *ACS Omega* **2019**, *4*, 7330–7343. [CrossRef]
21. Hoang, T.M.C.; van Eck, E.R.H.; Bula, W.P.; Gardeniers, J.G.E.; Lefferts, L.; Seshan, K. Humin Based By-Products from Biomass Processing as a Potential Carbonaceous Source for Synthesis Gas Production. *Green Chem.* **2015**, *17*, 959–972. [CrossRef]

22. Filiciotto, L.; Balu, A.M.; Romero, A.A.; Rodríguez-Castellón, E.; Van der Waal, J.C.; Luque, R. Bening-by-Design Preparation of Humin-Based Iron Oxide Initiatoric Nanocomposites. *Green Chem.* **2017**, *17*, 4423–4434. [CrossRef]

23. Kang, S.; Fu, J.; Deng, Z.; Jiang, S.; Zhong, G.; Xu, Y.; Guo, J.; Zhou, J. Valorization of Biomass Hydrolysis Waste: Activated Carbon from Humins as Exceptional Sorbent for Wastewater Treatment. *Sustainability* **2018**, *10*, 1795. [CrossRef]

24. Chernysheva, D.V.; Chus, Y.A.; Klushin, V.A.; Lastovina, T.A.; Pudova, L.S.; Smirnova, N.V.; Kravchenko, O.A.; Chernyshev, V.M.; Ananikov, V.P. Sustainable Utilization of Biomass Refinery Wastes for Accessing Activated Carbons and Supercapacitor Electrode Materials. *ChemSusChem* **2018**, *11*, 3599–3608. [CrossRef] [PubMed]

25. Tosi, P.; Van Klink, G.P.M.; Celzard, A.; Fierro, V.; Vincent, L.; de Jong, E.; Mija, A. Auto-Crosslinked Rigid Foams Derived from Biorefinery Byproducts. *ChemSusChem* **2018**, *11*, 2797–2809. [CrossRef] [PubMed]

26. Mija, A.; de Jong, E.; van der Waal, J.C.; van Klink, G. Humins Containing Foam. WO 2017074183 A1 20170504, 4 May 2017.

27. Blank, W.J.; He, Z.A.; Picci, M. Catalyssis of the Epoxy-Carboxyl Reaction. *J. Coat. Technol.* **2002**, *74*, 33–41. [CrossRef]

28. Vidil, T.; Tournilhac, F.; Musso, S.; Robisson, A.; Leibler, L. Control of Reactions and Network Structures of Epoxy Thermosets. *Prog. Polym. Sci.* **2016**, *62*, 126–179. [CrossRef]

29. Ellis, B. Introduction to the Chemistry, Synthesis, Manufacture and Characterization of Epoxy Resins. In *Chemistry and Technology of Epoxy Resins*; Ellis, B., Ed.; Springer Science Business Media: Berlin, Germany, 1993; pp. 1–36.

30. Pin, J.-M.; Guigo, N.; Vincent, L.; Sbirrazzuoli, N.; Mija, A. Copolymerization as a Strategy to Combine Epoxidized Linseed Oil and Furfuryl Alcohol: The Design of a Fully Bio-based Thermoset. *ChemSusChem* **2015**, *8*, 4149–4161. [CrossRef]

31. Dinu, R.; Mija, A. Polyfuranic Frame Networks with Elastomeric Behaviour Based on Humins Biorefinery By-Products. *Green Chem.* **2019**. [CrossRef]

32. Rozenberg, B.A. Thermodynamics and Mechanism of Reactions of Epoxy Oligomers with Amines. *Adv. Polym. Sci.* **1986**, *75*, 113–165.

33. Ashcroft, W.R. Curing Agents for Epoxy Resins. In *Chemistry and Technology of Epoxy Oligomers with Amines*; Ellis, B., Ed.; Springer Science + Business Media: Berlin, Germany, 1993; pp. 37–71.

34. Braun, M. Fundamentals and Transition-State Models. Aldol Additons of Group 1 and 2 Enolates. In *Modern Aldol Reactions*; Mahrwald, R., Ed.; Wiley-VCH Verlag GmbH & Co. KGaA: Baden-Württemberg, Germany, 2008; Volume 1, pp. 1–61.

35. Patil, S.K.R.; Lund, C.R.F. Formation and Growht of Humins via Aldol Addition and Condensation during Acid-Catalyzed Conversion of 5-Hydroxymethylfurfural. *Energy Fuels* **2011**, *25*, 4745–4755. [CrossRef]

36. List, B. Amine-Catalyzed Aldol Reactions. In *Modern Aldol Reactions*; Mahrwald, R., Ed.; Wiley-VCH Verlag GmbH & Co. KGaA: Baden-Württemberg, Germany, 2008; Volume 1, pp. 161–200.

37. Markert, M.; Mulzer, M.; Schetter, B.; Mahrwald, R. Amine-Catalyzed Direct Aldol Addition. *J. Am. Chem. Soc.* **2007**, *129*, 7258–7259. [CrossRef] [PubMed]

38. Muralidhara, A.; Tosi, P.; Mija, A.; Sbirrazzuoli, N.; Len, C.; Engelen, V.; De Jong, E.; Marlair, G. Insights on Thermal and Fire Hazards of Humins in Support of Their Sustainable Use in Advanced Biorefineries. *ACS Sustain. Chem. Eng.* **2018**, *6*, 16692–16701. [CrossRef]

39. Heltzel, J.; Patil, S.K.R.; Lund, C.R.F. Humin Formation Pathways. In *Reactions and Mechanisms in Thermocatalytic Biomass Conversion II*; Schlaf, M.Z., Zhang, Z.C., Eds.; Springer Science + Business: Singapore, 2016; pp. 105–118.

40. Guzmán, D.; Ramis, X.; Fernández-Francos, X.; De la Flor, S.; Serra, A. Preparation of New Biobased Coatings from a Triglycidyl Eugenol Derivative through Thiol-Epoxy Click Reaction. *Prog. Org. Coat.* **2018**, *114*, 259–267. [CrossRef]

41. Pin, J.-M.; Guigo, N.; Mija, A.; Vincent, L.; Sbirrazzuoli, N.; Van der Waal, J.C.; De Jong, E. Valorization of Biorefinery Side-Stream Products: Combination of Humins with Polyfurfuryl Alcohol for Composite Elaboration. *ACS Sustain. Chem. Eng.* **2014**, *2*, 2182–2190. [CrossRef]

42. Munteanu, S.B.; Vasile, C. Spectral and Thermal Characterization of Styrene-Butadiene Copolymers with Different Architectures. *J. Optoelectron. Adv. Mater.* **2005**, *7*, 3135–3148.

43. Montero, A.L.; Montero, L.A.; Martínez, R.; Spange, S. Ab Initio Modelling of Crosslinking in Polymers. *J. Mol. Struct. THEOCHEM* **2006**, *770*, 99–106. [CrossRef]

44. Vandenberg, E.J. Polymerization of Glycidol and Its Derivatives: A New Rearrangement Polymerization. *J. Polym. Sci. Pol. Chem.* **1985**, *23*, 915–949. [CrossRef]

45. Sunder, A.; Hanselmann, R.; Frey, H.; Mülhaupt, R. Controlled Synthesis of Hyperbranched Polyglycerols by Ring-Opening Multibranching Polymerization. *Macromolecules* **1999**, *32*, 4240–4246. [CrossRef]

46. Gao, C.; Yan, D. Hyperbranched Polymers: From Synthesis to Applications. *Prog. Polym. Sci.* **2004**, *29*, 183–275. [CrossRef]

47. Voit, B.I.; Lederer, A. Hyperbranched and Highly Branched Polymer Architectures—Synthetic Strategies and Major Characterization Aspects. *Chem. Rev.* **2009**, *109*, 5924–5973. [CrossRef] [PubMed]

48. Chen, S.; Xu, Z.; Zhang, D. Synthesis and Application of Epoxy-Ended Hyperbranched Polymers. *Chem. Eng. J.* **2018**, *343*, 283–302. [CrossRef]

49. Huang, Y. Concerning the Solvent Effect in the Aldol Condensation. *Monatsh. Chem.* **2000**, *131*, 521–523. [CrossRef]

50. Kapoor, M.; Majumber, A.B.; Nath Gupta, M. Promiscuous Lipase-Catalyzed C-C Bond Formation Reactions Between 4 Nitrobenzaldehyde and 2-Cyclohexen-1-One in Biphasic Medium: Aldol and Morita-Baylis-Hillman Adduct Formations. *Catal. Lett.* **2015**, *145*, 527–532. [CrossRef]

51. Emrick, T.; Chang, H.; Fréchet, J.M.J. The Preparation of Hyperbranched Aromatic and Alipathic Polyether Epoxies by Chloride-catalyzed Proton Transfer Polymerization from ABn and A2 + B3 Monomers. *J. Polym. Sci. Pol. Chem.* **2000**, *38*, 4850–4869. [CrossRef]

52. Ahn, K.D.; Kim, M.H. Enhanced Cationic Photocuring of Epoxides with Styrene Oxide as a Reactive Diluent. *Prog. Org. Coat.* **2012**, *73*, 194–201. [CrossRef]

Sample Availability: Samples of the compounds HG40B5, HP40B5 and HP20G20B5 are available from the authors.

 *molecules*

Article

Photocuring of Epoxidized Cardanol for Biobased Composites with Microfibrillated Cellulose

Sara Dalle Vacche *, Alessandra Vitale and Roberta Bongiovanni

Department of Applied Science and Technology, Politecnico di Torino, Corso Duca degli Abruzzi 24, 10129 Torino, Italy; alessandra.vitale@polito.it (A.V.); roberta.bongiovanni@polito.it (R.B.)
* Correspondence: sara.dallevacche@polito.it; Tel.: +39-011-090-4565

Academic Editor: Sylvain Caillol
Received: 17 October 2019; Accepted: 24 October 2019; Published: 25 October 2019

Abstract: Cardanol is a natural alkylphenolic compound derived from Cashew NutShell Liquid (CNSL), a non-food annually renewable raw material extracted from cashew nutshells. In the quest for sustainable materials, the curing of biobased monomers and prepolymers with environmentally friendly processes attracts increasing interest. Photopolymerization is considered to be a green technology owing to low energy requirements, room temperature operation with high reaction rates, and absence of solvents. In this work, we study the photocuring of a commercially available epoxidized cardanol, and explore its use in combination with microfibrillated cellulose (MFC) for the fabrication of fully biobased composites. Wet MFC mats were prepared by filtration, and then impregnated with the resin. The impregnated mats were then irradiated with ultraviolet (UV) light. Fourier Transform InfraRed (FT-IR) spectroscopy was used to investigate the photocuring of the epoxidized cardanol, and of the composites. The thermomechanical properties of the composites were assessed by thermogravimetric analysis, differential scanning calorimetry, and dynamic mechanical analysis. We confirmed that fully cured composites could be obtained, although a high photoinitiator concentration was needed, possibly due to a side reaction of the photoinitiator with MFC.

Keywords: biobased epoxy; cardanol; cationic photocuring; microfibrillated cellulose; biobased composites; sustainable materials

---

1. Introduction

The growing environmental concerns and the foreseen depletion of fossil resources call for a more sustainable economy; this gives impulse to the development of materials derived from biomass, produced with environmentally friendly processes. The photo-induced curing of biobased monomers, although still scarcely documented, is particularly suited for the preparation of sustainable polymer-based materials [1,2]: in fact, photopolymerization is considered a green technology owing to low energy requirements, room temperature operation with high reaction rates, and no need for solvents [3].

Cardanol is an alkylphenolic compound derived from Cashew NutShell Liquid (CNSL), an annually renewable and non-edible raw material extracted from cashew nutshells, byproduct of the food industry. Epoxidized and acrylated derivatives of cardanol, which may be suitable for photopolymerization, have been recently synthesized [4–9]. Commercial epoxidized cardanol resins are available on the market, and their thermal curing with amine hardeners has been reported, the thermomechanical properties obtained varying widely depending on the hardener used [10,11]. Few studies though have been published on the photo-induced cationic polymerization of epoxidized cardanol derivatives, namely using electron beam and UV (254 nm) as radiation sources. Different epoxidized derivatives of cardanol, including also epoxide groups on the aliphatic chain, have been used in these works, where hexafluorophosphate and hexafluoroantimonate type cationic initiators were employed [12–14].

Photopolymerized epoxy cardanol showed however relatively weak mechanical properties, due to the low functionality of the monomers, leading to a low degree of crosslinking, and to the presence of the flexible aliphatic side chain. Therefore, its use as an additive has also been explored; as an example, cationic photopolymerization with an ultraviolet source in the UVA range (315–400 nm) of a cycloaliphatic epoxy-based materials containing 10 wt.% epoxidized cardanol was reported [15].

UV-curing biobased monomers in combination with natural reinforcements, for fully biobased composite materials, is another interesting possibility. Microfibrillated cellulose (MFC) may be derived by a variety of wood and annual plants biomass; MFC sheets have high mechanical and barrier properties which are very attractive for packaging and electronics applications. However, these excellent properties are highly affected by humidity. Therefore, MFC has been combined with crosslinked polymer matrices. A solvent exchange process has been reported for the fabrication of composites with microfibrillated cellulose to preserve a structure of individual nanofibers; composites with a thermally cured epoxy matrix, and UV curable composites with an acrylate matrix have been prepared [16,17]. To the best of our knowledge no reports are available at the moment on composites of microfibrillated cellulose with a biobased cardanol epoxy photocrosslinked matrix.

In this work we confirm the feasibility of the photo-induced curing with UV light of a commercial epoxidized cardanol with an iodonium hexafluorophosphate type cationic initiator. The results obtained by photopolymerization are compared with those obtained for the thermal curing of an epoxidized cardanol with the same molecular structure [10]. Although the structures of the monomers and prepolymers used in previous works on photo-induced curing of epoxidized cardanol derivatives [12–14] are different from the one used here, some comparisons are possible and will be discussed. We hereby also confirm the possibility of preparing fully biobased photocured composites of epoxidized cardanol with microfibrillated cellulose, although a fairly high concentration of photoinitiator, of as much as 15 wt.%, was needed for leading the crosslinking reaction to completion. The reason for this is attributed to a side reaction of the photoinitiator with the microfibrillated cellulose.

2. Materials and Methods

2.1. Materials

The epoxidized cardanol (EC) resin NC-514S, with an epoxy equivalent weight of 418 g/mol, whose structure is shown in Figure 1, was provided by the Cardolite Corp. (US). Omnicat 250, which is a 75% solution of iodonium, (4-methylphenyl) [4-(2-methylpropyl) phenyl]-hexafluorophosphate(1-) in propylene carbonate (IGM Resins, US), was used as photoinitiator (PI); its structure is also shown in Figure 1. The microfibrillated cellulose Exilva F01-V, in the form of a paste with 10 wt.% concentration of cellulose microfibrils in water, was provided by Borregaard (Norway). For solvent exchange, acetone ≥99.5%, by Sigma-Aldrich S.r.l. (Italy), was used.

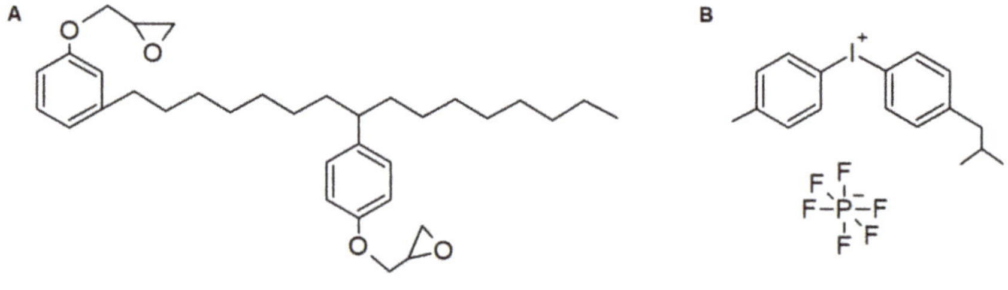

Figure 1. Structures of the epoxidized cardanol NC-514S (**A**) and of the photoinitiator (**B**).

2.2. Preparation of the Resin and of the Composites

The photocurable resins were prepared by mixing the EC with the desired amount of photoinitiator PI, and briefly stirring with a magnetic stirrer. The resins were kept in the dark and refrigerated, and used within short time from their preparation; they were used as prepared for the resin photopolymerization study, or diluted at a 30 wt.% concentration in acetone for impregnation of MFC mats.

For the preparation of the MFC mats, the cellulose microfibrils were diluted in distilled water at 0.75 wt.% concentration, and dispersed with a homogenizer (Ultraturrax T10, IKA®-Werke GmbH & Co. KG, Germany) at about 20 k rpm, for 3 min. The obtained suspension was then filtered with a Büchner funnel connected to vacuum, equipped with a 47 mm diameter Durapore® membrane filter made of hydrophilic poly(vinylidene fluoride) (PVDF) with 0.65 μm pore size. The microfibrillated cellulose thus formed a wet mat on the filter, containing approximately 85 wt.% of water. The water impregnating the MFC mat was then exchanged with acetone by soaking the mat in an acetone bath in a Petri dish, covered with a lid sealed with aluminum foil to prevent solvent evaporation. The acetone was refreshed three times over 24 h. To obtain dry MFC mats, used as a reference, acetone was replaced a fourth time, and was then allowed to slowly evaporate at room temperature with the lid of the Petri dish in place but not sealed. To obtain the composites, instead, once taken off the third acetone bath, the mat was directly soaked in the resin solution in acetone for 3 additional hours. The impregnated mat was then taken off the resin bath and dried under vacuum for 10 min, and then immediately irradiated with UV light. The final resin/microfibrillated cellulose weight ratio was controlled by the concentration of the resin in the acetone solution used for the impregnation of the MFC mat. An approximate resin content of 80 wt.% in the final composite was obtained when an acetone solution with 30 wt.% resin concentration was used as an impregnation bath.

A 5000-EC UV flood lamp system (Dymax Corporation, Torrington, CT, USA) with medium intensity mercury bulb (320–390 nm) was used for curing the resins, in the form of 10–50 μm-thick films on silicon wafers or glass slides, and the composites, in the form of 180 μm-thick self-standing films. The intensity was tuned at 100 ± 10 mW cm^{-2} by changing the distance between the specimen and the light source, and was measured by means of a UV Power Puck II radiometer (EIT, LLC., Leesburg, USA). The resin films were irradiated only on the free side, while the irradiation of the composites was carried out by turning the sample upside down every minute, to have homogeneous irradiation on the two sides; therefore, the irradiation time for the composites is indicated as $2 \times n$ minutes, where n is the number of minutes of irradiation per side.

In what follows the mixtures of EC and PI are referred to as "EC-X%PI" resins, where X indicates the PI weight percent in the resin mixture, and the corresponding composites are indicated as MFC-EC-X%PI.

2.3. Characterization Methods

Optical microscopy was performed in in reflection and transmission mode with a Nikon SMZ18 stereo microscope (Nikon Instruments Europe B.V, The Netherlands), and in reflection dark field mode with an Olympus BX53M microscope (Olympus Italia S.R.L., Italy).

Fourier Transform Infrared (FT-IR) analysis was performed with a Nicolet iS50 spectrometer (Thermo Fisher Scientific Inc., Waltham, MA, USA). For monitoring the photocuring reaction, the resins were spread on a silicon wafer with a 10 μm wire wound bar, and were analyzed in transmission mode, in the 400–4000 cm^{-1} range, with 32 scans and a resolution of 4 cm^{-1}. The dry MFC mat and the composites were analyzed by Attenuated Total Reflectance Fourier Transform Infrared (ATR FT-IR) spectroscopy analyses with an accessory equipped with a diamond crystal. The spectra were acquired in the 525–4000 cm^{-1} range, 32 scans per spectrum and a resolution of 4 cm^{-1}.

The insoluble fraction of the cured resins was assessed by measuring the weight of cured films, detached from the substrate and wrapped in a fine metallic mesh, before and after immersion in toluene or acetone for 24 h, and evaporation of the residual solvent at room temperature for 24 h followed by drying at 100 °C for 1 h.

Thermogravimetric analysis (TGA) was performed using a TGA/SDTA 851e apparatus by Mettler Toledo (Switzerland). Scans were made from 25 °C to 800 °C with a heating rate of 10 °C min^{-1}, under a 60 mL min^{-1} N$_2$ flux, to prevent thermo-oxidative processes. The first derivative of the weight profile was calculated to better resolve the main degradation steps of the analyzed materials.

Differential scanning calorimetry (DSC) was carried out using a DSC1 STARe instrument (Mettler Toledo, Switzerland) in the temperature range from −70 °C to 180 °C, using a heating rate of 10 °C min^{-1}, under N$_2$ flux.

Dynamic mechanical analysis (DMA) was performed with a TTDMA (Triton Technology Ltd., Keyworth, UK) equipment, in the tensile mode. The temperature increased from −100 °C to 180 °C at a 3 °C min^{-1} rate. The frequency was set at 1 Hz, and the strain was set at 0.1% for the composites, and 0.01% for the dry MFC mat. The specimens had a length of 10 mm between the clamps, and a width of 6 to 8 mm.

Wettability measurements were performed with a FTA 1000 C contact angle meter (First Ten Ångstroms, Portsmouth, VA, USA), equipped with video camera and image analysis software (FTA32 Software version 2.1, First Ten Ångstroms, Portsmouth, VA, USA), with the sessile drop technique. The testing liquid was water (γ =72 mN m^{-1}).

The water uptake was measured by immersing a dried specimen of known weight in distilled water at room temperature for 4 weeks and then weighing it again after lightly wiping the surface to remove excess water.

3. Results and Discussion

3.1. Morhology and Photocuring Behaviour

In this section, we report the photocuring of a commercial EC resin, with a cationic hexafluorophospate photoinitiator, and of its composites with microfibrillated cellulose. Dry microfibrillated cellulose mats prepared by solvent exchange, but without addition of resin, were also characterized for reference. As detailed below, while the resin containing 5 wt.% PI could reach full conversion within a few minutes of exposure to UV light, for the composites the curing reaction was somewhat hindered, and only by highly increasing the PI amount up to 15 wt.% a full cure could be obtained.

Photos of the dry MFC mat and of the cured composites are shown in Figure 2. The dry MFC mat obtained by solvent exchange was transparent, provided that a slow evaporation rate of acetone was ensured by keeping the Petri dish covered with his lid during drying. Otherwise, with fast evaporation rates a white non-transparent film was obtained (Figure S1); therefore, one can suggest that the porosity left in the mat upon evaporation depends highly on the solvent evaporation rate. The composites were relatively transparent; when uncured their color was yellowish, due to the color of the resin, and turned brownish upon curing, being darker with increasing PI concentration. Reaction of MFC with PI contributed to the color change, as also MFC impregnated with only PI was found to develop a brownish color upon irradiation (Figure S2).

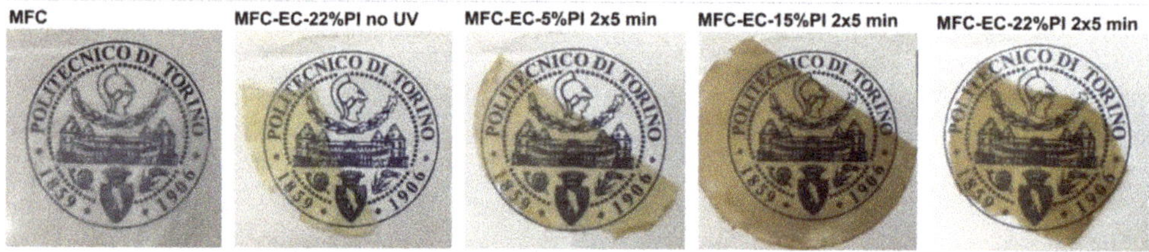

Figure 2. Photos of dry MFC mat, MFC-EC-22%PI composite before irradiation, and of composites irradiated for 2 × 5 min: MFC-EC-5%PI, MFC-EC-15%PI and MFC-EC-22%PI.

The dry MFC mat, and the uncured and cured composites were also observed by optical microscopy. Bright field images in both reflection and transmission mode are shown in Figure 3 and dark field images in reflection mode are shown in Figure 4. The homogeneous structure of entangled fibers shown by the dry MFC mat was maintained also in the composites, before and after curing. A few larger fibers were also visible. A rough surface structure, due to the cellulose microfibrils was highlighted by the images taken in reflection mode.

Figure 3. Bright field transmission (left) and reflection (right) images of: dry MFC mat, MFC-EC-15%PI before UV exposure, and MFC-EC-22%PI with 5 min of UV exposure per side.

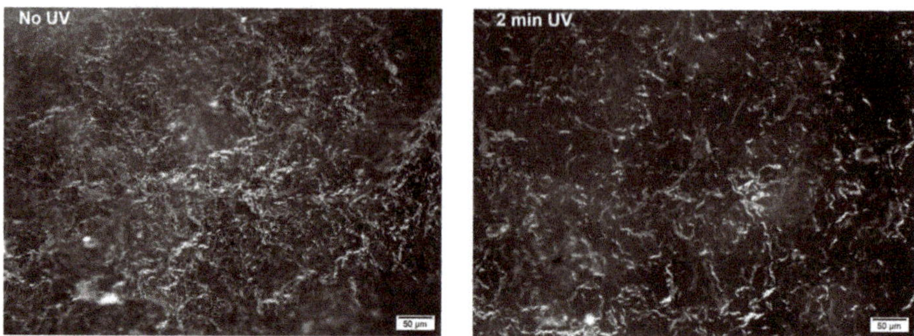

Figure 4. Reflection dark field images of MFC-EC-5%PI uncured (left) and with 2 min of UV exposure per side (right).

Both the dry MFC mat and the composites were flexible, and it was possible to roll them on a 4.9 mm diameter cylinder, as shown in Figure 5.

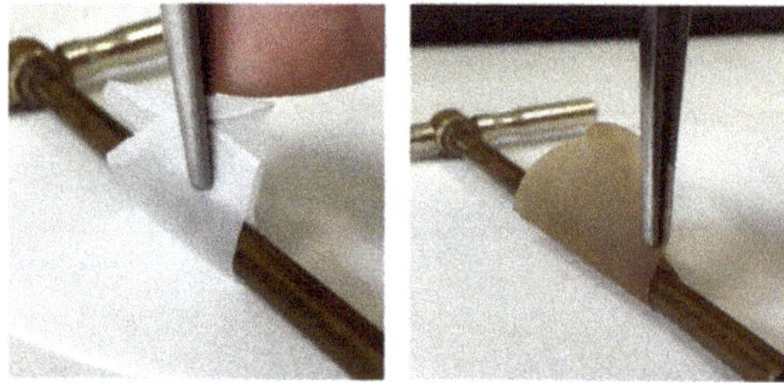

Figure 5. Dry MFC mat (left) and fully cured composite (right) rolled on a 4.9 mm diameter cylinder.

To explore the photo-induced curing of the cardanol resin and its composites, the FT-IR transmission spectra of the EC resin and of the PI, as well as the ATR FT-IR spectrum of the dry MFC mat were recorded (Figure 6) as reference.

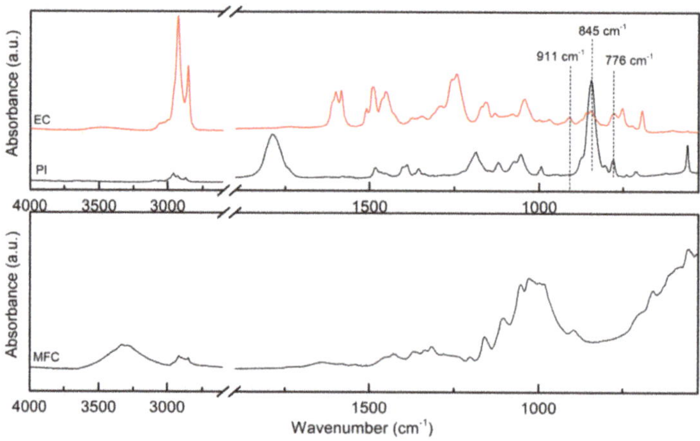

Figure 6. FT-IR transmission spectra of the epoxidized cardanol (EC) and of the photoinitiator (PI), and ATR FT-IR spectrum of the non-impregnated MFC mat (dry MFC).

The spectrum of the EC resin showed two intense peaks at 2927 and 2854 cm^{-1} corresponding to sp^3 C–H bond vibrations in the aliphatic chain of the epoxidized cardanol [4,5,12]. At higher wavenumber, some small peaks appeared; the one at 3056 cm^{-1} can be attributed to the stretching of the C–H bond of the methyl group attached to the epoxide ring [18,19]. As expected, the broad peak centered around 3350 cm^{-1} corresponding to the phenolic OH group of cardanol before epoxidation [5,20] was not present in the spectrum of EC, confirming that the hydroxyl groups are quantitatively replaced by epoxide groups. Also, the peak at about 3007–3009 cm^{-1} characteristic of C=C bonds in the aliphatic chain of cardanol [5,20] was not detectable in EC, indicating that the double bond content in this epoxidized cardanol is rather low [10]. The vibrations of the C=C aromatic bond were seen in the 1550–1650 cm^{-1} and 1400–1510 cm^{-1} regions [4,5,20]. A shoulder peak at 1258 cm^{-1} indicated the presence of the phenolic ether linkage [14,20]. The characteristic vibrations of the epoxide ring were found at 911 cm^{-1} [4,5,11,14,20], 860 cm^{-1} [4,12] and 776 cm^{-1} [4]. Among these, the most suitable to be followed to assess the degree of curing of the EC was the 911 cm^{-1} signal of the epoxide group [14], as the signals at 860 and 776 cm^{-1} overlapped with peaks characteristic of the meta substituted (670–710, 750–805, 870–900 cm^{-1}) and para substituted (845 cm^{-1} γCH) aromatic rings in the EC molecule [21–23], and with some photoinitiator's signals (845 cm^{-1} and 780 cm^{-1}).

Indeed, the spectrum of the photoinitiator showed peaks at 1790–1800 cm^{-1} and at 780 cm^{-1}, related to the propylene carbonate solvent, an intense peak at 845 cm^{-1} (stretching vibrations of the PF$_6$ anion), with shoulders at 876 cm^{-1} and 804 cm^{-1}, while the δ(PF$_6$) bending vibrations was observed as a narrow strong band at 558 cm^{-1} [24].

The spectrum of the MFC mat was characterized by a broad O–H stretching signal in the 3500–3000 cm^{-1} region, relatively weak C-H stretching peaks at 3000–2800 cm^{-1}, and in the fingerprint region intense bands attributed to the C–O stretching of the pyranose ring skeletal vibration in the 1150–1030 cm^{-1} range, and to the β-glycosidic bond vibration at 896 cm^{-1}; a weak and broad peak centered at 1639 cm^{-1} reflected the presence of water adsorbed into the cellulose fibrils [25–30].

The photocuring of the resin mixtures containing 5%, 15% and 22% photoinitiator was followed; the higher amounts of photoinitiator were used also for the resin as a comparison with the composites, which could not be cured with the lowest concentration. The formulations were exposed to UV radiation at 100 mW cm^{-2} with a medium intensity mercury bulb for intervals of 1 min, and the photo-induced reaction of ring opening photopolymerization was followed by FT-IR in transmission mode (Figure 7). The peak at 911 cm^{-1}, characteristic of the epoxide ring, decreased with increasing exposure time and eventually completely disappeared meaning the full conversion of the epoxide rings: for the EC-5%PI resin the peak disappeared within 3 min, while for the EC-15%PI and EC-22%PI resins less than 1 min of irradiation was sufficient. The peak at 3056 cm^{-1} and the shoulder at 860 cm^{-1} decreased as well with irradiation dose, although quantification was not possible due to overlap with other peaks. The peaks characteristic of the photoinitiator, i.e., 1800 cm^{-1}, 845 cm^{-1}, 780 cm^{-1} also decreased with increasing UV exposure. While the epoxide conversion proceeded, the broad peak centered around 3450 cm^{-1} increased, as expected from the formation of hydroxyls after the opening of the epoxide ring. The spectra revealed other interesting features of the reaction: upon irradiation, between 1760 cm^{-1} and 1700 cm^{-1} some weak peaks appeared, which could be attributed to the autooxidation of the residual double bonds in the aliphatic chain [5]. The intensity of the peaks at 2927 cm^{-1} and 2854 cm^{-1}, and at 1600 cm^{-1} and 1583 cm^{-1} did not change; however, the peaks became slightly broader. The FT-IR spectra of the cured resins of this work were similar to those present in the literature [14].

The high degree of curing of the EC-5%PI resin was confirmed by measuring the insoluble content, which was in the 86–89 % range both in acetone and in toluene.

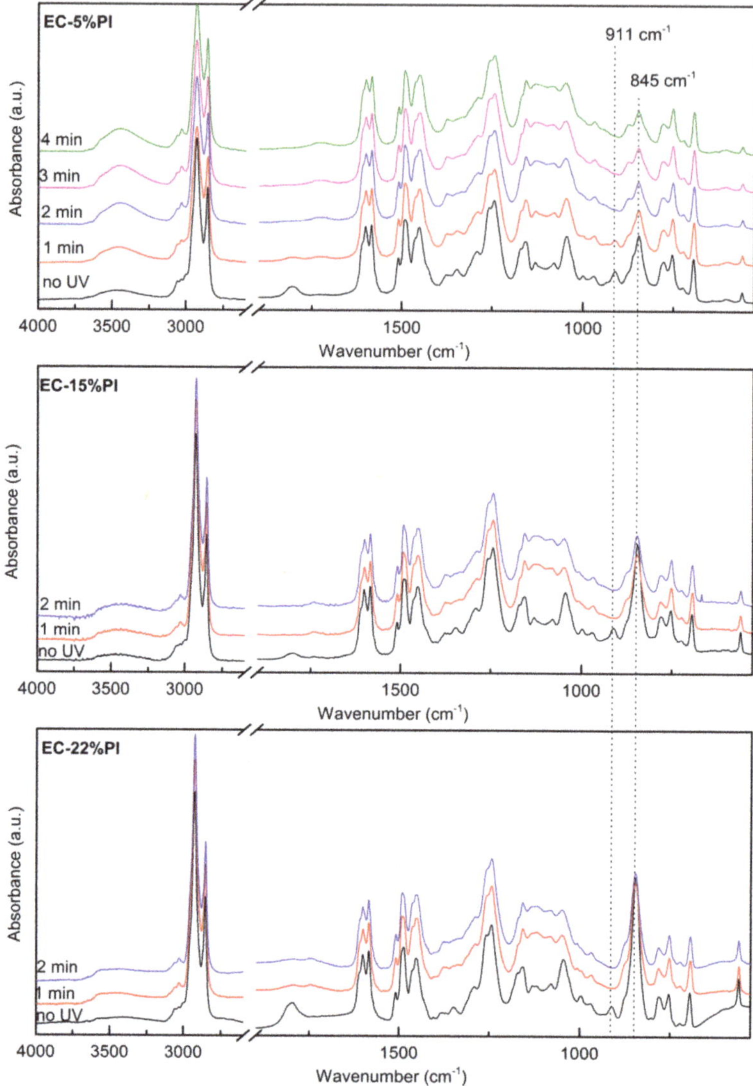

Figure 7. FT-IR spectra of the EC-5%PI, EC15% PI and EC-22%PI resins before and after different times of exposure to UV radiation with a 100 mW cm^{-2} intensity.

The photocuring of the composites upon exposure to UV radiation at 100 mW cm^{-2} for increasing time was followed by FT-IR analysis in ATR mode, as described in Section 3. The ATR FT-IR spectra of the composites with EC-5%PI, EC-15%PI and EC-22%PI resins before, and after different times of light exposure are shown in Figure 8. When the MFC mat was impregnated with the resins, the intensity of the O–H stretching signal diminished, and the characteristic peaks of the resins, as described in Section 2.1, appeared. When the 5% PI resin composites were irradiated, a moderate decrease of the epoxy signals was initially visible; however, after 2 min the conversion reaction did not seem to proceed further. When the composites with 15% and 22% PI resins were irradiated for increasing time, the same trends as for the resin alone were seen. The peak at 911 cm^{-1} diminished and eventually disappeared completely, after irradiating for more than 2 min per side, faster at the higher

PI concentration. The photocuring was however slower than for the resin alone. The consumption of the photoinitiator was confirmed by the decrease of the peaks at 1800 cm^{-1} and 845 cm^{-1}, which initial intensity increased with PI concentration in the resin. As the cationic curing is known to proceed also after the end of irradiation, partially cured samples with EC-5%PI and EC-15%PI were kept in the dark, and FT-IR spectra were taken again after 1 month. The MFC-EC-15%PI composite was eventually fully cured, while the MFC-EC-5%PI composite did not show a noticeable increase of the degree of curing, suggesting that for the latter all the available photoinitiator was consumed during the initial irradiation, and no further dark curing occurred.

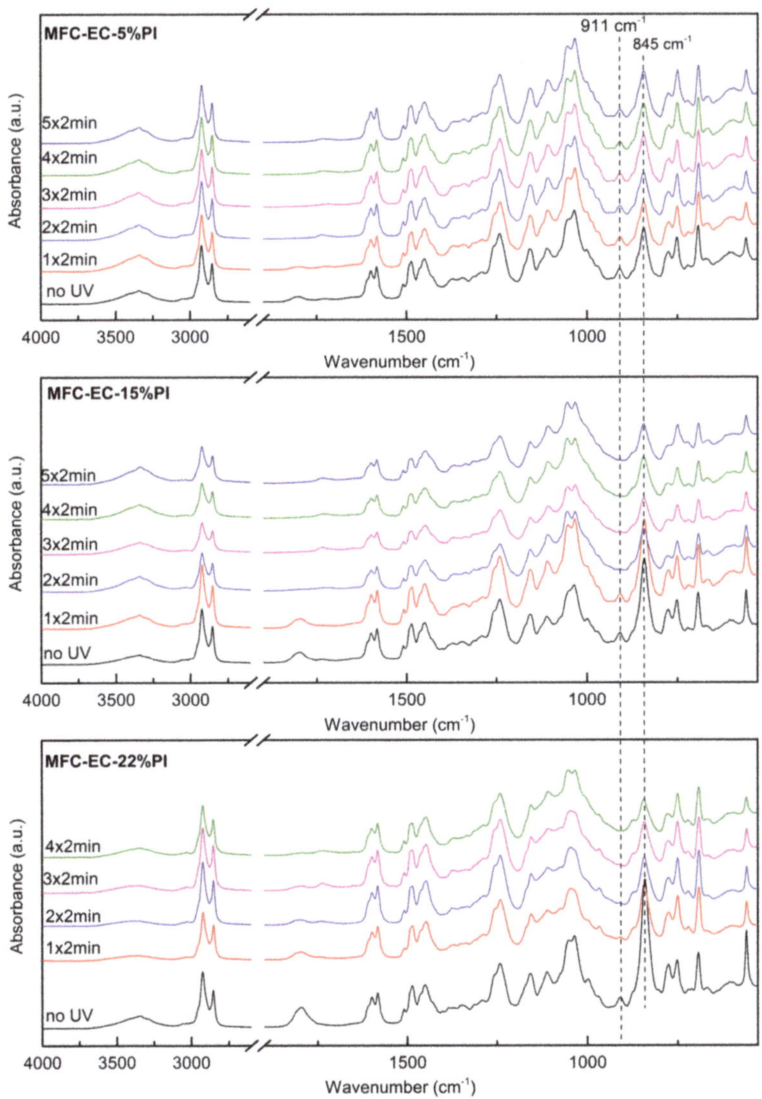

Figure 8. ATR FT-IR spectra composites MFC-EC-5%PI, MFC-EC-15%PI and MFC-EC-22%PI with different times of UV irradiation at 100 mW cm^{-2} ("n × 2 min" indicates that irradiation was carried out for n minutes on each side).

3.2. Characterization

The thermal stability of the dry MFC mat, of the EC monomer and of the fully cured EC-15%PI resin, as well as that of the uncured and cured composites was evaluated by thermogravimetric analysis in an inert atmosphere. The weight-loss curves, and their first derivatives, are reported in Figures 9 and 10. The thermogravimetric measurements were performed after 1 month from curing.

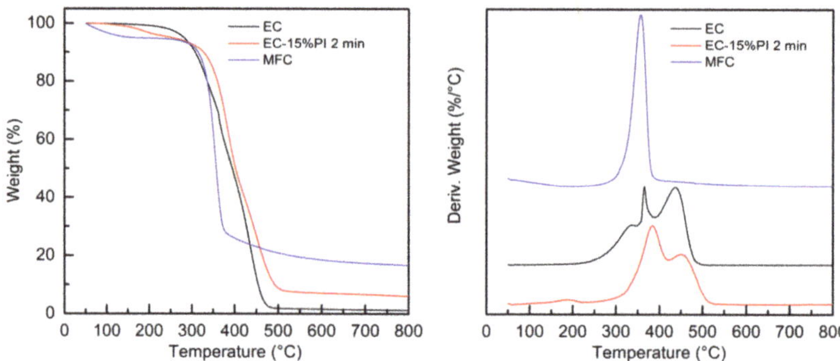

Figure 9. Weight (left), and first derivative (right) as function of temperature for the epoxidized cardanol resin EC, cured EC-15%PI resin, and dry MFC mat.

The dry MFC mat degraded in one step at around 355 °C, with a residue of about 17% at 750 °C. The initial weight loss below 100 °C can be attributed to adsorbed humidity. The EC presented its first weight loss well above 200 °C; the derivative curve highlighted three different degradation events, with maximum degradation rates at about 335 °C, 365 °C and 440 °C. The residual weight was of about 1%. For the cured EC-15%PI resin only two main degradation steps were present, with maximum rates at 380 °C and 450 °C, and a small degradation step appeared at about 190 °C attributable to the degradation of the photoinitiator. The residue was above 6%, confirming that crosslinking took place. As a comparison, the same epoxidized monomer, crosslinked by thermal curing, the temperature at which a weight loss of 30% had occurred was reported to be in the 350–366 °C range [10,31], and its char at 600 °C between 2% and 6.8% [10]. The thermal stability of our resin cured by UV light is also comparable to that of a different EC resin cured by electron beam with hexafluorophosphate photoinitiator [12].

The uncured and cured composites showed complex weight profiles, which are reported in Figure 10, together with the corresponding first derivatives that enable better visualization of the different weight-loss events.

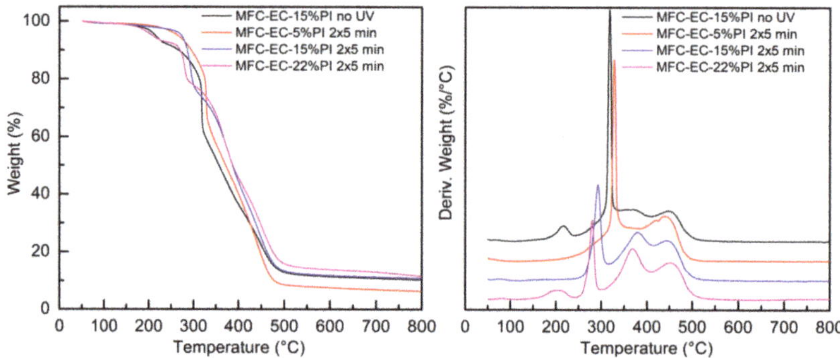

Figure 10. Weight (right) and derivative of the weight (left) as a function of temperature for uncured and cured composites.

Although a detailed understanding of all the degradation mechanisms involved is beyond the scope of this work, some general considerations can be done. The small initial weight loss before 100 °C can be attributed to evaporation of residual solvent, as seen also in the DSC analysis. The second weight loss for the uncured composite, around 200 °C, can be attributed to loss of the photoinitiator; this weight-loss step was not observed for the cured composites with 5% and 15% photoinitiator, while it was present in the cured composite with 22% of photoinitiator, suggesting that only at the highest concentration not all the photoinitiator was consumed even after dark curing. The sharp weight loss present for the uncured composite with maximum rate at about 320 °C was attributed to the degradation of the cellulose microfibrils; for the composite with 5% photoinitiator after exposure to UV light this weight loss was found at about 330 °C. In both cases the degradation temperature is slightly lower than for the dry MFC mat. Remarkably, for the cured composites with 15% and 22% PI a dramatic shift of MFC degradation to lower temperatures (280–290 °C) is detected. This effect is attributed to the action of the cationic photoinitiator, which upon irradiation generates a strong Brønsted acid. In the composites, the acid, which is meant to promote the cationic polymerization of the epoxidized monomer, is supposed to attack also the cellulose microfibrils, hydrolyzing them. Moreover, this mechanism may explain why a higher amount of photoinitiator is required to cure the composites with respect to the neat resin. In the Supplementary Information (Figure S3) is reported the thermogravimetric analysis of MFC impregnated with only photoinitiator and irradiated, showing a highly reduced thermal stability. For the uncured composite, the first degradation step characteristic of the uncured resin was present, overlapping with the MFC degradation, and the second and third degradation steps were visible with maximum rates at 370 °C and 445 °C. A similar pattern was shown for the UV irradiated MFC-EC-5%PI composite, demonstrating that the crosslinking reaction did not advance for this composite. For the cured MFC-EC-15%PI and MFC-EC-22%PI composites the first degradation step of the resin was not present, while the second and third steps became more marked; particularly the second step, which appeared at 370–380 °C, showed an increased intensity, as seen for the cured EC-15%PI resin. The residual weights of 7–12% shown by these composites is compatible with a presence of 20–30 wt.% of MFC.

The EC monomer, as well as the EC-5%PI and EC-15%PI fully cured resins, were analyzed by DSC. The DSC thermograms are shown in Figure 11.

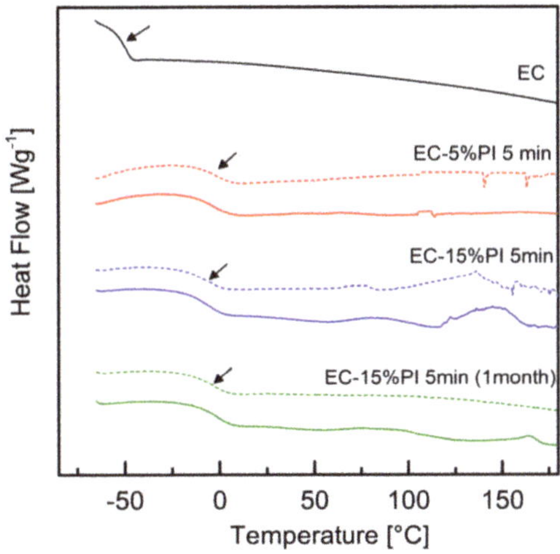

Figure 11. DSC thermograms of EC, of cured (5 min at 100 mW cm^{-2}) EC-5%PI and of cured EC-15%PI (5 min at 100 mW cm^{-2}) tested immediately and after one month from irradiation; full lines indicate the first heating scan, and dashed lines the second heating scan. The arrows indicate the glass transition region.

The curve of the uncured EC resin showed an inflection at −49 °C attributable to the glass transition. After curing, the glass transition temperature of the EC-5%PI resin increased to −4 °C; no other transitions could be remarked above this temperature. Furthermore, for both the uncured EC and the EC-5%PI cured resin no difference could be detected between the first and the second heating scan. On the other hand, for the EC-15%PI cured resin, a glass transition temperature of −7 °C was detected in the first heating scan, followed by two exothermic events that may indicate some residual reactivity. In the second heating scan the T_g increased to −3 °C, and the exothermic events were less intense. When this cured resin was tested one month after curing, the T_g was −3 °C both in the first and in the second heating scans; the exothermic events were very mild in the first heating scan, and disappeared in the second heating scan, indicating that the residual reactivity was almost completely suppressed. As a comparison, for EC thermally cured with amine hardeners, glass transition temperatures ranging from 9 °C to 158 °C were reported, depending on the type and amount of hardener used [10,31]. Furthermore, the maximum T_g obtained for an EC cured by electron beam with an hexafluorophosphate photoinitiator was −2.9 °C by DSC, while by DMA it was found to be around 13 °C; with an hexafluoroantimonate photoinitiator higher T_g were obtained [13]. The glass transition temperatures shown by a cardanol epoxy prepolymer cured by UVC with hexafluorophosphate photoinitiators combined with photosensitizers were around 25–30 °C by DSC and around 20 °C by DMA [14]. These higher results however must take into account the effect of the prepolymerization of the cardanol epoxy prepolymer on T_g.

The DSC thermograms of the composites are shown in Figure 12. The thermogram of the uncured MFC-EC-5%PI composite showed a step corresponding to the glass transition temperature of the uncured resin at −48 °C. The thermogram of the MFC-EC-5%PI composite irradiated for 5 min showed a T_g at −44 °C, confirming that negligible cure occurred. For the MFC-EC-15%PI and the MFC-EC-22%PI composites irradiated for 5 min the T_g was in the −3 °C to 1 °C range, similarly to what shown by the fully cured resin. The first heating run of all composites, before and after curing, also showed an endothermic peak around 60–75 °C, which disappeared in the second heating run, and which can be attributed to the evaporation of residual solvent.

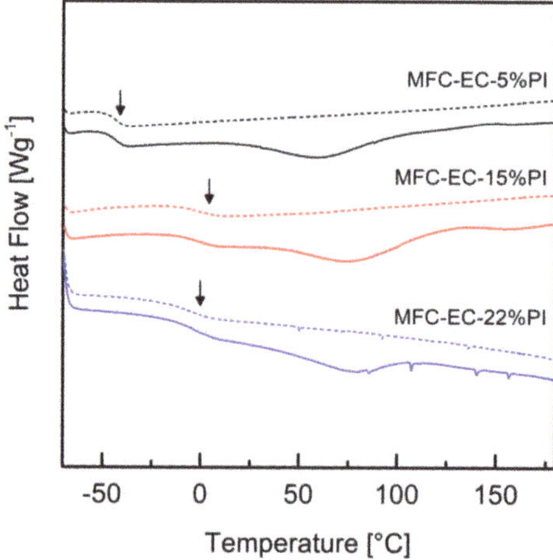

Figure 12. DSC thermograms of the cured composites with different photoinitiator concentrations. Full lines are for the first heating run and dashed lines for the second heating run in the DSC. The arrows indicate the glass transition region.

The storage modulus (E′) and the loss tangent (tan δ) as a function of temperature, measured for the uncured and cured composites, and for the dry MFC mat, are reported in Figure 13. The dry MFC mat did not show prominent thermal transitions in the explored temperature range, resulting in almost flat E′ and tan δ curves. The E′ was of about 10^9 Pa in the entire temperature range. At around 50 °C, a dip in the E′ curve, which may be due to the evaporation of residual solvent, appeared for the mat dried at 100 °C, but was not present for the mat dried at 180 °C. For the composites, the thermal transitions of the resin were detected during the temperature scan. When no UV-curing was performed on the composites, the E′ before glass transition was around 10^8 Pa, and a first large peak in the tan δ curve, related to the glass transition of the resin, was present with a peak temperature that varied in the −50 °C to −35 °C range, depending on the tested specimen. A smaller peak appeared in some cases at 10 °C. This suggests that although care was taken to shorten the time between specimen preparation and testing, keeping the sample in the dark and refrigerated, some curing of the resin may have happened even without UV irradiation, as also suggested by TGA. When the MFC-EC-15%PI and MFC-EC-22%PI composites were cured for 5 min per side at 100 mW cm^{-2} the peak of tan δ associated with the glass transition temperature shifted to about 12 °C. A very small peak was still visible at −52 °C, possibly due to some uncured resin residues. Above the glass transition temperature, the E′ dropped of about two orders of magnitude for uncured composites, while the decrease was smaller for the cured composites. The T_g obtained by DMA for the fully cured composite was similar to that obtained by DMA for the electron beam cured resin reported in [13], which also presented a similar difference between the T_g detected by DSC and that measured by DMA as that reported here.

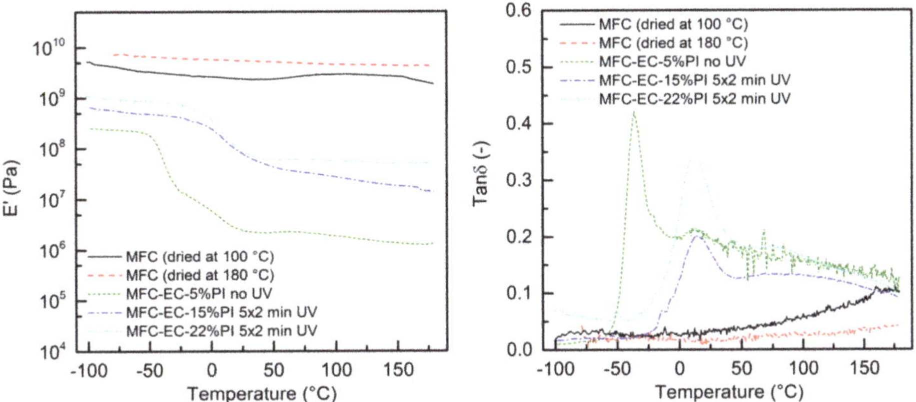

Figure 13. Storage modulus E′ and loss tangent (tan δ) as a function of temperature for uncured and cured composites, and of the MFC mat dried at 100 °C and 180 °C.

As mentioned above, MFC are hydrophilic and sensitive to water. As expected, the contact angle on the dry MFC mat was lower than 10° as soon as the drop touched the surface, while after a few seconds the drop was absorbed by the mat. The wetting by water of the MFC-EC-22%PI cured composites was hindered: the water contact angle was found to be in the 75° to 85° range, at different locations of a same specimen. This may be due to the microfibrils being covered by a thinner or thicker layer of resin at different locations of the surface, as the resin may fill the valleys of the rough MFC mat surface, as suggested also from the optical micrographs. Although the value is lower than 90° and the materials cannot be defined hydrophobic, the wettability of the composite is remarkably poor. In agreement with this result, the water uptake of the cured MFC-EC-15%PI composite after four weeks of immersion was found to be only 8%, so that the composites can be considered water resistant.

4. Conclusions

In this work we demonstrated that EC can be successfully cured by UV radiation, with a cationic photoinitiator, obtaining a transparent film at the rubbery state with a T_g below room temperature.

The resin was compatible with MFCs, the photocuring reaction of the composites proceeded to completion, although with a very high percentage of photoinitiator. The obtained films were transparent, rubbery, flexible, and water repellent. To fully exploit these materials, it might be interesting to increase the glass transition temperature, which can be easily done by copolymerization with other biobased polyfunctional epoxide resins.

Supplementary Materials: The following are available online: Figure S1: MFC dry mat obtained by solvent exchange followed by evaporation of the solvent in air (left), and its surface observed with a stereo microscope in reflection mode. Figure S2: Dry MFC mat (left) and MFC mat impregnated with photoinitiator, irradiated for 5 min at 100 mW cm^{-2} (right). Figure S3: Thermogravimetric analysis of MFC mat impregnated with photoinitiator, irradiated for 5 min at 100 mW cm^{-2}: weight profile and its first derivative as a function of temperature.

Author Contributions: Conceptualization, S.D.V. and R.B.; methodology, S.D.V.; formal analysis, S.D.V.; investigation, S.D.V. and A.V.; data curation, S.D.V.; writing—original draft preparation, S.D.V.; writing—review and editing, S.D.V., A.V. and R.B.; supervision, R.B.; project administration, S.D.V. and R.B.; funding acquisition, S.D.V. and R.B.; visualization, S.D.V.

Funding: The project ComBIOsites has received funding from the European Union's Horizon 2020 research and innovation programme under the Marie Skłodowska–Curie grant agreement No 789454.

Acknowledgments: The authors are grateful to Borregaard and Cardolite for donating materials; to Mauro Corrado and to Giulio Malucelli of Politecnico di Torino for use of the stereoscopic microscope and for the TGA measurements, respectively; to Vijayaletchumy Karunakaran and Deniz Ebeperi for their technical support.

Conflicts of Interest: The authors declare no conflict of interest. The funders had no role in the design of the study; in the collection, analyses, or interpretation of data; in the writing of the manuscript, or in the decision to publish the results.

References

1. Noè, C.; Malburet, S.; Bouvet-Marchand, A.; Graillot, A.; Loubat, C.; Sangermano, M. Cationic photopolymerization of bio-renewable epoxidized monomers. *Prog. Org. Coat.* **2019**, *133*, 131–138. [CrossRef]
2. Tehfe, M.-A.; Lalevée, J.; Gigmes, D.; Fouassier, J.P. Green chemistry: Sunlight-Induced cationic polymerization of renewable epoxy monomers under air. *Macromolecules* **2010**, *43*, 1364–1370. [CrossRef]
3. Tehfe, M.; Louradour, F.; Lalevée, J.; Fouassier, J.-P. Photopolymerization reactions: On the way to a green and sustainable chemistry. *Appl. Sci.* **2013**, *3*, 490–514. [CrossRef]
4. Chen, J.; Nie, X.; Liu, Z.; Mi, Z.; Zhou, Y. Synthesis and application of polyepoxide cardanol glycidyl ether as biobased polyepoxide reactive diluent for epoxy resin. *Acs Sustain. Chem. Eng.* **2015**, *3*, 1164–1171. [CrossRef]
5. Hu, Y.; Shang, Q.; Wang, C.; Feng, G.; Liu, C.; Xu, F.; Zhou, Y. Renewable epoxidized cardanol-based acrylate as a reactive diluent for UV-curable resins. *Polym. Adv. Technol.* **2018**, *29*, 1852–1860. [CrossRef]
6. Kanehashi, S.; Yokoyama, K.; Masuda, R.; Kidesaki, T.; Nagai, K.; Miyakoshi, T. Preparation and characterization of cardanol-based epoxy resin for coating at room temperature curing. *J. Appl. Polym. Sci.* **2013**, *130*, 2468–2478. [CrossRef]
7. Yuan, Y.; Chen, M.; Zhou, Q.; Liu, R. Synthesis and properties of UV-curable cardanol-based acrylate oligomers with cyclotriphosphazene core. *J. Coat. Technol. Res.* **2019**, *16*, 179–188. [CrossRef]
8. Kim, Y.H.; An, E.S.; Park, S.Y.; Song, B.K. Enzymatic epoxidation and polymerization of cardanol obtained from a renewable resource and curing of epoxide-containing polycardanol. *J. Mol. Catal. B Enzym.* **2007**, *45*, 39–44. [CrossRef]
9. Nguyen, T.K.L.; Livi, S.; Soares, B.G.; Barra, G.M.O.; Gérard, J.-F.; Duchet-Rumeau, J. Development of sustainable thermosets from cardanol-based epoxy prepolymer and ionic liquids. *ACS Sustain. Chem. Eng.* **2017**, *5*, 8429–8438. [CrossRef]
10. Jaillet, F.; Darroman, E.; Ratsimihety, A.; Auvergne, R.; Boutevin, B.; Caillol, S. New biobased epoxy materials from cardanol. *Eur. J. Lipid Sci. Technol.* **2014**, *116*, 63–73. [CrossRef]
11. Dworakowska, S.; Cornille, A.; Bogdał, D.; Boutevin, B.; Caillol, S. Formulation of bio-based epoxy foams from epoxidized cardanol and vegetable oil amine. *Eur. J. Lipid Sci. Technol.* **2015**, *117*, 1893–1902. [CrossRef]
12. Cheon, J.; Cho, D.; Song, B.K.; Park, J.; Kim, B.; Lee, B.C. Thermogravimetric and fourier-transform infrared analyses on the cure behavior of polycardanol containing epoxy groups cured by electron beam. *J. Appl. Polym. Sci.* **2015**, *132*. [CrossRef]

13. Cheon, J.; Cho, D. Effects of the electron-beam absorption dose on the glass transition, thermal expansion, dynamic mechanical properties, and water uptake of polycardanol containing epoxy groups cured by an electron beam. *J. Appl. Polym. Sci.* **2015**, *132*. [CrossRef]

14. Kanehashi, S.; Tamura, S.; Kato, K.; Honda, T.; Ogino, K.; Miyakoshi, T. Photopolymerization of bio-based epoxy prepolymers derived from cashew nut shell liquid (CNSL). *JFST* **2017**, *73*, 210–221. [CrossRef]

15. Chen, Z.; Chisholm, B.J.; Webster, D.C.; Zhang, Y.; Patel, S. Study of epoxidized-cardanol containing cationic UV curable materials. *Prog. Org. Coat.* **2009**, *65*, 246–250. [CrossRef]

16. Ansari, F.; Galland, S.; Johansson, M.; Plummer, C.J.G.; Berglund, L.A. Cellulose nanofiber network for moisture stable, strong and ductile biocomposites and increased epoxy curing rate. *Compos. Part A Appl. Sci. Manuf.* **2014**, *63*, 35–44. [CrossRef]

17. Galland, S.; Leterrier, Y.; Nardi, T.; Plummer, C.J.G.; Månson, J.A.E.; Berglund, L.A. UV-cured cellulose nanofiber composites with moisture durable oxygen barrier properties. *J. Appl. Polym. Sci.* **2014**, *131*. [CrossRef]

18. Luzuriaga, A.R.d.; Martin, R.; Markaide, N.; Rekondo, A.; Cabañero, G.; Rodríguez, J.; Odriozola, I. Epoxy resin with exchangeable disulfide crosslinks to obtain reprocessable, repairable and recyclable fiber-reinforced thermoset composites. *Mater. Horiz.* **2016**, *3*, 241–247. [CrossRef]

19. Nikolic, G.; Zlatkovic, S.; Cakic, M.; Cakic, S.; Lacnjevac, C.; Rajic, Z. Fast fourier transform IR characterization of epoxy GY systems crosslinked with aliphatic and cycloaliphatic EH polyamine adducts. *Sensors* **2010**, *10*, 684–696. [CrossRef]

20. Natarajan, M.; Murugavel, S.C. Thermal stability and thermal degradation kinetics of bio-based epoxy resins derived from cardanol by thermogravimetric analysis. *Polym. Bull.* **2017**, *74*, 3319–3340. [CrossRef]

21. Joseph, T.; Varghese, H.T.; Panicker, C.Y.; Viswanathan, K.; Dolezal, M.; Van Alsenoy, C. Spectroscopic (FT-IR, FT-Raman), first order hyperpolarizability, NBO analysis, HOMO and LUMO analysis of *N*-[(4-(trifluoromethyl) phenyl]pyrazine-2-carboxamide by density functional methods. *Arab. J. Chem.* **2017**, *10*, S2281–S2294. [CrossRef]

22. Coates, J. Interpretation of infrared spectra, a practical approach. In *Encyclopedia of Analytical Chemistry*; Meyers, R.A., Ed.; John Wiley & Sons Ltd.: Chichester, UK, 2006; pp. 10815–10837. ISBN 978-0-470-02731-8.

23. Roeges, N.P.G. *A Guide to the Complete Interpretation of Infrared Spectral of Organic Structures*; Wiley: New York, NY, USA,, 1994; ISBN 978-0-471-93998-6.

24. Logacheva, N.M.; Baulin, V.E.; Tsivadze, A.Y.; Pyatova, E.N.; Ivanova, I.S.; Velikodny, Y.A.; Chernyshev, V.V. Ni(II), Co(II), Cu(II), Zn(II) and Na(I) complexes of a hybrid ligand 4'-(4'''-benzo-15-crown-5)-methyloxy-2,2':6',2''-terpyridine. *Dalton Trans.* **2009**, 2482–2489. [CrossRef] [PubMed]

25. Mondragon, G.; Fernandes, S.; Retegi, A.; Peña, C.; Algar, I.; Eceiza, A.; Arbelaiz, A. A common strategy to extracting cellulose nanoentities from different plants. *Ind. Crop. Prod.* **2014**, *55*, 140–148. [CrossRef]

26. Rosa, M.F.; Medeiros, E.S.; Malmonge, J.A.; Gregorski, K.S.; Wood, D.F.; Mattoso, L.H.C.; Glenn, G.; Orts, W.J.; Imam, S.H. Cellulose nanowhiskers from coconut husk fibers: Effect of preparation conditions on their thermal and morphological behavior. *Carbohydr. Polym.* **2010**, *81*, 83–92. [CrossRef]

27. Lawther, J.M.; Sun, R. The fractional characterisation of polysaccharides and lignin components in alkaline treated and atmospheric refined wheat straw. *Ind. Crop. Prod.* **1996**, *5*, 87–95. [CrossRef]

28. Abraham, R.; Wong, C.; Puri, M. Enrichment of cellulosic waste hemp (cannabis sativa) hurd into non-toxic microfibres. *Materials* **2016**, *9*, 562. [CrossRef]

29. Alemdar, A.; Sain, M. Biocomposites from wheat straw nanofibers: Morphology, thermal and mechanical properties. *Compos. Sci. Technol.* **2008**, *68*, 557–565. [CrossRef]

30. Li, R.; Fei, J.; Cai, Y.; Li, Y.; Feng, J.; Yao, J. Cellulose whiskers extracted from mulberry: A novel biomass production. *Carbohydr. Polym.* **2009**, *76*, 94–99. [CrossRef]

31. Darroman, E.; Durand, N.; Boutevin, B.; Caillol, S. Improved cardanol derived epoxy coatings. *Prog. Org. Coat.* **2016**, *91*, 9–16. [CrossRef]

Sample Availability: Samples of the compounds reported in the paper are not available from the authors.

 *molecules*

Article

Synthesis of Pluri-Functional Amine Hardeners from Bio-Based Aromatic Aldehydes for Epoxy Amine Thermosets

Anne-Sophie Mora [1], Russell Tayouo [2], Bernard Boutevin [1], Ghislain David [1] and Sylvain Caillol [1,*]

[1] Institut Charles Gerhardt, UMR 5253—CNRS, Université de Montpellier, Ecole Nationale Supérieure de Chimie de Montpellier, 240 Avenue Emile Jeanbrau, 34296 Montpellier, France
[2] SAS Nouvelle Sogatra, 784 Chemin de la Caladette, 30350 Lezan, France
* Correspondence: sylvain.caillol@enscm.fr

Academic Editor: Sylvain Caillol
Received: 23 July 2019; Accepted: 4 September 2019; Published: 9 September 2019

Abstract: Most of the current amine hardeners are petro-sourced and only a few studies have focused on the research of bio-based substitutes. Hence, in an eco-friendly context, our team proposed the design of bio-based amine monomers with aromatic structures. This work described the use of the reductive amination with imine intermediate in order to obtain bio-based pluri-functional amines exhibiting low viscosity. The effect of the nature of initial aldehyde reactant on the hardener properties was studied, as well as the reaction conditions. Then, these pluri-functional amines were added to petro-sourced (diglycidyl ether of bisphenol A, DGEBA) or bio-based (diglycidyl ether of vanillin alcohol, DGEVA) epoxy monomers to form thermosets by step growth polymerization. Due to their low viscosity, the epoxy-amine mixtures were easily homogenized and cured more rapidly compared to the use of more viscous hardeners (<0.6 Pa s at 22 °C). After curing, the thermo-mechanical properties of the epoxy thermosets were determined and compared. The isophthalatetramine (IPTA) hardener, with a higher number of amine active H, led to thermosets with higher thermo-mechanical properties (glass transition temperatures (T_g and T_α) were around 95 °C for DGEBA-based thermosets against 60 °C for DGEVA-based thermosets) than materials from benzylamine (BDA) or furfurylamine (FDA) that contained less active hydrogens (T_g and T_α around 77 °C for DGEBA-based thermosets and T_g and T_α around 45 °C for DGEVA-based thermosets). By comparing to industrial hardener references, IPTA possesses six active hydrogens which obtain high cross-linked systems, similar to industrial references, and longer molecular length due to the presence of two alkyl chains, leading respectively to high mechanical strength with lower T_g.

Keywords: imine; epoxide; amine; thermoset; bio-based

1. Introduction

Amine is one of the most important functional groups in the chemical industry, highly present in various industrial fields, such as pharmaceuticals [1–3], agrochemicals [4,5], detergents [6,7], lubricants [8] and polymer industry [9,10]. Amines act as intermediates in the synthesis of different polymers including phenolic resins [11,12], polyimides [13], polyureas, polyurethane [14] and poly(hydroxy)urethanes [15], polyamide [16,17] and epoxy thermosets [18,19]. Our team has a long experience in the synthesis of epoxy thermosets [10,20] and has recently developed new efficient bio-based amine hardeners exhibiting high reactivity, by a direct amination method of epoxy monomers using an aqueous ammonia solution [21]. Due to the presence of many hydrogen bond sites on their structures, the obtained β-hydroxylamine hardeners showed high reactivity but also high viscosity.

Therefore, we worked on an alternative pathway to synthesize bio-based pluri-functional amine monomers, avoiding the formation of hydroxyl functions, in order to disfavor the hydrogen bonds effect and thus decrease viscosity. Hence, we synthesized amines from imine reduction, more generally called reductive amination. This methodology allows easily obtaining secondary or tertiary amine monomers with high yields and high reactivity [22–24]. Using aromatic aldehydes, amines monomers containing aromatic moieties can be synthesized from aliphatic amines which are less toxic than conventional aromatic amines. Moreover, benzyl amine monomers are much more reactive than aromatic ones. Aromaticity is really interesting in epoxy-amine formulation to improve the miscibility of amines with epoxy monomers, which are most of the time aromatic substances. Moreover, providing aromaticity tunes the hydrogen equivalent weight (HEW), decreases the volatility of amines and induces high thermo-mechanical properties for the final thermosets. Hence, such aromatic amines could be very highly desirable for epoxy thermosets synthesis.

Moreover, the imine reduction is a really simple method, using a variety of hydride source reducing agents [25,26]. The reduction step can also be performed by the aminocatalytic method using a catalyst [27], by one-step bio-catalyzed transformation using enzymes [28–30] or by the adaptation of the Haber–Bosch method used for the synthesis of ammonia [31]. This method involves the metal-catalyzed hydrogenation of unsaturated bonds as in the case of imine functions [32–36]. For instance, Jia et al. described the development of the ruthenium-catalyzed hydrogenation method for the reduction of imine intermediates to obtain amine monomers [37]. Exposito et al. recently developed the imine Pd-catalyzed hydrogenation using the industrial continuous-flow process, thanks to a Pd/C support, which was stable over multiple reaction–regeneration cycles [38].

For epoxy-amine thermosets formulation, the hardener requires a structure with at least three active hydrogens to form a cross-linked network when using a conventional diepoxy prepolymer, such as diglycidyl ether of bisphenol A (DGEBA). Most of the amine hardeners are petro-based diamines (four active H) such as 4,4′-Methylenebis(cyclohexylamine) (PACM) [39,40] and Triethyleneglycol diamine (EDR 148) [41,42], and there are only a handful of bio-based diamines, such as 1,6-hexamethylenediamine or 1,9-nonanediamine which can be obtained from renewable resources [43], 1,10-decanediamine from alcohols [44] and the diamine developed by Garrison et al. from renewable terpenoids [45]. However, beyond this functionality, very few polyfunctional amines are reported, and most of the time they are petro-based [10], such as Jeffamine T403, which exhibits a low reactivity [46]. Only a few poly-functional amine hardeners are bio-based, such as citric amido amine described by Bähr et al., which performed the aminolysis of citric acid to obtain a tri-functional primary amine [47]. Pluri-functional cardanol-based amines can also be obtained with formaldehyde and aliphatic di- or triamines to obtain the corresponding Mannich hardeners [48].

The reductive amination could be a solution to design polyfunctional amine hardeners since this is an easy method to tune and modify already existing amines. However, the imine reduction generally leads to secondary mono- or diamines as intermediates in pharmaceutical applications [49–51], except when using ammonia as the initial amine reactant [52,53], reaching primary amines which can be efficient hardeners [54,55]. This process was used for producing diamine monomers, which are useful as starting materials for various polyamides and polyurethanes but not for epoxy thermosets. Moreover, only a few amine monomers from this reductive amination method are bio-based such as 2,5-bis(aminomethyl)furan developed in 2015 by Le et al. [10,56,57]. Thanks to reductive amination, another possibility to synthesize hardeners is to slowly add an aldehyde in an excess of primary diamine monomers. Only a few examples are described in the literature. For instance, Micklitsch et al. synthesized an amine monomer exhibiting one secondary and one primary amine functions from ethylenediamine and benzaldehyde in the presence of $NaBH_4$ [58]. In 2012, Milelli et al. described the synthesis of a naphthalene diimide derivative, obtained from the addition of reduced benzylic imine on 1,4,5,8-naphthalenetetracarboxylic dianhydride monomer [59]. To the best of our knowledge, the reductive amination, thanks to imine formation, is hardly described in the literature for epoxy-amine applications. Only Kasemi et al. patented hardeners using aldehyde monomers and principally *m*-xylylenediamine (MXDA) or some derivatives of 1,2-propylenediamine as amine without any concerns about the biomass origin and reactant toxicity [60]. Moreover, this patent is very broad and imprecise and does not study the properties of hardeners.

Hence, our work proposes for the first time a reductive amination method for the synthesis of pluri-functional amine hardeners containing aromatic moieties, between three to six active H, forming a cross-linked system using diepoxy monomers. Moreover, this method respects green chemistry concepts due to the use of monomers that are or could be obtained from biomass and with a low toxicity. We have studied the impact of reaction conditions on the final amine functionality. Then, we have evaluated the impact of phenyl and furan moieties on the thermal resistance of monomers [61,62]. Finally, hardener properties were studied with the synthesis and the characterization of epoxy-amine thermosets.

2. Experimental

2.1. Materials and Methods

Benzaldehyde (purity $\geq$ 99.0%), 1,5-diamino-2-methylpentane as DYTEK®A (purity $\geq$ 99.0%), furfural (purity $\geq$ 99.0%), 1,6-hexanediamine (purity $\geq$ 98.0%) and terephthaldehyde (purity $\geq$ 99.0%) were purchased from Sigma-Aldrich (St. Quentin Fallavier, France). 2-methyltetrahydrofuran (purity $\geq$ 99.0%) was purchased from Alfa Aesar (Kandel, Germany). Isophthalaldehyde (purity $\geq$ 98.0%) was purchased from TCI Europe NV (Zwijndrecht, Belgium). Ethyl acetate (purity $\geq$ 99.9%) was purchased from VWR Chemicals (Fontenay-sous-Bois, France). Deuterated solvents were obtained from Sigma Aldrich for NMR study.

2.2. Characterization Techniques

^{1}H and ^{13}C NMR analyses were recorded on a 400 MHz Bruker Aspect NMR spectrometer at 23 °C (Rheinstetten, Germany) in deuterated solvents. Tetramethylsilane was used as a reference to the chemical shifts, which are given in parts per million (ppm).

Viscosities measurements were performed at 22 °C on the AR-1000 rheometer (TA Instruments, New Castle, DE, USA). A 60 mm diameter and 2° cone-plan geometry was used. The flow mode was used with a gradient from 1 to 0.01 rad·s^{-1}.

Fourier-transform infrared (FTIR) spectra were recorded using a Thermo Scientific Nicolet 6700 FTIR spectrometer with "diamond ATR" equipment (Waltham, MA, USA) in transmittance and with a band accuracy of 4 cm^{-1}. Additionally, 32 scans were performed in the range of 4000–650 cm^{-1} and with a resolution of 4. OMNIC software was used.

Thermogravimetric analyses (TGA) were recorded using a Netzsch F1-Libra analyzer (Selb, Germany) at a heating rate of 20 °C·min^{-1} from 25 to 600 °C (nitrogen stream). Each sample was placed in an alumina crucible and contained an amount between 9–10 mg. The moisture and volatile content, the percentage of residue at 600 °C, and the degradation temperature (T_d) were determined thanks to TGA analyses.

Differential scanning calorimetry (DSC) measurements were performed with the use of a Netzsch DSC200F3 calorimeter F3 (Selb, Germany, indium calibration, nitrogen stream). Pierced aluminum pans were used as crucibles with approximately 10 mg of the sample. A heating rate of 20 °C·min^{-1} from −100 °C to 120 °C was used to record the glass transition temperature (T_g). Right values were measured on a second heating ramp.

Dynamic mechanical analyses (DMA) were performed on Metravib DMA 25 with Dynatest 6.8 software (TA Instruments, New Castle, DE, USA). Uniaxial stretching of samples was carried out while heating at a rate of 3 °C·min^{-1} using a constant frequency of 1 Hz and a fixed strain (according to sample elastic domain). For T_g around 60 °C, DMA analyses were performed from −30 °C to + 160 °C. For T_g around 90 °C, DMA analyses were performed from 0 °C to +190 °C.

Cross-linking density: From rubber elasticity theory [63], uniaxial stretching was studied on the rubbery plateau at T = T_α +80 °C, and at very small deformations. According to these hypotheses, the cross-linking density (v'), was determined from Equation (1), where E' is the storage modulus, R is the universal gas constant and T_α is the vitreous transition temperature, in K. Calculated values are given for informational purposes only, and they can only be compared.

$$v' = \frac{E'_{T_{\alpha+80}}}{3RT_{\alpha+80}} \tag{1}$$

Swelling indices (SI) were measured with a sample of approximately 25 mg, which was placed in 25 mL of tetrahydrofuran (THF) for 24 h. This step was repeat three times for repeatability. The swelling index was calculated according to Equation (2), where m_1 is the mass of the material after swelling in THF during 24 h and m_2 is the initial mass of the material.

$$SI = \frac{m_1 - m_2}{m_2} \times 100 \tag{2}$$

Gel contents (GC) were measured after SI samples were dried in a ventilated oven at 70 °C for 24 h. The gel content was calculated according to Equation (3), where m_3 is the mass of the material after drying and m_2 is the initial mass of the material.

$$GC = \frac{m_3}{m_2} \times 100 \tag{3}$$

2.3. Synthesis of Amine Hardeners

First, 100 mL of H_2O-2-MeTHF mixture (70–30) were added to DYTEK®A (17.3 g, 149 mmol, 10 equivalents) in a 250 mL two-neck round-bottom flask. Then, isophthalaldehyde (2 g, 14.9 mmol, 1 equivalent) was solubilized in 30 mL of 2-MeTHF and then added dropwise using a dropping funnel. The reaction crude was stirred and heated until complete aldehyde conversion, at 110 °C. The solution was then cooled down to room temperature.

In the case of monoaldehyde use, only 5 equivalents of DYTEK®A were used for 1 equivalent of monoaldehyde.

In the case of one pot conditions for isophthalatetetramine (IPTA)2 synthesis:

Then, 100 mL of H_2O-2-MeTHF mixture (70–30) were added to DYTEK®A (17.3 g, 149 mmol, 10 equivalents) in a 250 mL round-bottom flask. Then, isophthalaldehyde (2 g, 14.9 mmol, 1 equivalent), was added. The reaction crude was stirred and heated until complete aldehyde conversion, at 110 °C. The solution was then cooled down to room temperature.

For each case, 2 equivalents of sodium borohydride were then added slowly to the theoretical amount of imine and solvent mixture, and the reaction crude was heated at reflux temperature until complete disappearance of imine signal in ^{1}H NMR. The solvent was removed under reduced pressure. The sodium borohydride was then neutralized by the addition of the reaction crude in water (200 mL). This aqueous solution was extracted with AcOEt (3 × 600 mL). Organic phase was washed with brine solution (400 mL). Then, organic phase was dried with $MgSO_4$, filtered and the solvent was removed under reduced pressure. Any trace of DYTEK®A was removed by distillation.

All the described structures corresponded to the attack of each amine group from the diamine compound onto dialdehyde.

Isophtalatediiminediamine (IPDIDA) ($C_{20}H_{34}N_4$) (Figure 1). Colorless liquid. IR (neat, v, cm^{-1}): 3381 and 3266 ($N_{primary}$-H), 2923 and 2845 (C-sp^3), 1643 (C = N), 1582 (δN-H), 1458 (C = C$_{aromatic}$), 1376, 1339, 1291, 1055, 1021, 967, 794. ^{1}H NMR (400 MHz, CDCl$_3$, ppm) δ: 8.26 (s, 1H, CH$_{imine}$), 8.23 (s, 1H, CH$_{imine}$), 7.99 (m, 1H, H$_{Ar}$), 7.75 (m, 2H, 2 × H$_{Ar}$), 7.40 (m, 1H, H$_{Ar}$), 3.58 (t, 2H, J = 6.0 Hz, HC = N-CH$_2$-CH$_2$), 3.46 (ddd as apparent dt, 2H, HC = N-CH$_2$-CH), 2.66 (t, 2H, J = 6.5 Hz, H$_2$N-CH$_2$-CH$_2$), 2.53 (ABX signal, 2 × dd, 2H, J = 12.6 and 6.9 Hz, 2 × H$_2$N-CHa-CH, 2 × H$_2$N-CHb-CH), 1.95–1.61 (m, 2H, 2 × CH), 1.55–1.07 (m, 12H, 4 × H$_{labiles}$, 4 × CH$_2$), 0.94–0.86 (m, 6H, 2 × CH$_3$). ^{13}C NMR (101 MHz, CDCl$_3$, ppm) δ: 160.5 and 160.4 (2 × CH$_{imine}$), 136.7 (2 × Cq$_{Ar}$), 129.8 and 129.7 (2 × CH$_{Ar}$), 128.9

(CH_{Ar}), 128.0 (CH_{Ar}), 68.2 (HC-CH_2-N = CH), 62.0 (H_2C-CH_2-N = CH), 48.4 (H_2N-CH_2-CH), 42.5 (H_2N-CH_2-CH_2), 36.2 (CH), 34.2 (CH), 32.5–28.4 (4 × CH_2), 18.2 and 17.5 (2 × CH_3).

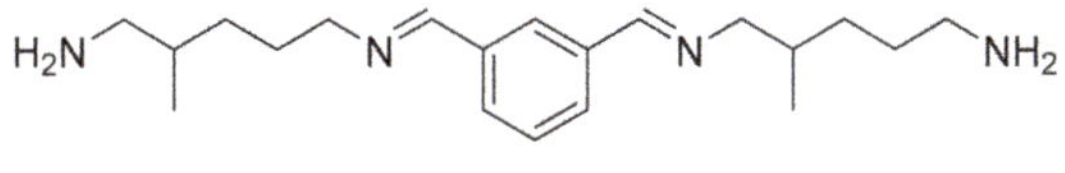

Figure 1. Isophtalatediiminediamine (IPDIDA).

IPTA ($C_{20}H_{38}N_4$) (Figure 2). Colorless liquid. IR (neat, ν, cm^{-1}): 3278 (broad N-H), 2920 and 2849 (C-sp^3), 1643–1454 (C = $C_{aromatic}$), 1558 (δN-H), 1376, 1291, 1155, 1003, 785. ^{1}H NMR (400 MHz, CDCl$_3$, ppm) δ: 7.26 (m, 4H, 4 × H_{ar}), 3.80 (s, 2H, Cq$_{ar}$-CH_2), 3.79 (s, 2H, Cq$_{ar}$-CH_2), 2.71–2.43 (m, 8H, 4 × H_2N-CH_2), 2.01–1.10 (m, 16H, 6 × $H_{labiles}$, 4 × CH_2, 2 × CH), 0.94 (d, 3H, J = 6.7 Hz, CH_3), 0.92 (d, 3H, J = 6.6 Hz, CH_3). ^{13}C NMR (101 MHz, CDCl$_3$, ppm) δ: 140.6 and 140.5 (2 × Cq$_{Ar}$), 128.4, 127.9, 127.8 and 126.7 (4 × CH_{Ar}), 55.9 (HN-CH_2-CH), 54.1 (2 × Cq$_{ar}$-CH_2), 49.9 (HN-CH_2-CH_2), 48.2 (H_2N-CH_2-CH), 42.2 (H_2N-CH_2-CH_2).

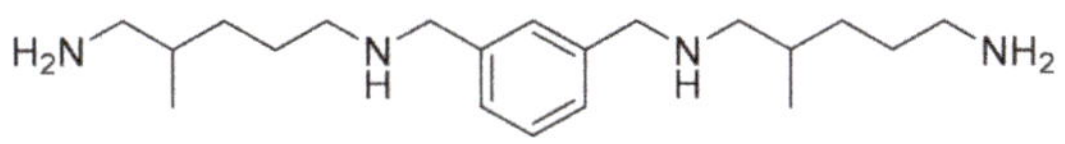

Figure 2. Isophtalatetetramine (IPTA).

Benzylimineamine (BIA) ($C_{13}H_{20}N_2$) (Figure 3). Colorless liquid. IR (neat, ν, cm^{-1}): around 3250 (N$_{primary}$-H, which disappear during reversibility), 3061 and 3026 (C-sp^2), 2926 and 2829 (C-sp^3), 1643 (C = N), 1579 (δN-H), 1493–1450 (C = $C_{aromatic}$), 1309, 1293, 1218, 1169, 1074, 1025, 968, 848. ^{1}H NMR (400 MHz, CDCl$_3$, ppm) δ: 8.30 or 8.27 (s, 1H, CH_{imine}), 7.76–7.71 (m, 2H, 2 × H_{Ar}), 7.45–7.40 (m, 3H, 3 H_{Ar}), 3.69–3.58 (m, 2H, HC = N-CH_2-CH_2) or 2.53 (ABX signal, 2 × dd, 2H, J = 11.4 and 7.1 Hz, HC = N-CHa-CH, HC = N-CHb-CH), 2.72 (t, 2H, J = 6.9 Hz, H_2N-CH_2-CH_2) or 2.59 (ABX signal, 2 × dd, 2H, J = 12.6 and 6.0 Hz, H_2N-CHa-CH, H_2N-CHb-CH), 2.02–1.54 (m, 7H, 2 × $H_{labiles}$, 2 × CH_2, CH), 1.00–0.92 (m, 3H, CH_3). ^{13}C NMR (101 MHz, CDCl$_3$, ppm) δ: 161.0 or 160.9 (CH_{imine}), 136.4 (Cq$_{Ar}$), 130.5 (CH_{Ar}), 128.6–128.1 (4 × CH_{Ar}), 68.2 (HC-CH_2-N = CH) or 62.0 (H_2C-CH_2-N = CH), 48.4 (H_2N-CH_2-CH) or 42.6 (H_2N-CH_2-CH_2), 34.3 (CH), 32.5–28.4 (2 × CH_2), 18.2 (CH_3). High Resolution Mass Spectrometry (HRMS) (*m/z*, ES+, [M + H$^+$]): $C_{13}H_{21}N_2$; calculated 205.1699; found 205.1706.

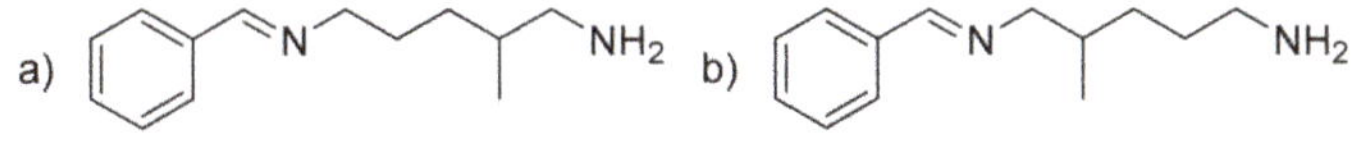

Figure 3. (**a**) Benzylimineamine (BIA) from A addition and (**b**) BIA from B addition.

Benzyldiamine (BDA) ($C_{13}H_{22}N_2$) (Figure 4). Colorless liquid. IR (neat, ν, cm^{-1}): 3375–3287 (broad N-H), 3093, 3069 and 3026 (C-sp^2), 2924 and 2850 (C-sp^3), 1667–1452 (C = $C_{aromatic}$ and δN-H), 1376, 1291, 1116, 1074, 1027, 806. ^{1}H NMR (400 MHz, CDCl$_3$, ppm) δ: 7.32–7.23 (m, 5H, 5 × H_{ar}), 3.78 or 3.77 (s, 2H, Cq$_{ar}$-CH_2), 2.68–2.40 (m, 4H, 2 × H_2N-CH_2), 1.98–1.07 (m, 8H, 3 × $H_{labiles}$, 2 × CH_2, CH), 0.92 (d, 3H, J = 6.7 Hz, CH_3) or 0.89 (d, 3H, J = 6.6 Hz, CH_3). ^{13}C NMR (101 MHz, CDCl$_3$, ppm) δ: 140.6 or 140.4 (2 × Cq$_{Ar}$), 128.4–126.9 (5 × CH_{Ar}), 55.8 (HN-CH_2-CH) or 49.7 (HN-CH_2-CH_2), 54.1 (Cq$_{ar}$-CH_2), 48.2 (H_2N-CH_2-CH) or 42.3 (H_2N-CH_2-CH_2), 36.1 or 33.2 (CH), 32.0 (CH_2), 27.5 (CH_2), 18.1 or 17.4 (CH_3). HRMS (*m/z*, ES+, [M + H$^+$]): $C_{13}H_{23}N_2$; calculated 207.1856; found 207.1862.

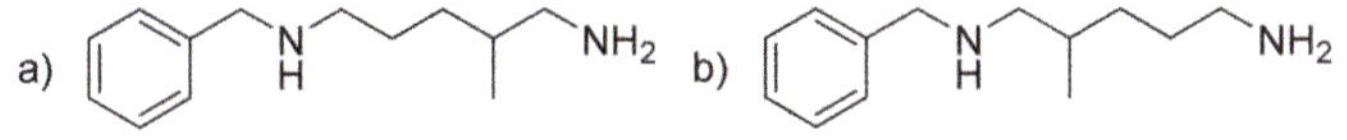

Figure 4. (**a**) Benzyldiamine (BDA) from A addition and (**b**) BDA from B addition.

Furfurylimineamine (FIA) ($C_{11}H_{18}N_2O$) (Figure 5). Brown liquid. IR (neat, ν, cm^{-1}): around 3250 ($N_{primary}$-H, which disappear during reversibility), 3110 (C-sp^2), 2927 and 2871 (C-sp^3), 1643 (C = N), 1579 (δN-H), 1483–1459 (C = C$_{aromatic}$), 1392, 1358, 1272, 1238, 1154, 1079, 1014, 931, 816. ^{1}H NMR (400 MHz, CDCl$_3$, ppm) δ: 8.30 or 8.27 (s, 1H, CH$_{imine}$), 7.51 (d, 1H, J = 1.7 Hz, O-CH$_{Ar}$), 6.72–6.71 (m, 1H, HC=CH-CH), 6.47 (dd, 1H, J = 3.4, 1.8 Hz, HC = CH-CH), 3.64–3.52 (m, 2H, HC = N-CH_2-CH$_2$) or 2.53 (ABX signal, 2 × dd, 2H, J = 11.5 and 7.1 Hz, HC = N-CHa-CH, HC = N-CHb-CH), 2.69 (t, 2H, J = 7.8 Hz, H$_2$N-CH_2-CH$_2$) or 2.57 (ABX signal, 2 × dd, 2H, J = 12.6 and 6.1 Hz, H$_2$N-CHa-CH, H$_2$N-CHb-CH), 2.00–1.19 (m, 7H, 2 × H$_{labiles}$, 2 × CH$_2$, CH), 0.95 (d, 3H, J = 6.7 Hz, CH$_3$). ^{13}C NMR (101 MHz, CDCl$_3$, ppm) δ: 151.5 (Cq$_{Ar}$), 149.7 or 149.5 (CH$_{imine}$), 144.5 (O-CH=CH), 113.7 (Cq$_{Ar}$-CH = CH), 111.5 (O-CH = CH), 68.4 (HC-CH$_2$-N = CH) or 62.0 (H$_2$C-CH$_2$-N=CH), 48.4 (H$_2$N-CH$_2$-CH) or 42.6 (H$_2$N-CH$_2$-CH$_2$), 34.1 (CH), 32.4 and 28.3 (2 × CH$_2$), 18.1 (CH$_3$). HRMS (*m/z*, ES+, [M + H$^+$]): $C_{11}H_{18}N_2O$; calculated 195.1492; found 195.1500.

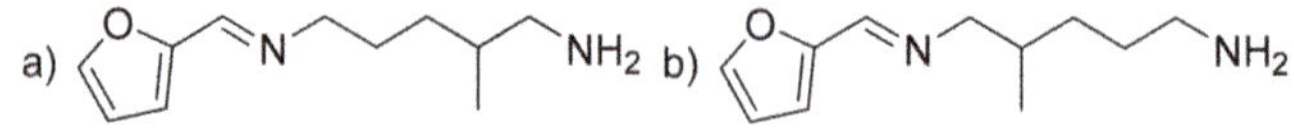

Figure 5. (**a**) Furfurylimineamine (FIA) from A addition and (**b**) FIA from B addition.

Furfuryldiamine (FDA) ($C_{11}H_{20}N_2O$) (Figure 6). Brown liquid. IR (neat, ν, cm^{-1}): 3667–3272 (broad N-H), 2924 and 2851 (C-sp^3), 1667–1454 (C = C$_{aromatic}$ and δN-H), 1377, 1336, 1147, 1110, 1075, 1008, 918, 802. ^{1}H NMR (400 MHz, CDCl$_3$, ppm) δ: 7.34 or 7.33 (d, 1H, J = 2 Hz, O-CH$_{ar}$), 6.29 or 6.28 (d, 1H, J = 3.2 Hz, HC = CH-CH), 6.15–6.14 or 6.14–6.13 (m, 1H, HC = CH-CH), 3.76 or 3.75 (s, 2H, Cq$_{ar}$-CH$_2$), 2.67–2.36 (m, 4H, HN-CH_2-CH$_2$ or HN-CH_2-CH, and H$_2$N-CH_2-CH$_2$ or H$_2$N-CH_2-CH), 1.97–1.07 (m, 8H, 3 × H$_{labiles}$, 2 × CH$_2$, CH), 0.89 (d, 3H, J = 6.8 Hz, CH$_3$). ^{13}C NMR (101 MHz, CDCl$_3$, ppm) δ: 154.0 (2 × Cq$_{Ar}$), 141.7 or 141.6 (O-CH = CH), 110.0 (Cq$_{Ar}$-CH = CH), 106.7 (O-CH = CH), 55.5 (HN-CH$_2$-CH) or 49.4 (HN-CH$_2$-CH$_2$), 48.3 (H$_2$N-CH$_2$-CH) or 42.5 (H$_2$N-CH$_2$-CH$_2$), 46.4 or 46.4 (Cq$_{ar}$-CH$_2$), 36.2 or 33.1 (CH), 32.0 (CH$_2$), 27.4 (CH$_2$), 18.1 or 17.4 (CH$_3$). HRMS (*m/z*, ES+, [M + H$^+$]): $C_{11}H_{21}N_2O$; calculated 197.1648; found 197.1656.

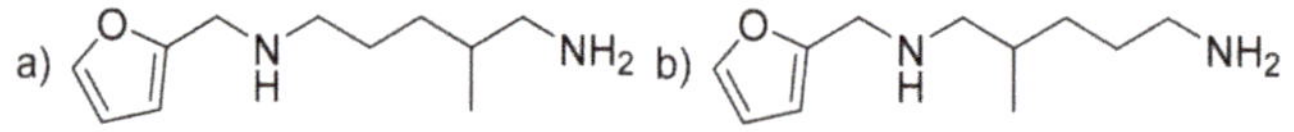

Figure 6. (**a**) Furfuryldiamine (FDA) from A addition and (**b**) FDA from B addition.

2.4. Amine Hydrogen Equivalent Weight (AHEW or HEW) Calculation

Each experimental HEW was determined by NMR ^{1}H titration using benzophenone as the intern reference. To this end, known weights of amine and benzophenone were poured into an NMR tube and 500 µL of deuterated chloroform were added. HEW values were determined according to Equation (4).

$$EEW = \frac{\int PhCOPh * H_{amine}}{\int amine * H_{PhCOPh}} * \frac{m_{amine}}{m_{PhCOPh}} * M_{PhCOPh} \tag{4}$$

	$\int_{PhCOPh}$:	integration of the benzophenone protons;
	$\int_{amine}$:	integration of the protons of the amine functions;
	H_{amine}:	number of protons of the amine functions;
where:	H_{PhCOPh}:	number of protons of the benzophenone;
	m_{amine}:	weight of the amine product;
	m_{PhCOPh}:	weight of benzophenone;
	M_{PhCOPh}:	molecular weight of benzophenone.

2.5. Synthesis of Epoxy Thermosets

The amount of hardener for 100 g of epoxy for a theoretical molar ratio of 1:2 between amine and epoxy functions was calculated according to Equations (5) and (6):

$$AHEW = \frac{M_{amine}}{n_{fonction\ NH}} \qquad (5)$$

$$m_{hardener} = \frac{100 \times HEW}{EEW} \qquad (6)$$

where AHEW (or HEW) is the amine hydrogen equivalent weight and EEW is the epoxy equivalent weight.

The optimal molar ratio was then determined with the adjustment of Equation (6) by multiplying the hardener mass by various ratio of amine/epoxy. Then, the optimal molar ratio, corresponding to the highest T_g, was determined by recording the T_gs using Differential Scanning Calorimetry (DSC) analysis.

Reactants were then mixed according to the previously determined optimal molar ratio and cured at 80 °C for 2 h to obtain the thermosets.

3. Results and Discussion

The aim of this study was to synthesize new pluri-functional and bio-based aminated hardeners containing aromatic moieties. To this end, the reductive amination method was applied to bio-based and non-toxic aromatic aldehydes with a non-toxic diamine (Figure 7). This method brings aromatic moieties, and thus to increase the thermo-mechanical properties of the final material, without having to synthesize aromatic amines which are generally toxic. The pluri-functionality of amine was easily studied by changing the active H number with the selection of the structure of the aldehyde monomer and the reaction conditions.

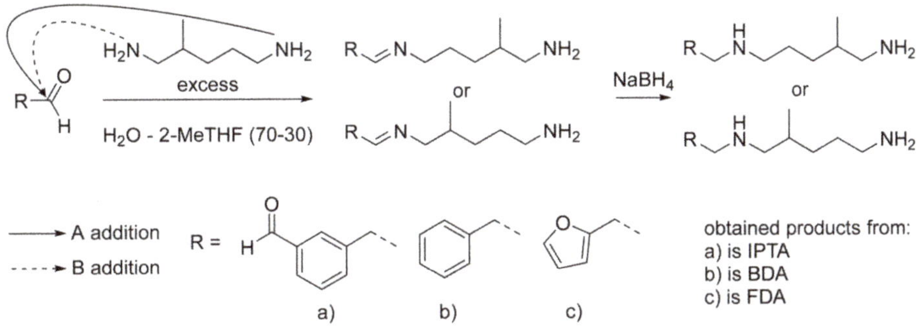

Figure 7. Amines syntheses from DYTEK®A as amine reactant and (**a**) isophthalaldehyde (IPA), (**b**) benzaldehyde and (**c**) furfural, as aldehyde monomers.

In this study, we studied the influence of the modification of the active H functionality on the thermo-mechanical properties. The use of dialdehyde monomers increases the active H functionality of the final hardener compared to the initial diamine (respectively six active H versus four), while monoaldehyde leads to three active H. In this view, isophthalaldehyde (IPA) was chosen as the dialdehyde monomer and benzaldehyde as the monoaldehyde reference. All previously cited aldehydes can be produced from biomass and are non-toxic [64–68]. Furfuraldehyde was also chosen, despite its toxicity, in order to compare the final thermoset properties with aromatic ones [69]. Due to its liquid aspect and its presence in the REACH (registration evaluation authorisation and restriction of chemicals) registration list, 2-methylpentane-1,5-diamine (DYTEK®A) was chosen as the amine reactant [70]. The presence of branched methyl in its structure decreases the viscosity thanks to the steric hindrance, providing a liquid aspect to the reactant. Moreover DYTEK®A could be synthesized by the methylation of natural glutamine [71,72].

3.1. Hardeners Synthesis

Each aldehyde reactant was respectively added dropwise, in an excess of initial amine reactant, in order to avoid the oligomer formation. To simplify the procedure, we chose $NaBH_4$ as a reducing agent for this study. However, this reducing step can be easily performed with an industrial process such as catalytic hydrogenation [31]. Only IPA-based amine was also synthesized in one-pot conditions, by mixing entirely aldehyde at the same time as amine, to favor the dimerization (i.e., in order to obtain a higher functionality). The aims of this other one-pot method was to study the influence of the increase of the active hydrogen functionality to six on the network properties. All amine characterizations are summarized in Table 1.

Table 1. Amine characterizations.

Amine Nomenclatures	A: B Addition Ratio [a]	HEW$_{th}$ (g·eq^{-1})	HEW$_{exp}$[b] (g·eq^{-1})	Amine Viscosity (Pa·s at 22 °C)	$T_d5\%$ (°C)	Char Yield[c] (%)	T_g[d] (°C)
IPTA1	51: 49	56	57	0.57	173	1	− 57
IPTA2	45: 55	56	59	1.40	204	4	− 35
BDA	48: 52	69	67	0.03	115	3	− 81
FDA	48: 52	65	66	0.02	126	2	− 84

[a] Determined at the end of imine synthesis reaction; [b] determined by NMR ^{1}H titration; [c] at 600 °C; [d] mid values. HEW: hydrogen equivalent weight.

The reductive amination method obtains full conversion of aldehyde during the imine synthesis step and then, full conversion of imine during the second step (spectra given in Supplementary Materials, parts 1 to 3). All amine hardeners from IPA, benzaldehyde and furfural were successfully synthesized.

The imines synthesis and their reduction may be easily monitored by the appearance and disappearance of the imine signal from both FTIR and ^{1}H NMR spectra (Figure 8). For instance, in the case of IPA-based hardener synthesis (IPTA), the reduction of imine moieties to amine functions may be observed with the disappearance of C = N stretching band at 1643 cm^{-1}. The ^{1}H NMR spectrum changes considerably after the reduction step. The disappearance of the –CH signal corresponding to the imine proton at 8.25 ppm is observed and then confirmed by the appearance of a singlet signal at 3.77 ppm, corresponding to –CH$_2$ signal of the reduced imine function. Moreover, the disappearance of the –CH$_2$ signal corresponding to the protons in α position of C = N bond designated as 5 and 5′ is shown in Figure 8. The reduction step induces the absence of conjugated system involving the imine bond. Consequently, the aromatic signals of the formed amines are shifted from the area 8.00–7.36 to the area 7.27–7.18 ppm.

^{1}H NMR spectra of the three synthesized amines from IPA (named IPTA1), benzaldehyde (named BDA) and furfural (named FDA) are displayed in Figure 9. The IPA-based hardener synthesized in one-pot conditions (IPTA2) shows a similar spectrum to IPTA1, with different integration values (spectra given in Supplementary Materials, part 1c). The signal corresponding to the Cq_{Ar}-CH$_2$ protons in α position of secondary amine appears at 3.75 ppm with a singlet (designated as 1 in Figure 9). Then, the signals of the other α-CH$_2$ of the secondary and primary amine moieties are overlapped at 2.51 ppm (designated as 2 in Figure 9). Moreover, the signals of the amine from the A addition are different from the signals of the amine from the B addition but are found in the same overlapped signal. Thus, there are at least four signals overlapped at 2.51 ppm. Due to this overlap, the A and B addition ratio could not have been determined thanks to the ^{1}H NMR spectra of the final amine monomers. However, before the reduction step, the α-CH$_2$ of the imine signal shifted to 3.50 ppm due to the conjugated system involving the imine bond. This signal is split into two different signals for A and B additions, determining the proportion of each other (spectra given in Supplementary Materials, parts 1 to 3).

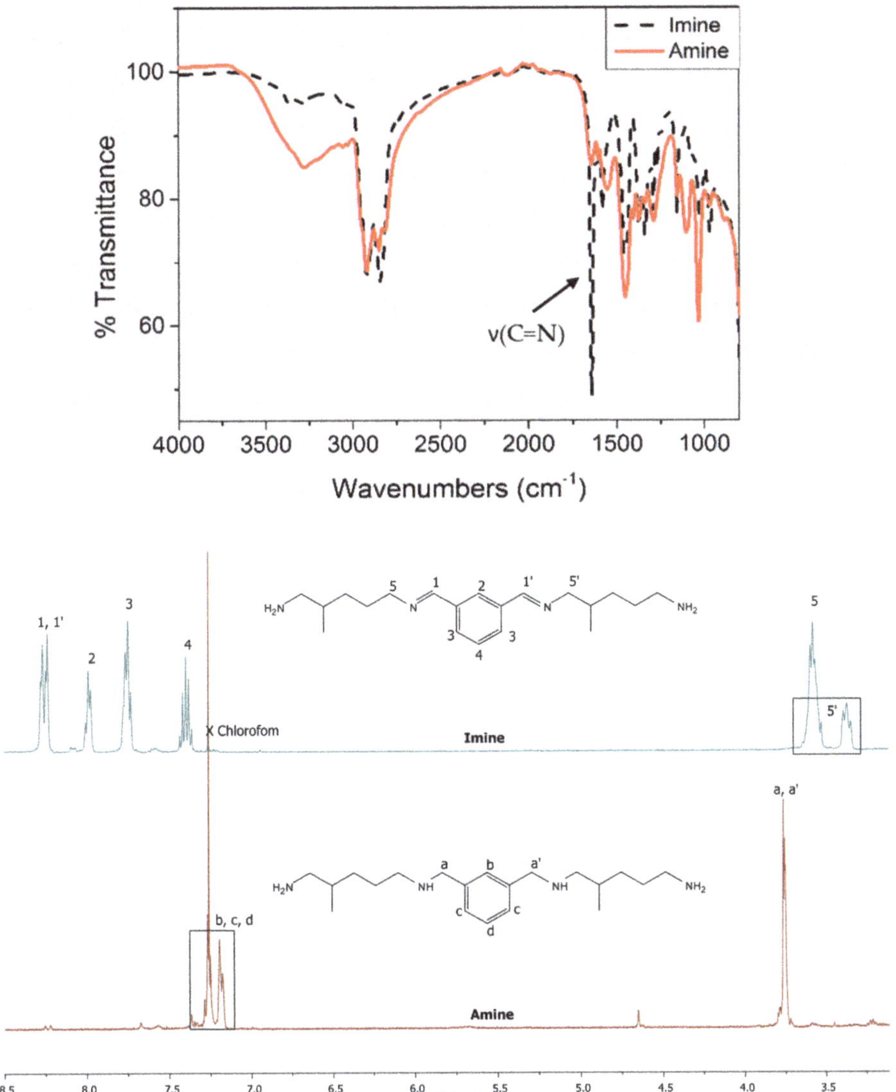

Figure 8. Imine signal disappearance of IPDIDA, the corresponding imine of IPTA amine monomer, followed by FTIR (up), and ^{1}H NMR in CDCl$_3$ (down).

The characterizations of the new bio-based amine hardeners named IPTA1, IPTA2, BDA and FDA are summarized in Table 1 (DSC and TGA thermograms are given in Supplementary Materials, parts 4 and 5). A and B additions were obtained in similar proportions, with an average ratio of 50:50 (determined using the ^{1}H NMR spectrum of imine intermediates). The experimental HEW was determined by NMR titration, as reported in the Materials and Methods section. Experimental and theoretical results were almost similar. The higher glass transition temperature (T_g) of IPTA2 compared to IPTA1 confirms the presence of more dimers in the IPTA2 hardener than in the IPTA1. The viscosity value follows the same trend with a higher value for IPTA2 than IPTA1. Due to the higher content of aromatic moieties in dimeric structure and so in IPTA2 hardener, T_g and $T_d5\%$ of IPTA2 are higher than IPTA1 monomeric structure. BDA and FDA both exhibit similar T_g and $T_d5\%$ values, showing similar behaviors for their thermomechanical properties. IPTA1, BDA and FDA exhibit a liquid aspect with low viscosities lower than 0.6 Pa s at 22 °C close to the water aspect, which are interesting properties for epoxy-amine formulations. The comparison of IPTA1 and IPTA2 hardeners shows an increase of viscosity with the number of dimers.

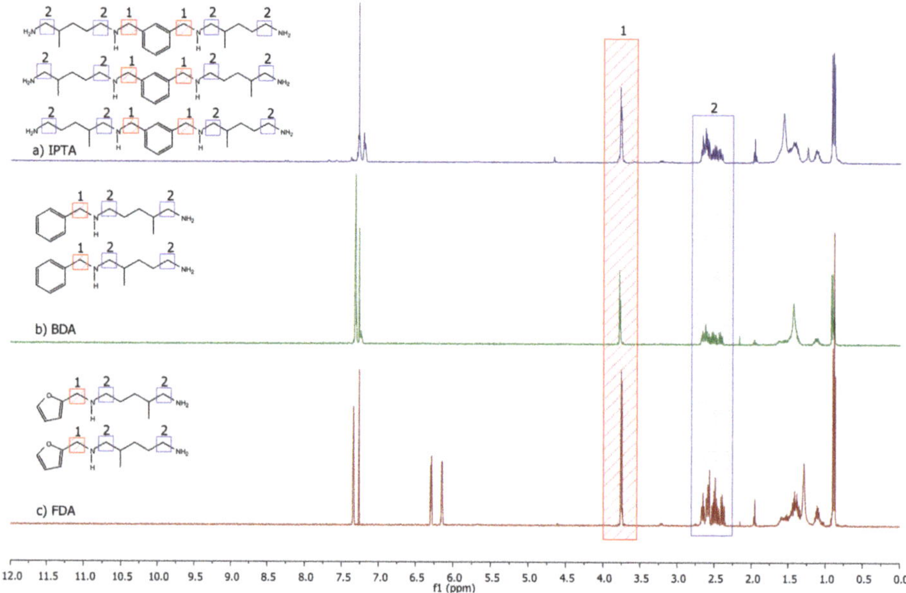

Figure 9. ^{1}H NMR spectra of (**a**) IPTA2, (**b**) BDA and (**c**) FDA in CDCl$_3$.

3.2. Thermoset Syntheses

Synthesized amines were then used to synthesize epoxy thermosets (also named P-materials) with different epoxy monomers. Hence, bulk materials (parallelepiped shape) were obtained by the curing of the synthesized amines with epoxy monomers, using previously-determined optimal ratios (method described in the Experimental section, DSC thermograms and optimal ratio given in Supplementary Materials, parts 6 and 7). First, hardeners were reacted with diglycidyl ether of bisphenol a (DGEBA), as a petro-sourced epoxy reference. These thermosets can be compared to literature results describing the networks' characteristics of MXDA- and DYTEK®A-based materials from DGEBA as an epoxy part [73–75]. *m*-Xylylenediamine (MXDA) is a petro-sourced arylamine hardener currently used in the industrial field of epoxy coatings due to its high reactivity [76]. It is interesting to compare MXDA and IPTA structures due to their similar dibenzyl center. We could observe the influence of the aliphatic chain addition provided by DYTEK®A using DYTEK®A-DGEBA thermoset results as data references. In the same way, thermosets (also named bio-materials) were then synthesized using the diglycidyl ether of vanillin alcohol (DGEVA) [77] as a bio-based epoxy derived from vanillin in order to increase the bio-based carbon content. Then, thermo-mechanical properties and chemical resistance in THF of each optimal bulk network are determined and summarized in Table 2.

Table 2. Thermoset characterizations.

Thermosets	Composition Amine-Epoxy	$T_d5\%$ (°C)	Char Yield [c] (%)	T_g[d] (°C)	T_α (°C)	ν' (mol·m^{-3})	E'_{glassy} (Pa)	$E'_{rubbery}$ (Pa)	SI[e] (%)	GC (%)
DYTEK®A-ref [a]	DYTEK®A-DGEBA	-	-	115	127	-	-	$2.90.10^7$	-	-
MXDA-ref [b]	MXDA-DGEBA	333	7	116	130	1 146	$1.2.10^8$	$1.80.10^7$	89	99
IPT1-P	IPTA1-DGEBA	351	7	99	101	1 311	$1.6.10^9$	$1.38.10^7$	32	100
IPT2-P	IPTA2-DGEBA	345	7	89	95	1 311	$2.14.10^9$	$1.38.10^7$	93	100
BD-P	BDA-DGEBA	354	4	77	88	460	$1.29.10^9$	$4.84.10^6$	123	97
FD-P	FDA-DGEBA	340	13	76	85	123	$2.71.10^8$	$1.30.10^6$	122	99
IPT1-Bio	IPTA1-DGEVA	312	16	59	57	997	$2.7.10^9$	$1.05.10^7$	41	100
IPT2-Bio	IPTA2-DGEVA	307	16	57	57	657	$1.72.10^9$	$6.92.10^6$	108	96
BD-Bio	BDA-DGEVA	325	12	45	50	407	$2.41.10^9$	$4.29.10^6$	134	95
FD-Bio	FDA-DGEVA	317	19	46	49	91	$5.45.10^8$	$9.59.10^5$	125	100

ref [a] literature results from Fu. et al. [78] and Meis et al. [79]; ref [b] literature results from Faye. et al. [77]; [c] at 600 °C; [d] mid values; [e] in THF.

Epoxy and amine reactants were mixed and then cured at 80 °C for 2 h, with reactant amounts corresponding to the respective optimal ratios. The end of the crosslinking reaction was confirmed by DSC analyses, with no residual enthalpy signal on each thermogram. Furthermore, high-gel content values (>90%), corresponding to a highly cross-linked material, confirmed the full conversion for each thermoset.

The thermal stabilities were determined using TGA under nitrogen steam. The 5% weight loss ($T_d5\%$) temperature and the char yields at 600 °C were recorded (Figure 10). By comparing IPTA1-based and IPTA2-based materials, results followed the same trend exhibiting similar $T_d5\%$ and char yield values. Furthermore, the P-materials showed higher thermal resistance with slightly higher $T_d5\%$ values around 350 °C against 315 °C for bio-materials, keeping however good thermal resistance. On the contrary, a higher char yield at 600 °C was observed for bio-based materials, meaning higher thermal resistance for high temperatures. This can be explained by the absence of the geminal dimethyl bridge on the bio-based epoxy structure, which has a low thermal stability. TGA results concluded that the slightly molecular weight difference between IPTA1 and IPTA2 has no impact on thermal stability. IPTA-based materials showed slightly higher char yields than BDA-based material with a residual mass of 3%–4% higher. FDA-based materials exhibited the highest char yield, meaning that furan moieties showed higher thermal resistance than benzyl moieties. All P-materials showed a higher thermal stability than MXDA-DGEBA material (MXDA-ref).

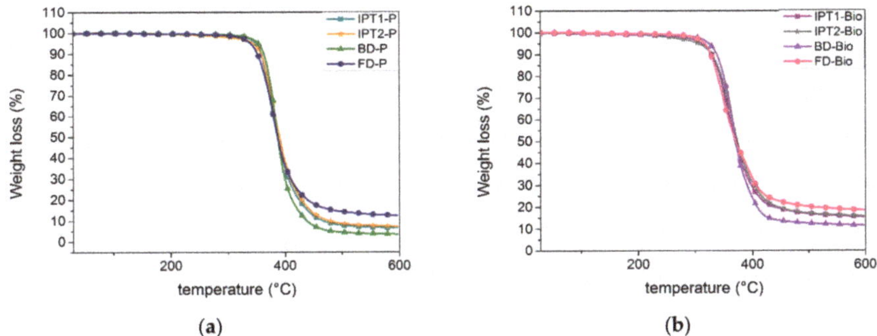

Figure 10. Thermogravimetry Analysis (TGA) measurements of the thermosets from new hardeners and (**a**) DGEBA or (**b**) DGEVA as epoxy.

The glass transition temperature values (T_g) were recorded by DSC and then compared to the alpha transition temperature values (T_α), which were determined by DMA analyses, corresponding to the mechanical manifestation of T_g (DMA in Figure 11, DSC thermograms given in Supplementary Materials, part 8). The transition from vitreous state to rubbery state induces a module loss, and thus the maximum value of the tan δ curve as a function of the temperature corresponds to the T_α. Results showed that T_gs and T_αs followed the same trend for each thermoset with each of narrowed tan δ peaks, suggesting that the materials are homogeneous. Overall, fully bio-based thermosets exhibited a lower T_g value than the P-material references, with a difference of 30 to 40 °C. This decrease of T_g value is due to the presence of methylene and methoxy moieties in DGEVA structure, which behaved as spacers and repulse polymeric chains for each other, thereby providing flexibility and thus lower T_g.

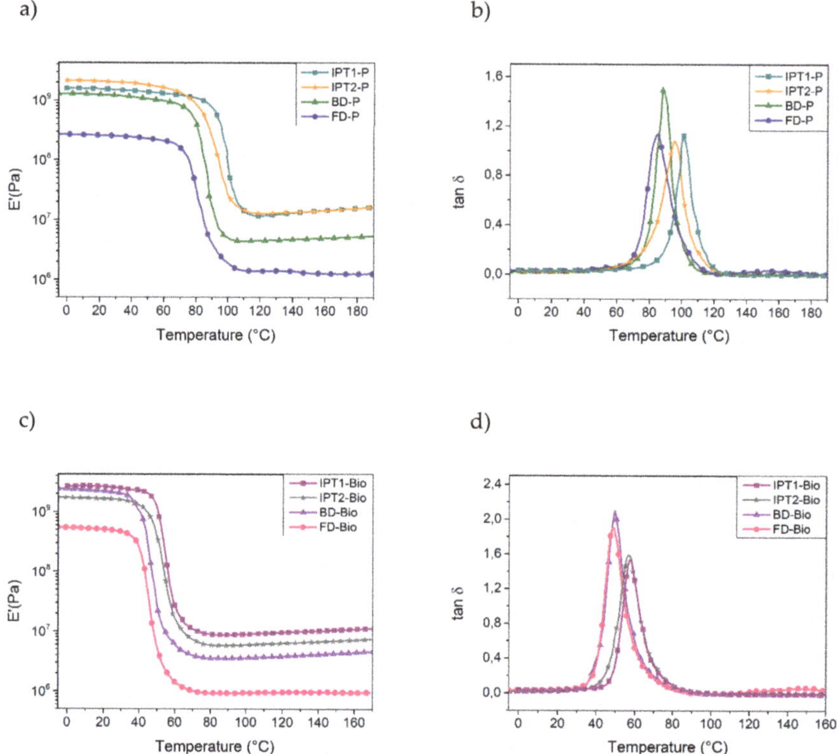

Figure 11. Dynamic Mechanical Analysis (DMA) thermograms (under air, 1 Hz, 3 K·min^{-1}): (**a**) E' curves of materials cured with DGEBA as epoxy; (**b**) Tan δ curves of materials cured with DGEBA as epoxy; (**c**) E' materials with DGEVA as epoxy; (**d**) Tan δ curves of materials cured with DGEVA as epoxy. All thermosets were cured with respective optimal ratios.

DSC results showed a difference of 10 °C between IPTA1- and IPTA2-based thermosets in favor to IPTA1-based ones due to the reduced molecule weight for the thermosets from IPTA1, that displayed shorter backbones than IPTA2. Then, the IPTA-based networks were compared to BD-P reference. BD-P is based from BDA which is a hardener with only three active hydrogen functions (one secondary and one primary amine) while IPTA-based hardeners show similar backbone structure containing at least two alkyl chains with six active hydrogens (two secondary and two primary amines). Due to the presence of more alkyl chains containing –NH functions in IPTA structure, the aromatic moieties were directly incorporated in the polymeric chain compared to the BDA structure, which induced alkyl polymeric chains with dandling aromatic moieties (Figure 12). Moreover, the six –NH reactive functions of IPTA increased the cross-linking density (v') values compared to BDA (respectively 1311 and 460 mol·m^{-3} for P-materials, and 997 and 659 mol·m^{-3} for bio-materials). Then, BDA-based and FDA-based thermosets exhibited similar T_g and T_α, without any distinction between furan and benzyl moieties. However, the cross-linking density values showed that thermosets from furan have more compact networks than benzyl-based thermosets (with respectively 460 and 123 $mol \cdot m^{-3}$ for P-materials, and 407 and 91 $mol \cdot m^{-3}$ for bio-materials). Moreover, those results are confirmed by the swelling index values, which are inversely proportional to the cross-linking density. A compact network involves a lower solvent ingress and thus a lower swelling index.

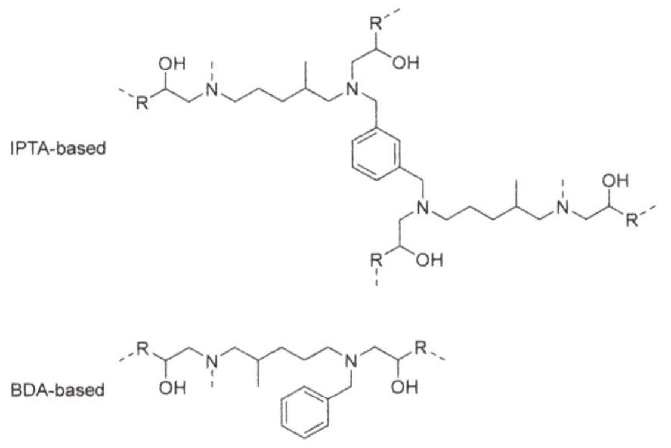

Figure 12. Theoretical structures of polymeric chain units for IPTA-based and BDA-based material.

Overall, the comparison of T_g and T_α for these four networks concluded that the higher the amount of active hydrogens, the higher T_g and T_α. However, by comparing IPT1-P, IPT2-P and BD-P with DYTEK®A-ref and MXDA-ref, the decrease of T_g and T_α could be noticed. In the case of BD-P, the addition of benzyl moiety induced a loss of one active hydrogen in the structure, reducing the cross-linking density and thus increasing the flexibility of the network (ν'_{BDA} = 460 $mol \cdot m^{-3}$ against 1146 for MXDA-ref). For IPTA-based thermosets, the active hydrogen functionality up to six, increased the cross linking-density ($\nu'_{IPTA1} = \nu'_{IPTA2}$ = 1311 $mol \cdot m^{-3}$). However, the presence of aliphatic chains induced a higher molecular length, allowing higher microscopic deformation and exhibiting lower T_g (in increasing order of molecular length: T_g of MXDA-ref = 116 °C ≈ T_g of DYTEK®A-ref, T_g of IPT1-P = 99 °C, and T_g of IPT2-P = 89 °C).

Finally, the storage modulus of vitreous (E'$_{glassy}$) and rubbery (E'$_{rubbery}$) domains were determined respectively at $T_{(\alpha-80)}$ and $T_{(\alpha+80)}$ using DMA analyses, allowing to obtain macroscopic deformation information. The storage modulus (E'$_{rubbery}$) is also linked to the cross-linking density, according to the rubber-elasticity theory [78]. The results showed similar storage modulus in the elastic domain for IPTA-based thermosets with an E'$_{rubbery}$ order of magnitude of 10^7 Pa, which corresponds to a high mechanical strength for thermosets compared to classical high performance thermosets such as MXDA-ref and DYTEK®A-ref materials [73–75]. It is interesting to note that similar mechanical strength was obtained with a lower T_g in the case of IPTA-based materials. These results were induced by the particular structure of IPTA compared to MXDA and DYTEK®A references. IPTA exhibits indeed six –N*H* active functions, increasing the cross-linking density and thus the mechanical strength at the macroscopic scale, while the longer molecular length, induced by the presence of two alkyl chains in IPTA backbone, decreases the T_g at the microscopic scale. In the case of BD-P and FD-P, which exhibits three –N*H* active functions, results showed a lower order of magnitude of 10^6 Pa with a lower value for FD-P (E'$_{rubbery}$ of BD-P = 4.84 × 10^6 Pa and E'$_{rubbery}$ of FD-P = 1.30 × 10^6 Pa), meaning a lower mechanical strength than IPT1-P, IPT2-P and BD-P. The slightly lower E'$_{rubbery}$ value could be assumed by the higher aromaticity of benzene moieties than furan ring [79] which induces higher π-stacking in BDA-based materials [80,81]. By comparing BD-P and FD-P to DYTEK®A-ref, it should be noted that the loss of one –N*H* active function induced a lower mechanical strength due to the decrease of cross-linking density. On the contrary, IPTA-based materials exhibited similar storage modulus thanks to higher cross-linking density despite the lower T_g value.

4. Conclusions

New pluri-functional amine hardeners based on various bio-based aldehydes were synthesized using the reductive amination method. Three amine monomers were obtained: IPTA, BDA and FDA from isophthalaldehyde, benzaldehyde and furfuraldehyde, respectively. IPTA exhibits six active amine hydrogens versus three for BDA and FDA. The molecular weight of IPTA and thus its properties could be modified by changing reaction conditions (dropwise addition or one-pot conditions). All hardeners exhibited no or slight color, without odor. As expected, theses hardeners are liquid with low viscosity (IPTA1, BDA and FDA < 0.6 Pa s at 22 °C), a lower viscosity than our previously synthetized β-hydroxylamine hardeners (>300 Pa s at 50 °C) [21]. Due to their low viscosities, the epoxy-amine mixtures were easily homogenized and rapidly cured with an optimal epoxy-amine ratio, using DGEBA as petro-sourced epoxy reference and DGEVA as bio-based epoxy monomer. Synthesized thermosets showed great thermo-mechanical properties with higher results for IPTA-based materials, which showed similar mechanical strength and cross-linking density than MXDA-ref and DYTEK®A-ref ($\nu'_{IPTA1} = \nu'_{IPTA2} = 1\ 311$ mol·m^{-3} against 1 146 for MXDA–ref and $E'_{rubbery}$ of IPTA1 = $E'_{rubbery}$ of IPTA2 = 1.38.10^7 Pa, $E'_{rubbery}$ of DYTEK®A-ref = 2.90.10^7 Pa, $E'_{rubbery}$ of MXDA-ref = 1.80.10^7 Pa). However, a lower T_g around 95 °C was observed for IPTA-based materials (against T_g = 116 °C for MXDA-ref and T_g = 115 °C for DYTEK®A-ref).

In fact, the functionality of six active hydrogens of IPTA led to unique, highly cross-linked systems exhibiting lower T_g due to the presence of two alkyl chains in the IPTA backbone, allowing microscopic scale deformation, and keeping high mechanical strength at the macroscopic scale, similar to industrial references.

Supplementary Materials: The following are available online, Figures in S1 to S3: monomer characterizations, Figures in S4: TGA measurements of monomers, Figures in S4: DSC measurements of monomers, Figures in S5: optimal ratio for DGEBA-based thermosets, Figures in S6: optimal ratio for DGEVA-based thermosets, Figures in S7: material characterizations.

Author Contributions: Conceptualization, A.-S.M., B.B., G.D., R.T. and S.C.; Methodology, A.-S.M.; Writing-Original Draft Preparation, A.-S.M. and S.C.

Funding: This research received no external funding.

Conflicts of Interest: The authors declare no conflict of interest.

Abbreviations

AHEW or HEW: amine hydrogen equivalent weight; BDA: benzyldiamine; BIA: benzylimineamine; DGEBA: diglycidyl ether of bisphenol A; DGEVA: diglycidyl ether of vanillyl alcohol; DMA: dynamic mechanical analysis; DSC: differential scanning calorimetry; EEW: epoxy equivalent weight; FDA: furfuryldiamine; FIA: furfurylimineamine; FTIR or IR: Fouriertransform infrared; GC: gel content; IPA: isophtalaldehyde; IPDIDA: isophlalatediiminediamine; IPTA: isophtalatetetramine; MXDA: *m*-xylylenediamine; NMR: nuclear magnetic resonance; PACM: 4,4'-methylenebis(cyclohexylamine); ppm: parts per million; SI: swelling index; THF: tetrahydrofuran; TGA: thermogravimetric analysis.

References

1. Graffius, G.C.; Jocher, B.M.; Zewge, D.; Halsey, H.M.; Lee, G.; Bernardoni, F.; Bu, X.; Hartman, R.; Regalado, E.L. Generic gas chromatography-flame ionization detection method for quantitation of volatile amines in pharmaceutical drugs and synthetic intermediates. *J. Chromatogr. A* **2017**, *1518*, 70–77. [CrossRef] [PubMed]

2. Liu, Y.-J.; Hu, C.-Y.; Lo, S.-L. Direct and indirect electrochemical oxidation of amine-containing pharmaceuticals using graphite electrodes. *J. Hazard. Mater.* **2019**, *366*, 592–605. [CrossRef] [PubMed]

3. Wang, X.; Yang, H.; Zhou, B.; Wang, X.; Xie, Y. Effect of oxidation on amine-based pharmaceutical degradation and N-Nitrosodimethylamine formation. *Water Res.* **2015**, *87*, 403–411. [CrossRef] [PubMed]

4. Peng, J.-B.; Wu, F.-P.; Xu, C.; Qi, X.; Ying, J.; Wu, X.-F. Direct synthesis of benzylic amines by palladium-catalyzed carbonylative aminohomologation of aryl halides. *Commun. Chem.* **2018**, *1*, 29. [CrossRef]

5. Legnani, L.; Prina-Cerai, G.; Delcaillau, T.; Willems, S.; Morandi, B. Efficient access to unprotected primary amines by iron-catalyzed aminochlorination of alkenes. *Science* **2018**, *362*, 434–439. [CrossRef] [PubMed]

6. Asokan, A.; Cho, M.J. Cytosolic Delivery of Macromolecules. 3. Synthesis and Characterization of Acid-Sensitive Bis-Detergents. *Bioconjug. Chem.* **2004**, *15*, 1166–1173. [CrossRef] [PubMed]

7. Hernández, A.; Martró, E.; Matas, L.; Jiménez, A.; Ausina, V. Mycobactericidal and tuberculocidal activity of Korsolex® AF, an amine detergent/disinfectant product. *J. Hosp. Infect.* **2005**, *59*, 62–66. [CrossRef] [PubMed]

8. Wolk, A.; Rosenthal, M.; Neuhaus, S.; Huber, K.; Brassat, K.; Lindner, J.K.N.; Grothe, R.; Grundmeier, G.; Bremser, W.; Wilhelm, R. A Novel Lubricant Based on Covalent Functionalized Graphene Oxide Quantum Dots. *Sci. Rep.* **2018**, *8*, 5843. [CrossRef]

9. Sarazen, M.L.; Sakwa-Novak, M.A.; Ping, E.W.; Jones, C.W. Effect of Different Acid Initiators on Branched Poly (propylenimine) Synthesis and CO_2 Sorption Performance. *ACS Sustain. Chem. Eng.* **2019**, *7*, 7338–7345. [CrossRef]

10. Froidevaux, V.; Negrell, C.; Caillol, S.; Pascault, J.-P.; Boutevin, B. Biobased Amines: From Synthesis to Polymers; Present and Future. *Chem. Rev.* **2016**, *116*, 14181–14224. [CrossRef]

11. Nair, C.P.R. Advances in addition-cure phenolic resins. *Prog. Polym. Sci.* **2004**, *29*, 401–498. [CrossRef]

12. Kiskan, B. Adapting benzoxazine chemistry for unconventional applications. *React. Funct. Polym.* **2018**, *129*, 76–88. [CrossRef]

13. Evsyukov, S.E.; Pohlmann, T.; Stenzenberger, H.D. m-Xylylene bismaleimide: A versatile building block for high-performance thermosets. *Polym. Adv. Technol.* **2015**, *26*, 574–580. [CrossRef]

14. Sarva, S.S.; Deschanel, S.; Boyce, M.C.; Chen, W. Stress–strain behavior of a polyurea and a polyurethane from low to high strain rates. *Polymer* **2007**, *48*, 2208–2213. [CrossRef]

15. Carré, C.; Ecochard, Y.; Caillol, S.; Averous, L. From the synthesis of biobased cyclic carbonate to polyhydroxyurethanes: A promising route towards renewable NonIsocyanate Polyurethanes. *ChemSusChem* **2019**, *12*, 3410–3430. [CrossRef] [PubMed]

16. Roy, M.; Wilsens, C.H.R.M.; Leoné, N.; Rastogi, S. Use of Bis(pyrrolidone)-Based Dicarboxylic Acids in Poly(ester–amide)-Based Thermosets: Synthesis, Characterization, and Potential Route for Their Chemical Recycling. *ACS Sustain. Chem. Eng.* **2019**, *7*, 8842–8852. [CrossRef]

17. Li, M.; Guan, Q.; Dingemans, T.J. High-Temperature Shape Memory Behavior of Semicrystalline Polyamide Thermosets. *ACS Appl. Mater. Interfaces* **2018**, *10*, 19106–19115. [CrossRef] [PubMed]

18. Lee, J.Y.; Jang, J. IR study on the character of hydrogen bonding in novel liquid crystalline epoxy resin. *Polym. Bull.* **1997**, *38*, 447–454. [CrossRef]

19. Ding, C.; Matharu, A.S. Recent Developments on Biobased Curing Agents: A Review of Their Preparation and Use. *ACS Sustain. Chem. Eng.* **2014**, *2*, 2217–2236. [CrossRef]

20. Auvergne, R.; Caillol, S.; David, G.; Boutevin, B.; Pascault, J.-P. Biobased Thermosetting Epoxy: Present and Future. *Chem. Rev.* **2014**, *114*, 1082–1115. [CrossRef]

21. Mora, A.-S.; Tayouo, R.; Boutevin, B.; David, G.; Caillol, S. Vanillin-derived amines for bio-based thermosets. *Green Chem.* **2018**, *20*, 4075–4084. [CrossRef]

22. Mićović, I.V.; Ivanović, M.D.; Piatak, D.M.; Bojić, V.D. A Simple Method for Preparation of Secondary Aromatic Amines. *Synthesis* **1991**, *1991*, 1043–1045. [CrossRef]

23. Alinezhad, H.; Tajbakhsh, M.; Zamani, R. Efficient and Mild Procedure for Reductive Methylation of Amines Using N-Methylpiperidine Zinc Borohydride. *Synth. Commun.* **2006**, *36*, 3609–3615. [CrossRef]

24. Ranu, B.C.; Majee, A.; Sarkar, A. One-Pot Reductive Amination of Conjugated Aldehydes and Ketones with Silica Gel and Zinc Borohydride. *J. Org. Chem.* **1998**, *63*, 370–373. [CrossRef]

25. McGonagle, F.I.; MacMillan, D.S.; Murray, J.; Sneddon, H.F.; Jamieson, C.; Watson, A.J.B. Development of a solvent selection guide for aldehyde-based direct reductive amination processes. *Green Chem.* **2013**, *15*, 1159–1165. [CrossRef]

26. Sato, S.; Sakamoto, T.; Miyazawa, E.; Kikugawa, Y. One-pot reductive amination of aldehydes and ketones with α-picoline-borane in methanol, in water, and in neat conditions. *Tetrahedron* **2004**, *60*, 7899–7906. [CrossRef]

27. Morales, S.; Guijarro, F.G.; García Ruano, J.L.; Cid, M.B. A General Aminocatalytic Method for the Synthesis of Aldimines. *J. Am. Chem. Soc.* **2014**, *136*, 1082–1089. [CrossRef] [PubMed]

28. Bornadel, A.; Bisagni, S.; Pushpanath, A.; Montgomery, S.L.; Turner, N.J.; Dominguez, B. Technical Considerations for Scale-Up of Imine-Reductase-Catalyzed Reductive Amination: A Case Study. *Org. Process Res. Dev.* **2019**, *23*, 1262–1268. [CrossRef]

29. France, S.P.; Howard, R.M.; Steflik, J.; Weise, N.J.; Mangas-Sanchez, J.; Montgomery, S.L.; Crook, R.; Kumar, R.; Turner, N.J. Identification of Novel Bacterial Members of the Imine Reductase Enzyme Family that Perform Reductive Amination. *ChemCatChem* **2018**, *10*, 510–514. [CrossRef]

30. Wetzl, D.; Gand, M.; Ross, A.; Müller, H.; Matzel, P.; Hanlon, S.P.; Müller, M.; Wirz, B.; Höhne, M.; Iding, H. Asymmetric Reductive Amination of Ketones Catalyzed by Imine Reductases. *ChemCatChem* **2016**, *8*, 2023–2026. [CrossRef]

31. Qian, J.; An, Q.; Fortunelli, A.; Nielsen, R.J.; Goddard, W.A. Reaction Mechanism and Kinetics for Ammonia Synthesis on the Fe(111) Surface. *J. Am. Chem. Soc.* **2018**, *140*, 6288–6297. [CrossRef] [PubMed]

32. Brenna, D.; Rossi, S.; Cozzi, F.; Benaglia, M. Iron catalyzed diastereoselective hydrogenation of chiral imines. *Org. Biomol. Chem.* **2017**, *15*, 5685–5688. [CrossRef] [PubMed]

33. He, Y.-M.; Feng, Y.; Fan, Q.-H. Asymmetric Hydrogenation in the Core of Dendrimers. *Acc. Chem. Res.* **2014**, *47*, 2894–2906. [CrossRef] [PubMed]

34. Zhang, B.; Guo, X.-W.; Liang, H.; Ge, H.; Gu, X.; Chen, S.; Yang, H.; Qin, Y. Tailoring Pt–Fe2O3 Interfaces for Selective Reductive Coupling Reaction To Synthesize Imine. *ACS Catal.* **2016**, *6*, 6560–6566. [CrossRef]

35. Kim, H.R.; Achary, R.; Lee, H.-K. DBU-Promoted Dynamic Kinetic Resolution in Rh-Catalyzed Asymmetric Transfer Hydrogenation of 5-Alkyl Cyclic Sulfamidate Imines: Stereoselective Synthesis of Functionalized 1,2-Amino Alcohols. *J. Org. Chem.* **2018**, *83*, 11987–11999. [CrossRef] [PubMed]

36. Wang, H.; Gao, Z.; Wang, X.; Wei, R.; Zhang, J.; Shi, F. Precise regulation of selectivity of supported nano-Pd catalysts using polysiloxane coatings with tunable surface wettability. *Chem. Commun.* **2019**, *55*, 8305–8308. [CrossRef] [PubMed]

37. Jia, L.; Makha, M.; Du, C.-X.; Quan, Z.-J.; Wang, X.-C.; Li, Y. Direct hydroxyethylation of amines by carbohydrates via ruthenium catalysis. *Green Chem.* **2019**, *21*, 3127–3132. [CrossRef]

38. Exposito, A.J.; Bai, Y.; Tchabanenko, K.; Rebrov, E.V.; Cherkasov, N. Process Intensification of Continuous-Flow Imine Hydrogenation in Catalyst-Coated Tube Reactors. *Ind. Eng. Chem. Res.* **2019**, *58*, 4433–4442. [CrossRef]

39. Webster, D.C.; Crain, A.L. Synthesis and applications of cyclic carbonate functional polymers in thermosetting coatings. *Prog. Org. Coat.* **2000**, *40*, 275–282. [CrossRef]

40. Chattopadhyay, D.K.; Webster, D.C. Hybrid coatings from novel silane-modified glycidyl carbamate resins and amine crosslinkers. *Prog. Org. Coat.* **2009**, *66*, 73–85. [CrossRef]

41. Demir, B.; Beggs, K.M.; Fox, B.L.; Servinis, L.; Henderson, L.C.; Walsh, T.R. A predictive model of interfacial interactions between functionalised carbon fibre surfaces cross-linked with epoxy resin. *Compos. Sci. Technol.* **2018**, *159*, 127–134. [CrossRef]

42. Ménard, R.; Caillol, S.; Allais, F. Chemo-Enzymatic Synthesis and Characterization of Renewable Thermoplastic and Thermoset Isocyanate-Free Poly(hydroxy)urethanes from Ferulic Acid Derivatives. *ACS Sustain. Chem. Eng.* **2017**, *5*, 1446–1456. [CrossRef]

43. Samanta, S.; Selvakumar, S.; Bahr, J.; Wickramaratne, D.S.; Sibi, M.; Chisholm, B.J. Synthesis and Characterization of Polyurethane Networks Derived from Soybean-Oil-Based Cyclic Carbonates and Bioderivable Diamines. *ACS Sustain. Chem. Eng.* **2016**, *4*, 6551–6561. [CrossRef]

44. Hibert, G.; Lamarzelle, O.; Maisonneuve, L.; Grau, E.; Cramail, H. Bio-based aliphatic primary amines from alcohols through the 'Nitrile route' towards non-isocyanate polyurethanes. *Eur. Polym. J.* **2016**, *82*, 114–121. [CrossRef]

45. Garrison, M.D.; Harvey, B.G. Bio-based hydrophobic epoxy-amine networks derived from renewable terpenoids. *J. Appl. Polym. Sci.* **2016**, *133*, 43621. [CrossRef]

46. Darroman, E.; Durand, N.; Boutevin, B.; Caillol, S. Improved cardanol derived epoxy coatings. *Prog. Org. Coat.* **2016**, *91*, 9–16. [CrossRef]

47. Bähr, M.; Bitto, A.; Mülhaupt, R. Cyclic limonene dicarbonate as a new monomer for non-isocyanate oligo- and polyurethanes (NIPU) based upon terpenes. *Green Chem.* **2012**, *14*, 1447–1454. [CrossRef]

48. Voirin, C.; Caillol, S.; Sadavarte, N.V.; Tawade, B.V.; Boutevin, B.; Wadgaonkar, P.P. Functionalization of cardanol: Towards biobased polymers and additives. *Polym. Chem.* **2014**, *5*, 3142–3162. [CrossRef]

49. Yang, J.-D.; Xue, J.; Cheng, J.-P. Understanding the role of thermodynamics in catalytic imine reductions. *Chem. Soc. Rev.* **2019**, *48*, 2913–2926. [CrossRef]

50. Peng, H.; Wei, E.; Wang, J.; Zhang, Y.; Cheng, L.; Ma, H.; Deng, Z.; Qu, X. Deciphering Piperidine Formation in Polyketide-Derived Indolizidines Reveals a Thioester Reduction, Transamination, and Unusual Imine Reduction Process. *ACS Chem. Biol.* **2016**, *11*, 3278–3283. [CrossRef]

51. Lv, J.; Wang, F.; Wei, T.; Chen, X. Highly Sensitive and Selective Fluorescent Probes for the Detection of HOCl/OCl–Based on Fluorescein Derivatives. *Ind. Eng. Chem. Res.* **2017**, *56*, 3757–3764. [CrossRef]

52. Yuan, Z.; Zhou, P.; Liu, X.; Wang, Y.; Liu, B.; Li, X.; Zhang, Z. Mild and Selective Synthesis of Secondary Amines Direct from the Coupling of Two Aldehydes with Ammonia. *Ind. Eng. Chem. Res.* **2017**, *56*, 14766–14770. [CrossRef]

53. Bódis, J.; Lefferts, L.; Müller, T.E.; Pestman, R.; Lercher, J.A. Activity and Selectivity Control in Reductive Amination of Butyraldehyde over Noble Metal Catalysts. *Catal. Lett.* **2005**, *104*, 23–28. [CrossRef]

54. Procede Pour La Preparation D'alpha, Omega-Diamines Aliphatiques, Et Produits Obtenus. French Patent FR1976003, 20 May 1977.

55. Nagareda, K.; Tokuda, Y.; Suzuki, S. Process for Producing Diamines from Dialdehydes. European Patent EP087, 18 November 1998.

56. Pelckmans, M.; Renders, T.; Van de Vyver, S.; Sels, B.F. Bio-based amines through sustainable heterogeneous catalysis. *Green Chem.* **2017**, *19*, 5303–5331. [CrossRef]

57. Le, N.-T.; Byun, A.; Han, Y.; Lee, K.-I.; Kim, H. Preparation of 2,5-Bis(Aminomethyl)Furan by Direct Reductive Amination of 2,5-Diformylfuran over Nickel-Raney Catalysts. *Green Sustain. Chem.* **2015**, *05*, 115–127. [CrossRef]

58. Micklitsch, C.M.; Yu, Q.; Schneider, J.P. Unnatural multidentate metal ligating α-amino acids. *Tetrahedron Lett.* **2006**, *47*, 6277–6280. [CrossRef]

59. Milelli, A.; Tumiatti, V.; Micco, M.; Rosini, M.; Zuccari, G.; Raffaghello, L.; Bianchi, G.; Pistoia, V.; Fernando Díaz, J.; Pera, B.; et al. Structure–activity relationships of novel substituted naphthalene diimides as anticancer agents. *Eur. J. Med. Chem.* **2012**, *57*, 417–428. [CrossRef]

60. Kasemi, E.; Kramer, A.; Stadelmann, U.; Burckhardt, U. Curing Agents for Low-Emission Epoxy Resin Products. US Patent US201533, 26 November 2015.

61. Shen, X.; Dai, J.; Liu, Y.; Liu, X.; Zhu, J. Synthesis of high performance polybenzoxazine networks from bio-based furfurylamine: Furan vs. benzene ring. *Polymer* **2017**, *122*, 258–269. [CrossRef]

62. Hu, F.; La Scala, J.J.; Sadler, J.M.; Palmese, G.R. Synthesis and Characterization of Thermosetting Furan-Based Epoxy Systems. *Macromolecules* **2014**, *47*, 3332–3342. [CrossRef]

63. Levita, G.; Petris, S.D.; Marchetti, A.; Lazzeri, A. Crosslink density and fracture toughness of epoxy resins. *J. Mater. Sci.* **1991**, *26*, 2348–2352. [CrossRef]

64. Widhalm, J.R.; Dudareva, N. A Familiar Ring to It: Biosynthesis of Plant Benzoic Acids. *Mol. Plant* **2015**, *8*, 83–97. [CrossRef] [PubMed]

65. Laurichesse, S.; Avérous, L. Chemical modification of lignins: Towards biobased polymers. *Prog. Polym. Sci.* **2014**, *39*, 1266–1290. [CrossRef]

66. Boukis, A.C.; Llevot, A.; Meier, M.A.R. High Glass Transition Temperature Renewable Polymers via Biginelli Multicomponent Polymerization. *Macromol. Rapid Commun.* **2016**, *37*, 643–649. [CrossRef] [PubMed]

67. Frost, J.W. Synthesis of Biobased Terephthalic Acids and Isophthalic Acids. US Patent WO201414, 18 September 2014.

68. Miller, K.K.; Zhang, P.; Nishizawa-Brennen, Y.; Frost, J.W. Synthesis of Biobased Terephthalic Acid from Cycloaddition of Isoprene with Acrylic Acid. *ACS Sustain. Chem. Eng.* **2014**, *2*, 2053–2056. [CrossRef]

69. Li, X.; Jia, P.; Wang, T. Furfural: A Promising Platform Compound for Sustainable Production of C_4 and C_5 Chemicals. *ACS Catal.* **2016**, *6*, 7621–7640. [CrossRef]

70. 2-Methylpentane-1,5-diamine-Substance Information-ECHA. Available online: https://echa.europa.eu/fr/substance-information/-/substanceinfo/100.035.945 (accessed on 28 May 2019).

71. Kagan, H.M.; Manning, L.R.; Meister, A. Stereospecific synthesis of alpha-methyl-L-glutamine by glutamine synthetase. *Biochemistry* **1965**, *4*, 1063–1068. [CrossRef]

72. Inoue, M.; Shinohara, N.; Tanabe, S.; Takahashi, T.; Okura, K.; Itoh, H.; Mizoguchi, Y.; Iida, M.; Lee, N.; Matsuoka, S. Total synthesis of the large non-ribosomal peptide polytheonamide B. *Nat. Chem.* **2010**, *2*, 280–285. [CrossRef] [PubMed]

73. Faye, I.; Decostanzi, M.; Ecochard, Y.; Caillol, S. Eugenol bio-based epoxy thermosets: From cloves to applied materials. *Green Chem.* **2017**, *19*, 5236–5242. [CrossRef]

74. Fu, B.X.; Namani, M.; Lee, A. Influence of phenyl-trisilanol polyhedral silsesquioxane on properties of epoxy network glasses. *Polymer* **2003**, *44*, 7739–7747. [CrossRef]

75. Meis, N.N.A.H.; van der Ven, L.G.J.; van Benthem, R.A.T.M.; de With, G. Extreme wet adhesion of a novel epoxy-amine coating on aluminum alloy 2024-T3. *Prog. Org. Coat.* **2014**, *77*, 176–183. [CrossRef]

76. Global M Xylylenediamine Market Report 2017—Market Reports World. Available online: https://www.marketreportsworld.com/global-m-xylylenediamine-market-report-2017-10576122 (accessed on 29 June 2019).

77. Fache, M.; Auvergne, R.; Boutevin, B.; Caillol, S. New vanillin-derived diepoxy monomers for the synthesis of biobased thermosets. *Eur. Polym. J.* **2015**, *67*, 527–538. [CrossRef]

78. Kaji, M.; Nakahara, K.; Endo, T. Synthesis of a bifunctional epoxy monomer containing biphenyl moiety and properties of its cured polymer with phenol novolac. *J. Appl. Polym. Sci.* **1999**, *74*, 690–698. [CrossRef]

79. Mucsi, Z.; Viskolcz, B.; Csizmadia, I.G. A Quantitative Scale for the Degree of Aromaticity and Antiaromaticity: A Comparison of Theoretical and Experimental Enthalpies of Hydrogenation. *J. Phys. Chem. A* **2007**, *111*, 1123–1132. [CrossRef] [PubMed]

80. Huber, R.G.; Margreiter, M.A.; Fuchs, J.E.; von Grafenstein, S.; Tautermann, C.S.; Liedl, K.R.; Fox, T. Heteroaromatic π-Stacking Energy Landscapes. *J. Chem. Inf. Model.* **2014**, *54*, 1371–1379. [CrossRef]

81. Matsumoto, K.; Hayashi, N. *Heterocyclic Supramolecules II*; Springer Science & Business Media: Berlin/Heidelberg, Germany, 2009; ISBN 978-3-642-02040-7.

Sample Availability: Samples of the compounds reported in the paper are not available from the authors.

 *molecules* 

Article

Synthesis and Characterization of a Bioartificial Polymeric System with Potential Antibacterial Activity: Chitosan-Polyvinyl Alcohol-Ampicillin

Andres Bernal-Ballen [1,*], Jorge Lopez-Garcia [2], Martha-Andrea Merchan-Merchan [3] and Marian Lehocky [2]

[1] Grupo de Investigación en Ingeniería Biomédica, Vicerrectoría de Investigaciones, Universidad Manuela Beltrán, Avenida Circunvalar No. 60-00, Bogotá 1101, Colombia
[2] Centre of Polymer Systems, University Institute, Tomas Bata University in Zlín, Trida Tomase Bati 5678, 76001 Zlín, Czech Republic; lopez@utb.cz (J.L.-G.); lehocky@utb.cz (M.L.)
[3] Maestría en Educación, Universidad Antonio Nariño, Conciencia, Calle 22 sur No. 12D-81, Bogotá 1101, Colombia; mmerchan30@uan.edu.co
*   Correspondence: andres_bernal9@hotmail.com; Tel.: +57-1-5460600

Academic Editors: Sylvain Caillol and Guillaume Couture
Received: 31 August 2018; Accepted: 30 October 2018; Published: 28 November 2018

Abstract: Bio-artificial polymeric systems are a new class of polymeric constituents based on blends of synthetic and natural polymers, designed with the purpose of producing new materials that exhibit enhanced properties with respect to the individual components. In this frame, a combination of polyvinyl alcohol (PVA) and chitosan, blended with a widely used antibiotic, sodium ampicillin, has been developed showing a moderate behavior in terms of antibacterial properties. Thus, aqueous solutions of PVA at 1 wt.% were mixed with acid solutions of chitosan at 1 wt.%, followed by adding ampicillin ranging from 0.3 to 1.0 wt.% related to the total amount of the polymers. The prepared bio-artificial polymeric system was characterized by FTIR, SEM, DSC, contact angle measurements, antibacterial activity against *Staphylococcus aureus* and *Escherichia coli* and antibiotic release studies. The statistical significance of the antibacterial activity was determined using a multifactorial analysis of variance with $p < 0.05$ (ANOVA). The characterization techniques did not show alterations in the ampicillin structure and the interactions with polymers were limited to intermolecular forces. Therefore, the antibiotic was efficiently released from the matrix and its antibacterial activity was preserved. The system disclosed moderate antibacterial activity against bacterial strains without adding a high antibiotic concentration. The findings of this study suggest that the system may be effective against healthcare-associated infections, a promising view in the design of novel antimicrobial biomaterials potentially suitable for tissue engineering applications.

Keywords: bio-artificial polymeric system; health-care associated infections; polyvinyl alcohol; chitosan; ampicillin

---

1. Introduction

Copious literature has evidenced the extraordinary applicability of polymers in medical fields [1–10]. In fact, these materials are used in tissue engineering, implantation of medical devices, artificial organs, prostheses, ophthalmology, dentistry, and bone repairing [7]. Nevertheless, natural polymers do not have appropriate mechanical properties whereas synthetics are deficient in terms of biocompatibility. In this matter, their combination has gathered growing interest from the scientific community in the recent past. Most of these systems, called bioartificial polymeric systems (BAPS), have attracting characteristics for biomedical applications, and were originally conceived with the aim of combining the features of synthetic polymers (good mechanical properties, easy processability, low production and transformation costs) with the specific tissue- and cel- compatibility of biopolymers [11,12].

The BAPS may be produced as hydrogels, films, scaffolds, and a great variety of potential applications has been reported including dialysis membranes, artificial skin, cardiovascular devices, implants, bandages, or even controlled drug-release systems [13–15]. In spite of the remarkable potential that the BAPS have revealed, they lack inherent attributes such as bactericidal or bacteriostatic properties, whereby the BAPS in the medical field might be focused on the appeasement of healthcare-associated infections which boosts health systems costs. As an approach for facing bacteria colonization, controlled released drug systems from polymeric matrixes might be catalogued as an auspicious solution [16,17].

One of the methods for reducing or preventing infections is the development of polymer-antibiotic systems [17]. BAPS might be included in this category because of the individual properties of both natural and synthetic components, as well as the interaction that might occur among them. In that respect, chitosan (CHI) as a natural polymer has been investigated and its antimicrobial properties against a wide range of target microorganisms (algae, bacteria, yeasts and fungi) have been reported in experiments involving *in vivo* and *in vitro* studies, and in different forms (solutions, films, and composites). Nonetheless, recent data in the literature have the tendency to characterize chitosan as bacteriostatic rather than bactericidal. This polymer has been extensively used in biomedical applications, as material for chemicals encapsulation and controlled release [18,19]. CHI is widely-used as an antimicrobial agent either alone or blended with other natural polymers (starch, gelatin, alginates), in food, pharmaceutical, textile, agriculture, water treatment, and cosmetics industries [20]. Furthermore, it is hydrophilic; hence, CHI films are poor barriers to moisture. For these reasons, some biodegradable synthetic polymers, such as poly (caprolactone), poly (lactic acid), and poly (vinyl alcohol) (PVA) were used to modify strength, water resistance, and thermal stability of CHI films [21–24].

As a synthetic polymer, PVA has centered the interest from scientists in diverse areas such as engineering, chemistry and medicine. In medical uses, properties such as a high capability of water absorption, chemical resistance, physical properties, biocompatibility, and complete biodegradability are responsible for its relevant uses. PVA has indeed unique features including excellent film-forming properties and is non-toxic, as well as its significance as a part of controlled drug delivery systems, dialysis membrane, wound dressing, artificial cartilage, and tissue engineering scaffold [25–27].

The combination of CHI and PVA may have beneficial effects on the biological characteristics of the blend. In fact, it has been reported that the chemical crosslinking between CHI and PVA may improve the mechanical strength and thermal stability, keeping the intrinsic properties of transparency and the swelling ability of CHI films [28,29]. Indeed, the use of plasticizers and crosslinkers imparts specific properties in the system and therefore, their mechanical drawbacks are diminished or attenuated.

On the other hand, it is generally assumed that the polycationic nature of CHI, conveyed by the positively charged -NH_3^+ groups of glucosamine, might be a fundamental factor contributing to its interaction with negatively charged surface components of many fungi and bacteria, causing substantial cell surface alterations, leakage of intracellular substances, and ultimately resulting in impairment of vital bacterial activities [30]. Consequently, it is expected that polymers with higher charge densities resulted in an improved antimicrobial activity. Indeed, there are successful studies related to PVA/CHI systems, which show adequate cell viability, non-toxicity, and suitable properties which can be tailored for prospective uses in tissue engineering [19,29,31,32].

Despite the remarkable properties that the individual polymers evince, reinforce such properties by adding an antibiotic might be deemed as an interesting approach for treating bacteria colonization. In this matter, sodium ampicillin (AMP) is used to treating urinary infections, salmonellosis, *Listeria* meningitis, and periodontitis among others which implies that AMP has a broad target for both Gram-positive and some Gram-negative bacteria [17].

For all the aforementioned reasons, this work deals with the synthesis and characterization of a BAPS as well as the evaluation of their antibacterial activity against *Staphylococcus aureus* and *Escherichia coli*.

2. Results and Discussion

2.1. Fourier Transform Infrared (FTIR)

Attenuated Total Reflectance Fourier Transform Infrared (ATR-FTIR) spectra of the assessed samples are shown in Figure 1. Pristine PVA (Figure 1A(a)) displays a large band observed between 3550 and 3200 cm^{-1}, which is linked to the stretching O–H from the intermolecular and intramolecular hydrogen bonds [33–36]. The band located at 2914 arises from saturated C–H stretching, whereas the band at 1423 cm^{-1} is related to –CH$_2$– bending [37]. An absence of a signal in the region of 1700 cm^{-1} indicates that only a small amount of acetate groups can be present in the polymer chain as the used PVA is highly hydrolyzed. Two peaks at 1659 and 1572 cm^{-1} may be associated with conjugated diones or single carbonyls in a conjugation with C=C double bonds in a solid state, where peaks over 1700 cm^{-1} are not manifested [38]. The peak at 1155 cm^{-1} is connected to C–O stretching modes and a strong dependency of its intensity on crystallinity degree of the solid PVA material was observed [39,40]. Other authors [41,42] connect this band with C–O–C in ether bridges, and crosslinking of PVA is deduced from the increase in absorbance at this wavenumber. The band centered at 1330 cm^{-1} may surge from combination frequencies of CH and OH [43]. A strong peak located at 1060–1030 cm^{-1} is assigned for a stretching C–O in a C–O–H group [41] but the band is shifted to 1096 cm^{-1} as a consequence of interactions with unsaturated bonds [37]. The band assigned to CH$_2$ rocking is located at 917 cm^{-1} whereas the signal at 854 cm^{-1} relies on C–O stretching [40].

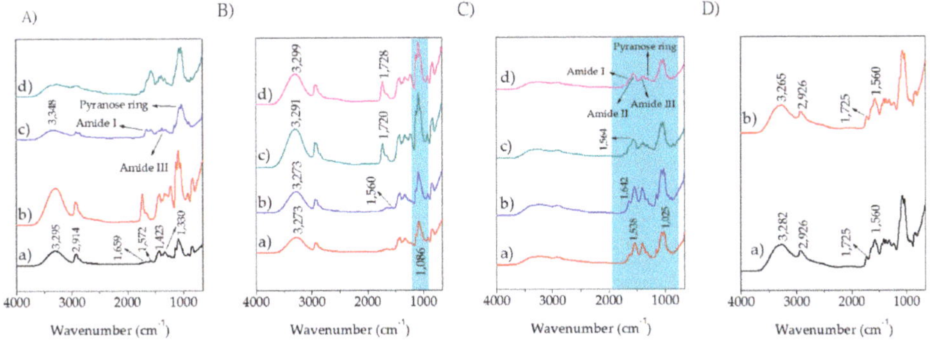

Figure 1. FTIR spectra for BAPS. (**A**). Polymers with and without lactic acid and glutaraldehyde (ADD): (a) PVA; (b) PVA/ADD; (c) CHI; (d) CHI/ADD. (**B**) PVA blends: (a) PVA/AMP 0.3%; (b) PVA/AMP 1%; (c) PVA/ADD/AMP 0.3%; (d) PVA/ADD/AMP 1.0%; (**C**) CHI blends: (a) CHI/AMP 0.3%; (b) CHI/AMP 1.0%; (c) CHI/ADD/AMP 0.3%; (d) CHI/ADD/AMP 1.0%; (**D**) PVA/CHI blends: (a) PVA/CHI/ADD/AMP 0.3%; (b) PVA/CHI/ADD/AMP 1.0%.

The FTIR spectrum for CHI exhibits a broad signal centered at 3348 cm^{-1} corresponding to -OH functional groups and -NH group stretching vibrations. The symmetric and asymmetric –CH$_2$– stretching occurred in the pyranose ring are located at 2920, 2880, 1430, 1320, 1275 and 1245 cm^{-1} [32,44,45]. The bands located at 1650, 1586 cm^{-1}, and 1322 cm^{-1} are associated with amide I, amide II, and amide III respectively as well as the saccharine-related signals at 1155 and 900 cm^{-1} [32,46–49]. The peak at 1030 and 1080 cm^{-1} indicate the C–O stretching vibration.

The addition of lactic acid (LA) as a plasticizer, and glutaraldehyde (GLU) as a crosslinker (hereinafter as ADD) does not alter significantly the spectra (Figure 1A(a,b)). A growth of OH peaks is a consequence of the presence of LA. The C=O band at approximately 1720 cm^{-1} may denote that the aldehyde groups of GA did not completely react with the -OH groups of the PVA chain or the contribution of remaining acetate groups of the PVA structure. In addition, the C–O stretching at 1088 cm^{-1} in pure PVA is replaced by a broader absorption band located between 1000 and 1140 cm^{-1}, which can be attributed to the ether (C–O) and the acetal ring (C–O–C) bands formed by the crosslinking reaction of PVA with GA. It can be assumed that GA has acted as a chemical crosslinker among PVA polymer chains [50]. By comparing CHI and CHI/ADD (Figure 1A(c,d)), a minor rise in the hydroxyl groups signal is evident. Nonetheless, the combination of both GLU and LA has contradictory effects; on one side, the crosslinker diminished the hydroxyl groups, but on the other, the plasticizer increased them. For that reason, there is no clear tendency depicted in the spectra. An increase of the signal for the imine bands as well as a drop on the amine are caused by the nucleophilic addition of the amine from CHI with the aldehyde [31].

AMP shows an absorption peak in the region of 1730–1720 cm^{-1} (Figure 1B–D), which is caused by C=O β-lactam stretching. The peaks at 1664 and 1560 cm^{-1} belong to C=O amide stretching and -NH amide groups respectively [51]. By virtue of no other modifications in the spectra, it is plausible to infer that no chemical interactions are occurring between PVA and CHI with AMP, which might imply that the antibiotic can be released from the matrix and even that the structure does not suffer significant modifications and no alterations on the antibacterial activity of AMP whatsoever. The same features are evidenced in the spectra for CHI/ADD/AMP as well as PVA/CHI/ADD/AMP.

Blends of PVA/CHI show two bands around 1640 and 1560 cm^{-1}, which are connected to symmetric stretching and bending of acetamide groups, respectively. The change in the characteristic shape of the CHI spectra, as well as the shifting of peak to a lower frequency range on account of hydrogen bonding between -OH of PVA and -OH or -NH$_2$ of chitosan were observed in the blended films [52]. These results suggested the formation of hydrogen bonds between the CHI and PVA molecules [53]. Special features are exhibited for the BAPS (Figure 1D). By comparing the CHI spectrum with that obtained for the BAPS, an increase of the signal for the imine peak as well as a drop on the amine signal caused by the nucleophilic addition of the amine from CHI with the aldehyde (crosslinking reaction with GLU) are presented [31]. Furthermore, a carboxylic acid dimer, which is manifested in the region of 1700–1725 cm^{-1} are originated from the acetic acid used for dissolving CHI and the absorption at 3300–3250 cm^{-1} related to -OH and -NH stretching vibrations broadened and shifted to a lower wavenumber suggests the formation of hydrogen bonds between CHI and PVA [54].

2.2. Scanning Electron Microscopy (SEM)

Figure 2 shows the surface and cross section of scanning electron microscopy (SEM) images of PVA/ADD/AMP, CHI/ADD/AMP and PVA/CHI/ADD. No obvious agglomeration particles were observed, suggesting well-dispersed components as well as a homogeneous structure with no phase separation. The surface anomalies may be probably accounted for manipulation or storage; likewise, samples based on CHI possess wavy areas, which may be ascribed to solvent evaporation.

With regard to the cross-section images, they depict that all of the specimens have gone through a homogeneous and effective mixing process since there are no special moieties in the sample and they look considerably uniform. It is well-established that LA plasticizes the material and its presence increases the toughness. On the contrary, GLU promotes brittleness. The SEM pictures are in total correspondence with the obtained results from the FTIR spectra where there is a defined intermolecular interaction between PVA and CHI; however, this mentioned interaction is not a physical barrier, which could interfere with the release of AMP from the obtained films.

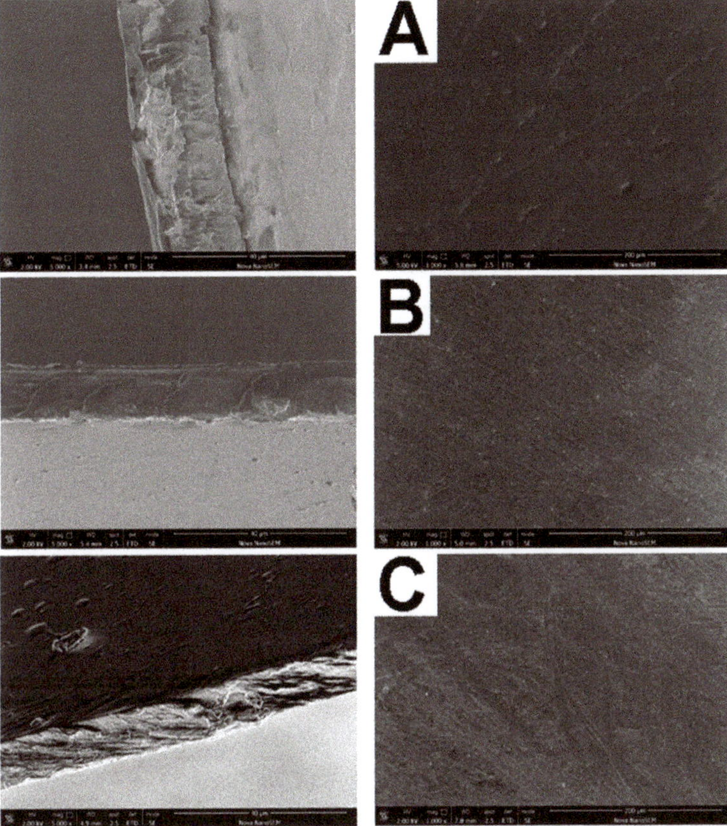

Figure 2. Cross section and surface SEM images of the Bioartificial polymeric systems (BAPS): (**A**) PVA/ADD/AMP. (**B**) CHI/ADD/AMP. (**C**) PVA/CHI/ADD/AMP.

2.3. Differential Scanning Calorimetry (DSC)

The thermograms for the prepared samples are depicted in Figure 3 and they correspond to the first heating scan. The glass transition temperature (T_g) (72 °C) and melting point (T_m) (218 °C) for PVA (Figure 3A(a)) have been reported in several publications and the results are in a good concordance with those obtained in the performed experiment [55,56]. The presence of LA and GLU influence the thermal features of the samples. Thus, the plasticizer reduces the number of lattices, causing a decrease in the endothermic peak compared with pristine PVA, whereas the low concentration of the crosslinker in comparison to LA produces a partial grafted hydroxyl group of PVA [1]. As a result, blends of PVA/ADD melted at lower temperatures (Figure 3A(d)).

The thermograms exhibit a second endothermic peak, which is ascribed to the melting of the other minor crystalline phases. Simultaneously, no evidence of transitions related to AMP is manifested in the thermograms due to its low concentration in the samples, and because the reported T_m of the antibiotic (220 °C) [57,58] is hidden by the endothermic manifestation for PVA.

CHI shows an endothermic transition between 100–120 °C that might be attributed to the evaporation of residual water [44] whereas a sharp peak at 180 °C does not correspond to T_m since a recrystallization peak neither appears in the second cooling nor in the second heating scan [59]. The effects of ADD are visible in the thermograms as the amine functional groups of CHI are more reactive to GLU than the hydroxyl groups of PVA [32,60]. Therefore, a rise in the crystalline region is manifested. Furthermore, the crystallinity decreases as the crosslinking degree by GLU increases, since crosslinks between two CHI units or pendant GLU with one aldehyde free may constitute an obstacle to chitosan molecule packing. As in the case of PVA samples, the T_m of AMP is not visible because of peak overlapping. BAPS exhibit endothermic peaks which started from room temperature and lasted at around 150 °C. This peak can be attributed to the presence of water molecules inside the PVA and CHI structure including tightly bound water which is released at about 100 °C, and the free loosely bound water, which exited at around 150 °C [61]. A sharp endothermic peak near to 200 °C belongs to T_m of the blend since the endothermic peaks for PVA shifted towards lower temperatures by the presence of CHI [53]. The reduction of T_m indicates a good miscibility of the polymers that affects the crystallization process of PVA. This drop of either the crystallization or melting temperature is considered as a measure of the blend compatibility [62].

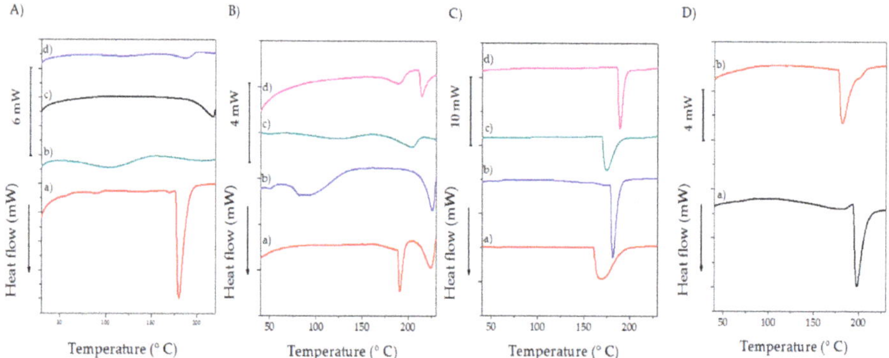

Figure 3. Thermogram for BAPS. (**A**). Polymers with and without ADD: (a) CHI; (b) CHI/ADD; (c) PVA; (d) PVA/ADD; (**B**) PVA blends: (a) PVA/AMP 0.3%; (b) PVA/AMP 1%; (c) PVA/ADD/AMP 0.3%; (d) PVA/ADD/AMP 1.0%; (**C**) CHI blends: (a) CHI/AMP 0.3%; (b) CHI/AMP 1.0%; (c) CHI/ADD/AMP 0.3%; (d) CHI/ADD/AMP 1.0%; (**D**) PVA/CHI blends: (a) PVA/CHI/ADD/AMP 0.3%; (b) PVA/CHI/ADD/AMP 1.0%.

2.4. Surface Wettability

In order to estimate the extent of hydrophilicity, contact angle measurements were carried out. The contact angle values of deionized water (θ_w), ethylene glycol (θ_e), and diiodomethane (θ_d) are reported in Table 1. Figure 4 exhibits the images taken for BAPS. Solid surfaces may be classified into two basic groups, hydrophilic (wettable with water and high surface energy) and hydrophobic (not wettable with water and low surface energy). Since a solid surface may be considered wettable if the contact angle is less than 90°, all the samples are in fact wettable. However, it is more reasonable to describe them as hydrophilic. The hydrophilicity of PVA stems from the hydrogen bonding between free -OH groups and water molecules. This feature might be reduced as the crosslinker reduces the availability of hydroxyl groups [50] as well as for the crystallinity (see DSC results).

In the case of CHI, the hydroxyl and amino groups increase the formed hydrogen bonding and, therefore, the material exhibits hydrophilic features. The plasticizer interrupts the sequence of hydrogen bonding, the intermolecular forces are affected reducing the availability of hydrophilic groups, and the material increases the contact angle. It may be observed that the samples with ADD are relatively more hydrophobic, having water contact angles above 50°. However, it seems that after some amount of AMP the hydrophilicity diminished again as a consequence of the intermolecular attractions that occur between the polymer and the antibiotic. Within this context, it is relevant to point out that a hydrophilic surface would offer a better contact with biological species and enhance the antimicrobial efficacy of the active ingredients [63]. It might explain that samples with ADD exhibit better antibacterial activity for *E. coli* than specimens that did not contain the plasticizer and crosslinker.

Table 1. Contact angle values for the prepared materials.

Description	Contact Angle (°)		
	Water	*Ethylene glycol*	*Diiodomethane*
PVA	<10 *	<10	34.93 ± 5.33
CHI	<10 *	<10	45.33 ± 1.71
PVA/AMP 0.3%	<10 *	***	35.24 ± 4.97
PVA/AMP 1.0%	<10 *	***	41.46 ± 3.82
CHI/AMP 0.3%	<10 *	55.76 ± 4.75	45.55 ± 3.34
CHI/AMP 1.0%	<10 *	47.73 ± 6.54	45.56 ± 3.02
PVA/ADD/AMP 0.3%	61.14 ± 4.13 **	***	22.94 ± 2.00
PVA/ADD/AMP 1.0%	43.12 ± 1.62 **	***	34.35 ± 2.59
CHI/ADD/AMP 0.3%	<10 *	***	45.51 ± 3.45
CHI/ADD/AMP 1.0%	<10 *	***	44.24 ± 5.69
PVA/CHI/ADD/AMP 0.3%	62.43 ± 3.82 **	44.36 ± 2.65	45.06 ± 2.61
PVA/CHI/ADD/AMP 1.0%	53.91 ± 2.67 **	43.22 ± 1.93	51.58 ± 3.88

* In these cases the samples are hydrophilic and the surface got wet very quickly. ** Although the samples have a water contact angle, the specimens got swollen in approximately 5 s. *** It reacted with the solvent.

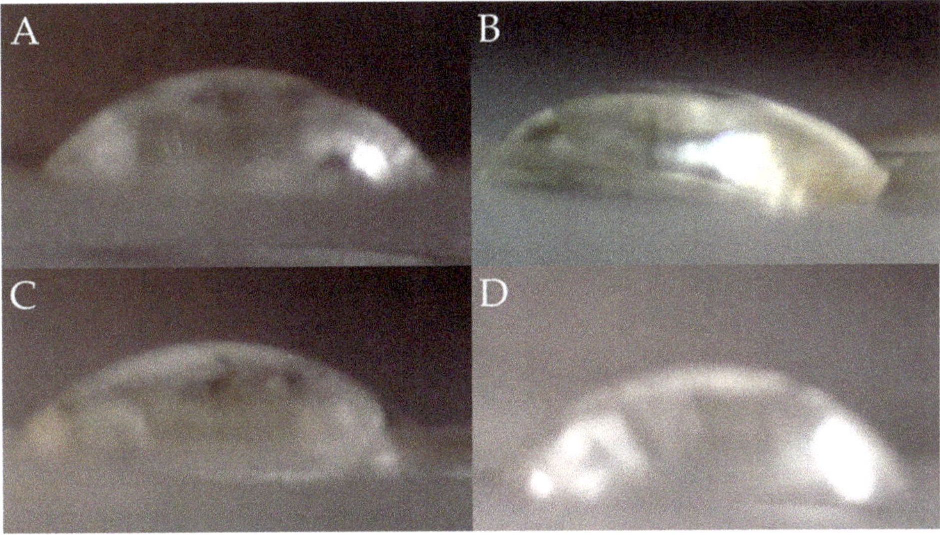

Figure 4. Images for contact angle measurements. (**A**) PVA/CHI/ADD/AMP 0.3% (ethylene glycol); (**B**) PVA/CHI/ADD/AMP 0.3% (diiodomethane); (**C**) PVA/CHI/ADD/AMP 1.0% (ethylene glycol) (**D**). PVA/CHI/ADD/AMP 1.0% (diiodomethane).

The presence of ADD has a higher impact on PVA than on CHI. As can be seen, a rise in the contact angle values is reported for samples containing PVA, including BAPS. It could indicate that the interactions between ADD and PVA are more effective in comparison to CHI.

2.5. Microbiological Testing

The method for evaluating the antibacterial activity was an agar diffusion test. The obtained results are reported in Table 2 and the inhibition halo of the tested materials are presented in Figure 5. As it can be expected, PVA did not show antibacterial activity whereas CHI requires a deeper sight. The antibacterial properties of this natural polymer depend on the molecular weight and generally show stronger bactericidal effects for Gram-positive than Gram-negative bacteria [64].

Furthermore, CHI is prepared in acetic acid, a substance which has antimicrobial activity, and bacteria in different growth stages have different sensitivity to CHI [65]. Indeed, it has been demonstrated that its antibacterial activity depends on the strain examined and on its growth phase, besides the concentration, degree of deacetylation, pH, temperature, and medium composition [66]. Then, it is plausible to affirm that the prepared samples of CHI do not have the appropriate conditions to be effective against the tested bacteria, although the presence of AMP might enhance its inherent bactericidal feature.

Table 2. Agar diffusion test of the studied sample against *Staphylococcus aureus* and *Escherichia coli*.

Sample	Inhibition Zone (mm) *Staphylococcus aureus*	Inhibition Zone (mm) *Escherichia coli*
PVA	-	-
CHI	-	-
PVA/AMP 0.3%	2.34 ± 0.69	-
PVA/AMP 1.0%	3.54 ± 0.60	1.26 ± 0.05
* CHI/AMP 0.3%	3.26 ± 0.87	-
* CHI/AMP 1.0%	4.36 ± 0.35	-
PVA/ADD/AMP 0.3%	-	1.27 ± 0.17
PVA/ADD/AMP 1.0%	1.30 ± 0.17	1.13 ± 0.05
* CHI/ADD/AMP 0.3%	3.61 ± 0.31	1.16 ± 0.21
* CHI/ADD/AMP 1.0%	4.18 ± 0.21	1.10 ± 0.16
* PVA/CHI/ADD/AMP 0.3%	1.18 ± 0.04	-
* PVA/CHI/ADD/AMP 1.0%	1.20 ± 0.10	-

* ANOVA test showed that CHI was the unique component which its significance was higher than $\rho < 0.05$.

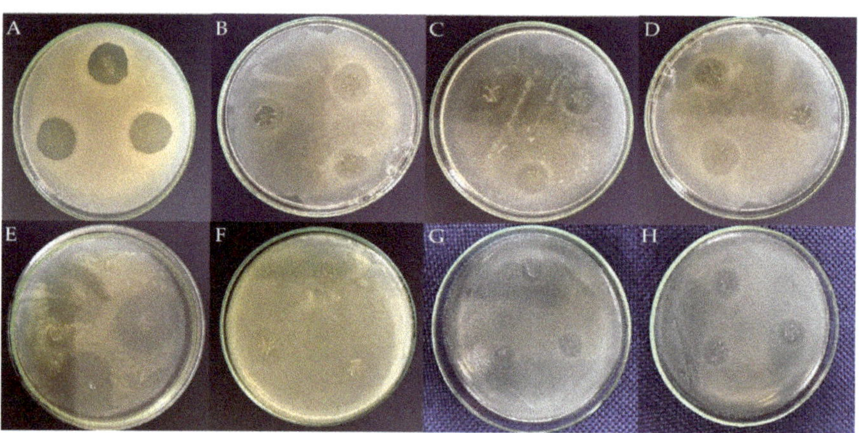

Figure 5. Inhibition zone for BAPS. (**A,B**) PVA/AMP 1.0% against *Staphylococcus aureus* and *Escherichia coli*. (**C,D**) PVA/ADD/AMP 1.0% against *Staphylococcus aureus* and *Escherichia coli*. (**E,F**) CHI/ADD/AMP 1.0% against *Staphylococcus aureus* and *Escherichia coli*. (**G**) PVA/CHI/ADD/AMP 0.3% against *Staphylococcus aureus* (**H**) PVA/CHI/ADD/AMP 1.0% against *Staphylococcus aureus*.

By comparing the presence of AMP in CHI and PVA against *Staphylococcus aureus* it is remarkable that at the same concentration of the antibiotic, the antibacterial activity of CHI is higher than PVA which might indicate that the bactericidal or bacteriostatic properties of CHI are reinforced by the presence of AMP. The combination of polymers with ADD increases the positive charges for CHI, which are responsible for its bactericidal features [67,68]. Therefore, CHI/ADD presents a higher antimicrobial activity than PVA/ADD. It may be also a consequence of synergic effects; for instance, this phenomenon is typically observed with phytochemicals, where certain molecules elicit greater influence as a group than as individual entities [67–69].

Three mechanisms have been proposed as an explanation for the antimicrobial properties of CHI. In the first one, the positive charges presented in the polymeric chain of CHI, on account of its amino group, interact with the negative charges from the residues of macromolecules in the membranes of microbial cells, interfering with a nutrient exchange between the exterior and interior of the cell. These charges may also compete with calcium for the electronegative sites in the membrane, compromising its integrity and causing a release of intracellular material and cellular death. The second mechanism states that CHI acts as a chelating agent, creating compounds from traces of metals essential to the cell, while the third mechanism asserts that CHI of low molecular weight is capable of entering into the nucleus of the cell, interacting with DNA, interfering with messenger RNA synthesis, affecting the synthesis of proteins, and inhibiting the action of various enzymes [18].

The molecular weight and the presence of ADD can alter the antimicrobial activity of CHI [24]. Nevertheless, the obtained results suggest that crosslinked CHI did not alter significantly their antibacterial capability. Furthermore, highly deacetylated CHI (75–85% in this research) exhibits a stronger antimicrobial activity than those with a higher proportion of acetylated amino groups, due to an increased solubility and a higher charge density [70].

It is relevant to emphasize that as far as inhibition is concerned, the highest amount of AMP exhibits the most effective antibacterial activity indicating that the antibiotic is released from the polymer matrix and it retains its antibacterial capacity. However, it is remarkably important that the prepared BAPS disclosed moderate antibacterial activity against bacterial strains without adding an extremely high concentration of antibiotic. In this matter, it has been established that AMP as a pure drug exhibits a minimum inhibitory concentration (MIC) of 3.13 and 0.20 µg/mL against *E. coli* and *S. aureus* strains [71]. For that reason, in samples with the same antibiotic concentration, the inhibition halo is higher for *S. aureus* than for *E. coli*. In the present study, the concentration of AMP was in the range of 30–100 µg/mL although it is plausible to infer that a minor amount of antibiotic is released from the matrix. That amount is located primarily on the surface of the film while the rest remains in the bulk, which can be delivered into the media once the sample dissolves as a consequence of the interaction with the solvent. In the same way, the samples that were crosslinked are diluted to a lesser extent, the antibiotic has a more complex pathway to get the exterior of the system. However, the effect of the plasticizer (an increase of the free spaces of the molecules) allows the diffusion of water and a greater amount of antibiotic can be released.

A difference in efficacy of Gram-positive and Gram-negative microorganism can be explained by the variance in the structure of the bacterial cell wall. The Gram-positive cell wall has a thicker peptidoglycan layer whereas the Gram-negative cell wall has one additional outer membrane [63]. That membrane impedes the action of the antibiotic. Although the results were consistent with the features for CHI and PVA, it is necessary to explain that AMP was effective against *E. coli* in PVA but not in CHI. *E. coli* cells are considered moderately hydrophilic whereas *S. aureus* cells exhibited more hydrophobic characteristics [72]. In this matter, the hydrophilic characteristics of PVA promote the liberation of AMP, which might interact ti a greater extent with this bacterium. On the contrary, CHI presents a lower hydrophilicity in comparison to PVA, resulting in a more complicated mechanism for AMP liberation. On the other hand, surface adherence is a natural tendency which is inherent to bacteria and other microorganisms. It has four basic steps: adhesion, colonization, formation and the subsequent bacterial biofilm growth. Biofilms act as a defense mechanism against external agents; in consequence, the aim of any antimicrobial materials is preventing bacterial adhesion and colonization, which are prerequisites to biofilm formation [73]. Therefore, any modification in the hydrophilicity of the surface will have repercussions for biofilm formation and the bacterial are more exposed to the action of the antibiotic. The use of ADD affected both PVA and CHI whether causing a reduction in the concentration of amino groups [48,74], or a lower content of hydroxyl groups [1]. Yet, interactions between the hydroxyl groups of PVA and the amino or hydroxyl groups of CHI might reduce the overall hydrophilicity of the system. Thus, the obtained results might be a consequence of the action of the released AMP, the hydrophilicity of the system, and the smoothness of the surface (see SEM images).

Two systems, PVA/CHI/ADD/AMP 0.3% and 1.0% did not reveal any inhibition over *E. coli*. As it might be foreseen, the difference between the expected and the obtained results may be ascribed to the fact that CHI hinders the adhesion of a Gram-positive strain but does not behave satisfactorily against Gram-negative bacteria, which may explain why the inhibition on PVA substrates is slightly better than on the CHI ones [75,76]. Moreover, the model of the wall of *E. coli* remains unsolved [77] although it has been established that the mechanism of AMP activity affects the cell wall synthesis as well as the acquisition of resistance of this microorganism against the mentioned antibiotic [78].

An ANOVA test shows that AMP reinforces the antibacterial activity of CHI, whereas PVA and ADD reduce slightly that property. Nonetheless, PVA and ADD enhance the poor mechanical properties of CHI and the combination of both consolidates the BAPS as a material with improved biological and physical features.

2.6. Antibiotic Release Studies

The mean cumulative mass of AMP released from the studied samples into distilled water as a function of elution time and the calculated constant of the Equation 1 are summarized in the Table 3.

Table 3. Release studies of AMP from the prepared samples.

% AMP	C_{MAX} (µg/g)	$-k \times 10^{-3}$ (1/h)	R^2
PVA/AMP 0.3%	26.69 ± 0.94	43.1	0.90
PVA/AMP 1.0%	70.54 ± 0.71	53.8	0.90
CHI/AMP 0.3%	27.11 ± 0.85	45.5	0.90
CHI/AMP 1.0 %	75.90 ± 2.09	70.5	0.90
PVA/ADD/AMP 0.3%	27.13 ± 0.40	0.40	0.90
PVA/ADD/AMP 1.0%	75.60 ± 1.63 *	0.45	0.90
CHI/ADD/AMP 0.3%	27.12 ± 0.44	57.9	0.90
CHI/ADD/AMP 1.0%	73.00 ± 2.16	68.7	0.95
PVA/CHI/ADD/AMP 0.3%	78.28 ± 0.23 *	0.33	0.95
PVA/CHI/ADD/AMP 1.0%	78.74 ± 2.58 *	0.41	0.95

* These values were obtained after 24 h. The others after 6 h.

According to the obtained data, the release profiles of PVA, CHI, and CHI/ADD showed a higher rate constant (k). These results might indicate that the elution of AMP from the systems occurred in the first part of the experiment. Then, as it is shown in Figure 6A,B, the released antibiotic levels off its concentration in the media, and after the third measurement no AMP was detected. These final results are consistent with the antibacterial test as well as the surface wettability which it might be related to high affinity to the polar media of the PVA, CHI and the β-lactamic antibiotic used in this study.

The relevant results from the antibiotic release profiles focused on the BAPS were compared to all batches of samples and exhibited the smallest rate constant. In this matter of fact, the participation of ADD in the systems is related to the controlled release of AMP from the BASP which corresponded to the long-lasting antibacterial activity of the system. The results are concomitant with the antibacterial activity as well as the water contact angle results, showing the crucial role of ADD in the BAPS system likewise the synergic performance of the PVA and CHI. The stability of the samples could be also taken as outstanding results, since it showed a fast swelling due to the presence of LA, allowing the entrance of water molecules in the bulk. However, the crosslinked PVA causes a decrease in hydrophilicity, and the sustainable activity of the AMP for longer periods of time from the BASP is achieved.

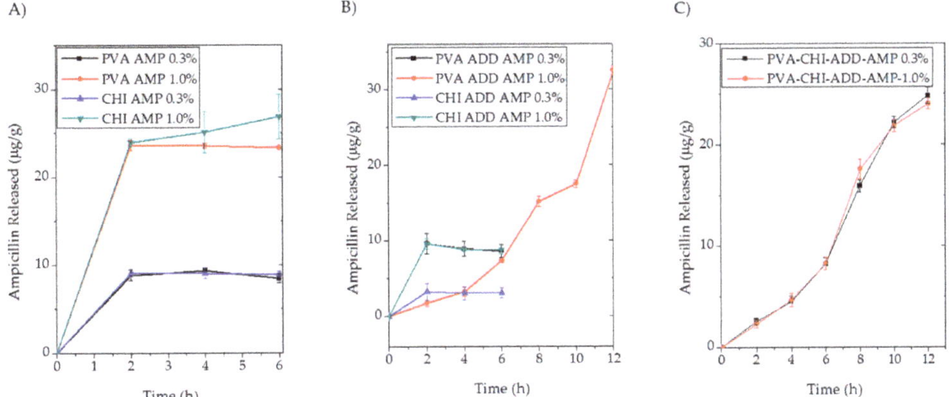

Figure 6. Release profile of AMP from BAPS measured by UV-VIS spectrophotometer.

3. Materials and Methods

Poly(vinyl alcohol) (PVA, M_w 130,000 g mol^{-1}) with a 99% hydrolysis, chitosan of low molecular weight and 75–85% deacetylation degree (CHI), and a 25% solution of glutaraldehyde (GLU) were provided by Sigma Aldrich, Bogotá, Colombia. Lactic acid (LA) was purchased at Fermont (Bogotá-Colombia), hydrochloric acid (HA) at Scharlau (Bogotá, Colombia), anhydrous ethylene glycol (99.8%) and diiodomethane (99%) were purchased from Sigma Aldrich (Saint Louis, MO, USA), and sodium ampicillin (AMP) was produced by Farmalógica, S.A. (Bogotá-Colombia) and donated to this research by The *Hospital Cardiovascular del Niño de Cundinamarca* (Soacha, Colombia). All of the reagents were used as they were received.

3.1. Sample Preparation

An aqueous solution of PVA at 1 wt.% was prepared by dissolving the polymer in distilled water for 6 h at 80 °C under continuous magnetic stirring (ThermoScientific—SP131325Q, Shangai, China). Thereupon, LA at 5 wt.%, HA at 1.2 wt.%, and GLU at 0.25% with respect to the total amount of the polymer were added as plasticizer and crosslinker agents and the solution was stirred for 20 more min. CHI was dissolved in acetic acid (0.5 M) at room temperature by the slow addition of the polymer to the solvent under mild magnetic stirring (ThermoScientific—SP131325Q, Shangai, China) for getting a 1 wt.% solution. Then, GLU (0.25 wt.%), HA (1.2 wt.%), and LA (5 wt.%) related to the total amount of the polymer were added as crosslinker and plasticizer agents. Blends of a 1:1 ratio of PVA and CHI were prepared and stirred until a homogeneous appearance with and without ADD followed by the addition of AMP (0 to 1 wt.%, related to the polymer). The solutions were stirred for 10 min and films were obtained using the casting method pouring 0.5 mL per cm^2 on plastic Petri dishes and the films were allowed to dry at 37 °C for a week in a no-air circulating oven.

3.2. Fourier Transform Infrared (FTIR)

FITR spectroscopy analysis was carried out on NICOLET 6700 FTIR spectrometer device (Thermo scientific, Waltham, MA, USA) equipped with attenuated total reflectance (ATR) accessory utilizing the Zn–Se crystal and software package OMNIC over the range of wavelengths from 4000 to 600 cm^{-1} at room temperature under a resolution of 4 cm^{-1}. Each spectrum represents 64 co-added scans referenced against an empty ATR cell spectrum.

3.3. Scanning Electron Microscopy (SEM)

Micrographs of the prepared samples were taken by the scanning electron microscope Nova NanoSEM 450 (FEI, Brno, Czech Republic) with a Schottky field emission electron source operated at an acceleration voltage ranging from 200 V to 30 kV and a low-vacuum SED (LVD) detector. A coating with a thin layer of gold was performed by a sputter coater SC 7640 (Quorum Technologies, Newhaven, East Sussex, UK).

3.4. Differential Scanning Calorimetry (DSC)

Calorimetric measurements were carried out in a DSC 1 calorimeter, Mettler Toledo (Greifensee, Zurich, Switzerland), under nitrogen flowing at a rate of 30 mL min^{-1}. The specimens were pressed in sealed aluminum pans. A heating cycle was performed in order to acquire the glass transition temperature (T_g) and melting temperature (T_m). The samples were cooled down by nitrogen at an exponentially decreasing rate. The heating of the cycle was performed from 25 to 250 °C at a rate of 20 °C/min. The T_g was determined as the midpoint temperature by standard extrapolation of the linear part of DSC curves using Mettler-Toledo Stare software and the T_m as the maximum value of the melting peak.

3.5. Surface Wettability

The wettability of the samples was evaluated by contact angle measurement. The sessile drop method was employed for this purpose on a Surface Energy Evaluation (SEE) system equipped with a CCD camera (Advex Instruments, Brno, Czech Republic). Deionized water, diiodomethane, and anhydrous ethylene glycol were used as testing liquids at 22 °C and 60% relative humidity. The droplets volume was set to 5 µL for all experiments. Every representative contact angle value was an average of 5 independent measurements.

3.6. Microbiological Tests

The antibacterial properties of the films were assessed by using the agar diffusion test. Round specimens (8 mm in diameter) were placed on the surface of an individual nutrient agar plate, where a bacterial solution of chosen microorganisms had been swabbed uniformly (*Staphylococcus aureus*-ATCC 9144, and *Escherichia coli*- ATCC 11775). After 24 h incubation at 37 °C, the dimensions of the inhibition zones were measured in four directions, and the average values were used to calculate the circle zone inhibition area.

3.7. Antibiotic Release Studies

To measure the AMP release, round shaped samples (15 mm in diameter) were immersed into 10 mL distilled water at 37 °C with continuous shaking (100 rpm). After defined periods of time, the samples were transferred into the fresh medium to reach perfect sink conditions. The released AMP in the elutions was detected by UV-vis spectrophotometer (Thermo Scientific, Helios Gamma, Waltham, MA, USA) at a wavelength of 210 nm. Calibration dependences of the absorbance (A) on AMP concentration (C_{AMP} in µg·L^{-1}) for release in distilled water (A = 0.0528 C_{AMP} + 0.0436, R^2 = 0.9984) were determined prior to the release investigation. Afterwards, the cumulative mass was calculated. The measurements were performed by triplicate. The observed data of the cumulative mass of the released AMP related to 1 g of the sample material were evaluated by using first-order kinetics (Equation (1)) and regression processed by the least squared where C_{REL} (µg/g) is the experimental concentration of AMP that was released at time t, C_{MAX} (µg/g), means the maximal theoretical concentration of the AMP released from 1 g of the sample, and $-k(h^{-1})$ represents the rate constant i.e., time needed to reach C_{MAX}.

$$C_{REL} = C_{MAX} \times (1 - e^{-kt}) \tag{1}$$

3.8. Statistical Analysis

Microbiological tests were performed in quintuplicate and the experimental values are reported in form of average $\pm$ standard deviation. Results were statistically compared applying a multifactorial analysis of variance (ANOVA) with $p < 0.05$ with SPSS software.

4. Conclusions

This contribution delved into the incorporation of sodium ampicillin to PVA/CHI system. The studied-referred to herein as BAPS were uniform and stable with interesting antibacterial results. It was observed throughout this study that the presence of AMP is different in each system, where under the same concentration of the antibiotic, the antibacterial activity of CHI was higher than PVA which might indicate that the bactericidal or bacteriostatic properties of CHI were reinforced by synergic effects with AMP. Indeed, it is plausible to infer that AMP might be released from the polymer matrix and the antibiotic keeps its antibacterial activity. The prepared BAPS endorse an interesting breakthrough in the development of novel antimicrobial biomaterials potentially suitable for tissue engineering applications, since the samples with only 0.3% of AMP showed effectiveness against both Gram-positive and Gram-negative bacterial strains without adding a high antibiotic concentration.

Apart from the microbiological test, succinct spectroscopic, topographical and thermal assessments have been made shedding light on the chemical and thermal variations which occur when any of the starting material is added. Polymer science and material innovation is undeniably a path towards creativity with eminent progress in the last years and with great challenges to overcome and the bio-artificial polymeric systems appraise here are indeed a promising approach for tissue engineering applications. Hence, the findings of this study open the door to more in-depth research both from scientific and medical standpoints.

Author Contributions: A.B.-B. and J.L.-G. designed the study, carried out all the experiments and analyzed the obtained information. M.L. collaborated in the data collection and analysis. M.-A.M.-M. carried out the biological test. All authors approved the final manuscript.

Funding: The authors would like to express his gratitude to the Czech Science Foundation (project 17-10813S) for financial support.

Acknowledgments: The authors are grateful to Jose Leonardo Cely Andrade from Hospital Cardiovascular del Niño de Cundinamarca (Soacha, Colombia) for providing the antibiotic for this study.

Conflicts of Interest: The authors declare that there are no conflicts of interest.

References

1. Bernal, A.; Kuritka, I.; Saha, P. Preparation and characterization of poly(vinyl alcohol)-poly(vinyl pyrrolidone) blend: A biomaterial with latent medical applications. *J. Appl. Polym. Sci.* **2013**, *127*, 3560–3568. [CrossRef]

2. Bernal, A.; Kuritka, I.; Kasparkova, V.; Saha, P. The effect of microwave irradiation on poly(vinyl alcohol) dissolved in ethylene glycol. *J. Appl. Polym. Sci.* **2013**, *128*, 175–180. [CrossRef]

3. Bernal, A.; Balkova, R.; Kuritka, I.; Saha, P. Preparation and characterisation of a new double-sided bio-artificial material prepared by casting of poly(vinyl alcohol) on collagen. *Polym. Bull.* **2013**, *70*, 431–453. [CrossRef]

4. Cascone, M.G. Dynamic-mechanical properties of bioartificial polymeric materials. *Polym. Int.* **1997**, *43*, 55–69. [CrossRef]

5. Luckachan, G.E.; Pillai, C.K.S. Biodegradable Polymers—A Review on Recent Trends and Emerging Perspectives. *J. Polym. Environ.* **2011**, *19*, 637–676. [CrossRef]

6. Sionkowska, A.; Płanecka, A.; Kozłowska, J.; Skopińska-Wiśniewska, J. Photochemical stability of poly(vinyl alcohol) in the presence of collagen. *Polym. Degrad. Stab.* **2009**, *94*, 383–388. [CrossRef]

7. Jagur-Grodzinski, J. Biomedical application of functional polymers. *React. Funct. Polym.* **1999**, *39*, 99–138. [CrossRef]

8. Langer, R.; Tirrell, D.A. Designing materials for biology and medicine. *Nature* **2004**, *428*, 487–492. [CrossRef] [PubMed]

9. Lazzeri, L. Progress in bioartificial polymeric materials. *Trends Polym. Sci.* **1996**, *8*, 249–252.

10. Wong, S.-S.; Altinkaya, S.A.; Mallapragada, S.K. Crystallization of poly (vinyl alcohol) during solvent removal: Infrared characterization and mathematical modeling. *J. Polym. Sci. Part. B Polym. Phys.* **2007**, *45*, 930–935. [CrossRef]

11. Sionkowska, A. Current research on the blends of natural and synthetic polymers as new biomaterials: Review. *Prog. Polym. Sci.* **2011**, *36*, 1254–1276. [CrossRef]

12. Mano, V.; Ribeiro e Silva, M.E.S. Bioartificial polymeric materials based on collagen and poly(*N*-isopropylacrylamide). *Mater. Res.* **2007**, *10*, 165–170. [CrossRef]

13. Scotchford, C.A.; Cascone, M.G.; Downes, S.; Giusti, P. Osteoblast responses to collagen-PVA bioartificial polymers in vitro: The effects of cross-linking method and collagen content. *Biomaterials* **1998**, *19*, 1–11. [CrossRef]

14. Vasheghani, F.B.; Rajabi, F.H.; Ahmadi, M.H. Influence of solvent on thermodynamic parameters and stability of some multicomponent polymer complexes involving an acrylic polymer, poly (ethylene imine) and poly (vinyl pyrrolidone). *Polym. Bull.* **2007**, *58*, 553–563. [CrossRef]

15. Ahmad, S.I.; Hasan, N.; Zainul Abid, C.K.V.; Mazumdar, N. Preparation and characterization of films based on crosslinked blends of gum acacia, polyvinylalcohol, and polyvinylpyrrolidone-iodine complex. *J. Appl. Polym. Sci.* **2008**, *109*, 775–781. [CrossRef]

16. Merchan, M.; Sedlarikova, J.; Friedrich, M.; Sedlarik, V.; Sáha, P. Thermoplastic modification of medical grade polyvinyl chloride with various antibiotics: Effect of antibiotic chemical structure on mechanical, antibacterial properties, and release activity. *Polym. Bull.* **2011**, *67*, 997–1016. [CrossRef]

17. Merchan, M.; Sedlarikova, J.; Sedlarik, V.; Machovsky, M.; Svobodová, J.; Sáha, P. Antibacterial polyvinyl chloride/antibiotic films: The effect of solvent on morphology, antibacterial activity, and release kinetics. *J. Appl. Polym. Sci.* **2010**, *118*, 2369–2378. [CrossRef]

18. Goy, R.C.; de Britto, D.; Assis, O.B.G. A review of the antimicrobial activity of chitosan. *Polimeros* **2009**, *19*, 241–247. [CrossRef]

19. Croisier, F.; Jérôme, C. Chitosan-based biomaterials for tissue engineering. *Eur. Polym. J.* **2013**, *49*, 780–792. [CrossRef]

20. Kong, M.; Chen, X.G.; Xing, K.; Park, H.J. Antimicrobial properties of chitosan and mode of action: A state of the art review. *Int. J. Food Microbiol.* **2010**, *144*, 51–63. [CrossRef] [PubMed]

21. Liang, S.; Liu, L.; Huang, Q.; Yam, K.L. Preparation of single or double-network chitosan/poly (vinyl alcohol) gel films through selectively cross-linking method. *Carbohydr. Polym.* **2009**, *77*, 718–724. [CrossRef]

22. Sarasam, A.; Madihally, S. V Characterization of chitosan—Polycaprolactone blends for tissue engineering applications. *Biomaterials* **2005**, *26*, 5500–5508. [CrossRef] [PubMed]

23. Suyatma, N.E.; Copinet, A.; Tighzert, L.; Coma, V. Mechanical and barrier properties of biodegradable films made from chitosan and poly (lactic acid) blends. *J. Polym. Environ.* **2004**, *12*, 1–6. [CrossRef]

24. Martinez-Camacho, A.P.; Cortez-Rocha, M.O.; Ezquerra-Brauer, J.M.; Graciano-Verdugo, A.Z.; Rodriguez-Félix, F.; Castillo-Ortega, M.M.; Yépiz-Gómez, M.S.; Plascencia-Jatomea, M. Chitosan composite films: Thermal, structural, mechanical and antifungal properties. *Carbohydr. Polym.* **2010**, *82*, 305–315. [CrossRef]

25. Kaczmarek, H.; Podgórski, A. The effect of UV-irradiation on poly (vinyl alcohol) composites with montmorillonite. *J. Photochem. Photobiol. A Chem.* **2007**, *191*, 209–215. [CrossRef]

26. Ding, B.; Kim, H.-Y.; Lee, S.-C.; Lee, D.-R.; Choi, K.-J. Preparation and characterization of nanoscaled poly(vinyl alcohol) fibers via electrospinning. *Fibers Polym.* **2002**, *3*, 73–79. [CrossRef]

27. Zhou, W.Y.; Guo, B.; Liu, M.; Liao, R.; Rabie, A.B.M.; Jia, D. Poly(vinyl alcohol)/Halloysite nanotubes bionanocomposite films: Properties and in vitro osteoblasts and fibroblasts response. *J. Biomed. Mater. Res. Part A* **2010**, *93*, 1574–1587. [CrossRef] [PubMed]

28. Hajji, S.; Chaker, A.; Jridi, M.; Maalej, H.; Jellouli, K.; Boufi, S.; Nasri, M. Structural analysis, and antioxidant and antibacterial properties of chitosan-poly (vinyl alcohol) biodegradable films. *Environ. Sci. Pollut. Res.* **2016**, *23*, 15310–15320. [CrossRef] [PubMed]

29. Pineda-Castillo, S.; Bernal-Ballén, A.; Bernal-López, C.; Segura-Puello, H.; Nieto-Mosquera, D.; Villamil-Ballesteros, A.; Muñoz-Forero, D.; Munster, L. Synthesis and Characterization of Poly (Vinyl Alcohol)-Chitosan-Hydroxyapatite Scaffolds: A Promising Alternative for Bone Tissue Regeneration. *Molecules* **2018**, *23*, 2414. [CrossRef] [PubMed]

30. Raafat, D.; Sahl, H.-G. Chitosan and its antimicrobial potential—A critical literature survey. *Microb. Biotechnol.* **2009**, *2*, 186–201. [CrossRef] [PubMed]

31. De Souza Costa-Júnior, E.; Pereira, M.M.; Mansur, H.S. Properties and biocompatibility of chitosan films modified by blending with PVA and chemically crosslinked. *J. Mater. Sci. Mater. Med.* **2009**, *20*, 553–561. [CrossRef] [PubMed]

32. Mansur, H.S.; Costa, E.D.S.; Mansur, A.A.P.; Barbosa-Stancioli, E.F. Cytocompatibility evaluation in cell-culture systems of chemically crosslinked chitosan/PVA hydrogels. *Mater. Sci. Eng. C* **2009**, *29*, 1574–1583. [CrossRef]

33. Mansur, H.S.; Sadahira, C.M.; Souza, A.N.; Mansur, A.A.P. FTIR spectroscopy characterization of poly (vinyl alcohol) hydrogel with different hydrolysis degree and chemically crosslinked with glutaraldehyde. *Mater. Sci. Eng. C* **2008**, *28*, 539–548. [CrossRef]

34. Yang, J.M.; Su, W.Y.; Leu, T.L.; Yang, M.C. Evaluation of chitosan/PVA blended hydrogel membranes. *J. Membr. Sci.* **2004**, *236*, 39–51. [CrossRef]

35. Holland, B.J.; Hay, J.N. The thermal degradation of poly(vinyl alcohol). *Polymer* **2001**, *42*, 6775–6783. [CrossRef]

36. Finch, C.A. *Polyvinyl Alcohol: Developments*; Wiley: New York, NY, USA, 1992.

37. Lambert, J.B.; Gronert, S.; Shurvell, H.F.; Lightner, D.; Cooks, R.G. *Organic Structural Spectroscopy*; Prentice Hall: New York, NY, USA, 2001.

38. Tayyari, S.F.; Zahedi-Tabrizi, M.; Laleh, S.; Moosavi-Tekyeh, Z.; Rahemi, H.; Wang, Y.A. Structure and vibrational assignment of 3,4-diacetyl-2,5-hexanedione. A density functional theoretical study. *J. Mol. Struct.* **2007**, *827*, 176–187. [CrossRef]

39. Wong, J.Y.; Bronzino, J.D. *Biomaterials*, 1st ed.; CRC Press: Boca Raton, FL, USA, 2007.

40. Bhat, N.V.; Nate, M.M.; Kurup, M.B.; Bambole, V.A.; Sabharwal, S. Effect of γ-radiation on the structure and morphology of polyvinyl alcohol films. *Nucl. Instrum. Methods Phys. Res. Sect. B Beam Interact. Mater. Atoms* **2005**, *237*, 585–592. [CrossRef]

41. Petrova, N.V.; Evtushenko, A.M.; Chikhacheva, I.P.; Zubov, V.P.; Kubrakova, I.V. Effect of microwave irradiation on the cross-linking of polyvinyl alcohol. *Russ. J. Appl. Chem.* **2005**, *78*, 1158–1161. [CrossRef]

42. Natalia, V.A.; Evtushenko, A.M.; Chikhacheva, I.P.; Zubov, V.P.; Kubrakova, I. V Effect of microwave irradiation on the structuring of polyvinyl alcohol. *Mendeleev Commun.* **2005**, *15*, 170–172.

43. Kumar, G.N.H.; Rao, J.L.; Gopal, N.O.; Narasimhulu, K.V.; Chakradhar, R.P.S.; Rajulu, A.V. Spectroscopic investigations of Mn^{2+} ions doped polyvinylalcohol films. *Plymer* **2004**, *45*, 5407–5415. [CrossRef]

44. Pawlak, A.; Mucha, M. Thermogravimetric and FTIR studies of chitosan blends. *Thermochim. Acta* **2003**, *396*, 153–166. [CrossRef]

45. Marsano, E.; Vicini, S.; Skopińska, J.; Wisniewski, M.; Sionkowska, A. Chitosan and poly(vinyl pyrrolidone): Compatibility and miscibility of blends. *Macromol. Symp.* **2004**, *218*, 251–260. [CrossRef]

46. Oliveira, J.M.; Rodrigues, M.T.; Silva, S.S.; Malafaya, P.B.; Gomes, M.E.; Viegas, C.A.; Dias, I.R.; Azevedo, J.T.; Mano, J.F.; Reis, R.L. Novel hydroxyapatite/chitosan bilayered scaffold for osteochondral tissue-engineering applications: Scaffold design and its performance when seeded with goat bone marrow stromal cells. *Biomaterials* **2006**, *27*, 6123–6137. [CrossRef] [PubMed]

47. Li, M.; Cheng, S.; Yan, H. Preparation of crosslinked chitosan/poly(vinyl alcohol) blend beads with high mechanical strength. *Green Chem.* **2007**, *9*, 894. [CrossRef]

48. Li, Y.; Liu, T.; Zheng, J.; Xu, X. Glutaraldehyde-crosslinked chitosan/hydroxyapatite bone repair scaffold and its application as drug carrier for icariin. *J. Appl. Polym. Sci.* **2013**, *130*, 1539–1547. [CrossRef]

49. Coates, J. Interpretation of infrared spectra, a practical approach. *Encycl. Anal. Chem.* **2000**, *12*, 10815–10837.

50. Bernal-Ballén, A.; Kuritka, I.; Saha, P. Preparation and characterization of a bioartificial polymeric material: Bilayer of cellulose acetate-PVA. *Int. J. Polym. Sci.* **2016**, *2016*. [CrossRef]

51. Hussein-Al-Ali, S.H.; El Zowalaty, M.E.; Hussein, M.Z.; Geilich, B.M.; Webster, T.J. Synthesis, characterization, and antimicrobial activity of an ampicillin-conjugated magnetic nanoantibiotic for medical applications. *Int. J. Nanomed.* **2014**, *9*, 3801. [CrossRef] [PubMed]

52. Srinivasa, P.C.; Ramesh, M.N.; Kumar, K.R.; Tharanathan, R.N. Properties and sorption studies of chitosan—Polyvinyl alcohol blend films. *Carbohydr. Polym.* **2003**, *53*, 431–438. [CrossRef]

53. Kumar, H.M.P.N.; Prabhakar, M.N.; Prasad, C.V.; Rao, K.M.; Reddy, T.V.A.K.; Rao, K.C.; Subha, M.C.S. Compatibility studies of chitosan/PVA blend in 2% aqueous acetic acid solution at 30 °C. *Carbohydr. Polym.* **2010**, *82*, 251–255. [CrossRef]

54. Zheng, H.; Du, Y.; Yu, J.; Huang, R.; Zhang, L. Preparation and characterization of chitosan/poly (vinyl alcohol) blend fibers. *J. Appl. Polym. Sci.* **2001**, *80*, 2558–2565. [CrossRef]

55. Bhajantri, R.F.; Ravindrachary, V.; Harisha, A.; Crasta, V.; Nayak, S.P.; Poojary, B. Microstructural studies on BaCl$_2$ doped poly(vinyl alcohol). *Polymer* **2006**, *47*, 3591–3598. [CrossRef]

56. Liang, S.; Yang, J.; Zhang, X.; Bai, Y. The thermal-electrical properties of polyvinyl alcohol/AgNO$_3$ films. *J. Appl. Polym. Sci.* **2011**, *122*, 813–818. [CrossRef]

57. Marciniec, B.; Plotkowiak, Z.; Wachowski, L.; Kozak, M.; Popielarz-Brzezinska, M. Analytical study of β-irradiated antibiotics in the solid state. *J. Therm. Anal. Calorim.* **2002**, *68*, 423–436.

58. Mundargi, R.C.; Shelke, N.B.; Rokhade, A.P.; Patil, S.A.; Aminabhavi, T.M. Formulation and in-vitro evaluation of novel starch-based tableted microspheres for controlled release of ampicillin. *Carbohydr. Polym.* **2008**, *71*, 42–53. [CrossRef]

59. Gonzalez-Campos, J.B.; Prokhorov, E.; Luna-Barcenas, G.; Fonseca-Garcia, A.; Sanchez, I.C. Dielectric relaxations of chitosan: The effect of water on the α-relaxation and the glass transition temperature. *J. Polym. Sci. Part B Polym. Phys.* **2009**, *47*, 2259–2271. [CrossRef]

60. Alhosseini, S.N.; Moztarzadeh, F.; Mozafari, M.; Asgari, S.; Dodel, M.; Samadikuchaksaraei, A.; Kargozar, S.; Jalali, N. Synthesis and characterization of electrospun polyvinyl alcohol nanofibrous scaffolds modified by blending with chitosan for neural tissue engineering. *Int. J. Nanomed.* **2012**, *7*, 25.

61. Cui, Z.; Zheng, Z.; Lin, L.; Si, J.; Wang, Q.; Peng, X.; Chen, W. Electrospinning and crosslinking of polyvinyl alcohol/chitosan composite nanofiber for transdermal drug delivery. *Adv. Polym. Technol.* **2017**, *37*, 1917–1928. [CrossRef]

62. Bonilla, J.; Fortunati, E.; Atarés, L.; Chiralt, A.; Kenny, J.M. Physical, structural and antimicrobial properties of poly vinyl alcohol—Chitosan biodegradable films. *Food Hydrocolloids* **2014**, *35*, 463–470. [CrossRef]

63. Turalija, M.; Bischof, S.; Budimir, A.; Gaan, S. Antimicrobial PLA films from environment friendly additives. *Compos. Part. B Eng.* **2016**, *102*, 94–99. [CrossRef]

64. No, H.K.; Park, N.Y.; Lee, S.H.; Meyers, S.P. Antibacterial activity of chitosans and chitosan oligomers with different molecular weights. *Int. J. Food Microbiol.* **2002**, *74*, 65–72. [CrossRef]

65. Liu, N.; Chen, X.-G.; Park, H.-J.; Liu, C.-G.; Liu, C.-S.; Meng, X.-H.; Yu, L.-J. Effect of MW and concentration of chitosan on antibacterial activity of Escherichia coli. *Carbohydr. Polym.* **2006**, *64*, 60–65. [CrossRef]

66. Tao, Y.; Qian, L.-H.; Xie, J. Effect of chitosan on membrane permeability and cell morphology of Pseudomonas aeruginosa and Staphyloccocus aureus. *Carbohydr. Polym.* **2011**, *86*, 969–974. [CrossRef]

67. Kowalczyk, M.C.; Kowalczyk, P.; Tolstykh, O.; Hanausek, M.; Walaszek, Z.; Slaga, T.J. Synergistic effects of combined phytochemicals and skin cancer prevention in SENCAR mice. *Cancer Prev. Res.* **2010**, 1940–6207. [CrossRef] [PubMed]

68. López-Garcia, J.; Kuceková, Z.; Humpolíček, P.; Mlček, J.; Sáha, P. Polyphenolic extracts of edible flowers incorporated onto atelocollagen matrices and their effect on cell viability. *Molecules* **2013**, *18*, 13435–13445. [CrossRef] [PubMed]

69. Palmer-Young, E.C.; Sadd, B.M.; Irwin, R.E.; Adler, L.S. Synergistic effects of floral phytochemicals against a bumble bee parasite. *Ecol. Evol.* **2017**, *7*, 1836–1849. [CrossRef] [PubMed]

70. Dutta, P.K.; Tripathi, S.; Mehrotra, G.K.; Dutta, J. Perspectives for chitosan based antimicrobial films in food applications. *Food Chem.* **2009**, *114*, 1173–1182. [CrossRef]

71. Giunchedi, P.; Genta, I.; Conti, B.; Muzzarelli, R.A.A.; Conte, U. Preparation and characterization of ampicillin loaded methylpyrrolidinone chitosan and chitosan microspheres. *Biomaterials* **1998**, *19*, 157–161. [CrossRef]

72. Mitik-Dineva, N.; Wang, J.; Truong, V.K.; Stoddart, P.; Malherbe, F.; Crawford, R.J.; Ivanova, E.P. Escherichia coli, Pseudomonas aeruginosa, and Staphylococcus aureus attachment patterns on glass surfaces with nanoscale roughness. *Curr. Microbiol.* **2009**, *58*, 268–273. [CrossRef] [PubMed]

73. Lynch, A.S.; Robertson, G.T. Bacterial and fungal biofilm infections. *Annu. Rev. Med.* **2008**, *59*, 415–428. [CrossRef] [PubMed]

74. Ma, L.; Gao, C.; Mao, Z.; Zhou, J.; Shen, J.; Hu, X.; Han, C. Collagen/chitosan porous scaffolds with improved biostability for skin tissue engineering. *Biomaterials* **2003**, *24*, 4833–4841. [CrossRef]

75. López-Garcia, J.; Lehocky, M.; Humpoliček, P.; Sáha, P. HaCaT keratinocytes response on antimicrobial atelocollagen substrates: Extent of cytotoxicity, cell viability and proliferation. *J. Funct. Biomater.* **2014**, *5*, 43–57. [CrossRef] [PubMed]
76. Asadinezhad, A.; Novak, I.; Lehocky, M. Polysaccharides coatings on medical-grade PVC: A probe into surface characteristics and the extent of bacterial adhesion. *Molecules* **2010**, *15*, 1007–1027. [CrossRef] [PubMed]
77. Dik, D.A.; Fisher, J.F.; Mobashery, S. Cell-Wall Recycling of the Gram-Negative Bacteria and the Nexus to Antibiotic Resistance. *Chem. Rev.* **2018**, *118*, 5952–5984. [CrossRef] [PubMed]
78. Tenover, F.C. Mechanisms of antimicrobial resistance in bacteria. *Am. J. Infect. Control.* **2006**, *34*, S3–S10. [CrossRef] [PubMed]

Sample Availability: Samples of the compounds are not available from the authors.

Article

Limonene-Based Epoxy: Anhydride Thermoset Reaction Study

Guillaume Couture, Lérys Granado, Florent Fanget, Bernard Boutevin and Sylvain Caillol *

Institut Charles Gerhardt Montpellier-UMR 5253-CNRS, Université Montpellier,
ENSCM–240 Avenue Emile Jeanbrau, CEDEX 5, 34296 Montpellier, France; Guillaume.coutur@enscm.fr (G.C.);
Lers.granado@enscm.fr (L.G.); flrorian.fanget@enscm.fr (F.F.); Bernard.boutevin@enscm.fr (B.B.)
* Correspondence: sylvain.caillol@enscm.fr; Tel.: +330-467-144-327

Received: 27 September 2018; Accepted: 22 October 2018; Published: 23 October 2018

Abstract: The development of epoxy thermosets from renewable resources is of paramount importance in a sustainable development context. In this paper, a novel bio-based epoxy monomer derived from limonene was synthesized without epichlorohydrine and characterized. In fact, this paper depicts the synthesis of bis-limonene oxide (bis-LO). However, intern epoxy rings generally exhibit a poor reactivity and allow reaction with anhydride. Therefore, we used a reaction model with hexahydro-4-methylphthalic anhydride to compare reactivity of terminal and interepoxy functions. We also studied the influence of methyl group on intern epoxy functions. Furthermore, the influence of epoxy:anhydride stoichiometry and initiator amount was studied. These studies allow to propose an optimized formulation of bis-LO. Finally, a bis-LO-based thermoset was obtained and characterized.

Keywords: limonene; biobased polymers; anhydride; epoxy; DSC

1. Introduction

With the decrease of easily available fossil resources, ensuing price volatility, and the increasing awareness regarding environmental and human health protection, a booming amount of research has been carried out to find alternative to fossil-based chemicals. This statement is especially true for thermosetting polymers (or thermosets), which represent about 20% of the global plastic production with a large number of applications. In fact, thermosets, as crosslinked nonfusible three-dimensional networks, cannot be recycled like most of thermoplastic materials [1]. Thus, the synthesis of thermosets from bio-based resources appears to be of upmost importance.

Among the various thermosets designed at an industrial scale, polyepoxides account for 70% of the market thanks to their high mechanical properties, chemical and moisture resistance or easy processability [2]. Unfortunately, around 90% of them are based on diglycidylether of bisphenol A (DGEBA), an oil-based molecule obtained from bisphenol A (BPA). BPA has been classified as carcinogen mutagen and reprotoxic (CMR). Moreover this is an endocrine disruptor which is submitted to a more and more restrictive regulation in numerous countries [3]. An increasing number of academic and industrial researches have therefore been dedicated to find bio-based, nontoxic alternatives to BPA for the synthesis of original epoxy networks [1,3,4] such as: vanillin and its derivatives [5–8], eugenol [9], ferulic acid [10], isosorbide [11,12], catechin [13], cardanol [14,15], and vegetable oils [16,17]. Similarly, most epoxy networks are synthesized from epichlorohydrin, which is a CMR and a highly toxic molecule. Alternatives have thus been considered, especially via the epoxidation of preexisting carbon–carbon double-bonds. Among potential bio-based molecules, limonene (Scheme 1) appears as an interesting candidate. It is a cyclic monoterpene derivative found in many citrus fruits [18]. The latter are the most abundant tree crops with a global production of 88 million tons per year and include oranges and lemons among others. Around 50% of theses fruits are processed into juice or marmalade, yielding an estimated 40 million tons of citrus waste world-wide [19]. Depending on the fruit variety, season and geographic origin, a variable amount of limonene can be extracted, generally accounting for 4% *w/w* of the citrus waste, for an estimated production of 77 thousand of tons per year [20].

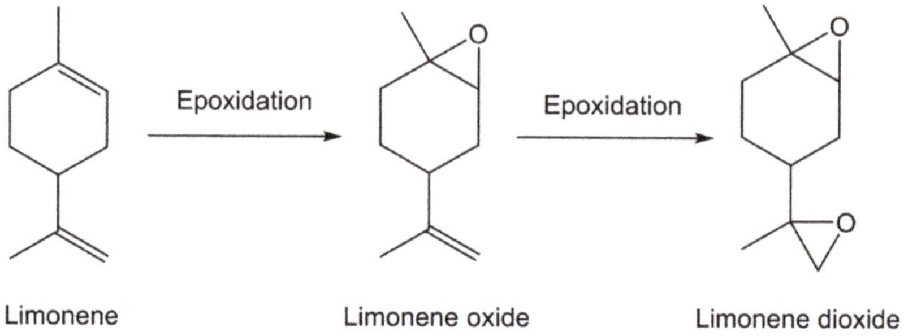

Scheme 1. Synthesis of limonene oxide and limonene dioxide from limonene.

The carbon–carbon double bond located on the aliphatic ring of limonene can be epoxidized to yield limonene oxide (Scheme 1) by various methods, generally involving organic peroxides, peroxy acids or recently hydrogen peroxide in the presence of enzymes [21]. The main advantage of this route lies in the absence of epichlorohydrin. Further epoxidation can also be carried out to obtain limonene dioxide. However, the difference in reactivity of its two epoxy groups and the limited tunability of limonene dioxide-based networks decrease its suitability for thermosets synthesis. More surprisingly, to the best of our knowledge, only one recent paper described the use of limonene oxide for this application. In fact, Morinaga and Sakamoto [22] synthesized multifunctional limonene-based epoxides using di-, tri- or tetra-thiols and crosslinked them with branched polyethyleneimine. They observed that a large excess of amine groups was required to obtain satisfactory crosslinking at 100 °C.

Very few other epoxy networks were synthesized with cycloaliphatic epoxides [23–25]. Recently, Wang and Schuman [25] crosslinked blends of DGEBA and vegetable-oil-derived epoxy monomers with hexahydro-4-methylphthalic anhydride (HMPA), creating the first partially bio-based epoxy thermoset obtained using cyclic anhydrides. Usually, epoxy monomers are crosslinked by toxic, oil-based polyamines, but according to Yang et al. [26], anhydrides provide epoxy thermosets with lower toxicity and higher glass transition temperature. Furthermore, Buchmeiser et al. proposed innovative approach by using protected *N*-heterocyclic carbenes as latent organocatalysts for the low-temperature curing of epoxy:anhydride thermosets [27,28]. Moreover, it is possible to obtain a one-component mixture with an increased pot-life with epoxy/anhydride systems [29] Anhydrides are potentially bio-based [30,31] and can provide the thermosets with tunable properties.

The mechanism of curing of epoxy with anhydrides differs from amines since it exhibits initiation, propagation and termination steps similar to a ring-opening polymerization. Conflicting initiation mechanisms have been reported, but most of them agree on the formation of a zwitterion [29,32–34]. For Park et al. [32], it occurs via the ring-opening of the epoxide by the catalyst (Scheme 2), while the nucleophilic attack can similarly occur on the anhydride according to Amirova et al. [29] (Scheme 3). Antoon and Koenig [35] describe an amine-catalyzed isomerization of the epoxy groups into unsaturated alcohols followed by the formation of a ternary complex reacting on another epoxy (Scheme S1 in Supporting Information). The formed anionic active centers then go through the propagation steps: alkoxide anions react with anhydrides to yield a mono-ester carboxylate that reacts with another epoxy ring. However, Park et al. mention that alkoxide anions can simultaneously react with other epoxides to yield ether instead of ester bonds (Scheme 2). Finally, termination step occurs by deactivation of the active center.

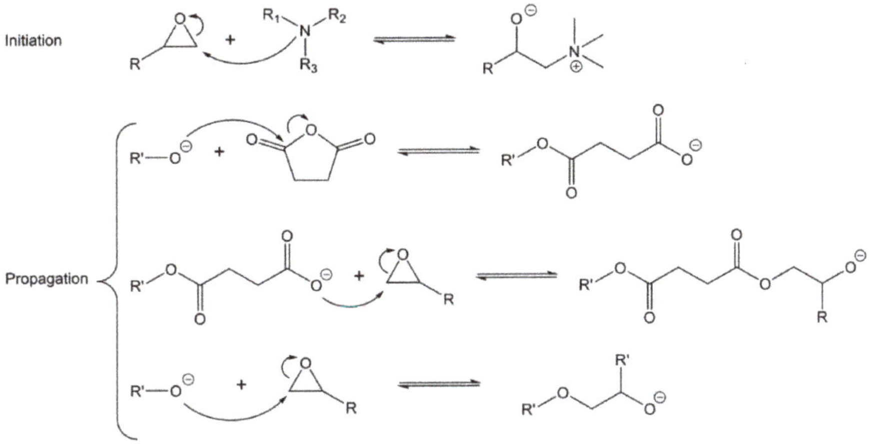

Scheme 2. Curing mechanism of succinic anhydride and epoxide initiated by a tertiary amine.

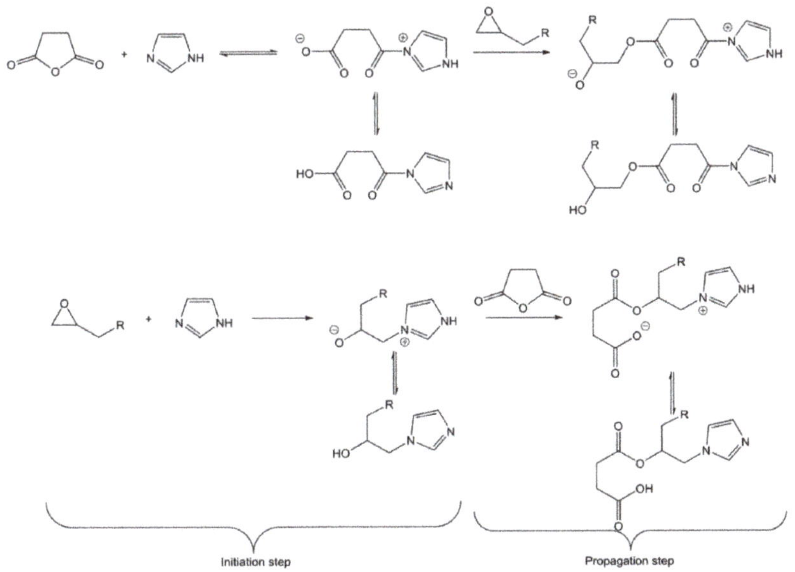

Scheme 3. Curing mechanism of succinic anhydride and epoxides initiated by imidazole.

Despite these complex and interesting mechanical studies, few information was given regarding the influence of stoichiometry and catalyst amount on thermosets, a trend also observed when DGEBA was cured with anhydrides [27,30]. Park et al. [30] mentioned a 1:1 epoxy:anhydride formulation ratio with epoxy in excess and observed slightly higher activation energy for the epoxy-rich formulations as well as a visible etherification peak on DSC curves. Matejka et al. [36] and Steinmann [37] studied the influence of various experimental parameters on epoxy-anhydride linear polymers. Steinmann [38] carried out the curing of DGEBA with phthalic anhydride and studied the influence of catalyst concentration on the T_g of the thermosets. She stated that the highest glass transition temperatures were obtained at different initiator concentrations depending on its nature, ranging between 0.05 and 0.1 mol%. Recently, Paramarta and Webster [39] studied the influence of stoichiometry and catalyst amount on the kinetics parameters of the curing reaction but did not provide thermal or mechanical properties. Finally, Kuncho et al. [40] noticed a maximum hardness for epoxidized linseed oil/HMPA/1,8-diazabicyclo[5.4.0]-undec-7-ene materials with 1:1 epoxy:anhydride ratio. They also observed an initial increase in hardness followed by a plateau when increasing the initiator amount. Additionally, Schultz et al. demonstrated by Matrix Assisted Laser Desorption Ionisation-Time of Flight (MALDI TOF) the perfect alternating structure of epoxy:anhydride thermosets [41].

Thus, the first aim of this paper is to further enhance the knowledge on cycloaliphatic epoxy/anhydride curing systems and the possibility to use bio-based cyclo-aliphatic epoxy monomers as precursors for thermosets materials. Indeed, cycloaliphatic epoxides could be directly obtained—without epichlorohydrin—from some terpenes, an important class of renewable resources. Hence, in a first part, the synthesis of bis-limonene oxide (bis-LO) is presented. Then, we compare the reactivity of epoxies DGEBA, bis-cyclohexene oxide (bis-CHO) and bis-LO with the help of model epoxies molecule phenyl glycidyl ether (PMO), cyclohexene oxide (CHO) and 3,4-epoxycyclohexylmethyl (MCHO), respectively (Figure 1). Noncrosslinking model molecules are chosen rather than thermosets to get free from the effect of formulations and to avoid unwanted variation of the kinetics due to the gelation and vitrification (occurring during crosslinked network formation). Hence, the influence of the various functions close to the epoxy moiety during the curing behavior can be clarified. The glycidyl epoxy position is compared to cycloaliphatic epoxy (PMO and CHO), the influence of methyl group on cycloaliphatic epoxy is evaluated (CHO and MCHO). The second part focuses on the influence of stoichiometry and initiator amount of thermo-mechanical behavior of model thermosets based on DGEBA and bis-CHO. This analysis is used to propose an optimized formulation of bis-LO with HMPA. Finally, the thermal resistance of the bis-LO is compared to model formulations by TGA.

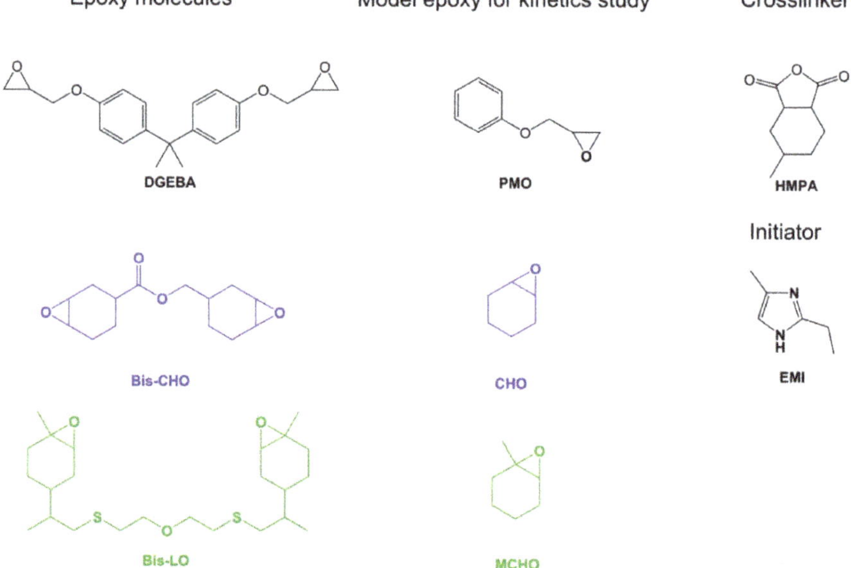

Figure 1. Chemical structure and abbreviations of the model molecules, diepoxides, curing agent and initiator catalyst.

2. Experimental Part

2.1. Materials

Cyclohexene oxide (98%), (+)-Limonene oxide, mixture of *cis* and *trans* (97%), 3,4-epoxycyclohexylmethyl 3,4-epoxycyclohexanecarboxylate, 2-ethyl-4-methylimidazole (95%), hexahydro-4-methylphthalic anhydride, mixture of *cis* and *trans* (96%), phenyl glycidyl ether (96%), diglycidyl ether of bisphenol A (DGEBA), and ethyl acetate were supplied by Sigma-Aldrich (Saint-Louis, MO, USA). Bis(2-mercaptoethyl)ether (95%) and 2,4,6-trimethylphenol (98%) were supplied by ABCR (Karlsruhe, Germany). Deuterated DMSO (DMSO-d6) (99.8%D) was supplied by Euriso-Top (Saint-Aubin, France). Azobisisobutyronitrile was supplied by Fluka and purified by recrystallization in methanol. All other reagents were used as received.

2.2. Synthesis of Bis-LO

In a 250 mL round-bottom flask with a magnetic stirrer, limonene oxide (20.00 g, 0.131 mol), azobisisobutyronitrile (2.00 g, 1.22×10^{-2} mol i.e., 10 wt.% compared to limonene), 40 mL of ethyl acetate and bis(2-mercaptoethyl) ether (8.90 g, 6.44×10^{-2} mol i.e., 0.49 molar equivalents of limonene) are inserted and heated at 60 °C for 16 hours. After cooling at room temperature, the crude product is poured into a separating funnel and 40 mL of a 2 M NaOH aqueous solution are added. The aqueous phase is washed with 40 mL of ethyl acetate and the total organic phase is washed with 40 mL of deionized water and 40 mL of brine. After drying over MgSO$_4$, the solvent was removed in a rotary evaporator and the product was dried overnight, in an oven at 40 °C under reduced pressure (1×10^{-1} mbar). Product appears as an uncolored, transparent, slightly viscous liquid.

2.3. Thermoset Curing

The epoxy monomer, HMPA and 2-ethyl-4-methylimidazole (EMI) were weighed, poured into an aluminum pan and stirred manually for 5 min prior to heating in an oven at the desired temperature. Bis-CHO-based thermosets were obtained after 4 h at 180 °C followed by 1 h at 190 °C, while their DGEBA-based counterparts were crosslinked at 120 °C for 4 h and 1 h at 140 °C.

2.4. Characterization Techniques

The Nuclear Magnetic Resonance (NMR) spectra were recorded on a Bruker AC 400 instrument (Billerica, MA, USA), using deuterated chloroform or dimethylsulfoxide-d_6 as solvents.

Attenuated total reflectance Fourier transform infrared absorption spectroscopy (ATR-FTIR) experiments were carried out on a NICOLET 6700 spectrometer from Thermo-Scientific (Waltham, MA, USA), equipped with a mercury-cadmium-tellurium (MCT) detector, in the middle infrared range with a resolution of 4 cm^{-1} and 32 scans were added to each spectrum.

Differential scanning calorimetry (DSC) experiments were carried out on a DSC200 F3 Maia from Netzsch (Selb, Germany), equipped with an intra-cooler system and under nitrogen environment. The DSC was calibrated with biphenyl, indium, bismuth and CsCl highly pure standards, at 10 °C/min. Samples were weighed into stainless-steel crucibles sealed with a gold seal for kinetics studies. DSC was also used to determine the glass transition temperatures of the cured thermosets (Al pan and pierced lids).

Thermo-gravimetric analysis (TGA) experiments were recorded on a TGA Q50 from TA instruments using a platinum crucible. The samples were heated up from 20 to 800 °C at 20 °C/min, under nitrogen atmosphere.

3. Results and Discussion

3.1. Synthesis of Bis-Limonene Oxide (Bis-LO)

Eugenol is a bio-based phenolic derivative exhibiting a carbon–carbon double-bond. Through functionalization with epichlorohydrin, it yields glycidylated eugenol that has been reacted by Zou et al. [42] with various di-thiols via a thiol-ene coupling to synthesize shape memory epoxy networks. Considering the similarities between glycidylated eugenol and limonene oxide, bis-limonene oxide derivatives have been synthesized following a similar pathway using azobisisobutyronitrile (AIBN) as a source of radicals (Scheme 4). Limonene oxide was used in slight excess to ensure a maximum yield of bis-epoxides. After purification steps, the obtained products were characterized by FTIR and NMR spectroscopies.

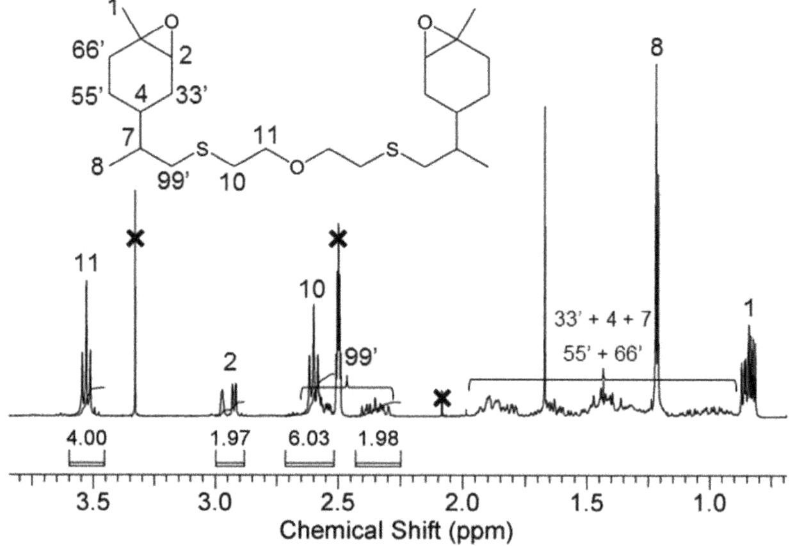

Scheme 4. Synthesis of bis-limonene oxide using thiol-ene coupling.

The analysis of the bis-LO by FTIR spectrum shows the disappearance of the C=C double bond signals at 884 cm^{-1} (C=C out-of-plane bending), 1645 cm^{-1} (C=C stretching) and 3072 cm^{-1} (=C-H stretching) when compared to limonene oxide (see Figure S1 in Supporting Information). All other signals of bis-LO correspond to a combination of the characteristic signals of both limonene oxide and bis(2-mercaptoethyl) ether such as C-H bonds stretching between 2780 and 3010 cm^{-1} and the C-O bond stretching of the ether groups at 1100 cm^{-1}. These encouraging results were confirmed by ^{1}H-NMR spectroscopy (Figure 2). Similarly to FTIR spectroscopic data, the signals of the limonene oxide C=C double bond protons at 4.6–4.7 ppm disappeared after the reaction, indicating a high conversion of limonene oxide towards the thiol-ene coupling. The signals shifted at 2.3–2.4 ppm and 2.5–2.6 ppm (**9** and **9′** on Figure 2) in the form of multiplets because of the asymmetric carbons C$_E$ and C$_H$ (see Figure S2 in Supporting Information). The latter also overlaps with the triplet of the -CH$_2$-S protons (**10**) from grafted bis(2-mercaptoethyl) ether at 2.6 ppm. A similar triplet can be observed at 3.5 ppm that corresponds to the -CH$_2$-O protons (**11**). The methanetriyl proton (**7**) that could also confirm the grafting unfortunately overlaps with all the other aliphatic signals between 0.8 and 2.0 ppm. Integrations highlight the efficiency of the synthesis and the absence of possible side-reactions such as epoxy ring opening from thiols. All these signal assignments were confirmed by ^{13}C, HSQC ^{1}H-^{13}C- and HMBC ^{1}H-^{13}C-NMR spectroscopies (Figures S2–S4, respectively).

Figure 2. ^{1}H-NMR spectrum of bis-LO recorded in deuterated DMSO.

For the purpose of limiting the number of costly purification steps regarding the potential applications of bis-limonene oxide derivatives, no chromatographic separation was carried out on bis-LO. Thus, the mixture might contain some traces of unreacted limonene oxide (although not observed by NMR spectroscopy) and isobutyronitrile from AIBN recombination. The epoxy content of bis-LO was therefore determined by NMR dosing using 2,4,6-trimethylphenol as an external standard. The signal of the methanetriyl proton (**2**) at 2.8–3.0 ppm was used as the reference in Equation (1).

$$Epoxy\ Equivalent\ (mol/g) = \frac{m}{\frac{m_S}{M_S} * \frac{2*\int_{2.8}^{3.0} CH_{epoxy}}{\int_{6.7}^{6.9} CH_S}} \tag{1}$$

With m and m_S the mass of the resin and the standard (in g), respectively, M_S the molecular weight of the standard (in g mol^{-1}). A value of 4.40 ± 0.08 mmol·g^{-1} (equivalent weight of epoxy mixture of EEW = 227 g/eq) were obtained for bis-LO and used for further materials formulations.

3.2. Kinetics Study on Model Molecules

The reaction between model epoxies (PMO, CHO and MCHO) and anhydride (2/1 n/n) in presence of initiator (1 wt%) is monitored by DSC, with nonisothermal programs, at constant heating rates ($\beta = 3, 5, 8, 10$ °C/min). Typical thermograms at 5 °C/min are presented in Figure 3, for the three studied systems. In all cases, exothermic signals are recorded, as expected for epoxy ring opening reactions. For PMO two peaks are observed (near 135 and 145 °C), whereas one single peak is present for CHO and MCHO. The first PMO peak is assigned to the reaction between epoxy and anhydride. The second peak corresponds to the reaction between two epoxy rings (i.e., homopolymerization). Notably, the reaction between HMPA and PMO lead to a fine and intense exothermic peak, whereas with CHO and MCHO the peak is much broader and less intense.

The reactivity of these model molecules can be readily appreciated by monitoring the onset of the exothermic peak (Table 1). For example, at 5 °C/min (Figure 3), the peak onsets are 126, 149 and 174 °C for PMO, CHO and MCHO, respectively. These significant differences lead to conclude in a trend of reactivity of the studied epoxy models with HMPA. Hence, glycidyl epoxy (PMO) is more reactive than cycloaliphatic epoxy (CHO) and even more thermal energy is required to enable the reaction ring opening of cycloaliphatic methyl-epoxy (MCHO).

In addition, it is interesting to compare the reaction enthalpy values (Table 1). CHO presents higher enthalpy per mass unit values (318.5 ± 6.8 J/g) compared to PMO (278.6 ± 12.9 J/g) and MCHO (182.0 ± 24.3 J/g). However, when the reaction enthalpy is calculated regarding the amount of epoxy, a different trend is observed. CHO and PMO present sensibly similar values (58 and 62 kJ/mol, respectively). These values are in accordance with literature data [43]. In contrast, the MCHO value of 36 kJ/mol is two times lower, suggesting again a lower reactivity of the methylated-epoxy groups.

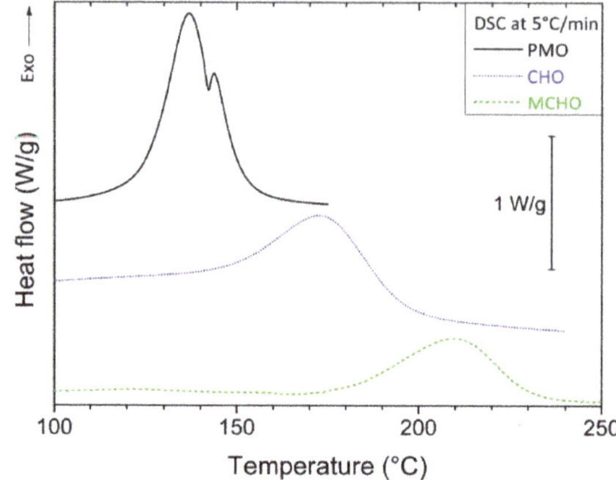

Figure 3. DSC thermograms of PMO, CHO and MCHO at 5 °C/min.

Table 1. Enthalpies of reaction and exothermic peak onsets as a function of the heating rates, for PMO, CHO and MCHO.

β (°C/min)	ΔH_{TOTAL} (J/g)			Peak Onset (°C)		
	PMO	**CHO**	**MCHO**	**PMO**	**CHO**	**MCHO**
3	287.9	312.8	214.4	118	137	170
5	284.9	320.8	170.0	126	149	174
8	281.9	327.1	185.3	136	160	193
10	259.5	313.1	158.1	137	166	201
Averages	278.6	318.5	182.0			
Deviations	12.9	6.8	24.3			

The conversion of the reaction (x) between model epoxies and anhydride is calculated by integrating DSC thermograms (with linear baseline approximation), with the following equation:

$$x(t, T) = \frac{\Delta H_{t,T}}{\Delta H_{TOTAL}} \tag{2}$$

where $\Delta H_{t,T} = \int_0^t \dot{q}(t, T)dt$ is the cumulative released heat of reaction at the time t and curing temperature T, as time integral of the instantaneous heat flow ($\dot{q}(t, T)$), and ΔH_{TOTAL} is the total enthalpy of the reaction (total area).

The nonisothermal kinetic profiles of PMO, CHO and MCHO are presented in Figure 4. The conversions (Equation (2)) are plotted against the temperatures, for different linear heating rates. The curves exhibit an expected sigmoidal shape. Notably, CHO and MCHO profiles are spread on a wide range of temperature (about 100–210 °C and 150–260 °C, respectively), whereas PMO profiles are restrained at lower temperature between 110 and 170 °C. The kinetic profiles follow the previous observation, PMO reaction occurs at lower temperature than CHO, and MCHO reaction proceeds at the highest temperature range.

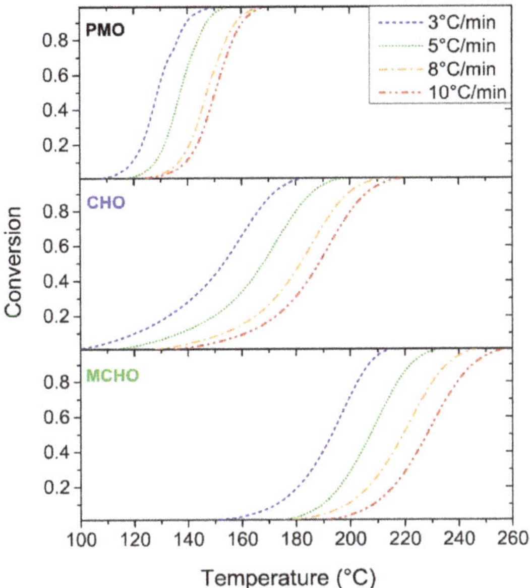

Figure 4. Nonisothermal kinetic profiles of PMO, CHO, and MCHO at 3, 5, 8 and 10 °C/min.

In order to evaluate the activation energy of the reaction between HMPA and PMO, CHO and MCHO, isoconversional analysis is performed on nonisothermal kinetic profile datasets (integral method). Isoconversional analysis is a model-free kinetics approach, which relies on the variable separation between conversion, x, and temperature, T, leading to the equation [44]:

$$\frac{dx}{dt} = k(T) \cdot f(x) \tag{3}$$

where dx/dt is the reaction rate, k is the rate constant and f is a function of the conversion (representing the reaction mechanisms). The rate constant is assumed to be described by the Arrhenius equation:

$$k(T) = A \cdot e^{-\frac{E}{RT}} \tag{4}$$

where A is the pre-exponential factor, E the activation energy, R the gas constant. Because x and T are considered independent variables (i.e., the temperature schedule does not influence the reaction mechanisms), the activation energy can be determined as follows:

$$\left[\frac{\partial \ln(dx/dt)}{\partial T^{-1}}\right]_x = \left[\frac{\partial \ln(k(T))}{\partial T^{-1}}\right]_x + \left[\frac{\partial \ln(f(x))}{\partial T^{-1}}\right]_x = -\frac{E_x}{R} \tag{5}$$

It is to be noted that one activation energy, E_x, is associated to one conversion, x. Hence, it is common to find variable activation energy with x, in isoconversional analysis.

In this study, isoconversional analysis is implemented using Vyazovkin integral method (VA). The detail of VA is very well described elsewhere [45]. In summary, VA relies on the following function minimization:

$$\min\Phi(E_x) = \sum_{i=1}^{n}\sum_{j\neq i} \frac{I_i(E_x, T_{x,i})}{I_j(E_x, T_{x,j})} \tag{6}$$

where indexes i and j represents two heating rates curves (Figure 3) and I are the temperature integrals:

$$I_i(E_x,\ T_{x,i}) = \beta_i^{-1} \int_{T_{x-\Delta x}}^{T_x} e^{-\frac{E_x}{RT_{x,i}}} \cdot dT \tag{7}$$

where β_i is the ith heating rate. In fine, E_x is the result of the function minimization associated to one conversion, x. The conversions are considered from $x = 0.01$ to $x = 0.99$, with $\Delta x = 0.01$, for the present implementation. It is important to stress that the activation energy values found with isoconversional analysis are apparent values. They may therefore arise from different contributing phenomena (e.g., competing reactions, diffusional restriction barriers).

The results of the VA method are presented in Figure 5. The activation energy is considered as a function of conversion. For the cases of PMO and MCHO reactions, the activation energy is constant for $x < 0.4$ ($E = 68$ kJ/mol). On the other hand, the activation energy increases slightly in a first regime for CHO. This increase is assigned to uncertainties linked to integration in early conversion data points. Then the trend of PMO curve differs from CHO and MCHO. CHO and MCHO curves exhibit a decreasing trend for $x > 0.4$. In contrast, the sharp increase of PMO curve is assigned to inaccuracies of calculations. Isoconversional analysis is unable to discriminate the reaction between HMPA and PMO, and PMO on itself, leading to an over-evaluation of the values of activation energy. Nonetheless, sufficiently far from $x \sim 0.7$ (conversion at which PMO starts to react on itself), the activation energies of HMPA/PMO (for $x < 0.4$) and homopolymerization PMO/PMO ($x > 0.8$) reactions can be respectively evaluated.

In overall, all curves are interestingly similar. These results allow concluding that the activation energies of glycidyl, methylated- and neat-cycloaliphatic epoxies are sensibly similar, near 64 ± 12 kJ/mol. This value is slightly above previous literature data [46,47]. Furthermore, the trends of CHO and MCHO curves offers interesting insights for the reaction between these cycloaliphatic epoxies and anhydride. The decreasing trend of activation energy suggests that the reaction is autocatalytic. The reaction products may self-catalyze the reaction, such as OH moieties as products of anhydride opening (as similarly described for epoxy-amine systems [48]).

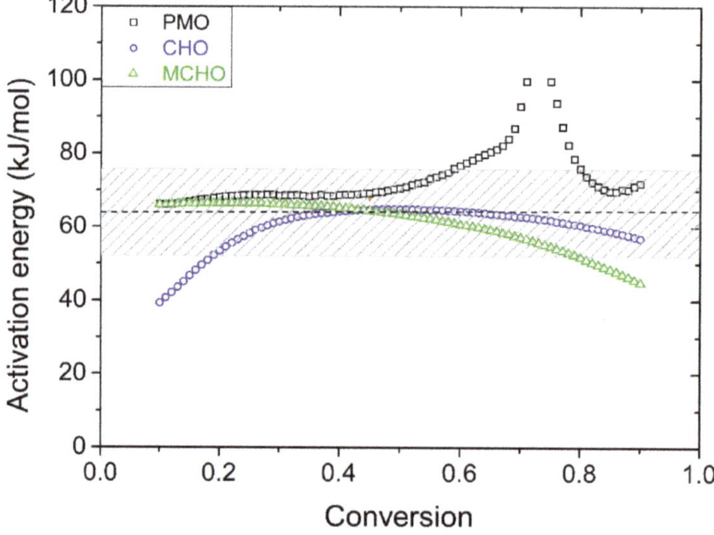

Figure 5. Activation energy as a function of the conversion between HMPA and PMO, CHO and MCHO, as calculated with VA method.

3.3. Influence of Stoichiometry and Initiator Amount on T_g

For epoxy/amine systems, the stoichiometry is known to display a crucial contribution in the three-dimensional network building and thus on the final glass transition temperature (T_g) of thermosets. Therefore the stoichiometry needs to be optimized to obtain a chemically and mechanically resistant material, with the highest glass transition temperature [5]. Nevertheless, to the best of our knowledge, no such rule has been established for epoxy/anhydride reaction. As follows, we propose to clarify the impact of the epoxy/anhydride ratio and the amount of the catalyst on the glass transition temperature, for two thermoset models bis-CHO/HMPA and DGEBA/HMPA, in presence of EMI as catalyst. All experimental data are gathered in Tables S1 and S2, and DSC thermograms are displayed on Figures S5 and S6 in Supporting Information.

The influence of the stoichiometry on T_g is reported in Figure 6 (top). The glass transition temperature of the materials was determined by DSC on cured formulations (Figure S5). T_g of DGEBA formulations are found near 158 °C, whereas glass transition temperatures are lower for bis-CHO formulations ($T_g \sim 120$ °C). The higher T_g of DGEBA formulations is explained by the presence of aromatic rings that confer higher rigidity to the three-dimensional polymer network and limit mobility of network. Additionally, the sulfur hinges confer added mobility to aliphatic chains, which contributes also to lower T_g value. Hence, the network needs less thermal energy to relax with cycloaliphatic epoxy, compare to aromatic-based ones. Interestingly, the glass transition temperatures are found to be constant over a rather wide range of anhydride/epoxy ratios, from 0.7 to 1.4, for both aromatic (DGEBA) and cycloaliphatic (bis-CHO) epoxies. Therefore, these results suggest that the stoichiometry does not strongly impact the glass transition temperature of final materials in epoxy/anhydride systems (unlike epoxy/amine).

In addition, the influence on the initiator amount is shown in Figure 6 (bottom). As previously described, the T_g of DGEBA formulations are globally higher than bis-CHO ones (ca. $+ 35$ °C). For initiator amounts higher than 2 wt.%, the T_g of both formulations linearly decreases from 158 to 142 °C for DGEBA and from 122 to 105 °C for bis-CHO. Interestingly, the slopes of the curves are rather similar. These decreases in glass transition is assigned to plasticizing effect of the EMI into the polymer network. Furthermore, the T_g of bis-CHO formulation decreases, for [EMI] < 2 wt.%. Hence, all reactive moieties are not activated at low amount of initiator, leading to incomplete crosslinking. This cannot be studied on DGEBA. The dangling chains, containing the unreacted epoxies and anhydride groups, act as plasticizers and contribute to lower the glass transition temperature.

In overall, it has been shown that stoichiometry has not a strong effect on the glass transition temperature of the final crosslinked materials. In opposition, the amount of initiator displays a greater role in the thermo-mechanical behavior. An optimum amount of EMI as initiator is determined from Figure 6, at ca. 2–2.5 wt.%. These findings would be of a precious help for further development of other epoxy/HMPA systems.

Consequently, a 1:1 ratio for epoxy:anhydride and an initiator amount of 2.5 wt.% were chosen for the formulation of bis-limonene oxide networks with HMPA (bis-LO). After curing (4 h at 180 °C) and post-curing (1 h at 190 °C), the material was solid and presented a moderate glass transition temperature ($T_g = 75$ °C).

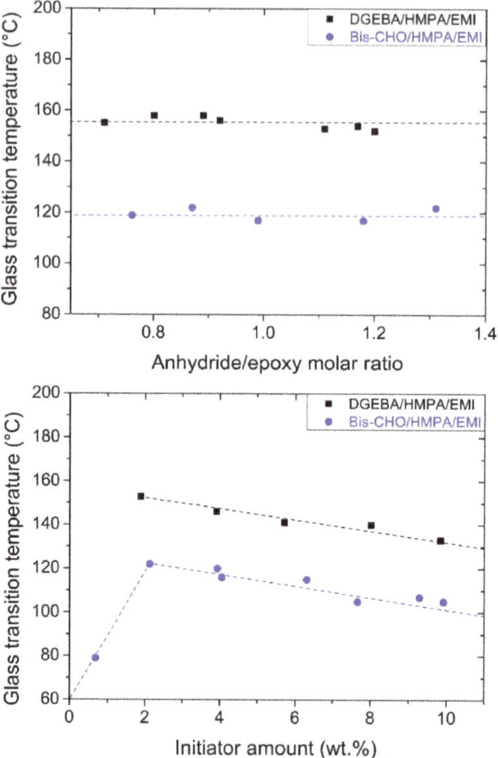

Figure 6. Influence of (**top**) thermoset stoichiometry and (**bottom**) initiator amount on glass transition temperature, for Bis-CHO/HMPA/EMI and DGEBA/HMPA/EMI thermosets.

3.4. Thermal Degradation Behavior

Thermal resistance of the bis-LO formulation was evaluated by TGA, in comparison to DGEBA and bis-CHO model formulations (Figure 7). As expected, DGEBA formulation exhibits higher thermal resistance than bis-CHO and bis-LO (temperature at 10% of degradation ($T_{d10\%}$) of DGEBA is 379 °C against 292 and 261 for bis-CHO and bis-LO, respectively). This thermal behavior of DGEBA is mainly due to the higher aromatic density. Bis-CHO exhibits rather better performances than bis-LO, probably because of the thioether-ether aliphatic chains between the epoxies in bis-LO. In fact, the relative small units in bis-CHO may results in denser crosslinked network, and thus enhancing the thermal resistance. In overall, the thermal behavior of bis-LO is relatively satisfying, as a green alternative of petroleum-based epoxy resins.

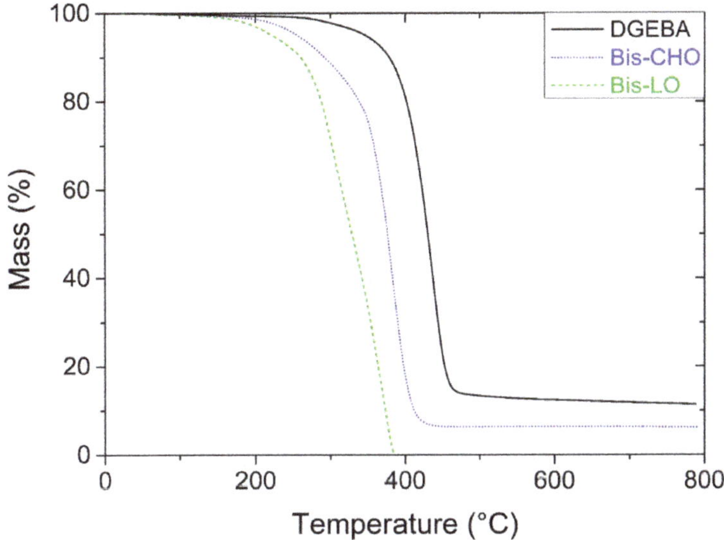

Figure 7. TGA thermograms of DGEBA, bis-CHO and bis-LO based networks with HMPA.

4. Conclusions

In this paper, we studied parameters that could influence reaction between anhydride and cycloaliphatic epoxide monomers. This reaction was much debated in literature and we showed the influence of reactants stoichiometry and initiator quantity on T_g value of network, which is an important parameter. Cycloaliphatic epoxides could be very interesting monomers directly obtained—without epichlorohydrin—from some terpenes, an important class of renewable resources. Hence, we extended our study to a novel approach to synthesize a limonene-based epoxy monomer, without the use of epichlorohydrin. We performed synthesis at the lab-scale with 20 g of limonene, but the epoxidation route could easily be upscaled, since this is the route which is currently performed at Industrial scale for the epoxidation of plant-oils. Formulation was performed with hexahydro-4-methylplphthalic anhydride to yield thermosets with high glass transition temperatures. This study allowed to compare reactivity of terminal and intern epoxy rings, as well as influence of methyl group on intern epoxy ring. This paper reports the influence of epoxy:anhydride stoichiometry and initiator amount that were previously much debated in literature. If the epoxy:anhydride stoichiometry has a low influence on a large range of ratios, the initiator amount has an impact on thermomechanical properties. The properties of the obtained materials were compared to the petroleum-based epoxy (DGEBA) cured with the same anhydride. This new bio-based thermoset exhibits good thermomechanical properties which allow to present limonene as a promising molecule for further studies for BPA replacement in some applications.

Supplementary Materials: The following are available online. Scheme S1: Curing mechanism of succinic anhydride and epoxides catalyzed by a tertiary amine as proposed by Antoon and Koenig. Figure S1: FTIR-ATR spectra of limonene oxide (in red) and Bis-LO (in blue). Figure S2: ^{13}C-NMR spectrum of Bis-LO recorded in DMSO *d*6. Figure S3: ^{1}H-^{13}C HSQC NMR spectrum of Bis-LO recorded in DMSO *d*6. Figure S4: ^{1}H-^{13}C HMBC NMR spectrum of Bis-LO recorded in DMSO *d*6. Table S1: Properties of the bis-CHO/HMPA/EMI-based and DGEBA/HMPA/EMI-based materials when changing the anhydride/epoxy molar ratio. Table S2: Properties of the bis-CHO/HMPA/EMI-based and DGEBA/HMPA/EMI-based materials when changing the initiator weight percentage. Figure S5: DSC thermograms of: i/ bis-CHO/HMPA/EMI-based thermosets on the left and ii/DGEBA/HMPA/EMI-based thermosets on the right, using varying stoichiometry. Figure S6: DSC thermograms of: i/ bis-CHO/HMPA/EMI-based thermosets on the left and ii/DGEBA/HMPA/EMI-based thermosets on the right, using varying amount of initiator.

Author Contributions: Conceptualization, B.B. and S.C.; Methodology, G.C., L.G. and F.F.; Writing-Original Draft Preparation, G.C., L.G. and S.C.

Funding: This research received no external funding.

Conflicts of Interest: The authors declare no conflict of interest.

References

1. Auvergne, R.; Caillol, S.; David, G.; Boutevin, B.; Pascault, J.-P. Biobased Thermosetting Epoxy: Present and Future. *Chem. Rev.* **2014**, *114*, 1082–1115. [CrossRef] [PubMed]
2. Pascault, J.-P.; Williams, R.J.J. *Epoxy Polymers: New Materials and Innovations*; Wiley-VCH Verlag GmbH Co. KGaA: Weinheim, Germany, 2010.
3. Ng, F.; Couture, G.; Philippe, C.; Boutevin, B.; Caillol, S. Bio-based aromatic epoxy monomers for thermoset materials. *Molecules* **2017**, *22*. [CrossRef] [PubMed]
4. Baroncini, E.A.; Kumar Yadav, S.; Palmese, G.R.; Stanzione, J.F. III. Recent advances in bio-based epoxy resins and bio-based epoxy curing agents. *J. Appl. Polym. Sci.* **2016**, *133*. [CrossRef]
5. Fache, M.; Auvergne, R.; Boutevin, B.; Caillol, S. New vanillin-derived diepoxy monomers for the synthesis of biobased thermosets. *Eur. Polym. J.* **2015**, *67*, 527–538. [CrossRef]
6. Fache, M.; Viola, A.; Auvergne, R.; Boutevin, B.; Caillol, S. Biobased epoxy thermosets from vanillin-derived oligomers. *Eur. Polym. J.* **2015**, *68*, 526–535. [CrossRef]
7. Fache, M.; Boutevin, B.; Caillol, S. Vanillin, a key-intermediate of biobased polymers. *Eur. Polym. J.* **2015**, *68*, 488–502. [CrossRef]
8. Fache, M.; Darroman, E.; Besse, V.; Auvergne, R.; Caillol, S.; Boutevina, B. Vanillin, a promising biobased building-block for monomer synthesis. *Green Chem.* **2014**, *16*, 1987–1998. [CrossRef]
9. Qin, J.; Liu, H.; Zhang, P.; Wolcott, M.; Zhang, J. Use of eugenol and rosin as feedstocks for biobased epoxy resins and study of curing and performance properties. *Polym. Int.* **2014**, *63*, 760–765. [CrossRef]
10. Maiorana, A.; Reano, A.F.; Centore, R.; Grimaldi, M.; Balaguer, P.; Allais, F.; Gross, R.A. Structure property relationships of biobased n-alkyl bisferulate epoxy resins. *Green Chem.* **2016**, *18*, 4961–4973. [CrossRef]
11. Chrysanthos, M.; Galy, J.; Pascault, J.-P. Influence of the Bio-Based Epoxy Prepolymer Structure on Network Properties. *Macromol. Mater. Eng.* **2013**, *298*, 1209–1219.
12. Chrysanthos, M.; Galy, J.; Pascault, J.-P. Preparation and properties of bio-based epoxy networks derived from isosorbide diglycidyl ether. *Polymer* **2011**, *52*, 3611–3620. [CrossRef]
13. Nouailhas, H.; Aouf, C.; Guerneve, C.L.; Caillol, S.; Boutevin, B.; Fulcrand, H. Synthesis and properties of biobased epoxy resins. part 1: Glycidylation of flavonoids by epichlorohydrin. *J. Polym. Sci. Part A Polym. Chem.* **2011**, *49*, 2261–2270. [CrossRef]
14. Jaillet, F.; Darroman, E.; Ratsimihety, A.; Auvergne, R.; Boutevin, B.; Caillol, S. New biobased epoxy materials from cardanol. *Eur. J. Lipid Sci. Technol.* **2014**, *116*, 63–73. [CrossRef]
15. Voirin, C.; Caillol, S.; Sadavarte, N.V.; Tawade, B.V.; Boutevin, B.; Wadgaonkar, P.P. Functionalization of cardanol: Towards biobased polymers and additives. *Polym. Chem.* **2014**, *5*, 3142–3162. [CrossRef]
16. Biermann, U.; Friedt, W.; Siegmund Lang, S.; Wilfried Lühs, W.; Guido Machmüller, G.; Metzger, J.O.; Klaas, M.R.G.; Schäfer, H.J.; Schneider, M.P. New syntheses with oils and fats as renewable raw materials for the chemical industry. *Angew. Chem. Int. Ed.* **2000**, *39*, 2206–2224. [CrossRef]
17. Zahradnik, L.; Tynova, E.; Kalouskova, H. Stable epoxy resins made from renewable nontraditional resources-economically and environmentally acceptable solution. *Koroze Ochr. Mater.* **2005**, *49*, 83–86.
18. Negro, V.; Mancini, G.; Ruggeri, B.; Fino, D. Citrus waste as feedstock for bio-based products recovery: Review on limonene case study and energy valorization. *Bioresour. Technol.* **2016**, *214*, 806–815. [CrossRef] [PubMed]
19. Sharma, K.; Mahato, N.; Cho, M.H.; Lee, Y.R. Converting citrus wastes into value-added products: Economic and environmently friendly approaches. *Nutrition* **2017**, *34*, 29–46. [CrossRef] [PubMed]
20. Harvey, B.G.; Guenthner, A.J.; Koontz, T.A.; Storch, P.J.; Reamsc, J.T.; Groshens, T.J. Sustainable hydrophobic thermosetting resins and polycarbonates from turpentine. *Green Chem.* **2016**, *18*, 2416–2423. [CrossRef]
21. Wiemann, L.O.; Faltl, C.; Sieber, V. Lipase-mediated epoxidation of the cyclic monoterpene limonene to limonene oxide and limonene dioxide. *Z. Naturforsch. B J. Chem. Sci.* **2012**, *67*, 1056–1060. [CrossRef]

22. Morinaga, H.; Sakamoto, M. Synthesis of multi-functional epoxides derived from limonene oxide and its application to the network polymers. *Tetrahedron Lett.* **2017**, *58*, 2438–2440. [CrossRef]
23. Yang, S.; Chen, J.S.; Körner, H.; Breiner, T.; Ober, C.K. Reworkable Epoxies: Thermosets with Thermally Cleavable Groups for Controlled Network Breakdown. *Chem. Mater.* **1998**, *10*, 1475–1482. [CrossRef]
24. Liu, W.; Wang, Z.; Xiong, L.; Zhao, L. Phosphorus-containing liquid cycloaliphatic epoxy resins for reworkable environment-friendly electronic packaging materials. *Polymer* **2010**, *51*, 4776–4783. [CrossRef]
25. Wang, R.; Schuman, T.P. Vegetable oil-derived epoxy monomers and polymer blends: A comparative study with review. *Exp. Polym. Lett.* **2013**, *7*, 272–292. [CrossRef]
26. Yang, T.; Zhang, C.; Zhang, J.; Cheng, J. The influence of tertiary amine accelerators on the curing behaviors of epoxy/anhydride systems. *Thermochim. Acta* **2014**, *577*, 11–16. [CrossRef]
27. Naumann, S.; Schmidt, F.G.; Speiser, M.; Bohl, M.; Epple, S.; Bonten, C.; Buchmeiser, M.R. Anionic Ring-Opening Homo- and Copolymerization of Lactams by Latent, Protected *N*-Heterocyclic Carbenes for the Preparation of PA 12 and PA 6/12. *Macromolecules* **2013**, *46*, 8426–8433. [CrossRef]
28. Altmann, H.J.; Naumann, S.; Buchmeiser, M.R. Protected *N*-heterocyclic carbenes as latent organocatalysts for the low-temperature curing of anhydride-hardened epoxy resins. *Eur. Polym. J.* **2017**, *95*, 766–774. [CrossRef]
29. Amirova, L.R.; Burilov, A.R.; Amirova, L.M.; Bauer, I.; Habicher, W.D. Kinetics and mechanistic investigation of epoxy-anhydride compositions cured with quaternary phosphonium salts as accelerators. *J. Polym. Sci. Part A Polym. Chem.* **2016**, *54*, 1088–1097. [CrossRef]
30. Pinazo, J.M.; Domine, M.E.; Parvulescu, V.; Petru, F. Sustainability metrics for succinic acid production: A comparison between biomass-based and petrochemical routes. *Catal. Today* **2015**, *239*, 17–24. [CrossRef]
31. Lin, Z.; Ierapetritou, M.; Nikolakis, V. Phthalic anhydride production from hemicellulose solutions: Technoeconomic analysis and life cycle assessment. *AIChE J.* **2015**, *61*, 3708–3718. [CrossRef]
32. Park, W.H.; Lee, J.K.; Kwon, K.J. Cure behavior of an epoxy-anhydride-imidazole system. *Polym. J.* **1996**, *28*, 407–411. [CrossRef]
33. Matejka, L.; Lovy, J.; Pokorny, S.; Bouchal, K.; Dusek, K. Curing epoxy resins with anhydrides. Model reactions and reaction mechanism. *J. Polym. Sci. Polym. Chem. Ed.* **1983**, *21*, 2873–2885. [CrossRef]
34. Chen, Y.-C.; Chiu, W.-Y.; Lin, K.-F. Kinetics study of imidazole-cured epoxy-phenol resins. *J. Polym. Sci. Part A Polym. Chem.* **1999**, *37*, 3233–3242. [CrossRef]
35. Antoon, M.K.; Koenig, J.L. Crosslinking mechanism of an anhydride-cured epoxy resin as studied by Fourier transform infrared spectroscopy. *J. Polym. Sci. Polym. Chem. Ed.* **1981**, *19*, 549–570. [CrossRef]
36. Matejka, L.; Pokorny, S.; Dusek, K. Acid curing of epoxy resins. A comparison between the polymerization of diepoxide-diacid and monoepoxide-cyclic anhydride systems. *Makromol. Chem.* **1985**, *186*, 2025–2036. [CrossRef]
37. Steinmann, B. Investigations on the curing of epoxy resins with hexahydrophthalic anhydride. *J. Appl. Polym. Sci.* **1989**, *37*, 1753–1776. [CrossRef]
38. Steinmann, B. Investigations on the curing of epoxides with phthalic anhydride. *J. Appl. Polym. Sci.* **1990**, *39*, 2005–2026. [CrossRef]
39. Paramarta, A.; Webster, D.C. Curing kinetics of bio-based epoxy-anhydride thermosets with zinc catalyst. *J. Therm. Anal. Calorim.* **2017**, *130*, 2133–2144. [CrossRef]
40. Kuncho, C.N.; Schmidt, D.F.; Reynaud, E. Effects of Catalyst Content, Anhydride Blending, and Nanofiller Addition on Anhydride-Cured Epoxidized Linseed Oil-Based Thermosets. *Ind. Eng. Chem. Res.* **2017**, *56*, 2658–2666. [CrossRef]
41. Woo, E.M.; Seferis, J.C. Cure kinetics of epoxy/anhydride thermosetting matrix systems. *J. Appl. Polym. Sci.* **1990**, *40*, 1237–1256. [CrossRef]
42. Zou, Q.; Ba, L.; Tan, X.; Tu, M.; Cheng, J.; Zhang, J. Tunable shape memory properties of rigid-flexible epoxy networks. *J. Mater. Sci.* **2016**, *51*, 10596–10607. [CrossRef]
43. Kretzschmar, K.; Hoffmann, K.W. Reaction enthalpies during the curing of epoxy resins with anhydrides. *Thermochim. Acta* **1985**, *94*, 105–112. [CrossRef]
44. Simon, P. Isoconversional methods-fundamentals, meaning and application. *J. Therm. Anal. Calorim.* **2004**, *76*, 123–132. [CrossRef]

45. Vyazovkin, S.; Burnham, A.K.; Criado, J.M.; Pérez-Maqueda, L.A.; Popescu, C.; Sbirrazzuoli, N. ICTAC Kinetics Committee recommendations for performing kinetic computations on thermal analysis data. *Thermochim. Acta* **2011**, *520*, 1–19. [CrossRef]
46. Harsch, M.; Karger-Kocsis, J.; Holst, M. Influence of fillers and additives on the cure kinetics of an epoxy/anhydride resin. *Eur. Polym. J.* **2007**, *43*, 1168–1178. [CrossRef]
47. Santiago, D.; Fernandez-Francos, X.; Ramis, X.; Salla, J.M.; Sangermano, M. Comparative curing kinetics and thermal-mechanical properties of DGEBA thermosets cured with a hyperbranched poly(ethyleneimine) and an aliphatic triamine. *Thermochim. Acta* **2011**, *526*, 9–21. [CrossRef]
48. Leukel, J.; Burchard, W.; Krüger, R.P.; Much, H.; Schulz, G. Mechanism of the anionic copolymerization of anhydride-cured epoxies–analyzed by matrix-assisted laser desorption ionization time-of-flight mass spectrometry (MALDI-TOF-MS). *Macromol. Rapid Commun.* **1996**, *17*, 359–366. [CrossRef]

Sample Availability: not available.

 *molecules*

Article

Synthesis and Characterization of Poly(Vinyl Alcohol)-Chitosan-Hydroxyapatite Scaffolds: A Promising Alternative for Bone Tissue Regeneration

Sergio Pineda-Castillo [1], Andrés Bernal-Ballén [1,*], Cristian Bernal-López [1], Hugo Segura-Puello [2], Diana Nieto-Mosquera [2], Andrea Villamil-Ballesteros [2], Diana Muñoz-Forero [2] and Lukas Munster [3]

[1] Grupo de Investigación en Ingeniería Biomédica, Vicerrectoría de Investigaciones, Universidad Manuela Beltrán, Avenida Circunvalar No. 60-00, Bogotá 110231, Colombia; sergiopinedac@outlook.com (S.P.-C.); bernalip10@hotmail.com (C.B.-L.)

[2] Laboratorio de Investigación en Cáncer. Universidad Manuela Beltrán, Avenida Circunvalar No. 60-00, Bogotá 110231, Colombia; hugo.segura@umb.edu.co (H.S.-P.); lorena.nieto@umb.edu.co (D.N.-M.); catalina.ballesteros@umb.edu.co (A.V.-B.); diana.munoz@umb.edu.co (D.M.-F.)

[3] Centre of Polymer Systems, University Institute. Tomas Bata University in Zlín, Trida Tomase Bati 5678, Zlin 76001, Czech Republic; munster@cps.utb.cz

* Correspondence: andres_bernal9@hotmail.com; Tel.: +57-3014192359

Academic Editors: Sylvain Caillol and Guillaume Couture
Received: 31 July 2018; Accepted: 10 September 2018; Published: 20 September 2018

Abstract: Scaffolds can be considered as one of the most promising treatments for bone tissue regeneration. Herein, blends of chitosan, poly(vinyl alcohol), and hydroxyapatite in different ratios were used to synthesize scaffolds via freeze-drying. Mechanical tests, FTIR, swelling and solubility degree, DSC, morphology, and cell viability were used as characterization techniques. Statistical significance of the experiments was determined using a two-way analysis of variance (ANOVA) with $p < 0.05$. Crosslinked and plasticized scaffolds absorbed five times more water than non-crosslinked and plasticized ones, which is an indicator of better hydrophilic features, as well as adequate resistance to water without detriment of the swelling potential. Indeed, the tested mechanical properties were notably higher for samples which were undergone to crosslinking and plasticized process. The presence of chitosan is determinant in pore formation and distribution which is an imperative for cell communication. Uniform pore size with diameters ranging from 142 to 519 μm were obtained, a range that has been described as optimal for bone tissue regeneration. Moreover, cytotoxicity was considered as negligible in the tested conditions, and viability indicates that the material might have potential as a bone regeneration system.

Keywords: scaffolds; chitosan; poly(vinyl alcohol); cell proliferation; cell differentiation

---

1. Introduction

Biomedical engineering is considered as an extension of chemical engineering towards biomaterials, and tissue engineering (TE) is one of its main branches [1]. TE aims to regenerate damaged tissues by combining cells from the body with highly porous scaffold biomaterials, which act as template for tissue regeneration and guide the growth of new tissue [2]. A very promising potential in creating biological alternatives for harvested tissues, implants, and prosthesis has been reported by TE [3] and the ability to aid treatment of numerous clinical situations, including spinal fusion, joint replacement, fracture nonunion, and pathological loss of bones, among other conditions [4]. Within this frame, bone tissue as a cornerstone of the muscle-skeletal system, is a hard, calcified connective tissue with functions of protection and mobility of the body, storing calcium as well as mesenchymal and hematopoietic stem cells [5]. Nonetheless, bones are susceptible to damage, specifically caused by fractures, tumors, osteoporosis, and other sources of trauma, and although bones have certain self-regenerative properties, there are conditions in which complete tissue restoration cannot be achieved exclusively by internal processes [6].

Over the last ten years, remarkable progress has been made in the development of surgical techniques for bone reconstruction. Synthesis of functional bone using combinations of cells and bioactive factors is considered as a promising approach towards bone regeneration, undoubtedly creating the possibility of tissue regeneration and repair [7]. However, TE has studied a technique able to supply the needs for supporting and proliferating regenerative cells: scaffolding [8]. This technique has been extensively studied in bone tissue regeneration and seeks to mimic bone extracellular matrix architecture.

Scaffolds aim at the creation of microenvironments for cell adhesion, proliferation and differentiation [9–11] and they play a key role in bone TE providing a 3-dimensional environment for cell seeding as well as filling bone defects while providing mechanical competence during bone regeneration [12]. Currently, there are different methods for synthesizing scaffolds and each method provides unique features to the sample and defines its final use [13–17]. Thus, polymers come as great candidates for scaffold fabrication due to their ability to be used in the aforementioned methods.

A large variety of polymers has been used in manufacturing scaffolds, both natural and synthetic. Natural ones provide outstanding properties to scaffolds, from fostering active interactions with cells to hampering immune response from the host, whereas synthetic polymers display better mechanical properties. Poly(L-lactic acid) (PLLA) [18], poly(glycolic acid) (PGA) [19], poly(caprolactone) (PCL) [20], and poly(lactic-co-glycolic) acid (PLGA) [21] are currently the most common materials used as scaffolds for bone regeneration [22,23]. Within natural polymers, chitosan (CH) deserves special attention because this polysaccharide exhibits antibacterial, antifungal, mucoadhesive and analgesic properties [24]. In addition, CH can be degraded by a set of human enzymes and is biocompatible under certain conditions [25]. Innumerable reports in which CH is used in scaffolds for bone regeneration can be found in literature [24,26–33] due to the fact that CH facilitates proliferation of osteoblast cells and helps to create mineralized bone matrix. While CH exhibits suitable biological properties for bone tissue regeneration, this type of material needs additional treatment, such as crosslinking, to be able to withstand the physical conditions required by a bone implant [25,34]. However, by blending synthetic polymers with natural ones, scaffolds might exhibit better mechanical stability [35]. In this frame, polyvinyl alcohol (PVA) represents an outstanding candidate for bone TE due to its biocompatibility and its low cytotoxicity as it has been demonstrated in previous studies [36–41]. Moreover, bioactive compounds are often used in TE to facilitate interactions between scaffolds and cells, promoting thus adhesion, proliferation and differentiation [42,43]. Therefore, hydroxyapatite, a major inorganic component in bone tissue [44], comes as a very promising bioactive compound as it may provide the material the ability to mimic natural bone functionality and ultimately promoting osteoconductivity [45]. Accordingly, several studies on CH-HA have been carried out, evidencing the ability of HA to improve bioactivity [44–48]. CH, PVA, and HA have been investigated into the bone regeneration frame separately and copious literature support the importance of the materials. Nonetheless, the combination of the mentioned materials appeared in only 5 publications in the last year according to Web of Science. Therefore, the system is still an unexplored niche, and it has a notably potential and innovative opportunity for bone tissue engineering.

Considering the aforementioned reasons, a crosslinked and plasticized PVA-CH-HA system rises as a potential candidate for bone tissue engineering and therefore, this research is aimed to determine the potential of the system in bone regeneration through biological, mechanical and physicochemical characterization.

2. Results and Discussion

Scaffold characterization consists of collecting information which might elucidate how the material will perform under specific conditions. In this matter, techniques such as degree of swelling, solubility degree, FTIR, DSC, SEM, mechanical characterization and cell culture will provide sufficient and relevant evidence about how the material behaves and how the material might interact with its own components as well as with the surroundings.

2.1. Degree of Swelling (DS) and Solubility Degree (SD)

As scaffolds are designed to be subject to high humidity conditions, swelling and solubility degree are fundamental parameters to be studied. Water absorption is an imperative feature for scaffolds as the ability to conduce and store water through a porous network is essential to achieve proper cell signaling and nutrition [45]. The mechanism of swelling requires the polymeric chains to be able to soak up water molecules across the matrix. To do so, reversible interactions such as hydrogen bonds must be fostered [25].

Significant differences were observed in both, crosslinked and plasticized (CPS) and non-crosslinked and non-plasticized samples (nCPS). Although considerable modifications in volume and fluctuations in shape were presented in both CPS and nCPS, the former were able to resist manipulation without fracture in contrast to nCPS which became fragile, yielding difficult manipulation and eventual fracture. CPS absorbed from 600 to 1700 times their own weight whereas nCPS presented notably higher values (Table 1). This behavior might be explained by three approaches which are depicted in Figure 1. In the first one, crosslinking between glutaraldehyde (GLU) and -NH$_2$ reduces the concentration of amino groups, leading to a reduction of interactions with water [49,50]. The second approach indicates that a lesser content of hydroxyl groups, due to PVA crosslinking, becomes available [51]. And finally, interactions between hydroxyl groups of PVA and amino or hydroxyl groups of CH reduces the number of interactions between -OH groups and water [52], reducing the overall hydrophilicity of the system. Since the hydrophilicity dominates the behavior of the scaffolds in this test, a higher concentration of CH increases DS. Although CH amine functional groups are more reactive to GLU than hydroxyl groups of PVA [53,54] the low concentration of the crosslinker did not dramatically reduce the presence of hydrophilic groups and the system exhibits a CH-like behavior, dominating the DS of the system. Furthermore, a reduction in the crystalline phase (see DSC) caused by CH produces a augmentation in the capacity of the material to absorb water and therefore, a rise in the DS [51].

On the other hand, nCPS did not reduce the availability of hydrophilic groups which allowed water molecules to create non-covalent bonds. In nCPS, PVA molecules are entangled with CH (physical interactions) whereas in the CPS ones, chemical crosslinking was achieved, forming covalent bonds among chains, fixing and reducing polymer mobility, which resulted in the lower DS [54].

Table 1. Degree of swelling and solubility degree for CPS and nCPS.

Type of Sample	Ratio (*wt/wt*)	Degree of Swelling after 24 h [%] (DS)	Average Weight Loss [%] (SD)
CPS	CH:PVA 1:1-GLU/GLY	1004.44 ± 17.70	9.44 ± 1.20
	CH:PVA 1:3-GLU/GLY	621.97 ± 61.12	9.38 ± 0.40
	CH:PVA 3:1-GLU/GLY	1766.18 ± 347.95	17.82 ± 16.21
nCPS	CH:PVA 1:1	5010.20 ± 158.78	49.57 ± 1.01
	CH:PVA 1:3	2633.24 ± 176.38	34.31 ± 3.24
	CH:PVA 3:1	9883.13 ± 182.25	N/A

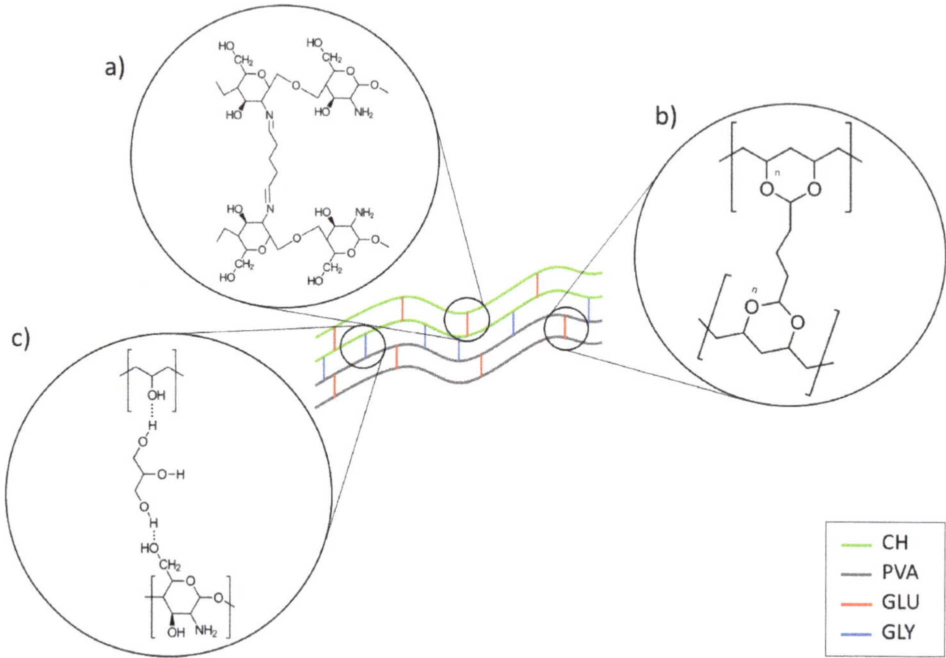

Figure 1. Schematic representation of interactions between CH, PVA, GLU, and GLY. (**a**) CH crosslinked by GLU. (**b**) PVA crosslinked by GLU. (**c**) Hydrogen bonding between CH and PVA with GLY.

There is an intimate relation between DS and SD. In both cases, the interactions with water are responsible for the observed behavior. It has been reported that solubility of PVA depends on the degree of hydrolysis, the molecular weight, and the tendency to form hydrogen bonding in aqueous solutions. In this matter, any alteration of those factors will have repercussions on the solubility [51]. Although CPS have more free space and therefore water could go through the system, the presence of GLU attaches the polymer chains and hydration might be difficult. Consequently, CPS exhibit lower SD than nCPS. The chemical bonds created during the treatment gives the material resistance and stability. Indeed, the lesser hydroxyl groups reduced the affinity for water leading to a reduction in the swelling as well as to the solubility. Although, CH is not soluble in water, its -NH$_2$ groups allow the polymeric chains to partially attach themselves to water molecules, increasing thus the solubility of the material. It can be expected that a higher PVA concentration supposes higher levels of solubility. Nonetheless, the combination of PVA with CH generates a system with lower SD than PVA alone, possibly due to intermolecular interactions between polymeric chains instead of water interactions [52]. It can be inferred then that CPS and nCPS with a high presence of CH lead to the material mechanical stability and a high capacity to absorb and transport water.

2.2. Fourier Transform Infrared Spectroscopy (FTIR)

FTIR is a crucial technique for polymer characterization and it provides information about crosslinking between polymeric chains. Figure 2 shows the obtained spectra for all the prepared samples and pure components. The spectra were vertically moved for better clarity. HA exhibits vibrations located at 630 and 3570 cm^{-1} which correspond to hydroxyl ions [55] as well as signals located at 1023 and 962 cm^{-1} that belong to PO$_4^{2-}$ groups, and one peak at 874 cm^{-1}, which is associated to CO$_3^{2-}$ [56,57].

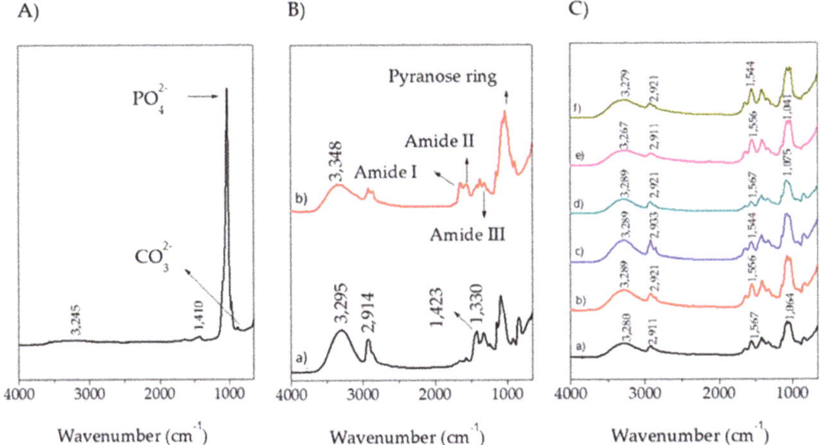

Figure 2. Collection of the obtained FTIR spectra. (**A**) Hydroxyapatite; (**B**) (a) PVA and (b) CH; (**C**) Prepared scaffolds (a) CH:PVA 1:1 GLU-GLY; (b) CH:PVA 1:1; (c) CH:PVA 1:3 GLU-GLY; (d) CH:PVA 1:3; (e) CH:PVA 3:1 GLU-GLY; (f) CH:PVA 3:1.

The spectrum for PVA shows a strong broad absorption centered at 3295 cm^{-1}, associated to hydroxyl functional groups. Saturated C–H stretching is manifested in the band at 2914 cm^{-1} and the band located at 1423 cm^{-1} corresponds to –CH$_2$– bending. No signals around 1700 cm^{-1} were visible, which indicates the absence of acetate groups, caused by the high percentage of hydrolysis of PVA. The bands at 1415 and 1330 cm^{-1} can be attributed to combination frequencies of CH and OH [58]. The CH$_2$ rocking band at 928 cm^{-1} and a peak at 848 cm^{-1} are associated with C–O stretching and these results are in a good agreement with other reports [59–62].

The CH spectrum exhibits a broad signal centered at 3348 cm^{-1} corresponding to OH functional groups and NH group stretching vibrations. The symmetric, and asymmetric –CH$_2$– stretching occurred in the pyranose ring are located at 2920, 2880, 1430, 1320, 1275 and 1245 cm^{-1} [54,63,64]. The bands located at 1650, 1586 cm^{-1}, and 1322 cm^{-1} are ascribed to amide I, amide II, and amide III respectively as well as the saccharine-related signals at 1155 and 900 cm^{-1} [49,54,65–67]. The peak at 1030 and 1080 cm^{-1} indicates the C–O stretching vibration.

The effect of the plasticizer (glycerol, GLY) on the samples can be evidenced in the slightly rise in the signal corresponding to hydroxyl groups. Nonetheless, the combination of both, GLY and GLU have contradictory consequences; on one side, the crosslinker diminished hydroxyl groups, but on the other side, the plasticizer increases them. For that reason, no clearly observable tendency is depicted in the spectra.

CPS presents an increase of the signal for imine bands as well as a drop on the amine caused by the nucleophilic addition of the amine from CH with the aldehyde [68]. Furthermore, blends of CH and PVA might produce carboxylic acid dimer which is manifested in the region of 1700–1725 cm^{-1} and are originated by the acetic acid used for dissolving CH. An evident decreasing of absorbance around 1550 cm^{-1} proves that crosslinking decreased the availability of NH$_2$ groups. Simultaneously, CPS spectra show lower absorbance at the –OH band because this group is consumed in the formation of C–O–C bonds [69]. Moreover, the absorption at 3300–3250 cm^{-1} related to OH and NH stretching vibrations broadened and shifted to a lower wave number with the increase of PVA, suggesting the formation of hydrogen bonds between CH and PVA [70].

A reduction of the intensities of bands in the range of 1500–1650 cm^{-1} was evidenced in samples where the CH content was diminished. Another observed feature is that the region from 3200 and 3500 cm^{-1} is wider in blends than in single components as a result of the overlapping signals from both hydroxyl and amino groups. The wavenumber of this peak has shifted toward lower values by increasing the PVA content in the samples, which indicates hydrogen bond formation between PVA and CH. The peak around 1560 cm^{-1} was a weaker N-H vibration and was depressed by increasing the PVA content in the samples. This is another indication of hydrogen bonding between PVA and CH chains [71].

2.3. Mechanical Characterization

Young's Modulus (*E*), tensile strength (*TS*) and yield strength (*YS*) were chosen for evaluating how the material behaves under mechanical loads and the results are reported in Table 2. The first remarkable feature of the studied samples is that differences in Young's Modulus are caused by the presence of the crosslinker and the plasticizer. As can be noticed, CPS display notably higher values than nCPS because GLU is chemically joined to both polymers reducing chain mobility. Nonetheless, as it was mentioned above (see swelling and solubility analysis), CH amine groups are more reactive to GLU than PVA, and therefore, the blend exhibits CH-like features. Indeed, by bonding polymer chains with GLU through Schiff bases and acetal bonds, the material improves its toughness. Therefore, CPS require more energy to be elongated and, ultimately, break. This behavior can be seen in Figure 3 where samples at >1% strain grouped in separate zones of the graph; CPS are found above nCPS along the Y-axis, enclosing a larger area under the curve than nCPS.

On the other hand, GLY can be considered as responsible for a higher *YS*. This may be caused by a reinforcement of the interactions between polymeric chains. GLY facilitates interactions between zones of the polymeric chains that were not crosslinked, increasing thus the number of interactions between them and its presence produces polymer-plasticizer interactions among the macromolecules to the detriment of polymer-polymer interactions and it causes a decrease of the hardness as well as Young's modulus [62]. Moreover, the presence of GLU and GLY has an important effect on *TS*. GLY is responsible for a relevant rise in intermolecular forces by creating interactions between polymeric chains through hydrogen bonds, as was explained previously (Figure 1c). These interactions might have the effect of preventing chitosan from crystallizing, although does not significantly destroy the crystalline component of PVA [72]. Simultaneously, some fractions of the polymeric chains are joined together chemically by GLU. Consequently, *TS* is affected by these intermolecular interactions and bonds, increasing thus the tensile strength of the sample [73].

Table 2. Mechanical properties for the prepared scaffolds.

Type of Sample	Proportion	Young's Modulus [MPa]	Tensile Strength [MPa]	Yield Strength [MPa]
CPS	CH:PVA 1:1-GLU/GLY	37.26 ± 10.35	0.96 ± 0.20	0.76 ± 0.34
	CH:PVA 1:3-GLU/GLY	27.29 ± 23.33	0.61 ± 0.41	0.57 ± 0.52
	CH:1 PVA 3:1-GLU/GLY	24.84 ± 9.83	0.67 ± 0.36	0.61 ± 0.35
nCPS	CH:PVA 1:1	4.70 ± 2.41	0.08 ± 0.04	0.03 ± 0.02
	CH:PVA 1:3	7.42 ± 1.10	0.23 ± 0.07	0.07 ± 0.02
	CH:PVA 3:1	2.61 ± 0.66	0.08 ± 0.05	0.02 ± 0.01

Polymer concentration is also a variable which affects the tested properties, *E*, *TS* and *YS*. As can be seen, the highest values were obtained for CH:PVA 1:1-GLU/GLY whereas a reduction of the tested parameters were obtained at the concentration of any of the two polymers changed. This behavior is in concordance with previous studies of CH and PVA interactions [74]. These evaluations indicate that CPS at equal concentrations of CH and PVA exhibit conditions to endure mechanical loads, which might indicate that under the tested conditions, the material will perform adequately as a candidate for bone tissue regeneration. Nonetheless, a higher concentration of CH is also appropriate for that purpose.

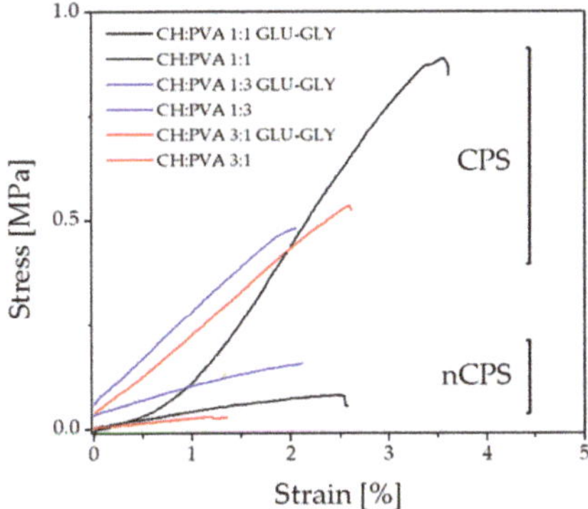

Figure 3. Tensile stress-strain curves for prepared samples.

2.4. Differential Scanning Calorimetry

Thermograms for pure components and prepared scaffolds are shown in Figure 4. Glass transition temperature (T_g) and melting point (T_m) (Table 3) for PVA are located at 72 °C and 218 °C respectively, results that have been reported for other authors [51,75,76]. CH, on the other hand, exhibits an endothermic transition between 100–120 °C which is attributed to evaporation of residual water [63]. A sharp peak centered at 161 °C is not a manifestation of T_m since a recrystallization peak did not appear neither in the second cooling nor in the second heating scan. Therefore, the peak presumably results from the dissociation process of interchain hydrogen bonding [77].

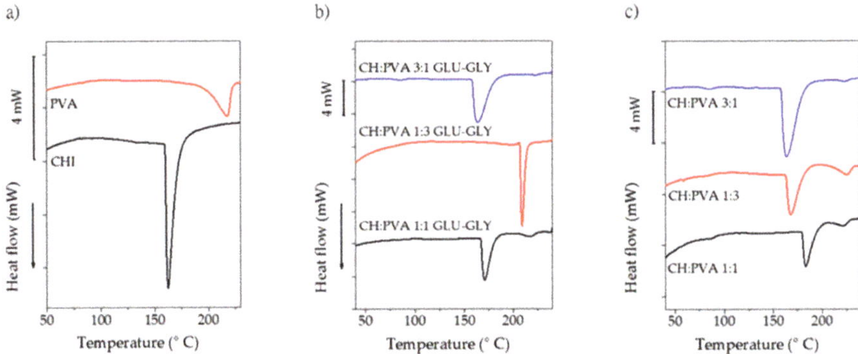

Figure 4. Thermograms for the studied materials. (**a**) Pure polymers; (**b**) CPS; (**c**) nCPS.

Table 3. Thermal characteristics of the studied samples.

Type of Sample	Ratio	Temperature (°C)		Enthalpy ΔH_f (mJ·g^{-1})	Crystallinity (%)
		T_g	T_m		
Pure polymers	CHITOSAN	-	-	-	-
	PVA	78	218	112	81
CPS	CH:PVA 1:1-GLU/GLY	-	170	67	48
	CH:PVA 1:3-GLU/GLY	-	210	125	90
	CH:PVA 3:1-GLU/GLY	-	160	266	-
nCPS	CH:PVA 1:1	-	185	69	49
	CH:PVA 1:3	-	165	108	78
	CH:PVA 3:1		162	224	-

Several attempts at obtaining T_g data were carried out and there was no unanimity in the reported values so DSC might not be an appropriate technique for detecting T_g since CH is a semi-crystalline polymer [78]. Nonetheless, it has been established that the degree of deacetylation (75–85% in this case) has not influence in the T_g, and α-relaxations occurred at 140–150 °C which might be a reliable value for T_g [79], transition which is hidden by the broad peak in the same region.

Important features are observable in the thermograms for the CPS and nCPS. Detecting T_g for CH is problematic, and the same difficulty was observed for that transition in the blends under the studied conditions, presumably for the overlapping of the transitions occurred in both CH and PVA. In the case of CPS, it can be assumed that the strong interactions such as inter-molecular hydrogen bonding inhibit the movement of the molecular chain sections and an increase of the T_g value might be achieved [80,81]. Furthermore, an increment in CH concentration causes higher values for T_g in the blend in comparison to PVA, because polymer interactions through the hydrogen bonding formation causes an effective increase in the mean molecular weight [82]. Despite the impossibility to detect T_g, all the prepared scaffolds should show a unique T_g which is a manifestation of the compatibility of PVA and CH [83]. On the other hand, the Tm of the blends is reduced as the concentration of CH rises and a depression of T_m is considered as a measure of the blend compatibility [42,83]. The addition of GLY led to more hydrogen bonds, which has an impact on the structural order of the matrix and therefore a reduction in T_g should be evident [84]. Nonetheless, apart from the low concentration of GLY and GLU, the presence of PVA is determinant in the thermal behavior of the scaffold. The T_m estimated for pure PVA is 218 °C which in combination with CH was notably reduced, yielding to a sharp and deep endothermic peak.

It might be expected that the crystalline nature of PVA was decreased by the addition of CH, which is in concordance to the DS and FTIR data [42]. Indeed, crosslinking limited the mobility of the polymers and suppressed the crystallites growing (in the case of PVA), disfavoring crystallization. Additionally, the reduction of crystallinity is an indicative of a lower extent of the confinement of molecular chain segments, and the interpenetration and entanglement of GLU-CH with PVA.

2.5. Scanning Electron Microscopy

SEM images provides information about structural and morphological features of the scaffolds. The first observable characteristic is that morphology varies depending on both the presence of GLU-GLY as well as the concentration of CH and PVA. As can be seen in Figure 5, there is a reduction in pore size as a consequence of chemical crosslinking in most examples, except in CH:PVA 3:1 in which average pore size increased four times when crosslinked. Conversely, a higher concentration of PVA led to a lower pore size [85]. Porosity decreased in CPS 1:3 compared to nCPS 1:3 suggesting that CH might act as a factor to stop the porosity size and void percentage of crosslinked-PVA-containing samples from decreasing. CH 3:1 PVA samples showed a higher pore size and void percentage than CH 1:3 PVA indicating that CH controls pore formation within the scaffold (Table 4). Equal concentration of each polymer showed a combination of both behaviors. Polymer concentration did not play a significant role in fiber formation. Lastly, it is reasonable to believe that void percentage and pore size is affected by freeze-drying and therefore, modifying vacuum pressure and time of might produce scaffolds with different pore size [40].

Another crucial parameter for pore formation is viscosity. The original solution of CH is too viscous, and the sample does not mix homogeneously, therefore the pore formation and size is reduced during the freeze-drying process. Consequently, the pore regularity is affected [86] and samples with lower CH concentration present less regular pores (Figure 6). Indeed, CPS 1:1 show higher pore size regularity and distribution while nCPS 1:1 exhibit a semi-porous surface, suggesting low pore interconnectivity. This behavior was consistently observed in all CPS and nCPS.

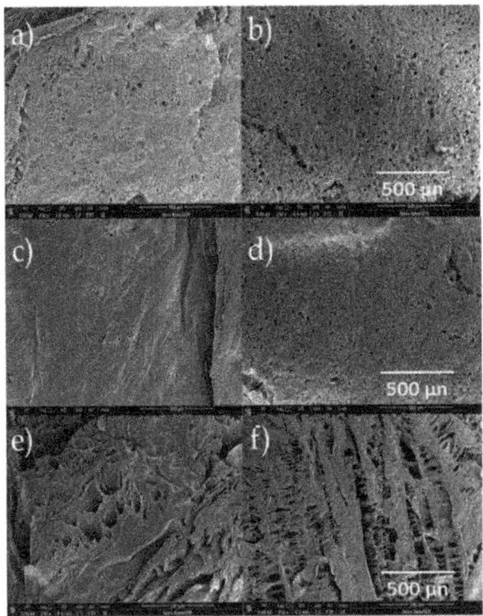

Figure 5. Surface SEM images for CPS (**left**) and nCPS (**right**) at 250×. (**a**) CH:PVA 1:1 GLU-GLY; (**b**) CH:PVA 1:1; (**c**) CH:PVA 1:3 GLU-GLY; (**d**) CH:PVA 1:3; (**e**) CH:PVA 3:1 GLU-GLY; (**f**) CH:PVA 3:1.

Table 4. Obtained pore size for the evaluated scaffolds.

Type of Sample	Ratio	Pore Size (µm)	Void Percentage (%)
CPS	CH:PVA 1:1-GLU/GLY	11.7 ± 5.1	40.86
	CH:PVA 1:3-GLU/GLY	0.6 ± 0.2	6.48
	CH:PVA 3:1-GLU/GLY	331.5 ± 111.8	70.22
nCPS	CH:PVA 1:1	25.5 ± 9.0	4.95
	CH:PVA 1:3	11.9 ± 4.3	30.22
	CH:PVA 3:1	74.5 ± 32.4	78.28

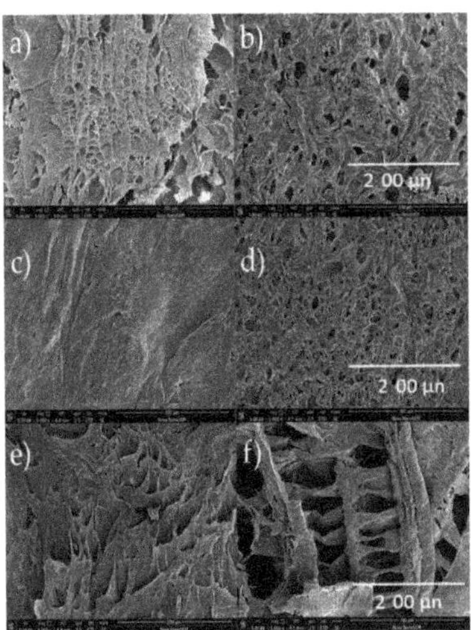

Figure 6. SEM images of surface for CPS (**left**) and nCPS (**right**) at 1000×. (**a**) CH:PVA 1:1 GLU-GLY; (**b**) CH:PVA 1:1; (**c**) CH:PVA 1:3 GLU-GLY; (**d**) CH:PVA 1:3; (**e**) CH:PVA 3:1 GLU-GLY; (**f**) CH:PVA 3:1.

Despite all the morphological features exhibited by CPS and nCPS, CH:PVA 3:1 GLU-GLU (Figure 7) seems to be adequate for bone tissue engineering. The data obtained during the evaluation of pore size revealed uniform pore formation with diameters ranging from 142 to 519 μm, a range that has been described as optimal for bone tissue regeneration [87] and the pore structure obtained is suitable for bone tissue engineering by exhibiting a homogeneous macro-porous network with open pores and some interconnectivity. Such hierarchical structure has potential application for future bone tissue in-growth, nutrient delivery and vascularization [88].

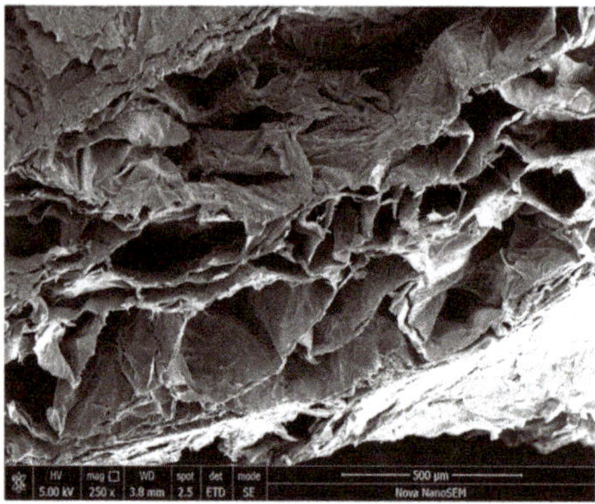

Figure 7. Bulk SEM images at 250× of CH:PVA 3:1 GLU-GLY.

2.6. Cell Culture

Cell culture was performed in triplicate for all the prepared scaffold during 17 days, and cell differentiation was evaluated through morphologic analysis. Initial conditions of trypsinized osteoblastic cells are shown in Figure 8. After three days of culture, a higher degree of differentiated cells were present in CPS, exhibiting the formation of osteocytic canaliculi (Figure 8b) which suggest proper cell signaling through osteocyte gap junctions [89]. Furthermore, the characteristic star-like morphology for osteocyte are visible in the image, suggesting that the process from osteoblast to osteocyte has been carried out. These cytoplasmic projections allow to the cell to communicate with surrounding osteocytes using gap join [90]. Proliferation continued during the first 5 days of the experiment. Later, it was not possible to evaluate it, as cells migrated inside the scaffolds and for observation it was needed to destroy the scaffold. Therefore, cell activity was evaluated through colorimetric observations of culture media and it was constant until 7 days of culture.

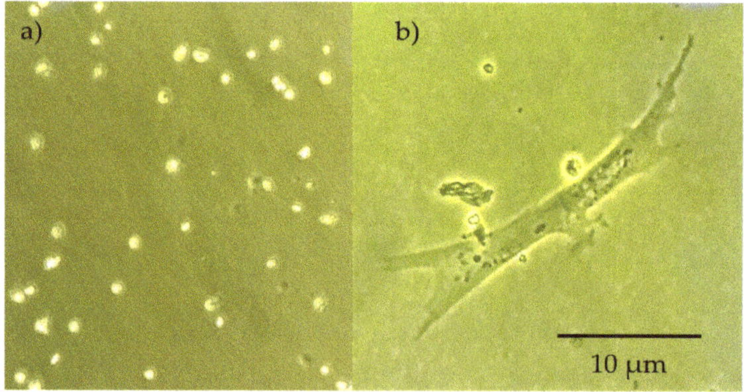

Figure 8. (**a**) Trypsinized osteoblastic cells. (**b**) Osteocyte cell found in the surroundings of sample CH 1:3 PVA + GLU/GLY.

Cell viability was assessed using the Trypan Blue Exclusion Method during the first days of the experiment. In this frame, Figure 9 exhibits cell with rigid membrane as well as translucid cells which might indicate that the material performed appropriately in terms of viability. On the other hand, dead cells appeared, and membrane fragmentation is presented, although it is important to emphasize that its number is considerably lower. Moreover, shattering did not affect substantially cell viability, and although turbidity might affect the adequate nutrients path, signaling was kept even in these adverse conditions. Another indicator for viability is that characteristic morphological changes occurs. In a viable state, the lacuna and the osteocyte inside are both star-shaped [91].

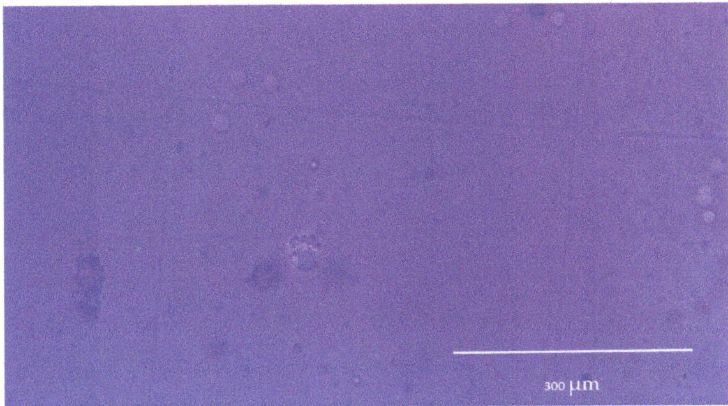

Figure 9. Cell culture after 5 days.

During the last days of the experiment, cell activity was significantly reduced until no change in culture media was observed, suggesting high mortality rate. Supporting this theory, cell detritus appeared in culture media. This fact might have been produced by early degradation of the sample, hindering continuous cell signaling. Nonetheless, the concentration of non-crosslinked GLU contained within the sample was too low and even if it was released as samples degraded, it might not have contributed to cytotoxicity.

3. Materials and Methods

3.1. General Information

Poly(vinyl alcohol) (PVA) (Mw = 130,000 g mol^{-1}) with 99% of hydrolysis, chitosan (CH) of medium molecular weight with deacetylation degree of 75–85%, glutaraldehyde (GLU) as aqueous solution grade II at 25%, hydroxyapatite (HA), ethanol 99.8%, Dulbecco's Modified Eagle's Medium–high glucose (DMEM high glucose culture medium), antibiotic solutions (penicillin-streptomycin, and amphotericin B), and Trypan blue were acquired from Sigma-Aldrich (Bogotá, Colombia). Glycerol (GLY) 99.5% was purchased from (Bogotá, Colombia). Fetal Calf Serum (FCS) was provided by Microgen (Bogotá, Colombia). None of the reagents were subject to further processing.

The scaffolds were prepared as follows: A 3% w/v acid solution of CH was prepared by the addition of the polymer to acetic acid at 0.2 M under mild manual stirring. Once the polymer was partially dissolved, the solution was stirred in a shaker for 24 h using a Unimax 1010DT instrument (Heidolph, Schwabach, Germany). On the other hand, an aqueous solution of PVA at 3% w/v was prepared in deionized water at 70 °C under vigorous magnetic stirred using a magnetic stirred (Model SP131325Q, Thermo-Scientific, Shangai, China). HA was then added to the previous solutions at 5% w/w related to the total amount of the polymer, and the new blends were stirred until a homogeneous system was formed. Two sets of polymeric solutions were prepared in three different volume ratios (1:1, 1:3 and 3:1). GLU and GLY at 0.001% w/w and 2.5% w/w related to the total amount of the polymer were added respectively to the first set as cross-linker and plasticizer respectively, and the blends were stored at 4 °C to allow crosslinking reactions. The second set was maintained as a control group. The prepared solutions were poured in molds to create disks and layers and frozen at −80 °C before freeze-drying process. Finally, the samples were freeze-dried at −40 °C and 3 mbar for 48 h (Super Modulyo manifold lyophilizer, Edwardsm, Wilmington, DE, USA) and kept at −4 °C until testing.

3.2. Degree of Swelling (DS) and Solubility Degree (SD)

Gravimetric method was used to obtain degree of swelling and solubility degree. Disk specimens were dried in a desiccator for 24 h to obtain dry weight (W1). Then, the samples were immersed in deionized water at 20 °C for several periods (2, 4, 6, 8, 10, 15, 20, 30, 60, 90 min and 24 h). At the end of each period excess of water was removed from the surface with filter paper and specimens were weighed again. The last measure (24 h) was considered as the swelling limit (W2) because in that period the equilibrium was reached. Finally, samples were dried until constant weight (W3). Degree of swelling (DS) and solubility degree (SD) were determined using Equations (1) and (2), respectively. The test was performed with three specimens for each sample (n = 3) to obtain statistically significant data:

$$DS = [(W_2 - W_1)/W_1] \times 100 \tag{1}$$

$$SD = [(W_1 - W_3)/W_3] \times 100 \tag{2}$$

3.3. FTIR-ATR Spectroscopy

Spectra for pure CH, PVA, HA and all blends with and without additives were obtained using a Nicolet iS5 spectrometer (Thermo Fisher, Waltham, MA, USA) equipped with attenuated total reflectance (ATR) accessory utilizing the Zn-Se crystal. Each spectrum represents 64 co-added scans referenced against an empty ATR cell spectrum. The spectra range was from 4000 to 650 cm^{-1} with a resolution of 1.92 cm^{-1}.

3.4. Mechanical Characterization

Young's modulus, tensile strength, and Yield strength were evaluated in tensile mode and tested on six specimens per sample using a rectangular test specimen specified in ASTM-F-2150-01/05 using a M350-CT Testometric tester (Lincoln Close, Rochdale, UK). The rate was 2 mm/min and the thickness of the samples was measured by a micrometer with the accuracy of 0.01 mm.

3.5. Differential Scanning Calorimetry (DSC)

Calorimetric measurements were carried out in a DSC 1 calorimeter, Mettler Toledo (Zurich, Switzerland), under nitrogen flowing at a rate 30 mL min^{-1}. The specimens were pressed in unsealed aluminum pans. Glass transition temperature (T_g) and melting temperature (T_m) were obtained, and the former was determined as the midpoint temperature by standard extrapolation of the linear part of DSC curves whereas the later as the maximum value of the melting peak. The samples were cooled down by air at an exponentially decreasing rate. The heating of the cycle was performed from 25 to 240 °C at a rate of 20 °C/min. The relative crystallinity (Xc) was estimated from the endothermic area using Equation (3), where ΔH_f is the measured enthalpy of the fusion from DSC thermograms and ΔH_f^0 is the enthalpy of fusion for 100% crystalline PVA (138.6 J g^{-1}) [92]:

$$Xc = \Delta H_f / \Delta H_f^0 \tag{3}$$

3.6. Scanning Electron Microscopy

Micrographs of the prepared samples were taken by a Nova NanoSEM 450 (FEI, Brno, Czech Republic) scanning electron microscope with Schottky field emission electron source operated at acceleration voltage ranging from 200 V to 30 kV and low-vacuum SED (LVD) detector. A coating with a thin layer of gold was performed by a sputter coater SC 7640 (Quorum Technologies, Newhaven, East Sussex, UK). Pore size and surface void percentage were determined using the Fiji biological image analysis software [93]. Average pore sizes were obtained by measuring ten randomly selected pores from SEM images and void percentage was calculated by processing SEM images to obtain 8-bit (black and white) micrographs of the surface of the scaffold and then calculating a percentage of void representing pixels using Equation (4), where V_p is the number of void pixels (black) and S_p (white) is the number of surface pixels:

$$\text{Void } \% = V_p / (V_p + S_p) \tag{4}$$

3.7. Cell Culture

Samples were sterilized using an ethanol series and neutralized in phosphate-buffered saline PBS at 4 °C which also worked as a progressive rehydration of the samples to avoid sudden and excessive swelling (see Section 2.1). Neutrality was confirmed by incubating the samples in DMEM culture media supplemented with 5% FBS, and 1% of antibiotic solution at 37 °C, 5% CO2 for 24 h to check color and pH changes of the media. Osteoblasts were obtained from a primary culture contained from an 8-day bird embryo; cells were washed in a PBS with antibiotics solution and subject to trypsinization for 30 min to separate them from the bottom of the flask to obtain a subculture. Once the subculture was confluent, cells were re-trypsinized and suspended for cultivation into the prepared scaffolds. Viability was assessed using the Trypan Blue Exclusion Method and cells were seeded at the same conditions of the initial culture. Cell culture was performed for 17 days, changing culture media and observing cells through the inverted microscope every three days.

3.8. Statistical Analysis

All experiments with quantitative measurements were performed in triplicate, except for mechanical testing where n = 6. All experimental values were expressed in form of average ± standard deviation. Results were statistically compared using two-way analysis of variance (ANOVA) with $p < 0.05$.

4. Conclusions

One of the main objectives for TE is the design of 3-dimensional materials able to mimic the bone extracellular matrix and ultimately allow bone cell adhesion, proliferation and differentiation. In this context, the prepared scaffolds were tested in order to elucidate how the material might behave in conditions which can be catalogued as similar for their possible performance as a candidate for bone regeneration. Individually, PVA and CH have singular properties which demonstrate benefits in biomedical science. However, the combination of both result in a new material with enhanced mechanical and biological properties. Within this framework, it is possible to claim that PVA-CH polymeric scaffolds with ulterior polymeric treatment, exhibit potential for bone tissue regeneration. The collected information suggest that CPS are considerably more pertinent for withstand moisture conditions with an appropriate solubility resistance, property which can be considered as desirable for bone tissue regeneration. Furthermore, the material proves to be suitable for osteoblastic colonization, differentiation, and proliferation within a range of two weeks, which is an indicative of negligible level of cytotoxicity. The addition of GLU, GLY and HA to the polymeric solutions produced a well porous interconnected material which favors cell growth; CH is determinant in pore formation as a higher concentration produces a higher uniformity and distribution of porosity. Variations in concentration of the polymers indicated that CPS with a CH:PVA 3:1 ratio exhibit a higher potential in tissue engineering and should be subjected to further analysis to determine to a full extent its limits in osteoblastic proliferation and ultimate bone regeneration.

Author Contributions: S.P.-C., C.B.-L., and A.B.-B. designed the study, and carried out all the experiments, wrote, and edited the manuscript. L.M. collaborated in the SEM images obtention. S.P.-C. performed statistical analysis.

H.S-P., D.N.-M., D.M.-F. and A.V.-B. performed biological tests and assisted biological data recollection and analysis. All authors approved the final manuscript.

Funding: This research received no external funding.

Acknowledgments: This work was supported by the Ministry of Education, Youth and Sports of the Czech Republic—Program NPU I (LO1504). Special thanks to Universidad Distrital Francisco José de Caldas, because of their cooperation in freeze-drying process.

Conflicts of Interest: The authors declare no conflict of interest.

References

1. Jagur-Grodzinski, J. Polymers for tissue engineering, medical devices, and regenerative medicine. Concise general review of recent studies. *Polym. Adv. Technol.* **2006**, *17*, 395–418. [CrossRef]
2. O'Brien, F.J. Biomaterials & scaffolds for tissue engineering. *Mater. Today* **2011**, *14*, 88–95.
3. Yang, S.; Leong, K.-F.; Du, Z.; Chua, C.-K. The design of scaffolds for use in tissue engineering. Part I. Traditional factors. *Tissue Eng.* **2001**, *7*, 679–689. [CrossRef] [PubMed]
4. Sabir, M.I.; Xu, X.; Li, L. A review on biodegradable polymeric materials for bone tissue engineering applications. *J. Mater. Sci.* **2009**, *44*, 5713–5724. [CrossRef]
5. Buckwalter, J.A.; Glimcher, M.J.; Cooper, R.R.; Recker, R. Bone biology. I: Structure, blood supply, cells, matrix, and mineralization. *Instruct. Course Lect.* **1996**, *45*, 371–386.
6. Dimitriou, R.; Jones, E.; McGonagle, D.; Giannoudis, P. V Bone regeneration: Current concepts and future directions. *BMC Med.* **2011**, *9*, 66. [CrossRef] [PubMed]
7. Henkel, J.; Woodruff, M.A.; Epari, D.R.; Steck, R.; Glatt, V.; Dickinson, I.C.; Choong, P.F.M.; Schuetz, M.A.; Hutmacher, D.W. Bone regeneration based on tissue engineering conceptions—A 21st century perspective. *Bone Res.* **2013**, *1*, 216–248. [CrossRef] [PubMed]
8. Available online: http://www.redalyc.org/articulo.oa?id=179214945008 (accessed on 27 June 2018).
9. Liu, X.; Ma, P.X. Polymeric scaffolds for bone tissue engineering. *Ann. Biomed. Eng.* **2004**, *32*, 477–486. [CrossRef] [PubMed]
10. Polo-Corrales, L.; Latorre-Esteves, M.; Ramirez-Vick, J.E. Scaffold design for bone regeneration. *J. Nanosci. Nanotechnol.* **2014**, *14*, 15–56. [CrossRef] [PubMed]
11. Yi, H.; Rehman, F.U.; Zhao, C.; Liu, B.; He, N. Recent advances in nano scaffolds for bone repair. *Bone Res.* **2016**, *4*, 16050. [CrossRef] [PubMed]
12. Motamedian, S.R.; Hosseinpour, S.; Ahsaie, M.G.; Khojasteh, A. Smart scaffolds in bone tissue engineering: A systematic review of literature. *World J. Stem Cells* **2015**, *7*, 657. [CrossRef] [PubMed]
13. Hutmacher, D.W.; Woodfield, T.B.F.; Dalton, P.D. Scaffold design and fabrication. In *Tissue Engineering*, 2nd ed.; Elsevier: London, UK, 2015; pp. 311–346.
14. Sears, N.A.; Seshadri, D.R.; Dhavalikar, P.S.; Cosgriff-Hernandez, E. A review of three-dimensional printing in tissue engineering. *Tissue Eng. Part B Rev.* **2016**, *22*, 298–310. [CrossRef] [PubMed]
15. Nishio, Y.; Suzuki, H.; Sato, K. Molecular orientation and optical anisotropy induced by the stretching of poly(vinyl alcohol) poly(N-vinyl pyrrolidone) blends. *Polymer* **1994**, *35*, 1452–1461. [CrossRef]
16. Sachlos, E.; Czernuszka, J.T. Making tissue engineering scaffolds work. Review: The application of solid freeform fabrication technology to the production of tissue engineering scaffolds. *Eur. Cell Mater.* **2003**, *5*, 39–40. [CrossRef]
17. Subia, B.; Kundu, J.; Kundu, S.C. Biomaterial scaffold fabrication techniques for potential tissue engineering applications. In *Tissue Engineering*; InTech: Rijeka, Croatia, 2010.
18. Lasprilla, A.J.R.; Martinez, G.A.R.; Lunelli, B.H.; Jardini, A.L.; Maciel Filho, R. Poly-lactic acid synthesis for application in biomedical devices—A review. *Biotechnol. Adv.* **2012**, *30*, 321–328. [CrossRef] [PubMed]
19. Shum, A.W.T.; Mak, A.F.T. Morphological and biomechanical characterization of poly(glycolic acid) scaffolds after in vitro degradation. *Polym. Degrad. Stab.* **2003**, *81*, 141–149. [CrossRef]
20. Lam, C.X.F.; Hutmacher, D.W.; Schantz, J.-T.; Woodruff, M.A.; Teoh, S.H. Evaluation of polycaprolactone scaffold degradation for 6 months in vitro and in vivo. *J. Biomed. Mater. Res. Part A* **2009**, *90*, 906–919. [CrossRef] [PubMed]

21. Zhang, W.; Yang, Y.; Zhang, K.; Luo, T.; Tang, L.; Li, Y. Silk-Poly(lactic-co-glycolic acid) Scaffold /Mesenchymal Stem Cell Composites for Anterior Cruciate Ligament Reconstruction in Rabbits. *J. Biomater. Tissue Eng.* **2017**, *7*, 571–581. [CrossRef]

22. Stratton, S.; Shelke, N.B.; Hoshino, K.; Rudraiah, S.; Kumbar, S.G. Bioactive polymeric scaffolds for tissue engineering. *Bioact. Mater.* **2016**, *1*, 93–108. [CrossRef] [PubMed]

23. Ramalingam, M.; Tiwari, A.; Ramakrishna, S.; Kobayashi, H. *Integrated Biomaterials for Biomedical Technology*; John Wiley & Sons: New York, NY, USA, 2012.

24. Di Martino, A.; Sittinger, M.; Risbud, M.V. Chitosan: A versatile biopolymer for orthopaedic tissue-engineering. *Biomaterials* **2005**, *26*, 5983–5990. [CrossRef] [PubMed]

25. Croisier, F.; Jérôme, C. Chitosan-based biomaterials for tissue engineering. *Eur. Polym. J.* **2013**, *49*, 780–792. [CrossRef]

26. Sultana, N.; Mokhtar, M.; Hassan, M.I.; Jin, R.M.; Roozbahani, F.; Khan, T.H. Chitosan-based nanocomposite scaffolds for tissue engineering applications. *Mater. Manuf. Process.* **2015**, *30*, 273–278. [CrossRef]

27. Wang, F.; Wang, M.; She, Z.; Fan, K.; Xu, C.; Chu, B.; Chen, C.; Shi, S.; Tan, R. Collagen/chitosan based two-compartment and bi-functional dermal scaffolds for skin regeneration. *Mater. Sci. Eng. C* **2015**, *52*, 155–162. [CrossRef] [PubMed]

28. Madihally, S.V.; Matthew, H.W.T. Porous chitosan scaffolds for tissue engineering. *Biomaterials* **1999**, *20*, 1133–1142. [CrossRef]

29. Suh, J.-K.F.; Matthew, H.W.T. Application of chitosan-based polysaccharide biomaterials in cartilage tissue engineering: A review. *Biomaterials* **2000**, *21*, 2589–2598. [PubMed]

30. Kim, I.-Y.; Seo, S.-J.; Moon, H.-S.; Yoo, M.-K.; Park, I.-Y.; Kim, B.-C.; Cho, C.-S. Chitosan and its derivatives for tissue engineering applications. *Biotechnol. Adv.* **2008**, *26*, 1–21. [CrossRef] [PubMed]

31. Nettles, D.L.; Elder, S.H.; Gilbert, J.A. Potential use of chitosan as a cell scaffold material for cartilage tissue engineering. *Tissue Eng.* **2002**, *8*, 1009–1016. [CrossRef] [PubMed]

32. Jafari, M.; Paknejad, Z.; Rad, M.R.; Motamedian, S.R.; Eghbal, M.J.; Nadjmi, N.; Khojasteh, A. Polymeric scaffolds in tissue engineering: A literature review. *J. Biomed. Mater. Res. Part B Appl. Biomater.* **2017**, *105*, 431–459. [CrossRef] [PubMed]

33. Bedian, L.; Villalba-Rodriguez, A.M.; Hernández-Vargas, G.; Parra-Saldivar, R.; Iqbal, H.M.N. Bio-based materials with novel characteristics for tissue engineering applications—A review. *Int. J. Biol. Macromol.* **2017**, *98*, 837–846. [CrossRef] [PubMed]

34. Arce Guerrero, S.; Valencia Llano, C.; Garzón-Alvarado, D.A. Obtención de un biocompuesto constituido por fosfato tricálcico y quitosana para ser usado como sustituto óseo en un modelo animal. *Rev. Cuba. Investig. Biomédicas* **2012**, *31*, 268–277.

35. Bernal, A.; Balkova, R.; Kuritka, I.; Saha, P. Preparation and characterisation of a new double-sided bio-artificial material prepared by casting of poly(vinyl alcohol) on collagen. *Polym. Bull.* **2013**, *70*, 431–453. [CrossRef]

36. Georgieva, N.; Bryaskova, R.; Tzoneva, R. New Polyvinyl alcohol-based hybrid materials for biomedical application. *Mater. Lett.* **2012**, *88*, 19–22. [CrossRef]

37. Pangon, A.; Saesoo, S.; Saengkrit, N.; Ruktanonchai, U.; Intasanta, V. Multicarboxylic acids as environment-friendly solvents and in situ crosslinkers for chitosan/PVA nanofibers with tunable physicochemical properties and biocompatibility. *Carbohydr. Polym.* **2016**, *138*, 156–165. [CrossRef] [PubMed]

38. Chahal, S.; Hussain, F.S.J.; Kumar, A.; Rasad, M.S.B.A.; Yusoff, M.M. Fabrication, characterization and in vitro biocompatibility of electrospun hydroxyethyl cellulose/poly(vinyl) alcohol nanofibrous composite biomaterial for bone tissue engineering. *Chem. Eng. Sci.* **2016**, *144*, 17–29. [CrossRef]

39. Kheradmandi, M.; Vasheghani-Farahani, E.; Ghiaseddin, A.; Ganji, F. Skeletal muscle regeneration via engineered tissue culture over electrospun nanofibrous chitosan/PVA scaffold. *J. Biomed. Mater. Res. Part A* **2016**, *104*, 1720–1727. [CrossRef] [PubMed]

40. Kanimozhi, K.; Basha, S.K.; Kumari, V.S. Processing and characterization of chitosan/PVA and methylcellulose porous scaffolds for tissue engineering. *Mater. Sci. Eng. C* **2016**, *61*, 484–491. [CrossRef] [PubMed]

41. Echeverri, C.E.; Vallejo, C.; Londoño, M.E. Síntesis y caracterización de hidrogeles de alcohol polivinílico por la técnica de congelamiento/descongelamiento para aplicaciones médicas. *Rev. EIA* **2009**, *12*, 59–66.

42. Kumar, H.M.P.N.; Prabhakar, M.N.; Prasad, C.V.; Rao, K.M.; Reddy, T.V.A.K.; Rao, K.C.; Subha, M.C.S. Compatibility studies of chitosan/PVA blend in 2% aqueous acetic acid solution at 30 C. *Carbohydr. Polym.* **2010**, *82*, 251–255. [CrossRef]

43. Rao, S.H.; Harini, B.; Shadamarshan, R.P.K.; Balagangadharan, K.; Selvamurugan, N. Natural and synthetic polymers/bioceramics/bioactive compounds-mediated cell signaling in bone tissue engineering. *Int. J. Biol. Macromol.* **2017**, *110*, 88–96. [CrossRef] [PubMed]

44. Thien, D.V.H.; Hsiao, S.W.; Ho, M.H.; Li, C.H.; Shih, J.L. Electrospun chitosan/hydroxyapatite nanofibers for bone tissue engineering. *J. Mater. Sci.* **2013**, *48*, 1640–1645. [CrossRef]

45. Mi Zo, S.; Singh, D.; Kumar, A.; Cho, Y.W.; Oh, T.H.; Han, S.S. Chitosan-hydroxyapatite macroporous matrix for bone tissue engineering. *Curr. Sci.* **2012**, *102*, 1438–1446.

46. Brun, V.; Guillaume, C.; Mechiche Alami, S.; Josse, J.; Jing, J.; Draux, F.; Bouthors, S.; Laurent-Maquin, D.; Gangloff, S.C.; Kerdjoudj, H.; et al. Chitosan/hydroxyapatite hybrid scaffold for bone tissue engineering. *Biomed. Mater. Eng.* **2014**, *24*, 63–73. [PubMed]

47. Chen, Y.; Yu, J.; Ke, Q.; Gao, Y.; Zhang, C.; Guo, Y. Bioinspired fabrication of carbonated hydroxyapatite/chitosan nanohybrid scaffolds loaded with TWS119 for bone regeneration. *Chem. Eng. J.* **2018**, *341*, 112–125. [CrossRef]

48. Tsiourvas, D.; Sapalidis, A.; Papadopoulos, T. Hydroxyapatite/chitosan-based porous three-dimensional scaffolds with complex geometries. *Mater. Today Commun.* **2016**, *7*, 59–66. [CrossRef]

49. Li, Y.; Liu, T.; Zheng, J.; Xu, X. Glutaraldehyde-crosslinked chitosan/hydroxyapatite bone repair scaffold and its application as drug carrier for icariin. *J. Appl. Polym. Sci.* **2013**, *130*, 1539–1547. [CrossRef]

50. Ma, L.; Gao, C.; Mao, Z.; Zhou, J.; Shen, J.; Hu, X.; Han, C. Collagen/chitosan porous scaffolds with improved biostability for skin tissue engineering. *Biomaterials* **2003**, *24*, 4833–4841. [CrossRef]

51. Bernal, A.; Kuritka, I.; Saha, P. Preparation and characterization of poly(vinyl alcohol)-poly(vinyl pyrrolidone) blend: A biomaterial with latent medical applications. *J. Appl. Polym. Sci.* **2013**, *127*, 3560–3568. [CrossRef]

52. Elt, O.; Gurny, R. Structure and interactions in chitosan hydrogels formed by complexation or aggregation for biomedical applications. *Eur. J. Pharm. Biopharm.* **2004**, *57*, 35–52.

53. Alhosseini, S.N.; Moztarzadeh, F.; Mozafari, M.; Asgari, S.; Dodel, M.; Samadikuchaksaraei, A.; Kargozar, S.; Jalali, N. Synthesis and characterization of electrospun polyvinyl alcohol nanofibrous scaffolds modified by blending with chitosan for neural tissue engineering. *Int. J. Nanomed.* **2012**, *7*, 25. [CrossRef]

54. Mansur, H.S.; de Costa, E.S.; Mansur, A.A.P.; Barbosa-Stancioli, E.F. Cytocompatibility evaluation in cell-culture systems of chemically crosslinked chitosan/PVA hydrogels. *Mater. Sci. Eng. C* **2009**, *29*, 1574–1583. [CrossRef]

55. Ramay, H.R.; Zhang, M. Preparation of porous hydroxyapatite scaffolds by combination of the gel-casting and polymer sponge methods. *Biomaterials* **2003**, *24*, 3293–3302. [CrossRef]

56. Berzina-Cimdina, L.; Borodajenko, N. Research of calcium phosphates using Fourier transform infrared spectroscopy. In *Infrared Spectroscopy-Materials Science, Engineering and Technology*; InTech: Rijeka, Croatia, 2012.

57. Fathi, M.H.; Hanifi, A.; Mortazavi, V. Preparation and bioactivity evaluation of bone-like hydroxyapatite nanopowder. *J. Mater. Process. Technol.* **2008**, *202*, 536–542. [CrossRef]

58. Kumar, G.N.H.; Rao, J.L.; Gopal, N.O.; Narasimhulu, K.V.; Chakradhar, R.P.S.; Rajulu, A.V. Spectroscopic investigations of Mn^{2+} ions doped polyvinylalcohol films. *Polymer* **2004**, *45*, 5407–5415. [CrossRef]

59. Holland, B.J.; Hay, J.N. The thermal degradation of poly(vinyl alcohol). *Polymer* **2001**, *42*, 6775–6783. [CrossRef]

60. Mansur, H.S.; Sadahira, C.M.; Souza, A.N.; Mansur, A.A.P. FTIR spectroscopy characterization of poly(vinyl alcohol) hydrogel with different hydrolysis degree and chemically crosslinked with glutaraldehyde. *Mater. Sci. Eng. C* **2008**, *28*, 539–548. [CrossRef]

61. Bernal, A.; Kuritka, I.; Kasparkova, V.; Saha, P. The effect of microwave irradiation on poly(vinyl alcohol) dissolved in ethylene glycol. *J. Appl. Polym. Sci.* **2013**, *128*, 175–180. [CrossRef]

62. Bernal-Ballén, A.; Kuritka, I.; Saha, P. Preparation and characterization of a bioartificial polymeric material: Bilayer of cellulose acetate-PVA. *Int. J. Polym. Sci.* **2016**, *2016*, 3172545. [CrossRef]

63. Pawlak, A.; Mucha, M. Thermogravimetric and FTIR studies of chitosan blends. *Thermochim. Acta* **2003**, *396*, 153–166. [CrossRef]

64. Marsano, E.; Vicini, S.; Skopińska, J.; Wisniewski, M.; Sionkowska, A. Chitosan and poly(vinyl pyrrolidone): Compatibility and miscibility of blends. *Macromol. Symp.* **2004**, *218*, 251–260. [CrossRef]

65. Oliveira, J.M.; Rodrigues, M.T.; Silva, S.S.; Malafaya, P.B.; Gomes, M.E.; Viegas, C.A.; Dias, I.R.; Azevedo, J.T.; Mano, J.F.; Reis, R.L. Novel hydroxyapatite/chitosan bilayered scaffold for osteochondral tissue-engineering applications: Scaffold design and its performance when seeded with goat bone marrow stromal cells. *Biomaterials* **2006**, *27*, 6123–6137. [CrossRef] [PubMed]

66. Li, M.; Cheng, S.; Yan, H. Preparation of crosslinked chitosan/poly(vinyl alcohol) blend beads with high mechanical strength. *Green Chem.* **2007**, *9*, 894. [CrossRef]

67. Coates, J. Interpretation of infrared spectra, a practical approach. *Encycl. Anal. Chem.* **2000**, *12*, 10815–10837.

68. De Souza Costa-Júnior, E.; Pereira, M.M.; Mansur, H.S. Properties and biocompatibility of chitosan films modified by blending with PVA and chemically crosslinked. *J. Mater. Sci. Mater. Med.* **2009**, *20*, 553–561. [CrossRef] [PubMed]

69. Figueiredo, K.C.S.; Alves, T.L.M.; Borges, C.P. Poly(vinyl alcohol) films crosslinked by glutaraldehyde under mild conditions. *J. Appl. Polym. Sci.* **2009**, *111*, 3074–3080. [CrossRef]

70. Zheng, H.; Du, Y.; Yu, J.; Huang, R.; Zhang, L. Preparation and characterization of chitosan/poly(vinyl alcohol) blend fibers. *J. Appl. Polym. Sci.* **2001**, *80*, 2558–2565. [CrossRef]

71. Koosha, M.; Mirzadeh, H. Electrospinning, mechanical properties, and cell behavior study of chitosan/PVA nanofibers. *J. Biomed. Mater. Res. Part A* **2015**, *103*, 3081–3093. [CrossRef] [PubMed]

72. Miya, M.; Iwamoto, R.; Mima, S. FT-IR study of intermolecular interactions in polymer blends. *J. Polym. Sci. Part B Polym. Phys.* **1984**, *22*, 1149–1151. [CrossRef]

73. Zhou, Y.S.; Yang, D.Z.; Nie, J. Effect of PVA content on morphology, swelling and mechanical property of crosslinked chitosan/PVA nanofibre. *Plast. Rubber Compos.* **2007**, *36*, 254–258. [CrossRef]

74. Yu, Z.; Li, B.; Chu, J.; Zhang, P. Silica in situ enhanced PVA/chitosan biodegradable films for food packages. *Carbohydr. Polym.* **2018**, *184*, 214–220. [CrossRef] [PubMed]

75. Bhajantri, R.F.; Ravindrachary, V.; Harisha, A.; Crasta, V.; Nayak, S.P.; Poojary, B. Microstructural studies on BaCl2 doped poly(vinyl alcohol). *Polymer* **2006**, *47*, 3591–3598. [CrossRef]

76. Liang, S.; Yang, J.; Zhang, X.; Bai, Y. The thermal-electrical properties of polyvinyl alcohol/AgNO$_3$ films. *J. Appl. Polym. Sci.* **2011**, *122*, 813–818. [CrossRef]

77. Chuang, W.-Y.; Young, T.-H.; Yao, C.-H.; Chiu, W.-Y. Properties of the poly(vinyl alcohol)/chitosan blend and its effect on the culture of fibroblast in vitro. *Biomaterials* **1999**, *20*, 1479–1487. [CrossRef]

78. Milosavljević, N.B.; Kljajević, L.M.; Popović, I.G.; Filipović, J.M.; Kalagasidis Krušić, M.T. Chitosan, itaconic acid and poly(vinyl alcohol) hybrid polymer networks of high degree of swelling and good mechanical strength. *Polym. Int.* **2010**, *59*, 686–694.

79. Gonzalez-Campos, J.B.; Prokhorov, E.; Luna-Barcenas, G.; Fonseca-Garcia, A.; Sanchez, I.C. Dielectric relaxations of chitosan: The effect of water on the α-relaxation and the glass transition temperature. *J. Polym. Sci. Part B Polym. Phys.* **2009**, *47*, 2259–2271. [CrossRef]

80. Hu, H.; Xin, J.H.; Hu, H.; Chan, A.; He, L. Glutaraldehyde–chitosan and poly(vinyl alcohol) blends, and fluorescence of their nano-silica composite films. *Carbohydr. Polym.* **2013**, *91*, 305–313. [CrossRef] [PubMed]

81. Cascone, M.G.; Barbani, N.P.; Giusti, C.C.; Ciardelli, G.; Lazzeri, L. Bioartificial polymeric materials based on polysaccharides. *J. Biomater. Sci. Polym. Ed.* **2001**, *12*, 267–281. [CrossRef] [PubMed]

82. Bonilla, J.; Fortunati, E.; Atarés, L.; Chiralt, A.; Kenny, J.M. Physical, structural and antimicrobial properties of poly vinyl alcohol—Chitosan biodegradable films. *Food Hydrocoll.* **2014**, *35*, 463–470. [CrossRef]

83. Lewandowska, K. Miscibility and thermal stability of poly(vinyl alcohol)/chitosan mixtures. *Thermochim. Acta* **2009**, *493*, 42–48. [CrossRef]

84. Dehnad, D.; Mirzaei, H.; Emam-Djomeh, Z.; Jafari, S.-M.; Dadashi, S. Thermal and antimicrobial properties of chitosan–nanocellulose films for extending shelf life of ground meat. *Carbohydr. Polym.* **2014**, *109*, 148–154. [CrossRef] [PubMed]

85. Ahmad, A.L.; Yusuf, N.M.; Ooi, B.S. Preparation and modification of poly(vinyl) alcohol membrane: Effect of crosslinking time towards its morphology. *Desalination* **2012**, *287*, 35–40. [CrossRef]

86. Hsieh, W.-C.; Chang, C.-P.; Lin, S.-M. Morphology and characterization of 3D micro-porous structured chitosan scaffolds for tissue engineering. *Colloids Surf. B Biointerfaces* **2007**, *57*, 250–255. [CrossRef] [PubMed]

87. Karageorgiou, V.; Kaplan, D. Porosity of 3D biomaterial scaffolds and osteogenesis. *Biomaterials* **2005**, *26*, 5474–5491. [CrossRef] [PubMed]

88. Mansur, H.S.; Costa, H.S. Nanostructured poly(vinyl alcohol)/bioactive glass and poly(vinyl alcohol)/chitosan/bioactive glass hybrid scaffolds for biomedical applications. *Chem. Eng. J.* **2008**, *137*, 72–83. [CrossRef]
89. Sasaki, M.; Hongo, H.; Hasegawa, T.; Suzuki, R.; Zhusheng, L.; de Freitas, P.H.L.; Yamada, T.; Oda, K.; Yamamoto, T.; Li, M.; et al. Morphological aspects of the biological function of the osteocytic lacunar canalicular system and of osteocyte-derived factors. *Oral Sci. Int.* **2012**, *9*, 1–8. [CrossRef]
90. Bonewald, L.F. The amazing osteocyte. *J. Bone Miner. Res.* **2011**, *26*, 229–238. [CrossRef] [PubMed]
91. Bloch, S.L.; Kristensen, S.L.; Sørensen, M.S. The viability of perilabyrinthine osteocytes: A quantitative study using bulk-stained undecalcified human temporal bones. *Anat. Rec. Adv. Integr. Anat. Evol. Biol.* **2012**, *295*, 1101–1108. [CrossRef] [PubMed]
92. Asran, A.S.; Henning, S.; Michler, G.H. Polyvinyl alcohol-collagen-hydroxyapatite biocomposite nanofibrous scaffold: Mimicking the key features of natural bone at the nanoscale level. *Polymer* **2010**, *51*, 868–876. [CrossRef]
93. Schindelin, J.; Arganda-Carreras, I.; Frise, E.; Kaynig, V.; Longair, M.; Pietzsch, T.; Preibisch, S.; Rueden, C.; Saalfeld, S.; Schmid, B.; et al. Fiji: An open-source platform for biological-image analysis. *Nat. Methods* **2012**, *9*, 676–682. [CrossRef] [PubMed]

Sample Availability: Samples of the compounds are not available from the authors.

Article

Enzymatic Synthesis of Amino Acids Endcapped Polycaprolactone: A Green Route Towards Functional Polyesters

Stéphane W. Duchiron [1], Eric Pollet [1,*], Sébastien Givry [2] and Luc Avérous [1,*]

[1] BioTeam/ICPEES-ECPM, UMR CNRS 7515, Université de Strasbourg, 25 rue Becquerel, 67087 Strasbourg CEDEX 2, France; stephane.duchiron@etu.unistra.fr

[2] J. SOUFFLET S. A., Centre de Recherche et d'Innovation Soufflet—Division Biotechnologies, Quai du Général Sarail, 10402 Nogent sur Seine CEDEX 2, France; sgivry@soufflet.com

* Correspondence: eric.pollet@unistra.fr (E.P.); luc.averous@unistra.fr (L.A.); Tel.: +33-368-852-786 (E.P.)

Received: 20 October 2017; Accepted: 23 January 2018; Published: 30 January 2018

Abstract: ε-caprolactone (CL) has been enzymatically polymerized using α-amino acids based on sulfur (methionine and cysteine) as (co-)initiators and immobilized lipase B of *Candida antarctica* (CALB) as biocatalyst. In-depth characterizations allowed determining the corresponding involved mechanisms and the polymers thermal properties. Two synthetic strategies were tested, a first one with direct polymerization of CL with the native amino acids and a second one involving the use of an amino acid with protected functional groups. The first route showed that mainly polycaprolactone (PCL) homopolymer could be obtained and highlighted the lack of reactivity of the unmodified amino acids due to poor solubility and affinity with the lipase active site. The second strategy based on protected cysteine showed higher monomer conversion, with the amino acids acting as (co-)initiators, but their insertion along the PCL chains remained limited to chain endcapping. These results thus showed the possibility to synthesize enzymatically polycaprolactone-based chains bearing amino acids units. Such cysteine endcapped PCL materials could then find application in the biomedical field. Indeed, subsequent functionalization of these polyesters with drugs or bioactive molecules can be obtained, by derivatization of the amino acids, after removal of the protecting group.

Keywords: enzymatic polymerization; caprolactone; amino acids; methionine; cysteine; ring opening polymerization; *Candida antarctica* lipase B; polyester functionalization

---

1. Introduction

During the last two decades, there has been a high focus in polymer for biomedical applications, for numerous purposes from drug delivery to tissue engineering [1,2]. All these fields require specific physical, chemical and biological properties. To reach such properties, several architectures have been developed from polymer chemistry by copolymerization, grafting or chemical functionalization [3–5]. There is a growing need for polymers with easily tunable properties which can combine different macromolecular architectures.

One of the most important properties required for most biomedical uses is the polymer biocompatibility [2], in connection with the non-toxicity of the degradation products and residual monomers [6]. Poly(ε-caprolactone) (PCL) is a well-known polyester that has been extensively studied in the last decades because of its non-toxicity, biocompatibility and bioresorbability [6–12]. PCL-based materials have been developed for numerous biomedical applications such as tissue engineering, wound dressing or single-use medical equipment [8–10]. For example, Park et al. demonstrated its great potential for bone-repair scaffolds [11] and several articles report its use for drug release and delivery systems [11,12].

Nowadays, the ring opening polymerization (ROP) of lactones by metal-based catalysts is the main way to obtain well-controlled polyester chains [13,14]. However, these residual catalysts may induce some toxicity (in the case of biomedical applications) [15], environmental pollution (in the case of compostability) and an increase in the polymer degradation kinetics [16] which could limit the material use. As potential alternatives, some common organocatalysts (like 4-dimethylaminopyridine, triflic acid or *N*-heterocyclic carbene) [17] show great potential for lactone ROP, but may also present some toxicity issues and drawbacks.

Enzymatic catalysis currently shows a great potential to substitute toxic metal-based catalysts or organocatalysts to limit the final toxicity and negative environmental impacts, in perfect agreement with a sustainable development and concepts of green chemistry [18]. By enzymatic catalysis, reactions can be performed under mild conditions (low temperature and pressure). Systems with high catalytic activity and very good reaction control of enantio-, chemo-, regio-, and stereo-selectivity can be expected. Owing to these advantages, enzymatic processes could provide precise control of the final polymer architectures, allowing the synthesis of polymers by a clean process, without by-products and with energy savings [19]. Enzymatic polymerization can thus be regarded as an environment-friendly synthetic process for polymeric materials, providing one of the best examples of "green polymer chemistry". Among the vast choice of enzymes, lipases (E.C. 3.1.1.3) can be found in most organisms from microbial, plant and animal kingdoms [20]. They are serine hydrolase that catalyze ester bond cleavage in aqueous medium (physiological action is cleavage of triglyceride) and they are also able to catalyze esters bond formation in organic medium (reverse reaction) [21]. Currently, lipases seem to be the most efficient or at least the most known enzymes for polyester synthesis by enzymatic ring opening polymerization (eROP).

Lactones are one of the most widely studied monomers in eROP catalyzed by lipases [22], especially by lipase B from *Candida antarctica* (CALB) which is the most efficient enzyme for such reaction. Some studies have shown that large lactones with low ring strain are more compatible with lipases binding site than shorter ones with higher ring strain [23]. Interestingly, chemical ROP shows the inverse reactivity trend with monomer ring strain.

Because of the chemical similarity, other heterocycles based on sulfur, and to a lesser extent on nitrogen, have been studied. Indeed, if some works have been published on the eROP of thiolactones [24–29], the enzymatic synthesis of polyamides by ROP seems much more difficult and there are only a few works on the eROP of lactam. These studies show that lactam unit integration is limited. Besides, it is very difficult to obtain long polyamide chains because of huge stability of large lactam compared to β-lactam [30–32]. However, lipases could catalyze the synthesis of lactams from five to nine membered cycles, and the formation of cyclic amino acids dimers and trimers [33]. Some other strategies have been used to synthesize polyamides or polyesteramides such as polycondensation reaction catalyzed by lipases (e.g., synthesis of low molar mass poly(aspartic acid)) [34] or eROP catalyzed by proteases instead of lipases or cutinases which are the most commonly used enzymes [34–37]. Some authors also tried to modify enzymes by mutagenesis, aiming at improving polyamide production, but with mixed results so far [38].

However, polyesters containing peptides or amino acids are of great interest for pharmaceutical and biomedical applications such as drug carriers and controlled drug delivery systems or as resorbable implants and tissue engineering materials. Indeed, polyester functionalization with amino acids has been reported to improve the biomedical properties compared to the parent polymer [39,40]. Among the various benefits arising from the introduction of amino acids into polyester chains, one may cite the improved interactions with proteins, cells or genes as well as the possibility for subsequent functionalization with bioactive molecules or drugs [39,41]. Such polyester–amino acid or polyester–peptide conjugates can be obtained by chemical modification of previously formed polyester chains, or by using a functionalized ester monomer but a more elegant way consists in using peptide or amino acids as (macro-)initiators, for instance, for the ring opening polymerization of lactones [42,43]. Various natural amino acids have been tested and reported in the literature for their ability to initiate the ring opening polymerization of ε-caprolactone (CL) or lactide, in bulk, without the use of metal-based catalysts [43–46]. However, the literature on the chemical polymerization of ε-caprolactone initiated by amino acids remains limited and, to the best of our knowledge, the enzyme-catalyzed ROP of CL initiated by natural amino acids has never been reported.

In the present work, we thus report assays of two distinct synthetic routes to enzymatically synthesize functional polyesters based on ε-caprolactone (CL) and two selected natural amino acids. Cysteine (Cys) and methionine (Met) have been selected since their respective side groups present adequate chemical functions: (i) a thiol for cysteine, which allows subsequent branching, grafting or cross-linking; and (ii) a thioether for methionine that could allow a control of the polymer thermal resistance, for example by sulfonation of the thioether [47]. A protection–unprotection approach of the chemical groups has been used and then two different strategies were tested: (i) preliminary tests based on direct polymerization, using methionine and cysteine amino acids; and (ii) an indirect approach, involving protected cysteine. The macromolecular architectures and the mechanisms to obtain the corresponding polymers have been analyzed through different techniques. The main goal of this study is to synthesize amino acids containing polyesters, and to compare and draw conclusions on the relative reactivity of acid, ester, amine and thiol functions, by enzymatic catalysis.

2. Results and Discussion

In a preliminary approach, we tested the direct enzyme-catalyzed polymerization of CL with the unmodified amino acids, as shown in Figure 1.

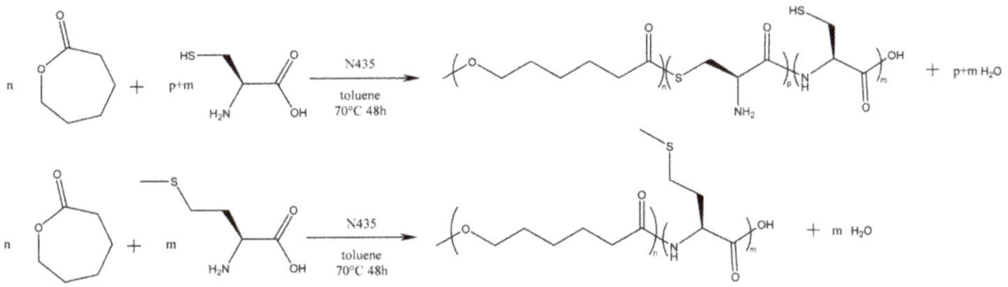

Figure 1. Preliminary direct approach for the polymerization of CL with unmodified amino acids.

Main results of the corresponding syntheses are summarized in Table 1, where the first line corresponds to the blank test without amino acid, the second to the fifth lines correspond to CL polymerization with cysteine and the last four to CL polymerization with methionine. These experiments allow to compare the two amino acids (AA) and the effect of the AA feed content.

Table 1. Results obtained from the direct polymerization of ε-caprolactone (CL) with the two unmodified amino acids (methionine and cysteine).

Amino Acid	AA Feed Content (mol %)	M_n [1] (g/mol)	Đ [1]	T_m (°C)	T_c (°C)	X (%)	T_d [2] (°C)	η [3] (%)
-	0	15 000	1.7	54	34	59	400	95
Cys	1	11 300	2.1	51	36	52	401	92
Cys	2	11 200	2.1	51	32	51	399	90
Cys	5	11 500	2.2	52	32	47	400	85
Cys	10	10 500	2.1	52	36	46	399	86
Met	1	11 200	2.2	52	36	45	395	93
Met	2	11 000	2.2	50	35	49	398	89
Met	5	10 800	2.2	52	36	46	394	87
Met	10	10 500	2.1	51	35	55	396	84

[1] Values determined by SEC and given as PS standards; [2] main degradation temperature at maximum degradation rate; [3] final yield obtained after vacuum drying.

Table 1 clearly shows that addition of small amounts of amino acids in the reaction medium does not lead to significant differences on the resulting polymers. The number average molar masses (Mn) are quite constant, ranging from 10,500 to 11,500 g·mol^{-1} with cysteine and from 10,500 to 11,200 g·mol^{-1} with methionine. These values are slightly lower than for the blank test that shows a molar mass around 15,000 g·mol^{-1}. This could be due to the initiation by amino acids or more likely by the water molecules brought by the amino acids that are very hygroscopic compounds. It is noteworthy that these values are given as polystyrene (PS) standards and are thus largely overestimated. The real Mn values are likely around half those presented here [46] and are in good agreement with those reported for amino acids initiated polymerization of CL in bulk [43–46]. One can also notice the relatively high dispersity (Đ), around 2, which could mean that there are two different initiation steps and/or competitive transesterification reactions. As reported in a previous work, the acrylic resin-immobilized form of CALB (N435) is an efficient transesterification catalyst [48]. This high dispersity is also linked to a long reaction time (48 h) compared to the more conventional 24 h often used for eROP of CL [49]. Such a high dispersity could also be explained by an initiation either by the amine of methionine or cysteine or by the thiol of cysteine, both being in competition with the conventional water initiation. Indeed, if initiation involves amine, thiol and hydroxyl (water) groups then their different reactivity may lead to different initiation rate constants that will generate different chains populations that could contribute to a higher dispersity. For the sake of comparison, dispersity values ranging from 1.3 to 1.9 have been reported for amino acid initiated PCL [43–46]. However, the occurrence of numerous transesterification reactions is more likely the main reason for the high dispersity observed here.

From the TGA results, one can observe a non-negligible weight loss before the main degradation peak (Td at around 400 °C) mainly due to water (below 150 °C) and residual CL monomer (below 250 °C) losses. These molecules have been chemically identified by FTIR analysis of the evolved gas from the degradation (Figure S1).

A representative DSC thermogram is presented in Figure S2 but all products display similar pattern. Indeed, one can notice from the DSC results (Table 1) that thermal properties seem not to be affected by the addition of amino acid during the synthesis and these values are very close to those expected and observed for PCL homopolymer. For example, melting temperature (Tm) ranges between 50 °C and 52 °C, which is comparable to, or slightly below, the melting temperature of PCL of such molar mass [43,50]. Only the crystallinity (X) shows some variations, without clear trend, but these values are very sensitive to the limits set for the peak integration and thus to the melting peak width. Nevertheless, these values are in good agreement with the crystallinity reported for PCL obtained from arginine-based macro-initiators [43]. In addition, all products show a double melting peak, which is also consistent with the thermal behavior of PCL chains of similar molar mass. This is due to crystals imperfections and some difficulties for the polymer to crystallize when molar masses are limited [50]. In the same way, degradation temperatures (Td around 400 °C) are very close to the one of neat PCL. These results tend to demonstrate that there is no, or a very limited, insertion of amino acids in the PCL polymer chain.

This has been confirmed by the ^{1}H- and ^{13}C-NMR analyses displaying only the characteristic peaks of PCL chains (Figure S3), highlighting that mainly PCL homopolymers have been obtained. The presence of residual CL has also been confirmed by ^{1}H-NMR spectroscopy (Figure S3a), with an estimated content ranging from 1% to 3% after the polymer precipitation recovery step, depending on the AA feed content. The highest value is obtained for the reaction involving 10 mol % Cys which also corresponds to the lowest CL monomer conversion. Such presence is somewhat expected since our precipitation protocol in cold methanol (−78 °C) is done to recover the smallest polymer chains but is also able to precipitate CL. One can notice that, initial amino acids do not appear on products analysis. They are probably precipitated at the end of the reaction when chloroform is added and then eliminated by filtration together with the enzyme.

The results of these preliminary tests show that the insertion of amino acid units in the polymer chain has not been clearly established and thus the corresponding products were not considered for further characterization. In a previously published study, Sobczak et al. claimed a successful initiation with both methionine and cysteine (at a content of ca. 2 mol %) and thus their incorporation in PCL chains but without giving any proof of such insertion [45]. In our case, it is assumed that the limited insertion of amino acids and their lack of reactivity are mainly due to their poor solubility in CL and toluene but that could also be due to a poor affinity with the lipase binding site. To improve the amino acid incorporation in the chain and to increase the solubility of the amino acid in the reaction medium, *N*-Boc cysteine hexyl ester (*N*-Boc Cys HE) (Figure 2) has been used. Indeed, tertiobutyl and hexyl group drastically increase the solubility of amino acids in organic solvent [51]. Besides, compared to the native carboxylic acid, hexyl ester could make the monomer closer to the lipase natural substrates (i.e., fatty esters) and could thus increase the monomer affinity with enzyme binding site, leading to higher reactivity.

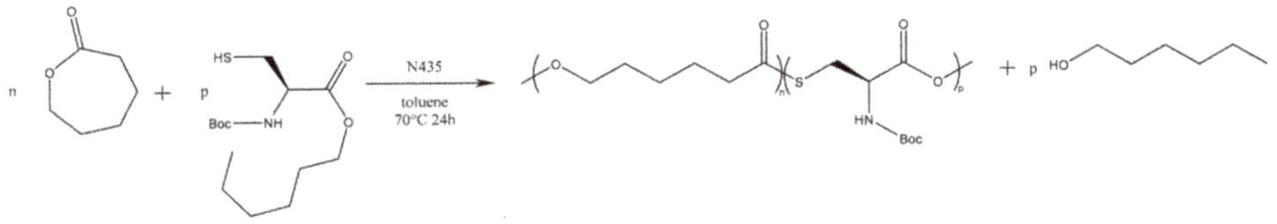

Figure 2. Second polymerization strategy based on the use of *N*-Boc protected cysteine hexyl ester.

However, since the first experiments did not show any reactivity of the thioether function of methionine, and since *N*-protected methionine would have only one reactive group left, making its polymerization impossible, this second strategy has been used only on cysteine derivatives.

Results of the polymerization reactions carried out on the modified cysteine are summarized in Table 2. At first, one can notice a significant decrease of the molar masses with the increase in amino acid feed content. This is consistent with the results reported by Liu et al. and Sobczak et al. showing a global decrease of the PCL molar mass with increasing amount of amino acid [44,45]. Besides, such decrease is not surprising since there are four different potential initiating species. Two of them are initially present in the reaction medium (residual water and *N*-Boc Cys HE) and the others are produced during the polymerization reaction (hexanol from transesterification of *N*-Boc Cys HE and water produced by reaction on thiol function of *N*-Boc Cys HE). Among these four potential initiators, three of them thus show an increasing content with the increment of the amino acid feed content. One can also notice that, for the 1% *N*-Boc Cys HE feed content, the final molar mass is higher than in preliminary step and close to the blank test (see Table 1). This is likely due to the lower hygroscopicity of the protected amino acid compared to native one.

Table 2. Results obtained from the polymerization of CL with *N*-Boc cysteine hexyl ester.

AA Feed Content (mol %)	Mn [1] (g/mol)	Đ [1]	Tm (°C)	X (%)	Td [2] (°C)	Mass Loss at 250 °C (%)	η [3] (%)	AA Final Content [4] (%)
1	14 300	2.3	54	54	405	1.1	96	0.7
2	11 400	2.5	54	54	404	1.4	93	0.8
5	8 000	1.6	51	61	402	2.0	94	1.6
10	6 500	1.6	52	54	405	3.3	96	1.9

[1] Values determined by SEC and given as PS standards; [2] main degradation temperature at maximum degradation rate; [3] final yield obtained after vacuum drying; [4] values determined by NMR.

According to these results, one can notice that there are on average only 1 to 1.3 amino acid units per chain. This means that the amino acid reacts mainly as initiator or as chain ending species. This is consistent with the results reported in the literature for similar systems and the amino acid content is also in agreement with contents reported for creatine endcapped PCL [45]. This low content in incorporated amino acid is also confirmed by TGA analyses with the mass loss below 250 °C which corresponds to the loss of the Boc protecting group, the residual water and the unreacted monomers as shown by the FT-IR coupling (Figure S1).

The ^{13}C-NMR characterization shows interesting pattern (Figure 3). Most of the characteristic peaks of protected amino acid could be easily identified except for the three peaks corresponding to the carbonyl of Boc protecting group, to the quaternary carbon of the tertiobutyl group and to the carbonyl of hexyl ester function. The absence of these peaks is not surprising because of the usually weak response of sp^4 carbons in ^{13}C-NMR and the small amount of amino acid in the polymers (with a maximum of 1.9 mol %, Table 2).

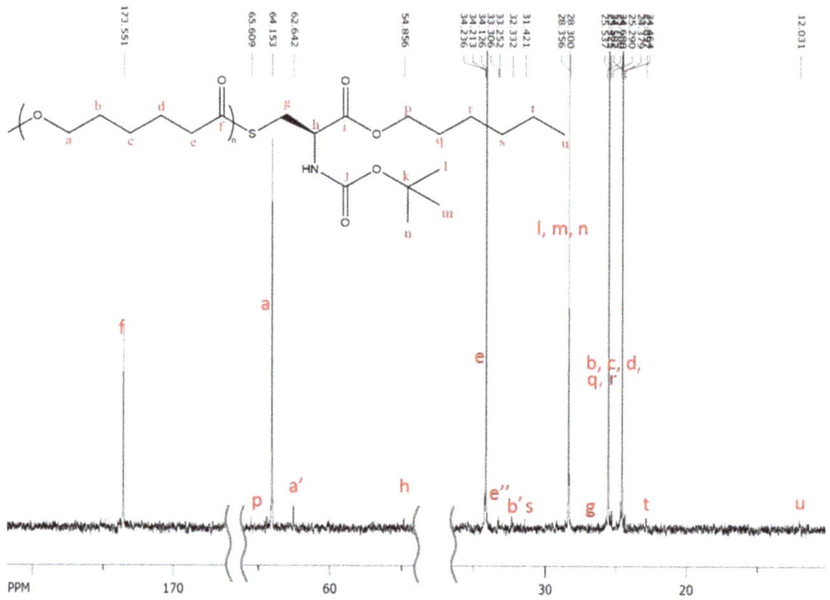

Figure 3. ^{13}C-NMR of polymerization product between CL and 10 mol % of *N*-Boc Cys HE.

Interestingly, one could also notice that the signals of hexyl ester are always observed in the polymer which could mean that at least part of *N*-Boc Cys HE did not react on its hexyl ester function or that hexanol released by transesterification of *N*-Boc Cys HE initiated an eROP side reaction. This second possibility is more probable due to the well-known ability of N435 to catalyze transesterification reactions.

One can also notice on the ^{13}C-NMR spectrum several small peaks at 63 ppm and between 30 ppm and 34 ppm (see assignment on Figure 3) with a′ and b′ corresponding to alcohol chain-end and e″ corresponding to thioester linkage.

These results seem to indicate that several distinct populations of polymer chains could be formed due to some side reactions, such as hexanol initiation, but also due to the multiple reactive sites on the *N*-Boc Cys HE monomer.

^{1}H-NMR spectra of the products obtained from CL polymerization with *N*-Boc Cys HE (Figure 4, Figures S4a, S5 and S6 for 10%, 1%, 2% and 5% amino acid feed content, respectively) show significantly different patterns than those observed with the first polymerization strategy.

From such spectrum, one can clearly see that the polymer chains are mainly composed of CL units. Here again low intensity signals corresponding to alcohol chain-end (a′) and to the thioester linkage (e″) can be identified. However, residual monomer content in the products cannot be determined precisely due to peaks overlapping at 4.1–4.2 ppm but is more likely lower than 5%. Besides, all characteristic peaks of *N*-Boc Cys HE can be clearly identified on the ^{1}H-NMR spectrum (Figure 4). This includes the hexyl ester signals corresponding to the chain population initiated by a thiol function and to the possible population initiated by hexanol resulting from the transesterification of cysteine hexyl ester. However, these analyses did not allow us to determine which reaction is more likely and to conclude on this specific point.

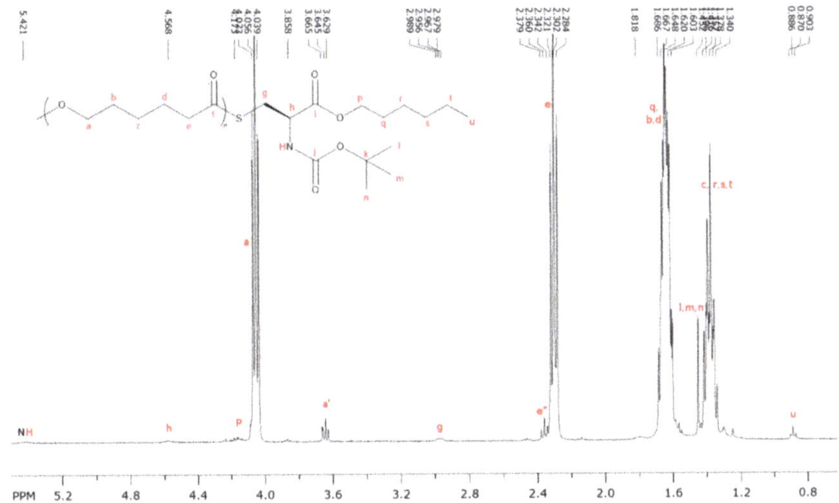

Figure 4. ^{1}H-NMR of polymerization product between CL and 10 mol % of *N*-Boc Cys HE.

To characterize more precisely those polymers, 2D NMR experiments were performed. Despite some absent (or too low intensity) signals on the ^{13}C-NMR, the proposed peaks assignment has been confirmed by HSQC (^{1}H–^{13}C), two-dimensional NMR experiment which directly shows the bonded protons and carbons. Figure 5 shows the confirmation of our assignment for the low intensity peaks in ^{13}C-NMR for carbons g and h, which make coupling spots at, respectively, 27 and 55 ppm.

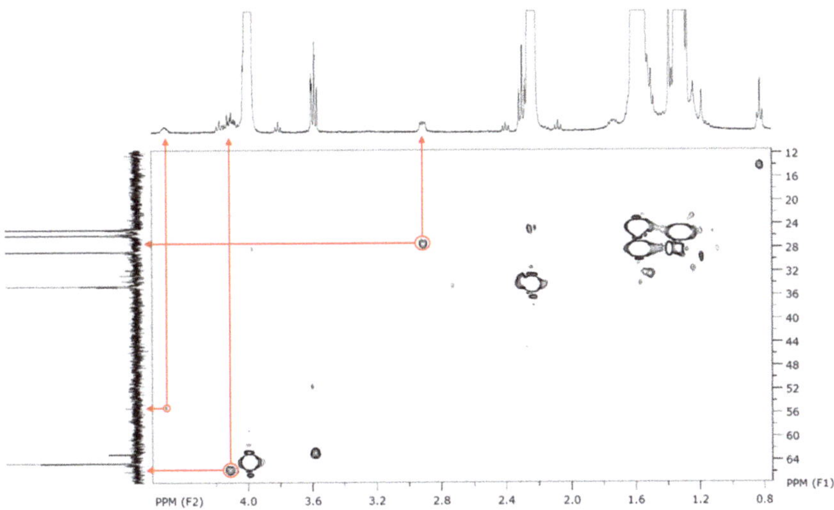

Figure 5. HSQC (^{1}H–^{13}C) 2D NMR of polymerization product between CL and 10 mol % of *N*-Boc Cysteine HE.

The NMR spectroscopy highlighted the possible presence of multiple polymer chain populations but, despite the deep analysis of the spectra, their different chemical structures remain only partially elucidated. Then, to clarify such complex polymer chains populations and structures, MALDI-ToF mass spectrometry analyses of the polymers of CL with *N*-Boc Cys HE were performed. Two examples of the resulting mass spectra are shown in Figure 6a,b, corresponding, respectively, to the polymerization products with 2% and 10% of *N*-Boc cysteine hexyl ester in the feed. The mass spectrum of the product obtained with 1% of *N*-Boc Cys HE in the feed is available in Figure S4b.

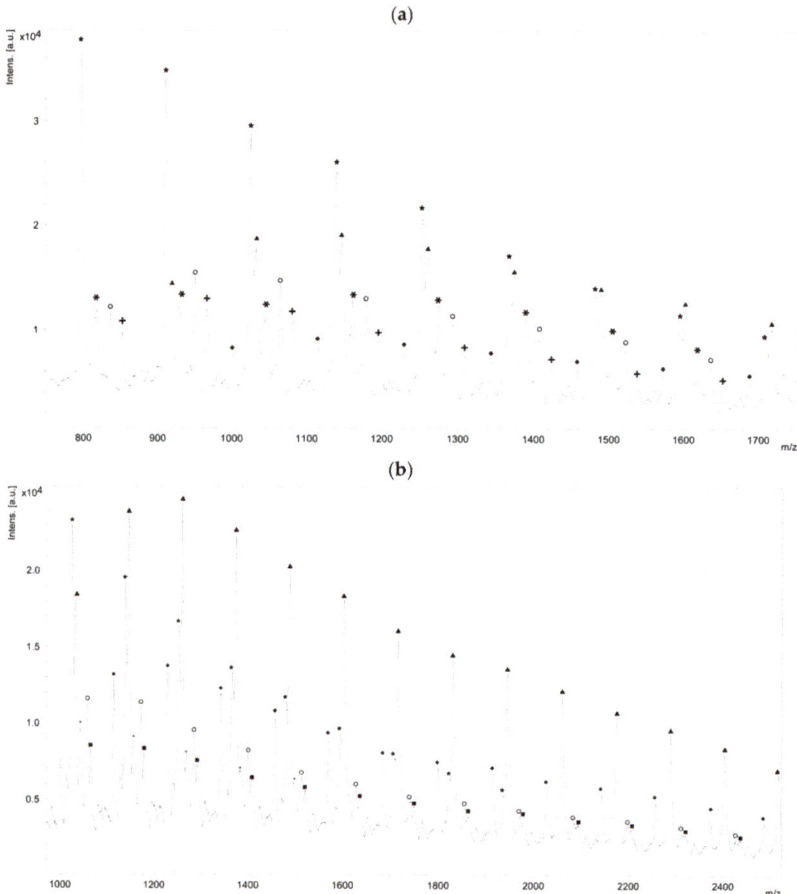

Figure 6. MALDI-ToF MS spectrum of polymerization product between CL and: (**a**) 2 mol %; and (**b**) 10 mol % of *N*-Boc Cys HE.

One can see in Figure 6a that there are at least six different distributions of polymer chains. These analyses also show that the m/z value of each peak logically verifies Equation (1).

$$m/z = M_{Na} - M_H + nM_{CL} + M_{chain-ends} \tag{1}$$

We can also notice in Figure 7a that, for all distributions, the interval between two successive peaks of a same distribution is 114 m/z, which exactly corresponds to one CL unit. This result validates our previous hypothesis made from the amino acid content and confirms that there is roughly only one amino acid per polymer chain. This is also in perfect agreement with MALDI-ToF results reported for PCL chains initiated by phenylalanine or arginine [45,46]. Regarding the m/z values distribution, one can observe a significant difference with the Mn values determined from SEC. This can be explained by the overestimation arising from the values given as PS standards but also from underestimated values given by MS analyses due to higher response usually observed for the lower molar mass compared to the longer chains.

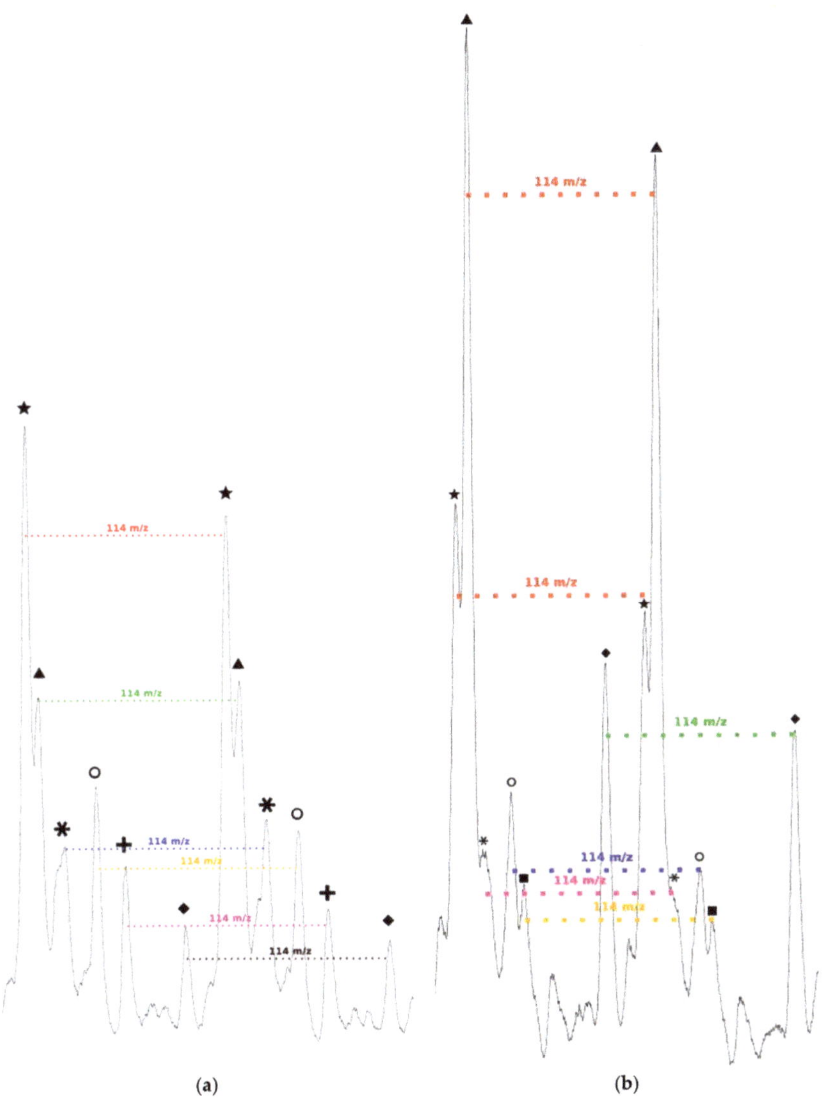

Figure 7. Interval between the successive peaks of a same polymer chain distribution for the polymerization of CL with: (**a**) 2 mol %; and (**b**) 10 mol % of *N*-Boc Cys HE.

A detailed analysis of these mass spectroscopy spectra allows to identify and propose the different types of chain-ends and/or polymer structure corresponding to each distribution. The main proposed structures and mechanisms for polymers obtained with 2 and 10 mol % of *N*-Boc Cys HE in the feed are summarized in Table 3.

Concerning the products obtained with 2 mol % of *N*-Boc Cys HE in the feed, as expected, the main distribution corresponds to chains initiated by the thiol function of cysteine and this initiation seems to be more important than the one from residual water. This could easily be explained by the very low water content since we minimize it for each product and at each step. One can also notice chains termination by an amino acid due to transesterification reaction involving the hexyl ester. Interestingly, this termination seems to become more important for longer polymer chains. That could indicate that after a first oligomerization/polymerization step, the chain growth is then mainly made by transesterification and chain coupling. As supposed from the NMR results, there is a chain distribution initiated by hexanol resulting from transesterification of the hexyl ester amino acid (Figure 2). From the MALDI-ToF mass spectra, one can also identify the presence of cyclic PCL chains which have already been reported in the literature [45,46,52]. All these chains populations resulting from transesterification reactions have already been reported for amino acids endcapped PCL [45,46].

Table 3. Proposed structures and mechanisms for CL polymerization with *N*-Boc Cys HE.

Marker	Structure	Supposed Mechanism
★	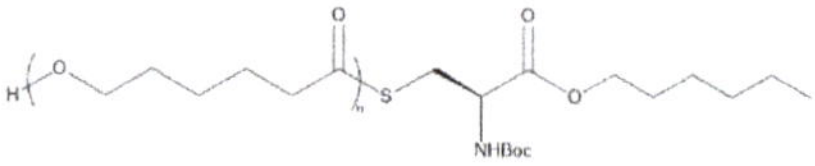	initiated by thiol of *N*-Boc Cys HE
○	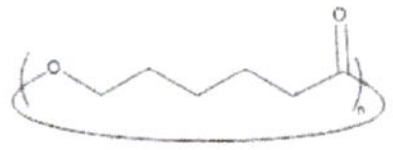	terminated by chain cyclization
▲	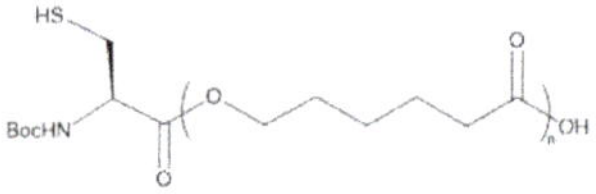	initiated by water and ending by transesterification of *N*-Boc Cys HE
*	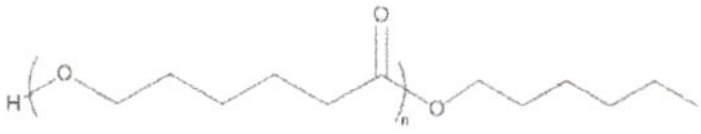	initiated by hexanol
+		initiated by residual water
◆	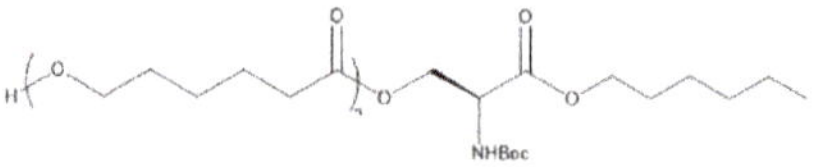	initiated by *N*-Boc Ser HE resulting from substitution of S by O in the binding site during the deacylation step
■	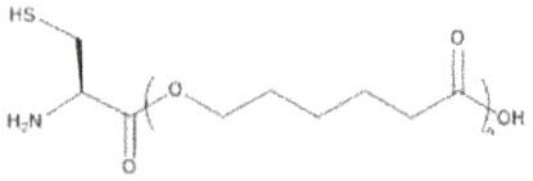	initiated by water and ending by transesterification of cysteine hexyl ester

Surprisingly, there is another distribution with a rather unexpected chain-end. According to the *m*/*z* values, a putative structure involving a *N*-Boc Serine hexyl ester (*N*-Boc Ser HE) is proposed. This hypothesized structure could originate from a substitution of the sulfur atom of the modified cysteine amino acid by an oxygen (from the serine residue of the active site) during the deacylation step of the reaction. A potential intermediate that could explain such mechanism is proposed (Figure 8) but there is neither experimental data nor existing literature references to support this statement.

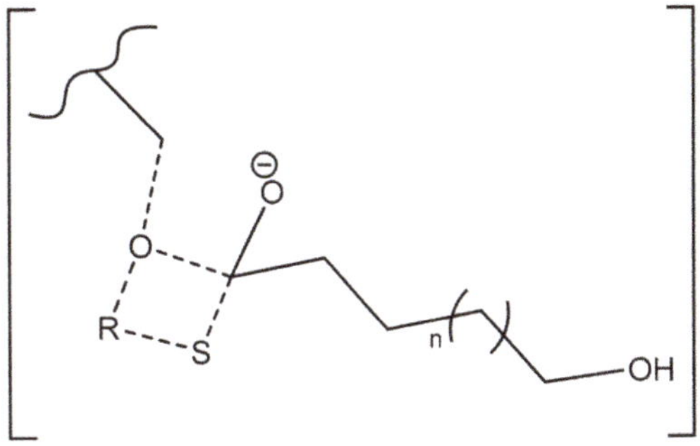

Figure 8. Potential intermediate formed in the enzymatic polymerization deacylation step that could lead to either a thioester- or an ester-initiated chain.

However, from the low peak intensity in the mass spectra, it seems that such substitution would be of very low occurrence and could explain why such products were not identified by NMR spectroscopy (also because their characteristic peaks could be masked by other signals).

Interestingly, for high amino acid contents in the feed (10 mol %), the MALDI-ToF mass spectrum is slightly different (Figure 6b). Indeed, six main distributions are also observed in this case but with two main differences in the identified structures (Table 3). One can first notice the absence of chains initiated by water molecules. This could be explained by the higher amount of amino acid which makes water molecules negligible compared to the others available initiators such as the thiol function of cysteine and hexanol released by transesterification of *N*-Boc Cys HE. As already mentioned, the occurrence of such transesterification reactions, that are promoted and catalyzed by the lipase, had been previously reported and observed from MALDI ToF spectra for similar systems [45,46]. As a second difference, one can also notice a surprising chain population that seems to be terminated by a cysteine. This distribution could result from a termination reaction by transesterification with unprotected cysteine hexyl ester. The latter could originate from a protected amino acid that would have been deprotected during the synthesis of *N*-Boc Cys HE. Indeed, amino acid esterification reaction is performed in acidic medium and this could weaken, and possibly remove, the *N*-Boc protecting group. However, according to the literature, lipases are unlikely to catalyze such Boc deprotection [53]. However, this chain distribution seems to be the least important.

One can also notice from the peaks intensities that the relative amounts of the various distributions are also different for polymers synthesized with larger quantity of amino acid in the feed. For instance, for the low amino acid content, the main distribution seems to be the chains initiated by the cysteine thiol whereas for higher feed content in amino acid, the chain terminated by transesterification of the hexyl ester seems to be major one. This latter behavior could be explained by the lower molar mass obtained when the feed content in protected amino acid is increased. As a result, protected amino acid chain ends are in greater ratio. The occurrence of lipase catalyzed transesterification reactions affecting these chain ends is thus statistically higher. It was also noticed that for longer chains, the sulfur substitution seems to be more important while this had not been observed for low amino acid feed content. However, as mentioned before, this trend is unexplained and the proposed mechanism and structure remain uncertain.

As seen in Figure 7b, and as already observed for the low amino acid content, for all distributions the interval between two successive peaks of a same distribution is 114 *m/z*, which corresponds exactly to one CL unit. This result confirms that, even if the amino acid content in the feed is increased, there is only around one amino acid unit incorporated in a polymer chain.

3. Materials and Methods

3.1. Materials

ε-Caprolactone (CL) was purchased from Sigma Aldrich (St. Louis, MO, USA) and distilled over CaH_2 under reduced pressure before each use. Methionine (Met) and cysteine (Cys) amino acids as well as Novozym®435 (N435, Novozymes, Bagsværd, Denmark), the acrylic resin-immobilized form of *Candida antarctica* lipase B (CALB), were purchased from Sigma Aldrich (St. Louis, MO, USA) and used after freeze drying. Anhydrous toluene was freshly distilled over sodium under nitrogen atmosphere prior each use. Hexanol was freshly distilled on molecular sieve under vacuum prior use. Other chemicals (*N*-Boc cysteine, *N*-Boc methionine, 4 Å molecular sieve, *p*-toluenesulfonic acid, and $NaHCO_3$) and solvents (methanol, diethyl ether, ethanol, chloroform, and dichloromethane) were purchased from Sigma Aldrich (St. Louis, MO, USA) and used without further purification or drying.

3.2. N-Boc Cysteine Hexyl Ester (N-Boc Cys HE) Synthesis

Typically, 1 g of *N*-Boc cysteine (4.5 mmol) was added in a round bottom flask with 20 equivalents of freshly distilled hexanol (11.4 mL; 90 mmol) under argon atmosphere. Then 2 mol % of *p*-toluenesulfonic acid were added to the reaction flask and the system was capped with a distillation bridge. Then the system was heated to 140 °C for 3 h, few droplets of water were collected by the distillation bridge. After cooling to 20 °C, the reaction medium was neutralized by $NaHCO_3$. After filtration, the organic phase was diluted in diethyl ether and washed several times with water. Then, residual alcohol and solvent were distilled to produce about 1.05 g (76% of yield) of a slightly yellow oil. ^{1}H-NMR ($CDCl_3$; 400 MHz): δ = 0.87–0.90 ppm (t, 3H, H_3C-CH2-); δ = 1.25–1.39 ppm (m, 6H, CH_3-C_3H_6-CH_2CH_2CO-); δ = 1.45 ppm (s, 9H, tertiobutyl); δ = 1.53–1.70 ppm (m, 2H, C_4H_6-CH_2CH_2O-); δ = 2.91–3.12 ppm (m, 2H, HS-CH_2CH); δ = 4.15–4.27 ppm(m, 2H, CH_2-CH_2O); δ = 4.35 ppm (s, 1H, NH); δ = 4.48–4.62 ppm (m, 1H, HS-CH_2-C*H*-).

3.3. Enzymatic Polymerization Setup

All reactions were carried out in dry toluene (2 mL), at 70 °C. For that, 3.95 mmol of CL (0.9017 g) and the adequate quantity of amino acid in molar equivalent (from 40 μmol to 0.395 mmol; e.g., for cysteine from 4.8 mg to 47.8 mg for 1% to 10 mol %, respectively) and 4 Å molecular sieve (0.05 g) were introduced into a previously dried Schlenck tube under an inert dry argon atmosphere. The tube was immediately capped with a rubber septum and then immersed in a heated oil bath at 70 °C. Toluene was transferred with a syringe through the rubber septum cap. A predetermined amount of N435 catalyst (50 mg) was quickly introduced in the tube under an inert dry argon atmosphere. The tube was immediately capped with a rubber septum. The enzyme addition marked the beginning (t0) of the polymerization. Reactions were terminated by dissolving the reaction mixture into chloroform and removing the catalyst by filtration. Part of the solvent in the filtrate was then stripped by rotary evaporation at 35 °C. The polymer in the resulting concentrated solution was precipitated in cold methanol (in a dry ice-ethanol bath at approximately −70°C) to ensure also the recovery of the shortest chains. The precipitated polymer was recovered by filtration and dried overnight at 30 °C under vacuum. For the polymerization involving *N*-Boc cysteine hexyl ester, similar procedure than for eROP of unprotected amino acids and CL was performed.

3.4. Characterization Techniques

NMR analyses were performed on a Bruker Ascend™ 400 spectrometer (Bruker, Wissembourg, France) operating at 400.13 MHz and 100.62 MHz for ^{1}H and ^{13}C NMR, respectively. Spectra were obtained by performing at least 64 scans for ^{1}H, 1024 scans for ^{13}C, 8 scans for COSY and 8 scans for HSQC analyses. 1D and 2D NMR spectra were exploited with SpinWork 4.1 freeware software (Dr K. Marat, University of Manitoba, Winnipeg, Canada). All samples were prepared in deuterated

chloroform with typically 8 mg for ^{1}H and 15 mg for ^{13}C and 2D NMR analyses. The determination of
N-Boc cysteine hexyl ester content in the polymer was calculated according to Equation (2):

$$N - \text{Boc cysteine hexyl ester } (\%) = 100 \times \frac{I_{Boc}/9}{(I_{Boc}/9) + (I_{PCL} + I_{PCL\ endchain})/2} \qquad (2)$$

using the integration of the singlet at 1.45 ppm, corresponding to the 9 protons of Boc tertiobutyl
group, compared to the sum of the integrals of the two triplets at 3.65 and 4.06 ppm corresponding to
the ROCH_2R in poly(ε-caprolactone) chain-end and main chain units, respectively. For the syntheses
with unprotected cysteine, the residual CL monomer content in the final product was calculated from
the ratio of the integrals of the CL and PCL characteristic peaks (respectively, at 4.1 and 4.0 ppm).

Size exclusion chromatography (SEC) measurements were performed in chloroform (HPLC grade)
in a Shimadzu liquid chromatograph equipped with a LC-10AD isocratic pump, a DGU-14A degasser,
a SIL-10AD automated injector, a CTO-10A thermostated oven with a 5 μm PLGel Guard column,
two PL-gel 5 μm MIXED-C and a 5 μm 100 Å 300 mm-columns, and three online detectors: a Shimadzu
RID-10A refractive index detector, a Wyatt Technologies MiniDAWN 3-angle-light scattering detector
and a Shimadzu SPD-M10A diode array (UV) detector. Samples were dissolved in chloroform
(concentration 4 mg·mL^{-1}) and filtered through a 0.45 μm PTFE membrane. For all analyses the
injection volume was 100 μL, the flow rate was 0.8 mL·min^{-1} and the oven temperature was set at
25 °C. The given molar masses and dispersity were calculated from a calibration with polystyrene
standards in RI detection. Refractive index increment values (dn/dc) were measured by injecting a
known concentration of polymer and assuming 100% mass recovery from the system.

MALDI-ToF analysis starts with the samples preparation. Matrix solutions were freshly
prepared with Super DHB (9:1 mixture of 2,5-dihydroxybenzoic acid and 2-hydroxy-5-methoxybenzoic
acid, from Sigma Aldrich (St. Louis, MO, USA)) which was dissolved till saturation in a
H_2O/CH$_3$CN/HCOOH (50%/50%/1%) solution. Typically, a 1:1 mixture of the sample solution
in CH_2Cl_2 was mixed with the matrix solution and 1 μL of the resulting mixture was deposited
on the stainless steel plate. Mass spectra were acquired on a time-of-flight mass spectrometer
(MALDI-ToF-ToF Autoflex II ToF-ToF, Bruker Daltonics, Bremen, Germany) equipped with a nitrogen
laser (λ = 337 nm). An external multi-point calibration was carried out before each measurement
using the singly charged peaks of a standard peptide mixture (0.4 μM, in water acidified with 1%
HCOOH). Scan accumulation and data processing were performed with FlexAnalysis 3.0 software
(Bruker Daltonics, Bremen, Germany).

The thermal stability and degradation of the samples was investigated by a thermal gravimetric
analyzer (TGA) coupled with a FTIR for evolved gas analysis. TGA measurements were conducted
under dry air (at a flow rate of 75 mL·min^{-1}) using a Hi-Res TGA Q5000 apparatus from TA
Instruments (New Castle, DE, USA). The samples (5–9 mg placed in a platinum pan) were heated up
to 450 °C at 5 °C·min^{-1}. FTIR spectra were recorded on a Nicolet 380 (Thermo Electron Corporation,
Waltham, MA, USA) by performing 16 scans with 4 cm^{-1} resolution.

DSC measurement were performed on a TA Q200 DSC (New Castle, DE, USA). in sealed
aluminum pan, typically on about 1 mg of purified sample, in a heat-cool-heat cycle at 10 and
5 °C·min^{-1} for heating and cooling, respectively. Results were exploited using TA Universal Analysis
software (New Castle, DE, USA). The crystallinity of the samples was determined using the neat PCL
common value of 139.5 J·g^{-1} for ΔH_m^0.

4. Conclusions

This work investigated the enzymatic polymerization of ε-caprolactone with unmodified or
modified sulfur-containing amino acids (cysteine and methionine). Two strategies were tested, a first
one based on a direct polymerization of CL with the native amino acids and a second route involving
the use of amino acids with protected functional groups. Even though lipases are catalysts that work
well at aqueous–organic interfaces and which could then be efficient for syntheses in heterogeneous

systems, the results from our first polymerization strategy did not show any significant reactivity of the amino acids. This lack of reactivity has been ascribed to a poor solubility of the amino acids in the reaction medium and to their low affinity with the lipase active site.

A second synthetic strategy based on the use of *N*-Boc and hexyl ester protected amino acid was thus established to enhance the solubility and improve the affinity with the lipase binding site. The significant conversion observed for the protected amino acids confirms that the solubility of these compounds and their affinity with the lipase binding site are key parameters for an efficient initiation of the CL enzymatic polymerization. This strategy produced various polymer chains populations that are mainly PCL homopolymer chains with distinct initiators and/or chain-ends, as highlighted by NMR and MALDI-ToF analyses. Each of the identified polymer chains distributions led us to formulate hypotheses on the mechanism of such enzymatic polymerization that once controlled could pave the way for the synthesis of new functionalized biocompatible polymers with potentially tunable macromolecular architectures and properties. Indeed, polyesters containing amino acids or peptides sequences are promising materials for biomedical applications such as controlled drug delivery systems or tissue engineering materials [39–41].

As far as cysteine-functionalized polymers are concerned, despite the different polymer backbones and synthesis strategies reported in the literature, some of the obtained materials have been described to form micelles showing enhanced cell adhesion or improved drug release properties [54,55]. In addition, the thiol group, which is present in some of the chains populations obtained with our strategy, and the carboxyl group can also be used for the grafting or conjugation of peptides sequences, antibodies, drugs or nanoparticles to obtain biomaterials with improved properties [56,57]. Regarding the polymer backbone functionalization using the thiol group, peptide conjugation using thiol-ene chemistry and native chemical ligation have been reported as efficient methods [56–58].

Thus, our polymerization strategy based on the use of modified amino acid opens perspectives for subsequent functionalization of the polymer chain by simply removing the protecting group that will make available a functional moiety on the polymer backbone. Then, further works would consist in investigating the polymer functionalization after deprotection and the potential biomedical applications of such cysteine-endcapped PCL chains.

Supplementary Materials: Supplementary materials are available online.

Acknowledgments: The authors would like to thanks European FEDER fund and Champagne-Ardenne Region (France) for their support to Soufflet S.A. on the Novopoly project. The authors also thank Ngov for here technical support.

Author Contributions: Stéphane W. Duchiron, Eric Pollet, Sebastien Givry and Luc Avérous conceived the experiments; Stéphane W. Duchiron performed the experiments; Stéphane W. Duchiron, Eric Pollet and Luc Avérous analyzed the data; and Stéphane W. Duchiron, Eric Pollet, Sebastien Givry and Luc Avérous wrote the paper.

Conflicts of Interest: The authors declare no conflict of interest.

References

1. Ricapito, N.G.; Ghobril, C.; Zhang, H.; Grinstaff, M.W.; Putnam, D. Synthetic biomaterials from metabolically derived synthons. *Chem. Rev.* **2016**, *116*, 2664–2704. [CrossRef] [PubMed]

2. Poole-Warren, L.A.; Patton, A.J. Introduction to biomedical polymers and biocompatibility. In *Biosynthetic Polymers for Medical Applications*; Poole-Warren, L., Martens, P., Green, R., Eds.; Woodhead Publishing Elsevier: Amsterdam, The Netherlands, 2016; Volume 1, pp. 3–31. ISBN 978-1-78242-105-4.

3. Lv, Z.; Chang, L.; Long, X.; Liu, J.; Xiang, Y.; Liu, J.; Liu, J.; Deng, H.; Deng, L.; Dong, A. Thermosensitive in situ hydrogel based on the hybrid of hyaluronic acid and modified PCL/PEG triblock copolymer. *Carbohydr. Polym.* **2014**, *108*, 26–33. [CrossRef] [PubMed]

4. Shi, S.; Shi, K.; Tan, L.; Qu, Y.; Shen, G.; Chu, B.; Zhang, S.; Su, X.; Li, X.; Wei, Y.; et al. The use of cationic MPEG-PCL-g-PEI micelles for co-delivery of Msurvivin T34A gene and doxorubicin. *Biomaterials* **2014**, *35*, 4536–4547. [CrossRef] [PubMed]

5. Li, Q.; Tan, W.; Zhang, C.; Gu, G.; Guo, Z. Novel triazolyl-functionalized chitosan derivatives with different chain lengths of aliphatic alcohol substituent: Design, synthesis, and antifungal activity. *Carbohydr. Res.* **2015**, *418*, 44–49. [CrossRef] [PubMed]

6. Ali, S.A.M.; Zhong, S.-P.; Doherty, P.J.; Williams, D.F. Mechanisms of polymer degradation in implantable devices: I. Poly(caprolactone). *Biomaterials* **1993**, *14*, 648–656. [CrossRef]

7. Khan, W.; Muntimadugu, E.; Jaffe, M.; Domb, A.J. Implantable medical devices. In *Focal Controlled Drug Delivery*; Domb, A.J., Khan, W., Eds.; Springer US: Boston, MA, USA, 2014; Volume 2, pp. 33–59. ISBN 978-1-4614-9433-1.

8. MacNeil, S. Biomaterials for tissue engineering of skin. *Mater. Today* **2008**, *11*, 26–35. [CrossRef]

9. Schlesinger, E.; Ciaccio, N.; Desai, T.A. Polycaprolactone thin-film drug delivery systems: Empirical and predictive models for device design. *Mater. Sci. Eng. C* **2015**, *57*, 232–239. [CrossRef] [PubMed]

10. Woodruff, M.A.; Hutmacher, D.W. The return of a forgotten polymer—Polycaprolactone in the 21st century. *Prog. Polym. Sci.* **2010**, *35*, 1217–1256. [CrossRef]

11. Park, K.E.; Kim, B.S.; Kim, M.H.; You, H.K.; Lee, J.; Park, W.H. Basic fibroblast growth factor-encapsulated PCL nano/microfibrous composite scaffolds for bone regeneration. *Polymer* **2015**, *76*, 8–16. [CrossRef]

12. Dash, T.K.; Konkimalla, V.B. Poly-ε-caprolactone based formulations for drug delivery and tissue engineering: A review. *J. Control. Release* **2012**, *158*, 15–33. [CrossRef] [PubMed]

13. Corma, A.; Iborra, S.; Velty, A. Chemical routes for the transformation of biomass into chemicals. *Chem. Rev.* **2007**, *107*, 2411–2502. [CrossRef] [PubMed]

14. Kricheldorf, H.R.; Berl, M.; Scharnagl, N. Poly(lactones). 9. Polymerization mechanism of metal alkoxide initiated polymerizations of lactide and various lactones. *Macromolecules* **1988**, *21*, 286–293. [CrossRef]

15. Tanzi, M.C.; Verderio, P.; Lampugnani, M.G.; Resnati, M.; Dejana, E.; Sturani, E. Cytotoxicity of some catalysts commonly used in the synthesis of copolymers for biomedical use. *J. Mater. Sci. Mater. Med.* **1994**, *5*, 393–396. [CrossRef]

16. Abe, H.; Takahashi, N.; Kim, K.J.; Mochizuki, M.; Doi, Y. Thermal degradation processes of end-capped poly(L-lactide)s in the presence and absence of residual zinc catalyst. *Biomacromolecules* **2004**, *5*, 1606–1614. [CrossRef] [PubMed]

17. Dubois, P.; Coulembier, O.; Raquez, J.-M. *Handbook of Ring-Opening Polymerization*; Wiley-VCH Verlag GmbH & Co. KGaA: Weinheim, Germany, 2009; ISBN 978-3-527-31953-4.

18. Shoda, S.; Uyama, H.; Kadokawa, J.; Kimura, S.; Kobayashi, S. Enzymes as green catalysts for precision macromolecular synthesis. *Chem. Rev.* **2016**, *116*, 2307–2413. [CrossRef] [PubMed]

19. Kobayashi, S. Recent developments in lipase-catalyzed synthesis of polyesters. *Macromol. Rapid Commun.* **2009**, *30*, 237–266. [CrossRef] [PubMed]

20. Schmid, R.D.; Verger, R. Lipases: Interfacial enzymes with attractive applications. *Angew. Chem. Int. Ed.* **1998**, *37*, 1608–1633. [CrossRef]

21. Albertsson, A.-C.; Srivastava, R.K. Recent developments in enzyme-catalyzed ring-opening polymerization. *Adv. Drug Deliv. Rev.* **2008**, *60*, 1077–1093. [CrossRef] [PubMed]

22. Uyama, H.; Kobayashi, S. Enzymatic ring-opening polymerization of lactones catalyzed by lipase. *Chem. Lett.* **1993**, *22*, 1149–1150. [CrossRef]

23. Van Buijtenen, J.; van As, B.A.C.; Verbruggen, M.; Roumen, L.; Vekemans, J.A.J.M.; Pieterse, K.; Hilbers, P.A.J.; Hulshof, L.A.; Palmans, A.R.A.; Meijer, E.W. Switching from S- to R-selectivity in the *Candida antarctica* lipase B-catalyzed ring-opening of ω-methylated lactones: tuning polymerizations by ring size. *J. Am. Chem. Soc.* **2007**, *129*, 7393–7398. [CrossRef] [PubMed]

24. Hedfors, C.; Hult, K.; Martinelle, M. Lipase chemoselectivity towards alcohol and thiol acyl acceptors in a transacylation reaction. *J. Mol. Catal. B* **2010**, *66*, 120–123. [CrossRef]

25. Iwata, S.; Toshima, K.; Matsumura, S. Enzyme-catalyzed preparation of aliphatic polyesters containing thioester linkages. *Macromol. Rapid Commun.* **2003**, *24*, 467–471. [CrossRef]

26. Kato, M.; Toshima, K.; Matsumura, S. Preparation of aliphatic poly(thioester) by the lipase-catalyzed direct polycondensation of 11-mercaptoundecanoic acid. *Biomacromolecules* **2005**, *6*, 2275–2280. [CrossRef] [PubMed]

27. Kato, M.; Toshima, K.; Matsumura, S. Enzyme-catalyzed preparation of aliphatic polythioester by direct polycondensation of diacid diester and dithiol. *Macromol. Rapid Commun.* **2006**, *27*, 605–610. [CrossRef]

28. Weber, N.; Klein, E.; Vosmann, K.; Mukherjee, K.D. Preparation of long-chain acyl thioesters—Thio wax esters—By the use of lipases. *Biotechnol. Lett.* **1998**, *20*, 687–691. [CrossRef]

29. Weber, N.; Klein, E.; Mukherjee, K.D. Long-chain acyl thioesters prepared by solvent-free thioesterification and transthioesterification catalysed by microbial lipases. *Appl. Microbiol. Biotechnol.* **1999**, *51*, 401–404. [CrossRef]

30. Stavila, E.; Arsyi, R.Z.; Petrovic, D.M.; Loos, K. *Fusarium solani pisi* cutinase-catalyzed synthesis of polyamides. *Eur. Polym. J.* **2013**, *49*, 834–842. [CrossRef]

31. Stavila, E.; Alberda van Ekenstein, G.O.R.; Woortman, A.J.J.; Loos, K. Lipase-catalyzed ring-opening copolymerization of ε-caprolactone and β-lactam. *Biomacromolecules* **2014**, *15*, 234–241. [CrossRef] [PubMed]

32. Elsässer, B.; Schoenen, I.; Fels, G. Comparative theoretical study of the ring-opening polymerization of caprolactam vs. caprolactone using QM/MM methods. *ACS Catal.* **2013**, *3*, 1397–1405. [CrossRef]

33. Stavila, E.; Loos, K. Synthesis of lactams using enzyme-catalyzed aminolysis. *Tetrahedron Lett.* **2013**, *54*, 370–372. [CrossRef]

34. Zhang, Y.; Xia, B.; Li, Y.; Wang, Y.; Lin, X.; Wu, Q. Solvent-free lipase-catalyzed synthesis: Unique properties of enantiopure D- and L- polyaspartates and their complexation. *Biomacromolecules* **2016**, *17*, 362–370. [CrossRef] [PubMed]

35. Ma, Y.; Sato, R.; Li, Z.; Numata, K. Chemoenzymatic synthesis of oligo(L-cysteine) for use as a thermostable bio-based material. *Macromol. Biosci.* **2016**, *16*, 151–159. [CrossRef] [PubMed]

36. Chen, Y.; Li, Y.; Gao, J.; Cao, Z.; Jiang, Q.; Liu, J.; Jiang, Z. Enzymatic PEGylated poly(lactone-co-β-amino ester) nanoparticles as biodegradable, biocompatible and stable vectors for gene delivery. *ACS Appl. Mater. Interfaces* **2016**, *8*, 490–501. [CrossRef] [PubMed]

37. Qin, X.; Xie, W.; Tian, S.; Ali, M.A.; Shirke, A.; Gross, R.A. Influence of N$_\varepsilon$-protecting groups on the protease-catalyzed oligomerization of L-lysine methyl ester. *ACS Catal.* **2014**, *4*, 1783–1792. [CrossRef]

38. Kawashima, Y.; Yasuhira, K.; Shibata, N.; Matsuura, Y.; Tanaka, Y.; Taniguchi, M.; Miyoshi, Y.; Takeo, M.; Kato, D.; Higuchi, Y.; et al. Enzymatic synthesis of nylon-6 units in organic solvents containing low concentrations of water. *J. Mol. Catal. B* **2010**, *64*, 81–88. [CrossRef]

39. Sun, H.; Meng, F.; Dias, A.A.; Hendriks, M.; Feijen, J.; Zhong, Z. α-amino acid containing degradable polymers as functional biomaterials: Rational design, synthetic pathway, and biomedical applications. *Biomacromolecules* **2011**, *12*, 1937–1955. [CrossRef] [PubMed]

40. Wang, B.; Ma, C.; Xiong, Z.C.; Bai, W.; Xiong, C.D.; Chen, D.L. Amino acid endcapped poly(p-dioxanone): Synthesis and crystallization. *J. Polym. Res.* **2013**, *20*, 116. [CrossRef]

41. Oledzka, E.; Sawicka, A.; Sobczak, M.; Nalecz-Jawecki, G.; Skrzypczak, A.; Kolodziejski, W. Prazosin-conjugated matrices based on biodegradable polymers and α-amino acids—Synthesis, characterization, and in vitro release study. *Molecules* **2015**, *20*, 14533–14551. [CrossRef] [PubMed]

42. Schappacher, M.; Soum, A.; Guillaume, S.M. Synthesis of polyester-polypeptide diblock and triblock copolymers using amino poly(ε-caprolactone) macroinitiators. *Biomacromolecules* **2006**, *7*, 1373–1379. [CrossRef] [PubMed]

43. Oledzka, E.; Sobczak, M.; Kolakowski, M.; Kraska, B.; Kamysz, W.; Kolodziejski, W. Development of creatine and arginine-6-oligomer for the ring-opening polymerization of cyclic esters. *Macromol. Res.* **2013**, *21*, 161–168. [CrossRef]

44. Liu, J.; Liu, L. Ring-opening polymerization of ε-caprolactone initiated by natural amino acids. *Macromolecules* **2004**, *37*, 2674–2676. [CrossRef]

45. Sobczak, M.; Oledzka, E.; Kolodziejski, W. Polymerization of cyclic, esters using aminoacid initiators. *J. Macromol. Sci. A* **2008**, *45*, 872–877. [CrossRef]

46. Oledzka, E.; Sokolowski, K.; Sobczak, M.; Kolodziejski, W. α-Amino acids as initiators of ε-caprolactone and L,L-lactide polymerization. *Polym. Int.* **2011**, *60*, 787–793. [CrossRef]

47. Haddadi, H.; Korani, E.M.; Hafshejani, S.M.; Farsani, M.R. Highly selective oxidation of sulfides to sulfones by H$_2$O$_2$ catalyzed by porous capsules. *J. Clust. Sci.* **2015**, *26*, 1913–1922. [CrossRef]

48. Duchiron, S.W.; Pollet, E.; Givry, S.; Avérous, L. Mixed systems to assist enzymatic ring opening polymerization of lactide stereoisomers. *RSC Adv.* **2015**, *5*, 84627–84635. [CrossRef]

49. Córdova, A.; Iversen, T.; Hult, K.; Martinelle, M. Lipase-catalysed formation of macrocycles by ring-opening polymerisation of ε-caprolactone. *Polymer* **1998**, *39*, 6519–6524. [CrossRef]

50. Huang, Y.P.; Xu, X.; Luo, X.L.; Ma, D.Z. Molecular weight dependence of the melting behavior of poly(ε-caprolactone). *Chin. J. Polym. Sci.* **2002**, *20*, 45–51.

51. Hojo, K.; Hara, A.; Kitai, H.; Onishi, M.; Ichikawa, H.; Fukumori, Y.; Kawasaki, K. Development of a method for environmentally friendly chemical peptide synthesis in water using water-dispersible amino acid nanoparticles. *Chem. Cent. J.* **2011**, *5*, 49. [CrossRef] [PubMed]

52. Huijser, S.; Staal, B.B.P.; Huang, J.; Duchateau, R.; Koning, C.E. Topology characterization by MALDI-ToF-MS of enzymatically synthesized poly(lactide-co-glycolide). *Biomacromolecules* **2006**, *7*, 2465–2469. [CrossRef] [PubMed]

53. Bornscheuer, U.T.; Kazlauskas, R.J. *Hydrolases in Organic Synthesis: Regio- and Stereoselective Biotransformations*, 2nd ed.; Wiley-VCH Verlag GmbH & Co. KGaA: Weinheim, Germany, 2006; ISBN 978-3-52760-754-9.

54. Sun, J.; Chen, X.; Lu, T.; Liu, S.; Tian, H.; Guo, Z.; Jing, X. Formation of reversible shell cross-linked micelles from the biodegradable amphiphilic diblock copolymer poly(L-cysteine)-block-poly(l-lactide). *Langmuir* **2008**, *24*, 10099–10106. [CrossRef] [PubMed]

55. Ulkoski, D.; Scholz, C. Synthesis and application of aurophilic poly(cysteine) and poly(cysteine)-containing copolymers. *Polymers* **2017**, *9*, 500. [CrossRef]

56. Jana, N.R.; Erathodiyil, N.; Jiang, J.; Ying, J.Y. Cysteine-functionalized polyaspartic acid: A polymer for coating and bioconjugation of nanoparticles and quantum dots. *Langmuir* **2010**, *26*, 6503–6507. [CrossRef] [PubMed]

57. Schmitz, M.; Kuhlmann, M.; Reimann, O.; Hackenberger, C.P.R.; Groll, J. Side-chain cysteine-functionalized poly(2-oxazoline)s for multiple peptide conjugation by native chemical ligation. *Biomacromolecules* **2015**, *16*, 1088–1094. [CrossRef] [PubMed]

58. Kuhlmann, M.; Reimann, O.; Hackenberger, C.P.R.; Groll, J. Cysteine-functional polymers via thiol-ene conjugation. *Macromol. Rapid Commun.* **2015**, *36*, 472–476. [CrossRef] [PubMed]

Sample Availability: Not Available.

Article

Characterization of Non-Derivatized Cellulose Samples by Size Exclusion Chromatography in Tetrabutylammonium Fluoride/Dimethylsulfoxide (TBAF/DMSO)

Jérémy Rebière [1,2], Antoine Rouilly [1], Vanessa Durrieu [1,*] and Frédéric Violleau [2]

[1] Laboratoire de Chimie Agro-industrielle (LCA), Université de Toulouse, INRA, INPT, 31030 Toulouse, France; jeremy.rebiere@ensiacet.fr (J.R.); antoine.rouilly@ensiacet.fr (A.R.)

[2] Laboratoire de Chimie Agro-industrielle (LCA), Université de Toulouse, INRA, INPT, INP-EI PURPAN, 31062 Toulouse, France; frederic.violleau@purpan.fr

* Correspondence: vanessa.durrieu@ensiacet.fr; Tel.: +33-(0)-534-323-509; Fax: +33-(0)-534-323-597

Received: 28 September 2017; Accepted: 15 November 2017; Published: 16 November 2017

Abstract: This paper deals with the use of tetrabutylammonium fluoride/dimethylsulfoxide (TBAF/DMSO) to characterize the molar mass distribution of non-derivatized cellulosic samples by size exclusion chromatography (SEC). Different cellulose samples with various average degree of polymerization (DP) were first solubilized in this solvent system, with increasing TBAF rates, and then analyzed by SEC coupled to a refractive index detector (RID), using DMSO as mobile phase. The Molar Masses (MM) obtained by conventional calibration were then discussed and compared with suppliers' data and MM determined by viscosimetry measurements. By this non-classic method, molar mass of low DP samples (Avicel® and cotton fibers) have been determined. For high DP samples (α-cellulose and Vitacel®), dissolution with TBAF concentration of 10 mg/mL involved elution of cellulose aggregates in the exclusion volume, related to an incomplete dissolution or the dilution of TBAF molecules in elution solvent, preventing the correct evaluation of their molar mass.

Keywords: cellulose; molar mass; SEC analysis; conventional calibration

1. Introduction

Cellulose is a natural, biodegradable and biocompatible resource, and represents about 50% of the associated carbon on earth. Its interest as a source of fibers in paper industry, textiles and materials, and more recently as a source of glucose for bio-ethanol production or a novel sugar-based chemistry, is constantly growing. For all these applications, the determination of molar masses, both average value and distribution is more and more expected, as they are key parameters to control and improve the industrial processes as well as the final product properties.

Size Exclusion Chromatography (SEC) is nowadays the favorite technique for this type of characterization, and several analytical methods have been established since the early 70's. They mainly differ from each other with respect to the cellulose dissolution step prior to SEC analysis. During the 70's and 80's, indirect dissolution was preferred, through the synthesis of non-polar cellulose derivatives, such as cellulose tricarbanilates (CTC) [1–5]. From the 90's, direct dissolution methods, without any derivatization, using polar solvent mixtures like lithium chloride/*N,N*-dimethylacetamide (LiCl/DMAc) have emerged [6–8]. McCormick et al. [6] have used lithium chloride/*N,N*-dimethylacetamide (LiCl/DMAc) as solvent system for cellulose characterization by SEC, and reported the stability, solubility and dissolving states of cellulose in this solvent system. Subsequently, LiCl/DMAc has been favorably used for cellulose molar mass distribution (MMD) characterization particularly on birch wood Kraft pulp, hardwood or softwood

Kraft pulps or aged papers [9–15]. Several works [16–18] reported cellulose dissolution in LiCl/DMAc without degradation, despite a slight decrease of the viscosity of cellulose solutions after 30 days has been observed [6]. However, some relevant problems concerning the use of LiCl/DMAc have been described: (i) the aggregation of cellulose in the solution depending on dissolution conditions and concentrations of cellulose or LiCl [19]; (ii) incomplete dissolution of several cellulose samples [20,21] and (iii) detrimental degradation of cellulose by heating in DMAc or LiCl/DMAc during the dissolution process [22], especially for high molecular weight samples. Consequently, cellulose characterization by SEC analyses using LiCl/DMAc as solvent is still currently a topic a numerous researches [23]. Other solvent systems have also been studied, such as LiCl/DMI described by Yanagisawa and Isogai [24,25].

TBAF/DMSO has been reported to be a good solvent for various cellulose samples, without affecting significantly their physical structure [26,27]. Even if the dissolution mechanism was not clearly explicated yet, it has been accepted that: (i) DMSO is an excellent swelling agent for cellulose, increasing the accessibility to hydrogen bonds; (ii) the fluoride ion, due to its strong hydrogen bond acceptor character, can then reduce the intermolecular attraction between the polysaccharide chains and disrupt the hydrogen bonds network [27]. For this reason, this system appears as a suitable candidate for SEC analysis of cellulose samples. Additionally, the replacement of a carcinogenic solvent such as DMAc by DMSO could be of interest for the health of the operators and for the environment. DMSO is classified as nontoxic with no risk for the human health according to US Environmental Protection Agency (EPA). Finally, DMSO is present in some industrial wastewaters and is readily biodegradable [28], and could be synthesized from renewable raw materials in the future.

The size exclusion separation mechanism is based on differences in the solute hydrodynamic volumes. Different methods can be used to obtain the molar mass of the components of chromatographed samples [29]. However, the most common and simplest method to define the relationship between the molar mass and elution volume is to calibrate the columns with narrowly distributed polymeric standards, using a refractive index detector. Pullulan standards are frequently used to obtain the MMD of cellulosic samples. Indeed, due to their linearity and the chemical structure (polymaltotriose units linked together by α-(1,6)-linkages), it is considered that the relation between molar mass and hydrodynamic volume is close to what is expected for cellulose [15,30–32].

In this study, TBAF/DMSO solvent was tested as a new greener solvent system for SEC analysis of four cellulose samples having various average degrees of polymerization (DP): Avicel® and cotton fibers known to have low DP (<300), and α-cellulose and Vitacel® chosen for their high DP (>1000). SEC was coupled to a refractive index detector and DMSO was used as mobile phase. The results obtained from pullulans calibration were then compared with suppliers' data and viscosimetric DP measurements.

2. Results and Discussion

A preliminary study in which TBAF/DMSO was directly used as elution solvent was performed with various TBAF concentrations (2.5, 5 and 10 mg/mL). However, the column pressure limit (30 bars) was rapidly reached, indicating column saturation. The higher the TBAF concentration was, the faster the saturation occurred. TBAF molecules seem to interact with PS-DVB co-polymer composing the stationary phase of the column. Consequently, DMSO alone was used as elution solvent and cellulose and pullulan samples were dissolved in TBAF/DMSO solutions, before SEC analysis. Detection was performed only with a refractive index detector using pullulans calibration.

2.1. Influence of TBAF Concentration on Pullulan Calibration Curves

Firstly, the influence of the TBAF concentration was studied on the pullulan standards. Three calibration curves with various TBAF concentrations 0, 5 and 10 mg/mL (named respectively CC DT 0, CC DT 0.5 and CC DT 1.0) were obtained (Figure 1) by polynomial regression, with satisfying correlation coefficients ($R^2 > 0.99\%$).

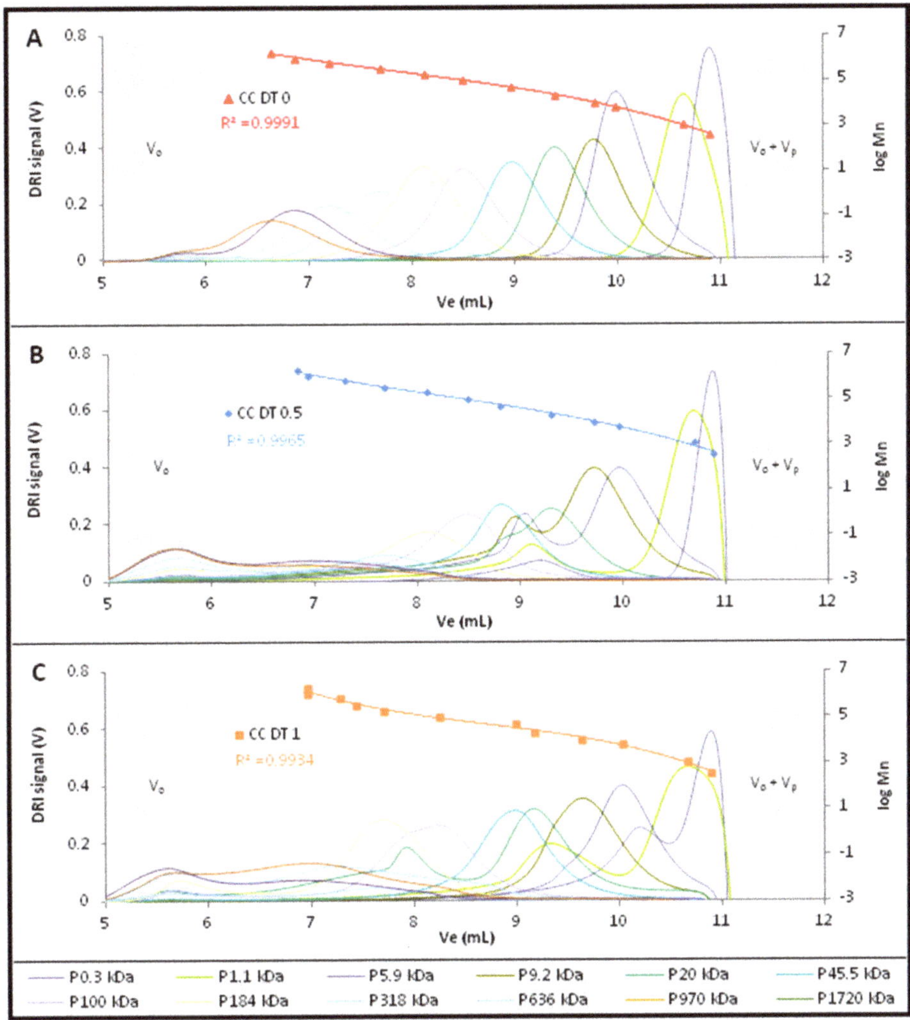

Figure 1. SEC calibration curves in DMSO (60 °C and 0.5 mL/min) of pullulan standards with different concentrations of TBAF: (**A**) 0 mg/mL; (**B**) 5 mg/mL; (**C**) 10 mg/mL.

However, increasing TBAF concentration involved significant modifications of the elution peaks shape and position (Figure 2). In the case of the low molecular weight samples (Mn lower than 100 kDa), for each standard a new elution peak appeared in addition to the one observed without TBAF. This new peak had a lower elution volume, suggesting the apparition of a second population with a higher hydrodynamic volume in the sample (Figure 2B). The increase of the hydrodynamic volume with TBAF concentration in solution can be explained by the formation of TBAF aggregates on pullulan chains in the presence of TBAF/DMSO due to the formation of the complex {DMSO-TBAF-pullulan}.

For high molecular weight pullulan samples (higher than 100 kDa), this aggregation phenomenon had a slightly different consequence. The elution peaks were flattened, widened and displaced to lower elution volume, and an exclusion peak (elution volume between 5 mL and 6 mL) appeared and increased in intensity with the increase of pullulan molecular weight as indicated by the red arrows (Figure 2A).

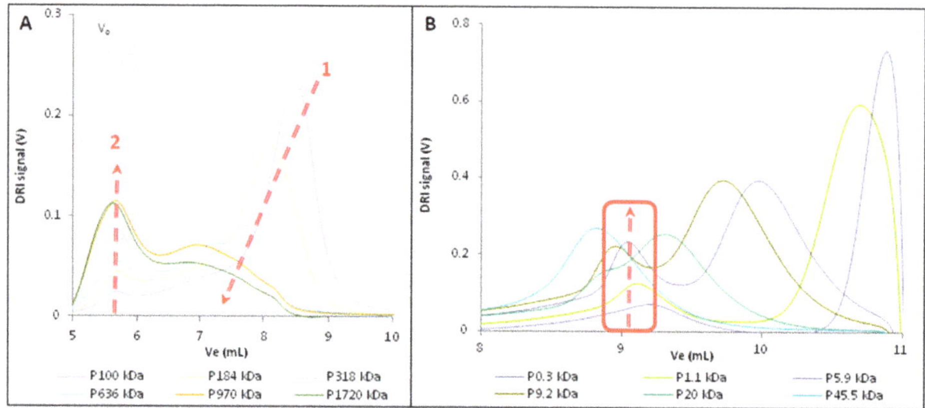

Figure 2. Increase of hydrodynamic volumes of pullulan standards prepared with 5 mg/mL of TBAF. (**A**) High molecular weight samples; (**B**) low molecular weight samples.

The exclusion rate of higher molar mass pullulans (i.e., 970 kDa & 1720 kDa) increased with TBAF concentration, but without affecting the major elution volume. The same phenomenon, i.e., the non-uniform effect on the higher- and lower-molecular weight macromolecules has been previously observed with pullulan standard solutions prepared in LiCl/DMAc, with an increased amount of LiCl in the standards solutions [33,34].

Despite these observations, the calibration curves obtained in these different conditions could be satisfactorily superimposed. As the one obtained in DMSO alone gave the most defined peaks and the higher correlation coefficient, it will be used as calibration curve for all the rest of the study.

2.2. Effect of TBAF Cencentration on AVICEL® Chromatograms

It has been proven that, in organic salts solvent systems, the concentration of the solid salt in the organic solvent can affect the cellulose dissolution. Indeed, Strlič et al. [34] demonstrated that the increase of LiCl content in LiCl/DMAc solutions improves cellulose dissolution. Considering this, several Avicel® samples in DMSO/TBAF have been prepared with different TBAF concentrations: from 2.5 mg/mL to 100 mg/mL. The obtained chromatograms are presented in Figure 3A. They demonstrated a strong influence of the TBAF concentration on the samples fractionation and hydrodynamic volumes.

For the lowest TBAF concentration (2.5 mg/mL) the elution profile presented two different peaks. The first one was eluted inside the exclusion volume and accounted for more than 50% of the signal. The second one, at high elution volume, was relatively large. Increasing the TBAF concentration (5 mg/L, 10 mg/L and 20 mg/L) a symmetric peak appeared between 7.5 mL and 9.5 mL, for 10 mg/L and 20 mg/mL this peak was particularly well defined and reproducible. Increasing further the TBAF concentration (50 mg/mL and 100 mg/mL), the chromatograms showed a widened peak indicating a large hydrodynamic volume dispersion inside this sample. For a TBAF concentration of 100 mg/L, even a second peak appeared for high elution volume.

Several hypotheses could explain these phenomena (Figure 3B). A too low TBAF concentration (below 5 mg/mL) does not seem to totally dissolve cellulosic chains, meaning one part of the sample may only be swollen. This swollen fraction, having a too high hydrodynamic volume to be retained in the column system, corresponds to the peak eluted in the exclusion volume as mentioned previously.

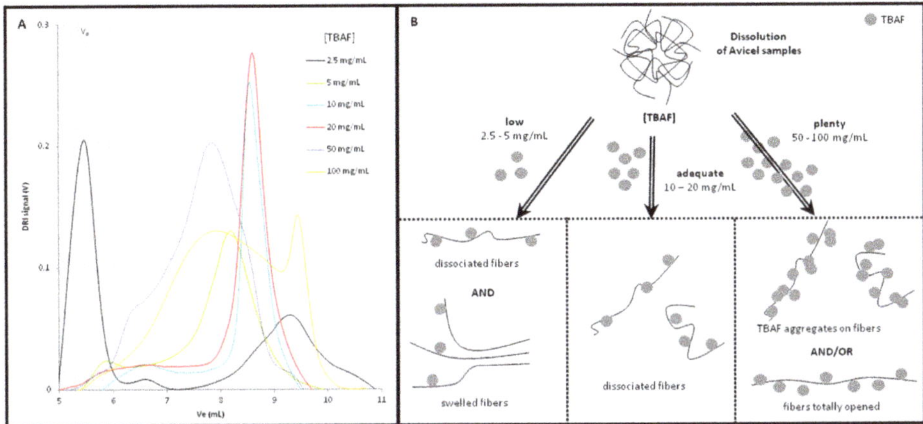

Figure 3. (**A**) Chromatograms of Avicel® samples in different TBAF/DMSO solutions (DMSO; 60 °C; 0.5 mL/min) and (**B**) Representation of different TBAF concentration related phenomena affecting the elution volume, among to TBAF concentration.

In ideal conditions, cellulose dissolution should be total, allowing a complete dissociation of polymeric chains, without influence of TBAF concentration on elution volumes. This seems to be the case for 10 mg/mL and 20 mg/mL, as both elution peaks were very similar.

Finally, for higher concentrations (50 mg/mL and 100 mg/mL), the excess of TBAF molecules involved a dispersion on hydrodynamic volume, that may be related either to their aggregation on dissociated macromolecules, or to total disruption of the intramolecular hydrogen bonds leading to the complete unfolding of cellulosic chains.

The molar mass obtained by conventional calibration with the calibration curve CC DT 0 confirms the phenomenon (Table 1).

Table 1. Influence of TBAF concentrations on MM distribution analyses of Avicel® samples.

Sample	(TBAF) (mg/mL)	%Excl. [a]	Dispersity	M_n (kDa)	M_w	M_v (kDa) Untreated	M_v (kDa) Dissolved *	Supplier MM (kDa) Inferior	Supplier MM (kDa) Superior
	100	2.4	2.7	82.1	221.3				
			1.1	12.6	14.2				
	50	3.5	2.1	110.0	233.4				
Avicel®	20	0.9	1.1	43.6	48.4	36.6	47.8	24.3	56.7
	10	0.8	1.1	43.8	46.9				
	5	1.3	1.3	68.2	87.0				
	2.5	64.5	nd	nd	nd				

[a] %Excl. is the no-retention percent of the sample; * After dissolution in DMSO/TBAF as described in our previous work [26]; nd: not determined.

In case of Avicel® samples dissolved with 2.5, 5, 50 and 100 mg/mL, the obtained molar masses were higher than the ones given by the suppliers or measured by viscosimetry [26]. On the contrary, the values obtained for the samples 10 mg/mL and 20 mg/mL are consistent with the reference ones, confirming that with such TBAF concentrations, the solvent system completely dissolves Avicel®, without any apparent modification of its hydrodynamic volume.

Usually, the molar mass distribution determination using standard calibration curve requires working with the exact same dissolution conditions for both standards and samples. Accordingly, for the

Avicel® sample dissolved with a 10 mg/mL TBAF concentration, the average molar masses calculated using two different calibration curves (CC DT 0 and CC DT 1) were compared (Table 2).

Table 2. Determination of molar masses among two different calibration curves of Avicel® samples dissolved with 10 mg/mL of TBAF concentration in solution.

	%Excl. [a]	Dispersity	Mn (kDa)	Mw (kDa)
C DT 0	0.8	1.1	43.8	46.9
CC DT 1	0.9	1.1	45.7	49.8

[a] %Excl. is the no-retention percent of the sample.

Some caution is required when considering these values, because pullulan were used as standards as no cellulose standards were available on the market. As described by Berggren et al. [35], molar mass of cellulose samples determined by conventional calibration should be corrected by a factor determined by an absolute method based on light scattering. Nonetheless, the similarity of the obtained values supports the suitability of the CC DT 0 calibration curve for all samples dissolved in DMSO/TBAF system.

2.3. Chromatogram and Molar Mass Distribution of Cellulose Samples

The four different cellulose samples were first dissolved into the solution of 10 mg/mL TBAF concentration. The chromatograms obtained presented two different behaviors depending on the celluloses nature (Figure 4A).

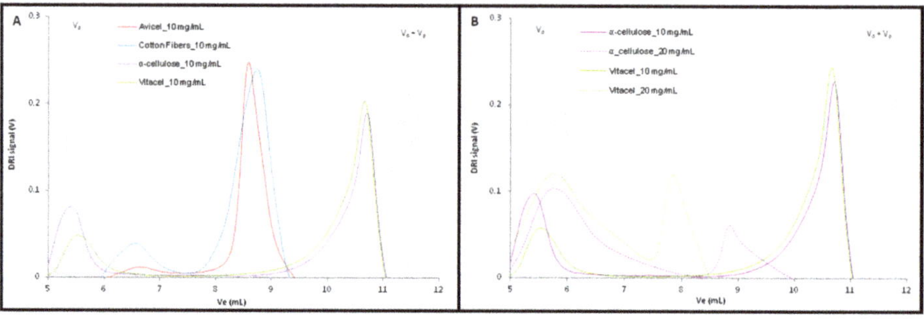

Figure 4. SEC chromatograms of cellulose samples dissolved with [TBAF] of 10 mg/mL and 20 mg/mL.

Avicel® and cotton fibers, the lowest molar mass samples, had similar chromatograms displaying a well-defined peak between 8 mL and 9 mL, and a smaller one (especially for cotton fibers) between 6 mL and 7 mL, which is in accordance with their molar masses being close to each other.

The well-defined peaks indicated that the dissolution was completed and the fibers were totally dissociated during the process. For both samples, the peaks were in the range of the calibration curve and the excluded volume percentage was negligible, allowing the calculation of the average molecular weight (Table 3).

The obtained values were comparable to the viscosimetric ones and those obtained from the suppliers. Vitacel® and α-cellulose, the highest molar mass samples, were also characterized by a similar DRI profile with two distinguished peaks identified on their respective chromatogram. But, the first peak was in the exclusion volume, whereas the second one, corresponding to very low hydrodynamic volume molecules, may be attributed to TBAF molecules complexed with DMSO, instead of cellulosic macromolecules (Figure 4A). Consequently, the average molecular weight were not calculated for these two samples.

Table 3. Molar masses of cellulose samples dissolved with 10 mg/mL and 20 mg/mL TBAF concentration.

Sample	[TBAF] (mg/mL)	%Excl. [a]	Dispersity	Mn (kDa)	Mw (kDa)	Mv (kDa)		Supplier MM (kDa)	
						Untreated	*Dissolved* *	*Inferior*	*Superior*
Avicel	10	0.8	1.1	43.8	46.9	36.6	47.8	24.3	56.7
Cotton Fibers	10	1.2	1.2	38.4	47.6	43.9	44.4	30.8	48.6
α-cellulose	20	62.3	nd	nd	nd	196.3	190.2	113.1	194.2
Vitacel	20	52.6	nd	nd	nd	174.1	195.2	71.3	365.2

[a] %Excl. is the no-retention percent of the sample; * After dissolution in DMSO/TBAF as described in our previous work [26]; nd: not determined.

A higher TBAF concentration (20 mg/mL) was then tested for α-cellulose and Vitacel®, in order to facilitate their dissolution and improve their fractionation. The obtained chromatograms (Figure 4B) presented again two distinct peaks. The first one was still eluted in the exclusion volume, but the second one was displaced to more reasonable elution volume (between 7.5 mL and 10 mL) corresponding probably to the totally dissolved cellulosic fraction. Unfortunately, the first peak accounted for more than 50% of the injected molecules (Table 2), preventing a correct calculation of the average molar masses, so the 20 mg/mL TBAF concentration did not seem sufficient to completely dissolve the high molecular weight cellulosic samples. For further investigation on analyses of high molecular weight cellulosic sample, it would hence be pertinent to test higher TBAF concentrations, keeping in mind that too high a concentration may lead to other artefacts during the fractionation.

3. Materials and Methods

3.1. Materials

Dimethyl sulfoxide (DMSO, 99.7%, anhydrous), tetrabutylammonium fluoride trihydrate (TBAF, 97%) and bis(ethylenediamine)copper(II) hydroxide solution (CED solution, 1 M in copper, molar ratio of ethylenediamine/copper of 2:1) were purchased from Sigma-Aldrich (Darmstadt, Germany). Four different samples of cellulose (one single batch each) have been studied with different average of DP (Table 1). Avicel® (PH-101, Ph Eur, cellulose microcrystalline, batch n°BCBK2051V), α-cellulose (powder, batch n°BCBH3503V) and cotton fibers (cotton linters, medium fibers, batch n°MKBQ8042V) were purchased from Fluka (Sigma Aldrich, Darmstadt, Germany), and Vitacel® (L600/30, Ph Eur, powdered cellulose, batch n°7120891215 X) from JRS Pharma (Rosenberg, Germany). Pullulan standards P 1720 kDa, P 970 kDa, P 636 kDa, P 318 kDa, P 184 kDa, P 100 kDa, P 45.5 kDa, P 20 kDa, P 9.2 kDa, P 5.9 kDa, P 1.1 Da and P 0.3 Da were purchased from Polymer Standards Service (PSS, Mainz, Germany).

3.2. Viscosimetric Measurements

Viscosimetric molar masses (Mv) of the different cellulosic samples, before and after dissolution in TBAF/DMSO solvent system, were determined indirectly by viscosimetric DP measurements of cellulose samples dissolved in the cupridiethylenediamine (CED). The samples were prepared and analyzed according to ASTM D-1795 [36].

3.3. Standard Procedure for Pullulan Dissolution in TBAF/DMSO

Pullulan standards were used to calibrate the column system. For each standard, the initial solution was prepared with 1.5 mg of pullulan samples placed under continuous magnetic stirring in 1.0 mL of DMSO during 30 min at ambient temperature.

To study the influence of the TBAF concentration on the molar mass distribution, three different pullulan standard ranges were prepared. Then, into each initial standard solution, 0.5 mL of TBAF/DMSO solution (containing 0 mg/mL, 15 mg/mL or 30 mg/mL of TBAF) were added. Finally, three SEC calibration curves with various TBAF concentrations (0, 5 and 10 mg/mL) were obtained (see Section 3.5).

3.4. Standard Procedure for Cellulose Dissolution in TBAF/DMSO

Two mg of cellulose samples with 10.0 g of DMSO were placed under continuous magnetic stirring during 2 h at 60 °C to swell the cellulosic fibers. Then, 10.0 g of TBAF/DMSO with 20 mg/mL and 40 mg/mL of TBAF were added. Finally, different solutions were obtained with 1 mg/mL of cellulose concentration and 10 mg/mL and 20 mg/mL of TBAF concentration, under the same stirring during 30 min. According to our previous study, the reaction temperature varied between 60 °C and 100 °C depending on the type of cellulose [26].

3.5. Size Exclusion Chromatography

The SEC system used consisted of a high-pressure isopump (G1310A, Agilent 1100 series, Agilent Technologies, Waldbronn, Germany), a stainless steel in-line filter with a 0.1 μm poly(tetrafluoroethylene) (PTFE) membrane (Millipore, Burlington, MA, USA), an automatic injector (G1313A, Agilent 1100 series, Agilent Technologies, Waldbronn, Germany), a column oven (G1316A, Agilent 1100 series, Agilent Technologies, Waldbronn, Germany), a SEC column packed with styrene-divinylbenzene (PS-DVB) copolymer gel (Mixed-B KD-806M, Shodex, München, Germany) preceded by a guard column (Mixed-B KD, Shodex) and a refractive index detector (RID-10A, Shimadzu, Kyoto, Japan). Data acquisition and processing were carried out using ASTRA software (Wyatt Technologies, Solvang, CA, USA).

SEC conditions were as follows: 1 mg/mL as sample concentration, 100 μL of injection volume, 0.5 mL/min flow rate and the column temperature set to 60 °C. The detector cell of RID was kept at 40 °C. The elution solvent was pure DMSO.

4. Conclusions

SEC analyses of various cellulosic samples were performed using TBAF/DMSO as solvent. Calibration curves from pullulan standards were successfully obtained, with different TBAF concentrations. The results underlined the impact of TBAF concentration on chromatographic profiles for both pullulan and cellulose samples.

For low molar mass cellulosic samples (Avicel® and cotton fibers), this method allowed the determination of the average molecular weight by conventional calibration. For higher molar mass cellulosic samples, the tested TBAF concentration range was not optimal which prevented the correct calculation of their average molecular weight. The saturation of the column by TBAF molecules did not allow to use TBAF/DMSO solutions as elution solvent. Improvements are still needed to apply a correction factor to the calibration curves and/or to consider the use of absolute method based on light scattering (MALS detector).

Author Contributions: J.R., A.R., V.D. and F.V. conceived and designed the experiments; J.R. performed the experiments; J.R., A.R., V.D. and F.V. analyzed the data; and wrote the paper.

Conflicts of Interest: The authors declare that they have no potential conflict of interest.

References

1. Lauriol, J.-M.; Froment, P.; Pla, F.; Robert, A. Molecular weight distribution of cellulose by on-line size exclusion chromatography—Low angle laser light scattering. Part I—Basic experiments and treatment of data. *Holzforschung* **1987**, *41*, 109–113. [CrossRef]

2. Evans, R.; Wearne, R.H.; Wallis, A.F.A. Molecular weight distribution of cellulose as its tricarbanilate by high performance size exclusion chromatography. *J. Appl. Polym. Sci.* **1989**, *37*, 3291–3303. [CrossRef]

3. Evans, R.; Wearne, R.H.; Wallis, A.F.A. Effect of amines on the carbanilation of cellulose with phenylisocyanate. *J. Appl. Polym. Sci.* **1991**, *42*, 813–820. [CrossRef]

4. Sjöholm, E. Size exclusion chromatography of cellulose and cellulose derivatives. In *Handbook of Size Exclusion Chromatography and Related Techniques: Revised and Expanded*; Chromatographic Science Series; Wu, C.-S., Ed.; Dekker: New York, NY, USA, 2003; Volume 91.

5. Henniges, U.; Kloser, E.; Patel, A.; Potthast, A.; Kosma, P.; Fischer, M.; Fischer, K.; Rosenau, T. Studies on DMSO-containing carbanilation mixtures: Chemistry, oxidations and cellulose integrity. *Cellulose* **2007**, *14*, 497–511. [CrossRef]

6. McCormick, C.L.; Callais, P.A.; Hutchinson, B.H. Solution studies of cellulose in lithium chloride and *N,N*-dimethylacetamide. *Macromolecules* **1985**, *18*, 2394–2401. [CrossRef]

7. Dawsey, T.R.; McCormick, C.L. The lithium chloride/dimethylacetamide solvent for cellulose: A literature review. *J. Macromol. Sci. Part C* **1990**, *30*, 405–440. [CrossRef]

8. Timpa, J.D. Application of universal calibration in gel permeation chromatography for molecular weight determinations of plant cell wall polymers: Cotton fiber. *J. Agric. Food Chem.* **1991**, *39*, 270–275. [CrossRef]

9. Striegel, A.M.; Timpa, J.D. Size Exclusion Chromatography of Polysaccharides in Dimethylacetamide-Lithium Chloride. In *Strategies in Size Exclusion Chromatography*; ACS Symposium Series; American Chemical Society: Washington, DC, USA, 1996; Volume 635, pp. 366–378.

10. Strlič, M.; Kolar, J. Size exclusion chromatography of cellulose in LiCl/*N,N*-dimethylacetamide. *J. Biochem. Biophys. Methods* **2003**, *56*, 265–279. [CrossRef]

11. Sjöholm, E.; Gustafsson, K.; Colmsjo, A. Size exclusion chromatography of lignins using lithium chloride/*N,N*-dimethylacetamide as mobile phase. I. Dissolved and residual birch kraft lignins. *J. Liq. Chromatogr. Relat. Technol.* **1999**, *22*, 1663–1685. [CrossRef]

12. Dupont, A.-L. Cellulose in lithium chloride/*N,N*-dimethylacetamide, optimisation of a dissolution method using paper substrates and stability of the solutions. *Polymer* **2003**, *44*, 4117–4126. [CrossRef]

13. Kennedy, J.F.; Rivera, Z.S.; White, C.A.; Lloyd, L.L.; Warner, F.P. Molecular weight characterization of underivatized cellulose by GPC using lithium chloride-dimethylacetamide solvent system. *Cellul. Chem. Technol.* **1990**, *24*, 319–325.

14. Silva, A.A.; Laver, M.L. Molecular weight characterization of wood pulp cellulose: Dissolution and size exclusion chromatographic analysis. *Tappi J.* **1997**, *80*, 173–180.

15. Westermark, U.; Gustafsson, K. Molecular size distribution of wood polymers in birch kraft pulps. *Holzforschung* **1994**, *48*, 146–150. [CrossRef]

16. Turbak, A.F.; El-Kafrawy, A.; Snyder, F.W.; Auerbach, A.B. Solvent System for Cellulose. U.S. Patent 4302252 A, 24 November 1981.

17. El-Kafrawy, A.; Turbak, A.F. The dissolution of cellulose in anhydrous chloral/aprotic solvents. *J. Appl. Polym. Sci.* **1982**, *27*, 2445–2456. [CrossRef]

18. Turbak, A.F. Newer cellulose solvent systems. *Tappi J.* **1983**, 105–110.

19. Röder, T.; Morgenstern, B.; Schelosky, N.; Glatter, O. Solutions of cellulose in *N,N*-dimethylacetamide/lithium chloride studied by light scattering methods. *Polymer* **2001**, *42*, 6765–6773. [CrossRef]

20. Tamai, N.; Tatsumi, D.; Matsumoto, T. Rheological properties and molecular structure of tunicate cellulose in LiCl/1,3-dimethyl-2-imidazolidinone. *Biomacromolecules* **2004**, *5*, 422–432. [CrossRef] [PubMed]

21. Sjöholm, E.; Gustafsson, K.; Pettersson, B.; Colmsjö, A. Characterization of the cellulosic residues from lithium chloride/*N,N*-dimethylacetamide dissolution of softwood kraft pulp. *Carbohydr. Polym.* **1997**, *32*, 57–63. [CrossRef]

22. Potthast, A.; Rosenau, T.; Sixta, H.; Kosma, P. Degradation of cellulosic materials by heating in DMAc/LiCl. *Tetrahedron Lett.* **2002**, *43*, 7757–7759. [CrossRef]

23. Henniges, U.; Vejdovszky, P.; Siller, M.; Jeong, M.-J.; Rosenau, T.; Potthast, A. Finally Dissolved! Activation Procedures to Dissolve Cellulose in DMAc/LiCl Prior to Size Exclusion Chromatography Analysis—A Review. *Curr. Chromatogr.* **2014**, *1*, 52–68. [CrossRef]

24. Yanagisawa, M.; Shibata, I.; Isogai, A. SEC–MALLS analysis of cellulose using LiCl/1,3-dimethyl-2-imidazolidinone as an eluent. *Cellulose* **2004**, *11*, 169–176. [CrossRef]

25. Yanagisawa, M.; Isogai, A. SEC-MALS-QELS study on the molecular conformation of cellulose in LiCl/amide solutions. *Biomacromolecules* **2005**, *6*, 1258–1265. [CrossRef] [PubMed]

26. Rebière, J.; Heuls, M.; Castignolles, P.; Gaborieau, M.; Rouilly, A.; Violleau, F.; Durrieu, V. Structural modifications of cellulose samples after dissolution into various solvent systems. *Anal. Bioanal. Chem.* **2016**, *408*, 8403–8414. [CrossRef] [PubMed]

27. Östlund, A.; Lundberd, D.; Nordstierna, L.; Holmberg, K.; Nydén, M. Dissolution and Gelation of Cellulose in TBAF/DMSO Solutions: The Roles of Fluoride Ions and Water. *Biomacromolecules* **2009**, *10*, 2401–2407. [CrossRef] [PubMed]

28. Marti, M.; Molina, L.; Aleman, C.; Armelin, E. Novel Epoxy Coating Based on DMSO as a Green Solvent, Reducing Drastically the Volatil Organic Compound Content and Using Conducting Polymers As a Nontoxic Anticorrosive Pignment. *ACS Sustain. Chem. Eng.* **2013**, *1*, 1609–1618. [CrossRef]

29. Jackson, C.; Barth, H. Concerns regarding the practice of multiple detector size-exclusion chromatography. In *Chromatographic Characterization of Polymers: Hyphenated and Multidimensional Techniques*; Provder, T., Barth, H.G., Urban, M.W., Eds.; American Chemical Society: Washington, DC, USA, 1995; pp. 59–68.

30. Sjöholm, E.; Gustafsson, K.; Berthold, F.; Colmsjö, A. Influence of the carbohydrate composition on the molecular weight distribution of kraft pulps. *Carbohydr. Polym.* **2000**, *41*, 1–7. [CrossRef]

31. Sjöholm, E.; Gustafsson, K.; Eriksson, B.; Brown, W.; Colmsjö, A. Aggregation of cellulose in lithium chloride/*N,N*-dimethylacetamide. *Carbohydr. Polym.* **2000**, *41*, 153–161. [CrossRef]

32. Strlič, M.; Kolar, J.; Zigon, M.; Pihlar, B. Evaluation of size-exclusion chromatography and viscometry for the determination of molecular masses of oxidised cellulose. *J. Chromatogr. A* **1998**, *805*, 93–99. [CrossRef]

33. Striegel, A.M.; Timpa, J.D. Gel permeation chromatography of polysaccharides using universal calibration. *Int. J. Polym. Anal. Charact.* **1996**, *2*, 213–220. [CrossRef]

34. Strlič, M.; Kolenc, J.; Kolar, J.; Pihlar, B. Enthalpic interactions in size exclusion chromatography of pullulan and cellulose in LiCl–*N,N*-dimethylacetamide. *J. Chromatogr. A* **2002**, *964*, 47–54. [CrossRef]

35. Berggren, R.; Berthold, F.; Sjöholm, E.; Lindström, M. Improved Methods for Evaluating the Molar Mass Distributions of Cellulose in Kraft Pulp. *J. Appl. Polym. Sci.* **2003**, *88*, 1170–1179. [CrossRef]

36. ASTM D 1795. In *Standard Test Method for Intrinsic Viscosity of Cellulose*; ASTM: West Conshohocken, PA, USA, 2013.

Sample Availability: Samples of the compounds are not available from the authors.

Article

Characteristics of Multifunctional, Eco-Friendly Lignin-Al$_2$O$_3$ Hybrid Fillers and Their Influence on the Properties of Composites for Abrasive Tools

Łukasz Klapiszewski [1,*], Artur Jamrozik [1,2], Beata Strzemiecka [1], Iwona Koltsov [3], Bartłomiej Borek [4], Danuta Matykiewicz [5], Adam Voelkel [1] and Teofil Jesionowski [1]

[1] Institute of Chemical Technology and Engineering, Faculty of Chemical Technology, Poznan University of Technology, Berdychowo 4, PL-60965 Poznan, Poland; artur.robert.jamrozik@gmail.com (A.J.); beata.strzemiecka@put.poznan.pl (B.S.); adam.voelkel@put.poznan.pl (A.V.); teofil.jesionowski@put.poznan.pl (T.J.)

[2] Wielkopolska Centre of Advanced Technologies, Umultowska 89 C, PL-61614 Poznan, Poland

[3] Polish Academy of Sciences, Institute of High Pressure Physics, Sokolowska 29/37, PL-01142 Warszawa, Poland; iwona.koltsov@gmail.com

[4] RHL-Service, Budziszynska 74, PL-60179 Poznan, Poland; bartek@rhl.pl

[5] Institute of Materials Technology, Faculty of Mechanical Engineering and Management, Poznan University of Technology, Piotrowo 3, PL-61138 Poznan, Poland; danuta.matykiewicz@put.poznan.pl

* Correspondence: lukasz.klapiszewski@put.poznan.pl; Tel.: +48-61-665-37-48

Received: 25 September 2017; Accepted: 3 November 2017; Published: 7 November 2017

Abstract: The main aim of the present study was the preparation and comprehensive characterization of innovative additives to abrasive materials based on functional, pro-ecological lignin-alumina hybrid fillers. The behavior of lignin, alumina and lignin-Al$_2$O$_3$ hybrids in a resin matrix was explained on the basis of their surface and application properties determined by inverse gas chromatography, the degree of adhesion/cohesion between components, thermomechanical and rheological properties. On the basis of the presented results, a hypothetical mechanism of interactions between lignin and Al$_2$O$_3$ as well as between lignin-Al$_2$O$_3$ hybrids and phenolic resins was proposed. It was concluded that lignin compounds can provide new, promising properties for a phenolic binder combining the good properties of this biopolymer as a plasticizer and of alumina as a filler improving mechanical and thermal properties. The use of such materials may be relatively non-complicated and efficient way to improve the performance of bonded abrasive tools.

Keywords: lignin-Al$_2$O$_3$ hybrid materials; abrasive tools; lignin; thermomechanical properties; rheological studies

---

1. Introduction

Resin-bonded abrasive products are complex composites consisting of abrasive, wetting agent (e.g., resole), binder (e.g., novolac), and fillers (e.g., pyrite, cryolite) [1]. Several factors influence the properties of the final abrasive tool during the production process and exploitation. The first stage during production of abrasive tools is covering the abrasive grains by resole. Appropriate covering of grains by resole is crucial for homogeneity of the semi-product and the final product [2]. In the next stage novolac mixed with filler is added. Then, the semi-product is pressed and hardened according to a specific temperature program. The appropriate hardening of the semi-product is highly important for the efficiency of the final product [3]. Hardening of resins in the final product also depends on the fillers used. Some of them can accelerate the hardening rate and some of them can modify this process [3]. Moreover, the fillers play a very important role in the work of the grinding tools, as they collect heat and prevent melting of the resin [4,5]. Inorganic compounds are broadly used as fillers. Conjugation of these fillers, characterized by polar surface properties, with a non-polar polymer matrix is difficult [6]. The use of an organic-inorganic hybrid filler may overcome this problem. Moreover, such hybrid fillers may increase the thermal resistance and mechanical strength. This effect may result from the possible reactions between active groups present in the inorganic and organic components.

Alumina is one of the most commonly used abrasive materials. In the present study, alumina was used as a filler. An Al_2O_3 filler can act as an additional abrasive and can collect heat. In order to increase the functionality of the final product, alumina was combined with lignin. Lignin is a natural polymer with a similar structure to phenolic resins used as binders in abrasive articles. Today, interest in natural resource polymers is growing due to the depletion of conventional petrochemical resources [7]. There are already known cases of successful use of biopolymers such as cellulose in advanced applications [8,9]. Lignin is the most available material in nature after cellulose [10]. Modified lignin is a polarographically active material and in recent years this biopolymer has also found interesting applications in electrochemistry [11–14]. As an aromatic biopolymer, it is a potential substitute for the polymers obtained from petroleum, due to its comparable or improved physicochemical properties and lower manufacturing cost. The presence of numerous hydroxyl groups in aromatic rings enables its use as a starting material for the synthesis of a wide range of polymers (such as polyethers, polyesters, polyethylene and polyurethane) [15]. Literature reports also suggest the potential use of lignocellulosic materials, including pure lignin and/or lignosulfonate, as fillers in a large group of polymers [16–22]. The problem of application of lignin to polyolefins was described in [19–22]. In case of mixing lignin with phenolic resins, the problem is not associated with the homogeneity of the polymer-lignin system, but insufficient mechanical properties [23,24]. Thus, it is expected that the application of a lignin-alumina hybrid as a filler may improve the mechanical properties of the final product. A very important aspect of the use of lignin-alumina hybrid as a filler is the reduction of emissions of harmful compounds into the atmosphere, due to the increased thermal stability of such a system in comparison with phenolic resins and/or lignin systems [24]. The described biopolymer is also one of the potential low-cost and readily available sorbents of environmentally harmful metal ions [25,26]. In order to be used as a sorbent, lignin can be obtained chiefly as a waste product from the paper industry and subjected to chemical modification to increase the number of functional groups [27].

There is a limited number of reports which describe attempts to use lignin and/or lignosulfonate in the preparation of advanced inorganic-organic hybrid materials. The concern is mainly the combination of biopolymers with the widely used and well-established silica [13,14,21,22,26,28–30]. Direct linking of natural polymers (lignin and lignosulfonates) with alumina has not been previously described.

The aim of our study was the preparation of new hybrid lignin-alumina fillers, which have not yet been described in the literature. The next step will be to test applications of the model composites in the abrasive industry. Lignin-alumina hybrid fillers were preliminarily tested to establish whether they may serve as new, promising, eco-friendly fillers for abrasive tool production. It is expected that such hybrid fillers should: (i) reinforce the final composite and (ii) possess higher thermal stability than lignin itself.

2. Results

2.1. Dispersive-Morphological Properties of Lignin-Al₂O₃ Hybrids

Aluminum oxide exhibited the presence of primary particles with diameters close to 100 nm, which showed a tendency to form aggregates (<1 µm). Al_2O_3 had different dispersive-morphological properties (see Table 1 and Figure 1a).

The particle size distribution of Al_2O_3 is very broad (from 142 nm to 955 nm, data from a Zetasizer Nano ZS apparatus). Addition of lignin slightly increased the particle size distribution. As follows from the data presented in Table 1, the increased lignin content in the hybrid filler resulted in a shift of the size distribution of particles (including primary particles and agglomerates, respectively) to larger sizes.

It should be noted that the commercial Kraft lignin used in the study contains particles of a wide range of sizes, which indicates the possibility to form large agglomerate structures. The presence of primary particles and secondary agglomerates was also confirmed by SEM images. Figure 1a,b present the SEM images of Al_2O_3 and lignin, respectively, while Figure 1c,d show images of lignin-alumina hybrids obtained with the use of different ratios of lignin to Al_2O_3 (8:1 *wt/wt* and 8:6 *wt/wt* respectively). It can be observed that 50% by volume of the lignin-alumina (8:1 *wt/wt*) hybrid system was occupied by particles with diameters smaller than 3.6 μm, while 90% of the sample volume was taken up by particles with diameters smaller than 5.3 μm. The average particle size in the hybrid system was 3.3 μm (see Table 1).

Figure 1. SEM images of alumina (**a**); lignin (**b**) and lignin-alumina hybrid materials (8:1, *wt/wt*) (**c**) and (8:6, *wt/wt*) (**d**).

Table 1. Dispersive characteristic of pure alumina and lignin-alumina hybrid fillers.

Sample	Dispersive Properties				
	Particle Size Distribution from Zetasizer Nano ZS (nm)	Particle Diameter from Mastersizer 2000 (μm)			
		d(0.1) *	d(0.5) **	d(0.9) ***	D(4.3) ****
Al_2O_3	142–955	2.5	3.7	5.5	3.7
Lignin-Al_2O_3 (8:1, *wt/wt*)	531–1106	2.1	3.6	5.3	3.3
Lignin-Al_2O_3 (8:2, *wt/wt*)	396–825	2.5	3.7	5.2	3.4
Lignin-Al_2O_3 (8:4, *wt/wt*)	295–955	2.4	3.5	5.2	3.2
Lignin-Al_2O_3 (8:6, *wt/wt*)	142–825	2.4	3.6	5.3	3.4

* d(0.1)—10% of the volume distribution is below this diameter value; ** d(0.5)—50% of the volume distribution is below this diameter value; *** d(0.9)—90% of the volume distribution is below this diameter value; **** D(4.3)—average particle size in examined system.

2.2. Fourier Transform Infrared Spectroscopy

FTIR analysis was performed in order to identify the functional groups present in the structure of alumina, lignin (Figure 2a) and lignin-Al_2O_3 hybrid fillers (Figure 2b). The most important bands are summarized in Table 2.

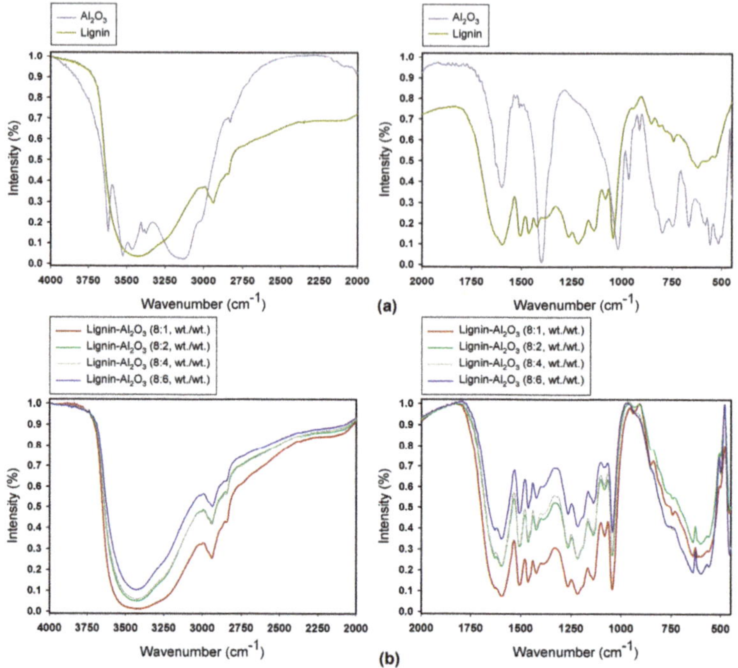

Figure 2. FTIR spectra of pure alumina and lignin (**a**) and of lignin-Al₂O₃ hybrid fillers (**b**).

Table 2. Vibrational frequency wavenumbers (cm^{-1}) for lignin, Al_2O_3 and lignin-Al_2O_3 fillers.

Lignin	Alumina	Lignin-Al₂O₃ Hybrid	Vibrational Assignment
-	3635, 3543 and 3473	overshadowed	Al–OH stretching
3432	3145	3430	O–H stretching, absorbed water
2935, 2877	-	2937, 2879	CH_x stretching
1648	-	1646	C=O stretching
1618	1620	1619	O–H bending of water
1602 not visible	-	1602 not visible	C–C, C=C (aromatic skeleton), stretching
1508	-	1508	
1471	-	1470	C–H (CH₃ + CH₂), bending
1419	-	1418	C–C, C=C (aromatic skeleton), stretching
-	1390	1389	Al–O as Si cage (TO₄)
1271	-	1271	C–O (guaiacyl unit) stretching
1226	-	1226	C–OH (phenolic OH) stretching
1139	-	1140	Aromatic C–H (guaiacyl unit), stretching
1080	-	1077	C–O stretching
1045	-	1039	C–OH + C–O–C (aliphatic OH + ether) stretching
-	1035	1039	Al–OH symmetric bending
-	970, 893	969, 893	–OH deformation linked to Al^{3-}
856, 751	-	858, 751	Aromatic C–H (guaiacyl unit), bending
-	788, 750, 693, 564 and 512	788, 751, 695, 565 and 512	Al–O in which aluminum ions are in both tetrahedral and octahedral sites
534	-	534	CH_x bending

The spectrum obtained for pure alumina (Figure 2a) revealed the presence of physically bound water, confirmed by the band at 3145 cm^{-1}, which results from O–H group stretching vibrations. Additionally, the band at 1620 cm^{-1} is caused by bending vibrations of the same group [31]. The bands at 3635 cm^{-1}, 3543 cm^{-1} and 3473 cm^{-1} are attributed to Al–OH stretching vibrations [32]. Symmetric bending vibrations of Al–OH produce a band at 1035 cm^{-1}, while the bands at 788 cm^{-1}, 750 cm^{-1}, 693 cm^{-1}, 564 cm^{-1} and 512 cm^{-1} are attributed to Al–O vibrations, in which aluminum ions occupy both tetrahedral and octahedral sites [33].

Figure 2a also shows the spectrum of pure lignin. The results show the presence of stretching vibrations of O–H groups (phenolic O–H and aliphatic O–H) at 3432 cm^{-1}, and stretching vibrations of C–H (–CH$_2$ and –CH$_3$) at 2965–2830 cm^{-1}. Stretching vibrations from ketone groups (C=O) are associated with the band at 1648 cm^{-1}, while those at 1602 cm^{-1}, 1508 cm^{-1} and 1419 cm^{-1} are attributed to stretching vibrations of the C–C, C=C bonds in the aromatic skeleton. Stretching vibrations of ether groups (C–O–C) appear at 1095–1000 cm^{-1}, and further bands in the range 1345–1250 cm^{-1} correspond to C–O stretching vibrations (C–O(H), C–O(Ar)). Below a value of 1000 cm^{-1} the spectrum contains bands attributed to in-plane and out-of-plane bending vibrations of aromatic C–H bonds. These results are in full agreement with earlier work [29,30,34,35].

The FTIR spectra of lignin-Al$_2$O$_3$ hybrid materials are presented in Figure 2b. The spectra revealed the presence of characteristic bonds for alumina: Al–O stretching vibrations at 1389 cm^{-1} (Al–O as Si cage (TO$_4$)) and Al–OH symmetric bending vibrations at 1039 cm^{-1}. The bands at 788 cm^{-1}, 751 cm^{-1}, 695 cm^{-1}, 565 cm^{-1} and 512 cm^{-1} are attributed to bending vibrations of Al–O. An important broad band in the range 3600–3200 cm^{-1} comes from stretching vibrations of O–H groups, which occur in the structure of both lignin and alumina. Functional groups which were observed in pure lignin are also present: C–H bonds at 2937 cm^{-1} and 2879 cm^{-1}, and different types of carbon atom bonds in the 1650–1000 cm^{-1} range.

Based on the FTIR spectra for the pure precursors (alumina and lignin) and organic-inorganic hybrid fillers, it can be observed that the intensity of bands in the hybrids increased in comparison with pure materials. This confirms the effectiveness of the proposed method of preparation. It is associated with an increase in the intensity of the bands attributed to particular functional groups. Furthermore, the intensity of the bands increased with increasing content of lignin relative to alumina.

2.3. Thermogravimetric Analysis–Mass Spectrometry

The thermal decomposition of the pure components is presented in Figure 3a,b. The alumina used in the preparation of lignin composites did not show any transition in the DSC curve during thermal treatment (Figure 3a). However, the decomposition of pure lignin produced four events which are clearly visible on the TG and DTG curves (Figure 3b). Degradation of lignin is influenced by its nature and by the reaction temperature, heating rate and degradation atmosphere [36]. Sample mass loss while heating occurs due to release of water (between RT and 200 °C) and other lignin decomposition products such as CH$_3$ ($m/z = 15$), CO ($m/z = 28$), HCHO ($m/z = 30$), and CO$_2$ ($m/z = 44$) (see Figure 3c) [37]. The decomposition of the polymer structure in lignin begins at 200 °C and continues up to 700 °C. These observations are in agreement with [36]. The DSC-TG-MS results for all compositions presented in Figures 4 and 5 show that there is no difference in terms of the trends of curves between samples. However, DTG curves for lignin-Al$_2$O$_3$ composites show lack of signal above 800 °C characteristic for pure lignin and CO$_2$ release.

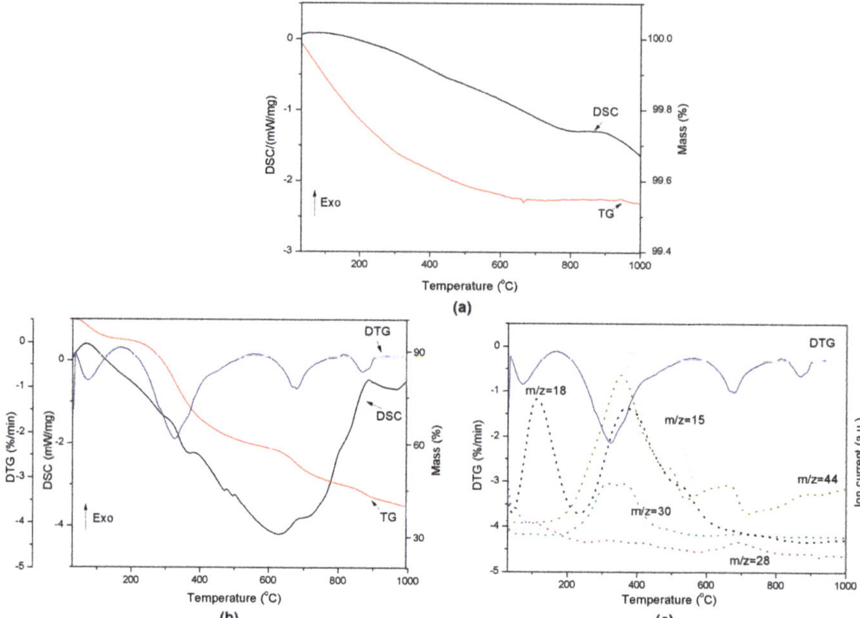

Figure 3. Thermal decomposition of Al$_2$O$_3$ represented by DSC-TG curves (**a**) and thermal decomposition of lignin represented by DSC-TG-DTG (**b**) and DTG-MS curves (**c**).

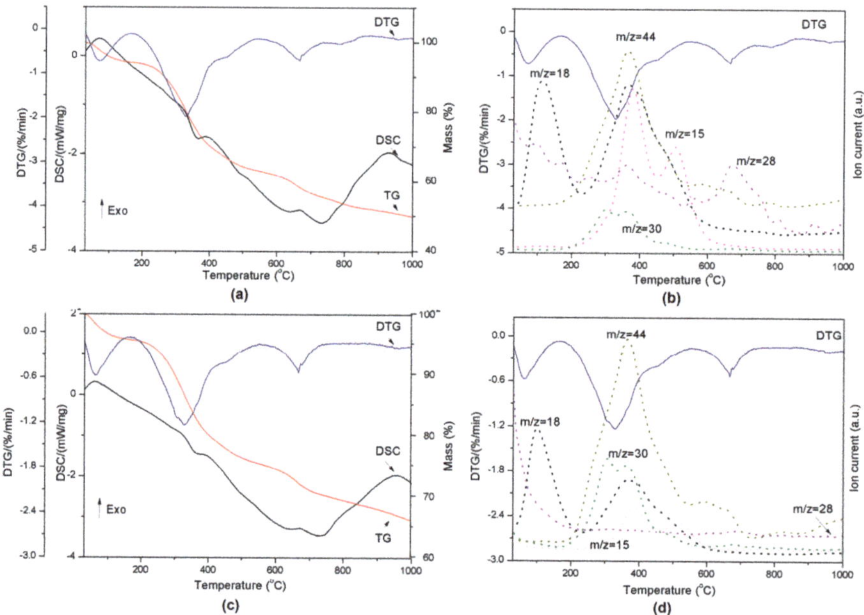

Figure 4. Thermal decomposition of lignin-Al$_2$O$_3$ (8:1, *wt/wt*) represented by DSC-TG-DTG (**a**) and DTG-MS curves (**b**) and lignin-Al$_2$O$_3$ (8:6, *wt/wt*) represented by DSC-TG-DTG (**c**) and DTG-MS curves (**d**).

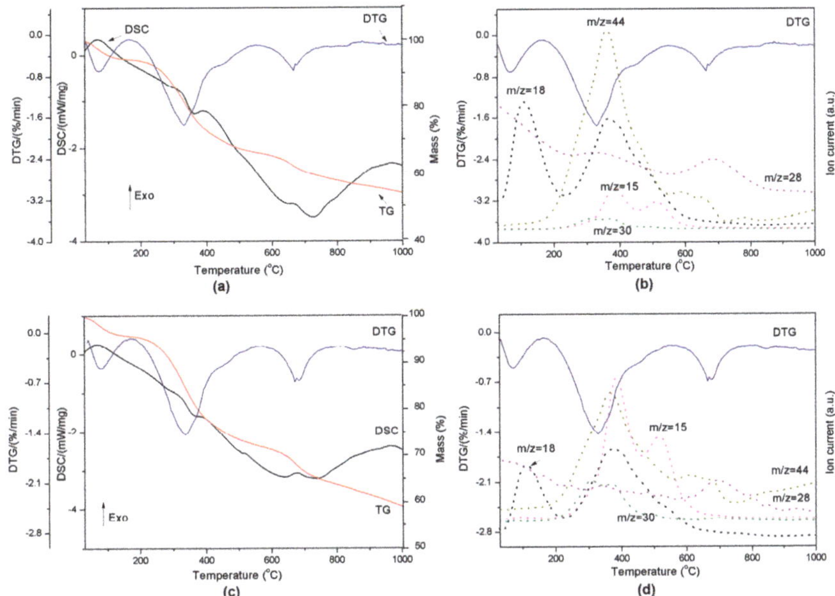

Figure 5. Thermal decomposition of lignin-Al$_2$O$_3$ (8:2, *wt/wt*) represented by DSC-TG-DTG (**a**) and DTG-MS curves (**b**) and lignin-Al$_2$O$_3$ (8:4, *wt/wt*) represented by DSC-TG-DTG (**c**) and DTG-MS curves (**d**).

The presence of Al$_2$O$_3$ in a composite slightly reduced the onset temperature of H$_2$O release from the material. This fact is visible especially in case of compositions from 8:2 *wt/wt* to 8:6 *wt/wt* (Table 3). The second transition started at higher temperatures for all composites than for pure lignin. This confirms that such material is more stable than Kraft lignin at the temperatures under 200 °C, which are most common for the grinding process in the presence of coolants. The third endothermic event related to CO$_2$/N$_2$ release varied between compositions and the highest onset temperature was found for the 8:2 mixture. The decomposition rate of materials presented in Table 3 increased with Al$_2$O$_3$ amount. The exception was composition 8:6 *wt/wt*, probably due to the smallest amount of sample mass loss.

Table 3. Comparison of lignin and different lignin-Al$_2$O$_3$ composition collected during thermal decomposition of samples. DTG and Tp are first derivative of sample mass loss signal and peek temperature, respectively.

Sample Composition	Tonset from DSC (°C)			
Lignin	Lignin-Al$_2$O$_3$ Hybrid Fillers (*wt/wt*)			
	8:1	8:2	8:4	8:6
34	32	27	28	28
324	330	329	328	327
650	645	678	646	640
Speed of Decomposition from DTG (%/min)				
Lignin	Lignin-Al$_2$O$_3$ Hybrid Fillers			
	8:1	8:2	8:4	8:6
0.85	0.72	0.69	0.50	0.58
2.14	1.98	1.76	1.43	1.23
0.65	0.72	0.70	0.71	0.20
Sample Mass Loss (%), Where Temperature Ranges Are, a: RT-160 °C, b: 160–550 °C, c: 550–1000 °C				
6.8 [a]	5.9 [a]	5.6 [a]	4.4 [a]	4.1 [a]
33.8 [b]	31.0 [b]	28.1 [b]	23.6 [b]	20.4 [b]
20.2 [c]	13.2 [c]	13.0 [c]	12.2 [c]	9.2 [c]

In addition, the remaining results of thermal analysis are presented in Figure 5 and Table 3. All powders released the same gases as pure lignin during heating. However, in contrast to pure lignin, the lignin-alumina samples produced only three events during heating, at ~30, ~325 and ~650 °C. They did not exhibit a transition at 865 °C. The results indicate that the increase of Al_2O_3 quantity slows down the reaction at approximately 325 °C (see Table 3), when the gases are released.

2.4. Inverse Gas Chromatography

IGC analysis was used to evaluate the surface properties of the fillers (see Table 4). All of the studied materials demonstrated medium surface activity (γ_s^d about 35–40 mJ/m^2). Such values of γ_s^d are in agreement with data published in the literature, e.g., for phenol-formaldehyde-lignin resin (41.9 mJ/m^2 for a system with 17% of lignin in place of phenol) [38]. Organic-inorganic hybrid fillers with higher lignin content exhibit similar surface properties to lignin, but they are slightly less active, as some active groups may be connected to hydroxyl groups on the aluminum oxide surface.

Table 4. Dispersive, (γ_s^d) donor-acceptor ($\gamma+$, $\gamma-$) components of the free surface energy of studied hybrid fillers and comparison with alumina and lignin.

Sample	γ_s^d (mJ/m^2)	γ^+ (mJ/m^2)	γ^- (mJ/m^2)	K_A (-)	K_D (-)	K_A/K_D (-)
Al_2O_3	37.2 ± 0.5	29.0 ± 0.1	177.2 ± 5.0	0.100 ± 0.010	0.260 ± 0.003	0.385
Lignin	35.2 ± 0.6	15.2 ± 0.2	46.4 ± 1.0	0.112 ± 0.005	0.161 ± 0.002	0.702
Lignin-Al_2O_3 (8:1, *wt/wt*)	33.2 ± 0.4	11.5 ± 0.1	38.2 ± 1.1	0.071 ± 0.001	0.140 ± 0.004	0.507
Lignin-Al_2O_3 (8:2, *wt/wt*)	36.4 ± 0.3	9.8 ± 0.2	35.8 ± 0.8	0.067 ± 0.001	0.143 ± 0.007	0.469
Lignin-Al_2O_3 (8:4, *wt/wt*)	35.6 ± 0.1	10.5 ± 0.1	39.3 ± 0.5	0.068 ± 0.001	0.146 ± 0.003	0.466
Lignin-Al_2O_3 (8:6, *wt/wt*)	35.7 ± 0.1	27.1 ± 0.3	148.1 ± 3.0	0.101 ± 0.002	0.220 ± 0.003	0.459

This is in agreement with the FITR spectra, where it can be seen that the signal from OH groups decreases for lignin-Al_2O_3 hybrids compared to pure lignin (see Figure 2) and OH groups for Al_2O_3 are not visible for hybrid fillers. Interestingly, the hybrid fillers with the highest amount of alumina have acid-base surface properties similar to those of alumina (products have a similar agglomeration behavior). Thus, a hybrid filler with a lignin-to-alumina ratio of 8:6 *wt/wt* has different surface properties than the other studied hybrid materials, and can behave differently in the abrasive article. It is reflected for example in the different rheological properties for composite with a lignin-to-Al_2O_3 ratio of 8:6 *wt/wt* (Table 5). Moreover, this is in agreement with particle size distribution results: the hybrid with a lignin-to-Al_2O_3 ratio of 8:6 was characterized by a similar size distribution to that of alumina (see Section 2.1).

Table 5. Characteristic points for the curing process.

Sample	Softening Point		Cross Over Point	
	Temperature (°C)	Viscosity (Pa·s)	Temperature (°C)	G' = G'' (Pa)
Al_2O_3	136.4	1.029	158	27,400
Lignin	136.1	49.68	157	898,000
Lignin-Al_2O_3 (8:1, *wt/wt*)	135.5	3950	158	430,000
Lignin-Al_2O_3 (8:2, *wt/wt*)	135.7	2913	160	533,000
Lignin-Al_2O_3 (8:4, *wt/wt*)	136.7	1294	158	680,000
Lignin-Al_2O_3 (8:6, *wt/wt*)	135.6	633	160	326,000

All of the studied fillers are more likely to act as electron donors than acceptors. As regards the Al_2O_3 surface, O atoms with a free electron pair can act as electron donors, and the higher value of K_D than K_A indicates that the access of the test compounds to O atoms on the alumina surface is easier than to Al atoms with an electron gap. In the case of the hybrid fillers, the values of the K_A and K_D parameters are lower, as some active groups are involved in the linking between lignin and alumina. In the case of the hybrid with a lignin-to-silica ratio of 8:6 *wt/wt*, the surface has acid-base properties similar to those of alumina. The hypothetical interactions between lignin and alumina are presented in Figure 6.

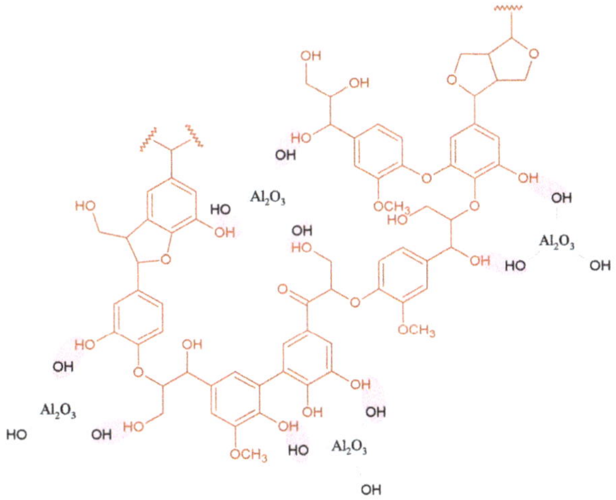

Figure 6. Hypothetical interactions between lignin and alumina.

2.5. Rheological Studies

All samples after the first preparation steps still had a powder form after rotation. For the 8:1 and 8:2 *wt/wt* lignin-alumina hybrid samples, a problem was encountered in reaching a measuring gap of 1.2 mm. For these samples the maximum normal force achieved was 50 N, because they had almost two times lower bulk densities and different thermal conductivity than the other samples. It is possible that the samples were not sufficiently softened. When the first step of sample preparation is analyzed (in terms of the relative change of the gap in time/temperature), conclusions can be drawn about the softening process, namely if the gap starts changing earlier (at lower temperature), the studied sample has a lower softening point; for example, for pure resin it is 84.2 °C, but for the resin with lignin-alumina additive (8:1, *wt/wt*) it is 86.4 °C. The relative change of the gap from the initial position provides information about the degree of softening and partly about thermal conductivity. For a larger relative change, it can be assumed that the sample is more plastic; e.g., for pure resin the size of the relative change of gap is 0.37 mm, while for the resin with the 8:6 *wt/wt* additive the change is 0.15 mm.

When it is observed, the G' and G'' results for the pure novolac, 9% + lignin and 9% alumina samples in the first preparation steps are very noisy. This is caused by the pure contact of the rotor with the sample. For the pure novolac and 9% + alumina samples, it is noticed that the normal force is very low—which means weak contact with the rotor, because the volume of the samples changes during the softening process. For the 9% + lignin sample a big force it is obtained—this means weak contact with the rotor, because the sample is still a powder and slippage effects are observed. These conditions can't be changed because the 1.2 mm gap should be constant and it is a compromise between a liquid sample and a solid.

During the experimental curing process, the pure resin is always in liquid state, even after 15 min at 160 °C, as it is thermoplastic without the addition of a cross-linking agent (urotropine). The other samples were solids at the end of the measurements. Table 5 summarizes the characteristic points for the curing process. The softening point was defined as the lowest value of the complex viscosity. At this point the sample has its most liquid form. After this point the curing process begins, where the storage modulus G' and loss modulus G'' increase. The second characteristic point is the crossover with the same value $G' = G''$. Prior to this crossover point, the measured material acts as a fluid more than elastic solid; afterwards storage modulus starts growing up faster than loss modulus and material acts as an elastic solid more than fluid.

For four lignin-alumina samples (8:1, 8:4, 8:2 and 8:6 *wt/wt*) the modulus of complex viscosity was at the same level at the end of the measurements. The results of the curing process are presented in Figure 7 as complex viscosity at measuring points, to show the influence of the additives on the curing process. Examples of phase transformation are shown in Figure 8a for resole with the 8:2 *wt/wt* lignin-alumina hybrid, and in Figure 8b for resole with the 8:6 *wt/wt* hybrid.

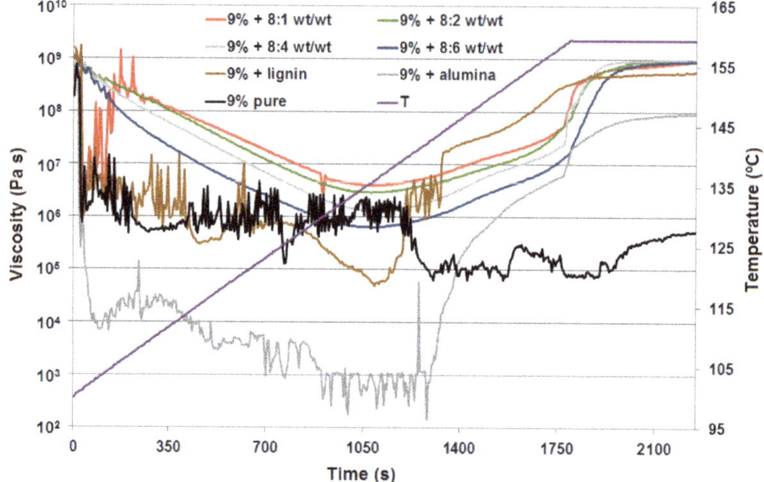

Figure 7. The curing process as complex viscosity.

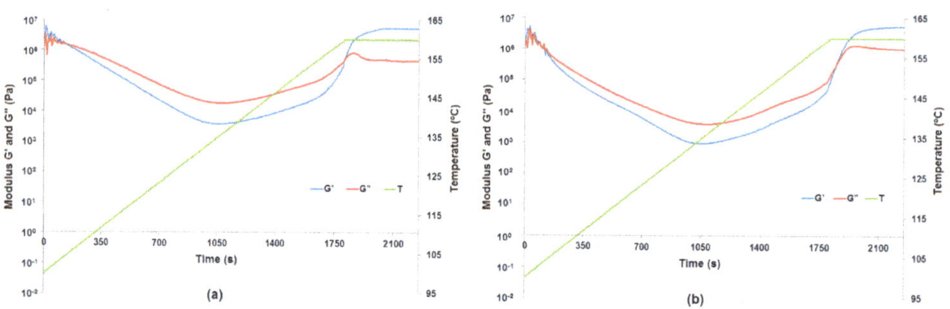

Figure 8. Phase transformation of lignin-Al$_2$O$_3$ (8:2, *wt/wt*) represented by the moduli G' and G" (**a**) and phase transformation of lignin-Al$_2$O$_3$ (8:6, *wt/wt*) represented by the moduli G' and G" (**b**).

At the beginning of the curing process up to 104 °C, the samples of all hybrid materials (8:1, 8:2, 8:4 and 8:6 *wt/wt*) were more elastic solids (powder-like) because the storage modulus G' is greater than the loss modulus G". Next, a softening process was observed, where the sample acted as a fluid more than an elastic solid up to a temperature of 136 °C. After that the curing process starts, but the result differ depending on the rate of temperature change. For pure resin and resin with the addition of Al$_2$O$_3$ (Figure 9a), lignin-Al$_2$O$_3$ hybrid material (8:6 *wt/wt*) and pure lignin (Figure 9b), the measurements up to 140 °C appear disrupted because the softening process was different. Additionally, an increase in the normal force at the measuring point is observed in the graph above 140 °C. The sample expanded, the normal force increased, and the measurement conditions were better to preserve the moduli G' and G". For all samples apart from pure resin, a characteristic change in normal force was observed. The range of change in the normal force provides information about the internal dynamic process: if the force is greater, the process is more turbulent. For the resin with pure lignin, the normal force reached 15 N, but it was lower for systems with lignin-Al$_2$O$_3$ additives (8:1, 8:2, 8:4 and 8:6 *wt/wt*), and reached the lowest value (1.6 N) for systems with Al$_2$O$_3$.

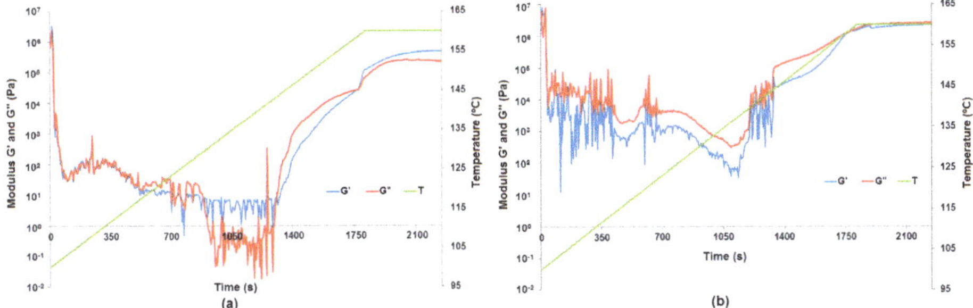

Figure 9. Phase transformation of resin with pure Al$_2$O$_3$ represented by the moduli G$'$ and G$''$ (**a**) and phase transformation of resin with pure lignin represented by the moduli G$'$ and G$''$ (**b**).

2.6. Dynamic-Mechanical Properties

DMTA analysis is often used to assess the interaction between materials and provides information about the viscoelastic behavior of the composites, described by the storage modulus G$'$ and glass transition temperature Tg [39–41]. The values of Tg and G$'$ for the composites determined at various temperatures are given in Table 6. The glass transition temperature Tg is described as a single number representing a wide temperature region. The position of tan δ at its maximum was taken as the glass transition temperature of the composites [42]. Plots of the storage modulus (G$'$) and mechanical loss factor tan δ versus temperature T are shown in Figure 10a,b. The G$'$ values of the modified composites decrease with an increase in the lignin-Al$_2$O$_3$ content. The highest value of G$'$ (2750 MPa) was observed for the reference sample. The sample with lignin-Al$_2$O$_3$ (8:1, *wt/wt*) gave the highest value of G$'$ among the modified composites. This may be the result of the presence of bulky lignin particles in the phenol matrix. Moreover, all modified samples had a lower glass transition temperature than the reference sample, which may be caused by the plasticizing properties of lignin [43,44], which can facilitate the preparation of resin blends and their processing during the formation of finished products. The lignin chains introduced into the matrix can increase the flexibility of the composites and may contribute to energy dissipation through internal friction [45]. The plasticizing effect of lignin-alumina fillers can decrease the fragility of composites. These phenomena may have a positive impact on the efficiency of the final abrasive tool [46].

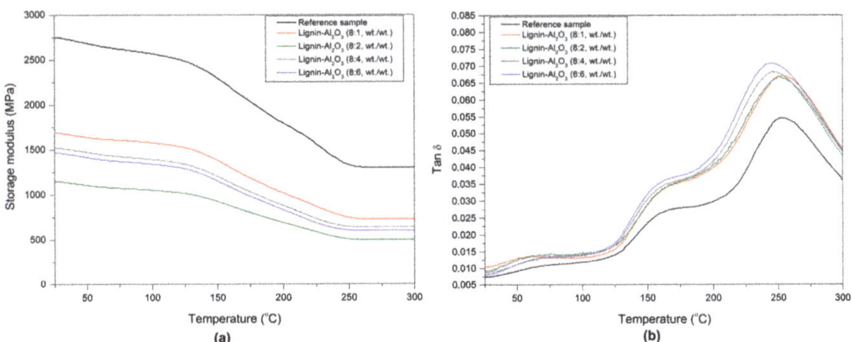

Figure 10. Storage modulus G$'$ (**a**) and tan δ (**b**) versus temperature for the composites obtained.

Table 6. Values of storage modules and glass transition temperature of composites obtained.

Sample	G′ 25 °C (MPa)	G′ 50 °C (MPa)	G′ 300 °C (MPa)	Tan δ_{max}	T_g (°C)
Reference sample	2750	2680	1350	0.055	252
Lignin-Al$_2$O$_3$ (8:1, wt/wt)	1690	1640	776	0.076	244
Lignin-Al$_2$O$_3$ (8:2, wt/wt)	1150	1110	530	0.067	250
Lignin-Al$_2$O$_3$ (8:4, wt/wt)	1520	1470	691	0.068	247
Lignin-Al$_2$O$_3$ (8:6, wt/wt)	1470	1420	652	0.071	244

2.7. Scanning Electron Microscopy Analysis of Composites

The structure of composites with hybrid lignin-alumina fillers was fairly homogeneous. The abrasive grains were well-bounded in all of the studied fillers. Only some small filler agglomerates can be seen. There were no essential differences in the homogeneity of the composites depending on the ratio of lignin to alumina in the fillers. Particularly noteworthy are the SEM images of the composite without the organic-inorganic hybrid filler, consisting exclusively of novolac, corundum and resole (Figure 11a,b). The characteristic structures shown in the images demonstrate the homogeneity of the resulting mixture. The addition of organic-inorganic materials with appropriate ratios of lignin to alumina did not significantly deteriorate the morphological and microstructural properties (Figure 11c,d). Only small differences arise from the variation in the quantity of biopolymer relative to inorganic material (Table 6).

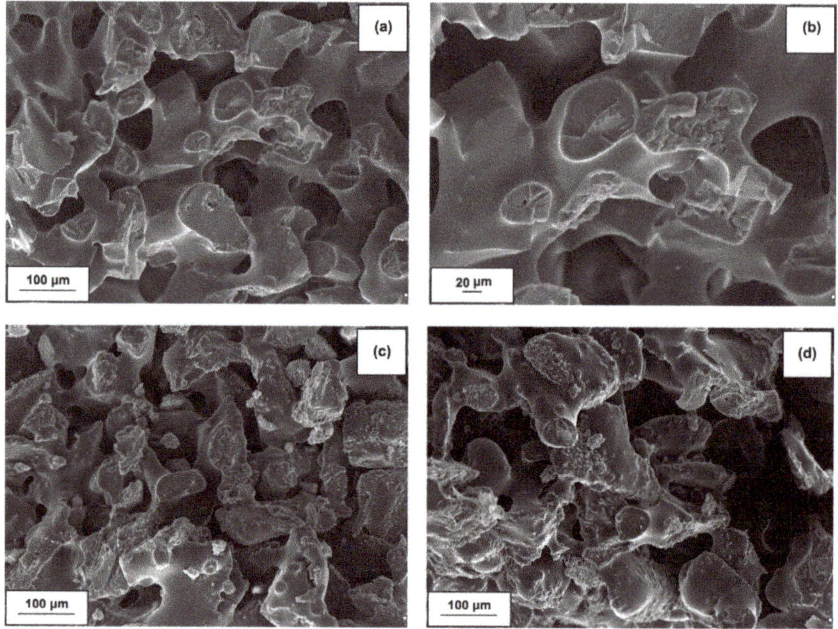

Figure 11. SEM images of novolac + corundum + resole composite (**a,b**) and novolac + corundum + resole + lignin-Al$_2$O$_3$ systems with ratios of organic-inorganic filler equal to 8:1 wt/wt (**c**) and 8:6 wt/wt (**d**).

2.8. Assessment of Emission of Phenol and Formaldehyde by Means of HS-GC Analysis

The emission of phenol and formaldehyde was measured by HS-GC analysis. The peak area values for each tested sample were compared, which allowed to determine the influence of hybrid filler addition on the amount of two main volatile organic compounds released from the mixture. The peak with retention time equal to 1.07 min was attributed to formaldehyde and another with retention time equal to 1.21 min was attributed to phenol.

The values of the peak area for emitted formaldehyde are presented in Table 7 and for phenol in Table 8. The amount of released formaldehyde is slightly lower for sample with hybrid lignin-Al_2O_3 than for sample with Kraft lignin only. Moreover, the composition with kraft lignin emitted slightly more formadehyde than samples with zeolite micro 20, pure resol or pure novolak. This results from the fact that the Kraft lignin in temperatures above 180 °C undergoes thermal decomposition and emits formaldehyde among others [36]. Generally, no significant impact of studied fillers addition on formaldehyde emission can be observed.

Table 7. The value of peak area of the formaldehyde emitted from tested samples during HS analysis.

Sample	Peak Area, S_{peak} (μV·s) *
Novolac	$1.21 \times 10^6 \pm 0.11 \times 10^6$
Resol	$1.90 \times 10^6 \pm 0.28 \times 10^6$
Kraft lignin	$2.52 \times 10^6 \pm 0.09 \times 10^6$
Lignin-Al_2O_3 (8:4, *wt/wt*)	$2.30 \times 10^6 \pm 0.08 \times 10^6$
Resol + novolac + Kraft lignin	$2.73 \times 10^6 \pm 0.18 \times 10^6$
Resol + novolac + lignin-Al_2O_3 (8:4, *wt/wt*)	$2.24 \times 10^6 \pm 0.18 \times 10^6$
Resol + novolac + zeolite micro 20	$2.05 \times 10^6 \pm 0.20 \times 10^6$

* The sum of seven injections from the same vial ± standard deviation for three repetitions of the whole analysis for three independent vials.

Table 8. The value of peak area of the phenol emitted from tested samples during HS analysis.

Sample	The Peak Area, S_{peak} (μV·s) *
Novolac	$0.32 \times 10^6 \pm 0.05 \times 10^6$
Resol	$10.53 \times 10^6 \pm 0.60 \times 10^6$
Kraft lignin	$0.02 \times 10^6 \pm 0.00 \times 10^6$
Lignin-Al_2O_3 (8:4, *wt/wt*)	$0.01 \times 10^6 \pm 0.00 \times 10^6$
Resol + novolac + Kraft lignin	$5.49 \times 10^6 \pm 0.40 \times 10^6$
Resol + novolac + lignin-Al_2O_3 (8:4, *wt/wt*)	$3.47 \times 10^6 \pm 0.31 \times 10^6$
Resol + novolac + zeolite micro 20	$4.86 \times 10^6 \pm 0.45 \times 10^6$

* The sum of seven injections from the same vial ± standard deviation for three repetitions of the whole analysis for three independent vials.

In case of phenol emission, addition of lignin-Al_2O_3 hybrid caused a significant decrease of the amount of released phenol in comparison to Kraft lignin or zeolite micro 20. Addition of all studied fillers notably decreased the emission of phenol by approximately 2–3 times and the highest decrease of phenol emission was observed for composition with lignin-Al_2O_3 hybrid.

3. Materials and Methods

3.1. Preparation of Novel Lignin-Al_2O_3 Hybrid Filler

The novel, functional lignin-Al_2O_3 hybrid materials were prepared by a mechanical method from commercial alumina (Sigma-Aldrich, St. Louis, MO, USA) and Kraft lignin (Sigma-Aldrich). Hybrid additives were produced using 8 parts by weight of lignin with 1, 2, 4 and 6 parts of Al_2O_3, respectively. To combine the Al_2O_3 and lignin, a mechanical process was used whereby the initial powders were ground and simultaneously mixed using a Pulverisette 6 Classic Line planetary ball mill (Fritsch, Idar-Oberstein, Germany). The vessel with the materials for grinding was placed eccentrically on the mill's rotating base. The direction of rotation of the base is opposite to that of the vessel, with a speed ratio of 1:2. The three agate balls inside the vessel move due to the Coriolis force. To obtain suitably homogeneous final materials, grinding was continued for 6 h. To prevent possible overheating of the material due to continuous grinding, every 2 h the mill automatically switched off for 5 min, after which it began operating again. Immediately after grinding, the lignin-Al_2O_3 hybrid materials were sifted using a sieve with a mesh diameter of 40 μm.

3.2. Preparation of Abrasive Composites with Lignin-Al_2O_3 Hybrids

The model abrasive composites were prepared by mixing resole, filler, novolac and abrasive grains, in a ratio of 3:5:12:80 by weight. The proportions of the components were chosen as the standard values used in the abrasive industry. The components were mixed using a mechanical mixer at a slow rate of 200 rpm for a short time (about 3 min)—the process was carried out at room temperature. White fused alumina with a 120 mesh granulation was used as an abrasive. Novolac contains 9% hexamethylenetetramine (hexamine). Firstly, the abrasive grains were covered by resole, then the mixture of novolac and filler was added and homogenized. The composites prepared this way were formed into cuboids. The samples were then hardened according to the following temperature program: heating from 50 °C up to 180 °C, heating rate 0.2 °C/min, then heating at 180 °C for 10 h.

3.3. Physicochemical and Dispersive-Morphological Characteristics of Lignin-Alumina Hybrids

3.3.1. Particle Size Distribution

The dispersive properties of the products were evaluated using Mastersizer 2000 (0.2–2000 μm) and Zetasizer Nano ZS (0.6–6000 nm) instruments (Malvern Instruments Ltd., Malvern, UK), employing the laser diffraction and non-invasive back scattering (NIBS) techniques respectively. During the experiments, no pre-treatment was used for breaking down the agglomerates of the investigated products.

3.3.2. Scanning Electron Microscopy

The surface morphology and microstructure of the lignin-alumina products and precursors were examined on the basis of SEM images recorded by an EVO40 scanning electron microscope (Zeiss, Jena, Germany). Before testing, the samples were coated with Au for a time of 5 s using a PV205P coater (Oerlikon Balzers Coating SA, Brügg, Switzerland).

3.3.3. Fourier Transform Infrared Spectroscopy

Fourier transform infrared spectroscopy (FTIR) measurements were performed on a Vertex 70 spectrophotometer (Bruker, Mannheim, Germany) at room temperature (RT). The sample was analyzed in the form of pellets, made by pressing a mixture of anhydrous KBr (approximately 0.25 g) and 1.5 mg of the tested substance in a special steel ring under a pressure of approximately 10 MPa. FTIR spectra were obtained in the transmission mode between 4000 and 450 cm^{-1}. The analysis was performed at a resolution of 0.5 cm^{-1}.

3.3.4. Thermogravimetric Analysis—Mass Spectrometry

TG-DSC analysis was carried out using a Jupiter STA 449 F1 instrument (Netzsch, Selb, Germany). The analysis was performed with a heating rate of 10 °C/min and a maximum temperature of 1000 °C. Measurements were conducted under a constant flow of helium (40 cm^3/min). The sample mass was approximately 30 mg. The volatile products evolved during heating were detected by a 403C Aëolos mass spectrometer (QMS, Selb, Germany) coupled online to the STA instrument. The QMS was operated with an electron impact ionizer with an energy of 70 eV. During the measurements, the m/z ratio was recorded in the range of 2–150 amu, where m is the mass of the molecule and z its charge.

3.3.5. Inverse Gas Chromatography

Surface properties of the hybrid fillers as well as alumina and lignin were tested by inverse gas chromatography (IGC). IGC experiments were carried out using a SEA Advanced apparatus (Surface

Energy Analyzer produced by Surface Measurement System Ltd., London, UK) equipped with a flame ionization detector. The studied hybrid fillers were applied to inert glass beads in a quantity of 1% (200 mg), placed in a glass chromatographic column (30 cm length, 0.4 cm inner diameter). The column oven temperature was 30 °C, and the temperature of the detector and injector was 150 °C. Dead time was determined by means of methane injection. Helium (flow rate 15 cm^3/min) was used as the carrier gas. The following test compounds were used: nonpolar—hexane, heptane, octane, nonane, decane; and polar—ethyl acetate, dichloromethane, ethanol, dioxane, acetonitrile, acetone.

The free surface energy, γ_s^{total}, and its dispersive (γ_s^d) and specific components (acid, γ_s^+ and basic, γ_s^-) were determined. The γ_s^d parameter was calculated according to the Schultz–Lavielle method, using Equation (1) [47]:

$$R \cdot T \cdot \ln V_N = 2 \cdot N \cdot a \cdot \sqrt{\gamma_s^d \cdot \gamma_l^d} + C \tag{1}$$

where: R is the gas constant, 8.314 J/mol·K; T is the temperature of measurement (K); V_N is the net retention volume (m^3); N is Avogadro's constant, 6.022×10^{23} L/mol; a is the cross-sectional area of the adsorbate (m^2); γ_s^d is the dispersive component of surface free energy (mJ/m^2); γ_l^d is the dispersive component of the surface tension of the probe molecule in liquid state (mJ/m^2); C is a constant.

Retention data for polar and nonpolar test compounds are necessary to quantify the acidic and basic properties of the examined surface. These are described by the parameters γ_s^+, γ_s^-, which were estimated according to the Good–van Oss concept [48] described by Equation (2):

$$\Delta G^{sp} = 2 \cdot N_A \cdot a \cdot \left(\left(\gamma_l^+ \cdot \gamma_s^- \right)^{1/2} + \left(\gamma_l^- \cdot \gamma_s^+ \right)^{1/2} \right) \tag{2}$$

In Equation (2), γ_l^+, γ_l^- are the electron acceptor and donor parameters of the probe molecules, respectively, and ΔG^{sp} is the specific component of the free energy of adsorption of the polar compound. The method of determination of ΔG^{sp} is described in many publications, e.g., [47,48]. For the calculation of γ_s^+, γ_s^- dichloromethane (DM) and ethyl acetate (EA) were used as test compounds. DM is a monopolar acid, and $\gamma_{DM}^- = 0.0$ mJ/m^2. Equation (2) is reduced to:

$$\gamma_S^- = \Delta G_{DM}^{sp} / (4 \cdot N_A^2 \cdot a_{DM}^2 \cdot \gamma_{DM}^+) \tag{3}$$

and γ_S^- can be easily calculated. The value of γ_{DM}^+ was established as 5.2 mJ/m^2 on the basis of [48]. Similarly, EA is a monopolar base, $\gamma_{EA}^+ = 0.0$ mJ/m^2, and the γ_S^+ parameter for the examined solid can be calculated from Equation (4):

$$\gamma_S^+ = \Delta G_{EA}^{sp} / \left(4 \cdot N_A^2 \cdot a_{EA}^2 \cdot \gamma_{EA}^- \right) \tag{4}$$

The value of γ_{EA}^- was established as 19.2 mJ/m^2 [46].

The acid-base properties of the studied fillers, as well as lignin and alumina, were assessed in terms of the parameters K_A and K_D, describing respectively the acid and base properties of the surface. These parameters were calculated from the straight line (Equation (5)):

$$\frac{\Delta G_{sp}}{AN^*} = K_A \cdot \frac{DN}{AN^*} + K_D \tag{5}$$

where: K_A is the parameter expressing the acidic properties of the solid surface; K_D is the parameter expressing the basic properties of the solid surface; ΔG_{sp} is the specific component of the free energy of adsorption of the polar compound; DN is the donor number of the polar test solute; AN^* is the acceptor number of the polar test solute.

3.4. Rheological Studies

Samples for rheological measurements were prepared by a mechanical method in a closed container with simultaneous mixing. The following composites were studied: novolac and lignin; novolac and Al_2O_3; novolac and lignin-Al_2O_3 hybrid (8:1, 8:2, 8:4 and 8:6 *wt/wt*).

Pure novolac was measured as a reference sample, as well as novolac with additive (constant proportion 3:1.25 *wt/wt*). To measure samples in a powder state in the rotational rheometer, it was decided to divide the measurements into two steps: firstly sample preparation, and secondly monitoring of the curing process. The rheological behavior of a sample was tested using an RS6000 Thermo Scientific rheometer (HAAKE, Vreden, Germany) with disposable plate-plate rotor with diameter 20 mm and disposable lower plate. The disposable measuring system enables monitoring of the cross-linking of the resin up to total curing.

The rheometer was set to an initial temperature of 80 °C. Next, the sample was loaded onto the rheometer with a sample loading tool. With the border around the lower plate, the geometry can easily be filled with powders. Next, the automatic lift, controlled by a RheoWin device (HAAKE), moves the upper geometry into the measuring position with a force of 20 N for 90 s. After this procedure, the sample forms a puck with the same geometry, independently of the bulk density. The measuring position was reached when the rheometer touched the sample with a normal force of 5 N. Next, a temperature module with the Peltier system started to change the temperature of the sample from 80 to 100 °C over 30 min. This process allowed to soften the sample.

In the next step, the sample prepared as described above was cross-linked in a temperature sweep from 100 to 160 °C over 30 min, and the cross-linking process was further monitored at 160 °C for 15 min. For protection against heating loss, a solvent trap made from teflon was used. Because in the first step of the sample preparation process the samples were observed to have different structures, the rheometer achieved a gap of 1.2 mm. When the rheometer rotor moved down to the measuring position, the normal force was recorded. Attainment of the normal force can be characterized by the degree of softening of the sample and its thermal conductivity. The classical rheological measurement to monitor the curing process is the oscillation test. The oscillation test is nondestructive when a small deformation or stress is involved. Because the viscoelastic properties of the sample vary over a large range in the curing process, the method with controlled deformation $\gamma = 0.01$ was chosen. For characterization of viscoelastic properties, storage modulus represents the elastic nature of the sample, while the loss modulus represents its viscous nature. When the sample forms a network structure in the curing process, it is expected that changes in the nature of the sample from more viscous to more elastic will be observed. Equilibrium of elastic properties means that a full cure state was reached in the curing process.

3.5. Dynamic-Mechanical Properties

The dynamic-mechanical properties of samples with dimensions of $10 \times 4 \times 50$ mm were investigated by dynamic mechanical thermal analysis (DMTA) in torsion mode using an Anton Paar MCR 301 apparatus (Ashland, VA, USA) operating at a frequency of 1 Hz. The temperature range was from 25 to 300 °C, with a heating rate of 2 °C/min. The position of tan δ at its maximum was taken as the glass transition temperature.

3.6. Headspace Gas Chromatography

An automatic headspace sampler (TurboMatrix HS 40, PerkinElmer Waltham, MA, USA) and a gas chromatograph system (Clarus 580, PerkinElmer) were used for HS-GC measurement. The GC system was equipped with a flame ionization detector and an Elite-5 capillary column (30 m × 0.25 mm i.d., with 0.25 µm film thickness, PerkinElmer) operating at temperatures of: vial 180 °C, transfer line 200 °C, column 210 °C with helium transfer gas (flow rate 2 mL/min) was employed. All examined samples consisted of: 0.95 g of organic resin binder (MD 1/11 novolak resin, LERG S. A., Pustków-Osiedle,

Poland), 0.2 g of wetting agent (Rezol S resole resin, LERG S. A., Pustków-Osiedle, Poland) and 0.35 g of tested filler (pure lignin or lignin-Al_2O_3 hybrid (8:4 wt/wt) as well as zeolite micro 20—filler commercially used in abrasive industry). Also the samples of: (i) 0.35 g Kraft lignin; (ii) 0.95 g novolak and (iii) 0.2 g Rezol S were tested. All the components were precisely mixed together to achieve a homogeneous mixture. The vial thermostating time was equal to 5 min. The number of injections for each sample was equal to 7: the analysis was performed 7 times for the same vial as multiple headspace till the peak area decreased significantly (near the limit of detection). The repetition of the method was determined by performance of the analysis for the same composition three times by preparing new vial as described above. The volume of headspace vials was 20 mL. The qualitative analysis of phenol and formaldehyde was performed on the basis of comparison of the retention time of the tested materials with standards: pure phenol and formaldehyde (paraformaldehyde). Phenol and formaldehyde were of analytical grade (purity > 99%, Sigma Aldrich, Steinheim am Albuch, Germany). Each compound was placed in the vial separately and the HS-GC analysis was performed as for other studied materials.

4. Conclusions

The results presented in the framework of this study demonstrate that novel lignin-alumina hybrid fillers, which were not previously described in the literature, can be obtained in a relatively simple way by intensive mechanical mixing of the biopolymer with Al_2O_3. The use of lignin-alumina hybrids makes it possible to obtain final composite abrasive articles with higher plasticity due to the lignin part, and also better heat conductivity due to the Al_2O_3. Moreover, it turns out that the addition of even a small quantity of alumina (lignin-to-alumina ratio 8:1 wt/wt) can increase the thermal conductivity of lignin, and thus improve the thermomechanical properties of the final composite used for abrasive tool production. The inorganic-organic hybrid fillers added to the composition of the abrasive tool have the most influence on the dynamics of cross-linking at temperatures of approx. 160 °C and the "internal" turbulence process at temperatures of approx. 140 °C, by changing the normal force. It is worth noticing that the addition of lignin-Al_2O_3 hybrids notably decreased phenol emission and slightly limited formaldehyde emission in comparison to commercially used filler natural zeolite micro 20 as well as pure Kraft lignin. The thorough physicochemical analysis of the new hybrid fillers has shown that chemical bonds may be formed between the hydroxyl groups present in both lignin and alumina. Further research will certainly be continued in this direction, especially with the use of kraft lignin derivatives combined with alumina using mechanical and chemical methods to increase the interaction between the precursors. Additionally, the study of the durability properties of the adhesives, using natural and/or QUV accelerated tests to prevent the ageing effects of temperature, humidity and UV exposure on the coating will be particularly important in the near future.

Acknowledgments: The study was financed within the National Centre of Science Poland funds according to decision No. DEC-2014/15/B/ST8/02321.

Author Contributions: Ł.K. Planning studies. Preparation and characterization of hybrid materials. Results development. Manuscript preparation. Coordination of all tasks in the paper. A.J. Characterization of hybrid materials using FTIR. Participation in HS-GC, IGC analysis performance. B.S. Planning studies. Characterization of hybrids using IGC. HS-GC analysis performance. Results development. I.K. Analysis and interpretation of thermogravimetric analysis—mass spectrometry. B.B. Analysis and interpretation of rheological studies. D.M. Analysis and interpretation of dynamic mechanical properties (DMTA). A.V. Experimental investigation. Research discussion. Elaboration of the obtained results. T.J. Planning studies. Experimental investigation. Results development.

Conflicts of Interest: The authors declare no conflict of interest.

References

1. Voelkel, A.; Strzemiecka, B. Characterization of fillers used in abrasive articles by means of inverse gas chromatography and principal component analysis. *Int. J. Adhes. Adhes.* **2007**, *27*, 188–194. [CrossRef]
2. Voelkel, A.; Strzemiecka, B.; Jesionowski, T. The examination of the degree of coverage of the fused alumina abrasive by resol wetting agent by Inverse GC. *Chroma* **2009**, *70*, 1393–1397. [CrossRef]

3. Strzemiecka, B.; Voelkel, A.; Hinz, M.; Rogozik, M. Application of inverse gas chromatography in physicochemical characterization of phenolic resin adhesives. *J. Chromatogr. A* **2014**, *1368*, 199–203. [CrossRef] [PubMed]

4. Laza, J.M.; Alonso, J.; Vilas, J.L.; Rodríguez, M.; León, L.M.; Gondra, K.; Ballestero, J. Influence of fillers on the properties of a phenolic resin cured in acidic medium. *J. Appl. Polym. Sci.* **2008**, *108*, 387–392. [CrossRef]

5. Strzemiecka, B.; Voelkel, A.; Chmielewska, D.; Sterzyński, T. Influence of different fillers on phenolic resin abrasive composites. Comparison of inverse gas chromatographic and dynamic mechanical-thermal analysis characteristics. *Int. J. Adhes. Adhes.* **2014**, *51*, 81–86. [CrossRef]

6. Dang, A.; Ojha, S.; Hui, C.M.; Mahoney, C.; Matyjaszewski, K.; Bockstaller, M.R. High-transparency polymer nanocomposites enabled by polymer-graft modification of particle fillers. *Langmuir* **2014**, *30*, 14434–14442. [CrossRef] [PubMed]

7. Imre, B.; Pukánszky, B. From natural resources to functional polymeric biomaterials. *Eur. Polym. J.* **2015**, *68*, 481–487. [CrossRef]

8. Cataldi, A.; Deflorian, F.; Pegoretti, A. Microcrystalline cellulose filled composites for wooden artwork consolidation: Application and physic-mechanical characterization. *Mater. Des.* **2015**, *83*, 611–619. [CrossRef]

9. Cataldi, A.; Dorigato, A.; Deflorian, F.; Pegoretti, A. Thermo-mechanical properties of innovative microcrystalline cellulose filled composites for art protection and restoration. *J. Mater. Sci.* **2014**, *49*, 2035–2044. [CrossRef]

10. Bozsódi, B.; Romhányi, V.; Pataki, P.; Kun, D.; Renner, K.; Pukánszky, B. Modification of interactions in polypropylene/lignosulfonate blends. *Mater. Des.* **2016**, *103*, 32–39. [CrossRef]

11. Milczarek, G.; Inganäs, O. Renewable cathode materials from biopolymer/conju-gated polymer interpenetrating networks. *Science* **2012**, *335*, 1468–1471. [CrossRef] [PubMed]

12. Milczarek, G. Lignosulfonate-modified electrodes: Electrochemical properties and electrocatalysis of NADH oxidation. *Langmuir* **2009**, *25*, 10345–10353. [CrossRef] [PubMed]

13. Jesionowski, T.; Klapiszewski, Ł.; Milczarek, G. Kraft lignin and silica as precursors of advanced composite materials and electroactive blends. *J. Mater. Sci.* **2014**, *49*, 1376–1385. [CrossRef]

14. Jesionowski, T.; Klapiszewski, Ł.; Milczarek, G. Structural and electrochemical properties of multifunctional silica/lignin materials. *Mater. Chem. Phys.* **2014**, *147*, 1049–1057. [CrossRef]

15. Hatakeyama, H.; Hatakeyama, T. Lignin structure, properties, and application. *Adv. Polym. Sci.* **2010**, *232*, 1–63. [CrossRef]

16. Thakur, V.K.; Thakur, M.K.; Raghavan, P.; Kessler, M.R. Progress in green polymer composites from lignin for multifunctional applications: A review. *ACS Sustain. Chem. Eng.* **2014**, *2*, 1072–1092. [CrossRef]

17. Thakur, V.K.; Thakur, M.K.; Gupta, R.K. Review: Raw natural fiber-based polymer composites. *Int. J. Polym. Anal. Charact.* **2014**, *19*, 256–271. [CrossRef]

18. El-Zawawy, W.K.; Ibrahim, M.M.; Belgacem, M.N.; Dufresne, A. Characterization of the effects of lignin and lignin complex particles as filler on a polystyrene film. *Mater. Chem. Phys.* **2011**, *131*, 348–357. [CrossRef]

19. Faludi, G.; Hári, J.; Renner, K.; Móczó, J.; Pukánszky, B. Fiber association and network formation in PLA/lignocellulosic fiber composites. *Compos. Sci. Technol.* **2013**, *77*, 67–73. [CrossRef]

20. Faludi, G.; Dora, G.; Renner, K.; Móczó, J.; Pukánszky, B. Biocomposite from polylactic acid and lignocellulosic fibers: Structure-property correlations. *Carbohydr. Polym.* **2013**, *92*, 1767–1775. [CrossRef] [PubMed]

21. Klapiszewski, Ł.; Pawlak, F.; Tomaszewska, J.; Jesionowski, T. Preparation and characterization of novel PVC/silica-lignin composites. *Polymers* **2015**, *7*, 1767–1788. [CrossRef]

22. Bula, K.; Klapiszewski, Ł.; Jesionowski, T. A novel functional silica/lignin hybrid material as a potential bio-based polypropylene filler. *Polym. Compos.* **2015**, *36*, 913–922. [CrossRef]

23. Kharade, A.Y.; Kale, D.D. Effect of lignin on phenolic novolak resins and moulding powder. *Eur. Polym. J.* **1998**, *34*, 201–205. [CrossRef]

24. Hattali, S.; Benaboura, A.; Dumarçay, S.; Gérardin, P. Evaluation of alfa grass soda lignin as a filler for novolak molding powder. *J. Appl. Polym. Sci.* **2005**, *97*, 1065–1068. [CrossRef]

25. Guo, X.; Zhang, S.; Shan, X. Adsorption of metal ions on lignin. *J. Hazard. Mater.* **2008**, *151*, 134–142. [CrossRef] [PubMed]

26. Klapiszewski, Ł.; Bartczak, P.; Wysokowski, M.; Jankowska, M.; Kabat, K.; Jesionowski, T. Silica conjugated with kraft lignin and its use as a novel 'green' sorbent for hazardous metal ions removal. *Chem. Eng. J.* **2015**, *260*, 684–693. [CrossRef]

27. Ge, Y.; Li, Z.; Kong, Y.; Song, Q.; Wang, K. Heavy metal ions retention by bi-functionalized lignin: Synthesis, applications, and adsorption mechanisms. *J. Ind. Eng. Chem.* **2014**, *20*, 4429–4436. [CrossRef]

28. Qu, Y.; Tian, Y.; Zou, B.; Zhang, J.; Zheng, Y.; Wang, L.; Li, Y.; Rong, C.; Wang, Z. A novel mesoporous lignin-silica hybrid from rice husk produced by a sol-gel method. *Bioresour. Technol.* **2010**, *101*, 8402–8405. [CrossRef] [PubMed]

29. Klapiszewski, Ł.; Nowacka, M.; Milczarek, G.; Jesionowski, T. Physicochemical and electrokinetic properties of silica-lignin biocomposites. *Carbohydr. Polym.* **2013**, *94*, 345–355. [CrossRef] [PubMed]

30. Klapiszewski, Ł.; Rzemieniecki, T.; Krawczyk, M.; Malina, D.; Norman, M.; Zdarta, J.; Majchrzak, I.; Dobrowolska, A.; Czaczyk, K.; Jesionowski, T. Kraft lignin/silica-AgNPs as a functional material with antibacterial activity. *Coll. Surf. B* **2015**, *134*, 220–228. [CrossRef] [PubMed]

31. Hassanzadeh-Tabrizi, S.A.; Taheri-Nassaj, E. Economical synthesis of Al_2O_3 nanopowder using a precipitation method. *Mater. Lett.* **2009**, *63*, 2274–2276. [CrossRef]

32. Costa, T.M.H.; Gallas, M.R.; Benvenutti, E.V.; da Jornada, J.A.H. Study of nanocrystalline γ-Al_2O_3 produced by high-pressure compaction. *J. Phys. Chem. B* **1999**, *103*, 4278–4284. [CrossRef]

33. Naskar, M.K. Hydrothermal synthesis of petal-like alumina flakes. *J. Am. Ceram. Soc.* **2009**, *92*, 2392–2395. [CrossRef]

34. Rahman, M.A.; de Santis, D.; Spagnoli, G.; Ramorino, G.; Penco, M.; Phuong, V.T.; Lazzeri, A. Biocomposites based on lignin and plasticized poly(L-lactic acid). *J. Appl. Polym. Sci.* **2013**, *129*, 202–214. [CrossRef]

35. Mancera, C.; Ferrando, F.; Salvadó, J.; El Mansouri, N.E. Kraft lignin behavior during reaction in an alkaline medium. *Biomass Bioenerg.* **2011**, *35*, 2072–2079. [CrossRef]

36. Brebu, M.; Vasile, C. Thermal degradation of lignin—A review. *Cellul. Chem. Technol.* **2010**, *44*, 353–363.

37. Jakab, E.; Faix, O.; Till, F. Thermal decomposition of milled wood lignins studied by thermogravimetry/mass spectrometry. *J. Anal. Appl. Pyrol.* **1997**, *40–41*, 171–186. [CrossRef]

38. Matsushita, Y.; Wada, S.; Fukushima, K.; Yasuda, S. Surface characteristics of phenol-formaldehyde-lignin resin determined by contact angle measurement and inverse gas chromatography. *Ind. Crop. Prod.* **2006**, *23*, 115–121. [CrossRef]

39. Faulstich de Paiva, J.M.; Frollini, E. Unmodified and modified surface sisal fibers as reinforcement of phenolic and lignophenolic matrices composites: Thermal analyses of fibers and composites. *Macromol. Mater. Eng.* **2006**, *291*, 405–417. [CrossRef]

40. Menard, K.P. *Dynamic Mechanical Analysis*, 2nd ed.; CRC Press: Boca Raton, FL, USA, 2008; ISBN 9781420053128.

41. Brostow, W.; Hagg Lobland, H.E. *Materials: Introduction and Applications*; John Wiley & Sons: New York, NY, USA, 2017; ISBN 978-0-470-52379-7.

42. Kalogeras, I.M.; Hagg Lobland, H.E. The nature of the glassy state: Structure and glass transitions. *J. Mater. Educ.* **2012**, *34*, 69–94.

43. Kelley, S.; Rials, T.G.; Glasser, W.G. Relaxation behaviour of the amorphous components of wood. *J. Mater. Sci.* **1987**, *22*, 617–624. [CrossRef]

44. Wang, J.; Laborie, M.P.G.; Wolcott, M.P. Kinetic analysis of phenol-formaldehyde bonded wood joints with dynamical mechanical analysis. *Thermochim. Acta* **2009**, *491*, 58–62. [CrossRef]

45. Guigo, N.; Mija, A.; Vincent, L.; Sbirrazzuoli, N. Eco-friendly composite resins based on renewable biomass resources: Polyfurfuryl alcohol/lignin thermosets. *Eur. Polym. J.* **2010**, *46*, 1016–1023. [CrossRef]

46. Gardziella, A.; Pilato, L.A.; Knop, A. *Phenolic Resins-Chemistry, Applications, Standardization, Safety and Ecology*; Springer: Berlin, Germany, 2000; ISBN 3540655174.

47. Schultz, J.; Lavielle, L.; Martin, C. Proprietes de surface des fibres de carbone dèterminèes par chromatographie gazeuse inverse. *J. Chim. Phys.* **1987**, *84*, 231–237. [CrossRef]
48. Oss, C.; Good, R.; Chaudhury, M. Additive and nonadditive surface tension components and the interpretation of contact angles. *Langmuir* **1988**, *4*, 884–891. [CrossRef]

Sample Availability: Samples of the compounds are not available from the authors.

Article

Single Actin Bundle Rheology

Dan Strehle [1,†], Paul Mollenkopf [1,2,†], Martin Glaser [1,2], Tom Golde [1], Carsten Schuldt [1,2], Josef A. Käs [1] and Jörg Schnauß [1,2,*]

[1]   Faculty of Physics and Earth Sciences, Peter Debye Institute, Leipzig University, Linnéstr. 5,
      04103 Leipzig, Germany; dan.strehle@gmx.net (D.S.); paul.mollenkopf@uni-leipzig.de (P.M.);
      martin.glaser@uni-leipzig.de (M.G.); tom.golde@uni-leipzig.de (T.G.); schuldt@physik.uni-leipzig.de (C.S.);
      jkaes@physik.uni-leipzig.de (J.A.K.)
[2]   Fraunhofer Institute for Cell Therapy and Immunology (IZI), DNA Nanodevices Group, Perlickstraße 1,
      04103 Leipzig, Germany
*   Correspondence: joerg.schnauss@uni-leipzig.de; Tel.: +49-341-973-2753
†   These authors contributed equally to this work.

Received: 20 September 2017; Accepted: 19 October 2017; Published: 24 October 2017

Abstract: Bundled actin structures play an essential role in the mechanical response of the actin cytoskeleton in eukaryotic cells. Although responsible for crucial cellular processes, they are rarely investigated in comparison to single filaments and isotropic networks. Presenting a highly anisotropic structure, the determination of the mechanical properties of individual bundles was previously achieved through passive approaches observing bending deformations induced by thermal fluctuations. We present a new method to determine the bending stiffness of individual bundles, by measuring the decay of an actively induced oscillation. This approach allows us to systematically test anisotropic, bundled structures. Our experiments revealed that thin, depletion force-induced bundles behave as semiflexible polymers and obey the theoretical predictions determined by the wormlike chain model. Thickening an individual bundle by merging it with other bundles enabled us to study effects that are solely based on the number of involved filaments. These thicker bundles showed a frequency-dependent bending stiffness, a behavior that is inconsistent with the predictions of the wormlike chain model. We attribute this effect to internal processes and give a possible explanation with regard to the wormlike bundle theory.

Keywords: biopolymers; actin; bundles; optical tweezers; rheology; mechanical properties; dynamics

---

1. Introduction

The cytoskeleton is a meshwork of a variety of biopolymers, providing eukaryotic cells with mechanical stability and dynamic functions, which affect a cell's structure and function [1,2]. A dominating cytoskeletal component is the semiflexible polymer actin, which, for instance, builds up the actin cortex and is involved in crucial cellular processes such as wound healing, embryonic development, tissue engineering, and immune response [1,3–8].

Being one of the most relevant polymers in terms of mechanical impact on cell properties, actin has been subject to numerous studies, both theoretical and experimental [4,9–18]. Remarkably, individual filaments can be arranged in different structures such as networks or bundles and their interplay with crosslinking molecules enables a rich phase space of mechanical responses against external stimuli. This vast phase space is further enriched by the transience of these crosslinkers such as fascin, fibrin, and α-actinin, which bind and unbind with rates in the order of tenths to tens of seconds [19–21]. These varying structural isoforms are ubiquitous in the cytoskeleton, rendering it a composition of diversified structures [11]. To decouple the effects of the differing structures, it is essential to study them in a bottom-up, isolated fashion. Following this approach, single actin filaments and networks have been investigated rigorously with respect to their mechanical properties [4,11–18]. The mechanics of bundled structures and the molecular design principles responsible for their mechanical properties, however, still remain poorly understood. A major drawback is that bundles are highly anisotropic impeding studies employing established conventional approaches, such as bulk rheology, that rely on isotropic networks. The mechanical characterization of bundles relies on elaborate experiments on the μm-scale to determine their properties. Pioneering examples for these investigations are the determination of bundles' bending stiffness measured by observing passive contour undulations induced by thermal fluctuations [21], and the evaluation of time-dependent mechanical responses of crosslinked actin bundles after deformations actively induced via optical tweezers [20,22]. These experimental studies are accompanied by new theoretical approaches within the so-called worm-like bundle model aiming to develop an understanding of the microscopic picture of bundle mechanics. However, a complete rheological characterization has not been feasible so far, since main characteristics such as frequency-dependent mechanics could not be tested.

Here, we resolve this methodological limitation and present a novel approach based on optical tweezers technology. Optical tweezers were used to actively deform actin bundles to rheologically characterize these structures and their responses for different frequencies. The concept is based on the method previously described by Riveline et al. [23], and the theoretical framework introduced by Wiggins et al. [24] that describes elastic rods under the influence of hydrodynamic drag, thereby allowing the deduction of the filaments' stiffness from its contour shape. We adapted and extended this approach to actin filaments bundled by depletion forces, using methyl cellulose as a depletion agent. Compared to the work by Riveline et al., we measured bundles with larger spatial extensions and observed a total decay of the induced oscillation, as illustrated in Figure 1. We found that thin bundles obey the theoretical predictions of the worm-like chain model [25]. To investigate the impact of bundle thickness, we introduced an approach to thicken bundles by merging them in a controlled manner (see Materials and Methods). These thickened bundles exhibited a differing mechanical behavior, in particular, they showed a frequency-dependent stiffness, which is reminiscent of viscoelastic materials. Treating bundling as an unspecific type of crosslinking, this result may be explained by the consensus of the wormlike bundle (WLB) theory [15,26]. As our system does not contain specific crosslinking proteins, which have been assumed by the theoretic model, this result is not inherent with respect to the corresponding theory. We suppose that previously reported mechanisms such as velocity-dependent inter-polymer sliding friction [27] and bundle relaxation [7] play a significant and non-negligible role, which is not yet completely understood. Transitions between systems dominated by crosslinking and systems which are determined by friction and relaxation may be rather indistinct.

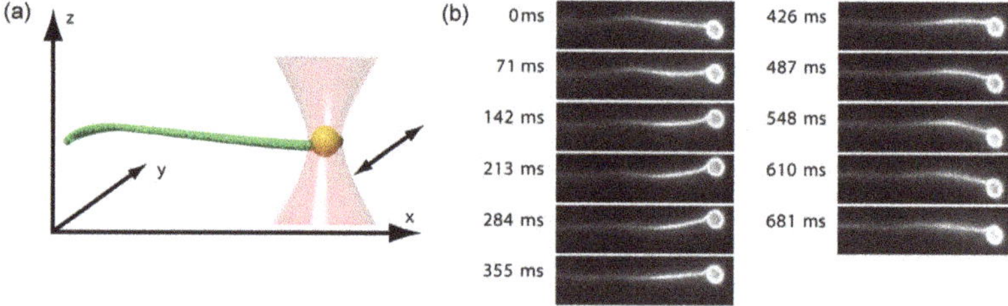

Figure 1. (**a**) A 2-μm polystyrene bead coated with streptavidin is attached to an actin bundle enriched with biotinylated actin monomers. This bead is trapped by optical tweezers, and an oscillatory movement of the trap in the *xy*-plane induces oscillations in the bundle. The oscillation amplitudes subsequently decay when traveling through the bundle, a process which can be captured by fluorescence microscopy; (**b**) Fluorescence images of the experimental procedure. The 11 images represent half a period at a frequency of 0.7 Hz.

2. Results

After an appropriate constellation of a bundle attached to a bead was found and adjusted with respect to its angle to the oscillation direction, the system was exposed to excitations spanning a frequency range between 0.04 Hz and 2.5 Hz. The experimental performance was limited by the sensitivity of the camera and the sample's degradation caused by bleaching effects. The minimum exposure time of 50 ms delimitated the temporal resolution for high frequencies (50 ms extended over one fifth of one period at 4 Hz). On the other side, 0.04 Hz already corresponded to an oscillation period of 25 s. As several periods must be recorded to minimize transient effects, photo-bleaching of the sample was a crucial limitation. As a first test, we verified that bundles did not change their behavior with time by exposing bundles to a defined and fixed driving frequency over a time interval of 600 s. We observed no overall tendency to diverge from an average value for the measured hydrodynamic length (data not shown). For the frequency tests, bundles were excited by a standard set of driving amplitudes with predefined frequencies, a method that had been introduced by Riveline et al. to study single filaments [23]. The unique combination with the evaluation processing, based on the theoretical framework by Wiggins et al. [24], allowed us to rheologically study bundle structures for the first time. The acquired hydrodynamic length shows a scaling with frequency $l_\omega \propto \omega^{-1/4}$, as predicted by the theory for a wormlike chain (WLC) (see Figure 2a) translating to a constant bending rigidity $\kappa_{eff} = \frac{\kappa}{\zeta} = l_\omega^4 \cdot \omega$ (see Figure 2b). The drag coefficient ζ is a function of the Reynolds number and scales with the ratio L/d, where L denotes the contour length and d the bundle diameter. In our experiments this ratio was only subject to minimal variations since the length of the bundles exceeded their according diameters by orders of magnitudes (L >> d). Consequently, the impact of thickness variations (Δd) to the drag coefficient are negligible. The amplitude decay of tested bundles yielded a bending stiffness in the order of $10^{-21} \frac{J\,m}{Pa \cdot s}$. For comparison, for a bundle of 50 tightly crosslinked filaments measured in a medium with a viscosity of $\eta = 50$ mPa·s, a bending rigidity of $1.5 \times 10^{-19} \frac{J\,m}{Pa \cdot s}$ was observed [28]. While most of the bundles obeyed the prediction of the WLC model and displayed a stiffness invariance over the whole frequency range, it was conspicuous that rather thick bundles (see Figure 2b: blue dotted) showed a softening behavior for low frequencies. This indicated stress relaxation mechanisms, which are not accounted for by the wormlike chain model.

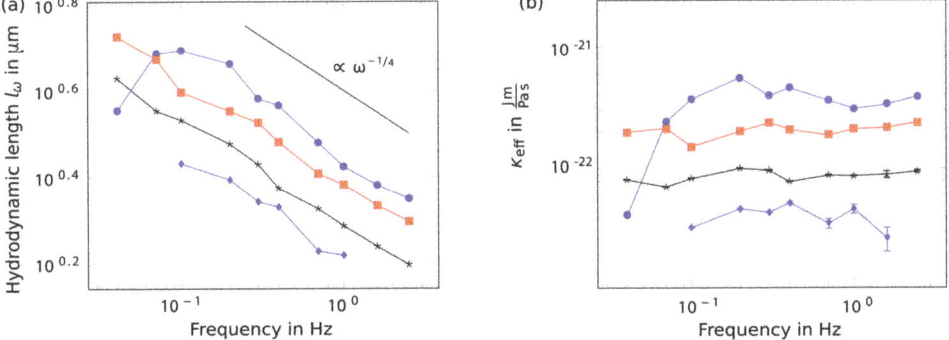

Figure 2. Hydrodynamic length and persistence length. Different colors and markers depict different individual bundles of varying thicknesses (blue dotted: thin bundle, blue circles: thick bundle, others: intermediate). (**a**) Regarding the frequency dependence of the hydrodynamic length, wiggling bundles with a bead attached to their ends revealed a power law exponent of $-1/4$; (**b**) which translated into a constant bending stiffness. Few bundles, however, also showed a softening behavior towards low frequencies.

The thickness of an individual bundle cannot be determined sufficiently, as emphasized previously [20]. A theoretically possible approach to extrapolate the bundle thickness from thermal fluctuation [29] by observation of the attached microbead is impeded by the fact that bundle twist and bending contributions cannot be separated in an exact manner. A further constraint is photo-bleaching effects, rendering experiment times too short to gain an adequate dataset for statistically significant values. Comparing the fluorescence intensities of bundles to those of single filaments at least enabled a rough estimation of the number of filaments in a bundle. This approach yielded bundle sizes of 20 to 70 filaments per bundle. Using the technique to merge bundles described above, the thickness of one specific bundle was increased successively between measurements, verifying that indeed a thinner and a thicker bundle, i.e., different numbers of filaments, are compared. The bundle stiffness increased with increasing thickness (see Figure 3a). However, in contrast to the measurements on single bundles with lower thicknesses, the bending stiffness—instead of being constant—showed a scaling with the frequency according to $\kappa \propto \omega^{-1/2}$. The bending stiffness gradually increases for lower frequencies, which is in contrast to some of the a priori thick bundles that showed softening behavior for low frequencies. Inherently, different frequencies agitated bundles on different length scales, which were governed by the hydrodynamic length. Thus, thickening a bundle led to a stiffening, and probing on an extended length scale resulting in a higher bending stiffness for longer bending modes (see Figure 3b).

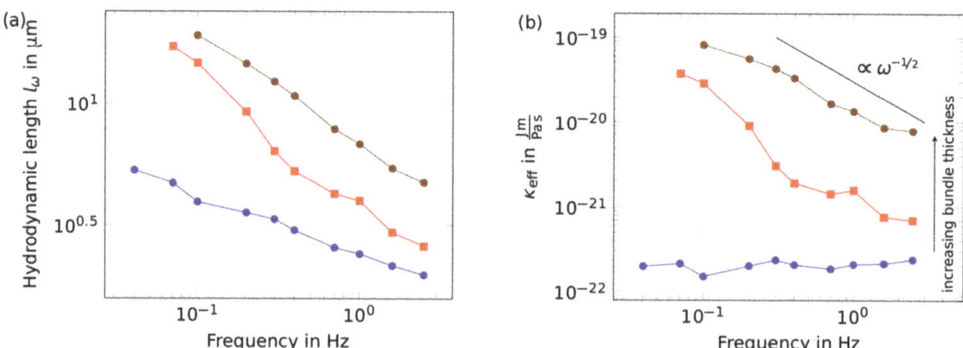

Figure 3. Rheological response of a successively thickened bundle. (**a**) While comparably thin bundles (blue circles) obey the predictions determined by the wormlike chain model, thicker bundles (intermediate: red squares, thick: brown circles) deviated from the $\omega^{-1/4}$-scaling, which translates into a frequency-dependent bending stiffness (**b**). With more and more bundles merged, the bending stiffness increased. At the same time, the bundles stiffened markedly for lower frequencies.

3. Discussion

Within this manuscript, we introduce a novel method to study highly anisotropic bundled filament structures and investigate the frequency dependence of their mechanic response. By monitoring the propagation of an oscillatory motion excited by amplitudes induced by an optical tweezers setup, the bending stiffness of the bundle was determined for varying oscillation frequencies between 0.04 Hz and 2.5 Hz. The hydrodynamic length was derived from the decay of the oscillation amplitude, which was proven to be constant for long time measurements over 600 s, excluding mechanically exited degradation effects and bundle instabilities. Bundles were formed and stabilized by depletion forces, allowing us to study the mechanical effects exclusively caused by the filaments and their inter-filament interactions. We would like to note that bundles cannot be treated as homogenous rods since the contour of the bundle highly depends on the involved filaments, which are not equally long but rather reveal a length distribution, leading to a slightly varying diameter along the entire structure. These indeterminable uncertainties surely influence the experiments and may result in small deviations in measured hydrodynamic lengths. However, we consider these deviations to be negligible. In preliminary experiments, we made sure that defects, kinks, or homogeneities were measurable only if distinguishable under the microscope. Mechanical responses of filament bundles exposed to wiggling excitations varied in magnitude, but the majority of the tested bundles obeyed the predictions derived from the WLC model. The scaling of the hydrodynamic length with the frequency revealed a power law exponent of $-1/4$. With $\kappa_{eff} \propto l_\omega^4$, this translates into a constant effective bending stiffness over the whole frequency range, which was also assumed by the WLC model. Consequentially, these bundles can be described as semiflexible wormlike chains with bending stiffness as their defining material property.

However, a number of bundles showed a softening behavior at very low frequencies. This was conspicuous at frequencies corresponding to oscillation periods larger than 5 s. These observations coincide with the theory for viscoelastic materials, which release the strain-induced stress over time. As a consequence, these bundles appeared stiffer when exposed to high frequencies, since the stress relaxation becomes increasingly dominant at distinctively larger time scales [4,30,31]. Conversely, they showed a softening behavior at low frequencies, when stresses can be relaxed due to a restructuring of the material, a mechanism which was also indicated by earlier experiments [20]. Actively bending actin bundles and holding them in a deformed state for a sufficient time facilitated a reconstruction of the material, illustrating an internal relaxation [20]. At low frequencies and large time scales, individual filaments within a bundle are able to slide against each other. This is captured in the theoretical framework provided by the worm like bundle theory, which explains the softening of a bundle with a decreasing contribution of crosslinking to the bending stiffness [15,32,33]. In the spirit of this theory, bundle cohesiveness is less pronounced under these circumstances.

Composite bundles were formed by iteratively merging individual bundles simply by converging two of them, a process we denoted as the zipping effect. Although we were not able to produce bundles of defined thicknesses (since the measurement of this quantity was not feasible for the aforementioned reasons), this method allowed us to compare the mechanical properties of one particular bundle with increasing thickness. The mechanical responses of these thickened bundles were remarkably different from the responses of simple a priori thick bundles. For comparably thin bundles, the scaling of the hydrodynamic length with frequency was the same as that for individual bundles and followed the relation $l_\omega \propto \omega^{-1/4}$. However, a significant increase in thickness yielded a change of the entire mechanical appearance. These bundles displayed a stiffness increase towards lower frequencies, which is in stark contrast to the previous rheological measurements in this study, which showed a softening of bundles probed for low frequencies.

This counterintuitive behavior becomes more evident when having a closer look at the expression of the hydrodynamic length, which sets the length scale over which the bending of the driven bundle decays. The length of the bundles' deformation mode is determined by the hydrodynamic length. Hence, probing bundles at lower frequencies is equivalent to measurements on longer lengths, leading to the interpretation that thick bundles appear stiffer at longer lengths (see Figure 4). A similar result has been reported by Taute et al. [34] as well as Pampaloni et al. [35], who observed a length-dependent stiffness of microtubules. Microtubules serve as a model system for wormlike bundles due to their protofilament architecture [15,26]. Composed by multiple protofilaments laterally connected by highly reversible tubulin-tubulin bonds, they react elastically to small deformations while undergoing continuous deformation characteristics for long wavelength modes [36]. For contour lengths $L > 5$ μm, fluctuation measurements revealed a relaxation time scaling $\tau \propto L^4$, as predicted by the wormlike chain model for semiflexible polymers [34]. For shorter contour lengths, the relaxation time scaled with the microtubule length as $\tau \propto L^2$. Measured persistence lengths significantly increased with filament length, but exhibited a plateau value for filaments smaller than 5 μm, which was also predicted by the wormlike bundle theory, proposing that different stiffness scalings are inherently connected with different regimes [15,26]. For actin bundles induced by the depletion agent polyethylene glycol (PEG), a behavior was observed that has been associated with a regime describing highly coupled bundles [37]. Interestingly, we observed a scaling behavior that fits the predictions made by the WLB theory for a regime where the bundle response to bending is dominated by the shearing of crosslinkers. This is rather surprising, since we formed bundles exclusively by depletion forces without any crosslinking proteins. However, another recent study on bundles that also did not involve specific crosslinking proteins revealed that molecular crowding and electrostatic interactions lead to an elastic coupling between filaments [22]. The stiffness of the relevant bending modes was predicted to scale with the wave vector in the form of $\kappa_n \propto q_n^{-2}$, according to the WLB theory. Crosslinker shearing as well as filament bending and stretching were assumed to contribute approximately equally to the bending energy of the bundle [15]. Since, in our experiments, one mode was predominantly excited due to wiggling at the fixation point with a wave length of $q^{-1} \propto l_\omega$, the effective bending stiffness for thick bundles was not constant, but rather scaled with the hydrodynamic length as $\kappa_{eff} \propto l_\omega^2$ and with the frequency as $\kappa_{eff} \propto \omega^{-1/2}$. The consensus of these predictions shows that mechanisms contributing to the mechanical appearance of anisotropic bundled structures are versatile and still not completely understood. One of these contributions might be inter-filament friction. Ward et al. [27] recently reported unexpectedly large inter-polymeric forces due to friction in bundled systems scaling logarithmically with the sliding velocity. Additionally, Schnauß et al. [7,38] found contractile forces acting against the lateral extraction of single filaments out of a bundle, which facilitates an exponential relaxation after stresses (e.g., caused by motion in a viscous medium) are released. It is very likely that both friction and bundle relaxation influence the frequency-dependent response of bundles. These types of interactions in a fully coupled regime are not yet covered by the WLB model. The phenomenological distinction between these contributions and the comparison to crosslinked systems is not trivial; however, all of these processes are based on inter-filament interactions, which can be captured in the frame of stick-slip models [39].

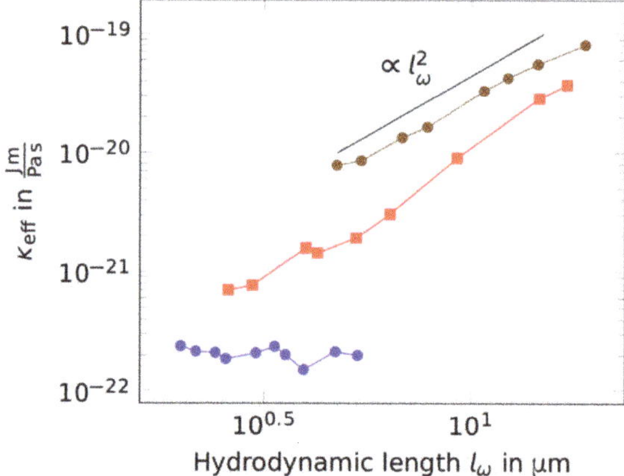

Figure 4. Stiffness scaling with the hydrodynamic length for increasing bundle diameters (thin bundles: blue circles, intermediate bundles: red squares, thick bundles: brown circles). For thickened bundles, the bending stiffness shows a scaling behavior with the excitation frequency. This scaling is more evident when looking at the bending stiffness with respect to the hydrodynamic length. While probing the bundle with different frequencies, the length scale on which the bundle was deformed changed. This translates into a higher bending stiffness for longer deformation modes.

Further measurements on depletion force-induced bundles, as well as on bundles crosslinked by specific proteins, are necessary to separate these diverse contributions to the mechanical response of a bundle. The method presented here is well-suited since it provides the possibility to actively excite filamentous bundles and thereby covers the investigation of frequency-dependent frictional effects.

The comparison of depletion force-induced bundles to bundles crosslinked by α-actinin should isolate the influences of crosslinker shearing and friction in combination with bundle relaxation, respectively.

In conclusion, we introduce a new and powerful tool providing remarkable insights into the rheological nature of filamentous bundles. Our initial data revealed perceptions of frequency-dependent mechanical responses of such bundles, which have not been experimentally accessible before.

4. Materials and Methods

4.1. Protein Preparation, Bead Functionalization, and Bundle Formation

G-actin was prepared from rabbit skeletal muscle as described previously by Smith et al. [40] according to the protocols established by Humphrey et al. [41]. Monomeric actin was polymerized at a concentration of 5 μM in F-buffer (0.1 M KCl, 1 mM $MgCl_2$, 0.1 mM $CaCl_2$, 0.5 mM TrisHCl, pH 7.8) in the presence of phalloidin-Tetramethylrhodamine B isothiocyanate (TRITC) (Sigma-Aldrich, St. Louis, MO, USA) overnight [42]. Biotinylated G-actin (Cytoskeleton, Inc., Denver, CO, USA) was added to one tenth of the actin concentration and subsequently incubated at room temperature. Biotin-actin was thereby incorporated at the ends of the already-formed filaments.

Polystyrene beads, 2 μm in diameter and functionalized by a covalently bound streptavidin coat (Polysciences, Inc., Warrington, PA, USA), were added in an additional incubation period. Streptavidin and its counterpart biotin form a highly selective ligand-receptor bond with a very low dissociation constant. This facilitates the attachment of the functionalized beads predominantly to the ends of the filaments enriched with the accumulation of biotinylated actin.

The composition of the final solution in F-buffer was diluted to a total actin concentration of 200 nM in the presence of polystyrene beads and glucose/glucose oxidase as an anti-bleaching agent. The F-buffer contained 0.8% methyl cellulose (400 or 4000 cP, Sigma-Aldrich, St. Louis, MO, USA) acting as a bundling agent by inducing depletion forces.

About 10 μL of the final solution was deposited on a Sigmacote-treated (Sigma-Aldrich, St. Louis, MO, USA) squared coverslip with a 22-mm edge length and vacuum grease-lined edges. A 24 mm × 50 mm coverslip was placed on top of the drop and then pressed flat, with care taken to remove trapped air pockets. The sample chamber was pressed together between two aluminum plates in a controlled fashion with a micrometer screw, as previously described by Strehle et al. [20]. The 2-μm beads were free in solution and trappable in the potential of a focused laser beam pertaining to an optical tweezers setup. In the best-case scenario, a bead was already attached to an actin bundle; however, this attachment could have been achieved by approaching the bead to the end of a free bundle in the solution.

4.2. Optical Tweezers Setup and Fluorescence Microscopy

For the presented experiments, an optical tweezers setup was used, as previously described in [7]. Sample observation and simultaneous recording of fluorescence microscopy images during experimental procedures were performed with a Hamamatsu Orca ER digital CCD (charge-coupled device) camera (Hamamatsu Photonics Deutschland GmbH, Herrsching am Ammersee, Germany).

For data acquisition, all components were controlled and integrated by a self-written LabView (National Instruments, Munich, Germany) program that allowed visualization of fluorescent beads and rhodamine-phalloidin labeled F-actin [42]. Fluorescence images of beads and bundles were recorded, and bead positions were determined in evaluation by cross-correlation analysis.

4.3. Experimental Procedure

Bundling F-actin can be achieved by different approaches. One possibility is to introduce additional crosslinking proteins such as α-actinin in a sufficiently high concentration to facilitate a parallel alignment of filaments to increase the number of available binding sites [8,43]. Here, the investigated anisotropic filament structures were depletion force-induced bundles in order to avoid additional interactions between the components.

In the presence of non-interacting polymers, an attractive force is generated between colloidal filaments. In the spirit of the model by Asakura and Osaka [44], polymers are treated as freely interpenetrating hard spheres excluded from the colloid surface by a thin layer. This shell creates a positive free energy difference, which is lowered if two colloidal particles share this excluded volume. Consequently, the total entropy of the system is increased and thus the free energy of the system is decreased [45]. These bundles were found to be able to merge easily when approaching each other solely due to the described entropic effect. Exploiting this behavior, a rather simple method to attain increasingly thick bundles was conceived. This "zipping"- process (see Figure 5) was used to increase the bundle thickness successively between measurements.

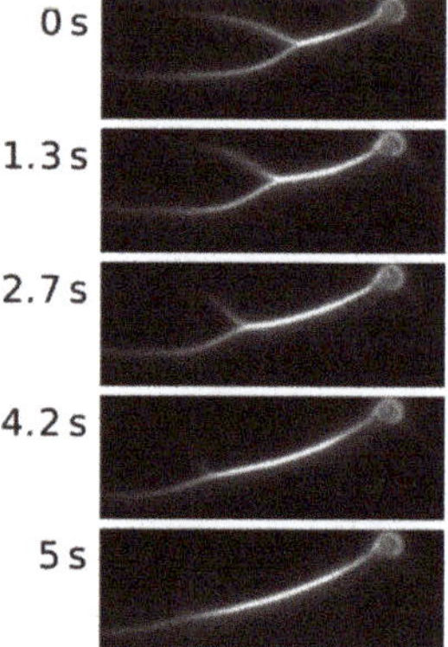

Figure 5. Time lapse of the zipping process. Bundles were thickened by merging with other bundles. This was achieved by dragging the bundle with the optically trapped bead into the vicinity of another, freely floating bundle.

Bundles were mechanically excited by moving the attached polystyrene beads via optical tweezers with defined amplitudes in the y-direction at varying frequencies. The position of the optical trap was controlled via a self-written LabView software enabling the wiggling procedure in a controlled and reproducible fashion (see Figure 1). Phalloidin-TRITC-labeled filaments were observed with fluorescence microscopy and recorded with a CCD camera, which allowed us to trace the bundle contours over time.

Obtained images were processed for evaluation in several steps. First, the image was preset geometrically for a better handling in evaluation. The respective direction of deflection was rotated and set as the y-direction, while the bundle contour in the undisturbed state was rotated and set as the x-direction. Subsequently, the largest anticipated amplitude was chosen as the boundary and the image was cropped to the corresponding width. Afterwards, a filter was applied, marginally smoothing the images in the x-direction averaging with a Gaussian profile while also enhancing edges (emphasizing horizontal lines) in the y-direction. This enabled the application of a parabola fit to the contour and, by the determination of the parabola's maximum, the bundle deflection was defined.

4.4. Analytical Tools

Elastic rods under the influence of hydrodynamic drag [23] were described theoretically by Wiggins et al. [24] by solving a differential equation for the motion of a filament under deformation by hydrodynamic flow, which was derived from the Hamiltonian for a wormlike chain. In the weakly-bending rod limit and for small deformations, the bending force for a WLC can be written as:

$$f_{bend} = -\kappa y_{xxxx}\hat{e}_y,$$

where κ denotes the bending stiffness and subscripts the partial derivatives with respect to the subscripted variable. The experiments took place at low Reynolds numbers, where the bending force is balanced by the hydrodynamic drag, which is represented by a velocity-dependent force:

$$f_{drag} = \zeta(y_t - u)\hat{e}_y,$$

with background velocity u and friction coefficient ζ. This gives the equation of motion:

$$\zeta(y_t - u) = -\kappa y_{xxxx}.$$

Rescaling the distance along the filament with the hydrodynamic length scale, we get:

$$l_\omega = \left(\frac{\kappa}{\omega\zeta}\right)^{1/4} = \left(\frac{k_B T l_p}{\omega\zeta}\right)^{1/4},$$

which sets the scale on which the filaments' displacement amplitude decays. Using a product ansatz for appropriate border conditions gives the solution for a filament driven at one end as

$$y(\eta, t) = \frac{y_0}{2}\left[e^{-\tilde{S}\eta}\cos\left(\tilde{C}\eta - \omega t\right) + e^{-\tilde{C}\eta}\cos\left(\tilde{S}\eta + \omega t\right)\right].$$

with $\eta = \frac{x}{l_\omega}$, $\tilde{C} \simeq 0.92$ and $\tilde{S} \simeq 0.38$ this solution describes two waves decaying on different time scales; the excitation wave in the first term and its reflection by the medium in the second term. A similar ansatz for a filament driven at its middle point yields the solution for one side of the contour:

$$y(\eta, t) = \frac{y_0}{2}\left\{e^{-\tilde{S}\eta}\left(\cos\left(\tilde{C}\eta - \omega t\right) + \sin\left(\tilde{C}\eta - \omega t\right)\right) + e^{-\tilde{C}\eta}\left(\cos\left(\tilde{S}\eta + \omega t\right) + \sin\left(\tilde{S}\eta + \omega t\right)\right)\right\}.$$

Figure 6a illustrates the recorded bundle positions over time. Eminent spikes are a direct result of noisy bundle recognition, especially in the bead vicinity. Non-orthogonal bundles, i.e., bundles that were not aligned orthogonal to the driving direction for their entire length, were excluded from further analysis since they did not show substantial oscillations. To eliminate random contributions due to thermal undulations as well as poor bundle detections, a band pass filter was used to isolate the oscillatory bundle movement with the driving frequency.

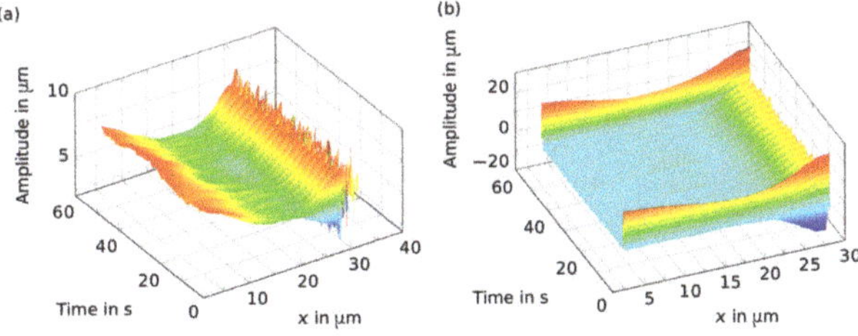

Figure 6. Wiggling bundle position data. Panel (**a**) shows the amplitude of the movement of the bundle contour in the *y*-direction with respect to its point along the bundle. The bead that is controllably deflected by the optical tweezers setup to induce oscillations is near position *x* = 27 μm, and is attached to one end of the bundle. Bundles that are not orthogonal to the oscillation direction were excluded from further analysis. Panel (**b**) shows the data after band passing. The first and last periods need to be excluded from l_ω -fitting because of the border effects of filtering

Acknowledgments: We would like to thank Klaus Kroy for fruitful discussions. We gratefully acknowledge support from the German Research Foundation (DFG) and Universität Leipzig within the program of Open Access Publishing. Furthermore we acknowledge funding from the Deutsche Forschungsgemeinschaft (DFG-1116/17-1) as well as the European Research Council (ERC-741350). P.M. acknowledges funding from the European Social Fund (ESF—100316844). J.S. acknowledges financial support through the Fraunhofer Attract project 601 683.

Author Contributions: D.S., J.A.K., and J.S. designed the project. D.S., C.S., P.M., M.G., T.G., and J.S. performed experiments. D.S., P.M., and J.S. analysed data. D.S., P.M., M.G., T.G., C.S., J.S., and J.A.K. wrote the manuscript.

Conflicts of Interest: The authors declare no conflict of interest.

References

1. Huber, F.; Schnauß, J.; Rönicke, S.; Rauch, P.; Müller, K.; Fütterer, C.; Käs, J. Emergent complexity of the cytoskeleton: From single filaments to tissue. *Adv. Phys.* **2013**, *62*, 1–112. [CrossRef] [PubMed]
2. Smith, D.; Gentry, B.; Stuhrmann, B.; Huber, F.; Strehle, D.; Brunner, C.; Koch, D.; Steinbeck, M.; Betz, T.; Käs, J. The cytoskeleton: An active polymer-based scaffold. *Biophys. Rev. Lett.* **2009**. [CrossRef]
3. Lodish, H.F. *Molecular Cell Biology*, 6th ed.; W.H. Freeman: New York, NY, USA, 2008.
4. MacKintosh, F.C.; Kas, J.; Janmey, P.A. Elasticity of Semiflexible Biopolymer Networks. *Phys. Rev. Lett.* **1995**, *75*, 4425–4428. [CrossRef] [PubMed]
5. Backouche, F.; Haviv, L.; Groswasser, D.; Bernheim-Groswasser, A. Active gels: Dynamics of patterning and self-organization. *Phys. Biol.* **2006**, *3*, 264–273. [CrossRef] [PubMed]
6. Hotulainen, P.; Lappalainen, P. Stress fibers are generated by two distinct actin assembly mechanisms in motile cells. *J. Cell Biol.* **2006**, *173*, 383–394. [CrossRef] [PubMed]
7. Schnauß, J.; Golde, T.; Schuldt, C.; Schmidt, B.U.S.; Glaser, M.; Strehle, D.; Händler, T.; Heussinger, C.; Käs, J.A. Transition from a Linear to a Harmonic Potential in Collective Dynamics of a Multifilament Actin Bundle. *Phys. Rev. Lett.* **2016**, *116*, 108102. [CrossRef] [PubMed]
8. Wachsstock, D.H.; Schwartz, W.H.; Pollard, T.D. Affinity of alpha-actinin for actin determines the structure and mechanical properties of actin filament gels. *Biophys. J.* **1993**, *65*, 205–214. [CrossRef]
9. Hinner, B.; Tempel, M.; Sackmann, E.; Kroy, K.; Frey, E. Entanglement, Elasticity, and Viscous Relaxation of Actin Solutions. *Phys. Rev. Lett.* **1998**, *81*, 2614–2617. [CrossRef]
10. Semmrich, C.; Storz, T.; Glaser, J.; Merkel, R.; Bausch, A.R.; Kroy, K. Glass transition and rheological redundancy in F-actin solutions. *Proc. Natl. Acad. Sci. USA* **2007**, *104*, 20199–20203. [CrossRef] [PubMed]
11. Gardel, M.L.; Shin, J.H.; MacKintosh, F.C.; Mahadevan, L.; Matsudaira, P.; Weitz, D.A. Elastic Behavior of Cross-Linked and Bundled Actin. *Science* **2004**, *304*, 1301–1305. [CrossRef] [PubMed]
12. Isambert, H.; Maggs, A.C. Dynamics and Rheology of Actin Solutions. *Macromolecules* **1996**, *29*, 1036–1040. [CrossRef]
13. Käs, J.; Strey, H.; Tang, J.X.; Finger, D.; Ezzell, R.; Sackmann, E.; Janmey, P.A. F-actin, a model polymer for semiflexible chains in dilute, semidilute, and liquid crystalline solutions. *Biophys. J.* **1996**, *70*, 609–625. [CrossRef]
14. Morse, D.C. Tube diameter in tightly entangled solutions of semiflexible polymers. *Phys. Rev. E* **2001**, *63*, 31502. [CrossRef] [PubMed]
15. Heussinger, C.; Bathe, M.; Frey, E. Statistical mechanics of semiflexible bundles of wormlike polymer chains. *Phys. Rev. Lett.* **2007**, *99*, 48101. [CrossRef] [PubMed]
16. Koenderink, G.H.; Dogic, Z.; Nakamura, F.; Bendix, P.M.; MacKintosh, F.C.; Hartwig, J.H.; Stossel, T.P.; Weitz, D.A. An active biopolymer network controlled by molecular motors. *Proc. Natl. Acad. Sci. USA* **2009**, *106*, 15192–15197. [CrossRef] [PubMed]
17. Fabry, B.; Maksym, G.N.; Butler, J.P.; Glogauer, M.; Navajas, D.; Fredberg, J.J. Scaling the microrheology of living cells. *Phys. Rev. Lett.* **2001**, *87*, 148102. [CrossRef] [PubMed]
18. Falzone, T.T.; Blair, S.; Robertson-Anderson, R.M. Entangled F-actin displays a unique crossover to microscale nonlinearity dominated by entanglement segment dynamics. *Soft Matter* **2015**, *11*, 4418–4423. [CrossRef] [PubMed]
19. Tang, J.X.; Janmey, P. The Polyelectrolyte Nature of F-actin and the Mechanism of actin bundle formation. *J. Biol. Chem.* **1996**, *271*, 8556–8563. [CrossRef] [PubMed]
20. Strehle, D.; Schnauss, J.; Heussinger, C.; Alvarado, J.; Bathe, M.; Käs, J.; Gentry, B. Transiently crosslinked F-actin bundles. *Eur. Biophys. J. EBJ* **2011**, *40*, 93–101. [CrossRef] [PubMed]

21. Claessens, M.M.A.E.; Bathe, M.; Frey, E.; Bausch, A.R. Actin-binding proteins sensitively mediate F-actin bundle stiffness. *Nat. Mater.* **2006**, *5*, 748–753. [CrossRef] [PubMed]

22. Rückerl, F.; Lenz, M.; Betz, T.; Manzi, J.; Martiel, J.-L.; Safouane, M.; Paterski-Boujemaa, R.; Blanchoin, L.; Sykes, C. Adaptive Response of Actin Bundles under Mechanical Stress. *Biophys. J.* **2017**, *113*, 1072–1079. [CrossRef] [PubMed]

23. Riveline, D.; Wiggins, C.H.; Goldstein, R.E.; Ott, A. Elastohydrodynamic study of actin filaments using fluorescence microscopy. *Phys. Rev. E* **1997**, *56*, R1330–R1333. [CrossRef]

24. Wiggins, C.H.; Riveline, D.; Ott, A.; Goldstein, R.E. Trapping and Wiggling: Elastohydrodynamics of Driven Microfilaments. *Biophys. J.* **1998**, *74*, 1043–1060. [CrossRef]

25. Aragon, S.R.; Pecora, R. Dynamics of wormlike chains. *Macromolecules* **1985**, *18*, 1868–1875. [CrossRef]

26. Bathe, M.; Heussinger, C.; Claessens, M.M.A.E.; Bausch, A.R.; Frey, E. Cytoskeletal bundle mechanics. *Biophys. J.* **2008**, *94*, 2955–2964. [CrossRef] [PubMed]

27. Ward, A.; Hilitski, F.; Schwenger, W.; Welch, D.; Lau, A.W.C.; Vitelli, V.; Mahadevan, L.; Dogic, Z. Solid friction between soft filaments. *Nat. Mater.* **2015**, *14*, 583–588. [CrossRef] [PubMed]

28. Shin, J.H.; Mahadevan, L.; So, P.T.; Matsudaira, P. Bending stiffness of a crystalline actin bundle. *J. Mol. Biol.* **2004**, *337*, 255–261. [CrossRef] [PubMed]

29. Wilhelm, J.; Frey, E. Radial Distribution Function of Semiflexible Polymers. *Phys. Rev. Lett.* **1996**, *77*, 2581–2584. [CrossRef] [PubMed]

30. Claessens, M.M.A.E.; Tharmann, R.; Kroy, K.; Bausch, A.R. Microstructure and viscoelasticity of confined semiflexible polymer networks. *Nat. Phys.* **2006**, *2*, 186–189. [CrossRef]

31. Janmey, P.A.; Euteneuer, U.; Traub, P.; Schliwa, M. Viscoelastic Properties of Vimentin Compared with Other Filamentous Biopolymer Networks. *J. Cell Biol.* **1991**, *113*, 155–160. [CrossRef] [PubMed]

32. Heussinger, C.; Schüller, F.; Frey, E. Statics and dynamics of the wormlike bundle model. *Phys. Rev. E* **2010**, *81*, 21904. [CrossRef] [PubMed]

33. Lieleg, O.; Claessens, M.M.A.E.; Heussinger, C.; Frey, E.; Bausch, A.R. Mechanics of bundled semiflexible polymer networks. *Phys. Rev. Lett.* **2007**, *99*, 88102. [CrossRef] [PubMed]

34. Taute, K.M.; Pampaloni, F.; Frey, E.; Florin, E.-L. Microtubule dynamics depart from the wormlike chain model. *Phy. Rev. Lett.* **2008**, *100*, 28102. [CrossRef] [PubMed]

35. Pampaloni, F.; Lattanzi, G.; Jonás, A.; Surrey, T.; Frey, E.; Florin, E.-L. Thermal fluctuations of grafted microtubules provide evidence of a length-dependent persistence length. *Proc. Natl. Acad. Sci. USA* **2006**, *103*, 10248–10253. [CrossRef] [PubMed]

36. Kononova, O.; Kholodov, Y.; Theisen, K.E.; Marx, K.A.; Dima, R.I.; Ataullakhanov, F.I.; Grishchuk, E.L.; Barsegov, V. Tubulin bond energies and microtubule biomechanics determined from nanoindentation in silico. *J. Am. Chem. Soc.* **2014**, *136*, 17036–17045. [CrossRef] [PubMed]

37. Broedersz, C.P.; MacKintosh, F.C. Modeling semiflexible polymer networks. *Rev. Mod. Phys.* **2014**, *86*, 995–1036. [CrossRef]

38. Schnauß, J.; Händler, T.; Käs, J. Semiflexible Biopolymers in Bundled Arrangements. *Polymers* **2016**, *8*, 274. [CrossRef]

39. Bullerjahn, J.T.; Kroy, K. Analytical catch-slip bond model for arbitrary forces and loading rates. *Phys. Rev. E* **2016**, *93*, 12404. [CrossRef] [PubMed]

40. Smith, D.; Ziebert, F.; Humphrey, D.; Duggan, C.; Steinbeck, M.; Zimmermann, W.; Käs, J. Molecular motor-induced instabilities and cross linkers determine biopolymer organization. *Biophys. J.* **2007**, *93*, 4445–4452. [CrossRef] [PubMed]

41. Humphrey, D.; Duggan, C.; Saha, D.; Smith, D.; Käs, J. Active fluidization of polymer networks through molecular motors. *Nature* **2002**, *416*, 413–416. [CrossRef] [PubMed]

42. Cooper, J.A. Effects of Cytochalasin and Phalloidin on Actin. *J. Cell Biol.* **1987**, *105*, 1473–1478. [CrossRef] [PubMed]

43. Liu, J.; Taylor, D.W.; Taylor, K.A. A 3-D reconstruction of smooth muscle alpha-actinin by CryoEm reveals two different conformations at the actin-binding region. *J. Mol. Biol.* **2004**, *338*, 115–125. [CrossRef] [PubMed]

44. Asakura, S.; Oosawa, F. On Interaction between Two Bodies Immersed in a Solution of Macromolecules. *J. Chem. Phys.* **1954**, *22*, 1255–1256. [CrossRef]
45. Hosek, M.; Tang, J.X. Polymer-induced bundling of F actin and the depletion force. *Phys. Rev. E* **2004**, *69*, 51907. [CrossRef] [PubMed]

Sample Availability: Samples of the compounds are not available from the authors.

Article

Development, Optimization and In Vitro/In Vivo Characterization of Collagen-Dextran Spongious Wound Dressings Loaded with Flufenamic Acid

Mihaela Violeta Ghica [1], Mădălina Georgiana Albu Kaya [2,*], Cristina-Elena Dinu-Pîrvu [1], Dumitru Lupuleasa [3] and Denisa Ioana Udeanu [4]

[1] Department of Physical and Colloidal Chemistry, Faculty of Pharmacy, University of Medicine and Pharmacy "Carol Davila", Bucharest 020956, Romania; mihaela.ghica@umfcd.ro (M.V.G.); ecristinaparvu@yahoo.com (C.-E.D.-P.)

[2] Department of Collagen Research, Division of Leather and Footwear Research Institute, National Research & Development Institute for Textiles and Leather, Bucharest 031215, Romania

[3] Department of Pharmaceutical Technology and Biopharmacy, Faculty of Pharmacy, University of Medicine and Pharmacy "Carol Davila", Bucharest 020956, Romania; office@colegfarm.ro

[4] Department of Clinical Laboratory and Food Safety, Faculty of Pharmacy, University of Medicine and Pharmacy "Carol Davila", Bucharest 020956, Romania; denisaudeanu@gmail.com

* Correspondence: madalina.albu@icpi.ro; Tel.: +40-723-395-108

Received: 7 August 2017; Accepted: 13 September 2017; Published: 15 September 2017

Abstract: The aim of this study was the development and optimization of some topical collagen-dextran sponges with flufenamic acid, designed to be potential dressings for burn wounds healing. The sponges were obtained by lyophilization of hydrogels based on type I fibrillar collagen gel extracted from calf hide, dextran and flufenamic acid, crosslinked and un-crosslinked, and designed according to a 3-factor, 3-level Box-Behnken experimental design. The sponges showed good fluid uptake ability quantified by a high swelling ratio. The flufenamic acid release profiles from sponges presented two stages—burst effect resulting in a rapid inflammation reduction, and gradual delivery ensuring the anti-inflammatory effect over a longer burn healing period. The resistance to enzymatic degradation was monitored through a weight loss parameter. The optimization of the sponge formulations was performed based on an experimental design technique combined with response surface methodology, followed by the Taguchi approach to select those formulations that are the least affected by the noise factors. The treatment of experimentally induced burns on animals with selected sponges accelerated the wound healing process and promoted a faster regeneration of the affected epithelial tissues compared to the control group. The results generated by the complex sponge characterization indicate that these formulations could be successfully used for burn dressing applications.

Keywords: collagen sponges; flufenamic acid; in vitro drug delivery; experimental design; taguchi technique; in vivo burn healing

1. Introduction

Cutaneous burns are acute wounds characterized by an immediate and increased production of specific proinflammatory mediators both at systemic and local levels, resulting in tissue regeneration delay and other systemic disorders at cardiac, respiratory, renal and intestinal levels [1–6]. In addition, tissue injury induces pain signals thus having a direct effect on patients' quality of life [5,7–9]. For these reasons, a first step in burns healing improvement is the control of inflammation and associated pain at the affected tissue.

The use of non-steroidal anti-inflammatory drugs (NSAIDs) ease the post-burn inflammatory response by inhibiting the normal inflammatory phase, successfully reducing both persistent and temporary pain and inflammation near the wound surface without delaying the epithelial healing [7,9–13]. Topical NSAIDs delivery systems offer several benefits compared to the systemic administration route, mainly the avoidance of systemic toxicity (such as gastro-intestinal bleeding) and side effects at hepatic and renal levels, as well as increased patient compliance [7,11,14–16]. Also, NSAIDs release directly to the burn level at a sustained rate, ensuring a sufficient and effective drug concentration reaches the affected tissue, which is important in combating and controlling the inflammation and pain that occurs during the healing process [3,9,10].

Among the large group of classical NSAIDs, flufenamic acid (FFA) was selected as the model compound in this research. FFA is an anthranilic acid derivative that inhibits both forms of cyclooxygenase izoenzymes, and is accountable for prostaglandin biosynthesis [17–20]. FFA demonstrated its anti-inflammatory and analgesic potential in various topical marketed formulations intended for use in different rheumatic disorders and soft tissue injuries [21,22].

NSAIDs-based local burn management currently relies upon the use of biodegradable and non-biodegradable biopolymers as potential carriers for controlled drug delivery systems (also called drug dressings) [23]. Such NSAIDs-based drug delivery systems (DDS) with non-biodegradable polymers support are to be avoided as they need frequent replacements that affect a patient's comfort and can cause additional trauma to the new epithelial tissue formed, the risk of secondary infection being very high [7,9,24–27]. Therefore, biodegradable polymers are considered to be ideal vehicles for drugs delivery [28], collagen being preferred due to its properties such as biodegradability, bioresorbability, biocompatibility, hemostasis ability, well-known structure, and reduced manufacturing cost [29–35]. The use of collagen as drug release support is however limited by its mechanical properties and resistance to in vivo enzymatic degradation, but these characteristics could be enhanced by chemical reaction using various crosslinkers such as glutaraldehyde [31,32,36,37], or by blending with other biopolymers. The biopolymer selected in this study was dextran, a polysaccharide with bacterial origins that has a high tissular biocompatibility and hydrophylicity. Moreover, it increases in vitro stability, decreases in vivo imonogenicity of proteins such as collagen and different enzymes, and is a potential candidate for wound healing stimulation [38–41].

Collagen can be used in various forms as hydrogels, membranes, fibers, matrices and composite materials, made from undenatured collagen [31–33,42]. Due to their ability to absorb large amounts of exudate, to adhere to wet wounds, to maintain a moist healing environment and to favour the regeneration of granular and epithelial tissue at wound level, the porous forms (spongious matrices) obtained by lyophilization of collagen solutions/gels are recommendable for treating burns with different tissue damage degrees [31,42,43]. Collagen spongious matrices have significant potential for burn tissue restoration and can prevent, in association with an anti-inflammatory drug, other further tissue injury caused by the inflammation process [13,44–46].

To minimize the overall burn wound inflammation and the unwanted influence of the prolonged inflammatory response, the design of drug controlled delivery systems that ensure both a rapid and gradual drug release is needed. In this context, the proper selection of the drug vehicle is of paramount importance to obtain DDS with desired release characteristics in accordance with the therapeutic indication and application site.

Thus, the aim of this paper was the development, analysis and optimization of some spongious matrices loaded with flufenamic acid, based on a support of collagen and dextran, un- and cross-linked with glutaraldehyde, obtained by lyophilization of hydrogels and designed according to a 3-factor, 3-level Box-Behnken factorial experiment. As drug release from collagen matrices is governed by the combined effect of sponge swelling (and consecutively to drug diffusion through its network) and the hydrolytic activity of the enzymes that exist in the biological fluids (determining the collagen degradation) [30,47–50], we explored the influence of the formulation factors on the sponges swelling ability and the percent of FFA released from the sponges, as well as on their resistance to enzymatic degradation.

The experimental statistic design was combined with response surface methodology (RSM), this method proving to be efficient in the development and optimization of the formulation of different pharmaceutical systems and technological processes [31,32,51–56] as it allows the researcher to investigate the simultaneous effects of formulation factors, to evaluate the interactions effects on system responses, to provide predictive mathematical models, and also to reduce the number of experiments saving in this way both time and resources [52,55–59]. The optimization technique was complemented with the Taguchi approach in order to set—through the Signal-to-Noise indicator (S/N ratio)—the adequate variation conditions for the formulation factors in order to ensure the products' quality and process robustness [31,32,60–63].

The selected optimal formulations of the spongious matrices were further subject to preliminary in vivo experiments carried out on Wistar rats in order to test the therapeutic potential in burn healing and to identify any potential major side effect induced by the treatment [64–66].

2. Results and Discussion

2.1. Design of Experiments

A Box-Behnken factorial experiment with 3-factors and 3-levels was used to determine the influence of the formulation factors on some of the physical-chemical and biological sponges parameters considered to be affecting the flufenamic acid delivery mechanism from the designed systems.

The formulation factors (X_i) selected as independent variables were collagen (C), dextran (DX) and glutaraldehyde (GA) concentrations (g%) (Table 1). The coded levels for each input variable were 1 for low, 2 for middle and 3 for high level respectively. The physical-chemical and biological parameters involved in the drug delivery process from collagen sponges—considered as dependent variables (measured responses—Y_i)—were swelling ratio, SR (g/g), released percent, RP (%) and weight loss, WL (%) The coded forms of these variables are listed in Table 1 along with the specific constraints imposed by the application of these porous forms in the burn wounds.

Table 1. Process variables and experimental conditions in 3-factor, 3-level Box-Behnken experimental design.

Independent Variables	Coded Symbol	Coded and Uncoded Variation Levels		
		Low (1)	Middle (2)	High (3)
Collagen, C (g%)	X_1	0.8	1.0	1.2
Dextran, DX (g%) *	X_2	0.0	0.6	1.2
Glutaraldehyde, GA (g%) *	X_3	0.0	0.0060	0.012
Dependent Variables	**Coded Symbol**	**Constraints**		
Swelling ratio, SR (g/g)	Y_1	Maximize		
Released percent, RP (%)	Y_2	Maximize		
Weight loss, WL (%)	Y_3	Minimize		

* the amounts of DX and GA are reported to collagen dry substance.

Box-Behnken design is a rotatable quadratic design [51,67] where the highest or lowest levels of all factor formulations do not appear simultaneously [68], the trials combinations being placed at the mid-points of each edge and at the center of the cube [51]. Thus, the number of experiments is reduced from 27 (corresponding to the full factorial design) to 13, summarized in Table 2. The spongious matrices were coded as M1 to M13.

Table 2. Formulation factors values in different Box-Behnken experimental trials for the hydrogels used to prepare the corresponding spongious matrices; observed and predicted responses for the spongious matrices.

Trials No.	Exp. Name	Independent Variables (Coded Level/Physical Level)			Responses					
		X_1—C	X_2—DX	X_3—GA	Y_1—SR (g/g)		Y_2—RP (%)		Y_3—WL (%)	
					Obs.	Pred.	Obs.	Pred.	Obs.	Pred.
1	M1	1 (0.8)	1 (0.0)	2 (0.006)	45.65	45.81	75.13	75.41	38.02	36.77
2	M2	3 (1.2)	1 (0.0)	2 (0.006)	28.41	28.91	76.51	75.60	32.49	31.37
3	M3	1 (0.8)	3 (1.2)	2 (0.006)	40.14	39.81	86.81	87.43	34.20	33.54
4	M4	3 (1.2)	3 (1.2)	2 (0.006)	36.16	36.21	87.28	87.62	26.62	26.53
5	M5	1 (0.8)	2 (0.6)	1 (0.000)	38.77	39.34	88.38	87.65	36.42	37.21
6	M6	3 (1.2)	2 (0.6)	1 (0.000)	26.52	27.24	84.81	85.66	30.66	31.01
7	M7	1 (0.8)	2 (0.6)	3 (0.012)	31.56	31.34	79.12	78.94	27.32	28.97
8	M8	3 (1.2)	2 (0.6)	3 (0.012)	24.05	22.94	81.61	81.31	21.48	22.77
9	M9	2 (1.0)	1 (0.0)	1 (0.000)	37.23	36.12	83.38	82.11	35.36	36.40
10	M10	2 (1.0)	3 (1.2)	1 (0.000)	42.88	42.35	93.51	94.13	31.56	32.37
11	M11	2 (1.0)	1 (0.0)	3 (0.012)	35.13	35.55	73.69	75.58	28.40	28.15
12	M12	2 (1.0)	3 (1.2)	3 (0.012)	29.83	30.62	89.19	87.60	25.83	24.12
13	M13	2 (1.0)	2 (0.6)	2 (0.006)	33.59	33.61	95.09	95.43	33.24	32.32

2.2. Swelling Study

The sponge swelling behavior is an important parameter that needs to be taken into consideration in the formulation of burn dressings. It is a critical parameter for burn healing control, being influenced by the biological properties of the wound and the sponge physical-chemical composition [69].

A swelling analysis was performed to assess the spongious matrices' absorption medium ability. This is a very important feature as the absorption medium was phosphate buffer pH = 7.4, which was also used as the release medium in the kinetic experiments. The fluid absorption was initiated after sponge immersion in a moist environment and the high retention of fluid within the porous structure could be attributed to the binding of the absorption medium to the collagen and its increased hydrophilicity [9]. These properties also favor the formation of a gel layer after sponge contact with lesion exudate, ensuring the desired wet environment [70].

The fluid absorption characteristics of the designed spongious matrices were quantified using the swelling ratio (SR—g/g), evaluated from the modification of support weight before and after swelling with buffer solution, according to the following formula (Equation (1)):

$$SR = \frac{(W_t - W_0)}{W_0} \tag{1}$$

where W_0 is the dried sponge weight at the initial time, and W_t is the sponge weight after immersion at time t [31].

The swelling ratios of the collagen spongious matrices following 10 h of immersion in phosphate buffer pH 7.4 are summarized in Table 2. The designed sponges exhibited a high swelling ability, the values recorded for the swelling ratio being between 24.05 and 45.65 g/g, proving their capacity for absorbing a large amount of burn exudate, adhering to the wet lesion and favoring the formation of new regenerated tissue. A high absorption capacity is also a valuable property required in sponge design as drug diffusion depends on fluid penetration of the porous structure [31].

2.3. In Vitro Drug Release Kinetics Study and Data Modeling

The cumulative percentage of flufenamic acid released throughout the in vitro experiments is graphically presented in Figure 1a,b.

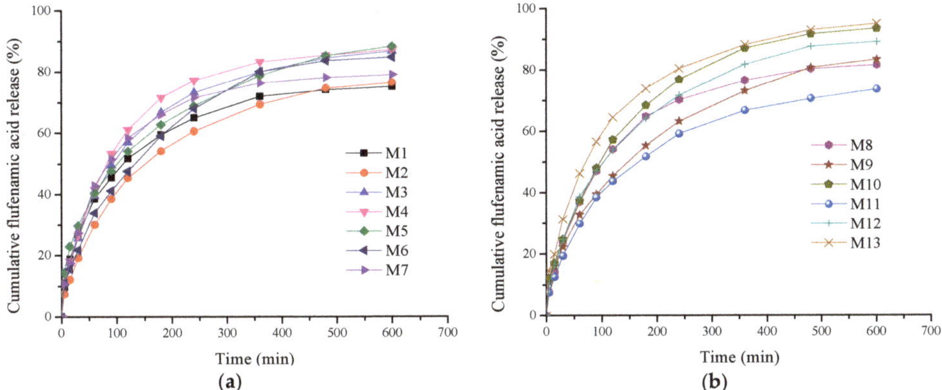

Figure 1. Time-dependent cumulative release profiles of FFA from collagen sponges: (**a**) experiments 1 to 7; (**b**) experiments 8 to 13.

The charts presented in Figure 1 indicate that the designed formulations present similar kinetic profiles. The flufenamic acid is markedly released in the first 60 min and the released drug percentage ranges from 29.97% (M11) to 46.37% (M13). This behavior—known as burst effect—is followed by a progressive anti-inflammatory drug delivery over a longer period of time set to 9 h. At the end of 10 h, a stationary drug release is reached.

This biphasic pattern of release involving the two stages is targeted to control the local inflammation and pain associated with a burn injury. Thus, the burst release effect ensures both a rapid reduction of painful sensation and the management of the pro-inflammatory mediators' cascade released at the burn level, and is needed immediately after lesion occurrence. The gradual drug delivery phase offers an anti-inflammatory and analgesic local effect over the longer period needed for burn healing.

The drug is released at a cumulative percentage higher than 73.69 and attaining 95.09% after 10 h of experiments (Table 2), resulting in an advantage for burn lesions treatment. This release profile is desirable for burn treatment as the first 12 h are critical and correspond to the peak of the inflammatory phase [70].

The flufenamic acid release profiles from the spongious matrices were assessed to determine the kinetic mechanism.

The drug mass transfer mechanism was evaluated by fitting the in vitro release experimental data to the Power law model (Equation (2)) and its particular case, the Higuchi kinetic model (Equation (3)) [71–73]:

$$\frac{m_t}{m_\infty} = k \cdot t^n \tag{2}$$

$$\frac{m_t}{m_\infty} = k \cdot t^{0.5} \tag{3}$$

where m_t/m_∞ is the fractional drug released in time t, k—the kinetic constant—and n—the release exponent characteristic—to the drug transport mechanism.

As noticed in Table 3, the values recorded for the determination coefficients (R^2) are higher for the Power law model (superior to 0.95) indicating a non-Fickian mechanism for FFA release from collagen spongious matrices, in line with our previous studies. The release exponent and kinetic constant values specific to the Power law model are listed in Table 3.

Table 3. Determination coefficients for flufenamic acid release from collagen sponges determined by application of Higuchi and Power law models; kinetic parameters for Power law model.

Formulation	Higuchi Model	Power Law Model	Release Exponent	Kinetic Constant (1/min^n)
M1	0.9296	0.9727	0.33	0.094
M2	0.9698	0.9803	0.41	0.056
M3	0.9452	0.9730	0.37	0.089
M4	0.9206	0.9577	0.36	0.099
M5	0.9679	0.9945	0.35	0.095
M6	0.9739	0.9830	0.42	0.062
M7	0.9006	0.9531	0.33	0.108
M8	0.9355	0.9618	0.38	0.079
M9	0.9851	0.9932	0.41	0.060
M10	0.9637	0.9779	0.40	0.077
M11	0.9672	0.9807	0.41	0.057
M12	0.9659	0.9822	0.39	0.076
M13	0.9375	0.9740	0.34	0.110

As we previously reported, and in accordance with other authors, this mechanism can be considered specific to drug delivery from swellable hydrophilic matrices like sponges and depends on many processes. Thus, a burn wound initially has a high content of exudate absorbed by, and penetrating, the porous spongious structure. A gel layer is formed at the contact between the wound and the sponge dressing, favoring the rapid diffusion of the drug retained at, and close to, the surface. The sponge's higher capacity for absorbing a large amount of fluid immediately after contact with the "fresh" wound is correlated with the drug's fast release in the first 60 min of experiments, reducing the burn inflammation and the associated pain at the application site.

The absorption process is followed by the hydration of the polymer network and simultaneous swelling of the relaxed polymer, and diffusion through the swollen matrix of drug entrapped in the polymer network during the lyophilization process.

The complex drug release mechanism from spongious forms previously described determines the non-respect of the Higuchi kinetic profiles characterized by a smaller drug diffusion rate than polymer relaxation rate.

Currently, it is considered that the potential erosion of the spongious support could also affect the drug delivery [47–50]. Therefore, an enzymatic study is required as a further step in tailoring the drug release kinetics.

2.4. In Vitro Enzymatic Degradation Analysis

In vitro biodegradation of collagen spongious matrices by collagenase solution was needed to simulate the in vivo behavior of the designed wound dressings. The degradation rate control is an important aspect in burn healing as in vivo resorption influences the tissue regeneration capacity [42]. Thus, a high degradation rate induces a very fast drug release due to the rapid support destructuring, which results in a loss of treatment efficiency, while a small degradation rate leads to a delay in affected tissue healing due to the necessity to successively remove and replace a wound dressing. Also, in this last case there is the possibility of producing new lesions and additional suffering, affecting patients' comfort.

Enzymatic degradation of collagen materials is dependent on, and determined by, their triple helical integrity, the crosslinking degree, and the availability of cleavage sites [74].

The resistance to enzymatic degradation was monitored through a weight loss parameter determined using the following equation (Equation (4)):

$$\% \text{ Weight Loss} = \frac{W_i - W_t}{W_i} \times 100 \qquad (4)$$

where W_i is the sponge initial weight, and W_t—the weight of the samples after a time t [75].

The weight loss values up to 10 h varied from 21.48 to 38.02 (%) as indicated in Table 2.

2.5. Optimization Technique

The experiments resulting from the Box-Behnken factorial design were subjected to an optimization technique based on experimental design, response surface methodology, and the Taguchi approach—presented in our previous works [31,32,60]—in order to determine the influence of the formulation factors on some responses involved in drug delivery from collagen spongious matrices. The physical levels of the independent variables to be optimized are considered for the hydrogels lyophilized to obtain the spongious matrices.

In order to set the reduced quadratic polynomial equations for each response, a stepwise regression analysis with backward elimination subroutine—which automatically eliminated the insignificant terms—was applied to the experimental data obtained. These regression models showing the quantitative effect of the formulation factors (X_i) and their interactions on the responses $(Y_i$, Table 2) are presented as Equations (5)–(7), with only the significant terms $(p < 0.05)$ being kept. The interactions between the formulation factors are the result of the terms containing two such parameters while each variable square indicate the quadratic relationship. The coefficients value for each independent variable, the combination of two independent variables, or the squared independent variables show their effects on each response. For the responses to be maximized—the swelling ratio and released drug percentage—the coefficients' positive sign signifies a synergistic effect and their negative sign signifies an antagonist effect. For the response to be minimized—weight loss—the signs' meaning is reversed.

From Equation (5) a positive impact on Y_1 response is noticed for X_2 (quadratic effect), as well as for the interactions between X_1 and X_2 and the interaction between X_1 and X_3, while X_1 (quadratic effect), X_2 (linear effect), X_3 (quadratic effect) and the interaction between X_2 and X_3 all have an antagonist effect on this response. Equation (6) indicates that the released drug percentage is positively influenced by X_1 and X_2 (linear effect) and by the interaction between X_1 and X_3, and negatively influenced by the quadratic effect of all formulation factors. Also, a positive quadratic effect of variables X_1 and X_3 and also of the interaction between X_1 and X_2 on the resistance to enzymatic degradation can be observed from Equation (7).

It can be noticed that the resistance to enzymatic degradation is influenced by the smallest number of factors.

The above equations presented in terms of coded factors are given below:

$$Y_1 = 59.58 - 39.22X_2 - 23.45X_1^2 + 13.92X_2^2 - 68353.76X_3^2 + 27.70X_1X_2 + 772.57X_1X_3 - 775.26X_2X_3 \quad (5)$$

$$Y_2 = -105.15 + 379.47X_1 + 30.77X_2 - 192.21X_1^2 - 17.29X_2^2 - 120827.46X_3^2 + 905.61X_1X_3 \quad (6)$$

$$Y_3 = 43.15 - 6.75X_1^2 - 57253.15X_3^2 - 3.36X_1X_2 \quad (7)$$

The predictive power of the regression models obtained was evaluated by way of goodness of fit (correlation—R, and determination—R^2 coefficients), analysis of variance (ANOVA) and residual analysis. The correlation coefficients values are 0.9951, 0.9899 and 0.9740 respectively, showing a high correlation degree between the observed and predicted values. The determination coefficients are 0.9902, 0.9799 and 0.9487 and indicate that only 0.98%, 2.01% and 5.13% of the total system response variation is not explained by the corresponding model.

The summary of variance analysis is presented in Table 4.

Table 4. Analysis of variance (ANOVA) for reduced regression polynomial models.

Responses	Sources of Variation	Sum of Squares	df	Mean Squares	F-Value	p-Value
Y_1	Regression	491.473	7	70.210	72.862	0.000099
	Residual	4.8180	5	0.963		<0.001
	Total	496.291	12			
Y_2	Regression	538.351	6	89.725	48.907	0.000078
	Residual	11.007	6	1.834		<0.001
	Total	549.358	12			
Y_3	Regression	258.315	3	86.105	55.54	0.000004
	Residual	13.953	9	1.550		<0.001
	Total	272.268	12			

It can be seen that the application of ANOVA proves the statistical significance of each regression model.

The good correlation between observed and predicted values is also sustained by the residual analysis presented in Table 2 and Figure 2, in which a linear distribution is noticed.

Moreover, the design normality was validated by normal probability plots of residuals (difference between observed and predicted responses). The experimental values for the responses are distributed near a straight line, confirming the robustness of design (Figure 3).

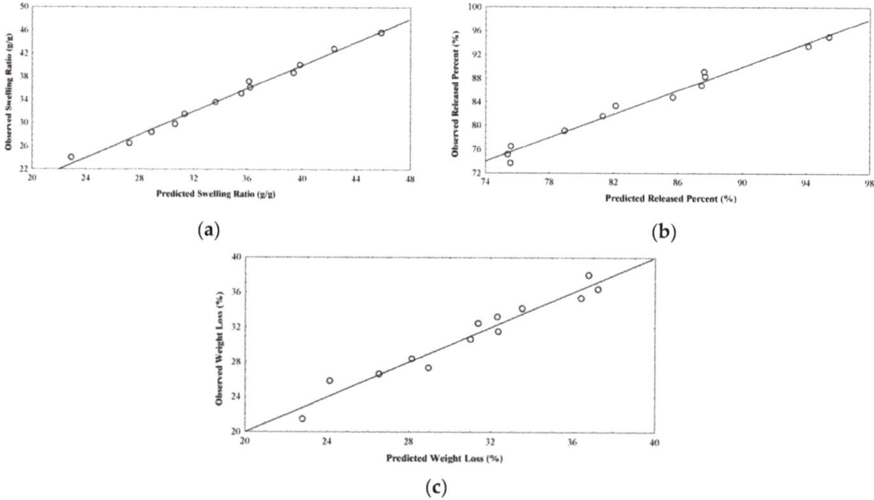

Figure 2. Plot showing correlation between observed and predicted values for: (**a**) swelling ratio (g/g); (**b**) released percent (%); (**c**) weight loss (%).

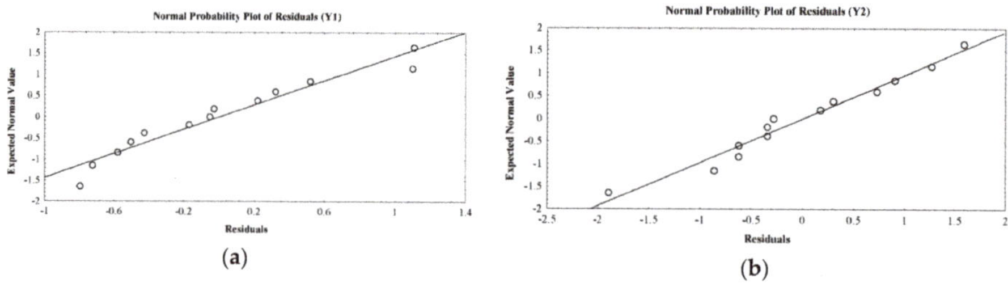

Figure 3. *Cont.*

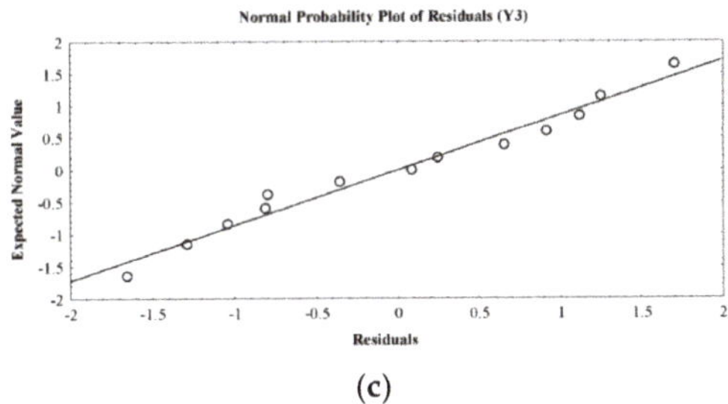

(**c**)

Figure 3. Plot showing correlation between expected normal values and residuals for: (**a**) swelling ratio (g/g); (**b**) released percent (%); (**c**) weight loss (%).

All statistical tests performed highlighted the high predictive power of the selected reduced regression polynomial models and the possibility of their use in correlating and validating the experimental results.

The influence of the formulation factors on sponge responses can be further visualized by plotting three-dimensional surface graphs. This technique—known as response surface methodology—is applied in order to assess the relationship between various independent variable combinations and for the determination of the conditions needed to obtain the targeted drug delivery kinetics.

The response surfaces are presented in Figures 4–6.

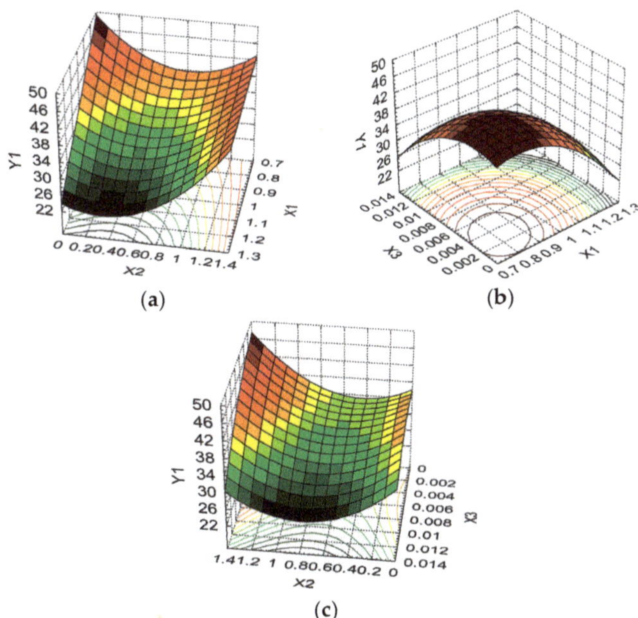

Figure 4. 3D response surface and contour plot showing the effect of various formulation factors on swelling ratio (Y_1): (**a**) collagen concentration (X_1) and dextran concentration (X_2); (**b**) collagen concentration (X_1) and glutaraldehyde concentration (X_3); (**c**) dextran concentration (X_2) and glutaraldehyde concentration (X_3).

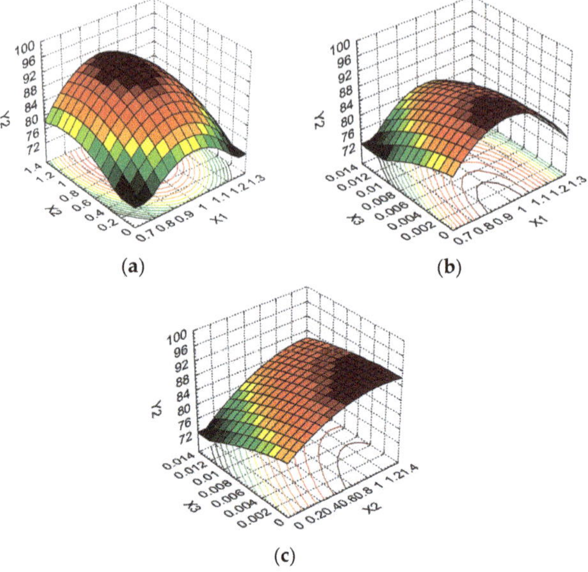

Figure 5. 3D response surface and contour plot showing the effect of various formulation factors on released percent (Y_2): (**a**) collagen concentration (X_1) and dextran concentration (X_2); (**b**) collagen concentration (X_1) and glutaraldehyde concentration (X_3); (**c**) dextran concentration (X_2) and glutaraldehyde concentration (X_3).

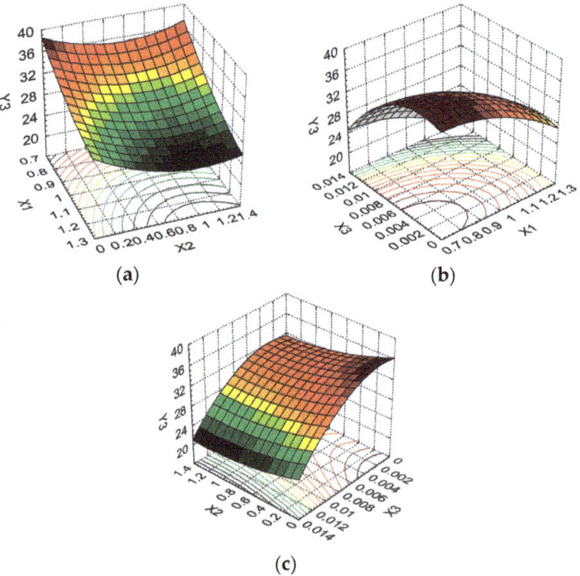

Figure 6. 3D response surface and contour plot showing the effect of of various formulation factors on weight loss (Y_3): (**a**) collagen concentration (X_1) and dextran concentration (X_2); (**b**) collagen concentration (X_1) and glutaraldehyde concentration (X_3); (**c**) dextran concentration (X_2) and glutaraldehyde concentration (X_3).

From Figure 4a we can see that the swelling ratio is strongly influenced by the collagen content for small amounts of dextran, this variation being attenuated by the addition of dextran. The swelling ratio decreases from 45.65 g/g (maximum observed value) to 28.41 g/g when collagen content increases and dextran is not present in the formulation (37.76% decrease) and from 31.56 g/g to 24.05 g/g (23.80% decrease) for dextran at the maximum level. It is suggested that the addition of collagen to the formulation increases the number of sponge pores, favoring the swelling process.

Figure 4b shows the dependence of the swelling ratio on the concentration of collagen and glutaraldehyde. A high swelling ratio is favored by both formulation factors at the lowest level. Y_1 has a more pronounced increase from 26.52 g/g to 38.77 g/g (46.19%) with a decrease in collagen concentration when the glutaraldehyde level is kept at a minimum, while the increase recorded for Y_1 is from 31.56 g/g to 38.77 g/g (22.84%) when the glutaraldehyde concentration decreases and the collagen level is kept at a minimum.

Concerning the dextran and glutaraldehyde influence on the swelling ratio, one can conclude from Figure 4c that the variation pattern is similar for different concentrations of glutaraldehyde, but small concentrations of glutaraldehyde and high quantities of dextran favor high values of Y_1. This variation is exemplified by a Y_1 increase of 15.17% (from 37.23 g/g to 42.88 g/g) when no glutaraldehyde is present in the formulation and dextran varies from the minimum to maximum level. Also, the smallest values for the swelling ratio are the result of dextran at the middle level and glutaraldehyde at the maximum level.

Figure 5a highlights the dependence of released drug percentage on collagen and dextran concentrations. The collagen variation influences Y_2 response, the highest released drug percentage being reached for mid to high amounts of dextran and mid values of collagen. Y_2 displays a more important variation from 73.69% to 95.09% (29.04% increase) with dextran concentration increase when collagen is at the middle level. Similar evolutions, but with smaller Y_2 increases, are seen when collagen is at the low and high levels and dextran varies between low to high levels.

The same variation pattern is also recorded for the influence of collagen and glutaraldehyde concentrations on released drug percentage, as it results from Figure 5b. The glutaraldehyde crosslinking properties are obvious as its absence results in high released drug percentage when collagen is at the middle level, the addition of this substance involving a reduction for Y_2 values.

Figure 5c shows the improvement of released drug percentage with the addition of dextran in the presence of glutaraldehyde. The drug release is favored by high amounts of dextran and by the lack of glutaraldehyde, as already mentioned. The released drug percentage increases by 26.90%—from 73.69%, in the absence of dextran and glutaraldehyde at the high level, to 93.51% when there is no glutaraldehyde but the maximum amount of dextran is included in the formulation.

According to the shape of Figure 6a, the upper level of collagen and the middle level of dextran determine a small weight loss value, the collagen variation having more influence on the Y_3 value. Thus, the Y_3 value decreases from 31.37% to 26.53% (15.43% decrease) for a dextran variation from minimum to maximum level at a maximum level of collagen and from 27.32% to 21.48% (21.38% decrease) for a collagen variation from minimum to maximum level for a middle level of dextran.

From Figure 6b we can conclude that the resistance to enzymatic degradation is increased for high collagen and glutaraldehyde values, both parameters having a similar effect on Y_3. This suggests that a higher concentration of collagen induces a denser matrix that is resistant to enzymatic degradation. Y_3 values decrease from 36.42% to 21.48% when collagen is at a minimum level and glutaraldehyde is absent from the formulation. When collagen and glutaraldehyde are at maximum levels, Y_3 values decrease by 41.02%.

According to Figure 6c, the analysis of dextran and glutaraldehyde formulation parameters indicates that the former has a small influence, and the latter an important influence, on Y_3 response, underlining the importance of glutaraldehyde concentration for resistance to enzymatic degradation.

As the hypothesis of this study is that drug release is controlled by both swelling degree and drug diffusion through the swollen polymeric matrix, as well as by the sponge support degradation rate, it was noticed that a small to medium degree of cross-linking determined higher values for the swelling ratio and released drug percentage simultaneously, with enhanced mobility of the polymeric chains resulting in macromolecular mobility and favoring drug diffusion. Meanwhile, a higher crosslinking degree is related to the incorporation of the drug into the sponge structure, reducing the polymeric chain hydration and consequently restricts the drug diffusion through the network matrix, but at the same time increases the resistance to enzymatic degradation.

In order to determine the sponge formulations that lead to responses that are minimally affected by noise factors, the final stage of our research involved the assessment of the effect of control factors on each response using the Signal-to-Noise ratio (S/N ratio), a performance indicator proposed by Taguchi.

The signal represents the desired value of the control factor (independent variable) while the noise represents the unwanted influence, the selection of a defined S/N ratio [32,60] depending on the constraints imposed on system responses. Thus, taking into consideration the constraints mentioned in Table 1b, the "larger-the-better" criterion has to be applied for S/N ratios related to swelling ability and the percentage of FFA released from the sponges, and the "smaller-the-better" criterion for the weight loss, which reflects resistance to enzymatic degradation.

The control factors' influence on the S/N ratio for each response, resulting in the optimal combination of formulation factors, is given in Table 5.

Table 5. Optimal combinations of independent variables coded levels, their effect size on S/N ratio for the dependent variables, expected and observed S/N value.

Control Factors (Independent Variables)	Y_1		Y_2		Y_3	
	"Larger—the—Better"	Effect Size	"Larger—the—Better"	Effect Size	"Smaller—the—Better"	Effect Size
X_1	1	1.112	2	0.272	3	0.918
X_2	3	0.702	3	0.523	2	0.343
X_3	2	0.563	1	0.354	3	1.526
S/N ratio expected (dB)	33.018		39.629		−26.909	
S/N ratio observed (dB)	32.072		39.417		−26.641	

The responses variation when the formulation factors are modified from low to high level is illustrated in Figure 7.

From Figure 7, and taking into account the differences between the variation levels for all responses, we can conclude that collagen has a primary influence on the swelling ratio and a moderate influence on released drug percentage and weight loss. The optimum coded level for this formulation factor are 1 for swelling ratio (corresponding to a X_1 concentration of 0.8), 2 for released drug percentage (corresponding to a X_1 concentration of 1.0) and 3 for the weight loss (corresponding to a X_1 concentration of 1.2).

The dextran has a primary influence on released drug percentage, a moderate influence on the swelling ratio, and a small influence on resistance to enzymatic degradation. The optimum coded levels of dextran for reducing noise factors to a minimum extent are 3 (maximum concentration of dextran, 1.2) for Y_1 and Y_2 responses, and 2 for resistance to enzymatic degradation.

The most important difference detected in the variation levels for glutaraldehyde is for the weight loss, while the other formulation factors are moderately affected by X_3. The reduction of noise factor effects occurs for the following glutaraldehyde optimum coded levels: 2 for Y_1 (middle level of X_1), 1 for Y_2 (absence of glutaraldehyde from the formulation) and 3 for Y_3 (maximum level of glutaraldehyde).

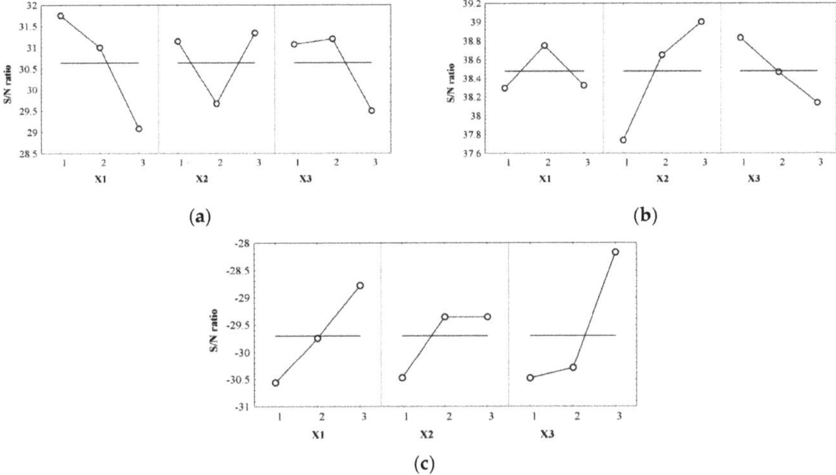

Figure 7. Control factors effects on S/N ratio for: (**a**) swelling ratio (Y_1); (**b**) released percent (Y_2); (**c**) weight loss (Y_3).

Following the analysis of the formulation factors' effect size on the responses, the results show that the main variable affecting the swelling ratio is collagen concentration (effect size 1.58 times higher than for the dextran concentration and 1.97 times higher than for the glutaraldehyde concentration).

The main effect of dextran is on the released drug percent, the effect size being 1.92 times higher in comparison with collagen concentration and 1.48 times higher than the glutaraldehyde concentration.

The glutaraldehyde strongly influences resistance to enzymatic degradation, and the collagen also has an important effect on this response. Thus, the glutaraldehyde effect size for the weight loss is 1.66 times superior to that of collagen concentration and is 4.49 times superior to that of dextran concentration.

By application of the Taguchi technique, three optimal stable and robust formulations were selected that least affect the swelling ratio, released drug percentage and sponge weight loss, see Table 6 for their compositions. All these formulations are included in the experimental design, their compositions being coded as 1:3:2 (sponge M3) for swelling ratio, 2:3:1 for released drug percentage (sponge M10) and 3:2:3 (sponge M8) for the weight loss which expresses resistance to enzymatic degradation.

Table 6. Composition of optimal formulations and the observed and predicted values of response variables.

Spongious Matrices	Composition of Optimal Formulation $X_1{:}X_2{:}X_3$ C:DX:GA (g%:g%:g%)	Response Variable	Observed Value	Predictive Value	Predicted Error (%)
M3	0.8:1.2:0.006	Y_1 (g/g)	40.14	39.81	+0.82
		Y_2 (%)	86.81	87.43	−0.71
		Y_3 (%)	34.20	33.54	+1.97
M8	1.2:0.6:0.012	Y_1 (g/g)	24.05	22.94	+4.84
		Y_2 (%)	81.61	81.31	+0.37
		Y_3 (%)	21.48	22.77	−5.67
M10	1.0:1.2:0.000	Y_1 (g/g)	42.88	42.35	+1.25
		Y_2 (%)	93.51	94.13	−0.65
		Y_3 (%)	31.56	32.37	−2.50

The regression models used proved to have strong predictive power according to the values presented in Table 6, the errors recorded being in the range ($-5.67 \div 4.84$)%.

It can be remarked that the best combination of formulation factors is not necessarily made by the highest/smallest values for the responses, but by the combination that leads to responses that are stable, robust, and insensitive to the noise factors, because the selection of an optimal product or of its manufacturing conditions consists not only in obtaining the best values of the responses, but also in finding the conditions for which the characteristics vary to the minimum extent [60].

2.6. Evaluation of the Collagen Sponges' Performance during the Wound Healing Process

The burn healing and anti-inflammatory action of the optimal sponges previously selected was confirmed by in vivo experiments performed on Wistar rats.

The animals were distributed in 4 groups of 5 individuals each as follows: M3 group was treated with the M3 FFA-sponge, M8 group with the M8 FFA-sponge, M10 group with the M10 FFA-sponge, and the Control group was not treated. M3 is a cross-linked sponge with a smaller concentration of collagen, but with a higher dextran concentration, M8 is a cross-linked sponge with a higher concentration of GA and a medium dextran concentration, and M10 is an uncross-linked sponge with a medium collagen concentration and a higher dextran concentration (Table 6).

The macroscopic images of burn wounds experimentally induced in rats are presented in Figure 8. The wounds were initially characterized by a white eschar with an affected epidermis and dermis and a hyperaemic zone in the periphery. During the first day, the wound area became fully hyperaemic due to red blood cell extravasation and a post-traumatic inflammatory process.

After inducing the burns, the application of the sponges accelerated the healing process of the lesions in comparison with the control group. After 24 h, the sponges were completely absorbed at the wound area and the epithelial tissue had started to cicatrize.

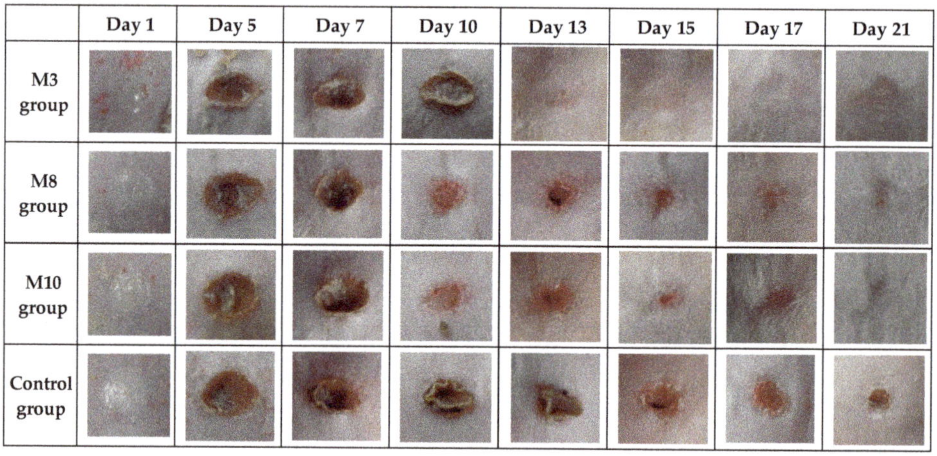

	Day 1	Day 5	Day 7	Day 10	Day 13	Day 15	Day 17	Day 21
M3 group								
M8 group								
M10 group								
Control group								

Figure 8. Evolution of wound healing process in rats without treatment (Control) and treated with collagen sponges M3, M8, M10 at different time intervals (day 1 is considered t = 0).

The healing process was evaluated according to the size profile of the wound as described by the following equation (Equation (8)):

$$\text{Healing } \% = \frac{W_i - W_t}{W_i} \times 100 \tag{8}$$

where the wound size (W) is an average measurement from the longest and shortest dimensions of the affected tissue, W_i is the initial wound size and W_t is the wound size at different time intervals [13,27,64–66]. The wound was considered healed when the scab fell off the experimental lesion.

The evolution of the wound diameter (mm) is presented in Table 7. The healing process was calculated according to Equation (8) and is presented in Figure 9.

The wound size was reduced by 36% in the case of treatment with the M3 sponge after 5 days, followed by the groups treated with the M8 and M10, which showed a decrease in wound size of 32% and 26% respectively in comparison with the control group (Table 7, Figure 9, ANOVA $p < 0.05$). After 7 days, the wound diameter in case of the treated groups decreased significantly compared to the mean diameter of the control group ($p < 0.05$), corresponding to a faster healing process. In the case of some animals from groups treated with M8 and M10 sponges, the scab fell off after 10 days and the healing process was completed after 15 days from the start of treatment. In the case of M3 the wound mean diameter was slightly decreased in the first 10 days in comparison with M8 and M10 groups ($p > 0.05$) but healing was completed for all animals after 17 days. The first 10 days after a burn are critical for tissue regeneration and the results demonstrated a significantly improved healing process after comparison of the treated groups with the control group ($p < 0.05$, Table 7).

Table 7. Evolution of the wound diameter (mm) after treatment with collagen sponges (M3, M8, M10).

Wound Diameter (mm)	M3 Group		M8 Group		M10 Group		Control Group	
	Mean	SEM	Mean	SEM	Mean	SEM	Mean	SEM
Day 1	10	0	10	0	10	0	10	0
Day 5	6.4 *	1.03	6.8 **	0.58	7.4 **	0.51	9.4	0.24
Day 7	6 **	0.89	6.8 **	0.37	7.4 ***	0.24	9.2	0.20
Day 10	5.4 **	0.51	6.8 *	0.58	7 *	0.32	8.2	0.37
Day 13	1.8 *	0.80	3.4	1.40	2.2	1.36	5.6	0.24
Day 15	1.4 *	0.87	0	0	0	0	1.6	0.98
Day 17	0	0	0	0	0	0	0.6	0.60
Day 21	0	0	0	0	0	0	0	0

SEM = standard error of the mean, student's t-test (vs. Control) * $p < 0.05$, ** $p < 0.01$, *** $p < 0.001$.

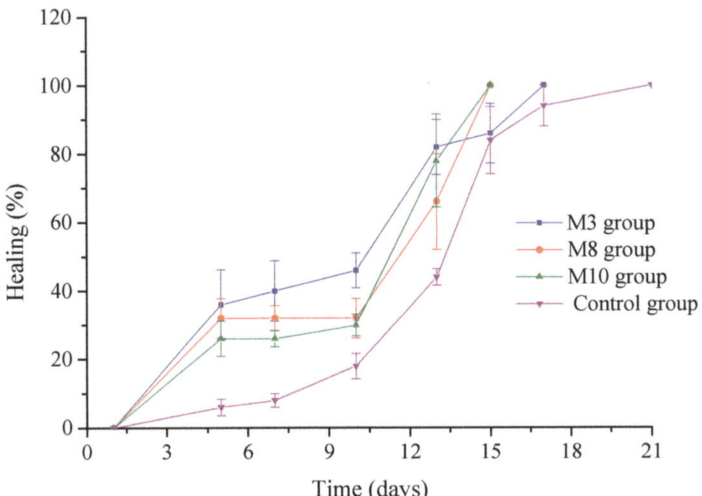

Figure 9. Evolution of healing process after treatment with collagen sponges (M3, M8, M10) and not-treated control group.

Our preliminary pharmacological study identified no significant differences in the healing process after treated groups comparison ($p > 0.05$). We observed that in case of the M3 group, after the scab fell off, the scar left was less visible compared to other groups (Figure 8). Further biological studies will be required to elucidate the skin repair mechanism and to explain how this cross-linked sponge with a smaller collagen concentration and a higher dextran concentration may offer better support for epithelial regeneration.

3. Materials and Methods

3.1. Materials

Type I collagen gel with a concentration of 2.54% (w/w) and pH 2.5 was extracted from calf hide by a technology currently used at the Collagen Department of Division Leather and Footwear Research Institute, Bucharest, Romania [42]. The chemicals were purchased from commercial suppliers as follows: flufenamic acid from MP Biomedicals (Solon, OH, USA), dextran from *Leuconostoc* spp. (Mw = 15,000–25,000) from Fluka (Steinheim, Germany), glutaraldehyde from Merck (Darmstadt, Germany), type I collagenase obtained from *Clostridium histolyticum* from Sigma-Aldrich (St. Louis, MO, USA), sodium hydroxide, monobasic potassium phosphate and disodium hydrogen phosphate from Merck (Germany), ether ethylic from Sigma Aldrich (USA). The water used was distilled and all other chemicals used for analysis were of analytical grade. The animals used for the pharmacological experiments were Wistar male rats and were supplied by the Animal Biobase of the "Carol Davila" University of Medicine and Pharmacy, Bucharest, Romania.

3.2. Methods

3.2.1. Preparation of Spongious Matrices

Collagen gels with concentrations of 0.8%, 1.0% and 1.2% and pH 7.4 were obtained from the initial collagen gel (2.54% and pH 2.5) using an NaOH 1 M solution under mechanical stirring. 0.6% and 1.2% dextran and 0.5% flufenamic acid solutions were added to the collagen gels according to the compositions presented in Table 1. Some of the hydrogels were cross-linked with 0.006% and 0.012% glutaraldehyde for 24 h at 4 °C. The amounts of dextran, flufenamic acid and glutaraldehyde were reported to collagen dry substance. The hydrogels were designed according to the 3^3 Box-Behnken design and their final composition is presented in Table 2. The hydrogels were then lyophilized using the Delta LSC 2-24 Martin Christ lyophilizer (Osterode am Harz, Germany) by the method previously described [37] and following the program presented in Figure 10. In this way, the corresponding spongious matrices were obtained and coded as M1 to M13. Briefly, the process started with freezing at −40 °C for 10 h and continued with main freeze drying at 0.1 mbar and the temperatures of +10 °C for 8 h, +25 °C for 12 h and +35 °C for 8 h. The final freeze drying lasted 10 h at the pressure of 0.001 mbar and the temperature of 35 °C.

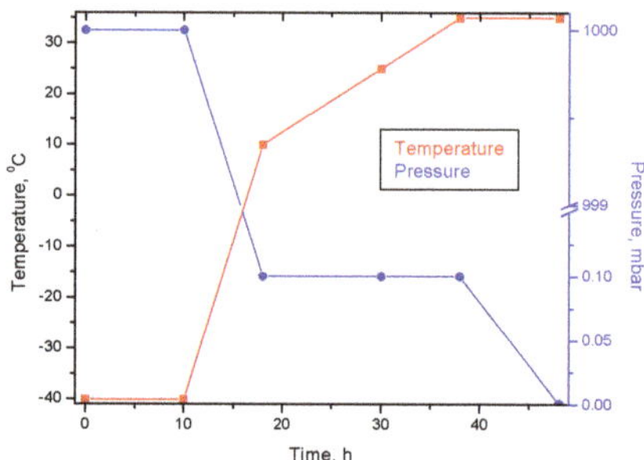

Figure 10. Freeze-drying chart of collagen-flufenamic acid spongious matrices.

3.2.2. Swelling Study

The spongious matrices' swelling ability was assessed using a general gravimetric method, at 37 °C. Collagen sponges were carefully weighed when in a dry condition (W_0) and were then immersed in the absorption medium represented by phosphate buffer solution pH 7.4 which simulates the wound physiological liquid pH [31]. At specific time intervals over a 10 h period, the swollen sponges were taken out of the swelling medium, hung to dry for one minute until no more drops were formed, then weighed (W_t) with a four decimal point electronic microbalance, and the swelling ratio was determined. Each experiment was conducted in triplicate and the results were averaged.

3.2.3. In Vitro Drug Release Kinetics Study and Data Modeling

The studies of flufenamic acid in vitro release kinetics from the collagen sponges were performed using dissolution equipment in conjunction with paddle stirrers (Essa Dissolver, city, Italy), as previously reported [76]. The disc shaped sponge samples had a diameter of 3 cm and a predetermined weight, and were put in a transdermal sandwich device before being placed in apparatus dissolution vessels. The determinations were carried out at 37 °C $\pm$ 0.5 °C with a rotational speed of 50 rpm. The release medium used was a phosphate buffer solution of pH 7.4. At pre-established time intervals over a 10 h period, samples of 5 mL were collected from the receiving medium and replaced with an equal volume of fresh phosphate buffer solution, maintained at 37 °C $\pm$ 0.5 °C, to keep a constant volume in the release vessel. The concentration of FFA in each sample was monitored by UV spectroscopy (Perkin-Elmer UV-Vis spectrophotometer, Überlingen, Germany) at its maximum absorbance corresponding to a wavelength of 288 nm, using the standard curve ($A_{1\%}^{1cm} = 534$), and the cumulative released drug percentage was determined. The release studies were performed in triplicate.

3.2.4. In Vitro Enzymatic Degradation Analysis

The enzymatic degradation of spongious matrices was investigated in a phosphate buffer solution with pH 7.4 containing 10^{-6} mg/mL collagenase and quantified using weight loss (WL) as a function of time. Sponge samples of 1 cm $\times$ 1 cm, accurately weighed (W_i), were immersed in the collagenase solution and incubated at 37 °C. At fixed time intervals over a 10 h period, the sponge pieces were extracted, weighed (W_t) and the degradation percentage was evaluated. Three determinations were performed for each sample and the results were averaged.

3.2.5. Design of Experiments and Optimization Technique

The preparation of the collagen hydrogels used to obtain the corresponding spongious matrices through lyophilization was performed in accordance with a 3-factor, 3-level Box-Behnken factorial design (Tables 1 and 2). The experiments resulting from the Box-Behnken factorial design were performed randomly and in triplicate to minimize the errors due to systematic trends in the factors. The statistical data analysis was performed using different routines of Statistica StatSoft Release software package. In order to set the quadratic polynomial equations for each response, a stepwise regression analysis with a backward elimination subroutine supposing automatic elimination of insignificant terms was applied to the experimental data obtained (only the significant terms with $p < 0.05$ were kept). To confirm the adequacy of the reduced quadratic models, the correlation (R) and determination (R^2) coefficients were assessed and the analysis of variance (ANOVA) and residual analysis were performed. Three-dimensional response surface graphs were further plotted to investigate the independent variables' effect on the system responses. The Taguchi approach, expressed as Signal/Noise ratio, was further applied to complement the optimization technique and to establish the adequate formulation factor combinations that would ensure the product's quality and process robustness.

3.2.6. Evaluation of the Collagen Sponges Performance during the Wound Healing Process

The in vivo experiment was performed on 20 Wistar male rats each weighing 180 ± 10 g. All animals used in the study were kept in standard laboratory conditions, were fed twice a day and received water *ad libitum*. The experiment was performed in compliance with the European Communities Council Directive 2010/63/UE and Law No. 43 of the Romanian Parliament from 11 April 2014 (ethic approval number 1030 from 24 May 2017).

The animals were anesthetized with ether ethylic and the hair was removed from the dorsal area. The experimental wound was induced using a special metallic device of 10 mm diameter. The device was heated in a boiling physiological serum and applied on the shaved dorsal area for 10 seconds. The severe burns measuring 10 mm diameter were sterilized and the collagen sponges were applied and fixed with a silk plaster. A control group was used in the study and the wounds were covered with sterile gauze.

The surface morphology of the wounds was recorded using a digital camera (Olympus SP-590UZ) and the wound diameter was measured every two days for 21 days. Any aspects of inflammation or infection of the wound, as well as any change in the animal's health status were also monitored.

Statistical analyses were performed using the GraphPad Prism 6 software. Error bars reported within the charts denote the standard errors of the mean ($n = 5$). The experimental data were evaluated using the student's t-test and analyses of variance. The results were considered significant at $p < 0.05$.

4. Conclusions

Collagen spongious matrices loaded with flufenamic acid intended to reduce the progression of burn lesions were designed according to the hypothesis that drug release at the affected site could be controlled both by the degree of sponge swelling and by drug diffusion through the swollen polymeric matrix, as well as by the topical support degradation rate.

The analysis and optimization of sponge formulations were performed based on a Box-Behnken experimental design and response surface methodology, followed by the Taguchi technique to select the formulations which were the most stable, robust and insensitive to noise factors. The formulations selected were three flufenamic acid sponges with the following ratio between collagen (%), dextran (%) and glutaraldehyde (%): 0.8:1.2:0.006 (M3), 1.2:0.6:0.012 (M8) and 1.0:1.2:0.000 (M10), the drug concentration of 0.5% being correlated with collagen dry substance.

The treatment of experimental burns on animals with the above selected sponges accelerated the wound healing process and promoted faster regeneration of the affected epithelial tissues compared to the control group. Further biological studies will be necessary to elucidate the mechanism of action in the tissue restoration stage of the burn healing process.

Acknowledgments: This study was financially supported by UEFISCDI, PN-III-Experimental Demonstration Project, project number PED 160/03.01.2017, code PN-III-P2-2.1-PED-2016-0813, Romania.

Author Contributions: Mihaela Violeta Ghica, Mădălina Georgiana Albu Kaya and Denisa Ioana Udeanu conceived and designed the experiments; Mihaela Violeta Ghica, Mădălina Georgiana Albu Kaya and Denisa Ioana Udeanu performed the experiments; Mihaela Violeta Ghica, Mădălina Georgiana Albu Kaya, Denisa Ioana Udeanu and Cristina-Elena Dinu-Pîrvu analyzed the data; Dumitru Lupuleasa contributed analysis tools; Mihaela Violeta Ghica, Mădălina Georgiana Albu Kaya and Denisa Ioana Udeanu wrote the paper.

Abbreviations

The following abbreviations are used in this manuscript:

FFA	Flufenamic acid
NSAID	Non-steroidal anti-inflammatory drug
DDS	drug delivery systems

C	collagen
DX	dextran
GA	glutaraldehyde
SR	swelling ratio
RP	released percent
WL	weight loss

References

1. Frieri, M.; Kumar, K.; Boutin, A. Wounds, burns, trauma, and injury. *Wound Med.* **2016**, *13*, 12–17. [CrossRef]
2. Farina, J.A., Jr.; Junqueira Rosique, M.; Rosique, R.G. Curbing inflammation in burn patients. *Int. J. Inflam.* **2013**, *2013*. [CrossRef] [PubMed]
3. Salibian, A.A.; Del Rosario, A.T.D.; Severo, L.A.M.; Nguyen, L.; Banyard, D.A.; Toranto, J.D.; Evans, G.R.D.; Widgerow, A.D. Current concepts on burn wound conversion—A review of recent advances in understanding the secondary progressions of burns. *Burns* **2016**, *42*, 1025–1035. [CrossRef] [PubMed]
4. Chang, K.C.; Ma, H.; Liao, W.C.; Lee, C.K.; Lin, C.Y.; Chen, C.C. The optimal time for early burn wound excision to reduce pro-inflammatory cytokine production in a murine burn injury model. *Burns* **2010**, *36*, 1059–1066. [CrossRef] [PubMed]
5. Widgerow, A.D.; Kalaria, S. Pain mediators and wound healing—Establishing the connection. *Burns* **2012**, *38*, 951–959. [CrossRef] [PubMed]
6. Ekenseair, A.K.; Kurtis Kasper, F.; Mikos, A.G. Perspectives on the interface of drug delivery and tissue engineering. *Adv. Drug Deliv. Rev.* **2013**, *65*, 89–92. [CrossRef] [PubMed]
7. Boateng, J.; Catanzano, O. Advanced therapeutic dressings for effective wound healing—A review. *J. Pharm. Sci.* **2015**, *104*, 3653–3680. [CrossRef] [PubMed]
8. Hanafiah, Z.; Potparic, O.; Fernandez, T. Addressing pain in burn injury. *Curr. Anaesth. Crit. Care* **2008**, *19*, 287–292. [CrossRef]
9. Shemesh, M.; Zilberman, M. Structure-property effects of novel bioresorbable hybrid structures with controlled release of analgesic drugs for wound healing applications. *Acta Biomater.* **2014**, *10*, 1380–1391. [CrossRef] [PubMed]
10. Morgado, P.I.; Miguel, S.P.; Correia, I.J.; Aguiar-Ricardo, A. Ibuprofen loaded PVA/chitosan membranes: A highly efficient strategy towards an improved skin wound healing. *Carbohydr. Polym.* **2017**, *159*, 136–145. [CrossRef] [PubMed]
11. Chang, Y.C.; Wang, J.D.; Hahn, R.A.; Gordon, M.K.; Joseph, L.B.; Heck, D.E.; Heindel, N.D.; Young, S.C.; Sinko, P.J.; Casillas, R.P.; et al. Therapeutic potential of a non-steroidal bifunctional anti-inflammatory and anti-cholinergic agent against skin injury induced by sulfur mustard. *Toxicol. Appl. Pharmacol.* **2014**, *280*, 236–244. [CrossRef] [PubMed]
12. Norman, A.T.; Judkins, K.C. Pain in the patient with burns. *Contin. Educ. Anaesth. Crit. Care Pain* **2004**, *4*, 57–61. [CrossRef]
13. Ghica, M.V.; Kaya, D.A.; Albu, M.G.; Popa, L.; Dinu-Pîrvu, C.E.; Cristescu, I.; Udeanu, D.I. Ibuprofen-collagen sponges for wound healing. In Proceedings of the 5th International Conference on Advanced Materials and Systems, Bucharest, Romania, 23–25 October 2014; pp. 213–218.
14. Rasekh, M.; Karavasili, C.; Soong, Y.L.; Bouropoulos, N.; Morris, M.; Armitage, D.; Li, X.; Fatouros, D.G.; Ahmad, Z. Electrospun PVP-indomethacin constituents for transdermal dressings and drug delivery devices. *Int. J. Pharm.* **2014**, *473*, 95–104. [CrossRef] [PubMed]
15. El-Badry, M.; Fetih, G. Preparation, characterization and anti-inflammatory activity of celecoxib chitosan gel formulations. *J. Drug Deliv. Sci. Technol.* **2011**, *21*, 201–206. [CrossRef]
16. Ferreira, H.; Matamá, T.; Silva, R.; Silva, C.; Gomes, A.C.; Cavaco-Paulo, A. Functionalization of gauzes with liposomes entrapping an anti-inflammatory drug: A strategy to improve wound healing. *React. Funct. Polym.* **2013**, *73*, 1328–1334. [CrossRef]
17. Ibolya, F.; Gyéresi, A.; Szabó-Révész, P.; Aigner, Z. Solid dispersions of flufenamic acid with PEG 4000 and PEG 6000. *Farmacia* **2011**, *59*, 60–69.
18. Mohamed, A.A.; Matijević, E. Preparation and characterization of uniform particles of flufenamic acid and its calcium and barium salts. *J. Colloid. Interface Sci.* **2012**, *381*, 198–201. [CrossRef] [PubMed]

19. Rubio, L.; Alonso, C.; Rodríguez, G.; Cócera, M.; López-Iglesias, C.; Coderch, L.; De la Maza, A.; Parra, J.L.; López, O. Bicellar systems as new delivery strategy for topical application of flufenamic acid. *Int. J. Pharm.* **2013**, *444*, 60–69. [CrossRef] [PubMed]

20. Baek, J.S.; Yeo, E.W.; Lee, Y.H.; Tan, N.S.; Loo, S.C.J. Controlled-release nanoencapsulating microcapsules to combat inflammatory diseases. *Drug Des. Dev. Ther.* **2017**, *11*, 1707–1717. [CrossRef] [PubMed]

21. Malinovskaja-Gomez, K.; Labouta, H.I.; Schneider, M.; Hirvonen, J.; Laaksonen, T. Transdermal iontophoresis of flufenamic acid loaded PLGA nanoparticles. *Eur. J. Pharm. Sci.* **2016**, *89*, 154–162. [CrossRef] [PubMed]

22. Mahrhauser, D.S.; Kählig, H.; Partyka-Jankowska, E.; Peterlik, H.; Binder, L.; Kwizda, K.; Valenta, C. Investigation of microemulsion microstructure and its impact on skin delivery of flufenamic acid. *Int. J. Pharm.* **2015**, *490*, 292–297. [CrossRef] [PubMed]

23. Huang, S.; Fu, X. Naturally derived materials-based cell and drug delivery systems in skin regeneration. *J. Control. Release* **2010**, *142*, 149–159. [CrossRef] [PubMed]

24. Elsner, J.J.; Zilberman, M. Novel antibiotic-eluting wound dressings: An in vitro study and engineering aspects in the dressing's design. *J. Tissue Viabil.* **2010**, *19*, 54–66. [CrossRef] [PubMed]

25. Hu, D.; Qiang, T.; Wang, L. Quaternized chitosan/polyvinyl alcohol/sodium carboxymethylcellulose blend film for potential wound dressing application. *Wound Med.* **2017**, *16*, 15–21. [CrossRef]

26. Brichacek, M.; Ning, C.; Gawaziuk, J.P.; Liu, S.; Logsetty, S. In vitro measurements of burn dressing adherence and the effect of interventions on reducing adherence. *Burns* **2017**, *43*, 1002–1010. [CrossRef] [PubMed]

27. Khalid, A.; Khan, R.; Ul-Islam, M.; Khan, T.; Wahid, F. Bacterial cellulose-zinc oxide nanocomposites as a novel dressing system for burn wounds. *Carbohydr. Polym.* **2017**, *64*, 214–221. [CrossRef] [PubMed]

28. Vedakumari, W.S.; Ayaz, N.; Karthick, A.S.; Senthil, R.; Sastry, T.P. Quercetin impregnated chitosan-fibrin composite scaffolds as potential wound dressing materials—Fabrication, characterization and in vivo analysis. *Eur. J. Pharm. Sci.* **2017**, *97*, 106–112. [CrossRef] [PubMed]

29. Puoci, F.; Piangiolino, C.; Givigliano, F.; Parisi, O.I.; Cassano, R.; Trombino, S.; Curcio, M.; Iemma, F.; Cirillo, G.; Spizzirri, U.G.; et al. Ciprofloxacin-collagen conjugate in the wound healing treatment. *J. Funct. Biomater.* **2012**, *3*, 361–371. [CrossRef] [PubMed]

30. Albu, M.G.; Titorencu, I.; Ghica, M.V. Collagen-based drug delivery systems for tissue engineering. In *Biomaterials Applications for Nanomedicine*; Pignatello, R., Ed.; InTech Open Acces Publisher: Rijeka, Croatia, 2011; pp. 333–358. ISBN 978-953-307-661-4.

31. Ghica, M.V.; Albu, M.G.; Popa, L.; Moisescu, S. Response surface methodology and Taguchi approach to assess the combined effect of formulation factors on minocycline delivery from collagen sponges. *Pharmazie* **2013**, *68*, 340–348. [PubMed]

32. Ghica, M.V.; Albu, M.G.; Leca, M.; Popa, L.; Moisescu, S. Design and optimization of some collagen-minocycline based hydrogels potentially applicable for the treatment of cutaneous wounds infections. *Pharmazie* **2011**, *66*, 853–861. [PubMed]

33. Antoniac, I.V.; Albu, M.G.; Antoniac, A.; Rusu, L.C.; Ghica, M.V. Collagen-bioceramic smart composites. In *Handbook of Bioceramics and Biocomposites*; Antoniac, I.V., Ed.; Springer: Basel, Switzerland, 2016; pp. 301–324, ISBN 978-3-319-12459-9.

34. Abou Neel, E.A.; Bozec, L.; Knowles, J.C.; Syed, O.; Mudera, V.; Day, R.; Hyun, J.K. Collagen-emerging collagen based therapies hit the patient. *Adv. Drug Deliv. Rev.* **2013**, *65*, 429–456. [CrossRef] [PubMed]

35. Nunes, P.S.; Rabelo, A.S.; Souza, J.C.; Santana, B.V.; da Silva, T.M.; Serafini, M.R.; Dos Passos Menezes, P.; Dos Santos Lima, B.; Cardoso, J.C.; Alves, J.C.; et al. Gelatin-based membrane containing usnic acid-loaded liposome improves dermal burn healing in a porcine model. *Int. J. Pharm.* **2016**, *513*, 473–482. [CrossRef] [PubMed]

36. Zhang, L.; Li, K.; Xiao, W.; Zheng, L.; Xiao, Y.; Fan, H.; Zhang, X. Preparation of collagen-chondroitin sulfate-hyaluronic acid hybrid hydrogel scaffolds and cell compatibility in vitro. *Carbohydr. Polym.* **2011**, *84*, 118–125. [CrossRef]

37. Lungu, A.; Titorencu, I.; Albu, M.G.; Florea, N.M.; Vasile, E.; Iovu, H.; Jinga, V.; Simionescu, M. The effect of BMP-4 loaded in 3D collagen-hyaluronic acid scaffolds on biocompatibility assessed with MG 63 osteoblast-like cells. *Dig. J. Nanomater. Biostruct.* **2011**, *6*, 1897–1908.

38. Coelho, J.F.; Ferreira, P.C.; Alves, P.; Cordeiro, R.; Fonseca, A.C.; Góis, J.R.; Gil, M.H. Drug delivery systems: Advanced technologies potentially applicable in personalized treatments. *EMPA* **2010**, *1*, 164–209. [CrossRef] [PubMed]

39. Unnithan, A.R.; Barakat, N.A.; Tirupathi Pichiah, P.B.; Gnanasekaran, G.; Nirmala, R.; Cha, Y.S.; Jung, C.H.; El-Newehy, M.; Kim, H.Y. Wound-dressing materials with antibacterial activity from electrospun polyurethane-dextran nanofiber mats containing ciprofloxacin HCl. *Carbohydr. Polym.* **2012**, *90*, 1786–1793. [CrossRef] [PubMed]

40. Liao, N.; Unnithan, A.R.; Joshi, M.K.; Tiwari, A.P.; Hong, S.T.; Park, C.H.; Kima, C.S. Electrospun bioactive poly (ε-caprolactone)—Cellulose acetate-dextran antibacterial composite mats for wound dressing applications. *Colloids Surf. A Physicochem. Eng. Asp.* **2015**, *469*, 194–201. [CrossRef]

41. Kim, M.K.; Kwak, H.W.; Kim, H.H.; Kwon, T.R.; Kim, S.Y.; Kim, B.J.; Park, Y.H.; Lee, K.H. Surface modification of silk fibroin nanofibrous mat with dextran for wound dressing. *Fiber Polym.* **2014**, *15*, 1137–1145. [CrossRef]

42. Albu, M.G. *Collagen Gels and Matrices for Biomedical Applications*; Lambert Academic Publishing: Saarbrücken, Germany, 2011; ISBN 978-01-9850-970-7.

43. Kumar, M.S.; Kirubanandan, S.; Sripriya, R.; Sehgal, P.K. Triphala incorporated collagen sponge—A smart biomaterial for infected dermal wound healing. *J. Surg. Res.* **2010**, *58*, 162–170. [CrossRef] [PubMed]

44. Lee, C.H.; Lee, Y. Collagen-based formulations for wound healing applications. In *Wound Healing Biomaterials*; Ågren, M., Ed.; Woodhead Publishing, Elsevier: Cambridge, UK, 2016; Volume 2, pp. 135–149. ISBN 978-1-78242-456-7.

45. De Almeida, E.B.; Cardoso, J.C.; de Lima, A.K.; de Oliveira, N.L.; de Pontes-Filho, N.T.; Lima, S.O.; Souza, I.C.L.; de Albuquerque-Júnior, R.L.C. The incorporation of Brazilian propolis into collagen-based dressing films improves dermal burn healing. *J. Ethnopharmacol.* **2013**, *147*, 419–425. [CrossRef] [PubMed]

46. Akturk, O.; Tezcaner, A.; Bilgili, H.; Deveci, M.S.; Gecit, M.R.; Keskin, D. Evaluation of sericin/collagen membranes as prospective wound dressing biomaterial. *J. Biosci. Bioeng.* **2011**, *112*, 279–288. [CrossRef] [PubMed]

47. Metzmacher, I.; Radu, F.; Bause, M.; Knabner, P. A model describing the effect of enzymatic degradation on drug release from collagen minirods. *Eur. J. Pharm. Biopharm.* **2007**, *67*, 349–360. [CrossRef] [PubMed]

48. Radu, F.A.; Bause, M.; Knabner, P.; Friess, W.; Metzmacher, I. Numerical simulation of drug release from collagen matrices by enzymatic degradation. *Comput. Vis. Sci.* **2009**, *12*, 409–420. [CrossRef]

49. Boateng, J.S.; Matthews, K.H.; Stevens, H.N.E.; Eccleston, G.M. Wound healing dressings and drug delivery systems: A review. *J. Pharm. Sci.* **2008**, *97*, 2892–2923. [CrossRef] [PubMed]

50. Siafaka, P.I.; Zisi, A.P.; Exindari, M.K.; Karantas, I.D.; Bikiaris, D.N. Porous dressings of modified chitosan with poly(2-hydroxyethyl acrylate) for topical wound delivery of levofloxacin. *Carbohydr. Polym.* **2016**, *143*, 90–99. [CrossRef] [PubMed]

51. Abdelbary, A.A.; AbouGhaly, M.H.H. Design and optimization of topical methotrexate loaded niosomes for enhanced management of psoriasis: Application of Box-Behnken design, in vitro evaluation and in vivo skin deposition study. *Int. J. Pharm.* **2015**, *485*, 235–243. [CrossRef] [PubMed]

52. Dinarvand, M.; Rezaee, M.; Foroughi, M. Optimizing culture conditions for production of intra and extracellular inulinase and invertase from *Aspergillus niger* ATCC 20611 by response surface methodology (RSM). *Braz. J. Microbiol.* **2017**, *48*, 427–441. [CrossRef] [PubMed]

53. Koca, S.; Aksoy, D.O.; Cabuk, A.; Celik, P.A.; Sagol, E.; Toptas, Y.; Oluklulu, S.; Koca, H. Evaluation of combined lignite cleaning processes, flotation and microbial treatment, and its modelling by Box Behnken methodology. *Fuel* **2017**, *192*, 178–186. [CrossRef]

54. Moghddam, S.R.M.; Ahad, A.; Aqil, M.; Imam, S.S.; Sultana, Y. Formulation and optimization of niosomes for topical diacerein delivery using 3-factor, 3-level Box-Behnken design for the management of psoriasis. *Mater. Sci. Eng. C Mater. Biol. Appl.* **2016**, *69*, 789–797. [CrossRef] [PubMed]

55. Tan, Y.H.; Abdullah, M.O.; Nolasco-Hipolito, C.; Zauzi, N.S.A. Application of RSM and Taguchi methods for optimizing the transesterification of waste cooking oil catalyzed by solid ostrich and chicken-eggshell derived CaO. *Renew. Energy* **2017**, *114*, 437–447. [CrossRef]

56. Tumbas Šaponjac, V.; Čanadanović-Brunet, J.; Ćetković, G.; Jakišić, M.; Djilas, S.; Vulić, J.; Stajčić, S. Encapsulation of beetroot pomace extract: RSM optimization, storage and gastrointestinal stability. *Molecules* **2016**, *21*, 584. [CrossRef] [PubMed]

57. Li, H.Z.; Zhang, Z.J.; Hou, T.Y.; Li, X.J.; Chen, T. Optimization of ultrasound-assisted hexane extraction of perilla oil using response surface methodology. *Ind. Crops Prod.* **2015**, *76*, 18–24. [CrossRef]

58. Wang, H.; Liu, Y.; Wei, S.; Yan, Z. Application of response surface methodology to optimise supercritical carbon dioxide extraction of essential oil from *Cyperus rotundus* Linn. *Food Chem.* **2012**, *132*, 582–587. [CrossRef] [PubMed]

59. Nguyen, X.H.; Bae, W.; Gunadi, T.; Park, Y. Using response surface design for optimizing operating conditions in recovering heavy oil process, Peace River oil sands. *J. Pet. Sci. Eng.* **2017**, *17*, 37–45. [CrossRef]

60. Ghica, M.V.; Popa, L.; Saramet, G.; Leca, M.; Lupuliasa, D.; Moisescu, S. Optimization of the pharmaceutical products and process design applying Taguchi quality engineering principles. *Farmacia* **2011**, *59*, 321–328.

61. Apparao, K.C.; Birru, A.K. Optimization of die casting process based on Taguchi approach. *Mater. Today Proc.* **2017**, *4*, 1852–1859. [CrossRef]

62. Raza, Z.A.; Anwar, F. Fabrication of chitosan nanoparticles and multi-response optimization in their application on cotton fabric by using a Taguchi approach. *Nano Struct. Nano Obj.* **2017**, *10*, 80–90. [CrossRef]

63. Yang, Q.; Zhong, Y.; Zhong, H.; Li, X.; Du, W.; Li, X.; Chen, R.; Zeng, G. A novel pretreatment process of mature landfilleachate with ultrasonic activated persulfate: Optimization using integrated Taguchi method and response surface methodology. *Process Saf. Environ. Prot.* **2015**, *98*, 268–275. [CrossRef]

64. Ramli, N.A.; Wong, T.W. Sodium carboxymethylcellulose scaffolds and their physicochemical effects on partial thickness wound healing. *Int. J. Pharm.* **2011**, *403*, 73–82. [CrossRef] [PubMed]

65. Perumal, S.; Ramadass, S.K.; Madhan, B. Sol-gel processed mupirocin silica microspheres loaded collagen scaffold: A synergistic bio-composite for wound healing. *Eur. J. Pharm. Sci.* **2014**, *52*, 26–33. [CrossRef] [PubMed]

66. Liu, X.; Niu, Y.; Chen, K.C.; Chen, S. Rapid hemostatic and mild polyurethane-urea foam wound dressing for promoting wound healing. *Mater. Sci. Eng. C* **2017**, *71*, 289–297. [CrossRef] [PubMed]

67. Oyinade, A.; Kovo, A.S.; Hill, P. Synthesis, characterization and ion exchange isotherm of zeolite Y using Box–Behnken design. *Adv. Powder Technol.* **2016**, *27*, 750–755. [CrossRef]

68. Ba-Abbad, M.M.; Chai, P.V.; Takriff, M.S.; Benamor, A.; Mohammad, A.W. Optimization of nickel oxide nanoparticle synthesis through the sol-gel method using Box—Behnken design. *Mater. Des.* **2015**, *86*, 948–956. [CrossRef]

69. Tavakoli, J.; Tang, Y. Honey/PVA hybrid wound dressings with controlled release of antibiotics: Structural, physico-mechanical and in vitro biomedical studies. *Mater. Sci. Eng. C* **2017**, *77*, 318–325. [CrossRef] [PubMed]

70. Morgado, P.I.; Lisboa, P.F.; Ribeiro, M.P.; Miguel, S.P.; Simões, P.C.; Correia, I.J.; Aguiar-Ricardo, A. Poly(vinyl alcohol)/chitosan asymmetrical membranes: Highly controlled morphology toward the ideal wound dressing. *J. Membr. Sci. Technol.* **2017**, *159*, 262–271. [CrossRef]

71. Ritger, P.L.; Peppas, N.A. A simple equation for description of solute release II. Fickian and anomalous release from swellable devices. *J. Control Release* **1987**, *5*, 37–42. [CrossRef]

72. Ritger, P.L.; Peppas, N.A. A simple equation for description of solute release I Fickian and non-fickian release from non-swellable devices in the form of slabs, spheres, cylinders or discs. *J. Control. Release* **1987**, *5*, 23–36. [CrossRef]

73. Higuchi, W.I. Analysis of data on the medicament release from ointments. *J. Pharm. Sci.* **1962**, *51*, 802–804. [CrossRef] [PubMed]

74. Ghica, M.V.; Ficai, A.; Marin, Ş.; Marin, M.; Ene, A.M.; Pătraşcu, J.M. Collagen/bioactive glass ceramic/doxycycline composites for bone defects. *Rev. Rom. Mater.* **2015**, *45*, 307–314.

75. Paunică-Panea, G.; Ficai, A.; Marin, M.M.; Marin, S.; Albu, M.G.; Constantin, V.D.; Dinu-Pîrvu, C.; Vuluga, Z.; Corobea, M.C.; Ghica, M.V. New collagen-dextran-zinc oxide composites for wound dressing. *J. Nanomater.* **2016**, *2016*. [CrossRef]

76. Albu, M.G.; Ghica, M.V. Spongious collagen-minocycline delivery systems. *Farmacia* **2015**, *63*, 20–25.

Sample Availability: No samples of the compounds are available from the authors.

Article

Phase Behaviour and Miscibility Studies of Collagen/Silk Fibroin Macromolecular System in Dilute Solutions and Solid State

Ima Ghaeli [1,2,3,*], Mariana A. de Moraes [4,5], Marisa M. Beppu [4], Katarzyna Lewandowska [6], Alina Sionkowska [6], Frederico Ferreira-da-Silva [1,7], Maria P. Ferraz [8] and Fernando J. Monteiro [1,2,3,*]

[1]   i3S—Instituto de Investigação e Inovação em Saúde, Universidade do Porto, Rua Alfredo Allen, 208, 4200-135 Porto, Portugal; ffsilva@ibmc.up.pt
[2]   INEB—Instituto de Engenharia Biomédica, Universidade do Porto, Rua Alfredo Allen, 208, 4200-135 Porto, Portugal
[3]   FEUP, Faculdade de Engenharia, Departmento de Enginharia Metalurgia e Materiais, Universidade do Porto, 4200-465 Porto, Portugal
[4]   School of Chemical Engineering, University of Campinas, 13083-852 Campinas, Brazil; mamoraes@unifesp.br (M.A.d.M.); beppu@feq.unicamp.br (M.M.B.)
[5]   Department of Chemical Engineering, Federal University of São Paulo, 09913-030 Diadema, Brazil
[6]   Nicolaus Copernicus University in Toruń, Faculty of Chemistry, Department of Chemistry of Biomaterials and Cosmetics, ul. Gagarina 7, 87-100 Toruń, Poland; reol@chem.umk.pl (K.L.); as@chem.umk.pl (A.S.)
[7]   IBMC—Instituto de Biologia Molecular e Celular, Universidade do Porto, Rua Alfredo Allen, 208, 4200-135 Porto, Portugal
[8]   FP-ENAS/CEBIMED, University Fernando Pessoa Energy, Environment and Health Research Unit/Biomedical Research Center, 200-150 Porto, Portugal; mpferraz@ufp.edu.pt
*   Correspondence: ema.ghaeli@gmail.com (I.G.); fjmont@fe.up.pt (F.J.M.); Tel.: +351-220-408-800 (F.J.M.)

Received: 18 July 2017; Accepted: 16 August 2017; Published: 18 August 2017

Abstract: Miscibility is an important issue in biopolymer blends for analysis of the behavior of polymer pairs through the detection of phase separation and improvement of the mechanical and physical properties of the blend. This study presents the formulation of a stable and one-phase mixture of collagen and regenerated silk fibroin (RSF), with the highest miscibility ratio between these two macromolecules, through inducing electrostatic interactions, using salt ions. For this aim, a ternary phase diagram was experimentally built for the mixtures, based on observations of phase behavior of blend solutions with various ratios. The miscibility behavior of the blend solutions in the miscible zones of the phase diagram was confirmed quantitatively by viscosimetric measurements. Assessing the effects of biopolymer mixing ratio and salt ions, before and after dialysis of blend solutions, revealed the importance of ion-specific interactions in the formation of coacervate-based materials containing collagen and RSF blends that can be used in pharmaceutical, drug delivery, and biomedical applications. Moreover, the conformational change of silk fibroin from random coil to beta sheet, in solution and in the final solid films, was detected by circular dichroism (CD) and Fourier transform infrared spectroscopy (FTIR), respectively. Scanning electron microscopy (SEM) exhibited alterations of surface morphology for the biocomposite films with different ratios. Surface contact angle measurement illustrated different hydrophobic properties for the blended film surfaces. Differential scanning calorimetry (DSC) showed that the formation of the beta sheet structure of silk fibroin enhances the thermal stability of the final blend films. Therefore, the novel method presented in this study resulted in the formation of biocomposite films whose physico-chemical properties can be tuned by silk fibroin conformational changes by applying different component mixing ratios.

Keywords: biopolymers; protein-protein interaction; silk fibroin; miscibility; coacervation

---

1. Introduction

Miscible blending of two biopolymers with different physicochemical characteristics is an interesting route to produce new materials with unique properties that may present the advantages of each single polymer and compensate the disadvantages over each one. The films prepared from blends of natural polymers can potentially be used in wound healing and skin tissue engineering applications [1,2]. The presence of proteins as natural macromolecules in blends may improve cell adhesion, due to the presence of more protein binding sites [3]. However, native physical structures of proteins, such as collagen, with a linear triple helix, limits its possible intra- and interchain interactions in blends [4,5]. The forces found in protein interactions are electrostatic, Van der Waals, hydrogen bonds, and hydrophobic and steric interactions, of which the electrostatic interactions are predominant [6]. Parameters such as pH and ionic strength may affect electrostatic interactions, whereas temperature may have an influence on hydrophobic and hydrogen bindings [7,8]. However, temperature induces protein denaturation. For example, heating collagen induces the cleavage of the intermolecular hydrophobic and hydrogen bonds, transforming the collagen triple helix into a randomly coiled form, allowing fibril formation and interactions with other proteins [9]. This work has been focused on the blending of collagen and silk fibroin as two relevant biomaterials showing high potential to be used for production of protein-based biocomposite films.

Collagen, as a biomaterial, is a key player in biomedical applications. Collagen acts as a natural scaffold for cell proliferation and has adequate mechanical strength, good biocompatibility, biodegradability, and ability to promote cellular attachment and growth. In addition, different functional groups along the collagen backbone may promote the incorporation of growth factors and other biological molecules [10]. Preservation of collagen native structure in biomedical applications may be of importance, since the collagen triple helix network may withstand the mechanical stresses through transmitting the forces and dissipating energy [11]. Moreover, the triple helix characteristics, such as high stability in the biological environment, binding ligands for cell surface receptors, and essential signals to influence cell activity [12,13], highlight the importance of protecting such conformational integrity for functional applications.

On the other hand, silk fibroin presents several interesting properties, such as excellent toughness and stiffness combined with low density. The high mechanical strength of silk fibroin attracted the attention of several researchers. Even though silk fibroin has good mechanical properties, the biomedical applications require other desired properties, such as high water retention capability and biodegradability, which are absent in the native silk fibroin. Hence, it is adequate to increase the amount of amorphous structure of fibroin molecules by dissolving fibers, disrupting the hydrogen interactions and inducing the transitions of fibroin to random coil conformation that results in regenerated silk fibroin solution (RSF) [14]. Solubility of native silk fibroin depends on the organic salts, which participate in the disruption of hydrogen bonds. Foo et al. [15], revealed that the hydrophilic parts of silk fibroin, stabilized by Ca^{2+} or other ions, cause the aggregation of fibroin molecules into hydrophilic regions, forming a gel through ionic crosslinks. Water molecules absorbed by hydrophilic regions restrain the premature crystallization of the hydrophobic domains. The highly concentrated salt solutions of silk fibroin have to be dialyzed in order to remove the salts and make it adequate for the preparation of SF-based materials [16].

The mechanism of fibroin self-assembly during dialysis has been described by several researchers [17,18], and the influence of various parameters, such as concentration [19], temperature [20], and ethanol content [21] on RSF self-assembly have been analyzed. Jin et al. [17] suggested a micellar structure pattern for silk fibroin chains in water in which the small hydrophilic parts of silk fibroin remain hydrated, while the large termini hydrophilic parts are located at the outer edge of micelles and the hydrophobic parts are placed between those two hydrophilic blocks [17]. Hence, the RSF solution after dialysis is water-soluble and metastable until the hydrophobic parts of micelles bind and eventually form gels. Even though RSF is a promising material with specific biological and functional characteristics, its partially amorphous structure, along with its limited solubility, restrain the applications of this biomaterial. Hence, blending with other biomaterials is a useful way to improve the properties of RSF [19].

Several studies on collagen/silk fibroin blends showed not only the improvement in mechanical properties of the final materials, but also a favorable environment of the mixtures for cell attachment and proliferation. Nevertheless, the blends were restricted to low collagen concentration or COL/RSF ratios, or to be able to incorporate higher collagen ratios, high temperatures are required that denature the collagen triple helix and facilitate the hydrogen bonding between the two biopolymers [21–27]. However, inducing electrostatic interactions between these two biopolymers, which has not been used up to now, may avoid the risk of the undesirable denaturation of collagen as the result of increasing the temperature. Aiming at preserving the collagen natural structure, this study tries to present a new method based on electrostatic interactions, for templating protein-protein interactions between collagen and regenerated silk fibroin macromolecules, in dilute solutions and in solid thin films. Considering the importance of salt for electrostatic interactions, three single-phase compositions of COL/RSF, with 75/25, 50/50 and 25/75 volume ratios (v/v), were prepared according to their phase diagram and dialyzed against distilled water. Different blends were obtained and physico-chemically characterized after dialysis.

2. Results

2.1. Miscibility Study of Collagen/Silk Fibroin

In this study, ternary solvent containing salt ions was used in order to induce electrostatic interactions between collagen and RSF chains and obtain miscible or semi-miscible blends.

The borderline mixing ratio points of collagen/RSF/ternary solvent, were obtained by blending the different volumes of component, and the points were plotted in the ternary phase diagram as shown in Figure 1. The miscible solutions can be identified in the single phase region, while phase-separated solutions (which can be easily visually observed as the fibrils start to be formed) are indicated as the two phase region.

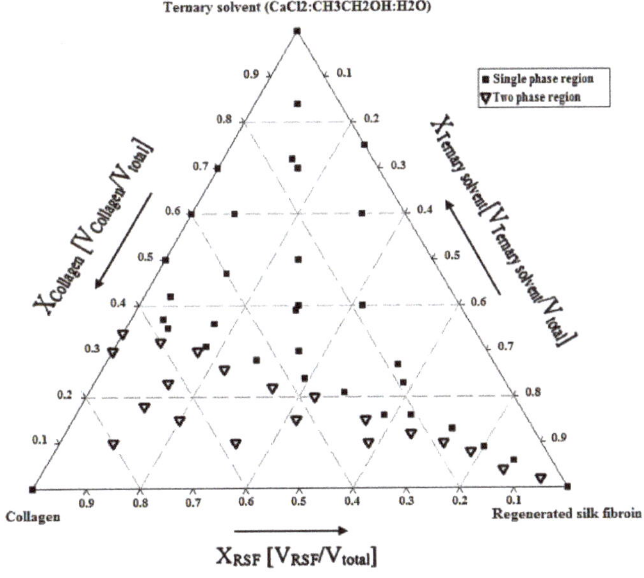

Figure 1. Ternary phase diagram of collagen/RSF/ternary solvent at 4 °C.

Regarding the polyelectrolyte nature of the involving proteins in this research, the electrostatic interaction between cationic amino groups of collagen and anionic groups of silk fibroin, and high calcium content, are the main cause of polyelectrolyte complex (PEC) formation.

The ζ potential measurements brought detailed understanding into the composition of charged groups in the mixture blends. Therefore, solutions with different mixing ratios according to the single-phase region in ternary phase diagram were measured and the results are shown in Table 1.

The net charge of all blends in the single phage region, is around zero, corresponding to almost electroneutrality of the solutions due to the high amount of salt. As shown in Table 1, the amount of ζ potential is negative for solutions containing more silk fibroin, which is attributed to higher amounts of silk chains with negative groups.

Table 1. ζ potential of collagen/RSF blends.

Sample	Zeta Potential (mV)
Collagen solution in acetic acid (0.5%)	39.06
Col/RSF: 75/25	1.41
Col/RSF: 50/50	3.12
Col/RSF: 25/75	−0.497

The quantitative analysis of miscibility by viscosimetry has been conducted by calculating the miscibility parameter (Δb_m) ,as well as relative and reduced viscosities, both theoretically (by methods of Krigbaum and Wall [28], and Garcia et al. [29]), and experimentally, and plotted against solution concentration.

The theoretical and experimental values for pure collagen, pure silk fibroin and their blends, according to the single-phase region of ternary phase diagram, are shown in Table 2. The positive miscibility parameter for all the blends indicates good miscibility for all prepared blends, which is due to the electrostatic interactions between chains and calcium ions, making the whole mixture more stable.

Table 2. Theoretical (by Krigbaum and Wall [28] and Garcia et al. [29] methods) and experimental values for pure collagen, pure silk fibroin and the mixtures.

W_{Collagen} (0.5%)	$[\eta]_m^{exp}$ (dL/g)	$[\eta]_m^{id}$ (dL/g)	$\Delta[\eta]_m$	b_m^{exp} (dL/g)^2	$b_m^{id^*}$ (dL/g)^2	$\Delta b_m{}^*$	$b_m^{id^{**}}$ (dL/g)^2	$\Delta b_m{}^{**}$
1 ($W_{\text{silk fibroin}}$:0)	2.48			38.44				
0.75	6.47	2.06	4.41	72.41	30.04	42.37	21.64	50.77
0.5	3.54	1.94	1.6	75.91	20.87	55.04	9.67	66.24
0.25	1.78	1.23	0.55	39.54	10.94	28.6	2.54	37.0
0 ($W_{\text{silk fibroin}}$:1)	0.81			0.2433				

$b_m^{id^*}$: determined according to Krigbaum and Wall [28]; $b_m^{id^{**}}$: determined according to Garcia et al. [29].

In a highly soluble polyelectrolyte mixture containing high amount of salt, the individual chains of each polyelectrolyte are separated from each other, with salt ions placed among them that yields to the rising of solution viscosity [30]. Increasing salt concentration in the complex system leads to the screening of the electrostatic interactions between two macromolecules, as well as the rearrangement of the polymer chains that may raise the viscosity of solution [31].

Figure 2 shows the reduced viscosity versus the concentration for pure collagen, pure silk fibroin and their blends. Silk fibroin solution has the lowest viscosity of all the solutions due to the solvation of fibroin chains by calcium ions of the ternary solvent. The viscosity of the primary mixtures with 25%, 50% and 75% of collagen (without adding the dilution solvent), is higher than the viscosity of pure silk fibroin and pure collagen.

Mixtures with more collagen content have higher ionic strength and viscosity. Therefore, initial mixtures with 75% and 50% collagen were more viscous than the others. Additionally, the reduced viscosities of mixtures with 50% and 75% collagen show close values, both above the values for other solutions, illustrating the high ionic strength in these solutions.

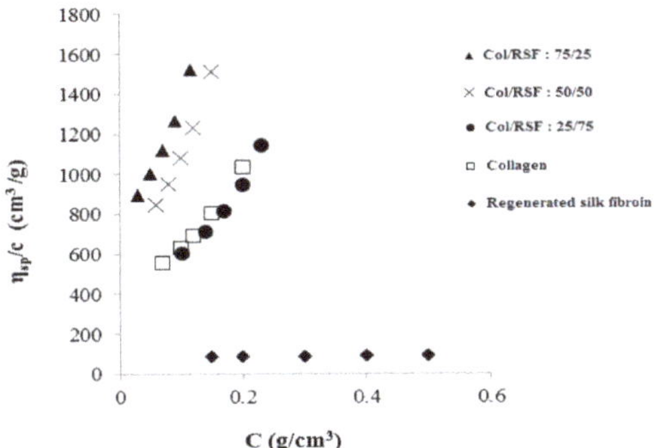

Figure 2. Reduced viscosity versus concentrations of collagen/RSF solutions.

However, adding NaCl solution as the diluting solvent to the aqueous system of proteins containing high calcium ions increases the probability of gradual precipitation after second time dilution. The imbalance of salt concentration among protein chains and the outside medium through dilution causes the calcium diffusion towards the medium by osmotic pressure. Moreover, the possibility of Ca^{2+} ion displacement by Na^+, in the sites requiring the ions in a dehydrated state, alters the conformation of binding sites, which is due to the kosmotropic behavior of NaCl, together with the prevention of binding sites by calcium ions [32]. In view of these considerations, the precipitation caused by dilution can be controlled by careful selection of the added volume of the NaCl solution.

2.2. Phase Change after Collagen/Silk Fibroin Dialysis

Slow diffusion of calcium salt through dialysis procedure, during three days, changes the phase behavior of mixtures from a homogeneous solution to a coacervate or a precipitate, depending on the degree of neutralization. The phase behavior after the last day of dialysis, when only residual salt amount may be present, shows a solid-liquid phase separation with formation of white solid complex coacervates or precipitates, for all the mixtures, which could be easily identified by naked-eye and optical microscope. This solid-liquid phase separation occurs because the removal of salt ions through the dialysis membrane increases the interaction of proteins COO^- and NH^{3+} ionic groups.

Figure 3 shows optical stereoscopic microscope images of blend mixtures after dialysis for three days. It should be noted that all the mentioned ratios correspond to those before dialysis procedure. As shown in Figure 3, the aggregates size increases with increasing amount of silk fibroin, while fibril formation occurred in all the mixtures.

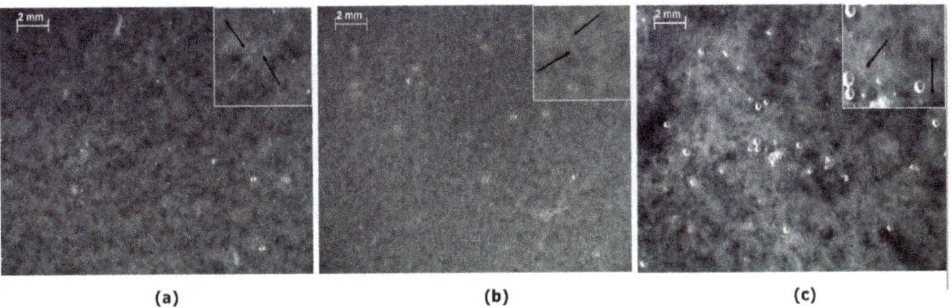

Figure 3. Optical microscope images of blend solutions after dialysis with the starting ratios (before dialysis) of: (**a**) Col/RSF: 75/25; (**b**) Col/RSF: 50/50; and (**c**) Col/RSF: 25/75. Arrows indicates fibril formation in the system.

To study the effect of dialysis (salt removal) on coacervate complexes, ζ potential has been used, and the results for the solutions after dialysis are presented in Table 3. The results show that dialysis procedure increased the charge density of solutions, leading to stronger attractive interactions among oppositely charged polyelectrolytes. The positive charge density for all the mixtures indicates the excess of NH^{3+} groups, which increases with increasing the collagen ratio. However, the ζ potential for silk fibroin rich mixture is lower than the other ones, which can be related to higher amounts of silk fibroin negative charges in the mixture.

Table 3. ζ potential of collagen/RSF blends after dialysis (the ratios are those before dialysis).

Sample	Zeta Potential (mV)
Silk fibroin after dialysis	−5.96
Col/RSF: 75/25 (after dialysis)	10.26
Col/RSF:50/50 (after dialysis)	9.03
Col/RSF:25/75 (after dialysis)	5.36

The ζ potential results are in agreement with optical images (Figure 3), proving that blend solutions of 75% and 50% collagen with high charge densities (Figure 3a,b), contain aggregates possessing charge-charge repulsion which inhibits their assembly. However, a lower charge density of solution with a higher silk fibroin ratio yields less repulsion and more aggregate assembly due to hydrophobic interactions (Figure 3c).

Finally, in order to confirm that the coacervate particles contain the complex of proteins, all the mixtures after dialysis procedure (as shown in Figure 3) have been passed through the gas and vacuum filters. Thereafter, the remaining solutions after filtration have been passed through two calibrated markers of the Ubbelhode viscosimeter, and the time has been measured. The RSF solution after dialysis with the reduced concentration from 0.5% to 0.17%, due to the dilution during three days of dialysis, showed a passing time of 111.16 s (without any filtration). Table 4 shows that, during three days of dialysis, all the aqueous solutions remaining after filtration present a passing time s close to that of distilled water, illustrating that the coacervate particles remaining behind the filters were protein complexes.

Table 4. Passing time (s) of the collagen/RSF solutions after filtrations (all the ratios are those before dialysis).

Dialysis Days	Col/RSF:75/25	Col/RSF:50/50	Col/RSF: 25/75	Water
Day 1	49.22 (s)	50.94 (s)	51.38 (s)	46.58 (s)
Day 2	48.59 (s)	49.32 (s)	49.95 (s)	46.58 (s)
Day 3	47.21 (s)	48.26 (s)	49.65 (s)	46.58 (s)

Slightly higher values of solution passing time than that of water may be due to the presence of remaining calcium ions in solution after interaction of collagen/RSF. Studies on the effect of lithium ions on silk fibroin films [33] confirmed that even after 72 days of dialysis, the salt ions could not be removed completely. The passing time is higher for the first day and decreases on the second and third days, respectively. After release of high salt levels during the first day, the gradual release during the second and third days is a combination of concentration gradient and the result of gradual electrostatic and hydrophobic interactions between the two polymers as well as the fibroin fibrillation. Due to the not-fully neutralized coacervates, these calcium ions may be electrostatically weakly bonded to the oppositely-charged free residues of the proteins inside the dialysis tube and, therefore, have more tendency to stay inside the dialysis tube rather than being removed to the water bath. However, more detailed studies are required to assess the amount of salt and non-mixed proteins in the remaining solutions after filtration.

2.3. Structural Characteristics of Collagen/Silk Fibroin Blend Solutions after Dialysis and Solid Films

After preparation of the blended films through drying of solutions at room conditions, their structures were analyzed via SEM, FTIR, DSC, and contact angle.

As can be seen in Figure 4, SEM images show a rough surface for collagen (Figure 4a) while a smooth surface for RSF (Figure 4e). According to the Col/SF ratios, significant changes in film surface morphology could be observed upon mixing RSF with collagen. Blended film with more collagen (Figure 4b) showed a surface similar to the collagen film (Figure 4a). However, the surface of the blended film with more silk fibroin (Figure 4d) has less roughness than the other ones. The fibrous-like structure could be observed for Col/RSF: 50/50 (Figure 4c), which confirms the previous results of optical microscopy (Figure 3).

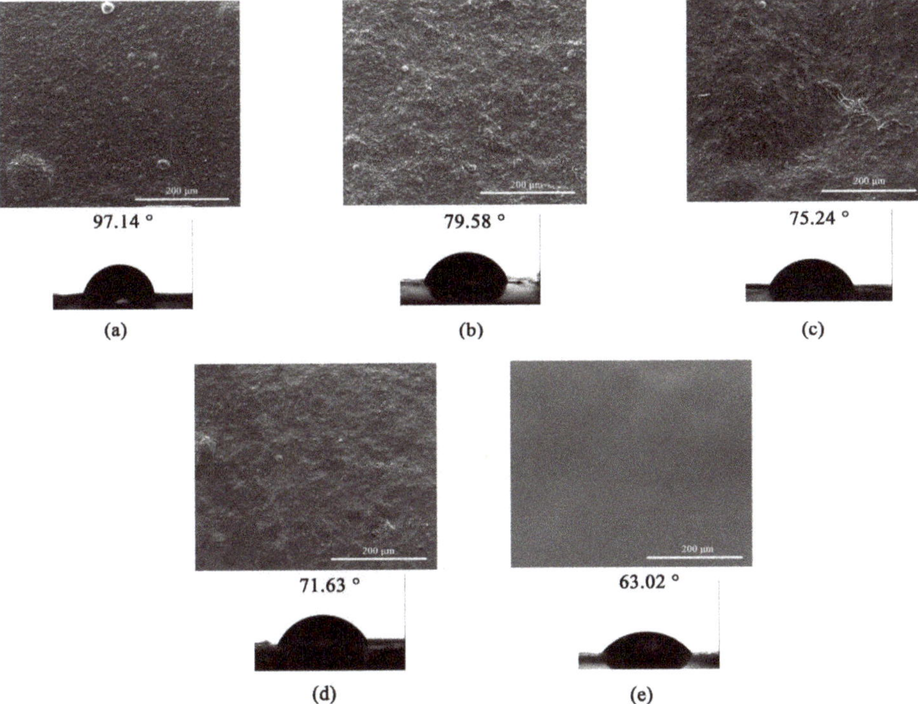

Figure 4. Scanning electron microscopy (SEM) and water contact angle images of blended films: (**a**) Collagen; (**b**) Col/RSF: 75/25; (**c**) Col/RSF: 50/50; (**d**) Col/RSF: 25/75; and (**e**) RSF at 500× magnification.

Moreover, the contact angle measurements of the films' surfaces shown in Figure 4 revealed that the RSF film had the highest hydrophilicity, and the hydrophilicity of blended films were in the range between pure silk fibroin and pure collagen. The results indicated that the hydrophilicity of the blended films improved with increasing the regenerated silk fibroin proportions. The observation demonstrated that the regenerated silk fibroin has significant influence on the wettability of the blended surfaces, which can be explained by the entanglement of RSF molecules and the exposure of their hydrophilic groups, which may be arranged on the surface of RSF chains [34].

Figure 5 present the DSC curves of the prepared films. Collagen presents an endothermic peak at around 57 °C, attributed to the evaporation of unbounded water molecules and the denaturation of collagen fibrils [35]. However, the very small endothermic peak at around 308 °C corresponds to the breaking of hydrogen bonds between alpha chains and to collagen transformation from the triple helix to the random coil structure [36]. Regenerated silk fibroin presents a tiny endothermal peak at around 62 °C, which is related to the loss of unbound water molecules. The second endothermal peak is at around 330 °C, attributed to the thermal degradation of silk fibroin chains.

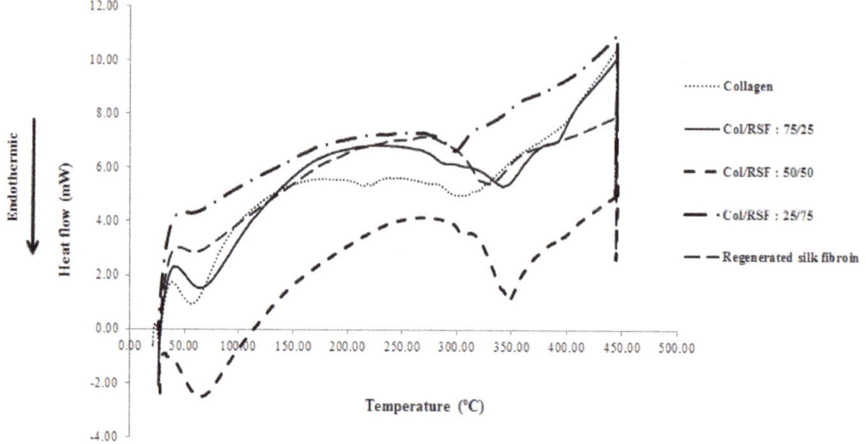

Figure 5. DSC curve of collagen/RSF blend films.

The absence of denaturation peak in the blend film containing higher silk fibroin ratio, shows the persistence of silk fibroin beta sheet structure in protection of collagen from denaturation through presumable covering of collagen chains while limiting the space for collagen triple helix motion. However, the existence of denaturation peak for other mixtures indicated that the collagen triple helix structure is maintained in those blend films. Thermal denaturation peaks in such blends shifted towards higher temperatures (65–67 °C) compared to collagen samples, demonstrating higher stability of the collagen triple helix. Thermal degradation of collagen/RSF mixtures with ratios of 75/25, 50/50, and 25/75 occurred at 342 °C, 348 °C (sharp peak), and 298 °C (small peak), respectively. The sample with higher silk fibroin ratio shows a tiny endothermic peak which is attributed to the molecular motions of alpha helix chains within the small amorphous regions. However, the absence of degradation peaks illustrates dominating silk fibroin beta sheet conformation. Nevertheless, mixtures with 50% and 75% collagen, show larger decomposition peak at higher temperatures, indicating that thermal decomposition is the sum of heat adsorbed to degrade the hydrogen bonds in both the collagen triple helix and the beta sheet structure of silk fibroin. In addition to the decomposition at higher temperature for the mixture with 50% collagen, the large observed decomposition peak may be the proof of the existence of more amorphous regions with alpha helix structures. Overall, decomposition of mixtures at higher temperatures than that of silk fibroin can be due to the beta sheet structures of silk fibroin in the blend films, along with higher thermal stability.

Figure 6 shows FTIR spectra of collagen/RSF blend films, as well as the individual polymer films. As reported in the literature, the spectral properties of silk fibroin showed two different structures of silk I and silk II, which are known to be rich in helical and beta-sheets, respectively. The spectral ranges of amide I (C=O and C–N stretching), amide II (N–H bending), and amide III (C–N stretching) are reported as 1655–1660 cm^{-1}, 1531–1542 cm^{-1}, and 1230 cm^{-1} for silk I, 1620–1630 cm^{-1}, 1515–1530 cm^{-1} and 1240 cm^{-1} for silk II, and 1640–1648 cm^{-1}, 1535–1545 cm^{-1}, and 1235 cm^{-1} for random coil structures [37,38]. Collagen amide I has been separated into three component peaks, including 1628–1633 cm^{-1} for hydroxyproline [39]. Additionally, the amide II bands [40] are presented at around 1550 cm^{-1}, and amide III bands [41] at around 1240 cm^{-1}.

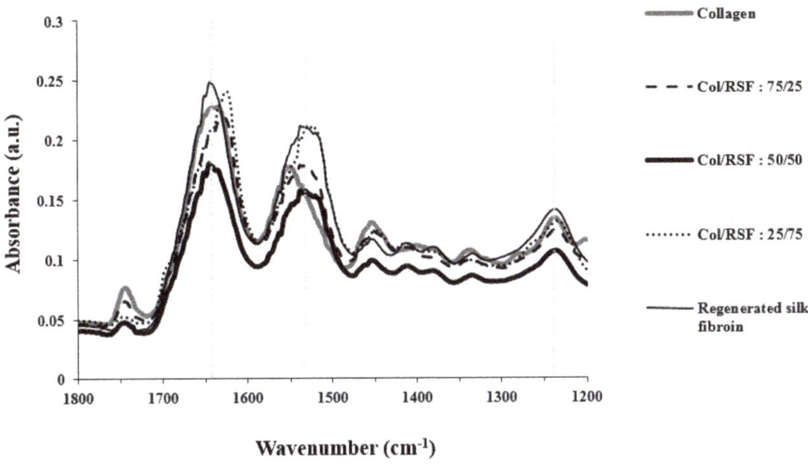

Figure 6. FTIR spectra of collagen/RSF films.

Amide I is the important peak in characterizing conformational changes. As shown in Figure 6, the position of amide I peaks of sample with 50/50 ratio remained unchanged for the fibroin molecules in the amide I region, showing the predominant contribution of silk fibroin with random coil chains in the mixture. Increasing collagen ratio to 75%, induced silk fibroin β-sheet structure as the amide I peak shifted to 1626 cm^{-1}. However, the unchanged position of amide II at 1531 cm^{-1} is consistent with the existence of some alpha helix structures of silk fibroin in the mixture. Blends with higher amounts of silk fibroin showed the amide I peak at 1623 cm^{-1}, while the amide II peak appeared near 1525 cm^{-1}, indicating that the blended films contain mostly crystalline beta sheets that may be attributed to the fibrillogenesis of silk fibroin. Table 5, summarizes the assignment of the major IR peaks for each polymer and their blend films.

Table 5. The FTIR band assignments of collagen/RSF blends.

	Amide I	Amide II	Amide III
Collagen	1634	1551	1238
Col/RSF: 75/25	1626	1531	1236
Col/RSF: 50/50	1644	1531	1237
Col/RSF: 25/75	1623	1525	1234
RSF	1644	1531	1237

Circular dichroism (CD) analysis was performed in order to investigate and affirm conformational transition of silk fibroin in the blended solutions after dialysis. As shown in Figure 7, CD spectrum of pure silk fibroin after dialysis showed typical random coil structure with a negative peak near 195 nm. For pure collagen, it was observed triple helix characteristic spectrum with a large negative peak near 197 nm, as well as a small positive peak centered at 220 nm.

Considering the persistence of natural collagen structure in the mixtures that was confirmed by DSC analysis and in order to obtain the spectra for regenerated silk fibroin when blended with collagen, the spectrum of the later was subtracted from the spectra measured for the blended solutions. As shown in Figure 7, the spectra of silk fibroin in the mixtures show a negative peak between 210 and 220 nm, and a positive peak between 195 and 200 nm, characteristic features of beta sheet conformations [42]. This indicates that silk fibroin structure changes toward beta sheet conformation due to the interaction with collagen.

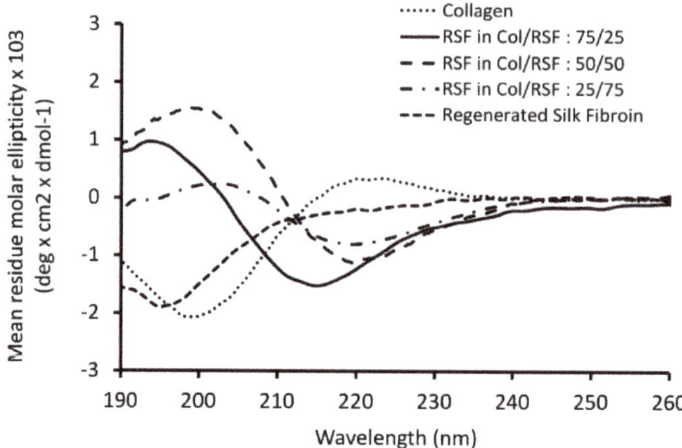

Figure 7. Circular dichroism (CD) spectrum of collagen/RSF mixtures after dialysis.

3. Discussion

Direct mixing of collagen and silk fibroin solutions resulted in phase separation, possibly due to the different pH values of silk fibroin solution (with pH of 7.17) and collagen solution (with pH of 2.74). The low collagen solution pH causes the protonation of carboxyl groups on silk fibroin to their non-ionic form, and amine groups to their cationic form. This is along with increasing the hydrophobicity of uncharged carboxyl groups with subsequent induction of silk fibroin morphological changes in the solution from the spherical micelles to nanofibrils. Additionally, conformational transition from random coil to β-sheet may occur [43–45]. Using higher temperatures (50–60 °C) as a way to prompt the interactions of collagen with other proteins through denaturating collagen molecules [22] is not the aim of this study, since the aim is the preservation of the collagen's native structure. Hence, collagen and RSF solutions were mixed, using calcium salt ions of the ternary solvent for inducing electrostatic interaction between biopolymer chains.

The chaotropic behavior of divalent calcium ions, under a phenomenon called "salting in", causes more ion-protein interactions than protein-protein interactions in the blend system, yielding a homogenous mixture [46]. The stability of the mixtures in the single-phase region is due to the quasi-equilibrium between oppositely-charged proteins (collagen and fibroin) and the introduction of a ternary solvent containing high amounts of salt which acts both as the third component of the phase diagram and a simple electrolyte.

In a system containing oppositely-charged proteins (considering their polyelectrolyte nature), increasing the ionic strength through adding salt to some extent, enhances the attraction between the oppositely-charged residues and causes a transition of an overcharged polyelectrolyte complex (PEC) to a neutral or uncharged complex [47,48]. Moreover, calcium ions with a radius of 4.1 Å in the hydrated state fit well to the distance of ~14 Å between adjacent triple helical molecules of collagen, probably interacting with the negatively-charged Asp or Glu side chains, forming salt bridges, and increasing the ionic strength of the solution [49–51].

Slow salt diffusion through the dialysis procedure changes the phase behavior of mixtures and causes coacervation or precipitation. Based on the results obtained from optical microscope images (Figure 3) and ζ potential analysis of the mixtures after dialysis, and considering the influence of proteins conformational structures on the aggregate formation, we hypothesized a model for mixtures with different ratios of collagen/silk fibroin (Figure 8).

In the mixtures containing collagen, it has been proved that the counterions were released to the media upon complexation with collagen [52]. Hence, in this research, at low ionic strength, the water and counterions may be released to the media upon the formation of hydrophobic and hydrogen bonds between the SF and collagen chains. Moreover, according to previous studies on silk fibroin mixtures [53], beta sheet conformation of silk fibroin showed by CD and FTIR analysis in this research may be due to the fact that upon complexation with collagen, silk fibroin chains may use collagen chains as a mold plate to stretch themselves. Through the process of the nucleation-dependent aggregation mechanism, once the beta sheet nucleus is formed, further growth of beta sheet units and beta sheet aggregation [54] will occur (Figure 8c).

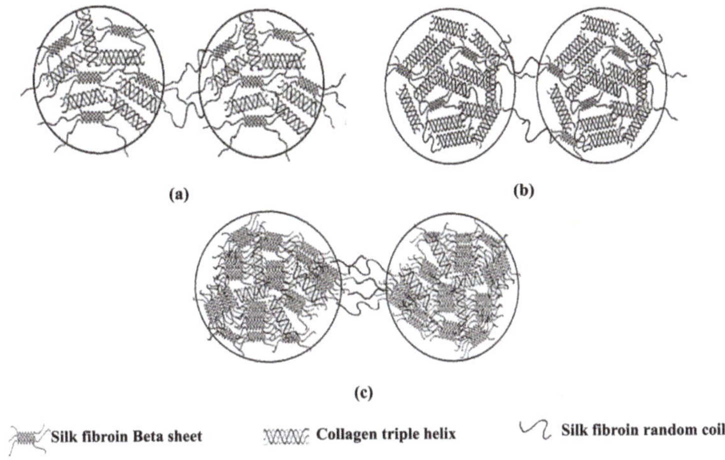

Figure 8. Schematic diagram of a hypothesized model for protein conformational changes in the two adjacent coacervate aggregates of collagen/RSF mixtures after dialysis with different starting ratios of: (a) Col/RSF: 75/25; (b) Col/RSF: 50/50; and (c) Col/RSF: 25/75.

On the other hand, the replacement of salt ions by water molecules (hydration) during dialysis causes the assembly of free collagen molecules, which probably happened in the mixtures with higher amounts of collagen (Figure 8b). Due to the specific hydroxyproline (Hyp) groups of collagen chains, the water molecules order around and between the collagen chains in a way that the triple helixes' organization forms a crystal packing. In these crystals, the triple helixes are bridged by ordered water molecules, causing the collagen assembly that induces collagen molecules aggregation or fibrillogenesis [55].

Overall, the characterization methods used in this study are simple techniques for primary assessment of phase behavior in collagen/silk fibroin blends. Further investigation can be done on ternary phase diagram of collagen/RSF/ternary solvent in order to obtain a boundary of miscibility/coacervate/precipitate, through application of other methods such as turbidimetry and potentiometric titration. Additionally, the size, shape, and inner composition characteristics of the precipitates can be further studied by small angle X-ray technique (SAXS) or WAXS (wide angle X-ray).

4. Materials and Methods

4.1. Sample Preparation

Collagen solution was prepared by dissolving type I bovine collagen (Bovine Achilles tendon, Sigma-Aldrich, St. Louis, Missouri, USA) in 0.5 mol/L acetic acid to the final concentration of 0.5% (*w/v* %) and stirring with high speed Turrax (T25D, IKA®, Janke and Kunkel IKA-Labortechnik, Staufen, Germany) at 10,000 rpm and 4 °C for 2–3 h.

B. mori silk fibroin was prepared by initially degumming by boiling silkworm cocoons in Na_2CO_3 solution at 1 g/L for 30 min at 85 °C. This procedure was repeated two times and, finally, the cocoons were boiled in distilled water for 30 min in order to separate the glue-like sericin from fibroin. After adequate washing with distilled water, the obtained silk fibroin fibers were dried at room temperature. Finally, silk fibroin fibers were milled to facilitate their dissolution process. Silk fibroin was dissolved in a solution of ternary solvent containing $CaCl_2$:CH_3CH_2OH:H_2O (1:2:8 molar ratio), at 85 °C in order to obtain the final concentration of 0.5% (*w/v* %). The fibroin solution was dialyzed against distilled water for 72 h.

4.2. Ternary Phase Diagram and Blend Preparation

Considering the key role of salt ions for inducing the ionic interactions between protein chains in solvent, the ternary phase diagram of collagen, RSF and ternary solvent at 4 °C (in order to prevent collagen denaturation) was analyzed. Aiming at obtaining different Col/RSF ratios of 100/0, 75/25, 50/50, 25/75 and 0/100, collagen/RSF/ternary solvent blends were prepared by selecting the relevant points in single-phase region of ternary phase diagram. Identifying volume fractions of each component ($X_{component}$ = $V_{component}$/V_{total}), the single-phase blends were prepared, at 4 °C. Later, the blended solutions were dialyzed against distilled water for three days. The resultant solutions were dried in polystyrene dishes at room temperature to obtain the blended films in the solid state.

4.3. Miscibility and ζ-Potential Analysis

In order to assess the miscibility, the viscosity behavior of mixtures in the single-phase region of ternary phase diagram, was analyzed at 25 ± 0.1 °C by Ubbelohde capillary viscometer (NCU Laboratory, Toruń, Poland). According to the intended blend ratios, different mass fractions of each polymer solution have been mixed. The intrinsic viscosity and the viscosity interaction parameter of each polymer solution as well as the ternary systems (collagen/RSF/ternary solvent) were obtained and used to estimate the miscibility of polymer mixtures through classical dilution method. Thus, each solution was prepared and diluted with NaCl (0.1 mol/L) to yield lower concentrations. The relative viscosities of blends were obtained by dividing solutions flow times by the value found for pure solvent (NaCl 0.1 mol/L). The intrinsic viscosity, the interaction parameter and Huggins coefficient values were determined according to the Huggins equation using solutions of several concentrations.

The values of experimental interaction parameter for all the blends, using diluted regime with solutions of 5 concentrations, were obtained from the plot of η_{sp}/c vs c (mass concentration, in g/mL) using Equation (1):

$$\frac{(\eta_{sp})_m}{c_m} = [\eta]_m^{exp} + b_m^{exp} c_m \tag{1}$$

where $(\eta_{sp})/c$ is the reduced viscosity, b_m^{exp} is the experimental viscosity interaction parameter of polymer mixture, $[\eta]_m^{exp}$ is the experimental intrinsic viscosity of the polymer blends, and c_m is the total concentration of solution.

The ideal values in this study were determined according to the Krigbaum and Wall [28], and Garcia et al. [29] techniques. The ideal interaction parameter of b_m^{id} have been determined by Krigbaum and Wall through Equation (2):

$$b_m^{id*} = b_A w_A^2 + b_B w_B^2 + 2b_{AB}^{id} w_A w_B \tag{2}$$

where w_A and w_B are the weight fractions of polymers A and B, respectively, and b_A and b_B are the interaction parameters of each individual polymer which were obtained from the slope of the plots of the reduced viscosity versus concentration. b_{AB}, the interspecific interaction parameter that was obtained by Equation (3):

$$b_{AB}^{id} = b_A^{1/2} b_B^{1/2} \tag{3}$$

The ideal interaction parameter by Garcia et al. [29], was calculated through Equation (4):

$$b_m^{id**} = b_A w_A^2 + b_B w_B^2 \tag{4}$$

The polymer mixture is miscible if $\Delta b_m = b_m^{exp} - b_m^{id} \geq 0$ and immiscible if $\Delta b_m = b_m^{exp} - b_m^{id} \geq 0$.

This viscosimetry analysis was done for all the mixtures after dialysis procedure. Thus, the solutions after dialysis have been passed through the gas and vacuum filters. The remaining solutions after filtration have been passed through two calibrated markers of the Ubbelhode viscometer, and the time has been measured.

The ζ-potential analysis was done for all solutions before and after dialysis through measuring the electrophoretic mobility by ZetaPALS (Brookhaven Instruments Corporation, Holtsville, NY, USA), as the temperature was maintained at 25 °C.

4.4. Light Stereoscopic Magnifier Microscope

The blend mixtures after dialysis procedure were observed and optical images were collected from the Leica EC3 stereo microscope equipped with Leica LAS software (Leica, Wetzlar, Germany). The mixtures in their containers were placed on a stage with a dark base and the pictures of the blend solutions were captured using the digital camera zoomed onto the solutions as closely as possible.

4.5. Structure of Collagen/Silk Fibroin Blended Solutions and Films after Dialysis

The structures of blended solutions after dialysis, were analyzed by circular dichroism (CD) using a J-815 (Jasco, Tokyo, Japan) spectrometer. Far-UV CD spectra were recorded between 190 and 260 nm using a 1 mm path length cuvette. CD spectra were acquired with a scanning speed of 100 nm/min, integration time of 1 s, and using a bandwidth of 1 nm. The spectra were averaged over eight scans and corrected by subtraction of the buffer signal. Spectra of silk fibroin in the mixtures were obtained by subtraction of pure collagen (Col/RSF 100/0 blend) spectrum from the mixture spectra. The results are expressed as the mean residue ellipticity θ_{MRW}, defined by Equation (5):

$$\theta_{MRW} = \theta obs(0.1MRW)/(lc) \tag{5}$$

where θobs is the observed ellipticity (mdeg), MRW is the mean residue weight (g/mol), c is the concentration (mg/mL), l is the light path length (cm), and θ_{MRW} is the mean residue ellipticity (deg.cm^2/dmol). MRW was calculated from data (MW and number of aminoacids) in the UniProt database as 76.1 for silk fibroin and 94.9 for collagen.

The structures of blended films were assessed by scanning electron microscopy (SEM), contact angle measurements, Fourier transform infrared spectroscopy (Perkin-Elmer 2000 FTIR spectrometer, Hopkinton, MA, USA), and differential scanning calorimetry (Setaram DSC 131, Caluire-et-Cuire, France).

For scanning electron microscopy (SEM), samples were coated with an Au/Pd thin film, by sputtering, using an SPI module sputter coater. The SEM analysis was performed using a high resolution (Schottky) environmental scanning electron microscope (FEI Quanta 400 FEG ESEM, Hillsboro, OR, USA). In order to analyze surface hydrophobicity, contact angle measurements were done using a digital imaging capture system (OCA 15, DataPhysics Instruments GmbH, Filderstadt, Germany). For this reason, the sessile drop method with distilled water at 25 °C was used and the contact angle was calculated using software version SCA 20. Thermal analysis of the prepared films has been performed using a Setaram DSC 131, from 25–450 °C at a scan rate of 10 °C/min with nitrogen flow of 50 mL/min. Moreover, FTIR analysis were carried out in the spectral range of 400–4000 cm^{-1} using a Perkin Elmer FTIR spectrophotometer model 2000, equipped with an ATR diamond cell accessory.

5. Conclusions

In this work, we studied the influence of salt ions on the phase behavior of collagen and silk fibroin mixtures. The results showed that ternary solvent containing calcium ions, as the third component in the ternary phase diagram, has a significant effect on the miscibility of mixtures. Its influence on implementing net charge density close to zero, making an almost electroneutral blend solution, was confirmed through the analysis of the observed ζ potential values. In such a system, the chains of collagen and silk fibroin should be in their most individual separated state with a high concentration of salt ions being placed among their chains. Moreover, the viscosity analysis reaffirmed the influence of ternary solvent to obtain high miscibility for all the blends. However, the maximum values were obtained for the mixture with the highest collagen ratio, due to the high ionic strength resulting from the higher amount of salt used in this mixture.

Removal of salt after the dialysis procedure yielded to complex coacervations (precipitation) with positive charge densities, demonstrating the importance of protein charge density and conformational structure. The protein-protein coacervate aggregates were formed as the oppositely charged protein chains got closer to each other and, after primary weak electrostatic interaction, the hydrogen and hydrophobic bonds were formed, leading to silk fibroin conformational changes from random coil to beta sheet. The conformational change of silk fibroin was reassured through CD spectra analysis of the solution after dialysis, as well as SEM, contact angle, DSC, and FTIR assessment of formed films after drying of dialyzed solutions.

Considering the physico-chemical changes of blended films, analyzed by SEM, contact angle, and DSC, it can be stated that the obtained blended films have tunable properties by varying the blend ratio. Overall, such preparation procedure for collagen/silk fibroin blends is a promising method for engineering of protein-protein interactions, which is important for the development of a wide range of biopharmaceutical applications of these materials, from drug delivery, to wound healing and tissue engineering.

Acknowledgments: This work was supported by FEDER funds through the Programa Operacional Factores de Competitividade (COMPETE) (POCI/01/0145/FEDER/007265) and by Portuguese funds through FCT (Fundação para a Ciência e a Tecnologia) under the Partnership Agreement PT2020 UID/QUI/50006/2013. Part of this work was funded by European Cooperation in Science and Technology (COST; Action MP1301). Furthermore, the support of Fernão Magalhães and LEPABE/ FEUP with DSC tests is greatly acknowledged.

Author Contributions: I.G., in cooperation with M.A.M., performed and analyzed the experiments and wrote the article; M.M.B. helped in planning the experiments and editing the manuscript; F.J.M. and M.P.F., supervised the work and executed the article editing; A.S. and K.L. provided access to the laboratory of viscometer and assisted with the experiments, analysis the data, and editing of the article; and F.F.-S. performed the CD measurements, analyzed the results, and edited the article.

References

1. Paichit, I.; Atchariya, F.; Anan, O.; Anuphan, S.; Jarupa, V. Effects of the blended fibroin/aloe gel film on wound healing in streptozotocin-induced diabetic rats. *Biomed. Mater.* **2012**, *7*, 035008.
2. Gu, Z.; Xie, H.; Huang, C.; Li, L.; Yu, X. Preparation of chitosan/silk fibroin blending membrane fixed with alginate dialdehyde for wound dressing. *Int. J. Biol. Macromol.* **2013**, *58*, 121–126. [CrossRef] [PubMed]
3. Silva, S.S.; Santos, M.I.; Coutinho, O.P.; Mano, J.F.; Reis, R.L. Physical properties and biocompatibility of chitosan/soy blended membranes. *J. Mater. Sci. Mater. Med.* **2005**, *16*, 575–579. [CrossRef] [PubMed]
4. Katz, E.P.; David, C.W. Energetics of intrachain salt-linkage formation in collagen. *Biopolymers* **1990**, *29*, 791–798. [CrossRef] [PubMed]
5. Katz, E.P.; David, C.W. Unique side-chain conformation encoding for chirality and azimuthal orientation in the molecular packing of skin collagen. *J. Mol. Biol.* **1992**, *228*, 963–969. [CrossRef]
6. Walker-Taylor, A.; Jones, D.T. *Computational methods for predicting protein-protein interactions, In Proteomics and Protein-Protein Interactions: Biology, Chemistry, Bioinformatics, and Drug Design*; Waksman, G., Ed.; Springer: Boston, MA, USA, 2005; pp. 89–114, ISBN 978-0-387-24531-7.

7. Zhang, J. Protein-protein interactions in salt solutions. In *Protein-Protein Interactions-Computational and Experimental Tools*; Cai, W., Hong, H., Eds.; INTECH Open Access: Rijeka, Croatia, 2012; ISBN 978-953-51-0397-4.

8. Ross, P.D.; Rekharsky, M.V. Thermodynamics of hydrogen bond and hydrophobic interactions in cyclodextrin complexes. *Biophys. J.* **1996**, *71*, 2144–2154. [CrossRef]

9. Rochdi, A.; Foucat, L.; Renou, J.P. Effect of thermal denaturation on water-collagen interactions: NMR relaxation and differential scanning calorimetry analysis. *Biopolymers* **1999**, *50*, 690–696. [CrossRef]

10. Gelse, K.; Pöschl, E.; Aigner, T. Collagens—Structure, function, and biosynthesis. *Adv. Drug Deliv. Rev.* **2003**, *55*, 1531–1546. [CrossRef] [PubMed]

11. Silver, F.H.; Landis, W.J. Viscoelasticity, energy Storage and transmission and dissipation by extracellular matrices in vertebrates. In *Collagen: Structure and Mechanics*; Fratzl, P., Ed.; Springer: Boston, MA, USA, 2008; pp. 133–154, ISBN 978-0-387-73906-9.

12. Knight, C.G.; Morton, L.F.; Peachey, A.R.; Tuckwell, D.S.; Farndale, R.W.; Barnes, M.J. The collagen-binding A-domains of integrins alpha(1)beta(1) and alpha(2)beta(1) recognize the same specific amino acid sequence, GFOGER, in native (triple-helical) collagens. *J. Biol. Chem.* **2000**, *275*, 35–40. [CrossRef] [PubMed]

13. Yannas, I.V.; Tzeranis, D.S.; Harley, B.A.; So, P.T.C. Biologically active collagen-based scaffolds: Advances in processing and characterization. *Philos. Trans. A Math. Phys. Eng. Sci.* **2010**, *368*, 2123–2139. [CrossRef] [PubMed]

14. Sah, M.; Pramanik, K. Regenerated silk fibroin from B. mori silk cocoon for tissue engineering applications. *Int. J. Environ. Sci. Dev.* **2010**, *1*, 404–408. [CrossRef]

15. Foo, C.W.P.; Bini, E.; Hensman, J.; Knight, D.P.; Lewis, R.V.; Kaplan, D.L. Role of pH and charge on silk protein assembly in insects and spiders. *Appl. Phys. A* **2006**, *82*, 223–233. [CrossRef]

16. Sashina, E.S.; Bochek, A.M.; Novoselov, N.P.; Kirichenko, D.A. Structure and solubility of natural silk fibroin. *Russ. J. Appl. Chem.* **2006**, *79*, 869–876. [CrossRef]

17. Jin, H.-J.; Kaplan, D.L. Mechanism of silk processing in insects and spiders. *Nature* **2003**, *424*, 1057–1061. [CrossRef] [PubMed]

18. Ochi, A.; Hossain, K. S.; Magoshi, J.; Nemoto, N. Rheology and dynamic light scattering of silk fibroin solution extracted from the middle division of bombyx mori silkworm. *Biomacromolecules* **2002**, *3*, 1187–1196. [CrossRef] [PubMed]

19. Yang, G.; Zhang, L.; Cao, X.; Liu, Y. Structure and microporous formation of cellulose/silk fibroin blend membranes: Part II. Effect of post-treatment by alkali. *J. Membr. Sci.* **2002**, *210*, 379–387. [CrossRef]

20. Tang, Y.; Cao, C.; Ma, X.; Chen, C.; Zhu, H. Study on the preparation of collagen-modified silk fibroin films and their properties. *Biomed. Mater.* **2006**, *1*, 242–246. [CrossRef] [PubMed]

21. Chomchalao, P.; Pongcharoen, S.; Sutheerawattananonda, M.; Tiyaboonchai, W. Fibroin and fibroin blended three-dimensional scaffolds for rat chondrocyte culture. *Biomed. Eng. Online* **2013**, *12*, 28–40. [CrossRef] [PubMed]

22. Lu, Q.; Feng, Q.; Hu, K.; Cui, F. Preparation of three-dimensional fibroin/collagen scaffolds in various pH conditions. *J. Mater. Sci. Mater. Med.* **2008**, *19*, 629–634. [CrossRef] [PubMed]

23. Hu, K.; Lv, Q.; Cui, F.Z.; Feng, Q.L.; Kong, X.D.; Wang, H.L.; Hunag, L.Y.; Li, T. Biocompatible fibroin blended films with recombinant human-like collagen for hepatic tissue engineering. *J. Bioact. Compat. Polym.* **2006**, *21*, 23–37. [CrossRef]

24. Lv, Q.; Hu, K.; Feng, Q.; Cui, F. Preparation of insoluble fibroin/collagen films without methanol treatment and the increase of its flexibility and cytocompatibility. *J. Appl. Polym. Sci.* **2008**, *109*, 1577–1584. [CrossRef]

25. Lv, Q.; Hu, K.; Feng, Q.; Cui, F. Fibroin/collagen hybrid hydrogels with crosslinking method: preparation, properties, and cytocompatibility. *J. Biomed. Mater. Res. A* **2008**, *84*, 198–207. [CrossRef] [PubMed]

26. Lv, Q.; Feng, Q.; Hu, K.; Cui, F. Three-dimensional fibroin/collagen scaffolds derived from aqueous solution and the use for HepG2 culture. *Polymer* **2005**, *46*, 12662–12669. [CrossRef]

27. Zhou, J.; Cao, C.; Ma, X.; Lin, J. Electrospinning of silk fibroin and collagen for vascular tissue engineering. *Int. J. Biol. Macromol.* **2010**, *47*, 514–519. [CrossRef] [PubMed]

28. Krigbaum, W.R.; Wall, F.T. Viscosities of binary polymeric mixtures. *J. Polym. Sci.* **1950**, *5*, 505–514. [CrossRef]

29. García, R.; Melad, O.; Gomez, C.M.; Figueruelo, J.E.; Campos, A. Viscometric study on the compatibility of polymer–polymer mixtures in solution. *Eur. Polym. J.* **1999**, *35*, 47–55. [CrossRef]

30. Wang, Q.; Schlenoff, J.B. The polyelectrolyte complex/coacervate continuum. *Macromolecules* **2014**, *47*, 3108–3116. [CrossRef]

31. van der Gucht, J.; Spruijt, E.; Lemmers, M.; Stuart, M.A.C. Polyelectrolyte complexes: Bulk phases and colloidal systems. *J. Colloid. Interface Sci.* **2011**, *361*, 407–422. [CrossRef] [PubMed]

32. Cramer, G.R.; Läuchli, A.; Polito, V.S. Displacement of Ca^{2+} by Na^+ from the plasmalemma of root cells: A primary response to salt stress? *Plant Physiol.* **1985**, *79*, 207–211. [CrossRef] [PubMed]

33. Yang, Y.; Kwak, H.W.; Lee, K.H. Effect of residual lithium ions on the structure and cytotoxicity of silk fibroin film. *Int. J. Ind. Entomol.* **2013**, *27*, 165–170. [CrossRef]

34. Yin, Z.; Wu, F.; Xing, T.; Yadavalli, V.K.; Kundu, S.C.; Lu, S. A silk fibroin hydrogel with reversible sol-gel transition. *RSC Adv.* **2017**, *7*, 24085–24096. [CrossRef]

35. Tiktopulo, E.I.; Kajava, A.V. Denaturation of type I collagen fibrils is an endothermic process accompanied by a noticeable change in the partial heat capacity. *Biochemistry* **1998**, *37*, 8147–8152. [CrossRef] [PubMed]

36. Bozec, L.; Odlyha, M. Thermal denaturation studies of collagen by microthermal analysis and atomic force microscopy. *Biophys. J.* **2011**, *101*, 228–236. [CrossRef] [PubMed]

37. Hu, X.; Kaplan, D.; Cebe, P. Determining beta-sheet crystallinity in fibrous proteins by thermal analysis and infrared spectroscopy. *Macromolecules* **2006**, *39*, 6161–6170. [CrossRef]

38. Chen, H.; Hu, X.; Cebe, P. Thermal properties and phase transitions in blends of Nylon-6 with silk fibroin. *J. Therm. Anal. Calorim.* **2008**, *93*, 201–206. [CrossRef]

39. Lazarev, Y.A.; Grishkovsky, B.A.; Khromova, T.B. Amide I band of IR spectrum and structure of collagen and related polypeptides. *Biopolymers* **1985**, *24*, 1449–1478. [CrossRef] [PubMed]

40. Rabotyagova, O.S.; Cebe, P.; Kaplan, D.L. Collagen structural hierarchy and susceptibility to degradation by ultraviolet radiation. *Mater. Sci. Eng. C Mater. Biol. Appl.* **2008**, *28*, 1420–1429. [CrossRef] [PubMed]

41. Susi, H.; Ard, J.S.; Carroll, R.J. The infrared spectrum and water binding of collagen as a function of relative humidity. *Biopolymers* **1971**, *10*, 1597–1604. [CrossRef] [PubMed]

42. Iizuka, E.; Yang, J.T. Optical rotatory dispersion and circular dichroism of the beta-form of silk fibroin in solution. *Proc. Natl. Acad. Sci. USA* **1966**, *55*, 1175–1182. [CrossRef] [PubMed]

43. Ayub, Z.; Arai, M.; Hirabayashi, K. Mechanism of the gelation of fibroin solution. *Biosci. Biotechnol. Biochem.* **1993**, *57*, 1910–1912. [CrossRef]

44. Zhou, P.; Xie, X.; Knight, D.P.; Zong, X.H.; Deng, F.; Yao, W.H. Effects of pH and calcium ions on the conformational transitions in silk fibroin using 2D raman correlation spectroscopy and ^{13}C solid-state NMR. *Biochemistry* **2004**, *43*, 11302–11311. [CrossRef] [PubMed]

45. Matsumoto, A.; Chen, J.; Collette, A.L.; Kim, U.J.; Altman, G.H.; Cebe, P.; Kaplan, D.L. Mechanisms of silk fibroin sol-gel transitions. *J. Phys. Chem. B* **2006**, *110*, 21630–21638. [CrossRef] [PubMed]

46. Curtis, R.A.; Prausnitz, J.M.; Blanch, H.W. Protein-protein and protein-salt interactions in aqueous protein solutions containing concentrated electrolytes. *Biotechnol. Bioeng.* **1998**, *57*, 11–21. [CrossRef]

47. Laos, K.; Brownsey, G.J.; Ring, S.G. Interactions between furcellaran and the globular proteins bovine serum albumin and β-lactoglobulin. *Carbohydr. Polym.* **2007**, *67*, 116–123. [CrossRef]

48. Seyrek, E.; Dubin, P.L.; Tribet, C.; Gamble, E.A. Ionic strength dependence of protein-polyelectrolyte interactions. *Biomacromolecules* **2003**, *4*, 273–282. [CrossRef] [PubMed]

49. Bianchi, E.; Conio, G.; Ciferri, A.; Puett, D.; Rajagh, L. The role of pH, temperature, salt type, and salt concentration on the stability of the crystalline, helical, and randomly coiled forms of collagen. *J. Biol. Chem.* **1967**, *242*, 1361–1369. [PubMed]

50. Freudenberg, U.; Behrens, S.H.; Welzel, P.B.; Müller, M.; Grimmer, M.; Salchert, K.; Taeger, T.; Schmidt, K.; Pompe, W.; Werner, C. Electrostatic interactions modulate the conformation of collagen I. *Biophys. J.* **2007**, *92*, 2108–2119. [CrossRef] [PubMed]

51. Li, S.T.; Katz, E.P. An electrostatic model for collagen fibrils. The interaction of reconstituted collagen with Ca++, Na+, and Cl. *Biopolymers* **1976**, *15*, 1439–1460. [CrossRef] [PubMed]

52. Chung, E.J.; Jakus, A.E.; Shah, R.N. In situ forming collagen-hyaluronic acid membrane structures: mechanism of self-assembly and applications in regenerative medicine. *Acta Biomater.* **2013**, *9*, 5153–5161. [CrossRef] [PubMed]

53. Chen, X.; Li, W.; Yu, T. Conformation transition of silk fibroin induced by blending chitosan. *J. Polym. Sci. Part B Polym. Phys.* **1997**, *35*, 2293–2296. [CrossRef]

54. Li, G.; Zhou, P.; Shao, Z.; Xie, X.; Chen, X.; Wang, H.; Chunyu, L.; Yu, T. The natural silk spinning process. A nucleation-dependent aggregation mechanism? *Eur. J. Biochem.* **2001**, *268*, 6600–6606. [CrossRef] [PubMed]
55. Bella, J.; Brodsky, B.; Berman, H.M. Hydration structure of a collagen peptide. *Structure* **1995**, *3*, 893–906. [CrossRef]

Sample Availability: Samples of the compounds are not available from the authors.

 *molecules*

Article

Polyurethane Foams for Thermal Insulation Uses Produced from Castor Oil and Crude Glycerol Biopolyols

Camila S. Carriço [1], Thaís Fraga [1], Vagner E. Carvalho [2] and Vânya M. D. Pasa [1,*]

[1] Laboratório de Produtos da Biomassa, Departamento de Química, Universidade Federal de Minas Gerais, Av. Antonio Carlos, 6627, 31270-901 Belo Horizonte, Minas Gerais, Brazil; cscarrico@ufmg.br (C.S.C.); tf.thais@hotmail.com (T.F.)

[2] Laboratório de Física de Superfícies, Departamento de Física, Universidade Federal de Minas Gerais, Av. Antonio Carlos, 6627, 31270-901 Belo Horizonte, Minas Gerais, Brazil; vagner@fisica.ufmg.br

* Correspondence: vmdpasa@terra.com.br; Tel.: +55-31-3409-6651

Received: 7 April 2017; Accepted: 15 June 2017; Published: 2 July 2017

Abstract: Rigid polyurethane foams were synthesized using a renewable polyol from the simple physical mixture of castor oil and crude glycerol. The effect of the catalyst (DBTDL) content and blowing agents in the foams' properties were evaluated. The use of physical blowing agents (cyclopentane and n-pentane) allowed foams with smaller cells to be obtained in comparison with the foams produced with a chemical blowing agent (water). The increase of the water content caused a decrease in density, thermal conductivity, compressive strength, and Young's modulus, which indicates that the increment of CO_2 production contributes to the formation of larger cells. Higher amounts of catalyst in the foam formulations caused a slight density decrease and a small increase of thermal conductivity, compressive strength, and Young's modulus values. These green foams presented properties that indicate a great potential to be used as thermal insulation: density (23–41 $kg \cdot m^{-3}$), thermal conductivity (0.0128–0.0207 $W \cdot m^{-1} \cdot K^{-1}$), compressive strength (45–188 kPa), and Young's modulus (3–28 kPa). These biofoams are also environmentally friendly polymers and can aggregate revenue to the biodiesel industry, contributing to a reduction in fuel prices.

Keywords: polyurethane foams; castor oil; crude glycerol; biopolyols; thermal insulator

1. Introduction

Rigid polyurethane foams are usually applied as thermal insulation in buildings, as well as in automobile and aerospace industries. The global rigid polyurethane foam market is expected to reach USD \$20.40 billion by 2020 in construction applications, such as in residential and commercial roofs, walls, panels, and doors, and in appliance applications [1]. For insulation purposes, thermal conductivity is the most important property to be studied. Typical thermal conductivity values for polyurethane foams are between 0.02 and 0.03 $W \cdot m^{-1} \cdot K^{-1}$ and the very efficient thermal insulation of these materials is due to the extremely low thermal conductivity of the blowing agent gas trapped in the closed porous structures [2,3]. Additives, as blowing agents, catalysts, and surfactants, are very important to adjust the final properties of the foam's synthesis from the main reaction of an isocyanate with a polyol. However, most of these reagents are derived from petrochemicals, increasing the petroleum dependence and, as a consequence, environmental problems. These aspects have encouraged the production of rigid foams from renewable materials [4].

In recent years, the uses of bio-based polyols have been investigated to produce sustainable and eco-friendly rigid polyurethane foams, such as lignin [5,6], biopitch [7,8], glycerol [9–11], and vegetable oils including castor [3,12–19], palm [20,21], soybean [11,22] starch [23,24], etc. Castor oil (CO) is a mixture of triglycerides, mainly ricinoleic acid, which is produced from the seed of the *Ricinus communis* plant. This low-cost renewable raw material has been usually used as a polyol for flexible material production, as polyurethane foams and elastomers. However, most of these materials are produced after chemical modifications of castor oil, such as transesterification, oxypropilation, hydroformilation, and ozonolysis, in order to introduce reactive hydroxyl groups, which involve oil pre-treatment with multiple steps and high costs.

Castor oil has high molecular weight, a low hydroxyl number, and poor reactivity, and also a slow cure time, low flame retardancy, and low miscibility with other components. These properties have limited the use of castor oil as a raw material polyol, especially to prepare rigid polyurethane foams, and the pre-treatment reaction seems to be efficient to minimize these disadvantages. Transesterification [16,18,19,25], thio-ene [26], and amidization [27] reactions have been employed to introduce reactive OH groups in castor oil in order to produce suitable polyols to synthesize rigid polyurethane foams. Recently, a work was published presenting a polyol production from the polymerization of glycerol followed by condensation of this polymerized material with castor oil. After this, rigid polyurethane foams were synthesized by partial substitution of a petrochemical polyol with this produced polyol [28]. These reactions are usually performed in the presence of an alcohol, and the use of glycerol has also been extensively studied. These foams presented good mechanical properties: densities of 35–50 kg·m^{-3}, compressive strengths of 127–475 kPa, and also great thermal stabilities, with a limit oxygen index of 20–30%, and a thermal conductivity of 0.021–0.029 W·mK^{-1}. However, the reactions to modify the oil and enhance its reactivity are conducted between 160 to 240 °C, for 2–6 h, involving multiple steps and some reactants, which increases the foam costs.

In the present study, polyurethane foams were prepared using a simple mixture of crude glycerol and pure castor oil. Binary mixtures with different contents of castor oil and crude glycerol were prepared in order to increase the functionality of this biopolyol without the extra costs of chemical modifications. The mixtures were achieved because the crude glycerol has three hydroxyl groups and a short chain, as well as good miscibility in the castor oil. The polyol was produced from the physical mixing of raw materials without any type of pre-treatment, and for direct use in polyurethane synthesis. This simple and innovative approach allowed for the production of renewable rigid polyurethane foams with good properties for thermal insulation uses. The effects of the catalyst and blowing agent in the mechanical and thermal conductivity properties of the foams were evaluated. This process is inexpensive, simple, and sustainable, and suitable to produce environmentally friendly materials with commercial applications.

2. Materials and Methods

2.1. Materials

Castor oil used was provided by PolyUrethane Company (Betim, Minas Gerais, Brazil). Crude glycerol, a co-product of biodiesel production, was kindly provided by Petrobrás (Usina Darcy Ribeiro—Montes Claros, Minas Gerais, Brazil). The crude glycerol has about 91% of glycerin, 5% of inorganic compounds (transesterification catalyst, acids, etc.), 1% of methanol, and about 3% of water and biodiesel, oleine, monoglicerides, and others. Pure glycerol (99.5%) was supplied by Synth. The isocyanate source (Desmodur 44 V 20), a mixture of 4,4′-diphenylmethane-diisocyanate, was supplied by Bayer Company. The surfactant was Tegostab 8460 (Evonik, Essen, Germany), a polyether-modified polysiloxane. The catalyst used in polyurethane foams synthesis was DBTDL (dibutyl tin dilaurate), an organometallic catalyst, manufactured by Evonik. The blowing agents used were cyclopentane (Sigma Aldrich, St. Louis, MO, USA), n-pentane (Sigma Aldrich, St. Louis, MO, USA), and distilled water.

2.2. Polyol Preparation

Polyols were obtained by binary mixtures of pure glycerol (P.A. glycerol) and castor oil in different proportions (20%, 40%, 50%, 60% and 80% of pure glycerol w/w). Biopolyols were also prepared from binary mixtures with different amounts of castor oil (Co) and crude glycerol (G), called GCo (10%, 20%, 30%, 40%, 50%, 60% and 70% of crude glycerol w/w). The raw materials were added in a beaker and homogenized for 1 min using a mechanical stirrer (Fisatom model 713 D, São Paulo, SP, Brazil).

2.3. Foam Synthesis

2.3.1. Study of the Best Binary Mixture to Produce Polyurethane Foams

A preliminary study to choose the best polyol formulation was carried out. For this purpose, rigid polyurethane foams were synthesized using the batch process method and binary polyols with different compositions. This comparative study was performed with the mixture of P.A. glycerol (20%, 40%, 50%, 60%, 80% w/w) and castor oil, and with crude glycerol (10%, 20%, 30%, 40%, 50%, 60%, 70%) and also castor oil. The NCO/OH molar ratio was kept equal to 2.0, the surfactant, 2% (w/w), and water was employed as the blowing agent (2%). These mixtures were kept under vigorous stirring for 1 min at 500 rpm. Afterwards, the isocyanate was added to the pre-mixture and stirred until complete homogenization. The formulations were poured into a wooden mold with the dimensions $7.0 \times 7.0 \times 20.0$ cm for the growth of the polymer foams and cured for 24 h at room temperature. After the curing time, the foams were demolded and visual inspections were performed. The best polyol formulation was chosen for the foam with the best properties (aspect, homogeneity, dimensional stability, and low friability).

2.3.2. Study of Catalyst and Blowing Agent Effect on the Foams Properties

After preliminary tests, the best binary polyol GCo with 10% crude glycerol, and 90% castor oil w/w was used to produce foams with good thermal stability, mechanical properties, and low thermal conductivity for insulation use. For this purpose, this best polyol was tested to prepare foams aiming to evaluate the effect of different catalyst (DBTDL) amounts (1% and 2% w/w); different blowing agent types (water, n-pentane, and cyclopentane), and contents. The tests to evaluate the different blowing agent amounts were carried out with: water = 2%, 4% and 6% w/w; n-pentane = 2% w/w; and cyclopentane = 2%, 4% and 6% w/w, but 6% was not suitable for use. These foam reactions were performed using the same methodology described for the preliminary tests (Section 2.3.1). These formulations are summarized in Table 1.

Table 1. Formulations of the foams prepared with GCo polyol (10% crude glycerol and 90% castor oil w/w), NCO/OH = 2.0, and 2% Tegostab (surfactant).

Formulations	Blowing Agent (Type)	Blowing Agent (%)	Catalyst DBTDL (%)
I	Water	2	1
II	Water	2	2
III	Water	4	1
IV	Water	4	2
V	Water	6	1
VI	Water	6	2
VII	n-pentane	2	2
VIII	ciclopentane	2	2
IX	ciclopentane	4	2

2.4. Characterization

2.4.1. FTIR Analyses

Fourier transform infrared (FTIR) spectra were recorded on an ABB Bomer spectrometer using 0.1% KBr pellets and an ATR accessory with resolution of 4 cm^{-1}, in the range of 4000–400 cm^{-1}.

2.4.2. Thermogravimetric Analyses (TGA and DTG)

The thermal behavior of the foams was evaluated by thermogravimetric analysis (TGA), which was carried out on a TA equipment model TA-50 Q at temperatures ranging from 30 °C to 800 °C with a heating rate of 10 °C·min^{-1} under a nitrogen flow of 40 mL·min^{-1}.

2.4.3. Morphology Analyses

The images obtained by optical microscopy were collected on an Olympus optical microscope, model BX41M, coupled to a TecVoz camera, model DNS 480. The morphology of polymers structure was characterized to observe the foam cellular structure by scanning electron microscope (SEM) images, which were obtained with a JSM–6360LV, a JEOL scanning microscope.

2.4.4. Determination of Apparent Density

Polyurethane foam densities were calculated according to the ASTM D-1622 standard by measuring three specimens of each sample with the dimensions of 50 × 50 × 25 mm.

2.4.5. Mechanical Properties

Mechanical testing (10% compressive strength) of polyurethane foams was performed on an Autograph Precision Universal Testing Machine AG-Xplus Series (Shimadzu, Kyoto, Japan), according to the ASTM D1621 standard method. At least five samples were analyzed to obtain the average value; the specimen size was 50 × 50 × 25 mm.

2.4.6. Thermal Conductivity

Thermal conductivity tests were performed by an instrument developed by Professor Vagner Carvalho in the Surface Laboratory at the Physics Department at Universidade Federal de Minas Gerais—Brazil [29]. This equipment has a unit that contains a tip, which is electronically controlled and has a temperature sensor. The measurements are based on the thermal comparison method by copper-constantan differential thermocouples that measure the contact temperature gradient and the changes in the sample temperature. The samples were tested six times and a simple average was taken as the comparator output. A calibration curve was constructed using various ranges of samples with a wide thermal conductivity and the data were plotted in order to obtain the relation between the thermal conductivity and the comparator output.

3. Results and Discussion

3.1. Study of the Best Binary Mixture to Produce Polyurethane Foams

The binary polyol production was first studied by the physical mixture of pure glycerol and castor oil, varying the glycerol content. Some foams did not present good dimensional stability (Figure 1a). Upon increasing the content of pure glycerol, it was observed that the foams became denser and softer. The formulations with 20% and 40% (*w*/*w*) of pure glycerol content did not grow as a typical foam, yielding a very rigid, solid material. The foams produced with the polyol containing 50% (*w*/*w*) of pure glycerol presented a high homogeneity, but upon increasing this content, the foams became very friable.

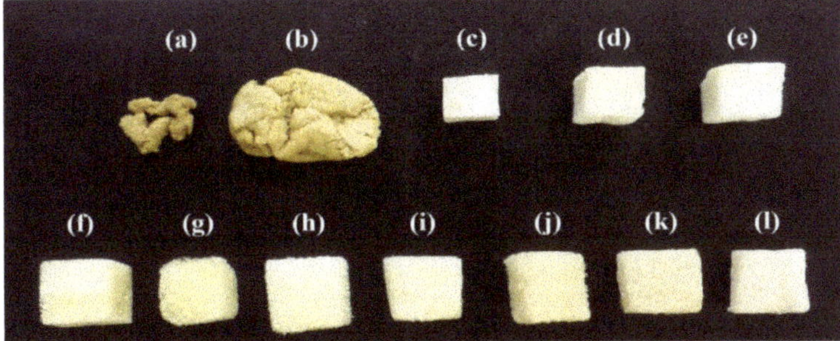

Figure 1. Foams produced varying the pure glycerol content: (**a**) 20%; (**b**) 40%; (**c**) 50%; (**d**) 60%; and (**e**) 80%, as well as varying the crude glycerol content: (**f**) 10%; (**g**) 20%; (**h**) 30%; (**i**) 40%; (**j**) 50%; (**k**) 60%; and (**l**) 70% of the polyols.

The pure glycerol was then replaced by crude glycerol, a co-product of biodiesel production, in order to synthesize new foams and the results were quite different. The foams with crude glycerol and castor oil polyol (named GCo, Figure 1f–l) were more homogeneous and presented good dimensional stability in comparison to those synthesized with pure glycerol (Figure 1a–e). Based on these experimental behaviors, we believe that the crude glycerol impurities (alkaline catalyst, methanol, methyl esters of fatty acids, fatty acid methyl esters) are responsible for the best properties of the foams. Further investigations can be made to understand this behavior. Similar behaviors have already been reported in the literature, evaluating the effects of the replacement of pure glycerol by crude glycerol for producing polyols from biomass liquefaction. These studies also affirm that these impurities of crude glycerol improved the polyols' and polyurethanes' properties [9,11,30].

It was observed that upon increasing the amount of crude glycerol, there was a decrease in the rigidity and the dimensional stability of the foams. For this reason, the foam produced from the polyol containing 10% of crude glycerol and 90% of castor oil (w/w) (Figure 1f) was chosen to perform further studies. The hydroxyl number (240 mg·KOH·g^{-1}) and viscosity (436.5 mm^2·s^{-1}) of this polyol were measured, indicating that these polyols are adequate to produce rigid foams [4]. Similar results have already reported in the literature for polyols from castor oil [26].

It is important to point out that the polyol used to produce our best foam, with 10% of glycerol and 90% of castor oil (w/w), has the molar ratio glycerol/castor oil approximately equal to 1 (considering the molar weight of glycerol and castor oil 92.09 and 895.33 g·mol^{-1}, respectively). Observing the structure of these molecules (Figure 2), there are three hydroxyl groups in each glycerol molecule and three instaurations of ricinoleic acid in the triglyceride structure that are suitable to be transformed into OH groups by pre-treatment reactions. Thus, we can consider that 1 mol of glycerol has the same number of OH groups as 1 mol of castor oil. Then, when we used a binary mixture 1:1, the OH amount is doubled. The same value of hydroxyl groups can be obtained by inserting an OH on each instauration of the castor oil ricinoleic chain. Our study was then performed using the binary mixture, without castor oil modification, to avoid extra costs in the process.

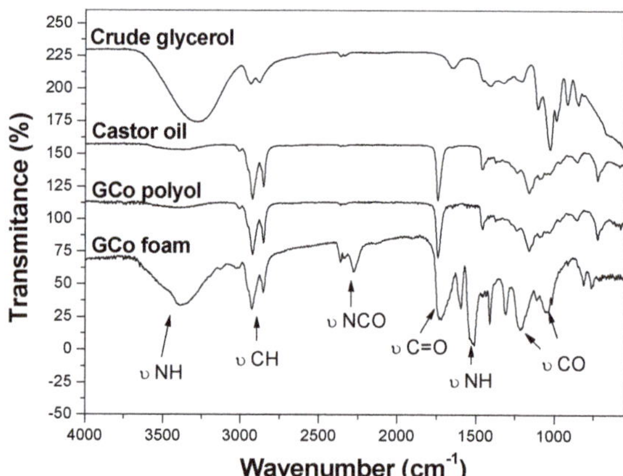

Castor oil

Glycerol

Figure 2. Structure of the castor oil (ricinoleic acid is the main component) and glycerol molecules.

3.2. Study of Catalyst and Blowing Agent Effect in the Foams Properties

The characterization data obtained for the different foams, which were prepared using the best binary polyol (10% crude glycerol and 90% castor oil w/w) will be discussed in this section. The formulations will be presented using Roman numerals, as shown in Table 1.

The FTIR spectra of the renewable raw materials used to produce GCo polyols are shown in Figure 3. The band corresponding to the hydroxyl group's vibration is observed at approximately 3700–3000 cm^{-1}. The characteristic stretches of double bonds in the castor oil groups C=C–H and C=C are observed at 3020 and 1740 cm^{-1}, respectively. The bands around 3018 and 2710 cm^{-1} are assigned to CH$_2$ and CH$_3$ stretches of aliphatic chains, which are quite pronounced in castor oil due to the 18-carbon chain. The characteristic band of carbonyl and carboxyl groups is observed to be centered at 1743 cm^{-1} in the castor oil spectrum. The alkenes deformation of CH$_2$ groups, present in the castor oil structure, is observed in a strong band at 1458 cm^{-1}. The bands around 1112–1000 cm^{-1} indicate the presence of primary and secondary hydroxyl groups. These bands are very pronounced in crude glycerol spectrum, due to the three hydroxyl groups present in its short chain [16,18].

Figure 3. FTIR spectra for the raw materials, GCo polyol, and GCo foam (formulation II of Table 1)

All of the foams' spectra produced from GCo polyols are very similar, while a typical foam spectrum is shown in Figure 3, which presents the characteristic polyurethane bands. The stretching and vibrations of NH groups were observed between 3808–3308, and 1512 and 1510 cm^{-1}, respectively. The deformation of CH$_2$ bonds was observed by the two thin bands at 2900 and 2890 cm^{-1}.

The vibration of N=C=N and N=C=O groups are attributed to bands between 2390 and 2150 cm^{-1}. Other vibration modes of the CH bond were also observed at 1464, 1418, 1364, and 1294 cm^{-1}. The band between 1730 and 1720 cm^{-1} corresponds to the stretch of the CO-free urethane bond, and around 1700 cm^{-1}, the hydrogen bond between the carbonyl and hydrogen atoms (from NH groups) from urethane is also observed. A band related to stretching asymmetric links OCONH was revealed at 1380 cm^{-1}. The bands between 1100 and 1000 cm^{-1} were attributed to primary and secondary hydroxyl groups [16,17].

The thermal behavior of the GCo foams containing different types and amounts of catalyst, shown in Figure 4, were evaluated by thermogravimetric analysis (TGA and DTG). The different foams presented similar thermal stability, and the DTG curves showed three regions of weight loss. The first event (around 300 °C) corresponds to urethane thermal degradation, free isocyanate, and alcohols; the second event is related to the degradation of rigid segments, at 370 °C; and the third event, approximately at 480 °C, is associated to the thermal degradation of the flexible segments and others segments of the remaining structure [31,32].

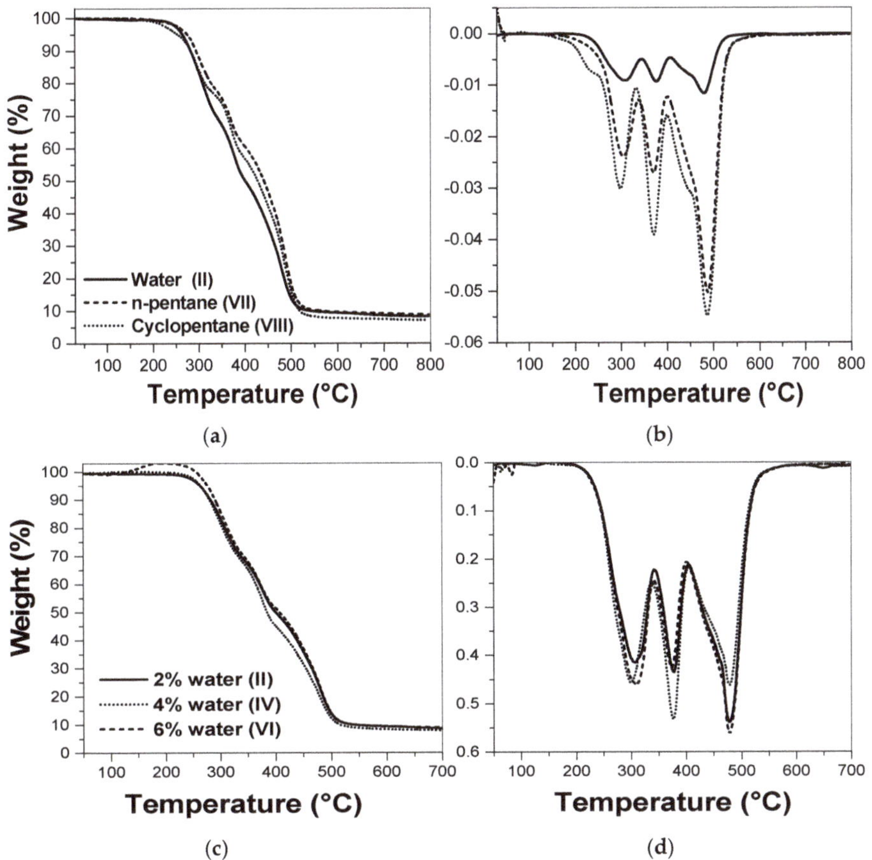

Figure 4. Thermogravimetric analysis: TGA (**a**,**c**) and DTG (**b**,**d**) curves of foams with GCo polyol with different types and amounts of blowing agents. (**a**,**b**) formulations II, VII, VIII; (**c**,**d**) formulations II, IV, VI of foams shown in Table 1.

The effect of different blowing agents on the thermal stability of the GCo foams was evaluated, as shown in Figure 4a,b. The results indicate that the blowing agent type did not significantly modify the thermal behavior of the foams, which is suggested by the similar curves of the foams synthesized with water, cyclopentane, and n-pentane.

The effect of the amount of blowing agent (water) in the formulations (Figure 4c,d) was also investigated. The results show that the amount of water as a blowing agent did not significantly affect the thermal stability of the foams produced with polyol GCo, taking into account that all of the curves presented the same profile, indicating a similar thermal stability.

Apparent density is an important parameter of cellular polymers. The effect of the blowing agent type on the apparent density of the foams produced from GCo polyols (Table 2) showed that the formulations with physical blowing agents (cyclopentane and n-pentane) produced foams with higher densities than those synthesized with the chemical blowing agent (water). Similar results have been reported in the literature [32–34], and this behaviour indicates that smaller cells are formed due to the rapid volatilization of physical blowing agents, which have a low boiling point, during the highly exothermic foam growth step in comparison with the CO_2, produced by the reaction of water with isocyanate [35].

Table 2. Density values of foams with different blowing agents.

Formulation	Blowing Agent	Apparent Density (kg·m^{-3})
II	Water	37.4
VII	n-pentane	61.3
VIII	Cyclopentane	99.3

The effect of the blowing agent (water) content on the foams' apparent densities was also evaluated, as shown in Figure 5a. In increasing the water amount, a decrease of density is observed, which suggests that higher cells are formed with the enhancement of CO_2 production from the water and isocyanate reaction [36].

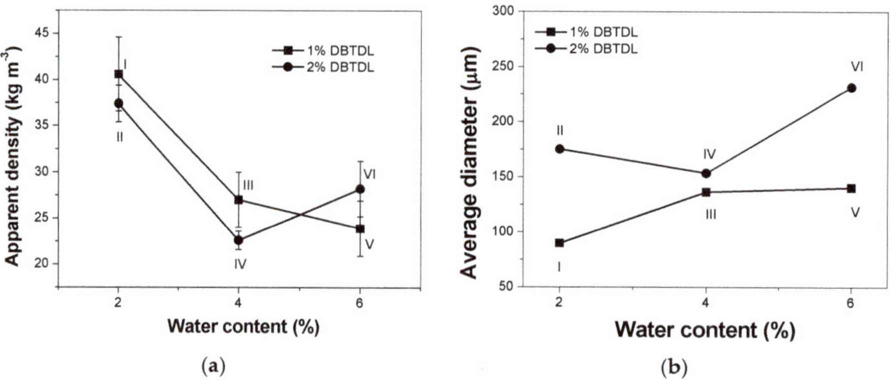

Figure 5. (**a**) Apparent density and (**b**) the average diameter of foams with different contents of the blowing agent (water) and catalyst. The numbers corresponding to the foam formulations (Table 1) are indicated in each point of these graphics.

Figure 5a also shows the influence of the catalyst content on the foams density. A decrease of apparent density was observed upon increasing the catalyst amount in the formulations. This behavior can be explained by the increment in the polymerization rate with the enhancement of the organometallic catalyst content in the formulation, avoiding the CO_2 releasing during the foam's cell formation [4]. As the reaction occurs with higher speed, the blowing agent is trapped in the structure and the cells presented higher diameters and lower densities (Figure 5a,b, respectively) [37]. This effect is more prominent in foams with higher water contents. These apparent density results are in agreement of the values measured for the same rigid polyurethane foams synthesized using castor oil polyols [19,26].

The effect of different blowing agents on the cellular foam structures can also be observed in Figure 6, which shows SEM images of the foams synthesized with water and cyclopentane. The foams prepared with water as the blowing agent showed the largest cell size, confirming the density data (Figure 5a). Pentane has a low boiling temperature (around 50 °C) and volatizes very quickly, as previously explained in the density data discussion. The foam with 6% cyclopentane presented a low dimensional stability and, for this reason, its SEM micrograph was not shown here.

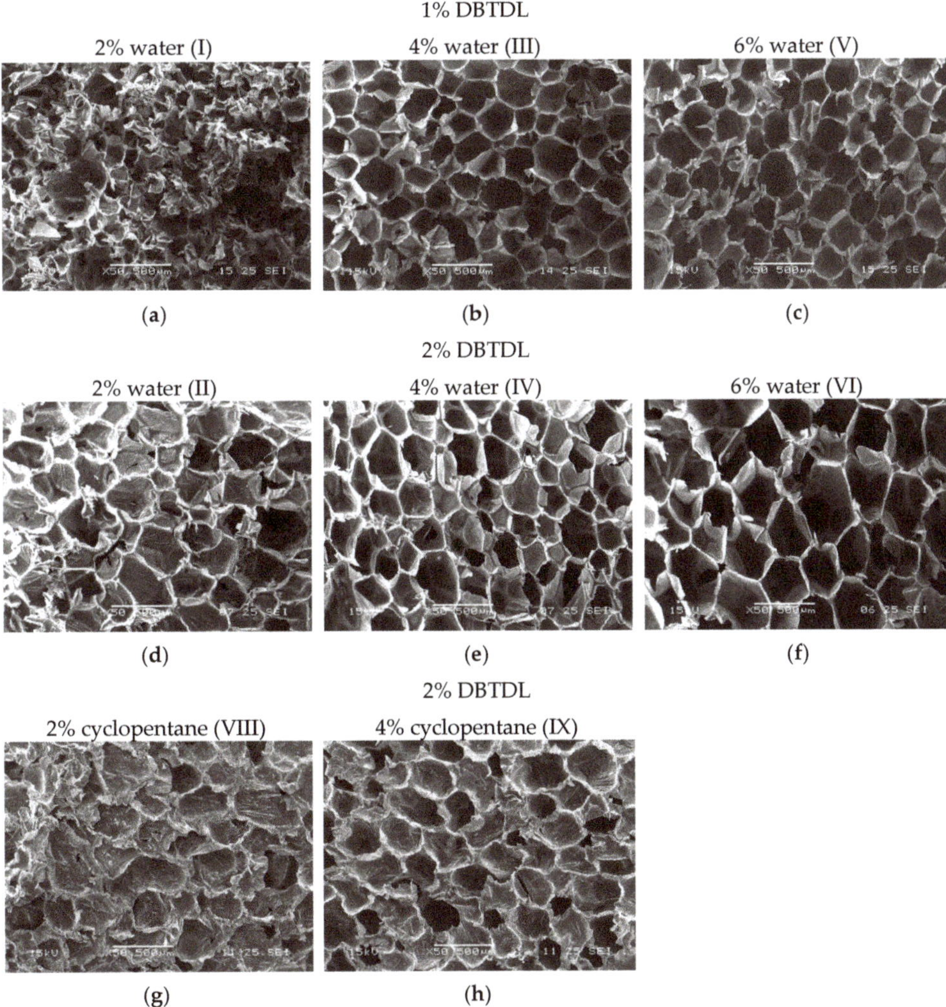

Figure 6. SEM micrographs of GCo foams with different types and contents of blowing agents and catalyst DBTDL (scale bar of 500 μm 50×). The numbers of the foam formulations (Table 1) are indicated in each micrograph.

The foams formulated with water as the blowing agent presented the best dimensional stability, the lowest apparent density, and the higher cells homogeneity. Based on these results, we chose this formulation, aiming to evaluate the effect of the catalyst amount on the mechanical and conductivity properties. Another important aspect to note is that the use of water as a blowing agent is considered a green and inexpensive option.

The effect of the water content as the blowing agent was also evaluated by SEM images, as shown in Figure 6. It was observed that the water concentration is directly proportional to the cell size (Figure 5b). These analyses are in agreement with the density data (Figure 5a). The foams produced with 4% water presented higher cell homogeneity in comparison to those containing 2% water. The foams formulated with 6% water produced larger and heterogeneous cells, indicating that 4% water is the optimal quantity to be used in the foam formulations.

The comparison of the catalyst amount in the foam cells (Figure 6) formulated with water showed that an increment of the catalyst content yielded cellular materials with higher average cell diameters, confirming the density values in Figure 5a. The foams synthesized with 2% DBTDL showed the best cell homogeneity, despite higher cell diameters, as observed in Figure 5b. The average diameter of the foams produced in this present process is smaller than the data reported in the literature (from 107 to 121 μm) of foams synthesized from pre-polymerized castor oil [28], which is an important result for our foam uses.

The main property for the application of foam as thermal insulation is its thermal conductivity. This parameter was measured for the rigid foams synthesized with water as a blowing agent, and the results are presented in Figure 7. It was observed that upon increasing the water amount in these formulations, there was a decrease in the thermal conductivity. This result can be explained by the decreasing density and the enhancement of the average cell diameter of the foams [38].

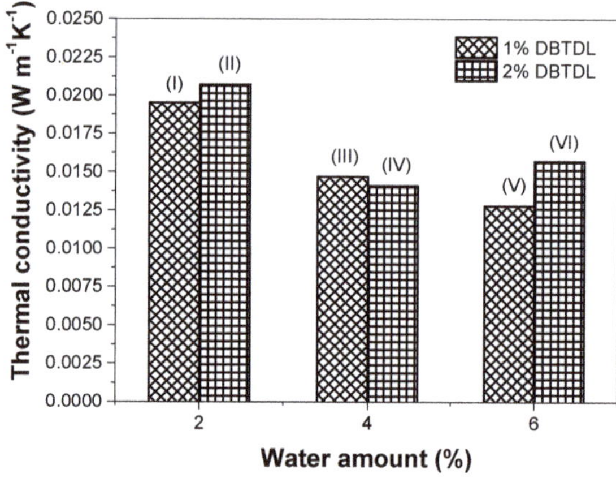

Figure 7. Thermal conductivity of foams with different contents of blowing agent (water) and catalyst (DBTDL). The numbers of foam formulations (Table 1) are indicated in each bar of this graphic.

The effect of the catalyst amount on this property is also presented in Figure 7. The use of a higher catalyst content in the formulations causes a slight thermal conductivity value enhancement, despite the density decrease as a consequence of cell size increase, as shown in Figure 5. The foams synthesized in this study presented better results in comparison to those reported in the literature for foams derived from renewable raw materials, whose values vary between 0.0233 and 0.0505 W·m^{-1}·K^{-1}, suggesting that these materials have a potential use as thermal insulation [22,39,40]. These thermal conductivity results are also better than those found for foams produced from pre-treated castor oil, especially if we consider the use of the very simple and inexpensive method of production [19,28].

The mechanical properties of the foams synthesized with different contents of blowing agent and catalyst were evaluated, and the results are shown in Figure 8. These results presented similar values to those reported in the literature of foams produced from castor oil polyols, which ranged from 125 to 220 kPa [16,19,25,26]. A significant decrease in the compressive strength and Young's modulus of the foams were observed with the addition of higher blowing agent amounts, which can be related with the decrease in density and the increase in cell size. As the cell structure becomes higher, less force is necessary to cause deformation in these foams [36].

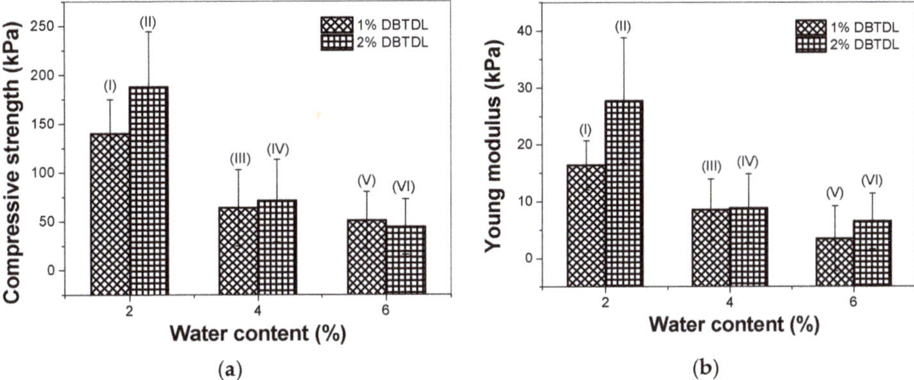

Figure 8. (**a**) Compressive strength and (**b**) Young's modulus of foams with different contents of blowing agent (water) and catalyst (DBTDL). The numbers corresponding to the foam formulations (Table 1) are indicated in each point of these graphics.

The results of the compressive strength and Young's modulus of foams with different amounts of catalyst in the formulations (Figure 8a,b) showed that there are no significant changes in the values upon increasing the catalyst amount, especially for the formulations with 4% and 6% of water as the blowing agent. The variations are within the experimental errors.

Comparing all of the formulations, it was observed that the foam with best thermal conductivity ($0.0141\ W \cdot m^{-1} \cdot K^{-1}$) was formulated with 1% of DBTDL and 6% of water, which also showed a low apparent density value ($23.9\ kg \cdot m^{-3}$). However, this sample showed a low compressive strength (51.01 kPa) and Young's modulus (3.44 kPa), suggesting its application as an insulator of places that do not receive high loads. The foam containing 2% of DBTDL and 2% of water possess a higher compressive strength (187.93 kPa) and Young's modulus (27.74 kPa), as well as a low apparent density value ($37.4\ kg \cdot m^{-3}$). On the other hand, the thermal conductivity value was the higher ($0.0207\ W \cdot m^{-1} \cdot K^{-1}$) in comparison with the other formulations; indeed, this insulation property value is in the range of typical commercial products [2].

4. Conclusions

Rigid polyurethane foams were synthesized from a polyol produced by the physical mixture of castor oil and crude glycerol, a co-product from the biodiesel industry. The polyol preparation method is simple and does not require any pre-treatment of the raw materials. The best formulation was obtained with 10% of crude glycerol and 90% of castor oil (*w/w*, molar ratio = 1:1) and water as the blowing agent. The addition of 10% glycerol increased the OH content of the biopolyol to the same extent as if the castor oil had been pre-treated in order to insert one hydroxyl group in each instauration (C=C) of ricinoleic acid.

The effect of the blowing agent on the foam synthesis was evaluated and the results showed that increasing the water amount caused a decrease in density, thermal conductivity, compressive strength, and Young's modulus. These behaviors are due to the cell size enhancement.

The evaluation of the effect of the catalyst (DBTDL) amount showed that increasing the catalyst content caused a decrease in density and an increase in thermal conductivity, however, there was only a slight influence on the compressive strength and Young's modulus.

These innovative foams presented properties that indicate their great potential to be used as thermal insulation. These materials are inexpensive, environmentally friendly, and can contribute to reducing the biodiesel prices in a biorefinery approach. These biopolymers can be synthesized by a simple process, which can be easily used in a large scale.

Acknowledgments: The authors would like to acknowledge the financial support of CNPq and Boeing Research and Technology, Brazil.

Author Contributions: Camila Carriço is a doctorate student that synthesize the polyols and foams. This paper is part of her thesis. Thaís Fraga is an undergraduate student, which collaborate with the characterizations analysis. Vagner Carvalho is the professor responsible to measure the foams thermal conductivity. Vanya Pasa is the professor that supervisor the work and revise the manuscript.

Conflicts of Interest: The authors declare no conflict of interest.

1. Grand View Research Global Rigid Polyurethane (PU) Foams Industry Trends And Market Segment Forecasts To 2020–Worldwide Rigid Polyurethane (PU) Foams Market, Product (Molded Foam Parts, Slabstock Polyether, Slabstock Polyester), and Segment Forecast to 2020. Available online: http://www.grandviewresearch.com/industry-analysis/rigid-polyurethane-pu-foams-industry (accessed on 3 October 2016).
2. Zhang, H.; Fang, W.-Z.; Li, Y.-M.; Tao, W.-Q. Experimental study of the thermal conductivity of polyurethane foams. *Appl. Therm. Eng.* **2017**, *115*, 528–538. [CrossRef]
3. Tibério Cardoso, G.; Claro Neto, S.; Vecchia, F. Rigid foam polyurethane (PU) derived from castor oil (*Ricinus communis*) for thermal insulation in roof systems. *Front. Archit. Res.* **2012**, *1*, 348–356. [CrossRef]
4. Vilar, W. *Química e Tecnologia dos Poliuretanos*, 3rd ed.; Vilar Consultoria: Rio de Janeiro, Brazil, 2004.
5. Xue, B.L.; Wen, J.L.; Sun, R.C. Lignin-based rigid polyurethane foam reinforced with pulp fiber: Synthesis and characterization. *ACS Sustain. Chem. Eng.* **2014**, *2*, 1474–1480. [CrossRef]
6. Cinelli, P.; Anguillesi, I.; Lazzeri, A. Green synthesis of flexible polyurethane foams from liquefied lignin. *Eur. Polym. J.* **2013**, *49*, 1174–1184. [CrossRef]
7. Araújo, R.C.S.; Pasa, V.M.D.; Melo, B.N. Effects of biopitch on the properties of flexible polyurethane foams. *Eur. Polym. J.* **2005**, *41*, 1420–1428. [CrossRef]
8. Melo, B.N.; Pasa, V.M.D. Thermal and morphological study of polyurethanes based on eucalyptus tar pitch and castor oil. *J. Appl. Polym. Sci.* **2004**, *92*, 3287–3291. [CrossRef]
9. Hu, S.; Li, Y. Polyols and polyurethane foams from acid-catalyzed biomass liquefaction by crude glycerol: Effects of crude glycerol impurities. *J. Appl. Polym. Sci.* **2014**, *131*, 9054–9062. [CrossRef]
10. Hu, S.; Li, Y. Two-step sequential liquefaction of lignocellulosic biomass by crude glycerol for the production of polyols and polyurethane foams. *Bioresour. Technol.* **2014**, *161*, 410–415. [CrossRef] [PubMed]
11. Hu, S.; Wan, C.; Li, Y. Production and characterization of biopolyols and polyurethane foams from crude glycerol based liquefaction of soybean straw. *Bioresour. Technol.* **2012**, *103*, 227–233. [CrossRef] [PubMed]
12. Narine, S.S.; Kong, X.; Bouzidi, L.; Sporns, P. Physical properties of polyurethanes produced from polyols from seed oils: I. Elastomers. *J. Am. Oil Chem. Soc.* **2007**, *84*, 55–63. [CrossRef]
13. Hatakeyama, H.; Matsumura, H.; Hatakeyama, T. Glass transition and thermal degradation of rigid polyurethane foams derived from castor oil-molasses polyols. *J. Therm. Anal. Calorim.* **2013**, *111*, 1545–1552. [CrossRef]
14. Cangemi, J.M.; dos Santos, A.M.; Neto, S.C.; Chierice, G.O. Biodegradation of polyurethane derived from castor oil. *Polímeros* **2008**, *18*, 201–206. [CrossRef]

15. Chethana, M.; Madhukar, B.S.; Siddaramaiah; Somashekar, R. Structure-property relationship of biobased polyurethanes obtained from mixture of naturally occurring vegetable oils. *Adv. Polym. Technol.* **2014**, *33*, 1–11. [CrossRef]

16. Li, Q.F.; Feng, Y.L.; Wang, J.W.; Yin, N.; Zhao, Y.H.; Kang, M.Q.; Wang, X.W. Preparation and properties of rigid polyurethane foam based on modified castor oil. *Plast. Rubber Compos.* **2016**, *45*, 16–21. [CrossRef]

17. Ristić, I.S.; Bjelović, Z.D.; Holló, B.; Mészáros Szécsényi, K.; Budinski-Simendić, J.; Lazić, N.; Kićanović, M. Thermal stability of polyurethane materials based on castor oil as polyol component. *J. Therm. Anal. Calorim.* **2013**, *111*, 1083–1091. [CrossRef]

18. Zhang, L.; Zhang, M.; Hu, L.; Zhou, Y. Synthesis of rigid polyurethane foams with castor oil-based flame retardant polyols. *Ind. Crops Prod.* **2014**, *52*, 380–388. [CrossRef]

19. Zhang, M.; Pan, H.; Zhang, L.; Hu, L.; Zhou, Y. Study of the mechanical, thermal properties and flame retardancy of rigid polyurethane foams prepared from modified castor-oil-based polyols. *Ind. Crops Prod.* **2014**, *59*, 135–143. [CrossRef]

20. Badri, K.H. Biobased polyurethane from palm kernel oil-based polyol: Polyurethane. In *Polyurethane*; InTech: Rijeka, Croatia, 2012; pp. 447–470.

21. Chuayjuljit, S.; Maungchareon, A.; Saravari, O. Preparation and Properties of Palm Oil-Based Rigid Polyurethane Nanocomposite Foams. *J. Reinf. Plast. Compos.* **2010**, *29*, 218–225. [CrossRef]

22. Tan, S.; Abraham, T.; Ference, D.; MacOsko, C.W. Rigid polyurethane foams from a soybean oil-based Polyol. *Polymer* **2011**, *52*, 2840–2846. [CrossRef]

23. Yoshioka, M.; Nishio, Y.; Saito, D.; Ohashi, H.; Hashimoto, M.; Shiraishi, N. Synthesis of biopolyols by mild oxypropylation of liquefied starch and its application to polyurethane rigid foams. *J. Appl. Polym. Sci.* **2013**, *130*, 622–630. [CrossRef]

24. Kim, D.; Kwon, O.; Yang, S.; Park, J.; Chun, B.C. Structural, Thermal, and Mechanical Properties of Polyurethane Foams Prepared with Starch as the Main Component of Polyols. *Fibers Polym.* **2007**, *8*, 155–162. [CrossRef]

25. Veronese, V.B.; Menger, R.K.; Madalena, M.; Forte, D.C.; Petzhold, C.L. Rigid Polyurethane Foam Based on Modified Vegetable Oil. *J. Appl. Polym. Sci.* **2011**, *120*, 530–537. [CrossRef]

26. Ionescu, M.; Radojči, D.; Wan, X.; Shrestha, M.L.; Petrovi, Z.S.; Upshaw, T.A. Highly functional polyols from castor oil for rigid polyurethanes. *Eur. Polym. J.* **2016**, *84*, 736–749. [CrossRef]

27. Stirna, U.; Lazdina, B.; Vilsone, D.; Lopez, M.J.; Vargas-Garcia, M.D.C.; Suarez-Estrella, F.; Moreno, J. Structure and properties of the polyurethane and polyurethane foam synthesized from castor oil polyols. *J. Cell. Plast.* **2012**, *48*, 476–488. [CrossRef]

28. Hejna, A.; Kirpluks, M.; Kosmela, P.; Cabulis, U.; Haponiuk, J.; Piszczyk, L. The influence of crude glycerol and castor oil-based polyol on the structure and performance of rigid polyurethane-polyisocyanurate foams. *Ind. Crops Prod.* **2017**, *95*, 113–125. [CrossRef]

29. Carvalho, V.E. *Contrução de um Comparador Térmico de Leitura Direta*; Universidade Federal de Minas Gerais: Belo Horizonte, Minas Gerais, Brazil, 1978.

30. Hu, S.; Li, Y. Polyols and polyurethane foams from base-catalyzed liquefaction of lignocellulosic biomass by crude glycerol: Effects of crude glycerol impurities. *Ind. Crop. Prod.* **2014**, *57*, 188–194. [CrossRef]

31. Corcuera, M.A.; Rueda, L.; Fernandez d'Arlas, B.; Arbelaiz, A.; Marieta, C.; Mondragon, I.; Eceiza, A. Microstructure and properties of polyurethanes derived from castor oil. *Polym. Degrad. Stab.* **2010**, *95*, 2175–2184. [CrossRef]

32. Carriço, C.S.; Fraga, T.; Pasa, V.M.D. Production and characterization of polyurethane foams from a simple mixture of castor oil, crude glycerol and untreated lignin as bio-based polyols. *Eur. Polym. J.* **2016**, *85*, 53–61. [CrossRef]

33. Modesti, M.; Adriani, V.; Simioni, F. Chemical and physical blowing agents in structural polyurethane foams: Simulation and characterization. *Polym. Eng. Sci.* **2000**, *40*, 2046–2057. [CrossRef]

34. Lim, H.; Kim, E.Y.; Kim, B.K. Polyurethane foams blown with various types of environmentally friendly blowing agents. *Plast. Rubber Compos.* **2010**, *39*, 364–369. [CrossRef]

35. Choe, K.H.; Lee, D.S.; Seo, W.J.; Kim, W.N. Properties of Rigid Polyurethane Foams with Blowing Agents and Catalysts. *Polym. J.* **2004**, *36*, 368–373. [CrossRef]

36. Thirumal, M.; Khastgir, D.; Singha, N.K.; Manjunath, B.S.; Naik, Y.P. Effect of Foam Density on the Properties of Water Blown Rigid Polyurethane Foam. *J. Appl. Polym. Sci.* **2008**, *108*, 1810–1817. [CrossRef]

37. Mills, N.J. *Polymer Foams Handbook: Engineering and Biomechanics Applications and Design Guide*; Butterworth Heinemann: Oxford, UK, 2007.
38. Jarfelt, U.; Ramnas, O. Thermal conductivity of polyurethane foam–best performance. In Proceedings of the 10th International Symposium on District Heating and Cooling, Göteborg, Sweden, 3–5 September 2006.
39. Ribeiro Da Silva, V.; Mosiewicki, M.A.; Yoshida, M.I.; Coelho Da Silva, M.; Stefani, P.M.; Marcovich, N.E. Polyurethane foams based on modified tung oil and reinforced with rice husk ash I: Synthesis and physical chemical characterization. *Polym. Test.* **2013**, *32*, 438–445. [CrossRef]
40. Gama, N.V.; Soares, B.; Freire, C.S.R.; Silva, R.; Neto, C.P.; Barros-Timmons, A.; Ferreira, A. Bio-based polyurethane foams toward applications beyond thermal insulation. *Mater. Des.* **2015**, *76*, 77–85. [CrossRef]

Sample Availability: Not Available.

 *molecules*

Review

Non-Conventional Features of Plant Oil-Based Acrylic Monomers in Emulsion Polymerization

Ananiy Kohut [1], Stanislav Voronov [1], Zoriana Demchuk [2], Vasylyna Kirianchuk [1], Kyle Kingsley [2], Oleg Shevchuk [1], Sylvain Caillol [3] and Andriy Voronov [2,*]

[1] Department of Organic Chemistry, Institute of Chemistry and Chemical Technologies, Lviv Polytechnic National University, 79000 Lviv, Ukraine; ananiy.kohut@gmail.com (A.K.); stanislav.voronov@gmail.com (S.V.); vasuluna411@ukr.net (V.K.); oshev2509@gmail.com (O.S.)
[2] Department of Coatings and Polymeric Materials, North Dakota State University, Fargo, ND 58108-6050, USA; zoriana.demchuk@ndsu.edu (Z.D.); kyle.kingsley.2@ndsu.edu (K.K.)
[3] ICGM, Univ Montpellier, CNRS, ENSCM, 34095 Montpellier, France; sylvain.caillol@enscm.fr
* Correspondence: andriy.voronov@ndsu.edu

Academic Editor: Dimitrios Bikiaris
Received: 8 June 2020; Accepted: 29 June 2020; Published: 30 June 2020

Abstract: In recent years, polymer chemistry has experienced an intensive development of a new field regarding the synthesis of aliphatic and aromatic biobased monomers obtained from renewable plant sources. A one-step process for the synthesis of new vinyl monomers by the reaction of direct transesterification of plant oil triglycerides with N-(hydroxyethyl)acrylamide has been recently invented to yield plant oil-based monomers (POBMs). The features of the POBM chemical structure, containing both a polar (hydrophilic) fragment capable of electrostatic interactions, and hydrophobic acyl fatty acid moieties (C_{15}-C_{17}) capable of van der Waals interactions, ensures the participation of the POBMs fragments of polymers in intermolecular interactions before and during polymerization. The use of the POBMs with different unsaturations in copolymerization reactions with conventional vinyl monomers allows for obtaining copolymers with enhanced hydrophobicity, provides a mechanism of internal plasticization and control of crosslinking degree. Synthesized latexes and latex polymers are promising candidates for the formation of hydrophobic polymer coatings with controlled physical and mechanical properties through the targeted control of the content of different POBM units with different degrees of unsaturation in the latex polymers.

Keywords: biobased polymers; plant oil-based monomers; mixed micelles; methyl-β-cyclodextrin inclusion complex; emulsion polymerization

---

1. Introduction

The problem of depletion of fossil raw materials has become of global importance and is complicated not only by economic but also environmental and political factors [1]. However, synthetic polymers are still one of the most widely used materials in everyday use. Environmental analysis shows that huge amounts of plastic waste are found in the environment, and their contribution to the ever-increasing amount of solid waste is a significant environmental threat [2]. Plant oils are low-cost raw materials for the manufacture of monomers and polymers [3–5], which have several advantages over conventional polymeric materials, namely biodegradability, non-toxicity, biocompatibility, and hydrophobicity. Therefore, an important task for synthetic chemists is to consider renewable raw materials as an alternative to raw counterparts of fossil origin. Due to the wide range of plant oils, the variety of their chemical compositions, and, in many cases, abandon of resources, they became an interesting object to be used in polymers synthesis as a renewable raw material.

The total production of plant oils in the world in 2019 amounted to more than 200 million tons, of which about 15% are used as raw materials for the synthesis of new chemical compounds and materials [6]. The presence of plant oils or their derivatives (fatty acids) in various compositions of polymeric materials improves their optical (gloss), physical (flexibility, adhesion) [7,8], and chemical (resistance to water and chemicals) properties [9]. The synthesis of biobased monomers from renewable resources is a promising platform for the synthesis and implementation of new environmentally friendly industrial polymer materials [10]. However, biomass molecules rarely possess suitable reactive functions for radical polymerization. Indeed, the double bonds of vegetable oils are poorly reactive in radical polymerization. Therefore, synthesizing radically polymerizable biobased monomers is a real challenge [11]. It should be noted that methods for the synthesis of vinyl monomers based on soybean, olive, and linseed oils through plant oil direct transesterification with *N*-(hydroxyethyl)acrylamide [10,12,13] has been recently developed along with methods of their copolymerization [14]. The ability of these plant oil-based monomers (POBMs) to undergo free radical (co)polymerization reactions has been confirmed and demonstrated [12,14,15]. The conversion, polymerization rate, and molecular weight of the polymers have been shown to depend on unsaturation of the plant oil chosen for POBMs synthesis [12,14].

2. Features of Synthesis of Vinyl Monomers from Plant Oils

Tang et al. [16] reported on the synthesis of *N*-hydroxyalkylamides and methacrylate hydrophobic monomers through the interaction of plant oil triglycerides with amino alcohols (Figure 1). They developed a new interesting approach that involves the use of a two-step process. In the first stage, the interaction of triglycerides with amino alcohols with the formation of *N*-hydroxyalkylamides by the amidation reaction:

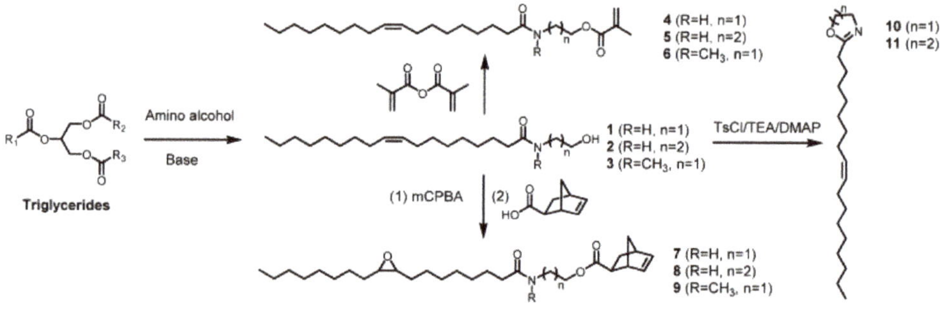

Figure 1. Transformation of triglycerides in N-hydroxyalkylamides and methacrylate monomers. Reprinted with permission from [16]. Copyright (2015) American Chemical Society.

Aminoalcohols, namely ethanolamine (R = H, n = 1), 3-amino-1-propanol (R = H, n = 2), and N-methylethanolamine (R = CH₃, n = 1), were used as chemicals which enables the synthesis of a wide range of new monomers. In the second stage, the corresponding monomers were synthesized by the interaction between N-hydroxyethylamides R′–C(O)–N(R)–(CH₂)ₙ–CH₂OH and methacrylic anhydride. In such a way, methacrylate monomer CH₂=C(CH₃)–C(O)–O–CH₂–(CH₂)ₙ–N(R)–C(O)–(CH₂)₇–CH=CH–(CH₂)₇–CH₃ based on high oleic soybean oil was synthesized.

A study of free radical polymerization of these monomers revealed that the double bonds in the fatty acid chains remain unaffected during the reaction. The synthesized methacrylate monomers based on high oleic soybean oil, depending on the nature of the branches in the side chain, have a wide range of glass transition temperatures [16]. In further studies, Tang and co-authors used more than twenty amino alcohols to establish the mechanism of amidation of plant oils, in particular high-oleic soybean oil [8]. Depending on the nature of the acyl fatty acid and amide structures, the polymers have a wide range of glass transition temperatures – from viscoelastic to thermoplastic materials. The possibility of controlled hydrogenation of unsaturated double bonds was also shown, which provides control over thermal and mechanical properties of polymers [8].

On the other hand, Harrison and Wheeler have shown for the first time [17] that the polymerization rate decreases with increasing unsaturation degree of acrylates containing acyl moieties of unsaturated fatty acids. They explained that the retarding of the polymerization rate and the low conversion are due to the interaction of radicals with the mobile hydrogen atoms in the acyl moieties. This leads to the formation of new allylic radicals of low activity which retards propagation of polymer chains (chain transfer reaction, retardation of the process) [17].

Through the esterification reaction of fatty alcohols with acryloyl chloride, Chen and Bufkin synthesized a range of acrylates, including linoleyl acrylate, oleyl acrylate, and lauryl acrylate, and studied their copolymerizability. They showed that the presence of fatty fragments in acrylate molecules determines their polarity, which leads to alternating copolymerization with methyl methacrylate [18].

In their next study [19], Chen and Bufkin confirmed Harrison's research results [17] and showed that the presence of fatty acid moieties in the structure of the new acrylates leads to a decrease in both the polymerization rate and polymer molecular weights because of chain transfer reactions. Increasing the content of unsaturated acyl groups in the copolymer also reduces its glass transition temperature. The authors suggested the mechanisms for crosslinking acrylate copolymers and studied their physical and mechanical properties [19].

Through amidation of triglycerides with amino alcohols, natural derivatives of surfactants and lubricants were obtained [20]. It was demonstrated that the high reactivity of the vinyl group of the acrylate monomers based on plant oils determines the course of their polymerization and copolymerization by a free radical mechanism [21]. Caillol et al. [22,23] proposed methods for the synthesis of a new monomer from cardanol (Figure 2), which is isolated from cashew nuts. Reactivity of the cardanol-based monomer in the copolymerization reactions with the POBMs was studied [24].

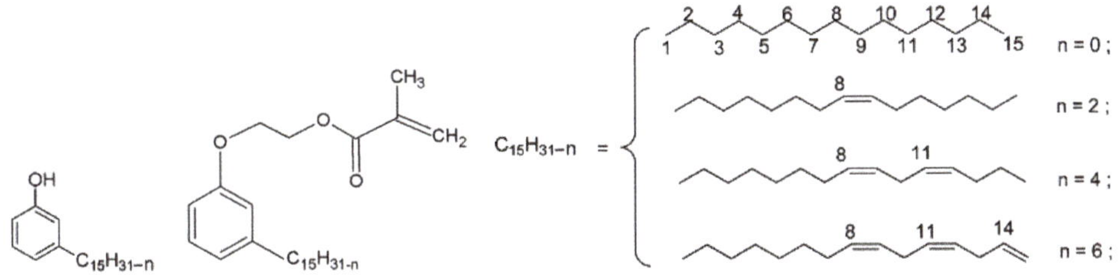

Figure 2. Structure of cardanol and a cardanol-based monomer. Adapted from [22] with permission from The Royal Society of Chemistry.

Cardanol-based polymers have a controlled flexibility due to "internal plasticization", which occurs due to the presence of a long side chain. The presence of these fragments also provides polymeric materials based on cardanol with hydrophobic properties. It should be noted that cardanol exhibits antimicrobial and anti-thermite properties and, under certain conditions, can provide them to copolymers based on it. It is compatible with a wide range of polymers, such as alkyds, melamines, polyesters, etc. [22].

In the study [23], synthetic pathways to molecular blocks from cardanol by one- or two-stage methods were reported. In particular: 1) dimerization/oligomerization of cardanol, to increase its functionality; 2) synthesis of reactive amines by thiolene coupling; 3) epoxidation and (meth)acrylation of cardanol for the synthesis of polymers and materials with oxirane or (meth)acrylate groups.

A feature of cardanol-based monomers and polymers is the presence of aromatic fragments and fatty moieties in their structure at the same time. The authors showed that cardanol-based polymeric materials have prospective properties for manufacturing coatings, in particular, with enhanced mechanical and thermal properties [23].

Recently, methacrylated ricinoleic acid monomer (Figure 3) was synthesized by the esterification reaction of ricinoleic acid with 2-hydroxyethyl methacrylate [25]:

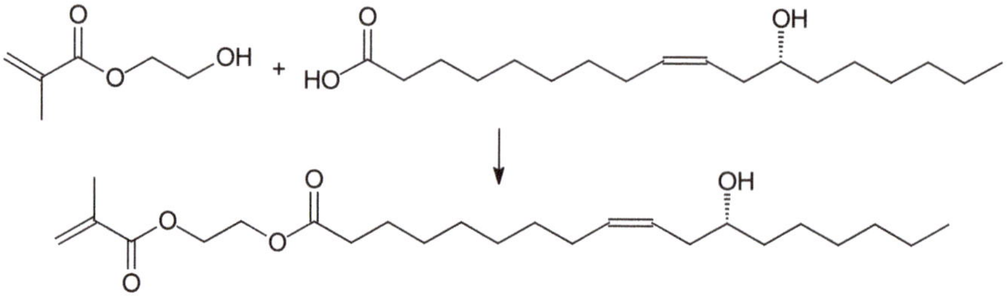

Figure 3. Schematic of synthesis of methacrylated ricinoleic acid monomer.

Other plant oil-based methacrylates have been widely studied in atom transfer radical polymerization (ATRP). A method for the ATRP of lauryl methacrylate was developed by Mandal et al. to obtain narrowly distributed polymers with predefined molecular weights [26]. Fatty alcohol derived methacrylates were used to prepare well defined main chain substituted polymethacrylates by copper-mediated ATRP [27].

It should be noted that all known methods for the synthesis of monomers from plant oil triglycerides are multi-stage and involve several steps, which makes their implementation in industry more challenging.

Recently, we developed a one-step process for the synthesis of new vinyl monomers by the direct transesterification reaction of plant oil triglycerides with N-(hydroxyethyl)acrylamide (Figure 4) [13,28].

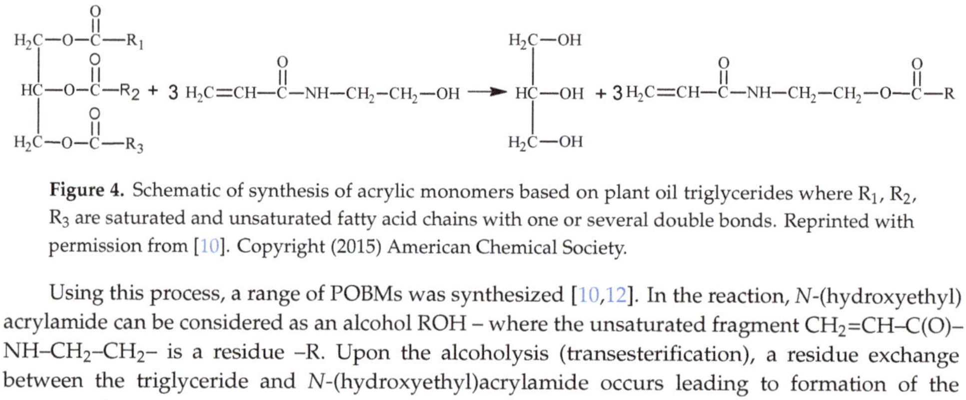

Figure 4. Schematic of synthesis of acrylic monomers based on plant oil triglycerides where R_1, R_2, R_3 are saturated and unsaturated fatty acid chains with one or several double bonds. Reprinted with permission from [10]. Copyright (2015) American Chemical Society.

Using this process, a range of POBMs was synthesized [10,12]. In the reaction, N-(hydroxyethyl) acrylamide can be considered as an alcohol ROH – where the unsaturated fragment CH_2=CH–C(O)–NH–CH_2–CH_2– is a residue –R. Upon the alcoholysis (transesterification), a residue exchange between the triglyceride and N-(hydroxyethyl)acrylamide occurs leading to formation of the corresponding monomers.

The POBMs synthesis was conducted in THF with an excess of N-(hydroxyethyl)acrylamide (molar ratio of N-(hydroxyethyl)acrylamide to triglyceride as 5.9: 1) in order to achieve complete transesterification of the triglycerides. The yield of the desired monomers was about 94–96%. The reaction by-product (glycerol) and the excessive N-(hydroxyethyl)acrylamide are easily removed by washing with brine after diluting the reaction mixtures with CH_2Cl_2. The POBMs sparingly soluble in water remain in the organic phase. To avoid free radical polymerization, 2,6-di-*tert*-butyl-p-cresol (0.05 to 0.1% by weight) was added as an inhibitor to the reaction mixtures prior the synthesis.

Upon the alcoholysis of olive oil with N-(hydroxyethyl)acrylamide, 2-N-acryloylaminoethyl oleate CH_2=CH–C(O)–NH–CH_2CH_2–O–C(O)–$(CH_2)_7$–CH=CH–$(CH_2)_7$–CH_3 is predominantly formed.

This monomer contains one double bond in the acryloylamide moiety and one double bond in the fatty acid chain. Such a structure, in comparison with the soybean oil-based monomer 2-*N*-acryloylaminoethyl linoleate (CH$_2$=CH–C(O)–NH–CH$_2$CH$_2$–O–C(O)–(CH$_2$)$_7$–CH=CH–CH$_2$–CH=CH–CH$_2$–(CH$_2$)$_3$–CH$_3$) reduces the extent of the chain transfer reaction. Accordingly, fewer less active radicals are formed and the polymerization reaction proceeds to higher conversion with a less pronounced retardation effect. This leads to the formation of polymers (copolymers) with a higher molecular weight and a lower polydispersity index.

The composition of plant oils is known to depend on the type of crop raw material. Therefore, an important step was to determine the content of different fatty acid chains in the mixture of monomers obtained upon the transesterification of olive oil with *N*-(hydroxyethyl)acrylamide. The content of fatty acid chains in the oil was determined from the integral intensities of the characteristic proton signals of each fatty acid chain and glycerol fragments using ^{1}H-NMR spectroscopy of the plant oil [29] (Figure 5).

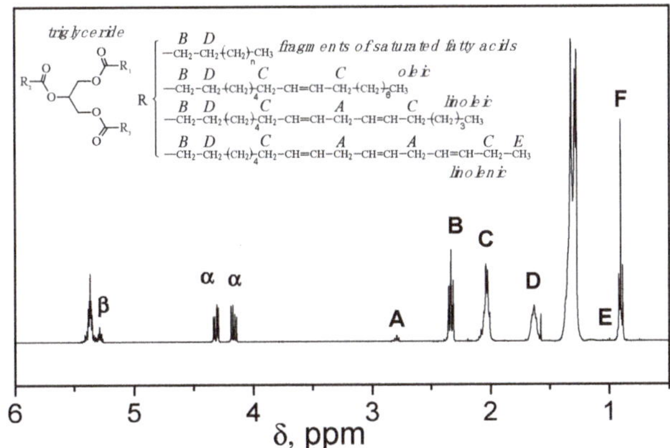

Figure 5. ^{1}H-NMR spectrum of olive oil. Adapted with permission from [12]. Copyright (2016) American Chemical Society.

The content of the linolenic acid units (linolenate) in olive oil was calculated by measuring the integral value of the signal at 1 ppm, which corresponds to the protons of the methyl group in a linolenic acid chain (signal E in Figure 5). Taking into account only one of the signals of α-protons of glycerol, i.e., the signal at 4.27 ppm (Figure 5), the ratio of the integral values is two α-glycerol protons to three protons of the linolenic acid methyl group. It should be kept in mind that three linolenic acid chains can undergo transesterification in the same glycerol molecule. A field correction factor is the ratio of two protons of glycerol to nine protons of the linolenic acid methyl groups. By converting the ratio of these areas to percentages, a ratio of 22.2% glycerol to 100% linolenic acid was obtained [29]. Calibration of the integral value of one of the glycerol α-proton signals to the value in the ^{1}H-NMR spectrum in the signal region at 0.98 ppm directly gives the percentage of linolenic acid in the sample −0.73%.

The percentage of linoleic acid chains (linoleates) in olive oil was determined from the ratio of the integral value of the signal at ~2.74 ppm, which corresponds to the methylene protons between two double bonds (signal A in Figure 5), to the integral value of one of glycerol α-protons. The field correction factor (33.3) is the ratio of two glycerol protons to six possible methylene protons in linoleate. The amount of linoleic acid is calculated by subtracting the twiced content of linolenic acid, determined from the peak at 2.74 ppm [26].

The percentage of oleic acid chains (oleates) was determined from the ratio of the integral value of the signal at ~2.02 ppm, which corresponds to the protons in the α-position to the double bond of all unsaturated fatty acids (signal C in Figure 5), to the integral value of one of glycerol α-protons.

Accordingly, the ratio of two glycerol protons to twelve possible protons in the α-position to the double bond of all unsaturated fatty acids is the field correction factor equal to 16.7. The percentage of oleic acid was calculated by subtracting the content of unsaturated linolenic and linoleic acid chains from the determined value [29].

The content of saturated fatty acid chains was determined from the fact that the total content of fatty acids is 100%, and the amount of unsaturated fatty acid chains was subtracted from 100%. The determined content of fatty acid chains is given in Table 1. These data are in a good agreement with the literature data obtained by gas chromatography of olive oil [30].

Table 1. Content of fatty acid chains in olive oil.

Fatty Acid Chains	Signal (Protons)	Field Correction Factor	%	Content of Fatty Acids in the Oil, % Calculated	Literature data
Linolenate (E)	0.95–1.05 ppm (-CH$_3$)	2H/9H	22.2	0.73	<1%
Linoleate (A)	2.75–2.85 ppm (-CH=CH-CH$_2$-CH=CH-)	2H/6H	33.3	$8.06 - 2 \cdot 0.73 = 6.6$	3.5–21
Oleate (C)	1.97-2.11 ppm (-CH$_2$-CH=CH-CH$_2$-)	2H/12H	16.7	$87.56 - (0.73 + 6.6) = 80.23$	55–83%
Content of saturated fatty acid chains:				$100 - (0.73 + 6.6 + 80.23) = 12.44$	1–20%

Thus, according to ^{1}H-NMR spectroscopy, the olive oil triglycerides include: saturated fatty acid chains (C18:0) −12.44%; oleic acid chains (C18:1) −80.23%; linoleic acid chains (C18:2) −6.60%; and linolenic acid chains (C18:3) −0.73%.

These results are consistent with the literature data obtained by gas chromatography of olive oil, where the saturated chains make up 1–20%, oleic acid chains 55–83%, linoleic acid chains 3.5–21%, and linolenic acid chains less than 1% [31]. Using this approach, monomers based on various plant oils were successfully synthesized and characterized (Table 2).

Table 2. Physico-chemical characteristics of POBMs.

Monomer from Oil	Content of Unsaturated Fatty Acids, % C$_{18:1}$	C$_{18:2}$	C$_{18:3}$	Iodine Value (Same for Oil), g/g	Molar Mass, g/mol
High Oleic Sunflower Oil (HOSFM)	86–89	3–6	0.5–1	105 (82)	379.0 [1]
Olive Oil (OVM)	65–85	3.5–20	0–1.5	110 (90)	379.3 [2]
High Oleic Soybean Oil (HOSBM)	70–73	13–16	0–1	124(105)	379.3 [2]
Canola Oil (CLM)	60–63	18–20	8–10	137 (96)	379.0 [2]
Corn Oil (COBM)	23–31	49–62	0–2	139 (120)	377.0 [1]
Sunflower Oil (SFM)	14–35	44–75	0–1	146 (128)	377.5 [2]
Soybean Oil (SBM)	22–34	43–56	7–10	149 (139)	377.3 [2]
Linseed Oil (LSM)	12–34	17–24	35–60	194 (177)	375.6 [1]

[1] Calculated. [2] Experimental.

The monomer structure was confirmed by FTIR and ^{1}H-NMR spectroscopy [12]. A ^{1}H-NMR spectrum of the olive oil-based monomer is shown in Figure 6. Similar spectra are recorded for other POBMs, which confirms the presence of the acryloylamide moiety in their molecules. This allows for predicting similar reactivity of the vinyl group in the monomers acryloylamide moiety in the radical polymerization [12].

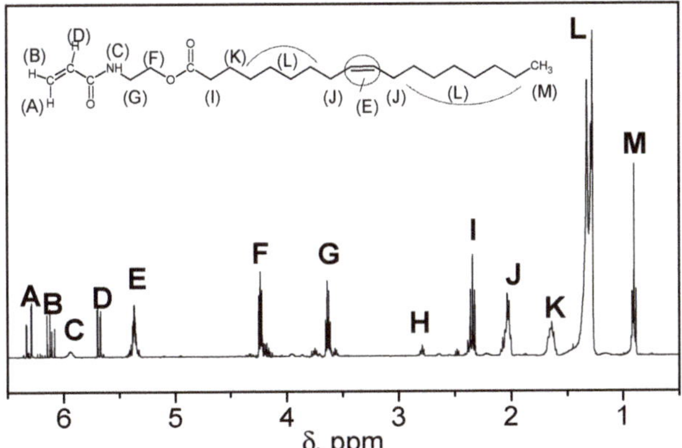

Figure 6. ^{1}H-NMR spectrum of the olive oil-based monomer. Adapted with permission from [12]. Copyright (2016) American Chemical Society.

The FTIR spectroscopy data indicate the addition of fatty acid acyl moieties to the acrylamide fragment (e.g., for the olive monomer—Figure 7). FTIR spectra of monomers based on other plant oils are pretty similar. Hence, the synthesized POBMs can be attributed to the conventional vinyl monomers because the acryloylamide moiety provides the participation of these monomers in the free radical polymerization.

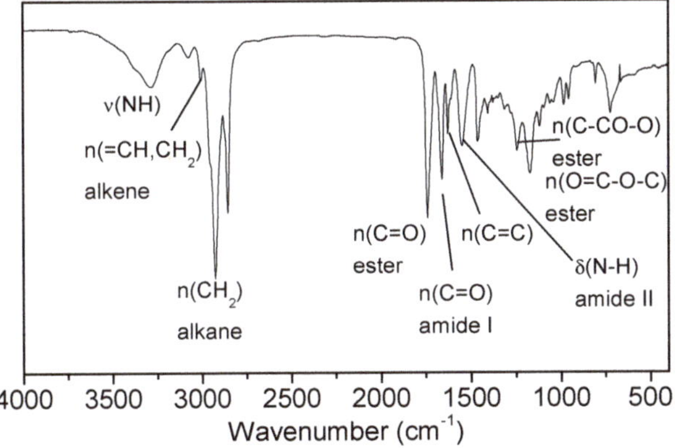

Figure 7. FTIR spectrum of the olive oil-based monomer. Adapted with permission from [12]. Copyright (2016) American Chemical Society.

The content of fatty acid chains in the monomer mixture produced by the transesterification of plant oils with *N*-(hydroxyethyl)acrylamide was determined on the example of the olive monomer by calculating the ratio of the integral values in the ^{1}H-NMR spectra of the commercial oil and the resulting monomer mixture (Figure 8).

According to the ^{1}H-NMR spectroscopy data, the monomer mixture consists of: saturated fatty acid chains (C18:0) −11.47% (stearic and palmitic); oleic acid chains (C18:1) −81.9%; linoleic acid chains (C18:2) −5.9%; and linolenic acid chains (C18:3) −0.73% after the olive oil transesterification.

Thus, the major component of the olive monomer is a monomer with oleic acid chains (81.9%) − 2-*N*-acryloylaminoethyl oleate. The soybean monomer is mainly a monomer with linoleic acid chains

(57.5%) – 2-*N*-acryloylaminoethyl linoleate whereas the linseed monomer predominantly consists of a monomer with linolenic acid chains (58%) – 2-*N*-acryloylaminoethyl linolenate.

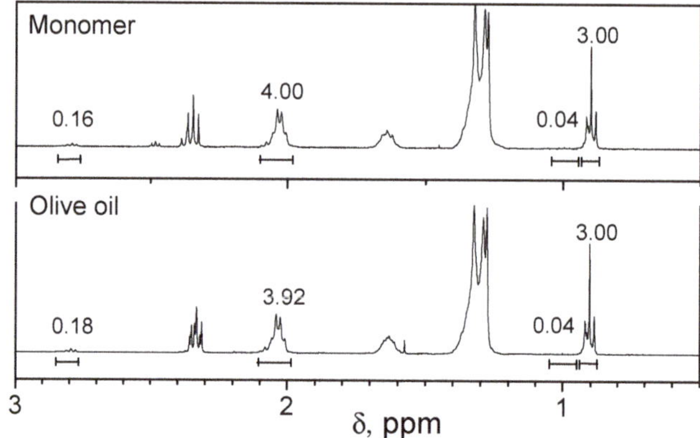

Figure 8. ^{1}H NMR spectra of the olive oil and the olive oil-based monomer.

One of the most important characteristics of monomers is the degree of unsaturation, which for POBMs is determined by the number of the double bonds in the fatty acid chains. To compare the monomers in terms of the unsaturation degree, their iodine values were determined and compared with those for the corresponding plant oils used for the synthesis of the POBMs (Table 2). The obtained results show that the iodine value for the monomers is higher than those for the oils, due to the presence of an unsaturated acryloylamide moiety. Depending on the type of oil, the iodine values for various monomers differ. For instance, the iodine value for 2-*N*-acryloylaminoethyl oleate (110 g/100 g) is significantly lower than that for 2-*N*-acryloylaminoethyl linoleate (149 g/100 g) due to the different unsaturation degree of the molecules. Besides the acryloylamide moiety, there are two double bonds in the fatty acid chain of the soybean monomer molecule [10]. The low solubility of these monomers in water implies their highly hydrophobic nature [12].

Analysis of the data obtained from the synthesis and characterization of plant oil-based acrylic monomers allows for drawing the POBM general formula (Figure 9) as:

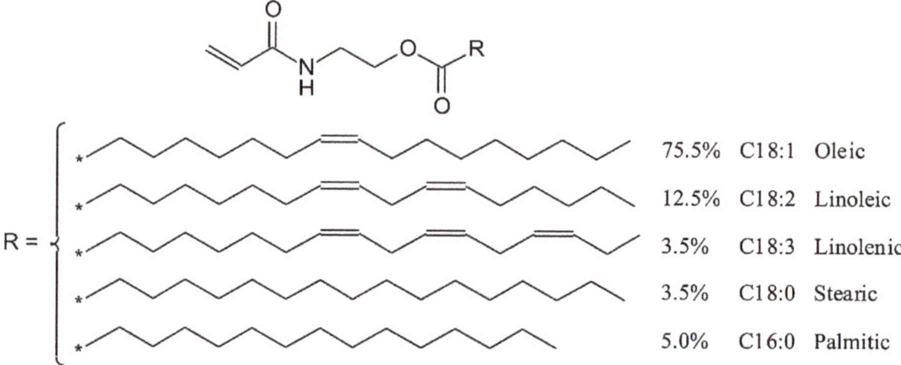

Figure 9. General formula of the plant oil-based monomers. Reprinted from [32]. Copyright (2018), with permission from Elsevier.

3. Features of Homo- and Copolymerization of Plant Oil-Based Acrylic Monomers

The reactivity of the POBMs in the free radical polymerization reactions (chain growth reaction) is determined by the presence of an acryloylamide moiety containing a vinyl group which is common for all monomers. However, the POBMs contain a certain amount of unsaturated fatty acid chains of various structures (the different numbers of both double bonds and hydrogen atoms in the α-position to the double bonds). This causes the monomers to participate in chain transfer reactions due to the abstraction of the allylic hydrogen atoms and the formation of less active radicals. Hence, the monomer chain transfer constants C_M clearly depend on monomer structure as follows: 0.033 (most unsaturated LSM) > 0.026 (SBM) > 0.023 (SFM) > 0.015 (least unsaturated OVM) with respect to decreasing number of C−H groups in the α-position of the fatty acid double bonds) [12]. This impacts the polymerization conversion and molecular weights of the resulting polymers (copolymers) from the POBMs.

The features of homopolymerization kinetics for 2-*N*-acryloylaminoethyl oleate revealed that the orders of reaction with respect to monomer and initiator are 1.06 and 1.2, respectively. The unsaturation degree of fatty acid chains in the POBM molecules was used as a criterion for comparing the kinetic data of homopolymerization of 2-*N*-acryloylaminoethyl oleate and 2-*N*-acryloylaminoethyl linoleate. The observed deviations of the orders of reaction are due to the specific mechanism of the homopolymerization reaction of 2-*N*-acryloylaminoethyl oleate which includes two simultaneous reactions (chain propagation and transfer) [12]. Although the propagating radicals might be very reactive, once the chain is transferred to the allylic C−H, the newly formed radical becomes more stable due to resonance stabilization and does not readily initiate new chains. In comparison with the soybean monomer, the olive monomer is less involved into the chain transfer reactions (the chain transfer constant $C_M = 0.015$ for OVM while for SBM $C_M = 0.026$). As a result, the homopolymerization of the olive monomer occurs at a higher rate of $12.2–45.3 \cdot 10^{-5}$ mol/(L·s) when compared with the more unsaturated soybean monomer $−4.3−11.3 \cdot 10^{-5}$ mol/(L·s). The molecular weights of the homopolymers were determined by gel permeation chromatography. The resulting homopolymers from OVM have higher number average molecular weights (16 800–23 200 g/mol for 2-*N*-acryloylaminoethyl oleate compared to 13 600–14 300 g/mol for 2-*N*-acryloylaminoethyl linoleate).

The reactivity of the POBMs in chain copolymerization was studied in their reactions with styrene and vinyl acetate. A characteristic feature of the POBMs is the presence of the acryloylamide fragment ($CH_2=CH-C(O)-NH-CH_2-CH_2-$) in their structure, which determines the monomer reactivity in copolymerization. The composition of the copolymers was determined from ^{1}H-NMR spectroscopy data. Radical copolymerization of the plant oil-based monomers is described with the classical Mayo–Lewis equation. The Alfrey–Price Q (0.41–0.51) and e (0.09–0.28) parameters are close for all POBMs and do not essentially depend on the monomer structure [13]. This is due to the presence of the same acryloylamide moiety in the POBM molecules which determines the monomer reactivity in polymerization reactions.

The features of copolymerization are determined by the structure of monomers based on plant oil triglycerides along with the degradative chain transfer by formation of less active radicals. Nevertheless, the growth of macrochains can be described with the conventional features of chain copolymerization, yielding copolymers with a potential to be used in a broad variety of applications.

Vinyl monomers from plant oils that have different degrees of unsaturation, soybean, and olive oils, were copolymerized in emulsion with styrene to investigate the kinetics features and feasibility of latex formation. The kinetics of emulsion copolymerization of styrene with the olive and soybean monomers agree with the Smith-Ewart theory since the number of nucleated latex particles is proportional to the surfactant and initiator concentration to the powers 0.58–0.64 and 0.39–0.46, respectively [33]. Copolymerization of styrene with POBMs follows the typical phenomenology for emulsion polymerization of hydrophobic monomers with a micellar nucleation mechanism. When the POBMs were copolymerized in emulsion with significantly more water soluble comonomers, methyl methacrylate and vinyl acetate, latex particle nucleation mainly occurred through homogeneous mode. For the emulsion copolymerization of methyl methacrylate with OVM and SBM, the reaction orders with respect to emulsifier and initiator are 0.33–0.67 and 0.56–0.69, respectively. It was shown that upon adding the highly hydrophobic POBMs into the monomer mixture, the latex polymer particles originating from micellar nucleation essentially increases [15].

Using emulsion and miniemulsion copolymerization of the olive and soybean monomers with styrene or methyl methacrylate, stable aqueous dispersions of polymers with latex particle sizes of 40–210 nm were produced. The content of the POBM units in the macromolecules of the latex polymers is 5–60 wt.%. The average molecular weight of the synthesized polymers varies in the range of 30,000–391,500. It was found that the molecular weight of the latex copolymers decreases with increasing unsaturation degree of the POBMs and their content in the reaction mixture, which is explained by degradative chain transfer to unsaturated fatty acid chains [15,33].

An analysis of the literature shows a rapid development of a new branch in polymer science related to the chemistry of aliphatic and aromatic biobased monomers obtained from renewable plant sources. The synthesis of such monomers allows for producing fully biobased polymers with biocompatible and biodegradable properties which do not pollute the environment. Copolymerization of monomers synthesized from various plant oils, which have different unsaturation, enables formation of copolymers with side branches of the macrochain with different unsaturation degrees. A large variety of the POBMs allows to synthesize polymers having moieties with different unsaturation and to make coatings with adjustable cross-linking degree thereof, including latex copolymers from fully biobased monomer mixtures [24]. Remarkably, copolymerization of the POBMs with commercial vinyl monomers enables synthesis of polymers with enhanced hydrophobicity along with the mechanism of internal plasticization.

Synthesized latex polymers and copolymers are prospective candidates for the formation of moisture-/water-resistant polymer coatings with controlled physical and mechanical properties using controlled content of incorporated POBM units with different unsaturation in the latex structure.

4. Preparation of Polymeric Coatings from Plant Oil-Based Monomers

As shown by Solomon [34], both crude and modified plant oils are common film-forming materials for the production of paints and varnishes. Despite the emergence of various synthetic polymers based on vinyl and acrylic monomers, crude plant oils are still the basis of some paints for painting the roofs of houses and other outdoor objects. One of the disadvantages of such materials is the slow drying and relatively low moisture resistance. The synthesis of plant oil-based acrylic monomers opens new avenues for producing homo- and copolymers thereof with adjustable physical and mechanical properties (e.g., flexibility, strength, *etc.*) due to formation of three-dimensional networks with controlled cross-linking density. The development of such coatings was shown to be carried out through oxidative cross-linking of the polymer network [35].

It should be noted that upon POBM application for the synthesis of latex polymers as highly hydrophobic film-forming protective coatings, they form three-dimensional networks with adjustable cross-linking density through the oxidative cross-linking mechanism.

Triglycerides of linseed oil contain about 52% of linolenic (*cis,cis,cis*-9,12,15-octadecatrienoic) acid chains which have three isolated carbon-carbon double bonds in their structure (Figure 10). Other fatty acids in linseed oil are oleic (*cis*-9-octadecenoic, 22%) and linoleic (*cis,cis*-9,12-octadecadienoic, 16%) acids. Hence, linolenate is a major constituent of LSM.

Oxidative cross-linking of the LSM-based copolymers is a free radical chain process consisting of chain initiation, propagation, and termination steps. Initiation, i.e., the formation of a fatty acid chain radical, can occur by thermal homolytic cleavage of a C–H bond or by a hydrogen atom abstraction from C–H by an initiator free radical. The bond dissociation energy of a bisallylic hydrogen is approximately 42 kJ/mol lower than that of an allylic hydrogen. As a result, linoleates and linolenates are more readily autooxidized and cross-linked in comparison with oleates.

Hydrogen atom abstraction from C-11 or C-13 of the linolenate moiety leads to pentadienyl radicals I and II (Figure 10). Subsequent oxygen addition to radicals I and II generates peroxyl radicals which are able of abstracting a hydrogen atom from a donor such as another linolenate moiety to form conjugated *trans,cis*-hydroperoxides I–IV (Figure 10). The hydroperoxides undergo decomposition reaction (Figure 11). The last three reactions lead to cross-linking of the macromolecules and formation of a polymer network.

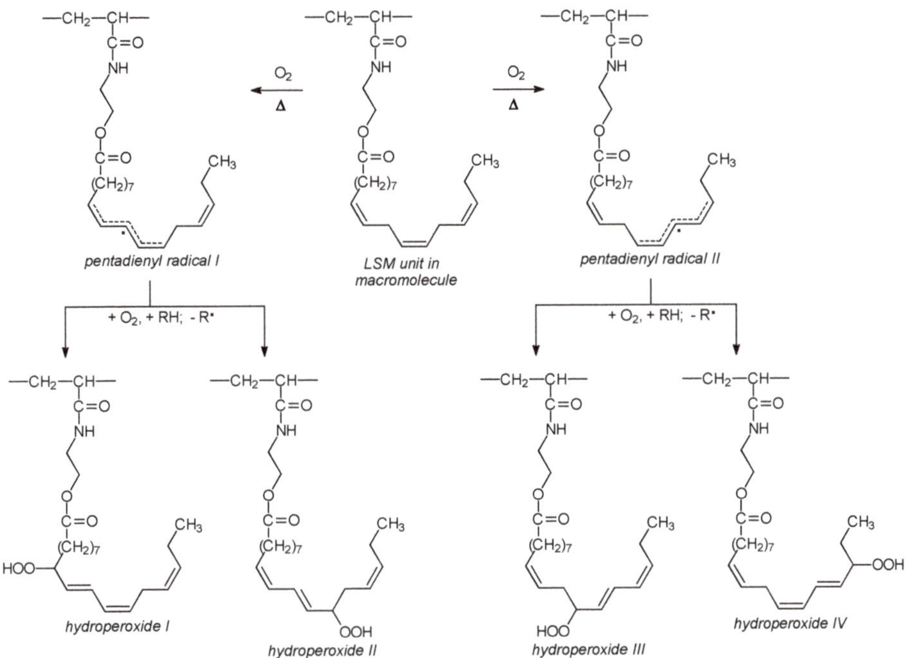

Figure 10. Formation of radicals and hydroperoxides during oxidative cross-linking of polymers from linseed oil-based monomer (LSM).

$$ROOH \longrightarrow RO^{\bullet} + {}^{\bullet}OH$$

$$2ROOH \longrightarrow RO^{\bullet} + ROO^{\bullet} + H_2O$$

$$ROOH + RH \longrightarrow RO^{\bullet} + R^{\bullet} + H_2O$$

$$RO^{\bullet} + R'H \longrightarrow ROH + R'^{\bullet}$$

$$RH + {}^{\bullet}OH \longrightarrow R^{\bullet} + H_2O$$

$$R^{\bullet} + R^{\bullet} \longrightarrow R{-}R$$

$$RO^{\bullet} + R^{\bullet} \longrightarrow R{-}O{-}R$$

$$RO^{\bullet} + RO^{\bullet} \longrightarrow R{-}O{-}O{-}R$$

where RH and R'H are unaltered fatty acid chains in macromolecules

Figure 11. Oxidative cross-linking of polymers from linseed oil-based monomer (LSM).

The main mechanical characteristics of films from latex polymers based on styrene or methyl methacrylate and the POBMs were determined. The composition of the copolymers was calculated from the ^{1}H-NMR spectroscopy data. The content of the olive and soybean monomers in the reaction mixture varied from 10 to 40 wt. % for copolymers with methyl methacrylate and from 25 to 60 wt. % for copolymers with styrene. A plasticizing effect and enhanced hydrophobicity were observed for the copolymers synthesized from 2-*N*-acryloylaminoethyl oleate and 2-*N*-acryloylaminoethyl linoleate as comonomers. The unsaturation degree of the olive and soybean monomers was used as an experimental parameter to control the latex properties. The effect of monomer unsaturation on the cross-linking density of the latex films and, thus, on the physical and mechanical properties of the coatings was demonstrated [35].

The decrease in the glass transition temperature of the latex copolymers (from 105 to 57 °C for the copolymers of the olive and soybean monomers with methyl methacrylate and from 100 to 5 °C for the copolymers with styrene) indicates that the presence of the OVM and SBM units in the macromolecules affects thermomechanical properties of the resulting latex copolymers. The olive and soybean monomer units impart flexibility to the macromolecules, improve the conditions of film formation, increase the strength in comparison with conventional polystyrene and poly(methyl methacrylate). Moreover, the OVM and SBM units in the macromolecules increase the hydrophobicity of polymeric latex films thus reducing the negative impact of water on the properties of the coatings.

Therefore, the incorporation of the hydrophobic fatty acid chains into the macromolecules of latex polymers allows for formation of polymer networks with a controlled cross-linking density along with enhancing the water resistance of the coatings. The copolymers based on the olive and soybean monomers enable the formation of coatings with low surface energy and water-repellent properties.

5. Formation of Micelles from the Complexes of Highly Hydrophobic Plant Oil-Based Monomers with Sodium Dodecyl Sulfate

The synthesized POBMs contain $CH_2=CH-C(O)-NH-CH_2CH_2-O-C(O)-$ as a hydrophilic fragment in their chemical structure and R– as a hydrophobic constituent in the fatty acid chain. Such a monomer structure imparts unique properties to the POBMs associated with the ability to participate in intermolecular (electrostatic and/or van der Waals) interactions to produce complexes (aggregates).

The influence of high oleic soybean oil-based (HO-SBM) and olive oil-based acrylic monomers on micellization of sodium dodecyl sulfate (SDS) was studied at different monomer and surfactant concentrations. The obtained data imply that SDS is able to solubilize highly hydrophobic monomer molecules. This leads to development of mixed (SDS/POBM) micelles (Figure 12). Surface activity of a SDS/POBM mixture varies by adding the monomer and, as a rule, is higher than for the surfactant alone [36]. It was also found that monomer molecules replace surfactant counterparts in the mixed micelles. Both the DLS and TEM data allow for conclusion that incorporation of POBM into the mixed micelles facilitates micellar association and development of 25–30 nm structures. Hence, solubilization of monomer molecules by SDS can affect mechanism of emulsion polymerization and reaction kinetics, as well as have an effect on development of latex particles and their morphology.

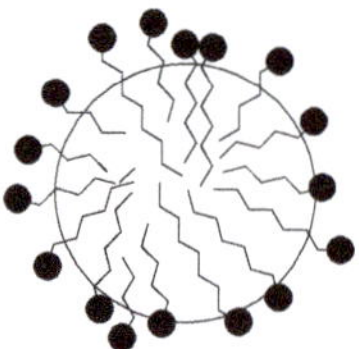

Figure 12. Possible location of the polar "heads" in a direct micelle from the SDS molecules and the POBM/SDS aggregate.

Surfactants form direct micelles due to the interactions between hydrocarbon parts of their molecules. Usually, the micelle has a layered structure depending on the packaging of surfactant molecules. A micelle contains a hydrocarbon nucleus, a water-hydrocarbon layer, which includes 2–3 methylene groups, and a layer of hydrated polar groups. It should be noted that surfactant molecules are freely localized in the micelle. They can leave the micelle or move to a different position; they can even be engulfed into the nucleus, which is determined by the hydrophobicity of the surfactant fragment [37,38].

The feature of micellization of sodium dodecyl sulfate (SDS) in the presence of the POBMs is surfactant ability to solubilize highly hydrophobic monomer molecules and promote development of mixed (SDS/POBM) micelles. As shown by Kingsley et al. [36], the packing of surface-active SDS/POBM complexes (aggregates) and the SDS molecules occurs upon micellization. In this event, the SDS molecules can be squeezed out of the micelle and localized at the interphase reducing the interfacial tension. Sodium dodecyl sulfate is known to have a micelle aggregation number of 60–70 [38]. It was found that the aggregation number decreases to 10–48 (Table 3) upon micellization of the SDS/POBM complexes. The micellar size depends on the packaging of the surfactants. Their length and packing density in the micelle determine the radius of the hydrocarbon nucleus. The formed colloidal solution simultaneously contains the SDS micelles and the mixed SDS/POBM micelles. It should be noted that their size is 18–38 nm (Table 3, Figure 13) whereas the SDS micellar size is 1.7–3.1 nm.

Table 3. Micellar parameters for SDS and POBM at different concentrations.

SDS + POBM (x,mol: y,mol)	N_{agg}	N_{POBM}	I_1/I_3	d,nm	PDI	σ, mN/m
SDS at 0.02 M	41	0	1.04	3.1	0.006	36.2
+ HO-SBM 0.01 M	25	20	0.94	18.2	0.02	34.1
0.02 M	15	25	0.94	23.3	0.05	31.2
0.04 M	12	39	0.92	25.4	0.04	30.4
SDS at 0.05 M	57	0	1.03	1.7	0.003	34.9
+ HO-SBM 0.02 M	46	22	0.96	27.3	0.03	31.5
0.04 M	38	34	0.95	28.5	0.05	29.8
SDS at 0.02 M	41	0	1.03	3.1	0.006	36.2
+ OVM 0.01 M	27	22	0.95	22.6	0.06	32.4
0.02 M	19	31	0.93	28.7	0.04	31.1
0.04 M	10	33	0.93	37.8	0.04	30.1
SDS at 0.05 M	57	0	1.03	1.7	0.003	34.9
+ OVM 0.02 M	48	23	0.95	28.2	0.02	32.6
0.04 M	31	28	0.94	33.9	0.04	30.2

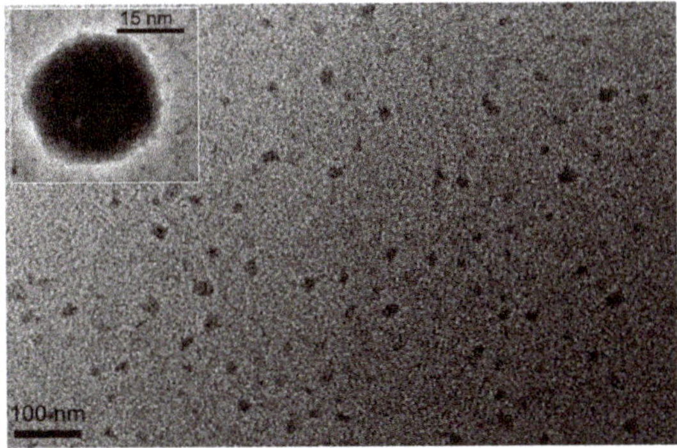

Figure 13. TEM micrograph of micelles prepared by mixing SDS (0.02M) and HO-SBM (0.01M) (inset shows morphology of selected individual micelle). Reprinted from Ref. 36, Copyright (2019), with permission from Elsevier.

If the fatty acid chain cannot leave the hydrocarbon nucleus, the polar group can be even drawn into the hydrophobic nucleus [38], which also affects the aggregation number. The authors explain the features of micellization (micellar parameters, size, structure, and surface tension) and the obtained results with the formation of surface-active SDS/POBM complexes (aggregates) of the following structure (Figure 14):

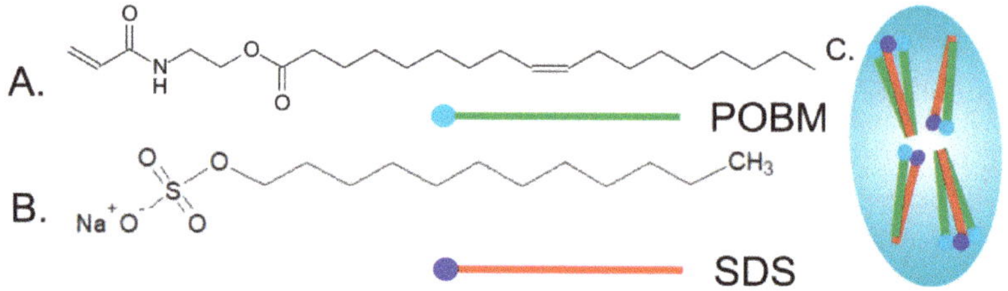

Figure 14. Chemical structure of plant oil-based monomer (**A**), surfactant (**B**), and schematic of POBM solubilization by SDS molecules (**C**). Reprinted from [36]. Copyright (2019), with permission from Elsevier.

The formation of a water-hydrocarbon layer of the spherical micelle leads to the localization of the acrylic groups of the POBM molecule at the interface. This opens up new possibilities for the formation of "core–shell" morphology, providing ability of POBM to undergo polymerization.

Chemical structure of POBM and SDS molecules is similar. They both have hydrophilic "head" and long non-polar "tail" with in average 17 (POBM) and 12 (SDS) carbon atoms. It was suggested that physical association of both polar "heads" and hydrophobic "tails" in water underlie intermolecular interactions. The results indicate that surface activity of a SDS/POBM mixture is usually higher than for the surfactant and varies by adding the monomer. Moreover, structure, size, and micellar parameters, observed changes upon the addition of the POBMs. The formed complexes can open up new opportunities for micellization and, thus, new approaches to conducting heterophase polymerization (e.g., conventional emulsion and miniemulsion polymerization).

6. Dual Role of Methyl-β-Cyclodextrin in the Emulsion Polymerization of Highly Hydrophobic Plant Oil-Based Monomers

The new acrylic monomers are also capable of complexation with cyclodextrins, which play a dual role in both protecting against allylic termination chain transfer as well as improving monomer polymerizability.

Methyl-β-cyclodextrin (M-β-CD), amphiphilic oligosaccharide, was utilized to enhance the polymerizability of high oleic soybean oil-based and linseed oil-based monomers in copolymerization with styrene. Interactions between M-β-CD and the POBMs leading to the development of 1:1 "host-guest" complex were revealed. In the presence of the oligosaccharide, polymer yield rises whereas the coagulum content decreases during the emulsion polymerization of POBMs with styrene implying their improved solubility in water. Latex polymers with a higher molecular weight were synthesized in the presence of inclusion complexes. Incorporation of POBM molecules into the M-β-CD cavities protects the fatty acid chains and reduces chain transfer as confirmed by ^{1}H-NMR spectroscopy. This effect is more pronounced for more unsaturated linseed oil-based monomer.

These results are explained by the specific structures of M-β-CD having a hydrophobic cavity and a hydrophilic exterior, and the POBM molecule, possessing a long hydrophobic fatty acid chain; this results in the formation of an inclusion complex (Figure 15).

To confirm the formation of the inclusion complex, it was used Powder X-ray diffractometry (PXRD) [39]. The PXRD spectra of H-SBM [monomer from hydrogenated (very low degree of unsaturation) soybean oil], M-β-CD, H-SBM/M-β-CD inclusion complex, and the physical mixture are depicted in Figure 16A. The diffractogram of the physical mixture of M-β-CD and H-SBM was the superposition of the monomer and the oligosaccharide implying that there is no interaction between them [40]. The diffraction pattern of the H-SBM/M-β-CD inclusion complex displayed no characteristic peaks that the pure monomer had. These data reveal that the self-lattice arrangement of the monomer was changed from a crystalline to amorphous state due to the H-SBM inclusion into the methyl-β-cyclodextrin cavity.

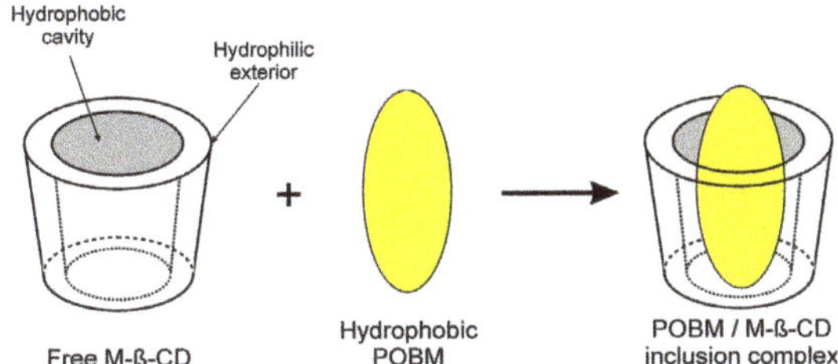

Figure 15. Schematic illustration of the association of free methyl-β-cyclodextrin and POBM to form a 1:1 inclusion complex. Reprinted from [32]. Copyright (2018), with permission from Elsevier.

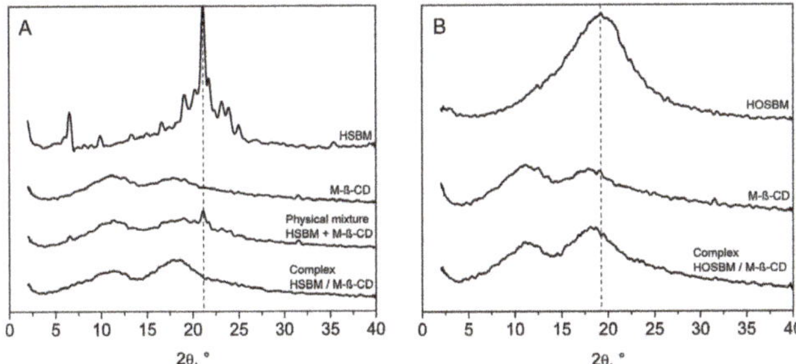

Figure 16. PXRD spectra of POBMs, methyl-β-cyclodextrin, their physical mixture, and inclusion complexes (**A**) – H-SBM, (**B**) – HO-SBM. Reprinted from [32]. Copyright (2018), with permission from Elsevier.

The diffractograms of HO-SBM, M-β-CD, and HO-SBM/M-β-CD inclusion complex are shown in Figure 16B. As both M-β-CD and high oleic soybean oil-based monomer are non-crystalline substances, the development of HO-SBM/M-β-CD inclusion complex cannot be proven by PXRD.

To this end, differential scanning calorimetry [41] was used to further confirm the complexation. The DSC thermograms of H-SBM, M-β-CD, H-SBM/M-β-CD inclusion complex, and the physical mixture are shown in Figure 17. Although the endothermic peaks of the monomer were detected in the DSC curve of the physical mixture, no peak in the melting range of the monomer was detected in the thermogram of H-SBM/M-β-CD. This confirms that the H-SBM molecules are entirely included into the oligosaccharide cavities and H-SBM/M-β-CD inclusion complex forms.

No peak in the melting range of high oleic soybean oil-based monomer was observed in the DSC curves of HO-SBM/M-β-CD inclusion complex and their physical mixture (data not shown). Obviously, this is because the HO-SBM molecules might be able to penetrate into the M-β-CD cavities since the monomer is an oily liquid. This phenomenon is known as the kneading method for the formation of cyclodextrin-guest complexes [42].

Even though PXRD and DSC studies confirmed the transitions from a crystalline to amorphous state of H-SBM upon its interaction with M-β-CD, these analyses neither definitely confirmed inclusion complexation (especially, for HO-SBM) nor determined the complex stoichiometry. Thus, electrospray ionization–mass spectrometry (ESI–MS) [42,43] was used to investigate the complexation of M-β-CD with POBM.

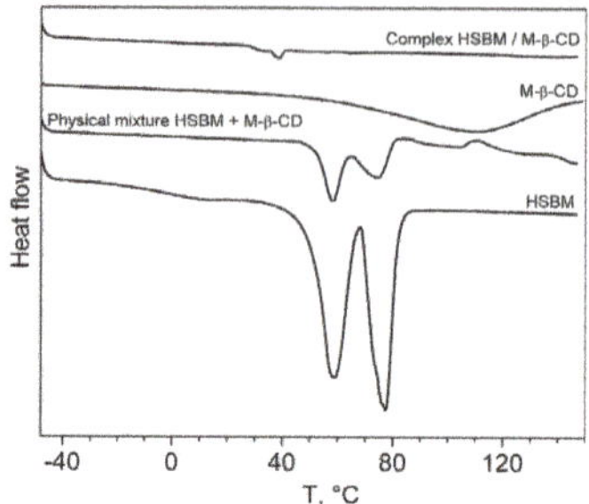

Figure 17. DSC thermograms of hydrogenated soybean monomer, methyl-β-cyclodextrin, their physical mixture, and inclusion complex. Reprinted from [32]. Copyright (2018), with permission from Elsevier.

The ESI mass spectra of inclusion complexes of the monomers with M-β-CD are presented in Figure 18. Ions detected in the spectrum of H-SBM/M-β-CD (Figure 18A) are attributed to sodium adducts of complexes of stearate-H-SBM with M-β-CD having 9–13 methyl groups. In the ESI mass spectrum of the HO-SBM/M–β–CD inclusion complex (Figure 18B), ions that correspond to the inclusion of oleate-HO-SBM by $M_{10\text{-}13}$-β-CD (sodium adducts) are revealed.

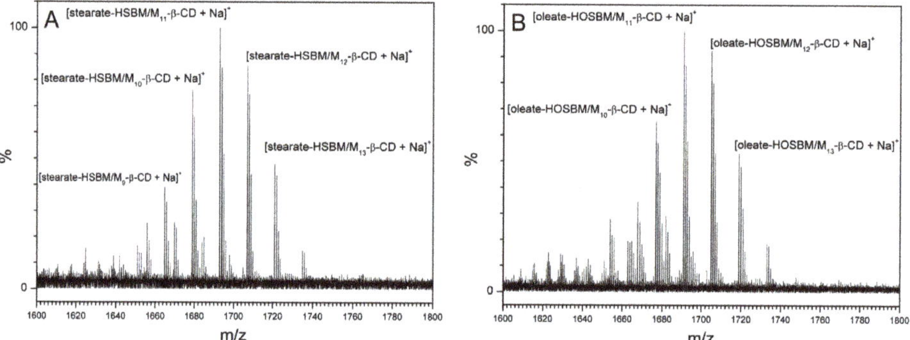

Figure 18. ESI mass spectra of H-SBM/M-β-CD (**A**) and HO-SBM/M-β-CD (**B**) inclusion complexes. Reprinted from [32]. Copyright (2018), with permission from Elsevier.

Thus, the M-β-CD complexation of monomer molecules leading to the 1:1 complex formation (Figure 15) was directly confirmed by the ESI–MS data. The emulsion copolymerization of the POBMs with styrene in the presence of methyl-β-cyclodextrin was found to allow for the preservation and protection of fatty acid double bonds (–CH=CH–) and the bisallylic hydrogen atoms (–CH=CH–CH₂–CH=CH–). The inclusion complexation decreases the chain transfer contribution by protecting the allylic moiety of the monomer fatty acid chains and simultaneously improves the aqueous solubility and availability of the monomers in the emulsion polymerization process. Therefore, a higher molecular weight of latex polymers from the highly hydrophobic POBMs was achieved with higher monomer conversion and lower coagulum formation.

7. Conclusions

Monomers based on various plant oils were synthesized through a two-step procedure. A one-step process for the synthesis of new vinyl monomers by the reaction of direct transesterification of plant oil triglycerides with *N*-(hydroxyethyl)acrylamide was recently patented. The features of homopolymerization kinetics for the POBMs were determined. The reactivity of the plant oil-based monomers was studied by radical copolymerization with vinyl monomers. The Alfrey–Price Q (0.41–0.51) and e (0.09–0.28) parameters are close for all POBMs and do not essentially depend on the monomer structure. The unique molecular structure of the plant oil-based acrylic monomers is that they simultaneously have a $CH_2=CH–C(O)–NH–CH_2CH_2–O–C(O)–$ polar (hydrophilic) fragment capable of electrostatic interactions and a hydrophobic fatty acid chain (C15–C17) capable of van der Waals interactions. This opens up new opportunities for formation of micelles from the complexes of highly hydrophobic plant oil-based monomers with sodium dodecyl sulfate and new approaches to conducting heterophase polymerization.

Preparation of polymeric coatings from the plant oil-based monomers through the incorporation of the hydrophobic fatty acid chains into the macromolecules of latex polymers allows for formation of polymer networks with a controlled cross-linking density along with enhancing the water resistance of the coatings. The copolymers based on the olive and soybean monomers enable the formation of coatings with low surface energy and water-repellent properties. The copolymers based on the olive and soybean monomers enable the formation of coatings with low surface energy and water-repellent properties.

The formation of inclusion complexes upon the interaction between the molecules of the plant oil-based acrylic monomers and methyl-β-cyclodextrin leads to a higher molecular weight of latex polymers from highly hydrophobic POBMs with higher monomer conversion and lower coagulum formation.

An analysis of the literature shows a rapid development of a new branch in polymer science related to the chemistry of aliphatic and aromatic biobased monomers from renewable plant sources. The synthesis of such monomers allows for producing fully biobased polymers with biocompatible and biodegradable properties which do not pollute the environment. Copolymerization of the biobased monomers with commercial monomers enables formation of copolymers with controlled physical, chemical, and mechanical properties.

8. Patents

Biobased Acrylic Monomers US 10,315,985 B2 June 11th, 2019.
Biobased Acrylic Monomers and Polymers Thereof US 10,584,094 B2 March 10th, 2020.

Author Contributions: Conceptualization, A.K., S.V., and A.V.; methodology, Z.D., V.K., K.K., and O.S.; investigation, A.K., Z.D., V.K., K.K., and O.S.; writing—original draft preparation, A.K., V.K., and S.V.; writing—review and editing, S.C. and A.V.; project administration, S.C. and A.V.; funding acquisition, A.V. All authors have read and agreed to the published version of the manuscript.

Funding: ND EPSCoR RII Track 1 ((IIA-1355466), North Dakota Department of Commerce, North Dakota Department of Agriculture, United Soybean Board.

Conflicts of Interest: The authors declare no conflict of interest. The funders had no role in the design of the study; in the collection, analyses, or interpretation of data; in the writing of the manuscript, or in the decision to publish the results.

References

1. Papageorgiou, G.Z. Thinking Green: Sustainable Polymers from Renewable Resources. *Polymers* **2018**, *10*, 952. [CrossRef] [PubMed]
2. Sharmin, E.; Zafar, F.; Akram, D.; Alam, M.; Ahmad, S. Recent advances in vegetable oils based environment friendly coatings: A review. *Ind. Crops Prod.* **2015**, *76*, 215–229. [CrossRef]
3. Xia, Y.; Larock, R.C. Vegetable oil-based polymeric materials: Synthesis, properties, and applications. *Green Chem.* **2010**, *12*, 1893–1909. [CrossRef]
4. Can, E.; Wool, R.P.; Kusefoglu, S. Soybean- and Castor-oil-based thermosetting polymers: Mechanical properties. *J. Appl. Polym. Sci.* **2006**, *102*, 1497–1504. [CrossRef]
5. Van De Mark, M.R.; Sandefur, K. Vegetable Oils in Paint and Coatings. In *Industrial Uses of Vegetable Oils*; Erhan, S.Z., Ed.; AOCS Press: Boca Raton, FL, USA, 2005; pp. 149–168.

6. Statista. Available online: https://www.statista.com/statistics/263933/production-of-vegetable-oils-worldwide-since-2000/ (accessed on 25 May 2020).
7. Moreno, M.; Miranda, J.I.; Goikoetxea, M.; Barandiaran, M. Sustainable polymer latexes based on linoleic acid for coatings applications. *Prog. Org. Coat.* **2014**, *77*, 1709–1714. [CrossRef]
8. Yuan, L.; Wang, Z.; Trenor, N.M.; Tang, C. Amidation of triglycerides by amino alcohols and their impact on plant oil-derived polymers. *Polym. Chem.* **2016**, *7*, 2790–2798. [CrossRef]
9. Moreno, M.; Lampard, C.; Williams, N.; Lago, E.; Emmett, S.; Goikoetxea, M.; Barandiaran, M.J. Eco-paints from bio-based fatty acid derivative latexes. *Prog. Org. Coat.* **2015**, *81*, 101–106. [CrossRef]
10. Tarnavchyk, I.; Popadyuk, A.; Popadyuk, N.; Voronov, A. Synthesis and Free Radical Copolymerization of a Vinyl Monomer from Soybean Oil. *ACS Sustain. Chem. Eng.* **2015**, *3*, 1618–1622. [CrossRef]
11. Molina-Gutierrez, S.; Ladmiral, V.; Bongiovanni, R.; Caillol, S.; Lacroix-Desmazes, P. Radical polymerization of biobased monomers in aqueous dispersed media. *Green Chem.* **2019**, *21*, 36–53. [CrossRef]
12. Demchuk, Z.; Shevchuk, O.; Tarnavchyk, I.; Kirianchuk, V.; Kohut, A.; Voronov, S.; Voronov, A. Free Radical Polymerization Behavior of the Vinyl Monomers from Plant Oil Triglycerides. *ACS Sustain. Chem. Eng.* **2016**, *4*, 6974–6980. [CrossRef]
13. Tarnavchyk, I.; Voronov, A. Biobased acrylic monomers. US patent 10,315,985 B2, 11 June 2019.
14. Demchuk, Z.; Shevchuk, O.; Tarnavchyk, I.; Kirianchuk, V.; Lorenson, M.; Kohut, A.; Voronov, S.; Voronov, A. Free Radical Copolymerization Behavior of Plant Oil-Based Vinyl Monomers and Their Feasibility in Latex Synthesis. *ACS Omega* **2016**, *1*, 1374–1382. [CrossRef] [PubMed]
15. Kingsley, K.; Shevchuk, O.; Kirianchuk, V.; Kohut, A.; Voronov, S.; Voronov, A. Emulsion Copolymerization of Vinyl Monomers from Soybean and Olive Oil: Effect of Counterpart Aqueous Solubility. *Eur. Polym. J.* **2019**, *119*, 239–246. [CrossRef]
16. Yuan, L.; Wang, Z.; Trenor, N.M.; Tang, C. Robust Amidation Transformation of Plant Oils into Fatty Derivatives for Sustainable Monomers and Polymers. *Macromolecules* **2015**, *48*, 1320–1328. [CrossRef]
17. Harrison, S.A.; Wheeler, D.H. The Polymerization of Vinyl and Allyl Esters of Fatty Acids. *J. Am. Chem. Soc.* **1951**, *73*, 839–842. [CrossRef]
18. Chen, F.B.; Bufkin, G. Crosslinkable Emulsion Polymers by Autoxidation. I. Reactivity Ratios. *J. Appl. Polym. Sci.* **1985**, *30*, 4571–4582. [CrossRef]
19. Chen, F.B.; Bufkin, G. Crosslinkable Emulsion Polymers by Autoxidation. II. *J. Appl. Polym. Sci.* **1985**, *30*, 4551–4570. [CrossRef]
20. Boshui, C.; Nan, Z.; Jiang, W.; Jiu, W.; Jianhua, F.; Kai, L. Enhanced biodegradability and lubricity of mineral lubricating oil by fatty acidic diethanolamide borates. *Green Chem.* **2013**, *15*, 738–743. [CrossRef]
21. Chen, R. Bio-based polymeric materials from vegetable oils. PhD Thesis, Iowa State University, Ames, IA, United States, 2014.
22. Voirin, C.; Caillol, S.; Sadavarte, N.V.; Tawade, B.V.; Boutevin, B.; Wadgaonkar, P.P. Functionalization of cardanol: Towards biobased polymers and additives. *Polym. Chem.* **2014**, *5*, 3142–3162. [CrossRef]
23. Jaillet, F.; Darroman, E.; Boutevin, B.; Caillol, S. A chemical platform approach on cardanol oil: From the synthesis of building blocks to polymer synthesis. *OCL - Oilseeds and fats crops and lipids* **2016**, *23*, 511–518. [CrossRef]
24. Demchuk, Z.; Li, W.S.J.; Eshete, H.; Caillol, S.; Voronov, A. Synergistic Effects of Cardanol- and High Oleic Soybean Oil Vinyl Monomers in Miniemulsion Polymers. *ACS Sustain. Chem. Eng.* **2019**, *7*, 9613–9621. [CrossRef]
25. Mhadeshwar, N.; Wazarkar, K.; Sabnis, A.S. Synthesis and characterization of ricinoleic acid derived monomer and its application in aqueous emulsion and paints thereof. *Pigment & Resin Technology* **2019**, *48*. [CrossRef]
26. Chatterjee, D.P.; Mandal, B.M. Facile atom transfer radical homo and block copolymerization of higher alkyl methacrylates at ambient temperature using CuCl/PMDETA/quaternaryammonium halide catalyst system. *Polymer* **2006**, *47*, 1812–1819. [CrossRef]
27. Çayli, C.; Meier, M.A.R. Polymers from renewable resources: Bulk ATRP of fatty alcohol derived methacrylates. *Eur. J. Lipid Sci. Technol.* **2008**, *110*, 853–859. [CrossRef]
28. Tarnavchyk, I.; Voronov, A. Biobased acrylic monomers and polymers thereof. US patent 10,584,094 B2, 10 March 2020.

29. Barison, A.; da Silva, C.W.; Campos, F.R.; Simonelli, F.; Lenz, C.A.; Ferreira, A.G. A simple methodology for the determination of fatty acid composition in edible oils through ^{1}H NMR spectroscopy. *Magn. Reson. Chem.* **2010**, *48*, 571–659. [CrossRef]

30. Bozell, J.J. Feedstocks for the future—biorefinery production of chemicals from renewable carbon. *Clean–Soil, Air, Water* **2008**, *36*, 641–647. [CrossRef]

31. Boskou, D. *Olive Oil Chemistry and Technology*, 2nd ed.; AOCS Publishing: New York, NY, USA, 2006; 288p.

32. Kohut, A.; Demchuk, Z.; Kingsley, K.; Voronov, S.; Voronov, A. Dual role of methyl-β-cyclodextrin in the emulsion polymerization of highly hydrophobic plant oil-based monomers with various unsaturations. *Eur. Polym. J.* **2018**, *108*, 322–328. [CrossRef]

33. Kingsley, K.; Shevchuk, O.; Demchuk, Z.; Voronov, S.; Voronov, A. The features of emulsion copolymerization for plant oil-based vinyl monomers and styrene. *Ind. Crops Prod.* **2017**, *109*, 274–280. [CrossRef]

34. Solomon, D.H. *The Chemistry of Organic Film Formers*; Robert, E., Ed.; Krieger Publishing Co.: Huntington, NY, USA, 1977; 412p.

35. Demchuk, Z.; Kohut, A.; Voronov, S.; Voronov, A. Versatile Platform for Controlling Properties of Plant Oil-Based Latex Polymer Networks. *ACS Sustain. Chem. Eng.* **2018**, *6*, 2780–2786. [CrossRef]

36. Kingsley, K.; Shevchuk, O.; Voronov, S.; Voronov, A. Effect of highly hydrophobic plant oil-based monomers on micellization of sodium dodecyl sulfate. *Colloids Surf. A Physicochem. Eng. Asp.* **2019**, *568*, 157–163. [CrossRef]

37. Mittal, K.L. *Micellization, Solubilization, and Microemulsions, Vols. 1 and 2*; Plenum: New York, NY, USA, 1977; 945p.

38. Rusanov, A.I. *Micellization in solutions of surfactants (in Russian)*; Khimiya: St. Petersburg, Russian Federation, 1992; 280p.

39. Michalska, P.; Wojnicz, A.; Ruiz-Nuño, A.; Abril, S.; Buendia, I.; León, R. Inclusion complex of ITH12674 with 2-hydroxypropyl-β-cyclodextrin: Preparation, physical characterization and pharmacological effect. *Carbohydr. Polym.* **2017**, *157*, 94–104. [CrossRef]

40. Liao, Y.; Zhang, X.; Li, C.; Huang, Y.; Lei, M.; Yan, M.; Zhou, Y.; Zhao, C. Inclusion complexes of HP-β-cyclodextrin with agomelatine: Preparation, characterization, mechanism study and in vivo evaluation. *Carbohydr. Polym.* **2016**, *147*, 415–425. [CrossRef] [PubMed]

41. Cabral Marques, H.M.; Hadgraft, J.; Kellaway, I.W. Studies of cyclodextrin inclusion complexes. I. The salbutamol-cyclodextrin complex as studied by phase solubility and DSC. *Int. J. Pharm.* **1990**, *63*, 259–266. [CrossRef]

42. Guo, M.Q.; Song, F.; Zhiqiang, L.; Shuying, L. Characterization of non-covalent complexes of rutin with cyclodextrins by electrospray ionization tandem mass spectrometry. *J. Mass Spectrom.* **2004**, *39*, 594–599. [CrossRef] [PubMed]

43. Marangoci, N.; Mares, M.; Silion, M.; Fifere, A.; Varganici, C.; Nicolescu, A.; Deleanu, C.; Coroaba, A.; Pinteala, M.; Simionescu, B.C. Inclusion complex of a new propiconazole derivative with β-cyclodextrin: NMR, ESI–MS and preliminary pharmacological studies. *Results Pharma Sci.* **2011**, *1*, 27–37. [CrossRef] [PubMed]

MDPI
St. Alban-Anlage 66
4052 Basel
Switzerland
Tel. +41 61 683 77 34
Fax +41 61 302 89 18
www.mdpi.com

Molecules Editorial Office
E-mail: molecules@mdpi.com
www.mdpi.com/journal/molecules